Communications
in Computer and Information Science   283

P. Balasubramaniam   R. Uthayakumar (Eds.)

Mathematical Modelling and Scientific Computation

International Conference, ICMMSC 2012
Gandhigram, Tamil Nadu, India, March 16-18, 2012
Proceedings

 Springer

Volume Editors

P. Balasubramaniam
Department of Mathematics
Gandhigram Rural Institute - Deemed University
Gandhigram 624302, Tamil Nadu, India
E-mail: balugru@gmail.com

R. Uthayakumar
Department of Mathematics
Gandhigram Rural Institute - Deemed University
Gandhigram 624302, Tamil Nadu, India
E-mail: uthayagri@gmail.com

ISSN 1865-0929   e-ISSN 1865-0937
ISBN 978-3-642-28925-5   e-ISBN 978-3-642-28926-2
DOI 10.1007/978-3-642-28926-2
Springer Heidelberg Dordrecht London New York

Library of Congress Control Number: 2012933232

CR Subject Classification (1998): G.1, G.2, G.3

Typesetting: Camera-ready by author, data conversion by Scientific Publishing Services, Chennai, India

Printed on acid-free paper

Springer is part of Springer Science+Business Media (www.springer.com)

Preface

Welcome to the Proceedings of the International Conference on Mathematical Modelling and Scientific Computation 2012 (ICMMSC 2012). It was the second international conference being organized by the Department of Mathematics, Gandhigram Rural Institute - Deemed University (GRI - DU), Gandhigram, Tamil Nadu, India.

Mathematical modelling and scientific computation are the promising, hot areas of current research and development, which can provide important advantages to users. Almost all mathematicians, industrialists, scientists, engineers and researchers in science disciplines are applying the techniques of mathematical modelling and computing to increasing effect. Scientific computation and information are the innovative thrust areas of research to study the stability analysis and control problems related to human needs.

To fulfill the needs of mathematicians, scientists, researchers, ICMMSC 2012 was organized by the Department of Mathematics, GRI - DU during March 16–18, 2012. This series contains collection of papers that have been peer reviewed by the researchers and presented in ICMMSC 2012.

ICMMSC 2012 intended to provide a common forum for researchers, scientists, engineers and practitioners throughout the world to share their ideas, latest research findings, developments and applications including their links to mathematics, computational sciences and information sciences.

In total, 332 papers were submitted to ICMMSC 2012, which were reviewed by the scientific peer-review committee consisting of 66 reviewers from various parts of the world. Based on the reviewers' comments, we accepted only 62 papers for publication in this volume.

In these proceedings, the accepted papers have been classified into two categories, namely, "Mathematical Modelling" and "Scientific Computation." Papers related to analysis, algebra, differential equations, dynamical systems, control theory, optimization techniques, fuzzy sets and logic, graph theory, coding theory are listed in Part 1. Finally, papers related to topics such as computational mathematics, image processing, pattern recognition, algorithms and cryptography are listed in Part 2.

As the Conference Chairs of ICMMSC 2012, we are very much thankful to all the national funding agencies for their grant support for the successful completion of this international conference.

ICMMSC 2012 was supported partially by the following funding agencies:

1. University Grants Commission (UGC) - Special Assistance Programme (SAP) - Departmental Research Support II (DRS II), Government of India, New Delhi

2. Council of Scientific and Industrial Research (CSIR), New Delhi, India

3. Department of Science and Technology (DST), Government of India, New Delhi, India
4. National Board for Higher Mathematics (NBHM), Department of Atomic Energy (DAE), Mumbai, India

We would like to express our sincere thanks to S.M. Ramasamy, Vice-Chancellor, GRI - DU, Gandhigram, Tamil Nadu, India, for his motivation and support. We also extend our profound thanks to all the authorities of GRI - DU as well as the faculty members and research scholars of the Department of Mathematics, GRI - DU, Gandhigram.

We are very much grateful to the Chairs and Members of the Technical Committee, Publication Committee, Financial Committee and Advisory Committee who worked as a team by investing their invaluable time and hard work to make this event a success.

We extend our hearty thanks to the keynote speakers who kindly accepted our invitation. Especially, we would like to thank the following professors:

N. Rudraiah (Emeritus Professor, Bangalore University, India)

P. Kandaswamy (Visiting Professor, GRI - DU, India)

S. Ponnusamy (Indian Institute of Technology, Madras)

Er. Meng Joo (Nanyang Technological University, Singapore)

K. Vajravelu (University of Central Florida, USA)

Kurunathan Ratnavelu, M.O. Wan Ainun, and *N. Kumaresan* (University of Malaya, Malaysia)

Santo Banerjee (Politecnico di Torino, Turin, Italy)

A total of 66 subject experts from around the world contributed to the peer-review process. We express our sincere gratitude to the reviewers for spending their valuable time to review the papers and sorting out the papers for presentation at ICMMSC 2012.

We thank Microsoft Research CMT for providing us with a wonderful "Conference Management Toolkit" for managing the papers. We also express our sincere thanks to Alfred Hofmann and his team at Springer for accepting our proposal to publish the papers in CCIS.

March 2012

P. Balasubramaniam
R. Uthayakumar

Organization

ICMMSC 2012 was organized by the Department of Mathematics, Gandhigram Rural Institute - Deemed University, Gandhigram, Tamil Nadu, India.

Organizing Committee

Patron	S.M. Ramasamy (Vice Chancellor, GRI - DU)
Conference Chairs	P. Balasubramaniam R. Uthayakumar
Technical Committee Chairs	S. Parthasarathy G. Nagamani
Publication Committee Chair	R. Rajkumar
Financial Committee Chair	P. Muthukumar
Organizing Team Members	G. Arockia Prabakar, R. Chandran, D. Easwaramoorthy, L. Jarina Banu, S. Jeeva Sathya Theesar, M. Kalpana, R. Krishnasamy, S. Lakshmanan, P. Muthukumar, M. Palanivel, M. Prakash, S. Priyan, C. Rajiv Ganthi, V.M. Revathi, J. Sahabtheen, R. Sathy, T. Senthil Kumar, V. Vembarasan, M. Vijayashree

Advisory Committee

R.P. Agarwal, USA
K. Balachandran, India
R.K. George, India
P. Kandaswamy, India
J.H. Kim, South Korea
N. Kumaresan, Malaysia
K. Ratnavelu, Malaysia
X. Mao, UK
E. Meng Joo, Singapore
A.K. Nandhakumaran, India
J.H. Park, South Korea

J.Y. Park, South Korea
S. Ponnusamy, India
S. Raviraja, Malaysia
N. Rudraiah, India
S. Banerjee, Italy
K. Somasundaram, India
N. Sugavanam, India
K. Vajravelu, USA
P. Veeramani, India
M.O. Wan Ainun, Malaysia
F. Zirilli, Italy

Referees

C. Alaca, Turkey
N.M.G. Al-Saidi, Iraq
A. Biswas, India
S. Arik, Turkey
T. Aravalluvan, India
A. Imre, Hungary
B. Prasad, India
K. Balachandran, India
H. Bao, China
G.L. Chia, Malaysia
M. Chris Monica, India
C. Claudio, Brazil
W. Feng, China
G. Ganesan, India
K.V. Geetha, India
G.V.D. Gowda, India
H. Yin, Urbana
G. Jayalalitha, India
T. Kalaiselvi, India
S. Karthikeyan, India
E. Karthikeyan, India
V. Kreinovich, USA
N. Kumaresan, Malaysia
O.M. Kwon, South Korea
N.F. Law, Hong Kong
G. Mahadevan, India
S. Marshal Anthoni, India
M. Nawi Nazri, Malaysia
S. Muralisankar, India
P. Muthukumar, India
G. Nagamani, India
V. Novak, Czech Republic
B.B. Pal, India
J.H. Park, South Korea

J.Y. Park, South Korea
J. Paulraj Joseph, India
V.N. Phat, Vietnam
R. Rajkumar, India
R. Rakkiyappan, India
R. Roopkumar, India
A. Roy, India
R. Sakthivel, South Korea
S. Banerjee, Italy
S. Saravanan, India
S. Mukherjee, India
P. Shanmugavadivu, India
S. Sankar Sana, India
S. Sivagurunathan, India
K. Somasundaram, India
Q. Song, China
S. Sivakumar, India
M. Sumathi, India
S. Prasath, USA
K. Shanti Swarup, India
S.P. Tiwari, India
I. Tsoulos, Greece
R.K. Upadhyay, India
M. Valliathal, India
V. Veeramani, Oman
C. Vidhya, India
S. Vimala, India
D. Vinayagam, India
A.V.A. Kumar, India
W. Wang, China
X. Li, China
X. Zhang, China

Table of Contents

Mathematical Modelling

Scientific Computation

Degree of Approximation of Function $f \in H_p^{(w)}$ Class in Generalized Hölder Metric by Matrix Means

Uaday Singh and Smita Sonker

Department of Mathematics,
Indian Institute of Technology Roorkee, Roorkee 247667 India
usingh2280@yahoo.co.in, smita.sonker@gmail.com

Abstract. In this paper, we shall compute the degree of approximation of function $f \in H_p^{(w)}$, a new Banach space introduced by Das, Nath and Ray ([2], [3]) through matrix means of Fourier series of f, which in turn generalizes most of the results of Liendler [5]. We use rows of the matrix $T \equiv (a_{n,k})$ without monotonicity and derive also some analogous results for the cases when rows of T are monotonic.

Keywords: Fourier series, Modulus of continuity, Hölder metric, Class $H_p^{(w)}$.

1 Introduction

Let $f \in L_p[0, 2\pi](p \geq 1)$ be a 2π-periodic function. Then, we write

$$s_n(f; x) = \frac{a_0}{2} + \sum_{k=1}^{n}(a_k cos(kx) + b_k sin(kx)) = \sum_{k=0}^{n} u_k(f; x)$$

the $(n + 1)$th partial sum of the Fourier series of f at a point x, which is a trigonometric polynomial of order (or degree) n. For $f \in L_p[0, 2\pi]$, $p \geq 1$ and a modulus of continuity w, Das, Nath and Ray ([2], [3]) have defined the following:

$$A(f; w) = sup_{t \neq 0}\frac{\|f(\cdot + t) - f(\cdot)\|_p}{w(|t|)}, \ H_p^{(w)} = \{f \in L_p[0, 2\pi], \ p \geq 1 : A(f, w) < \infty\};$$

and the norm in the space $H_p^{(w)}$ is defined by

$$\|f\|_p^{(w)} = \|f\|_p + A(f, w). \tag{1}$$

It can be verified that $\| \cdot \|_p^{(w)}$ is a norm on $H_p^{(w)}$, and called generalized Hölder metric. Given the spaces $H_p^{(w)}$ and $H_p^{(v)}$, if $w(t)/v(t)$ is non-decreasing then $H_p^{(w)} \subseteq H_p^{(v)} \subseteq L_p, p \geq 1$.

P. Balasubramaniam and R. Uthayakumar (Eds.): ICMMSC 2012, CCIS 283, pp. 1–10, 2012.

Approximation of $f \in L_p[0, 2\pi]$ by trigonometric polynomials $T_n(f; x)$ of degree n, obtained from the Fourier series of f is called Fourier approximation; and the degree of approximation $E_n(f)$ is given by

$$E_n(f) = Min_n \|f(x) - T_n(f; x)\|_p.$$

In the present paper, we shall consider approximation of $f \in H_p^{(w)}$ with the norm defined in (1) by trigonometric polynomials

$$\tau_n(f; x) = \tau_n(x) = \sum_{k=0}^{n} a_{n,k} s_k(f; x) \ \forall n \geq 0, \tag{2}$$

where $T \equiv (a_{n,k})$ is a lower triangular matrix with non-negative entries such that $a_{n,-1} = 0$, $A_{n,k} = \sum_{r=k}^{n} a_{n,k}$ so that $A_{n,0} = 1 \forall n \geq 0$. The $\tau_n(f; x)$ defined in (2) are also known as matrix means of Fourier series of f. The Fourier series of f is said to be T-summable to $s(x)$, if $\tau_n(f; x) \to s(x)$ as $n \to \infty$. In particular, if

$$a_{n,k} = \begin{cases} p_{n-k}/P_n, & 0 \leq k \leq n, \\ 0, & k > n, \end{cases}$$

and

$$a_{n,k} = \begin{cases} p_k/P_n, & 0 \leq k \leq n, \\ 0, & k > n, \end{cases} \tag{3}$$

where $P_n (= \sum_{k=0}^{n} p_k \neq 0) \to \infty$ as $n \to \infty$ and $P_{-1} = 0 = p_{-1}$, then the summability matrix T reduces to Nörlund and Riesz matrices, respectively, and $\tau_n(f; x)$ in (2) defines corresponding trigonometric polynomials $N_n(f; x)$ and $R_n(f; x)$. In case of $a_{n,k} = 1/(n+1)$ for $0 \leq k \leq n$ and $a_{n,k} = 0$ for $k > n$, $\tau_n(f; x)$ reduces to the Cesáro means of order one, denoted by $\sigma_n(f; x)$ and defined by

$$\sigma_n(f; x) = \frac{1}{n+1} \sum_{k=0}^{n} s_k(f; x). \tag{4}$$

If $a_{n,k} = 1$ for $k = n$ and $a_{n,k} = 0$ for $k \neq n$, then $\tau_n(f; x)$ reduces to partial sums $s_n(f; x)$.

Let $\lambda = \lambda_n$ be a non-decreasing sequence of positive numbers tending to ∞ for which $\lambda_0 = 1$ and $\lambda_{n+1} \leq \lambda_n + 1$. The generalized de la Vallée-Poussin means of Fourier series of f are defined by

$$V_n(x) = V_n(\lambda, f; x) = \frac{1}{\lambda_{n+1}} \sum_{k=n-\lambda_n}^{n} s_k(f; x). \tag{5}$$

We note that if $\lambda_n = n$, then the $V_n(x)$ means reduces to $\sigma_n(f; x)$, if $\lambda_n \equiv 1$, then to partial sums $s_n(f; x)$, and if $\lambda_n = [n/2]$, the integral part of $n/2$, then we get classical de la Vallée-Poussin means ([5], [8]). It is important to note that a summability method is consistent or regular, if for every convergent sequence $\{s_n\}$, $\lim_{n \to \infty} s_n = s \Rightarrow \lim_{n \to \infty} \tau_n = s$.

A positive sequence $\mathbf{a} = \{a_{n,k}\}$ is called almost monotonically decreasing with respect to k, if there exists a constant $K = K(\mathbf{a})$, depending on the sequence $\mathbf{a}$ only, such that $a_{n,p} \leq K a_{n,m}$ for all $p \geq m$ and we write that $\mathbf{a} \in AMDS$. Similarly $\mathbf{a} = \{a_{n,k}\}$ is called almost monotonically increasing with respect to k, if $a_{n,p} \leq K a_{n,m}$ for all $p \leq m$ and we write that $\mathbf{a} \in AMIS$ ([4], [5]). We shall also use the notation $\triangle_k a_{n,k} = a_{n,k} - a_{n,k+1}$ and $a_n << b_n$ if there is a positive constant K (may be different in different occurrences) such that $a_n \leq K b_n \, \forall n$.

2 Known Results

Das, Nath and Ray [2] have studied the degree of approximation of $f \in H_p^{(w)} (p \geq 1)$ through partial sums $s_n(f;x)$ of the Fourier series of f and proved the following:

Theorem 1. *If v and w are moduli of continuity such that $w(t)/v(t)$ is non-decreasing, furthermore, if $f \in H_p^{(w)}$, then*

$$\| f(x) - s_n(f;x) \|_p^{(v)} = O\left(\frac{w(\pi/n)}{v(\pi/n)} \log n + \frac{1}{n} \int_{\pi/n}^{\pi} \frac{w(t)}{t^2 v(t)} dt \right). \qquad (6)$$

Liendler [5] pointed out that the integral in the right hand side of (6) can be omitted, if there exists $\epsilon > 0$ such that $t^{-\epsilon} w(t)/v(t)$ is non-increasing; and proved the following theorems:

Theorem 2. *Let v and w be moduli of continuity such that $w(t)/v(t)$ is non-decreasing; moreover, for some $0 < \epsilon \leq 1$ let the function*

$$\gamma(t) = \gamma(v, w, \epsilon; t) = t^{-\epsilon} w(t)/v(t) \qquad (7)$$

be non-increasing. If $f \in H_p^{(w)} (p \geq 1)$, then

$$\| f(x) - s_n(f;x) \|_p^{(v)} << \frac{w(1/n)}{v(1/n)} \log n \, \forall n \geq 2. \qquad (8)$$

Theorem 3. *If the conditions of Theorem 2 are satisfied with $0 < \epsilon < 1$, and one of the following additional conditions is satisfied:*

(i) $\{p_n\} \in AMDS$,
(ii) $\{p_n\} \in AMIS$ and $np_n << P_n$,
(iii) $\sum_{m=1}^{n-1} m |\triangle p_m| << P_n$,

then

$$\| N_n(f;x) - f(x) \|_p^{(v)} << \frac{w(1/n)}{v(1/n)} \log n \, \forall n \geq 2. \qquad (9)$$

Theorem 4. *If the conditions of Theorem 2 are satisfied with $0 < \epsilon < 1$, $np_n << P_n$ and*

$$\sum_{m=0}^{n-1} |\triangle p_m| << P_n n^{-\epsilon},$$

then

$$\| R_n(f;x) - f(x) \|_p^{(v)} << \frac{w(1/n)}{v(1/n)} \log n \ \forall n \geq 2. \tag{10}$$

Theorem 5. *If the conditions of Theorem 2 are satisfied with $0 < \epsilon < 1$, then*

$$\| V_n(x) - f(x) \|_p^{(v)} << \frac{w(1/n)}{v(1/n)} \log n \ \forall n \geq 2. \tag{11}$$

We observe that the analogous of Theorems 3 and 4 for $f \in Lip(\alpha,p)$ can be found in [4] and their antecedents for monotonic $\{p_n\}$ in [1], which have been further generalized for $T \equiv (a_{n,k})$ in recent papers [7] and [6], respectively. We further observe that Theorem 2 is valid for $\epsilon = 1$, but in the subsequent results i.e., Theorems 3, 4 and 5 the author has excluded the case $\epsilon = 1$. It appears to be a typing error (cf. p-57, [5]).

3   Main Result

In this paper, we determine the degree of approximation of $f \in H_p^{(w)}(p \geq 1)$ through matrix means of the Fourier series of f and show that Theorems 3 and 4 are particular case of our theorem. More precisely, we prove:

Theorem 6. *Let v and w be moduli of continuity such that $w(t)/v(t)$ is non-decreasing; moreover, for some $0 < \epsilon \leq 1$ let the function*

$$\gamma(t) = \gamma(v, w, \epsilon; t) = t^{-\epsilon} w(t)/v(t) \tag{12}$$

be non-increasing. Let $T \equiv (a_{n,k})$ be a lower triangular infinite regular matrix. If one of the following additional conditions is satisfied:

(i) $\{a_{n,k}\} \in AMIS$,
(ii) $\{a_{n,k}\} \in AMDS$ and $(n+1)a_{n,0} << 1$,
(iii) $\sum_{k=0}^{n-1}(n-k)|\triangle_k a_{n,k}| << 1$,
(iv) $\sum_{k=0}^{n-1} k^{1-\epsilon}|\triangle_k a_{n,k}| << n^{-\epsilon}$ and $(n+1)a_{n,n} << 1$.

Then for $f \in H_p^{(w)}(p \geq 1)$

$$\| f(x) - \tau_n(f;x) \|_p^{(v)} << \frac{w(1/n)}{v(1/n)} \log n \ \forall n \geq 2. \tag{13}$$

Remark 1. If $T \equiv (a_{n,k})$ is Nörlund matrix, then conditions (i) to (iii) of our Theorem 6 reduces to conditions (i) to (iii) of Theorem 3, respectively, and (13) reduces to (9). Thus our Theorem 6 under conditions (i) to (iii) generalizes Theorem 3. Also if $T \equiv (a_{n,k})$ is Riesz matrix, then our Theorem 6 under condition (iv) reduces to Theorem 4.

4 Lemmas

We need the following lemmas to prove Theorem 6.

Lemma 1. *If $T \equiv (a_{n,k})$ satisfies condition (iii) of Theorem 6 i. e., $\sum_{k=0}^{n-1}(n-k)|\triangle_k a_{n,k}| \ll 1$, then*

$$\sum_{k=1}^{n} |\triangle_k k^{-1}(A_{n,0} - A_{n,k})| \ll n^{-1}. \tag{14}$$

Proof. We have

$$\triangle_k k^{-1}(A_{n,0} - A_{n,k}) = \frac{A_{n,0} - A_{n,k}}{k} - \frac{A_{n,0} - A_{n,k+1}}{k+1}$$

$$= k^{-1}(k+1)^{-1}(A_{n,0} - A_{n,k} - ka_{n,k}) = k^{-1}(k+1)^{-1}\left(\sum_{i=0}^{k-1} a_{n,i} - ka_{n,k}\right). \tag{15}$$

Next, we shall verify by induction that

$$|\sum_{i=0}^{k-1} a_{n,i} - ka_{n,k}| \leq \sum_{i=1}^{k} |a_{n,i-1} - a_{n,i}|. \tag{16}$$

For $k = 1$, we have

$$|\sum_{i=0}^{k-1} a_{n,i} - ka_{n,k}| = |a_{n,0} - a_{n,1}|$$

i.e.,(16) is true for $k = 1$. Let us assume that (16) is true for $k = m$, then for $k = m + 1$,

$$|\sum_{i=0}^{m} a_{n,i} - (m+1)a_{n,m+1}| = |\sum_{i=0}^{m-1} a_{n,i} + a_{n,m} + ma_{n,m} - (m+1)a_{n,m+1}|$$

$$\leq |\sum_{i=0}^{m-1} a_{n,i} - ma_{n,m}| + (m+1)|a_{n,m} - a_{n,m+1}|$$

$$= \sum_{i=1}^{m} i|a_{n,i-1} - a_{n,i}| + (m+1)|a_{n,m} - a_{n,m+1}|$$

$$= \sum_{i=1}^{m+1} i|a_{n,i-1} - a_{n,i}|.$$

Thus (16) is true for $k = m+1$, hence (16) is true for $1 \leq k \leq n$. Now, combining (15) and (16); then using Abel's transformation, we have

$$\sum_{k=1}^{n} |\triangle_k k^{-1}(A_{n,0} - A_{n,k})| \leq \sum_{k=1}^{n} k^{-1}(k+1)^{-1} \sum_{i=1}^{k} i|a_{n,i-1} - a_{n,i}|$$

$$= \sum_{k=1}^{n} \triangle(k^{-1}) \sum_{i=1}^{k} i|a_{n,i-1} - a_{n,i}| \leq \sum_{k=1}^{n+1} \left(\frac{1}{k} - \frac{1}{n+1}\right) k|a_{n,k-1} - a_{n,k}|$$

$$= \sum_{k=1}^{n+1} \left(\frac{n-k+1}{n+1}\right) |a_{n,k-1} - a_{n,k}| = \sum_{k=0}^{n} \left(\frac{n-k}{n+1}\right) |a_{n,k+1} - a_{n,k}|$$

$$\leq \frac{1}{n} \sum_{k=0}^{n-1} (n-k)|\triangle_k a_{n,k}| << n^{-1},$$

in view of condition (iii) of Theorem 6. Thus proof of Lemma 1 is complete. □

Lemma 2. *[5, p.56]: If the conditions of Theorem 6 are satisfied, then*

$$\| \sigma_n(f;x) - f(x) \|_p^{(v)} << \frac{w(1/n)}{v(1/n)} \log n \ \forall n \geq 2, \tag{17}$$

and

$$\| \sigma_n(f;x) - s_n(f;x) \|_p^{(v)} << \frac{w(1/n)}{v(1/n)} \log n \ \forall n \geq 2. \tag{18}$$

Lemma 3. *If*

$$\sum_{k=0}^{n-1} (n-k)|\triangle_k a_{n,k}| << 1,$$

and the conditions of Theorem 6 are satisfied, then

$$\| \tau_n(f;x) - s_n(f;x) \|_p^{(v)} << \frac{w(1/n)}{v(1/n)} \log n \ \forall n \geq 2. \tag{19}$$

Proof. By Abel's transformation and $a_{n,n+1} = 0$,

$$\tau_n(f;x) = \sum_{k=0}^{n} a_{n,k} s_k(f;x) = \sum_{k=0}^{n} a_{n,k} \left(\sum_{i=0}^{k} u_i(f;x)\right) = \sum_{k=0}^{n} A_{n,k} u_k(f;x),$$

and thus,

$$s_n(f;x) - \tau_n(f;x) = \sum_{k=0}^{n} (1 - A_{n,k}) u_k(f;x) = \sum_{k=1}^{n} k^{-1}(A_{n,0} - A_{n,k}) k u_k(f;x).$$

Hence, again by Abel's transformation and $A_{n,n+1} = 0$, we get

$$s_n(f;x) - \tau_n(f;x) = \sum_{k=1}^{n} \triangle_k k^{-1}(A_{n,0} - A_{n,k}) \sum_{i=1}^{k} i u_i(f;x) + (n+1)^{-1} \sum_{k=1}^{n} k u_k(f;x).$$

Therefore,

$$
\| s_n(f;x) - \tau_n(f;x) \|_p^{(v)} \le \sum_{k=1}^{n} |\triangle_k k^{-1}(A_{n,0} - A_{n,k})| \, \| \sum_{i=1}^{k} i u_i(f;x) \|_p^{(v)}
$$

$$
+ (n+1)^{-1} \| \sum_{k=1}^{n} k u_k(f;x) \|_p^{(v)} . \tag{20}
$$

Also,

$$
s_n(f;x) - \sigma_n(f;x) = (n+1)^{-1} \sum_{k=0}^{n}((n+1)u_k(f;x) - s_k(f;x)) = (n+1)^{-1} \sum_{k=1}^{n} k u_k(f;x),
$$

which implies that

$$
\| \sum_{k=1}^{n} k u_k(f;x) \|_p^{(v)} = (n+1) \| \sigma_n(f;x) - s_n(f;x) \|_p^{(v)} << \frac{w(1/n)}{v(1/n)} \log n \; \forall n \ge 2. \tag{21}
$$

in view of (18) of Lemma 2.

Combining (20) and (21), we obtain

$$
\| s_n(f;x) - \tau_n(f;x) \|_p^{(v)} << \sum_{k=1}^{n} |\triangle_k k^{-1}(A_{n,0} - A_{n,k})|(k+1) \frac{w(1/k)}{v(1/k)} \log k
$$

$$
+ \frac{w(1/n)}{v(1/n)} \log n << (n+1) \frac{w(1/n)}{v(1/n)} \log n \sum_{k=1}^{n} |\triangle_k k^{-1}(A_{n,0} - A_{n,k})| +
$$

$$
\frac{w(1/n)}{v(1/n)} \log n << (n+1) \frac{w(1/n)}{v(1/n)} \log n (n)^{-1} + \frac{w(1/n)}{v(1/n)} \log n
$$

$$
<< \frac{w(1/n)}{v(1/n)} \log n \; \forall n \ge 2.
$$

in view of Lemma 1 and increasing nature of $(k+1)w(1/k)logk/v(1/k)$. The proof of Lemma 3 is complete. $\square$

5 Proof of Theorem 6

Since the function $\gamma(t)$ given in (12) is non-increasing, the sequence $\gamma_n = n^\epsilon w(1/n) \log n / v(1/n)$ is nondecreasing, consequently we have

$$
w(1/n)/v(1/n) >> n^{-\epsilon} \ge 1/n \; (0 < \epsilon \le 1) \tag{22}
$$

and non-increasing nature of the sequence $w(1/n)/v(1/n)$ implies that

$$
w(1/\tilde{n})/v(1/\tilde{n}) << w(1/n)/v(1/n), \tag{23}
$$

where $\tilde{n} = [n/2]$. Now, we shall discuss the proof case by case.

Case I: Since $a_{n,k} \in AMIS$ implies that $1 = \sum_{k=0}^{n} a_{n,k} > \sum_{k=\tilde{n}}^{n} >> (n - \tilde{n} + 1)a_{n,\tilde{n}} >> na_{n,\tilde{n}}$, we have

$$\| \tau_n(f;x) - f(x) \|_p^{(v)} \le \sum_{k=0}^{n} a_{n,k} \| s_k(f;x) - f(x) \|_p^{(v)}$$

$$= \left(\sum_{k=0}^{\tilde{n}} + \sum_{k=\tilde{n}+1}^{n} \right) a_{n,k} \| s_k(f;x) - f(x) \|_p^{(v)} \tag{24}$$

$$<< a_{n,\tilde{n}}\{1 + \sum_{k=2}^{\tilde{n}} w(1/k) \log k / v(1/k)\} + \sum_{k=\tilde{n}+1}^{n} a_{n,k} w(1/k) \log k / v(1/k)$$

$$<< \frac{1}{n}\{1 + (\tilde{n}^{\epsilon} w(1/\tilde{n}) \log \tilde{n}/v(1/\tilde{n})) \sum_{k=2}^{\tilde{n}} k^{-\epsilon}\} + \{w(1/\tilde{n}) \log n/v(1/\tilde{n})\} \sum_{k=\tilde{n}+1}^{n} a_{n,k}$$

$$<< \frac{1}{n}\{1 + (\tilde{n}^{\epsilon} w(1/n) \log n/v(1/n))\tilde{n}^{1-\epsilon}\} + \{\log n \ w(1/n)/v(1/n)\}A_{n,0}$$

$$<< \frac{1}{n}\{1 + nw(1/n) \log n/v(1/n)\} + \log n \ w(1/n)/v(1/n)$$

$$<< \log n \ w(1/n)/v(1/n), \forall n \ge 2$$

in view of Theorem 2, increasing nature of γ_n, (22) and (23).

Case II: Since $\{a_{n,k}\} \in AMDS$ and $(n+1)a_{n,0} << 1$, thus by virtue of (23), we have

$$\| \tau_n(f;x) - f(x) \|_p^{(v)} \le \sum_{k=0}^{n} a_{n,k} \| s_k(f;x) - f(x) \|_p^{(v)}$$

$$= a_{n,0} \left(\sum_{k=0}^{\tilde{n}} + \sum_{k=\tilde{n}+1}^{n} \right) \| s_k(f;x) - f(x) \|_p^{(v)}$$

$$<< a_{n,0}\{1 + \sum_{k=2}^{\tilde{n}} w(1/k) \log k / v(1/k) + w(1/\tilde{n}) \log n/v(1/\tilde{n}) \sum_{k=\tilde{n}+1}^{n} 1\}$$

$$<< \frac{1}{n+1}\{1 + nw(1/n) \log n/v(1/n) + n \log n \ w(1/n)/v(1/n)\}$$

$$<< \log n \ w(1/n)/v(1/n), \ \forall n \ge 2$$

in view of arguments used in Case I.

Case III: Using Theorem 2 and Lemma 3, we have

$$\| \tau_n(f;x) - f(x) \|_p^{(v)} \le \| \tau_n(f;x) - s_n(f;x) \|_p^{(v)} + \| s_n(f;x) - f(x) \|_p^{(v)}$$

$$<< \log n \ w(1/n)v(1/n) \ \forall n \ge 2.$$

Case IV: Using Abel's transformation, we get

$$\tau_n(f;x) - f(x) = \sum_{k=0}^{n} a_{n,k}\{s_k(f;x) - f(x)\} = \sum_{k=0}^{n-1}(\triangle_k a_{n,k})\sum_{i=0}^{k}(s_i(f;x) - f(x))$$

$$+ a_{n,n}\sum_{k=0}^{n}\{s_k(f;x) - f(x)\}$$

$$= \sum_{k=0}^{n-1}(k+1)(\triangle_k a_{n,k})(\sigma_k(f;x) - f(x)) + (n+1)a_{n,n}(\sigma_n(f;x) - f(x)).$$

Hence,

$$\| \tau_n(f;x) - f(x) \|_p^{(v)} \leq \sum_{k=0}^{n-1}(k+1)|\triangle_k a_{n,k}| \, \| \sigma_k(f;x) - f(x) \|_p^{(v)}$$

$$+ (n+1)a_{n,n} \, \| \sigma_n(f;x) - f(x) \|_p^{(v)}$$

$$<< \sum_{k=2}^{n-1} k^{1-\epsilon+\epsilon}|\triangle_k a_{n,k}|w(1/k)\log k/v(1/k) + w(1/n)\log n/v(1/n)$$

$$<< (n^\epsilon w(1/n)\log n/v(1/n))\sum_{k=2}^{n-1} k^{1-\epsilon}|\triangle_k a_{n,k}|$$

$$+ w(1/n)\log n/v(1/n) << w(1/n)\log n/v(1/n) \; \forall n \geq 2.$$

in view of Lemma 2 and condition (iv) of our Theorem 6. Thus proof of Theorem 6 is complete.   □

6   Corollaries

Since every monotone sequence is almost monotone, the conditions (i) and (ii) of Theorem 6 are satisfied. Further, every sequence $\{a_{n,k}\}$ a non-decreasing with respect to k always satisfies condition (iii) of Theorem 6, e. g.,

$$\sum_{k=0}^{n-1}(n-k)|\triangle_k a_{n,k}| = \sum_{k=0}^{n-1}(n-k)(a_{n,k+1} - a_{n,k}) = A_{n,0} - (n+1)a_{n,0} << 1.$$

Thus, we have the following analogous result of Theorem 6 for monotone $\{a_{n,k}\}$:

Corollary 1. *Let v and w be moduli of continuity such that $w(t)/v(t)$ is non-decreasing; moreover, for some $0 < \epsilon \leq 1$ let the function*

$$\gamma(t) = \gamma(v, w, \epsilon; t) = t^{-\epsilon}w(t)/v(t),$$

be non-increasing. Let $T \equiv (a_{n,k})$ be lower triangular infinite regular matrix. If one of the following additional conditions is satisfied:

(i) $\{a_{n,k}\}$ is non-decreasing in k,
(ii) $\{a_{n,k}\}$ is non-decreasing in k and $(n+1)a_{n,0} << 1$.

Then (13) holds for $f \in H_p^{(w)}(p \geq 1)$.

Similarly, the analogues result of Corollary 1 for monotonic $\{p_n\}$ can be derived by replacing T with the Nörlund matrix.

References

1. Chandra, P.: Trigonometric Approximation of Functions in Lp–norm. J. Math. Anal. Appl. 275, 13–26 (2002)
2. Das, G., Nath, A., Ray, B.K.: An Estimate of the Rate of Convergence of Fourier Series in Generalized Hlder Metric, Analysis and Applications (Ujjain 1999), Narosa, New Delhi, pp. 43–60 (2002)
3. Das, G., Nath, A., Ray, B.K.: A New Trigonometric Method of Summation and its Application to the Degree of Approximation. Proc. Indian Acad. Sci. (Math. Sci.) 112(2), 299–319 (2002)
4. Leindler, L.: Trigonometric Approximation of Functions in Lp–norm. J. Math. Anal. Appl. 302, 129–136 (2005)
5. Leindler, L.: A Relaxed Estimate of the Degree of Approximation by Fourier Series in Generalized Hölder Metric. Anal. Math. 35, 51–60 (2009)
6. Mittal, M.L., Rhoades, B.E., Mishra, V.N., Singh, U.: Using Infinite Matrices to Approximate Functions of Class Lip(α, p) using Trigonometric Polynomials. J. Math. Anal. Appl. 326, 667–676 (2007)
7. Mittal, M.L., Rhoades, B.E., Sonker, S., Singh, U.: Approximation of Signals of Class Lip(α, p) by Linear Operators. Appl. Maths. Compu. 217, 4483–4489 (2011)
8. Mursaleen, M., Alotaibi, A.: Statistical Summability and Approximation by de la Valle-Poussin Mean. Appl. Math. Lett. 24, 320–324 (2011)

Angle Change of Plane Curve

Xiao Han, Baozeng Chu, and Liangping Qi

School of Information Engineering, China University of Geoscience (Beijing),
Xueyuan Rd.29, 100083, Beijing, China
hanxiao_8699@163.com

Abstract. In classical analysis, a curve's length can be defined as the supremum of the length of a polygonal line with turning points in the curve. To know the change of angles when one travles from one endpoint to another along the curve, a similar method can be taken. The analog is also treated in comlex analysis, but a more natural way to deal with such a problem exists, that is, define the change of angles to be the limit of polygonal line with turning points in the curve. Angle change between vectors and the sum of angle chage of a polygonal line are both well defined, then the way to find the angle change of a curve is showed here. A conjecture is posed. An abstract angle change function is also constructed. Further work is to solve the conjecture, to find the sufficient and necessary condition for a plane curve to be summable respect to total sum of angle change, and to study angle variation and the angle change function.

Keywords: angle change, summable respect to total sum of angle change, angle variation, angle change function.

1 Basic Concepts and Properties

A plane curve is a mapping from an interval in $\mathbb{R}$ into $\mathbb{R}^2$. For a closed interval I, let $\mathcal{P}[I]$ be the set of all the partitions of I.

Definition 1

$$\Delta(u, v) := \begin{cases} sign|u, v| \cdot \arccos \frac{u'v}{|u| \cdot |v|} & u, v \in \mathbb{R}^2/\{0\} \\ 0 & u, v \in \mathbb{R}^2 \wedge \ulcorner u = 0 \vee v = 0 \urcorner \end{cases} \quad (1)$$

is called the angel change of vectors $\{u, v\}$.

Definition 2. Given a plane curve $\alpha : I \mapsto \mathbb{R}^2, I = [a, b] \subset \mathbb{R}, P = \{t_i\}_0^{n+1} \in \mathcal{P}[I], t_0 = a, t_{n+1} = b, u_i = \alpha(t_i) - \alpha(t_{i-1}), i = 1 : n + 1$

$$\sum_1^n \Delta(u_i, u_{i+1}) := AC(P, \alpha) \quad (2)$$

P. Balasubramaniam and R. Uthayakumar (Eds.): ICMMSC 2012, CCIS 283, pp. 11–21, 2012.
© Springer-Verlag Berlin Heidelberg 2012

is called the (possitive oriented) sum of angle change of α on I respect to P
[1][2][4];

$$\sum_1^n |\Delta(u_i, u_{i+1})| := AAC(P, \alpha) = AV(P, \alpha) \qquad (3)$$

is called the absolute sum of angle change or angle variation of α on I respect
to P.

If $\exists\, A \in \mathbb{R} \ni \forall\, \epsilon > 0 \,\exists\, P_\epsilon \in \mathcal{P} \ni \forall\, P \supseteq P_\epsilon, P \in \mathcal{P}, |AC(P, \alpha) - A| <$
ϵ, α is said to be summable respect to total sum of angle change on I, $A :=$
$AC_\alpha[I] = AC_\alpha[a, b]$ is called the total sum of angle change of α on I, the set of
all plane curve summable respect to total sum of angle change on I is denoted
by $ACA[I][1]$.

If $\exists\, A \in \mathbb{R} \ni \forall\, \epsilon > 0 \,\exists\, \delta > 0 \ni \forall\, P \in \mathcal{P}, \|P\| < \delta, |AC(P, \alpha) - A| < \epsilon$,
α is said to be strongly summable respect to total sum of angle change on I,
$A := AC_\alpha^*[I]$ is called the total strong sum of angle change of α on I, the set of
all plane curve strongly summable respect to total sum of angle change on I is
denoted by $ACA^*[I]$.

If $\exists\, M > 0 \ni \forall\, P \in \mathcal{P}, AV(P, \alpha) \le M$, α is said to be of bounded angle
variation on I, denote the set of all plane curve of bounded angle variation on I
to be $AVB[I]$. If $\alpha \in AVB[I]$, the total angle variation of α on I is defined to
be [1][2][4]

$$\sup\{AV(P, \alpha) : P \in \mathcal{P}\} := AV_\alpha[I] = AV_\alpha[a, b]. \qquad (4)$$

Remark 1. (I) $\Delta(P_1, P_2) := \Delta(u, v).P_1 = \{s_i\}_0^{m+1} \in \mathcal{P}[I], P_2 = \{t_i\}_0^{n+1} \in$
$\mathcal{P}[J].I = [a, b], J = [b, c].u = \alpha(b) - \alpha(s_m), v = \alpha(t_1) - \alpha(b).\alpha : [a, c] \mapsto \mathbb{R}^2$.
(II) $\Delta(x, P) = \Delta(x, u), \Delta(P, x) = \Delta(v, x).P = \{t_i\}_0^{n+1} \in \mathcal{P}[I].I = [a, b].x \in$
$\mathbb{R}^2.u = \alpha(t_1) - \alpha(a), v = \alpha(b) - \alpha(t_n).\alpha : I \mapsto \mathbb{R}^2$.
(III) $A(u, P, v) = \Delta(u, P) + AC(P, \alpha) + \Delta(P, v)$.

Definition 3. *A plane curve $\alpha : I \mapsto \mathbb{R}^2$ is said to be seperable if there is a*
partition $P_0 \in \mathcal{P}[I]$ such that for any $P = \{t_i\}_0^{n+1} \supset P_0, P \in \mathcal{P}[I] : \alpha(t_{i+1}) -$
$\alpha(t_i) \ne 0$ $(i = 0 : n)$. That is, α has no adjacent points overlapped for sufficiently
finer partition.

Lemma 1

$$|u - v| \le \epsilon < |v|, u, v \in \mathbb{R}^2 \Rightarrow |\Delta(u, v)| \le \arcsin \frac{\epsilon}{|v|} = \arccos \sqrt{\frac{|v|^2 - \epsilon^2}{|v|^2}}. \qquad (5)$$

Proof. Let $|u - v| = r \le \epsilon$ next to prove $\frac{u'v}{|u| \cdot |v|} \ge \sqrt{\frac{|v|^2 - r^2}{|v|^2}}$.
First $u \ne 0$, otherwise $|v| \le \epsilon < |v|$; next $u'v > 0$, or $(u - v)'(u - v) = u'u -$
$2u'v + v'v > v'v \Rightarrow r^2 > |v|^2 > \epsilon^2$.

$$\therefore \frac{u'v}{|u| \cdot |v|} \geq \sqrt{\frac{|v|^2 - r^2}{|v|^2}} \Leftrightarrow \frac{(u'v)^2}{u'u} \geq v'v - (u-v)'(u-v)$$

$$= 2u'v - u'u \Leftrightarrow (u'u - u'v)^2 \geq 0,$$

$$\therefore 1 \geq \frac{u'v}{|u| \cdot |v|} \geq \sqrt{\frac{|v|^2 - r^2}{|v|^2}} \geq \sqrt{\frac{|v|^2 - \epsilon^2}{|v|^2}} \Rightarrow \arccos \frac{u'v}{|u| \cdot |v|}$$

$$\leq \arccos \sqrt{\frac{|v|^2 - \epsilon^2}{|v|^2}} = \arcsin \frac{\epsilon}{|v|}.$$

Remark 2. When the equality holds, $u'v = u'u \wedge r = \epsilon \Leftrightarrow (u-v)'u = 0 \wedge |u-v| = \epsilon \Leftrightarrow v'(u-v) = -\epsilon^2 \wedge |u-v| = \epsilon$, let $u = v + \epsilon(\cos\theta, \sin\theta)'$, then $v'(\cos\theta, \sin\theta)' = -\epsilon$, while this equation must have a solution for $\epsilon < |v|$. The inequality above also holds when $\leq$ is replaced by $<$.

Proposition 1. $u, v \in \mathbb{R}^2 \Rightarrow \Delta(u,v) + \Delta(v,u) = 0; \Delta(u,\lambda u) = 0 \ (\lambda \in \mathbb{R});$
$\Delta(\lambda u, \mu v) = \Delta(u,v) \ (\lambda, \mu > 0);$

$$\Delta(u,v) - \Delta(u,-v) = sign|u,v| \cdot \pi. \tag{6}$$

Proof. Only prove the last one. If $u, v \in \mathbb{R}^2/\{0\} \Delta(u,v) - \Delta(u,-v) = sign|u,v| \cdot (\arccos \frac{u'v}{|u| \cdot |v|} + \arccos \frac{-u'v}{|u| \cdot |v|}) = sign|u,v| \cdot \pi$; if $u = 0 \vee v = 0$, $\Delta(u,v) = \Delta(u,-v) = sign|u,v| = 0$.

Theorem 1. $u = r(\cos\alpha, \sin\alpha)', v = s(\cos\beta, \sin\beta)', r, s \in \mathbb{R}^+$ *then*

$$\beta - \alpha \in (-\pi, \pi) \Rightarrow \Delta(u,v) = \beta - \alpha; \tag{7}$$

$$\beta - \alpha \in [0, 2\pi) \Rightarrow \Delta(u,-v) = \beta - \alpha - \pi. \tag{8}$$

Proof. $\beta - \alpha \in (-\pi, \pi) \Rightarrow \Delta(u,v) = sign\sin(\beta - \alpha) \cdot \arccos\cos(\beta - \alpha) = \beta - \alpha; \beta - \alpha \in [0, 2\pi) \Rightarrow \Delta(u,-v) = -sign\sin(\beta - \alpha) \cdot \arccos - \cos(\beta - \alpha) =$
$$\begin{cases} -[\pi - (\beta - \alpha)] & \beta - \alpha \in [0, \pi) \\ 0 & \beta - \alpha = \pi \\ \beta - \alpha - \pi & \beta - \alpha \in (\pi, 2\pi) \end{cases} = \beta - \alpha - \pi.$$

Theorem 2. $u_i = r_i(\cos\alpha_i, \sin\alpha_i)', r_i \in \mathbb{R}^+, \alpha_i \in [0, 2\pi), i = 1 : n, u_{n+1} = u_1, r_{n+1} = r_1, \alpha_{n+1} = \alpha_1 \Rightarrow$

$$\sum_1^n \Delta(u_i, u_{i+1}) = k \cdot \pi \quad (k \in \mathbb{Z}). \tag{9}$$

Proof. First $\exists\, k_i \in \mathbb{Z} \ni \alpha_{i+1} + k_i \cdot 2\pi \in [\alpha_i, \alpha_i + 2\pi)$, if $\alpha_{i+1} \geq \alpha_i$, put $k_i = 0$, if $\alpha_{i+1} < \alpha_i$ let $k_i = 1$ $(i = 1 : n)$.

$$\therefore \Delta(u_i, -u_{i+1}) = \alpha_{i+1} + 2k_i\pi - \alpha_i - \pi,$$
$$\therefore \Delta(u_i, u_{i+1}) = \Delta(u_i, -u_{i+1}) + sign|u_i, u_{i+1}| \cdot \pi$$
$$= \alpha_{i+1} - \alpha_i + (2k_i - 1 + sign|u_i, u_{i+1}|) \cdot \pi,$$
$$\therefore \sum_1^n \Delta(u_i, u_{i+1}) = \sum_1^n (\alpha_{i+1} - \alpha_i) + k \cdot \pi = k \cdot \pi,$$
$$k = \sum_1^n (2k_i - 1 + sign|u_i, u_{i+1}|) \in \mathbb{Z}.$$

Specially, $\forall\, u, v, w, x, y \in \mathbb{R}^2/\{0\} \,\exists\, k_1, k_2 \in \mathbb{Z} \ni \Delta(u, v) + \Delta(v, w) + \Delta(w, u) = k_1 \cdot \pi, \Delta(u, x) + \Delta(x, y) + \Delta(y, w) + \Delta(w, u) = k_2 \cdot \pi \Rightarrow \exists\, k = k_1 - k_2 \in \mathbb{Z} \ni$

$$\Delta(u, v) + \Delta(v, w) = \Delta(u, x) + \Delta(x, y) + \Delta(y, w) + k \cdot \pi. \tag{10}$$

Theorem 3. $u, v, w \in \mathbb{R}^2, w \neq 0, \langle w, u \rangle \neq \pi, \langle w, v \rangle \neq \pi \Rightarrow$

$$|\Delta(u, v)| \leq |\Delta(u, w)| + |\Delta(v, w)|. \tag{11}$$

Proof. If $u = 0 \vee v = 0$, the conclusion is obvious; if $u \neq 0 \wedge v \neq 0$, let $u = (r\cos\alpha, r\sin\alpha)', v = (s\cos\beta, s\sin\beta)', w = (t\cos\gamma, t\sin\gamma)'$ $r, s, t \in \mathbb{R}^+, \alpha, \beta \in (\gamma - \pi, \gamma + \pi) \Rightarrow |\Delta(u, w)| = |\gamma - \alpha|, |\Delta(v, w)| = |\beta - \alpha| \Rightarrow |\Delta(u, v)| \leq |\arccos\cos(\beta - \alpha)| \leq |\beta - \alpha| \leq |\Delta(u, w)| + |\Delta(v, w)|.$

Proposition 2. $I = [a, b] \subset \mathbb{R}$, *seperable curve* $\alpha : I \mapsto \mathbb{R}^2 \in ACA[I] \Rightarrow \exists\, P_0 = \{t_i^{(0)}\}_0^{m+1} \in \mathcal{P}[I] \ni \forall\, P = \{t_i\}_0^{n+1} \in \mathcal{P}[I], P \supseteq P_0, t_1 = t_1^{(0)}, t_n = t_m^{(0)}$:

$$AC(P, \alpha) = AC(P_0, \alpha). \tag{12}$$

Proof. $\exists\, A \in \mathbb{R} \ni \forall\, 0 < \epsilon < \pi/2, \exists\, P_\epsilon \ni \mathcal{P} \ni \forall\, P \supseteq P_\epsilon, |AC(P, \alpha) - A| < \epsilon$. For α is seperable, choose P_ϵ so that α has no adjacent points overlapped for partition finer than P_ϵ. Then equality (9) or (10) applies. Put $P_0 = \{t_i\}_0^{m+1} \supseteq P_\epsilon, m \geq 2, P_0 \in \mathcal{P} \Rightarrow |AC(P_0, \alpha) - A| < \epsilon$, when add a new partition point t to (t_1, t_m), we may assume $t \in (t_1, t_2)$. Denote the new partition to be $P, u = \alpha(t_1) - \alpha(a), v = \alpha(t_2) - \alpha(t_1), w = \alpha(t_3) - \alpha(t_2), x = \alpha(t) - \alpha(t_1), y = \alpha(t_2) - \alpha(t) \Rightarrow AC(P_0, \alpha) - AC(P, \alpha) = \Delta(u, v) + \Delta(v, w) - \Delta(u, x) - \Delta(x, y) - \Delta(y, w) = k\pi, k \in \mathbb{Z}$. While $|AC(P_0, \alpha) - AC(P, \alpha)| < 2\epsilon < \pi \Rightarrow k = 0 \Rightarrow AC(P_0, \alpha) = AC(P, \alpha)$.

Theorem 4. $I = [a, b] \subset \mathbb{R}$, *seperable curve* $\alpha : I \mapsto \mathbb{R}^2 \in ACA[I], s_n, t_n \in (a, b] \to a, x_n = \alpha(s_n) - \alpha(a), y_n = \alpha(t_n) - \alpha(a), u \in \mathbb{R}^2, \lim_{n\to\infty} \Delta(u, x_n) = A, \lim_{n\to\infty} \Delta(u, y_n) = B \Rightarrow \exists\, k \in \mathbb{Z} \ni A - B = k\pi.$

Proof. If $\nexists\, k \in \mathbb{Z} \ni A - B = k\pi$, let $0 < \epsilon < \frac{1}{4}\min\{|A - B|, |\pi - |A - B||, 2\pi - |A - B|\}, \alpha \in ACA[I] \Rightarrow \exists\, C \in \mathbb{R} \ni \exists\, P_\epsilon = \{r_i\}_0^{m+1} \in \mathcal{P}[I] \ni \forall\, P \in$

$\mathcal{P}[I], P \supseteq P_\epsilon, |AC(P,\alpha) - C| < \epsilon$. For α is seperable, choose P_ϵ so that α has no adjacent points overlapped for partition finer than P_ϵ. Then equality (9) or (10) applies. Let $w = \alpha(r_1) - \alpha(a), v = \alpha(r_2) - \alpha(r_1), \xi_n = \alpha(r_1) - \alpha(s_n), \eta_n = \alpha(r_1) - \alpha(t_n), D = AC(P_\epsilon, \alpha) - \Delta(w, v). \exists N > 0 \ni \forall n > N, a < s_n < r_1, a < t_n < r_1 \Rightarrow |\Delta(x_n, \xi_n) + \Delta(\xi_n, v) + D - A| < \epsilon, |\Delta(y_n, \eta_n) + \Delta(\eta_n, v) + D - A| < \epsilon \Rightarrow |\Delta(x_n, \xi_n) + \Delta(\xi_n, v) - \Delta(y_n, \eta_n) - \Delta(\eta_n, v)| < 2\epsilon$. While $\exists k_{1,n}, k_{2,n} \in \mathbb{Z} \ni \Delta(u, x_n) + \Delta(x_n, \xi_n) + \Delta(\xi_n, v) + \Delta(v, u) = k_{1,n}\pi, \Delta(u, y_n) + \Delta(y_n, \eta_n) + \Delta(\eta_n, v) + \Delta(v, u) = k_{2,n}\pi \Rightarrow |k_n\pi - \Delta(u, x_n) + \Delta(u, y_n)| < 2\epsilon, k_n = k_{1,n} - k_{2,n} \in \mathbb{Z}. \lim_{n \to \infty} \Delta(u, x_n) = A, \lim_{n \to \infty} \Delta(u, y_n) = B \Rightarrow \exists N_1 > 0 \ni \forall n > N_1, |\Delta(u, x_n) - A| < \epsilon, |\Delta(u, y_n) - B| < \epsilon \Rightarrow |k_n - A + B| < 4\epsilon, n > \max\{N, N_1\}$, a contradiction.

Corollary 1. $I = [a, b] \subset \mathbb{R}$, *seperable curve* $\alpha : I \mapsto \mathbb{R}^2 \in ACA[I], S = \{A \in \mathbb{R} : \lim_{n \to \infty} \Delta(u, x_n) = A, u \in \mathbb{R}^2, x_n = \alpha(t_n) - \alpha(a), t_n \in (a, b] \to a\} \Rightarrow |S| \leq 3 \wedge \ulcorner |S| = 3 \Leftrightarrow S = \{-\pi, 0, \pi\} \urcorner$.

Theorem 5. $I = [a, b] \subset \mathbb{R}, \alpha : I \mapsto \mathbb{R}^2 \in ACA[I] \Leftrightarrow \forall \epsilon > 0 \exists P_\epsilon \in \mathcal{P}[I] \ni \forall P_1, P_2 \supseteq P_\epsilon, P_1, P_2 \in \mathcal{P}[I]$:

$$|AC(P_1, \alpha) - AC(P_2, \alpha)| < \epsilon. \tag{13}$$

Proof. $(\Rightarrow)\alpha \in ACA[I] \Rightarrow \exists A \in \mathbb{R} \ni \forall \epsilon > 0 \exists P_\epsilon \in \mathcal{P}[I] \ni \forall P \supseteq P_\epsilon, P \in \mathcal{P}[I] : |AC(P, \alpha) - A| < \epsilon/2 \Rightarrow \forall P_1, P_2 \supseteq P_\epsilon, P_1, P_2 \in \mathcal{P}[I] : |AC(P_i, \alpha) - A| < \epsilon/2 \ i = 1, 2 \Rightarrow |AC(P_1, \alpha) - AC(P_2, \alpha)| < \epsilon$.
$(\Leftarrow)\forall \epsilon > 0 \exists P_\epsilon \in \mathcal{P}[I] \ni \forall Q_1, Q_2 \supseteq P_\epsilon, Q_1, Q_2 \in \mathcal{P}[I] : |AC(Q_1, \alpha) - AC(Q_2, \alpha)| < \epsilon$, put $P_{n+1} = P_n \cup P_{\frac{1}{n}}, n = 1 : \infty \Rightarrow P_m \subseteq P_n, m < n \Rightarrow \forall \epsilon > 0, \exists N > 1/\epsilon \ni \forall m, n > N, |AC(P_m, \alpha) - AC(P_n, \alpha)| < 1/N < \epsilon \Rightarrow \{AC(P_n, \alpha)\}$ is a Cauchy sequence $\Rightarrow \exists A \in \mathbb{R} \ni AC(P_n, \alpha) \to A \Rightarrow \forall \epsilon > 0, \exists N_1 \ni \forall n > N_1 : |AC(P_n, \alpha) - A| < \epsilon/2$, let $n > \max\{N_1, 2/\epsilon\} \Rightarrow \forall P \supseteq P_n : |AC(P, \alpha) - AC(P_n, \alpha)| < 1/n < \epsilon/2 \Rightarrow |AC(P, \alpha) - A| < \epsilon$.

Corollary 2. $I = [a, b] \subset \mathbb{R}, \alpha : I \mapsto \mathbb{R}^2 \notin ACA[I] \Leftrightarrow \exists \epsilon > 0 \ni \forall P_\epsilon \in \mathcal{P}[I], \exists P_1, P_2 \supseteq P_\epsilon, P_1, P_2 \in \mathcal{P}[I]$:

$$|AC(P_1, \alpha) - AC(P_2, \alpha)| > \epsilon. \tag{14}$$

Theorem 6. $u + v = w \ u, v \in \mathbb{R}^2 \Rightarrow \Delta(u, v) \cdot \Delta(u, w) \geq 0, \Delta(u, v) \cdot \Delta(w, v) \geq 0 \wedge$

$$\Delta(u, v) = \Delta(u, w) + \Delta(w, v). \tag{15}$$

Proof. $u + v = w \Rightarrow |u, v| = |u, w| = |w, v| \Rightarrow \Delta(u, v) \cdot \Delta(u, w) \geq 0, \Delta(u, v) \cdot \Delta(w, v) \geq 0. |u, v| = 0 \Rightarrow \Delta(u, v) = \Delta(u, w) = \Delta(w, v) = 0$; if $|u, v| \neq 0$, let $u = (r\cos\alpha, r\sin\alpha)', v = (s\cos\beta, s\sin\beta)', w = (t\cos\gamma, t\sin\gamma)'$ where $r, s, t \in \mathbb{R}^+$, if $|u, v| > 0$, let $\beta, \gamma \in (\alpha, \alpha + \pi) \Rightarrow \Delta(u, v) = \beta - \alpha, \Delta(u, w) = \gamma - \alpha, |w, v| > 0 \Rightarrow \beta - \gamma \in (0, \pi) \Rightarrow \Delta(w, v) = \beta - \gamma \Rightarrow \Delta(u, v) = \Delta(u, w) + \Delta(w, v)$. The case when $|u, v| < 0$ is similar.

Lemma 2. $u, v \in \mathbb{R}^2 \Rightarrow$

$$|\Delta(u, v)| \geq |\Delta(u, u + v)|, |\Delta(u, v)| \geq |\Delta(u + v, v)|. \tag{16}$$

Proposition 3. $\lim_{n\to\infty} \Delta(u, x_n) = A \in (-\pi, 0) \cup (0, \pi)$ $u, x_n \in \mathbb{R}^2 \Rightarrow \exists v \in \mathbb{R}^2/\{0\} \ni$

$$\lim_{n\to\infty} \langle v, x_n \rangle = 0. \tag{17}$$

Proof. $A \in (0, \pi) \Rightarrow sign|u, x_n| \to 1 \Rightarrow \exists N \ni \forall n > N : |u, x_n| > 0, 0 <$ arccos $\frac{u' x_n}{|u| \cdot |x_n|} < \pi \therefore$ put $u = (r \cos\alpha, r \sin\alpha)', x_n = (r_n \cos\theta_n, r_n \sin\theta_n)'$ $r, r_n \in \mathbb{R}^+, \theta_n - \alpha \in (-\pi, \pi), n > N \Rightarrow \Delta(u, x_n) = \theta_n - \alpha \in (0, \pi) \Rightarrow \theta_n \to \alpha + A := \beta$, let $v = (r \cos\beta, r \sin\beta)' \Rightarrow \theta_n - \beta = \Delta(v, x_n) \to 0$; it is similar for the case $A \in (-\pi, 0)$.

Theorem 7. $I = [a, b] \subset \mathbb{R}$, *seperable curve* $\alpha : I \mapsto \mathbb{R}^2, \lim_{n\to\infty} \Delta(u, x_n) = A \neq 0, \lim_{n\to\infty} \langle v, x_n \rangle = 0, \lim_{t\to a} \Delta(v, x) \not\exists u, v \in \mathbb{R}^2, x_n = \alpha(t_n) - \alpha(a), x = \alpha(t) - \alpha(a)$ $t, t_n \in (a, b), t_n \to a \Rightarrow \alpha \notin ACA[I]$.

Proof. Suppose $\alpha \in ACA[I]$. $\lim_{n\to\infty} \langle v, x_n \rangle = 0 \Rightarrow \lim_{n\to\infty} \Delta(v, x_n) = 0$. $\lim_{t\to a} \Delta(v, x) \not\exists \Rightarrow \exists s_n \in (a, b) \to a, y_n = \alpha(s_n) - \alpha(a) \ni \lim_{n\to\infty} \Delta(v, y_n) \neq 0$, while $\{\Delta(v, y_n)\}_1^\infty$ is bounded, hence there is a subsequence of it converges, by Th.4 we may assume $\lim_{n\to\infty} \Delta(v, y_n) = \pi \Rightarrow sign|v, y_n| \to 1. \forall P = \{r_i\}_0^{m+1} \in \mathcal{P}[I], \forall 0 < \epsilon' < \pi/2 \exists N > 0 \ni \forall n > N : a < t_n < r_1, a < s_n < r_1, \Delta(u, x_n) \neq 0, \langle v, x_n \rangle < \epsilon'$. Put $n_1 > N, P_1 = P \cup \{t_{n_1}\}$.
(I) $\langle v, x_{n_1} \rangle = 0, x_{n_1} \neq 0. \forall \epsilon > 0 \exists N_1 > 0 \ni \forall n > N_1 : a < s_n < t_{n_1}, \pi - \epsilon < \Delta(v, y_n) < \pi$. Put $n_2 > N_1, P_2 = P_1 \cup \{s_{n_2}\}$, let $w = \alpha(r_1) - \alpha(t_{n_1}), \tau = x_{n_1} - y_{n_2} \Rightarrow \delta := AC(P_2, \alpha) - AC(P_1, \alpha) = \Delta(y_{n_2}, \tau) + \Delta(\tau, w) - \Delta(x_{n_1}, w) = \Delta(y_{n_2}, \tau) + \Delta(\tau, w) + \Delta(w, v)$. $\pi - \epsilon < \Delta(v, y_{n_2}) < \pi \Rightarrow |v, y_{n_2}| > 0, \pi - \epsilon < \langle v, y_{n_2} \rangle < \pi \Rightarrow \Delta(y_{n_2}, -v) = \Delta(y_{n_2}, -x_{n_1}) \in (0, \epsilon), y_{n_2} - x_{n_1} = -\tau \Rightarrow 0 < \Delta(y_{n_2}, -\tau) < \epsilon, 0 < \Delta(-\tau, -x_{n_1}) < \epsilon \Rightarrow -\pi < \Delta(y_{n_2}, \tau) < -\pi + \epsilon, 0 < \Delta(\tau, v) < \epsilon, -\pi < \Delta(\tau, -v) < -\pi + \epsilon, -\pi < \Delta(-\tau, v) < -\pi + \epsilon$.
(I, a) $\langle v, w \rangle = 0. \delta = \Delta(y_{n_2}, \tau) + \Delta(\tau, v) \in (-\pi, -\pi + 2\epsilon)$;
(I, b) $|v, w| < 0, |w, \tau| < 0. \exists \lambda, \mu > 0 \ni: w = \lambda v + \mu \tau \Rightarrow \Delta(\tau, v) = \Delta(\tau, w) + \Delta(w, v) \Rightarrow \delta = \Delta(y_{n_2}, \tau) + \Delta(\tau, v) \in (-\pi, -\pi + 2\epsilon)$;
(I, c) $\langle w, \tau \rangle = 0. \Rightarrow \delta = \Delta(y_{n_2}, \tau) + \Delta(\tau, v) \in (-\pi, -\pi + 2\epsilon)$;
(I, d) $|v, w| < 0, |w, \tau| > 0. \exists \lambda, \mu > 0 \ni: \tau = \lambda v + \mu w \Rightarrow \Delta(w, v) = \Delta(w, \tau) + \Delta(\tau, v) \Rightarrow \delta = \Delta(y_{n_2}, \tau) + \Delta(\tau, v) \in (-\pi, -\pi + 2\epsilon)$;
(I, e) $\langle v, w \rangle = \pi. \delta = \Delta(y_{n_2}, \tau) + \Delta(\tau, -v) \in (-2\pi, -2\pi + 2\epsilon)$;
(I, f) $|v, w| > 0, |w, \tau| > 0. \exists \lambda, \mu < 0 \ni: w = \lambda u + \mu \tau \Rightarrow \Delta(\tau, v) = \Delta(\tau, -w) + \Delta(-w, v). \because \Delta(w, \tau) = \pi - \Delta(\tau, -w), \Delta(v, w) = \pi - \Delta(-w, v) \therefore \Delta(\tau, v) = 2\pi + \Delta(\tau, w) + \Delta(w, v) \Rightarrow \delta = \Delta(y_{n_2}, \tau) + \Delta(\tau, v) - 2\pi \in (-3\pi, -3\pi + 2\epsilon)$;
(I, g) $\langle w, \tau \rangle = \pi. \delta = \Delta(y_{n_2}, \tau) + \Delta(-\tau, v) \in (-2\pi, -2\pi + 2\epsilon)$;
(I, h) $|v, w| > 0, |w, \tau| < 0. \exists \lambda, \mu > 0 \ni: v = \lambda w + \mu \tau \Rightarrow \Delta(\tau, w) = \Delta(\tau, v) + \Delta(v, w) \Rightarrow \delta = \Delta(y_{n_2}, \tau) + \Delta(\tau, v) \in (-2\pi, -2\pi + 2\epsilon)$;
$\therefore |\delta| > \pi - 2\epsilon \Rightarrow \alpha \notin ACA[I]$.
(II) $0 < \langle v, x_{n_1} \rangle < \epsilon'. \exists N_1 > 0 \ni \forall n > N_1 : a < s_n < t_{n_1}, \pi - \langle v, x_{n_1} \rangle < \Delta(v, y_n) < \pi$. Put $n_2 > N_1 \Rightarrow \exists N_2 > 0 \ni \forall n > N_2 : a < t_n < s_{n_2}, |\Delta(v, x_n)| < \pi - \Delta(v, y_n)$. Let $n_3 > N_2, P_2 = P_1 \cup \{s_{n_2}, t_{n_3}\}$, put $w = \alpha(r_1) - \alpha(t_{n_1}), \tau_1 = y_{n_2} - x_{n_3}, \tau_2 = x_{n_1} - y_{n_2} \Rightarrow \delta := AC(P_2, \alpha) - AC(P_1, \alpha) = \Delta(x_{n_3}, \tau_1) + \Delta(\tau_1, \tau_2) + \Delta(\tau_2, w) + \Delta(w, x_{n_1}). 0 < |\Delta(v, x_{n_1})| < \epsilon', \pi - \langle v, x_{n_1} \rangle < \Delta(v, y_{n_2}) <$

$\pi, |\Delta(v, x_{n_3})| < \pi - \Delta(v, y_{n_2}) < \epsilon' \Rightarrow \pi - 2\epsilon' < \Delta(x_{n_3}, y_{n_2}) < \pi, \pi - 2\epsilon' < \Delta(y_{n_2}, x_{n_1}) < \pi \Rightarrow \Delta(y_{n_2}, -x_{n_3}), \Delta(x_{n_1}, -y_{n_2}) \in (0, 2\epsilon') \Rightarrow \Delta(y_{n_2}, \tau_1), \Delta(\tau_1, -x_{n_3}), \Delta(x_{n_1}, \tau_2), \Delta(\tau_2, -y_{n_2}) \in (0, 2\epsilon') \Rightarrow \Delta(x_{n_3}, \tau_1), \Delta(y_{n_2}, \tau_2), \Delta(-\tau_2, x_{n_1}), \Delta(\tau_2, -x_{n_1}) \in (\pi - 2\epsilon', \pi) \Rightarrow \pi - 4\epsilon' < \Delta(\tau_1, \tau_2) < \pi.$

$(II, a)\ \langle w, \tau_2 \rangle = 0. \delta = \Delta(x_{n_3}, \tau_1) + \Delta(\tau_1, \tau_2) + \Delta(\tau_2, x_{n_1}) \in (2\pi - 8\epsilon', 2\pi);$

$(II, b)\ |\tau_2, w| < 0, |w, x_{n_1}| < 0. \exists\ \lambda, \mu > 0 \ni: w = \lambda\tau_2 + \mu x_{n_1} \Rightarrow \Delta(\tau_2, x_{n_1}) = \Delta(\tau_2, w) + \Delta(w, x_{n_1}) \Rightarrow \delta = \Delta(x_{n_3}, \tau_1) + \Delta(\tau_1, \tau_2) + \Delta(\tau_2, x_{n_1}) \in (2\pi - 8\epsilon', 2\pi);$

$(II, c)\ \langle w, x_{n_1} \rangle = 0. \delta = \Delta(x_{n_3}, \tau_1) + \Delta(\tau_1, \tau_2) + \Delta(\tau_2, x_{n_1}) \in (2\pi - 8\epsilon', 2\pi);$

$(II, d)\ |\tau_2, w| < 0, |x_{n_1}, w| < 0. \exists\ \lambda, \mu > 0 \ni: x_{n_1} = \lambda\tau_2 + \mu w \Rightarrow \Delta(\tau_2, w) = \Delta(\tau_2, x_{n_1}) + \Delta(x_{n_1}, w) \Rightarrow \delta = \Delta(x_{n_3}, \tau_1) + \Delta(\tau_1, \tau_2) + \Delta(\tau_2, x_{n_1}) \in (2\pi - 8\epsilon', 2\pi);$

$(II, e)\ \langle w, \tau_2 \rangle = \pi. \delta = \Delta(x_{n_3}, \tau_1) + \Delta(\tau_1, \tau_2) + \Delta(-\tau_2, x_{n_1}) \in (3\pi - 8\epsilon', 3\pi);$

$(II, f)\ |\tau_2, w| > 0, |w, x_{n_1}| > 0. \exists\ \lambda, \mu > 0 \ni: -w = \lambda\tau_2 + \mu x_{n_1} \Rightarrow \Delta(\tau_2, x_{n_1}) = \Delta(\tau_2, -w) + \Delta(-w, x_{n_1}). \because \Delta(\tau_2, w) = \pi + \Delta(\tau_2, -w), \Delta(w, x_{n_1}) = \pi + \Delta(-w, x_{n_1}) \therefore \Delta(\tau, x_{n_1}) = \Delta(\tau_2, w) + \Delta(w, x_{n_1}) - 2\pi \Rightarrow \delta = \Delta(x_{n_3}, \tau_1) + \Delta(\tau_1, \tau_2) + \Delta(\tau_2, x_{n_1}) + 2\pi \in (4\pi - 8\epsilon', 4\pi);$

$(II, g)\ \langle w, x_{n_1} \rangle = \pi. \delta = \Delta(x_{n_3}, \tau_1) + \Delta(\tau_1, \tau_2) + \Delta(\tau_2, -x_{n_1}) \in (3\pi - 8\epsilon', 3\pi);$

$(II, h)\ |\tau_2, w| > 0, |w, x_{n_1}| < 0. \exists\ \lambda, \mu > 0 \ni: \tau_2 = \lambda w + \mu x_{n_1} \Rightarrow \Delta(w, x_{n_1}) = \Delta(w, \tau_2) + \Delta(\tau_2, x_{n_1}) \Rightarrow \delta = \Delta(x_{n_3}, \tau_1) + \Delta(\tau_1, \tau_2) + \Delta(\tau_2, x_{n_1}) \in (2\pi - 8\epsilon', 2\pi);$

$\therefore |\delta| > 2\pi - 8\epsilon' \Rightarrow \alpha \notin ACA[I].$

Theorem 8. $I = [a, b] \subset \mathbb{R}$, seperable curve $\alpha : I \mapsto \mathbb{R}^2, \lim_{n\to\infty} \langle u, x_n \rangle = 0, \lim_{t\to a} \Delta(u, x) \not\exists\ u \in \mathbb{R}^2, x_n = \alpha(t_n) - \alpha(a) \neq 0, x = \alpha(t) - \alpha(a)\ t, t_n \in (a, b), t_n \to a \Rightarrow \alpha \notin ACA[I].$

Theorem 9. $I = [a, b] \subset \mathbb{R}$, seperable curve $\alpha : I \mapsto \mathbb{R}^2 \in ACA[I], \lim_{n\to\infty} \langle u, x_n \rangle = 0\ u \in \mathbb{R}^2, x_n = \alpha(t_n) - \alpha(a) \neq 0, x = \alpha(t) - \alpha(a)\ t, t_n \in (a, b), t_n \to a \Rightarrow$

$$\lim_{t\to a} \Delta(u, x) = 0. \tag{18}$$

Theorem 10. $I = [a, b] \subset \mathbb{R}$, seperable curve $\alpha : I \mapsto \mathbb{R}^2 \in ACA[I], x = \alpha(t) - \alpha(a), t \in (a, b) \Rightarrow \exists\ v \in \mathbb{R}^2/\{0\} \ni$

$$\lim_{t\to a} \Delta(v, x) = 0. \tag{19}$$

Proof. Let $u \in \mathbb{R}^2/\{0\}$. (I) If $\lim_{t\to a} \Delta(u, x) = A \in (-\pi, 0) \cup (0, \pi), (17) \Rightarrow (19)$; if $A = \pm\pi$, put $v = -u$; $A = 0$, done. (II) If $\lim_{t\to a} \Delta(u, x)$ does not exist, suppose $\lim_{n\to\infty} \Delta(u, x_n) = A, \lim_{n\to\infty} \Delta(u, y_n) = B\ x_n = \alpha(s_n) - \alpha(a), y_n = \alpha(t_n) - \alpha(a)\ s_n, t_n \in (a, b)\ s_n, t_n \to a$. We may assume $A \neq 0$. $A = \pm\pi \Rightarrow sign|u, x_n| \to \pm 1, \langle u, x_n \rangle \to \pi$, put $v = -u \Rightarrow \langle v, x_n \rangle \to 0$; if $A \in (-\pi, 0) \cup (0, \pi), (17) \Rightarrow \exists\ v \in \mathbb{R}^2/\{0\} \ni \langle v, x_n \rangle \to 0$. Hence $(18) \Rightarrow (19)$.

A similar conclusion remains true for the right endpoint.

Theorem 11. $I = [a, b] \subset \mathbb{R}, \alpha : I \mapsto \mathbb{R}^2 \in ACA[I] \Rightarrow \exists\ M > 0 \ni \forall\ P \in \mathcal{P}[I] : |AC(P, \alpha)| < M.$

Proof. $\alpha \in ACA[I] \Rightarrow \exists\ A \in \mathbb{R} \ni \forall\ \epsilon > 0\ \exists\ P_\epsilon = \{t_i\}_0^{n+1} \in \mathcal{P}[I] \ni: |AC(P, \alpha) - A| < \epsilon\ \forall\ P \supseteq P_\epsilon, P \in \mathcal{P}[I]. \because \forall\ u, v \in \mathbb{R}^2 : |\Delta(u, v)| < \pi$, while add or delete a partition point involes at most 5 angles $\Rightarrow \forall\ P \in \mathcal{P}[I], P' = P \cup P_\epsilon \Rightarrow |AC(P, \alpha)| < |AC(P', \alpha)| + 5n\pi < A + \epsilon + 5n\pi.$

Conjecture 1. $I = [a, b] \subset \mathbb{R}$, seperable curve $\alpha : I \mapsto \mathbb{R}^2 \in ACA[I] \Leftrightarrow \exists M > 0 \ni \forall P \in \mathcal{P}[I] : |AC(P, \alpha)| < M \wedge \exists u, v \in \mathbb{R}^2/\{0\} \ni : \lim_{s \to a} \Delta(u, x) = \lim_{t \to b} \Delta(y, v) = 0. x = \alpha(s) - \alpha(a), y = \alpha(b) - \alpha(t). s, t \in (a, b).$

For seperable curve α which has bounded sum of angle change, (9) and Th.10 give that $A(u, P, v)$ can attain at most finite many values in $\mathbb{R}$. Put $S = \{A = A(u, P, v) : P \in \mathcal{P}[I]\} = \{A_i\}_1^n$, then there exists $A \in S$ such that for any $P_0 \in \mathcal{P}[I]$, there is a partition $P \in \mathcal{P}[I], P \supseteq P_0 : A(u, P, v) = A$. For S may be divided into two classes: $S_1 = \{A : \forall P_0 \in \mathcal{P}[I], \exists P \in \mathcal{P}[I], P \supseteq P_0, A(u, P, v) = A\}; S_2 = S_1^c = \{A : \exists P_0 \in \mathcal{P}[I], \forall P \in \mathcal{P}[I], P \supseteq P_0, A(u, P, v) \neq A\}$, then S_1 must not be empty. If so, $\exists P_i \in \mathcal{P}[I], \forall P \in \mathcal{P}[I], P \supseteq P_i : A(u, P, v) \neq A_i, i = 1 : n$. Put $P_0 = \bigcup_1^n P_i$, then for any $P \in \mathcal{P}[I], P \supseteq P_0, A(u, P, v) \notin S$, a contradiction.

2   A Sufficient Condition

Theorem 12. $I = [a, b] \subset \mathbb{R}, f : I \mapsto \mathbb{R}^n$ *is differenciable,* f' *is continous at some* c *in* $I \Rightarrow \forall \epsilon > 0 \exists \delta > 0 \ni \forall x \neq y \in U(c, \delta) \cap I$

$$\left| \frac{f(x) - f(y)}{x - y} - f'(c) \right| < \epsilon. \qquad (20)$$

Proof. Let $f = (f_i)_1^{nT}, f'(c) = (a_i)_1^{nT} \Rightarrow f_i'$ is continous at $c \Rightarrow \forall \epsilon > 0 \exists \delta > 0 \ni \forall x \in U(c, \delta) \cap I$ $|f_i'(x) - a_i| < \frac{\epsilon}{\sqrt{n}} \Rightarrow \forall x < y \in U(c, \delta) \cap I \exists \xi_i \in (x, y) \ni$ $\left| \frac{f_i(x) - f_i(y)}{x - y} - a_i \right| = |f_i'(\xi_i) - a_i| < \frac{\epsilon}{\sqrt{n}} \Rightarrow |\frac{f(x) - f(y)}{x - y} - f'(c)|^2 = \sum_1^n [\frac{f_i(x) - f_i(y)}{x - y} - a_i]^2 < \epsilon^2$.

Theorem 13. $I = [a', b'] \subset \mathbb{R}$, *seperable curve* $\alpha : I \mapsto \mathbb{R}^2$ *had finite derivative* α' *everywhere on* I, α' *is continous at some* e *in* I *and* $\alpha'(e) = \tau \neq 0 \Rightarrow \exists \delta > 0 \ni \forall a < b < c < d \in U(e, \delta) \cap I, \forall t_1 < t_2 < \cdots < t_n \in (b, c)$

$$\Delta(u, v) + \Delta(v, w) = \Delta(u, x_1) + \sum_1^n \Delta(x_i, x_{i+1}) + \Delta(x_{n+1}, w), \qquad (21)$$

where $u = \alpha(b) - \alpha(a), v = \alpha(c) - \alpha(b), w = \alpha(d) - \alpha(c), x_1 = \alpha(t_1) - \alpha(b), x_i = \alpha(t_i) - \alpha(t_{i-1})$ $(i = 2 : n), x_{n+1} = \alpha(c) - \alpha(t_n)$.

Proof. It is sufficient to prove the case when $n = 1$, for the general case, add the additional points one by one. Let $t \in (b, c), x = \alpha(t) - \alpha(b), y = \alpha(c) - \alpha(t) \Rightarrow \forall 0 < \epsilon < |\tau| \exists \delta > 0 \ni \forall \xi \neq \eta \in U(e, \delta) \cap I \left| \frac{\alpha(\xi) - \alpha(\eta)}{\xi - \eta} - \tau \right| < \epsilon \Rightarrow |\frac{u}{b-a} - \tau|, |\frac{v}{c-b} - \tau|, |\frac{w}{d-c} - \tau|, |\frac{x}{t-b} - \tau|, |\frac{u}{c-t} - \tau| < \epsilon < |\tau| \Rightarrow |\Delta(u, \tau)|, |\Delta(v, \tau)|, |\Delta(w, \tau)|, |\Delta(x, \tau)|, |\Delta(y, \tau)| < \arcsin \frac{\epsilon}{|\tau|} \wedge u'\tau, v'\tau, w'\tau, x'\tau, y'\tau > 0 \Rightarrow |\Delta(u, v)|, |\Delta(v, w)|, |\Delta(u, x)|, |\Delta(x, y)|, |\Delta(y, w)| < 2 \cdot \arcsin \frac{\epsilon}{|\tau|}$. If ϵ is small enough, $\ni \arcsin \frac{\epsilon}{|\tau|} < \pi/12 \Rightarrow |\Delta(u, v) + \Delta(v, w)| < 4 \cdot \arcsin \frac{\epsilon}{|\tau|} < \pi/2, |\Delta(u, x) + \Delta(x, y) + \Delta(y, w)| < 6 \cdot \arcsin \frac{\epsilon}{|\tau|} < \pi/2$, hence the unique interger that makes the equality true is 0.

Likewith,

Theorem 14. $I = [a, d] \subset \mathbb{R}$, *seperable curve* $\alpha : I \mapsto \mathbb{R}^2$ *has finite derivative* α' *everywhere on* I, α' *is continous at* a *and* $\alpha'(a) = \tau \neq 0 \Rightarrow \exists\, \delta > 0 \ni \forall\, b < c \in \overset{\circ}{U}(a, \delta) \cap I, \forall\, t_1 < t_2 < \cdots < t_n \in (a, b)$

$$\Delta(\tau, u) + \Delta(u, v) = \Delta(\tau, x_1) + \sum_1^n \Delta(x_i, x_{i+1}) + \Delta(x_{n+1}, v), \qquad (22)$$

Where $u = \alpha(b) - \alpha(a), v = \alpha(c) - \alpha(b), x_1 = \alpha(t_1) - \alpha(a), x_i = \alpha(t_i) - \alpha(t_{i-1})$ $(i = 2 : n), x_{n+1} = \alpha(b) - \alpha(t_n)$.

Theorem 15. $I = [a, d] \subset \mathbb{R}$, *seperable curve* $\alpha : I \mapsto \mathbb{R}^2$ *has finite derivative* α' *everywhere on* I, α' *is continous at* d *and* $\alpha'(d) = \tau \neq 0 \Rightarrow \exists\, \delta > 0 \ni \forall\, b < c \in \overset{\circ}{U}(d, \delta) \cap I, \forall\, t_1 < t_2 < \cdots < t_n \in (c, d)$

$$\Delta(u, v) + \Delta(v, \tau) = \Delta(u, x_1) + \sum_1^n \Delta(x_i, x_{i+1}) + \Delta(x_{n+1}, \tau), \qquad (23)$$

where $u = \alpha(c) - \alpha(b), v = \alpha(d) - \alpha(c), x_1 = \alpha(t_1) - \alpha(b), x_i = \alpha(t_i) - \alpha(t_{i-1})$ $(i = 2 : n), x_{n+1} = \alpha(d) - \alpha(t_n)$.

Theorem 16. $I = [a, b] \subset \mathbb{R}$, *seperable curve* $\alpha : I \mapsto \mathbb{R}^2 \in C^1[I], \alpha' \neq 0 \ \forall\, t \in I \Rightarrow \alpha \in ACA^*[I] \subset ACA[I]$.

Proof. $\alpha \in C^1[I] \wedge \alpha' \neq 0 \ \forall\, t \in I$, then for any $t \in I$, there exists a $\delta > 0$ that makes (21) true. $\bigcup_{t \in I} U(t, \delta)$ is an open cover of compact set I, hence finite of the sets (let be) $\bigcup_1^n U(t_i, \delta_i) \supset I$. We may assume that any one of $U(t_i, \delta_i)$ do not contain another, and any one that is not covered by the others is necessary for the finite open cover, otherwise, that is contained by another or is redundant may be deleted. Let $a \leq t_1 < t_2 < \cdots < t_n \leq b$, then $t_1 - \delta_1 < a < t_2 - \delta_2 < t_1 + \delta_1 < t_3 - \delta_3 < t_2 + \delta_2 < \cdots < t_n - \delta_n < t_{n-1} + \delta_{n-1} < b < t_n + \delta_n$, thus the points above and a, b form a partition $P_0 = \{a, t_2 - \delta_2, \cdots, b\}$ of I. For the endpoints a and b, for any $0 < \epsilon < \min\{|\alpha'(a)|, |\alpha'(b)|\}$, there exists a $\delta > 0$ that makes (22) and (23) true. And $\forall\, \xi \neq \xi' \in U(a, \delta') \cap I, \eta \neq \eta' \in U(b, \delta') \cap I$ $\left| \frac{\alpha(\xi) - \alpha(\xi')}{\xi - \xi'} - \alpha'(a) \right| < \epsilon, \left| \frac{\alpha(\eta) - \alpha(\eta')}{\eta - \eta'} - \alpha'(b) \right| < \epsilon \Rightarrow |\Delta(\frac{\alpha(\xi) - \alpha(a)}{\xi - a}, \alpha'(a))| = |\Delta(\alpha'(a), \alpha(\xi) - \alpha(a))| < \arcsin \frac{\epsilon}{|\alpha'(a)|}, |\Delta(\frac{\alpha(\eta) - \alpha(b)}{\eta - b}, \alpha'(b))| = |\Delta(\alpha(b) - \alpha(\eta), \alpha'(b))| < \arcsin \frac{\epsilon}{|\alpha'(b)|} \ (\xi \neq a, \eta \neq b)$. $\forall\, P = \{s_i\}_0^{m+1} \in \mathcal{P}$, put

$$A(P) := \Delta(\alpha'(a), \alpha(s_1) - \alpha(a)) + AC(P, \alpha) + \Delta(\alpha(b) - \alpha(s_m), \alpha'(b)). \qquad (24)$$

Let $\delta < \frac{1}{4} \min\{\|P_0\|, \delta'\}$, hence $\forall\, P \in \mathcal{P}, \|P\| < \delta$, the inner of any partition interval of P_0 has at least four points of P, the interval $(a, a + \delta'), (b - \delta', b)$ also has at least four points of P [5]. Choose the largest , the secondly largest, and the leasts, the secondly least points form every partition interval of P_0 together with the endpoints a, b to form a new partition, called the simlified partition of P respect to P_0, denoted by $\hat{P}$. Hence $\forall P' \supset \hat{P}, A(P) = A(\hat{P})$.

$\forall\, P_1 \neq P_2 \in \mathcal{P}, \|P_1\| < \delta, \|P_2\| < \delta$, put $P = \hat{P}_2 \cap \hat{P}_2 \Rightarrow A(P) = A(\hat{P}_1) = A(\hat{P}_2) = A(P_1) = A(P_2) := A.$ $\therefore \forall\, P = \{s_i\}_0^{m+1} \in \mathcal{P}, \|P\| < \delta, |AC(P,\alpha) - A| = |AC(P,\alpha) - A(P)| = |\Delta(\alpha'(a), \alpha(s_1) - \alpha(a)) + \Delta(\alpha(b) - \alpha(s_m), \alpha'(b))| \leq |\Delta(\alpha'(a), \alpha(s_1) - \alpha(a))| + |\Delta(\alpha(b) - \alpha(s_m), \alpha'(b))| < \arcsin\frac{\epsilon}{|\alpha'(a)|} + \arcsin\frac{\epsilon}{|\alpha'(b)|}.$

3 Construction of Angle Change Function

Definition 4. *(Angle Change Function) A function $\Delta : \mathbb{R}^2 \times \mathbb{R}^2 \mapsto \mathbb{R}$ is called a angle change function of vectors in $\mathbb{R}^2$ if for any $u, v \in \mathbb{R}^2$: [3]*
(I) directivity: $\Delta(u,v) = \Delta(\lambda u, \mu v)$ $\lambda, \mu \in \mathbb{R}^+$;
(II) antisymmetry: $\Delta(u,v) = -\Delta(v,u)$;
(III) centrosymmetry: $\Delta(u,v) = \Delta(-u,-v)$;
(IV) periodicity: $\Delta(u, u+v) + \Delta(u+v, v) = \Delta(u,v)$.

Proposition 4. $\Delta(0,0) = \Delta(u,u) = \Delta(u,-u) = 0.$

Proof. $\Delta(0,0) + \Delta(0,0) = \Delta(0,0); \Delta(u,u) + \Delta(u,0) = \Delta(u,0); \Delta(u,-u) = -\Delta(-u,u) = -\Delta(u,-u).$

Theorem 17. $\exists\, \pi^\circ \in \mathbb{R} \ni \forall\, u, v \in \mathbb{R}^2 \wedge rank(u,v) = 2 :$

$$\Delta(u,v) + \Delta(v,-u) = sign|u,v| \cdot \pi^\circ. \tag{25}$$

Proof. (i)$\forall\, u, v \in \mathbb{R}^2 \wedge rank(u,v) = 2 : \pi^\circ := sign|u,v| \cdot [\Delta(u,v) + \Delta(v,-u)]$;
(ii)$\forall\, x, y \in \mathbb{R}^2 \wedge rank(x,y) = 2$. We may assume $rank(x,u) = 2, \Rightarrow \exists\, \lambda, \mu \in \mathbb{R} \wedge \mu \neq 0 \ni: x = \lambda u + \mu v.$ If $\lambda = 0, \mu < 0 \Rightarrow \Delta(u,x) + \Delta(x,-u) = \Delta(u,-v) + \Delta(-v,-u) = \Delta(-u,v) + \Delta(v,u) = -[\Delta(u,v) + \Delta(v,-u)] = sign|u,x| \cdot \pi^\circ$; if $\lambda > 0, \mu > 0 \Rightarrow \Delta(-u,v) + \Delta(v,x) = \Delta(-u,x), \Delta(v,x) + \Delta(x,u) = \Delta(v,u) \Rightarrow \Delta(u,x) + \Delta(x,-u) = \Delta(u,v) - \Delta(x,v) + \Delta(v,-u) + \Delta(x,v) = \Delta(u,v) + \Delta(v,-u) = sign|u,x| \cdot \pi^\circ.$ The others are similar : $\Delta(u,x) + \Delta(x,-u) = sign|u,x| \cdot \pi^\circ \Rightarrow \Delta(x,u) + \Delta(u,-x) = sign|x,u| \cdot \pi^\circ \Rightarrow \Delta(x,y) + \Delta(y,-x) = sign|x,y| \cdot \pi^\circ.$

Theorem 18. $u, v, v \in \mathbb{R}^2/\{0\} \Rightarrow \exists k \in \mathbb{Z} \ni: \Delta(u,v) + \Delta(v,w) + \Delta(w,u) = k\pi^\circ.$

Proof. Put $w = \lambda u + \mu v$ when $rank(u,v) = 2.$
(i)$\lambda > 0, \mu > 0. \Delta(\lambda u, w) + \Delta(w, \mu v) = \Delta(\lambda u, \mu v).$
$\therefore \Delta(u,w) + \Delta(w,v) = \Delta(u,v), \Delta(u,v) + \Delta(v,w) + \Delta(w,u) = 0.$
(ii)$\lambda = 0, \mu < 0. \Delta(u,v) + \Delta(v,w) + \Delta(w,u) = \Delta(u,v) + \Delta(v,-v) + \Delta(-v,u) = \Delta(u,v) + \Delta(v,-u) = sign|u,v| \cdot \pi^\circ.$
(iii)$\lambda < 0, \mu < 0. -w = -\lambda u - \mu v. \Delta(-\lambda u, -w) + \Delta(-w, -\mu v) = \Delta(-\lambda u, -\mu v).$
$\therefore \Delta(u,-w) + \Delta(-w,v) = \Delta(u,v) = sign|w,u| \cdot \pi^\circ - \Delta(w,u) + sign|v,w| \cdot \pi^\circ - \Delta(v,w).$
$\therefore \Delta(u,v) + \Delta(v,w) + \Delta(w,u) = (sign|w,u| + sign|v,w|) \cdot \pi^\circ.$
   Others are similar.

When $rank(u,v) = 1$, let $v = \lambda u$ with $\lambda \neq 0$. $\Delta(u,v) + \Delta(v,w) + \Delta(w,u)$
$= \Delta(u, \lambda u) + \Delta(\lambda u, w) + \Delta(w, u)$
$= \begin{cases} \Delta(u,w) + \Delta(w,u) = 0 \\ \Delta(-u,w) + \Delta(w,u) = \Delta(u,-w) + \Delta(w,u) = sign|u,v| \cdot \pi° \end{cases}$.

By induction,

Theorem 19. $u_i \in \mathbb{R}^2 / \{0\}, i = 1 : n+1, u_{n+1} = u_1 \Rightarrow \exists k \in \mathbb{Z} \ni \sum_1^n \Delta(u_i, u_{i+1})$
$= k \cdot \pi°$.

Hence when the angle change of vectors is replaced by a angle change function on vectors, similar results remain true.

References

1. Apostol, T.: Mathematical Analysis, English edn., 2nd edn. China Machine Press, Beijing (2004)
2. Rudin, W.: Principle of Mathematical Analysis, English edn., 3rd edn. China Machine Press, Beijing (2004)
3. Dudley, R.M.: Real Analysis and Probability, English edn., 2nd edn. China Machine Press, Beijing (2006)
4. Royden, H.L., Fitzpatrick, P.M.: Real Analysis, English edn., 4th edn. China Machine Press, Beijing (2010)
5. Zhou, M.: Real Function Theory, 2nd edn. (in Chinese). Peking University Press, Beijing (2008)

RETRACTED CHAPTER: On Faintly SP-θ (Semi-Pre-θ)-Continuous Functions

Alaa. M.F. AL. Jumaili and Xiao-Song Yang

Department of Mathematics,
Huazhong University of Science and Technology
Wuhan, 430074, China
alaa_mf1970@yahoo.com

Abstract. In this paper we introduce and study two new notions of faintly continuity which are called faintly sp-θ-Continuous functions and faintly semi-pre-θ-Continuous functions using the concept of b-θ open sets and β-θ open sets, the class of faintly sp-θ-Continuous functions is a generalization of faintly Pre-θ (Semi-θ)-Continuous functions due to A.A. El-Atik [1]. At the same time the class of faintly semi-pre-θ-Continuous functions is a generalization of faintly Pre-θ (Semi-θ)-Continuous functions and faintly sp-θ-Continuous functions. Some characterizations and several properties concerning faintly sp-θ-Continuous functions and faintly semi-pre-θ-Continuous functions are obtained. Furthermore the relationships among these notions and other well-known forms of faintly continuity are also given.

Keywords: Faintly b-θ-Continuity, Faintly β-θ Continuity, b-θ Open Sets, β-θ Open Sets.

1   Introduction

The notion of continuity is an important concept in general topology as well as all branches of mathematics. Of course its weak forms and strong forms of continuity are important, too. Semi-open sets, α-open sets, preopen sets, semi-preopen sets, b-open sets, play an important role in generalization of continuity in topological spaces. By using these sets many authors introduced and investigated various types of modifications of continuity. Long and Herrington [2] in 1982 defined a weak form of continuity called faintly continuous by making use of θ-open sets. T. Noiri, V. Popa, [3] in 1990 introduced and investigated three weaken forms of faintly continuity which are called faintly semi continuity and faintly precontinuity and faintly β-continuity. T. Noiri. [4] in 2003, defined β-θ open sets and J.H.Park, [5] in 2006, defined b-θ open sets in topological spaces and investigate some of their properties. Recently, A.A. El-Atik [1] in 2011, introduced and investigated two weaken forms of faint continuity which are called faintly Pre-θ (Semi-θ)-Continuous functions. The purpose of this paper is to introduce and investigate other forms of faint continuity namely faintly sp-θ-Continuous functions and faintly semi-pre-θ-Continuous functions. Some characterizations and basic properties of these new types of functions are obtained. And we discussed the relationships among these notions and several forms of faintly continuity.

The paper starting on page 22 of this volume has been retracted for reasons of plagiarism. In addition, the name of the second author was added without his knowledge or consent. The erratum to this chapter is available at DOI: 10.1007/978-3-642-28926-2_63

P. Balasubramaniam and R. Uthayakumar (Eds.): ICMMSC 2012, CCIS 283, pp. 22–31, 2012.
© Springer-Verlag Berlin Heidelberg 2012

2   Preliminaries

Throughout this paper, (X, T) and (Y, T^*) (or simply X and Y) mean topological spaces. For any subset A of X, The closure and interior of A are denoted by Cl(A) and Int(A), respectively.

Definitions 1. Let (X, T) be a topological space and A be a subset of X,

a)   A is said to be α -open [6] (resp. semi-open [7], preopen [8], β-open [9] or semi-preopen [10], γ-open [11] or b-open [12] or sp-open [13]) if $A \subset Int(Cl(Int(A)))$ (resp. $A \subset Cl(Int(A)), A \subset Int(Cl(A))$, $A \subset Cl(Int(Cl(A)))$, $A \subset Int(Cl(A)) \cup Cl(Int(A))$). The complement of a semi-open (resp. α -open, preopen, β-open, b-open) set is said to be semi-closed [14], (resp. α-closed [6], preclosed [15], β-closed [9], b-closed [11]).

b)   A point x of X is called a b-θ-cluster point of A if bCl $(U) \cap A \neq \emptyset$ for every b-open set U containing x. The set of all b-θ-cluster points of A is called the b-θ-closure of A and is denoted by $bCl_\theta(A)$ [5].A sub set A is said to be b-θ-closed if $A = bCl_\theta(A)$.The complement of a b-θ-closed set is called b-θ-open set. The union of all b-θ-open sets of X contained in A is called the b-θ interior of A and is denoted by $bInt_\theta(A)$. The family of all b-θ-open subsets of X containing a point $x \in X$ is denoted by $B\theta\Sigma(X, x)$. The family of all b-θ-open sets in X is denoted by $B\theta\Sigma(X, T)$. The family of all b-θ-closed sets in X is denoted by $BC_\theta(X, T)$.

c)   The β-θ-closure of A [4], denoted by $\beta Cl_\theta(A)$, is defined to be the set of all x $\in X$ such that $\beta Cl(V) \cap (A) \neq \emptyset$ for every $V \in \beta\Sigma(X, x)$ with $x \in V$. A subset A is said to be β-θ-closed [4] if $A = \beta Cl_\theta(A)$. The complement of a β-θ-closed set is said to be β-θ-open. The union of all β-θ open sets of X contained in A is called the β-θ-interior of A and is denoted by $\beta Int_\theta(A)$.The family of all β-θ-open subsets of X containing a point $x \in X$ is denoted by $\beta\theta\Sigma(X, x)$. The family of all β-θ-open sets in X is denoted by $\beta\theta\Sigma(X,T)$. The family of all β-θ-closed sets in X is denoted by $\beta C_\theta(X,T)$.

d)   A subset A is said to be b-regular [5] (resp. β-regular [4]) if it is both b-open And b-closed (resp. β-open and β-closed).

Remark 1. Since the notion of b-open (resp. β-open) sets and the notion of sp-open (resp. semi-preopen) sets are same, we will use the term b-θ-open (resp. β-θ-open) sets instead of sp-θ-open (resp. semi-pre-θ-open) sets.

Lemma 1. [5] For a subset A of a topological space X, the following properties hold:
a)   If $A \in B\Sigma(X, T)$, then $bcl(A) = bCl_\theta(A)$.
b) $A \in BR(X)$, if and only if A is b-θ-open and b-θ-closed.

Lemma 2. [4] For a subset A of a topological space X, the following properties hold:
a)   If $A \in \beta\Sigma(X, T)$, then $\beta Cl(A) = \beta Cl_\theta(A)$.
b) $A \in \beta R(X)$, if and only if A is β-θ-open and β-θ-closed.

Remark 2. It is obvious that b-regular $\Rightarrow$ b-θ-open $\Rightarrow$ b-open. But the Converses are not necessarily true as shown by the examples in [5].

Remark 3. It is obvious that β-regular ⇒ β-θ-open ⇒ β-open. But the Converses are not necessarily true as shown by the examples in [4].

Definitions 2. A function $f: (X, T) \rightarrow (Y, T^*)$ is said to be semi continuous [16] (resp. precontinuous [8], α-continuous [6], γ-continuous [11], β-continuous [9]) if $f^{-1}(V) \in S\Sigma(X,T)$ (resp. $f^{-1}(V) \in P\Sigma(X,T), f^{-1}(V) \in \alpha\Sigma(X, T), f^{-1}(V) \in \gamma\Sigma(X,T),$ $f^{-1}(V) \in \beta\Sigma(X,T))$ for every open set V of Y.

Definitions 3. A function $f: (X, T)$ and $\rightarrow (Y, T^*)$ is said to be faintly continuous [2] (resp. faintly semi-continuous [3], faintly precontinuous [3], faintly α-continuous [7], faintly b-continuous [18], faintly β-continuous [3], faintly m-continuous[19], quasi continuous [3], faintly pre-θ-continuous [1],faintly semi-θ-continuous[1]) if for each point x ∈ X and each θ-open set V containing $f(x)$, there exists an open (resp. semi-open, preopen, α-open, b-open, β-open, mx-open, θ-open, pre-θ-open, semi-θ-open) set U containing x such that $f(U) \subset V$.

Remark 4. For modifications of open sets defined in Definitions (1), the following relationships are known:

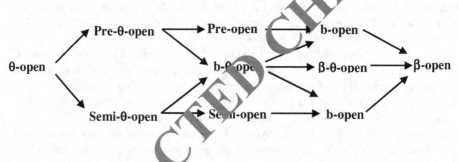

3   Characterizations of Faintly SP-θ(Semi-Pre-θ)-Continuous Functions

Definition 4. A function $f: (X, T) \rightarrow (Y, T^*)$ is said to be faintly b-θ-continuous functions (resp. faintly β-θ-Continuous functions) if for each point x ∈ X and each θ-open set V of Y containing $f(x)$, there exists $U \in B\theta\Sigma(X, x)$ (resp. $U \in \beta\theta\Sigma(X, x)$ such that $f(U) \subset V$.

Theorem 1. For a function $f:(X, T) \rightarrow (Y, T^*)$ the following statements are equivalent:

a) f is a faintly b-θ-continuous,
b) $f: (X, T) \rightarrow (Y, T^*_\theta)$ is b-θ-continuous,
c) $f^{-1}(V) \in B\theta\Sigma(X, x)$ for every $V \in T^*_\theta$,
d) $f^{-1}(F)$ is b-θ-closed in (X, T) for every θ- Closed subset F of (Y, T^*),
e) $bCl_\theta(f^{-1}(B)) \subseteq f^{-1}(Cl_\theta(B))$ for every subset B of Y,
f) $f^{-1}(Int_\theta(G)))) \subseteq bInt_\theta(f^{-1}(G))$.

Theorem 2. For a function $f: (X,T) \to (Y,T^*)$ the following statements are equivalent:

- a) f is a faintly β-θ-continuous,
- b) $f: (X,T) \to (Y,T^*_\theta)$ is β-θ-continuous,
- c) $f^{-1}(V) \in \beta\theta\Sigma(X, x)$ for every $V \in T^*_\theta$,
- d) $f^{-1}(F)$ is β-θ-closed in (X, T) for every θ- Closed subset F of (Y, T^*),
- e) $\beta Cl_\theta(f^{-1}(B)) \subseteq f^{-1}(Cl_\theta(B))$ for every Subset B of Y,
- f) $f^{-1}(Int_\theta(G))) \subseteq \beta Int_\theta(f^{-1}(G))$.

Proof. The proofs of Theorems (1) and (2) are straightforward and hence omitted.

Definition 5. A function $f: (X, T) \to (Y, T^*)$ is said to be strongly θ-continuous [20] (resp. b-θ-continuous [21], β-θ-continuous [22]), if $f^{-1}(V) \in (X, T_\theta)$ (resp. $B\theta\Sigma(X, T)$, $\beta\theta\Sigma(X, T)$) for every open set V of Y.

Remark 5. If (Y, T^*) is a regular space, we have $T^*=T^*_\theta$ and the next theorem follows immediately from the definition (5).

Theorem 3. Let Y be a regular space. Then a function $f: (X, T) \to (Y, T^*)$ is a b-θ-Continuous (resp. β-θ-continuous) if and only if it is faintly b-θ-continuous (resp. faintly β-θ-continuous).

Let $f: X \to Y$ be a function. A function $G: X \to X \times Y$ defined by $G(x) = (x, f(x))$ for every $x \in X$ is called the graph function of f.

Theorem 4. A function $f: (X, T) \to (Y, T^*)$ is faintly b-θ-continuous (resp. faintly β-θ-Continuous) if the graph function G: $X \to X \times Y$ is faintly b-θ-continuous (resp. faintly β-θ-Continuous).

Proof. Let $x \in X$ and V be a θ-open set of Y containing $f(x)$. Then $X \times V$ is θ-open in $X \times Y$ ([2, Theorem 5]) and contains G $(x) = (x, f(x))$. Therefore there exits $U \in B\theta\Sigma(X, x)$ (resp. U $\in \beta\theta\Sigma(X, x)$) such that G $(U) \subset X \times V$. This implies that $f(U) \subseteq V$. Thus f is faintly b-θ- continuous (resp. faintly β-θ-Continuous).

Comparison of Some Functions

In this section, we investigate the relationships among several weak forms of Continuity which are implied by weak continuity.

Theorem 5. A function $f: (X, T) \to (Y, T^*)$ is faintly b-θ-continuous (resp. Faintly β-θ-Continuous functions) if and only if $f^{-1}(V) \in B\theta\Sigma(X, T)$ (resp. $f^{-1}(V) \in \beta\theta\Sigma(X, T)$) for every θ-open set V of Y.

Proof. This proof is clear follows immediately from Definition (4).

Definition 6. A function $f : (X, T) \to (Y, T^*)$ is said to be:

a)   Weakly continuous [23] if $f^{-1}(V) \subset \text{Int}(f^{-1}(\text{Cl}(V)))$, for every $V \in T^*$.

b)   Almost weakly continuous [24] if $f^{-1}(V) \subset \text{Int}(\text{Cl}(f^{-1}(\text{Cl}(V))))$, for every $V \in T^*$.

Lemma 3. [12] Let A be a subset of a topological space X, then $x \in \text{bCl}(A)$ if and only if $A \cap U \neq \emptyset$ for every $U \in B\Sigma(X, T)$.

Lemma 4. [5] For a subset A of a topological space X, the following properties hold:

a)   If $A \in B\Sigma(X, T)$, then bcl (A) is b-regular and bcl (A) = bCl_θ (A).

b)   If A is b-θ-open, then A is the union of b-Regular sets.

From the definitions and Theorems (1) and (2) we have the following relationships:

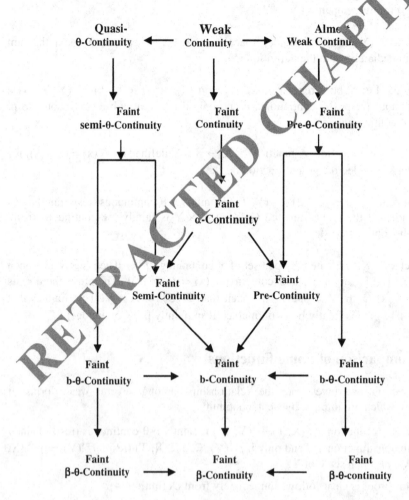

Lemma 5. [5] For a subset A of a topological space X, the following properties hold:
   a) A is b-regular if and only if A is b-θ-Closed and b-θ-open.
   b) If A is a b-open set, then bcl (A) is b-Regular.
   c) A is b-regular if and only if A = bInt(bCl(A)).

Lemma 6. [4] Let A be a subset of a topological space (X, T).
   (a) If A ∈ βΣ(X, T), then βCl(A) ∈ βR(X). (b) If A ∈ βR(X), then A ∈ βθΣ(X, T).

Theorem 6. For a function $f :(X, T) \rightarrow (Y, T^*)$ the following hold:
   a) Faintly b-θ- continuous implies faint b-Continuity.
   b) Faintly β-θ-Continuous implies faint β-Continuity.

Proof. The proof is clear follows directly from Lemma (5) and Lemma (6).

However the converses are not true in general by Examples 4.5 and 4.6 of [3], Example (2) of [2], Examples 4.2, 4.3, and 4.4 of [1], Examples 4.4 and 4.5 of [18], and the following examples.

Example 1. Let X = Y = {1, 2, 3, 4}. Define a topology T= {Ø, X, {1}, {2}, {1, 2}} on X and a topology T*= {Ø, Y, {4}, {3, 4}} on Y and let f: (X, T) → (Y, T*) be a function defined as follows: $f(1) = 4$, $f(2) = 3$, $f(3) = 2$, $f(4) =1$. Then f is faintly b-θ-Continuous (resp. β-θ-Continuous) but not faintly pre-θ-continuous because for a θ-open set {4} we have $f^{-1}(\{4\}) = \{1\}$ is b-θ-open (resp. β-θ-open) but not pre-θ-open.

Example 2. Let X = Y = {1, 2, 3, 4}. Define a topology T= {Ø, X, {1, 2}, {1, 2, 3}} on X and a topology T*= {Ø, Y, {1, 3, 4}, {2, 3, 4}, {3, 4}} on Y and let f: (X, T) → (Y, T*) be a function defined as follows: $f(1) = 3$, $f(2) = 4$, $f(3) =1$, $f(4) =2$. Then f is faintly b-θ-Continuous (resp. β-θ-Continuous) but not faintly semi-θ-continuous because for a θ-open set {3, 4} we have $f^{-1}(\{3, 4\}) = \{1, 2\}$ is b-θ-open (resp. β-θ-open) but not semi-θ-open.

Example 3. Let X = Y = {1, 2, 3}. Define a topology T= {Ø, X, {1}, {2}, {1, 2}, {1, 3}} on X and a topology T*= {Ø, Y, {1}, {2}, {1, 2}} on Y. Then the identity function f: X, T) → (Y, T*) is faintly b-Continuous (resp. faintly β- Continuous) but not faintly b-θ-continuous because for a θ-open set {1, 2} we have $f^{-1}(\{1, 2\}) \notin$ BθΣ(X, T).

Example 4. Let (X, T) and (Y, T*) be define as in Example (3). Then the identity function $f :(X, T) \rightarrow (Y, T^*)$ is faintly β-Continuous but not faintly β-θ-Continuous because for a θ-open set {1, 2} we have $f^{-1}(\{1, 2\}) \notin$ βθΣ(X, T).

Example 5. Let X = Y = {1, 2, 3, 4}. Define a topology T= {Ø, X, {1}, {2, 3}, {1, 2, 3}} on X and a topology T*= {Ø, Y, {1, 3, 4}, {2, 3, 4}, {3, 4}} on Y and let f: (X, T) → (Y, T*) be a function defined as follows: $f(1) = 2$, $f(2) = 3$, $f(3) = 1$, $f(4) = 4$. Then f is faintly β-θ-Continuous but not faintly b-θ-continuous because for a θ-open set {3, 4} we have $f^{-1}(\{3, 4\}) = \{2, 4\}$ is not b-θ-open in (X, T).

Park J.H. [5] has shown that if each subset A of a space X is b-regular, then $B\Sigma(X, T) = B\theta\Sigma(X, T)$. Also, T. Noiri [4] has shown that if each subset A of a space X is β-regular, then $\beta\Sigma(X, T) = \beta\theta\Sigma(X, T)$. Consequently we have the following result.

Theorem 7. Let every subset A in (X, T) is b-regular (resp. β-regular). Then a function $f: (X, T) \rightarrow (Y, T^*)$ is faintly b-θ- continuous (resp. Faintly β-θ-Continuous functions) if and only if faintly b-continuous (resp. faintly β-continuous).

Proof. The proof is straightforward. It follows from Lemma (3), Lemma (4) and Lemma (5) (resp. by Lemma (6)).

A space X is said to be submaximal if each dense subset of X is open in X and extremely disconnected (ED for short) if the closure of each open set of X is open in X.

Theorem 8. If (X, T) is submaximal ED, then the following are equivalent for a function $f: (X, T) \rightarrow (Y, T^*)$:
(a) f is faintly pre-θ-continuous; (b) f is faintly semi-θ-continuous;
(c) f is faintly b-θ- continuous; (d) f is faintly β-θ-continuous;
(e) f is faintly precontinuous; (f) f is faintly semi continuous;
(g) f is faintly α-continuous; (h) f is faintly b-continuous;
(i) f is faintly β-continuous; (j) f is faintly continuous.

Proof. This follows directly from the fact that if (X, T) is submaximal ED, then, T= $P\theta\Sigma(X, T) = S\theta\Sigma(X, T) = B\theta\Sigma(X, T) = \beta\theta\Sigma(X, T) = \alpha\Sigma(X, T) = P\Sigma(X, T) = S\Sigma(X, T) = B\Sigma(X, T) = \beta\Sigma(X, T)$.

Theorem 9. For a function $f: (X, T) \rightarrow (Y, T^*)$ the following properties hold:

  a) If every subset of X is b-open and closed, then each of faintly b-θ-continuous and faintly b-continuous are equivalent.
  b) If every subset A of X is β-open and closed, then each of faintly β-θ-continuous and faintly β-continuous is equivalent.
  c) If (X, T) is an indiscrete space, then faintly b-θ-continuous and faintly b-continuous are equivalent.
  d) If (X, T) is an indiscrete space, then faintly β-θ-continuous and faintly β-continuous are equivalent.

Proof. This proof follows from Theorem (4) in [25] and Lemma (2).

5   Some Basic Properties

In this section, using properties of $B\theta\Sigma(X, T)$ (resp. $\beta\theta\Sigma(X, T)$, we investigate Restrictions, compositions products for faintly b-θ-continuous (resp. faintly β-θ-Continuous) functions.

Theorem 10. If $f : (X, T) \to (Y, T^*)$ is faintly b-θ-continuous and $A \in B\theta\Sigma(X, T)$ then the restriction $f\mid_A : A \to Y$ is faintly b-θ-continuous.

Theorem 11. If $f : (X, T) \to (Y, T^*)$ is faintly β-θ-continuous and $A \in \beta\theta\Sigma(X, T)$ then the restriction $f\mid_A : A \to Y$ is faintly β-θ-continuous.

Proof. We prove only the Theorem (10). Since the order is shown similarly. Let V be any θ-open set of Y. By Theorem (1), we have $f^{-1}(V) \in B\theta\Sigma(X, T)$ and hence by definition of $B\theta\Sigma(X, T)$, $(f\mid_A)^{-1}(V) = f^{-1}(V) \cap A \in B\theta\Sigma(A)$. Therefore, it follows from Theorem (1) that $f\mid_A : A \to Y$ is faintly b-θ-continuous.

Theorem 12. If $f : (X, T) \to (Y, T^*)$ is faintly b-θ-continuous and $g : (Y, T^*) \to (Z, T^{**})$ is quasi θ-continuous, then $g \circ f$ is faintly b-θ-continuous.

Proof. Let A be any θ-open set in Z. Then $g^{-1}(A)$ is θ-open in Y and hence $(g \circ f)^{-1}(A) = f^{-1}(g^{-1}(A))$ is b-θ-open in X. Therefore $g \circ f$ is faintly b-θ-continuous.

Theorem 13. If $f : (X, T) \to (Y, T^*)$ is faintly β-θ-continuous and $g : (Y, T^*) \to (Z, T^{**})$ is quasi θ-continuous, then $g \circ f$ is faintly β-θ-continuous.

Proof. This proof is similar to that of Theorem (12) and is thus omitted.

From the implications between faintly continuous functions, we have the following Result.

Theorem 14. For functions $f : X \to Y$ and $g : Y \to Z$, the following hold:

   a) If f is faintly b-θ-continuous (resp. faintly β-θ-continuous) and g is strongly θ-continuous, then $g \circ f$ is faintly b-continuous (resp. faintly β-continuous)
   b) If f is b-continuous (resp. β-continuous) and g is faintly b-θ-continuous (resp. faintly β-θ-continuous), then $g \circ f$ is faintly b-continuous (resp. faintly β-continuous)
   c) If f quasi-b-irresolute (resp. quasi-β-irresolute) and g is faintly b-θ-Continuous (resp. faintly β-θ-continuous), then $g \circ f$ is faintly b-θ-continuous (resp. faintly β-θ-continuous)
   d) If f quasi-b-irresolute (resp. quasi-β-irresolute) and g is quasi θ-continuous.

Then $g \circ f$ is faintly b-θ-continuous (resp. faintly β-θ-continuous).

   Noiri and Popa [3] studied the product theorems for faintly semi-continuous (resp. faintly precontinuous, faintly β-continuous) functions. Also, Nasef [18] studied the product theorems for faintly α-continuous (resp. faintly γ-continuous) functions. Recently, A.A. El-Atik [1] studied the product theorems for faintly pre-θ-continuous (resp. faintly semi-θ-continuous) functions. In this paper we investigate the product theorems for faintly b-θ-continuous (resp. faintly β-θ-continuous) functions.

Let $\{X_\lambda: \lambda \in \Lambda\}$ and $\{Y_\lambda: \lambda \in \Lambda\}$ be two families of spaces with the same index set Λ. For each $\lambda \in \Lambda$, let $f_\lambda: X_\lambda \rightarrow Y_\lambda$ be a function. The Product space $\Pi\{X_\lambda: \lambda \in \Lambda\}$ will be denoted by $\Pi\{X_\lambda\}$ and the product function $\Pi f_\lambda: \Pi X_\lambda \rightarrow \Pi Y_\lambda$ is simply denoted by $f: \Pi X_\lambda \rightarrow \Pi Y_\lambda$.

Lemma 7. [26] Let $\{Y_\lambda: \lambda \in \Lambda\}$ be a family of spaces. Then $\Pi (Y_\lambda)_\theta \subset (\Pi Y_\lambda)_\theta$.

Since the class of $B\theta\Sigma(X, T)$ (resp. $\beta\theta\Sigma(X, T)$ is a topology on X, then the product of b-θ-open sets (resp. β-θ-open sets) is b-θ-open (resp. β-θ-open). Then we have the following result.

Theorem 15. Let $\{X_\lambda: \lambda \in \Lambda\}$ and $\{Y_\lambda: \lambda \in \Lambda\}$ be two a families of spaces with the same index set Λ. For each $\lambda \in \Lambda$, let $f_\lambda: X_\lambda \rightarrow Y_\lambda$ be a function. Then a function $f: \Pi X_\lambda \rightarrow \Pi Y_\lambda$ defined by $f(x_\lambda) = f_\lambda(x_\lambda)$ is b-θ-continuous (resp. β-θ-continuous) if and only if f_λ is b-θ-continuous (resp. β-θ-continuous) for each $\lambda \in \Lambda$.

Proof. We prove only the case of b-θ-continuity since the order is shown similarly. Let $V_\lambda \in T_{Y\lambda}$. Then by b-$\theta$-continuity of f_λ, we have $f^{-1}(\Pi V_\lambda) = \Pi f_\lambda^{-1}(V_\lambda) \in B\theta\Sigma(\Pi Y_\lambda)$. If $W \in T\Pi Y_\lambda$, then $W = \bigcup j \in \Lambda (\Pi V_{\lambda j})$ where $V_{\lambda j} \in B\theta\Sigma(\Pi Y_\lambda)$. Therefore $f^{-1}(W) = f^{-1}(\bigcup j \in \Lambda (\Pi V_{\lambda j})) = \bigcup j \in \Lambda \ f^{-1}(\Pi V_{\lambda j}) \in B\theta\Sigma(\Pi X_\lambda)$. Then f is b-θ-continuous.

Theorem 16. If a function $f: X \rightarrow \Pi Y_\lambda$ is faintly b-θ-continuous (resp. faintly β-θ-continuous) functions, then $P_\lambda f: X \rightarrow Y_\lambda$ is faintly b-θ-continuous (resp. faintly β-θ-continuous) for each $\lambda \in \Lambda$, where P_λ is the projection of ΠY_λ onto Y_λ.

Proof. Follows from the fact P_λ is continuous.

Theorem 17. If $f: \Pi X_\lambda \rightarrow \Pi Y_\lambda$ is faintly b-θ-continuous (resp. faintly β-θ-continuous functions, then $f_\lambda: X_\lambda \rightarrow Y_\lambda$ is faintly b-θ-continuous (resp. faintly β-θ-continuous) for each $\lambda \in \Lambda$.

Proof. Suppose that $f: \Pi X_\lambda \rightarrow \Pi Y_\lambda$ is faintly b-θ-continuous (resp. faintly β-θ-continuous). By Theorem (1) (resp. Theorem (2)) and Lemma (7), we have $f: \Pi X_\lambda \rightarrow (\Pi Y_\lambda)_\theta$ is b-θ-continuous (resp. β-θ-continuous). It follows from Theorem (15) that $f_\lambda: (X, T_\theta) \rightarrow (Y_\lambda)_\lambda$ is b-θ-continuous (resp. β-θ-continuous). Therefore by Theorem (1) (resp. Theorem (2)) $f_\lambda: X_\lambda \rightarrow Y_\lambda$ is faintly b-θ-continuous (resp. faintly semi β-θ-continuous).

Acknowledgments. The authors are thankful to the referees for giving the valuable comments to improve the paper.

References

1. El-Atik, A.A.: On Some Types of Faint Continuity. Thai Journal of Mathematics 9(1), 83–93 (2011)
2. Long, P.E., Herrington, L.L.: The T_θ-topology and Faintly Continuous Functions. Kyungpook Math. J. 22, 7–14 (1982)
3. Noiri, T., Popa, V.: Weak Forms of Faint Continuity. Bull. Math. Soc. Math. Roumanie. 34(82), 270–363 (1990)
4. Noiri, T.: Weak and Strong Forms of β-Irresolute Functions. Acta. Math. Hungar. 99, 315–328 (2003)
5. Park, J.H.: Strongly θ-b-Continuous Functions. Acta. Math. Hungar. 110(347–359 (2006)
6. Njástad: On Some Classes of Nearly Open Sets. Pacfic. J. Math. 15, 961– (1965)
7. Noiri, T., Popa, V.: Almost Weakly Continuous Multifunctions. Demonstra. Math. 26, 363–380 (1993)
8. Mashhour, A.S., Abd El-Monsef, M.E., El-Deeb, S.N.: On Precontinuous and Weak Precontinuous Mappings. Proc. Math. Phy. Soc. 53, 47–53 (1982)
9. Abd El-Monsef, M.E., El-Deeb, S.N., Mahmoud, R.A.: β-Open Sets and β-Continuous Mappings. Bull. Fac. Sci. Assiut. Univ. 12(1), 77–90 (1983)
10. Andrijevic, D.: Semi-Pre Open Sets. Mat. Vesnik. 38, 24–32 (1986)
11. El-Atik, A.A.: A Study of Some Types of Mappings on Topological Spaces. M. Sc. Thesis, Tanta Uni. Egypt (1997)
12. Andrijević, D.: On b-Open Sets. Mat. Bech. 48, 59–64 (1996)
13. Dontchev, J., Przemski, M.: On the Various Decompositions of Continuous and Some Weakly Continuous Functions. Acta. Math. Hungar. 71(1-2), 109–120 (1996)
14. Crossley, S.G., Hildebrand, S.K.: Semi-Closure. Texas. J. Sci. 22, 99–112 (1971)
15. El-Deeb, S.N., Hasanein, I.A., Mashhour, A.S., Noiri, T.: On p-Regulars Spaces. Bull. Math. Soc. Math. R. S. Roumanie. 27(75), 311–315 (1983)
16. Levine, N.: Semi-Open Sets and Semi-Continuity in Topological Spaces. Amer. Math. Monthly 70, 36–41 (1963)
17. Jafari, S., Noiri, T.: On Faintly α-Continuous Functions. Indian. J. Math. 42, 203–210 (2000)
18. Nasef, A.A.: Another Weak Forms of Faint Continuity. Chaos, Solitons & Fractals 12, 2219–2225 (2001)
19. Noiri, T., Popa, V.: Faintly m-Continuous Functions. Chaos, Solitons & Fractals 19, 1147–1159 (2004)
20. Noiri, T.: On δ-Continuous Functions. J. Korean. Math. Soc. 16, 161–166 (1980)
21. Zorlutuna, İ.: On b-Closed Space and θ-b-Continuous Functions. The Arabian Journal for Science and Engineering 34(2A), 205–216 (2009)
22. Noiri, T., Popa, V.: Strongly θ-β-Continuous Functions. J. Pure Math. 19, 31–39 (2002)
23. Levine, N.: A Decomposition of Continuity in Topological Spaces. Amer. Math. Monthly 63, 44–66 (1961)
24. Janković, D.: θ-Regular Spaces. Int. J. Math. Sci. 8, 615–624 (1985)
25. Rajesh, N., Salleh, Z.: Some More Results on b-θ-Open Sets. Buletinul Academiei De Stiinte A Reublicii Moldova. Matematica. 3(61), 70–80 (2009)
26. Rose, D.A.: Weak Continuity and Strongly Closed Sets. Int. J. Math. Sci. 7(4), 809–825 (1984)

RETRACTED CHAPTER: On e-Closed Spaces and θ-e-Continuous Functions

Alaa. M.F. AL. Jumaili and Xiao-Song Yang

Department of Mathematics,
Huazhong University of Science and Technology
Wuhan, 430074, China
alaa_mf1970@yahoo.com

Abstract. In this paper we introduce several characterizations of e-closed spaces which were introduced by Murad ÄOzkoc and GÄulhan Aslim [1]. We study and investigate a new class of continuous functions called θ-e-continuous functions which Transform e-closed spaces to quasi H-closed spaces and introduce new notions of regularity axioms by using notion of e-open sets and investigate the relationships between θ-e-continuous functions and separation axioms. Also we investigate strongly θ-e-closedness of graphs of functions.

Keywords: e-Open Set, e-θ-Closed Set, e-Regularity, e-Closed Space, θ-e-Continuity.

1 Introduction

It is well known that various types of functions play a significant role in the theory of classical point set topology. Also Generalized open sets play a very important role in General Topology and they are now the research topics of many topologists worldwide. Indeed a significant theme in General Topology and Real Analysis concerns the variously modified forms of Continuity, separation axioms, etc. In 1969, Porter and Thomas [2] introduced the class of quasi-H-closed spaces. In 1976, Thompson [3] introduced the class of s-closed spaces which is contained in the class of quasi-H-closed spaces. In 1987, Di Maio and Noiri [4] and in 1989 Abo-Khadra [5] introduced s-closed and p-closed spaces, respectively. On the other hand, the concepts of an e-open set and e-continuous functions are introduced by Erdal Ekici [6]. Recently, Murad ÄOzkoc and GÄulhan Aslim [1] developed the notions of e-θ-open sets, strongly θ-e-continuous functions and e-closed spaces. In this paper we introduce several characterizations of e-closed spaces including characterizations using nets and filter bases. These are parallel to characterizations of other generalizations of compactness such as s-closed, p-closed, s-closed and f-closed spaces in [7], [8], [4] and [9]. Also we introduce and investigate a new class of continuous functions, called θ-e-continuous functions, which contains the class of strongly θ-e-continuous functions. Moreover, we introduce and characterize new regularity axioms by using e-open sets and investigate the relationships between θ-e-continuous functions and separation axioms and the relationships among these notions and other well-known functions are also given. Finally, we investigate the graphs of θ-e-continuous functions.

P. Balasubramaniam and R. Uthayakumar (Eds.): ICMMSC 2012, CCIS 283, pp. 32–46, 2012.
© Springer-Verlag Berlin Heidelberg 2012

2   Preliminaries

Throughout this paper, (X, T) and (Y, T*) (or simply X and Y) mean topological spaces on which no separation axioms are assumed unless explicitly stated. For any subset A of X, the closure and interior of A are denoted by Cl(A) and Int(A), respectively. We recall the following definitions, which will be used often throughout this paper.

A point $x \in X$ is called a θ-cluster (resp. δ-cluster) point of A if Cl(V) ∩ A ≠ Ø (resp. Int(Cl(V)) ∩ A ≠ Ø) for every open subset V of X containing x. The set of all θ-cluster (resp. δ-cluster) points of A is called the θ-closure (resp. δ-closure) of A [10] and is denoted by $Cl_\theta(A)$ (resp. $Cl_\delta(A)$). If $A = Cl_\theta(A)$ (resp. $A = Cl_\delta(A)$), then A is said to be θ-closed (resp. δ-closed). The complement of a θ-closed (resp. δ-closed) set is said to be θ-open (resp. δ-open) set.

A subset A of a space X is called e-open [6] if $A \subset Cl(Int_\delta(A)) \cup Int(Cl_\delta(A))$, the complement of a e-open set is called e-closed. The intersection of all e-closed sets containing A is called the e-closure of A [6] and is denoted by e-Cl(A). The union of all e-open sets of X contained in A is called the e-interior [6] of A and is denoted by e-Int(A). A subset A is said to be regular open (resp. regular closed) if A = Int(Cl(A)) (resp. A=Cl(Int(A))). A subset A of a topological space X is e-regular [1] if it is e-open and e-closed.

A point x of X is called an e-θ-cluster point of A if e-Cl(V) ∩ A ≠ Ø for every e-open set V containing x. The set of all e-θ-cluster points of A is called e-θ-closure [1] of A and is denoted by e-Clθ(A). A subset A is said to be e-θ-closed if A = e-Clθ(A). The complement of an e-θ-closed set is said to be e-θ-open. The family of all regular open (resp. e-open, e-closed, e-regular, e-θ-open) subsets of X containing a point x ∈ X is denoted by RO(X, x) (resp. EΣ(X, x), EC(X, x), ER(X, x), EθΣ(X, x)). The family of all regular open (resp. e-open, e-closed, e-regular, e-θ-open) sets in X are denoted by RO(X, T) (resp. EΣ(X, T), EC(X, T), ER(X, T), EθΣ(X, T).

A subset A of X is called semiopen [11] (resp. preopen [12], if $A \subset Cl(Int(A))$ (resp. $A \subset Int(Cl(A))$) and the complement of a semiopen (resp. preopen) set are called semiclosed (resp. preclosed).

Remark 1. It is obvious that e-regular ⇒ e-θ-open ⇒ e-open. But the Converses are not necessarily true as shown by the examples in [1].

The following basic facts will be needed in the sequel.

Lemma 1. For an open set U, $e\text{-}Cl_\theta(U) \subseteq Cl(U)$.

Proof. Clear.

Lemma 2. [1] Let A and A_α ($\alpha \in \Lambda$) be any subsets of a space X. Then the following properties hold:

a) $A \in E\Sigma(X, T)$ if and only if e-Cl(A) $\in$ ER(X, T);

b)   If A_α is e-θ-open in X for each $\alpha \in \Lambda$, then $\bigcup_{\alpha \in \Lambda} A_\alpha$ is e-θ-open in X.

c) A is e-θ-open in X if and only if for each x∈ A, there exists H∈ ER(X, x), such that x ∈ H ⊂ A.

Lemma 3. [1] Let A be any subset of a space X. A is e-θ-open in X if and only if for each x ∈ A, there exists U∈ EΣ(X, T) containing x such that x ∈ U⊂ e-Cl(U)⊂A.

Lemma 4. Let X be a topological space and let A⊂X. Then,

a) e-Cl$_\theta$(X \ A) =X \ (e-Int$_\theta$(A)).
b) e-Int$_\theta$(X \ A) =X \ (e-Cl$_\theta$(A)).

Proof. Obvious.

3   e-Closed Spaces

Definition 1. A space X is called e-closed [1] if every cover of X by e-open sets has a finite subfamily whose e-closures cover X.

Definition 2. Let £ be a filter base on a topological space X and x_0 be a point of X, then:

a) £ is said to be e-θ-accumulate at x_0 if e-Cl(V) ∩ M ≠ Ø for every V ∈ EΣ(X,x) and every M∈ £.
b) £ is said to be e-θ-converge to x, if for each V ∈ EΣ(X, x), there exists M∈ £ Such that M⊂ e-Cl(V).

Theorem 1. For a topological space (X, T), the following properties are equivalent:

a) (X, T) is e-closed;
b) Every maximal filter base e-θ-converges to some point of X;
c) Every filter base e-θ-accumulates at some point of X;
d) For every family {V_λ: λ∈ Δ} of e-closed subsets of X such that ∩ {V_λ: λ ∈ Δ}= Ø, there exists a finite subset Δ_0 of Δ such that ∩ {e-Int (V_λ): λ ∈ Δ_0}= Ø.

Proof. (a) ⇒ (b): Let £ be a maximal filter base on X. Suppose that £ does not e-θ-converge to any point of X. Since £ is maximal, then £ does not e-θ-accumulate at any point of X. For each x ∈ X, there exist M_x ∈ £ and V_x ∈ EΣ(X, x) such that e-Cl(V_x) ∩ M_x = Ø. The family {V_x: x∈ X} is a cover of X by e-open sets of X. By (a), there exists a finite number of points $x_1, x_2, ..., x_n$ of X such that X =∪ {e-Cl(V_{xi}): i=1,2,...,n}. Since £ is a filter base on X, there exists M_0 ∈ £ such that M_0⊂ ∩ {M_{xi}: i=1, 2,..., n }.Therefore, we obtain M_0 = Ø .This is a contradiction.

(b) ⇒ (c): Let £ be any filter base on X. Then there exists a maximal filter base $£_0$ such that £⊂ $£_0$. By (b), $£_0$ e-θ-converges to some point x ∈ X. Then for every M∈ £ and every V∈ EΣ(X, x), there exists M_0 ∈ $£_0$ such that M_0 ⊂ e-Cl(V); hence M ∩ e-Cl(V) ⊃ M ∩ M_0≠ Ø. This shows that M is e-θ-accumulates at x.

(c) $\Rightarrow$ (d): Let $\{V_\lambda: \lambda \in \Delta\}$ be any family of e-closed subsets of X such that $\cap\{V_\lambda: \lambda \in \Delta\} = \emptyset$. Let $F(\Delta)$ denote the ideal of all finite subsets of Δ. Assume that $\cap\{$e-Int$(V_\lambda): \lambda \in \eta\} \neq \emptyset$ for every $\eta \in F(\Delta)$. Then the family $£ = \{\cap_{\lambda \in \eta}$ e-Int$(V_\lambda): \eta \in F(\Delta)\}$ is a filter base on X. By (c), $£$ e-θ-accumulates at some point $x \in X$. Since $\{X \setminus V_\lambda: \lambda \in \Delta\}$ is a cover of X, $x \in X \setminus V_{\lambda_0}$ for some $\lambda_0 \in \Delta$. Therefore, we obtain $X \setminus V_{\lambda_0} \in E\Sigma(X, x)$, e-Int$(V_{\lambda_0}) \in £$ and e-Cl$(X \setminus V_{\lambda_0}) \cap$ e-Int$(V_{\lambda_0}) = \emptyset$. This is a contradiction.

(d) $\Rightarrow$ (a): Let $\{V_\lambda: \lambda \in \Delta\}$ be a cover of X by e-open sets of X. Then $\{X \setminus V_\lambda: \lambda \in \Delta\}$ is a family of e-closed subsets of X such that $\cap\{X \setminus V_\lambda: \lambda \in \Delta\} = \emptyset$. By (d), there exists a finite subset Δ_0 of Δ such that $\cap \{$e-Int$(X \setminus V_\lambda): \lambda \in \Delta_0\} = \emptyset$; hence $X = \{$e-Cl $(V_\lambda): \lambda \in \Delta_0\}$. This shows that X is e-closed.

Definition 3. Let (x_i) be a net on a topological space X and x_0 be a point of X.

a) (x_i) is said to e-θ-accumulate at x_0 if for every $V \in E\Sigma(X, x)$ and every $i_0 \in I$, There is $i \in I$ such that $i \geq i_0$ and $x_i \in$ e-Cl(V).
b) (x_i) is said to e-θ-converge to x_0 if for every e-open set V containing x, there Exists i_0 such that $x_i \in$ e-Cl (V) for all $i \geq i_0$.

Theorem 2. For a topological space (X, T), the following properties are equivalent:

a) (X, T) is e-closed;
b) Every net of X has a e-θ-accumulation point;
c) Every maximal net of X is e-θ-converges.

Proof. (a) $\Rightarrow$ (b): Let (x_i) be a net of a e-closed space X. Then by (c) of Theorem (1) the derived filter from (x_i) has an e-θ-accumulation point which is, as can be clearly seen, an e-θ-accumulation point of (x_i), too.

(b) $\Rightarrow$ (c): Let (x_i) be a maximal net of X. By (a), (x_i) has an e-θ-accumulation point y_0. Then for each open set V containing y_0, (x_i) is residual either in e-Cl(V) or e-Cl$(X \setminus V) = X \setminus V$. But, since some terms of (x_i) are contained in e-Cl(V), it can not be residual e-Cl$(X \setminus V)$. Thus, (x_i) is residual in e-Cl(V), proving (x_i) is e-θ-converges to y_0.

(c) $\Rightarrow$ (a): By using (b) of Theorem (1), suppose $£$ is a maximal filter base on X and (x_i) is a derived net from $£$. Then (x_i) is maximal, hence by (c), (x_i) is e-θ-converges to y_0 for some point $y_0 \in X$. Now, clear that a point y_0 be an e-θ-accumulation point of $£$. But, since $£$ is maximal, then $£$ is e-θ-converges to y_0. Thus by Theorem (1), X is e-closed.

Theorem 3. A topological space X is e-closed if and only if every family of e-θ-closed subsets of X with the finite intersection property has a nonempty intersection.

Proof. Let $\{\Gamma_\lambda: \lambda \in \Delta\}$ be any family of e-θ-closed subsets of X with the finite intersection property. Suppose that $\cap \{\Gamma_\lambda: \lambda \in \Delta\} = \emptyset$. Then $\{X \setminus \Gamma_\lambda: \lambda \in \Delta\}$ is a cover of X by e-θ-open sets of X and for each $x \in X$, there exists a $\lambda(x) \in \Delta$ such that $x \in X \setminus \Gamma_{\lambda(x)}$. Since $X \setminus \Gamma_\lambda$ is e-θ-open, there exists a e-open set $V_{\lambda(x)}$ such that $x \in V_{\lambda(x)}$

$\subset$ e-Cl($V_{\lambda(x)}$) $\subset$ X \ $\Gamma_{\lambda(x)}$. The family $\{V_{\lambda(x)}: x \in X\}$ is a cover of X by e-open sets. By e-closedness of X, there exists a finite number of points $x_1, x_2, \ldots, x_n$ of X such that X=$\bigcup \{$e-Cl($V_{\lambda(xi)}$):i=1,2,...,n$\}$; hence X=$\bigcup \{$X \ $\Gamma_{\lambda(xi)}$ i=1,2,...,n$\}$. Therefore, we have $\cap \{\Gamma_{\lambda(xi):}$ i = 1, 2, ... , n$\}$ = $\emptyset$. This is a contradiction.

Conversely, let $\{V_\lambda: \lambda \in \Delta\}$ be a cover of X by e-open subsets of X. Suppose that X is not e-closed (i.e.) X $\neq \bigcup \{$e-Cl(V_λ): $\lambda \in \eta\}$ for every $\eta \in$ F(Δ) where F(Δ) denotes the ideal of all finite subsets of Δ. Then $\cap\{$X \ e-Cl(V_λ): $\lambda \in \eta\} \neq \emptyset$; hence the family $\{$X \e-Cl(V_λ): $\lambda \in \Delta\}$ has the finite intersection property and X\ e-Cl(V_λ) is an e-regular and so e-θ-closed subset of X for each $\lambda \in \Delta$. By hypothesis, $\cap\{$X \ e-Cl(V_λ): $\lambda \in \Delta\} \neq \emptyset$. Therefore, we have $\bigcup \{$e-Cl(V_λ): $\lambda \in \Delta\} \neq$ X. This is a contradiction to the fact that $\{V_\lambda: \lambda \in \Delta\}$ is a cover of X.

As an application of Theorem (3) we offer a "fixed set theorem" for multifunctions on e-closed spaces as analogous to Theorem (4) in [7].

Theorem 4. Let X be an e-closed space and let Φ: X$\rightarrow$X be a multifunction with the property that Φ (M) is e-θ-closed when M is e-θ-closed. Then there is a non empty e-θ-closed set H $\subset$ X with Φ (H) =H.

Proof. Let $\Omega = \{$M $\subset$ X: M $\neq \emptyset$, Φ (M) $\subset$ M and M is e-θ-closed in X$\}$ and apply Zorn's Lemma to choose a minimal element H of Ω under set inclusion. It follows that Φ (H) =H and the proof is complete.

Definition 4. A subset A of a topological space (X, T) is said to be e-closed relative to X if for every cover $\{$ $V_\lambda: \lambda \in \Delta$ $\}$ of A by e-open sets of X, there exists a finite subset Δ_0 of Δ such that A $\subset \bigcup \{$ e-Cl(V_λ): $\lambda \in \Delta_0$ $\}$.

Theorem 5. For a topological space (X, T), the following properties are equivalent:

   a) A is e-closed relative to X;
   b) Every maximal filter base on X which meets A e-θ-converges to some point of A;
   c) Every filter base on X which meets A e-θ-accumulates to some point of A;
   d) For every family $\{$(V_λ): $\lambda \in \Delta$ $\}$ of e-closed subsets of (X, T) such that $[\cap \{($V_\lambda$): $\lambda \in \Delta\}] \cap$A = $\emptyset$, there exists a finite subset Δ_0 of Δ such that $[\cap \{$e-Int(V_λ): $\lambda \in \Delta_0\}] \cap$A = $\emptyset$.

Proof. The proof is similar to that of Theorem (3) and hence is omitted.

A topological space (X, T) is called quasi-H-closed [2] (resp. s-closed [4], p-closed [5]) if every open (resp. semi-open, preopen) cover of X has a finite subfamily whose closures (resp. semi-closures, preclosures) cover X. It is clear that every e-closed space is s-closed, p-closed and quasi-H-closed.

Recall that, a topological space (X, T) is said to be T_e-space [13] if every e-open subset of X is open.

4   Characterizations θ-e-Continuous Functions

In this section, we obtain some characterizations of θ-e-Continuous functions.

Definition 5. A function $f: (X, T) \rightarrow (Y, T^*)$ is said to be θ-e-Continuous (briefly, θ-e-c.) if for each $x \in X$ and each open set V containing $f(x)$, there exists $U \in E\Sigma(X, x)$, such that $f(\text{e-Cl}(U)) \subset \text{Cl}(V)$.

Theorem 6. For a function $f: (X, T) \rightarrow (Y, T^*)$ the following statements are equivalent:

a) f is θ-e-Continuous;

b)   for every $x \in X$ and every open set V containing $f(x)$, there exists an e-regular set U in X containing x such that $f(U) \subset \text{Cl}(V)$;

c) $f(\text{e-Cl}_\theta(A)) \subset \text{Cl}_\theta(f(A))$ for every subset A of X;

d) $\text{e-Cl}_\theta(f^{-1}(B)) \subset f^{-1}(\text{Cl}_\theta(B))$ for every subset B of Y;

e) $f^{-1}(\text{Int}_\theta(B)) \subset \text{e-Int}_\theta(f^{-1}(B))$ for every subset B of Y;

f) $\text{e-Cl}_\theta(f^{-1}(V)) \subset f^{-1}(\text{Cl}(V))$ for every open subset V of Y;

g) $f^{-1}(V) \subset \text{e-Int}_\theta(f^{-1}(\text{Cl}(V)))$ for every open subset V of Y;

h) $\text{e-Cl}_\theta(f^{-1}(\text{Int}(M)) \subset f^{-1}(M)$ for every closed subset M of Y;

i) $\text{e-Cl}_\theta(f^{-1}(\text{Int}(\text{Cl}(V)))) \subset f^{-1}(\text{Cl}(V))$ for every open subset V of Y;

j) $\text{e-Cl}_\theta(f^{-1}(\text{Int}(\text{Cl}(B)))) \subset f^{-1}(\text{Cl}(B))$ for every subset B of Y;

k) $\text{e-Cl}_\theta(f^{-1}(V)) \subset f^{-1}(\text{Cl}(V))$ for every preopen subset V of Y;

l) $f^{-1}(\text{Int}(B)) \subset \text{e-Int}_\theta(f^{-1}(\text{Cl}(\text{Int}(B))))$ for every subset B of Y;

m) $f^{-1}(\text{Int}(M)) \subset \text{e-Int}_\theta(f^{-1}(M))$ for every preclosed subset M of Y.

Proof. The proofs of the above results are straightforward and are omitted.

Theorem 7. If $f: (X, T) \rightarrow (Y, T^*)$ is a θ-e-continuous function, then the following are hold:

a) $f^{-1}(V)$ is e-θ-open for every θ-open subset V of Y.

b) $f^{-1}(V)$ e-θ-closed for every θ-closed subset V of Y.

Proof. Let H be a θ-closed subset of Y. By (d) of Theorem (6), we have $\text{e-Cl}_\theta(f^{-1}(H)) \subset f^{-1}(H)$. Hence $f^{-1}(H)$ is e-θ-closed. It is obvious that (a) and (b) are equivalent.

Remark 3. The converse of Theorem (7) is not true in general as the following example shows.

Example 1. Let $X = \{1, 2, 3\}$. Define a topology $T = \{ \emptyset, X, \{1\}, \{2\}, \{1, 2\}\}$ and $T^* = \{ \emptyset, X, \{1\}, \{3\}, \{1, 3\}\}$ on X and let $f: (X, T) \rightarrow (X, T^*)$ be a function defined as follows: $f(1) = f(2) = 3$, $f(3) = 1$. Then f is satisfies (a) of Theorem (7) but is not θ-e-Continuous.

5   Some Basic Properties of θ-e-Continuous Functions

In this section, we obtain some properties of θ-e-Continuous functions.

Theorem 8. If $f: X \to Y$ is θ-e-continuous function and $g: Y \to Z$ is continuous function, then $g \circ f: X \to Z$ is θ-e-continuous function.

Proof. Let V be open subset of Z. Then, e-$Cl_\theta((g \circ f)^{-1}(V))$ = e-$Cl_\theta(f^{-1}(g^{-1}(V))) \subset f^{-1}(Cl(g^{-1}(V))) \subset f^{-1}(g^{-1}(Cl(V)))=(g \circ f)^{-1}(Cl(V))$, which proves that $g \circ f$ is θ-e-continuous.

Theorem 9. Let $f: X \to Y$ and $g: Y \to Z$ be functions. If $g \circ f$ is θ-e-continuous and f is surjective and satisfies the condition that f (e-$Cl_\theta(A)$) is e-θ-closed for each $A \subset X$, then g is θ-e-continuous.

Proof. Let V be open subset of Z. Since $g \circ f$ is θ-e-continuous. it follows by Theorem (6) that e-$Cl_\theta((g \circ f)^{-1}(V)) \subset (g \circ f)^{-1}(Cl(V)) = f^{-1}(g^{-1}(Cl(V)))$ and hence f (e-$Cl_\theta(f^{-1}(g^{-1}(V))))$ $\subset g^{-1}(Cl(V))$. By hypothesis, e-$Cl_\theta(f (f^{-1}(g^{-1}(V)))) \subset f$(e-$Cl_\theta(f^{-1}(g(V)))) \subset g^{-1}(Cl(V))$ and so e-$Cl_\theta(g^{-1}(V)) \subset g^{-1}(Cl(V))$, which proves that g is θ-e-continuous.

Theorem 10. Let $f: X \to Y$ and $g: Y \to Z$ be functions. If $g \circ f$ is θ-e-continuous and g is a clopen injection, then f is θ-e-continuous.

Proof. Let V be open subset of Z. Since g is clopen, g (V) is an open subset of Z. Since $g \circ f$ is θ-e-continuous, then by Theorem (6), e-$Cl_\theta((g \circ f)^{-1}(g(V))) \subset (g \circ f)^{-1}(Cl(g(V))) = f^{-1}(g^{-1}(Cl(g(V))))$. Furthermore, since g is closed and injective, e-$Cl_\theta(f^{-1}(V))$ = e-$Cl_\theta(f^{-1}(g^{-1}(g(V)))) \subset f^{-1}(g^{-1}(Cl(g(V)))) \subset f^{-1}(Cl(g^{-1}(g(V)))) = f^{-1}(Cl(V))$. This shows that f is θ-e-continuous.

Theorem 11. Let $f: X \to Y$ be a θ-e-continuous surjection. If X is e-closed, then Y is quasi-H-closed.

Proof. Let $\{V_\lambda :\}\lambda \in I$ be an open cover of Y. If $x \in X$, then $f(x) \in V_{\lambda(x)}$ for some λ $(x) \in I$. Since f is θ-e-continuous, there exists a e-open set U_x such that f (e-$Cl(U_x)$) $\subset Cl(V_{\lambda(x)})$. Thus $\{U_x\}_{x \in X}$ is a e-open cover of X. Since X is e-closed, there exist $x_1, x_2, ... , x_n \in X$ such that $X \subset \bigcup$ {e-$Cl(U_{xi})$:$x_i \in X$, i=1,2,...,n}. So we obtain $f(X) \subset f(\bigcup$ {e-$Cl(U_{xi})$:$x_i \in X$, i=1,2,...,n}) $\subset \bigcup$ {$Cl(V_{\lambda(xi)}$):$x_i \in X$, i=1,2,...,n}. Since f is surjective, then Y is quasi-H-closed.

6   Separation Axioms

In this section, we introduce new characterizations of e-regularity and investigate the relationships between θ-e-continuous functions and separation axioms.

Definition 6. A space X is called e-regular[1] if for each closed set $M \subset X$ and each point $x \in X \backslash M$, there exist disjoints e-open sets U and V such that $x \in U$ and $M \subset V$.

Theorem 12. The following are equivalent for a space X:

a) X is e-regular;

b) For each $x \in X$ and for each open set U of X containing x, there exists $V \in ER(X, T)$, such that $x \in V \subset U$;

c) For each $x \in X$ and for each open set U of X containing x, there exists $V \in E\theta\Sigma(X, T)$, such that $x \in V \subset U$;

d) Every open set is e-θ-open;

e) Every closed set is e-θ-closed;

f) Every closed set is the intersection of e-regular sets;

g) Every open set is the union of e-regular sets;

h) $e\text{-}Cl_\theta(A) \subset Cl(A)$ for every subset A of X.

Proof. The proof of (a) $\Rightarrow$ (b) $\Rightarrow$ (c) $\Rightarrow$ (d) $\Rightarrow$ (e) $\Rightarrow$ (f) $\Rightarrow$ (g) are obivious thus omitted.

(g) $\Rightarrow$ (h): Let A be any subset of X. Suppose that $x \notin Cl(A)$. Then there exists an open set U containing x such that $U \cap A = \emptyset$. By (g), there exists $F_x \in ER(X, T)$, such that $x \in F_x \subset U$. Therefore, we have $F_x \cap A = \emptyset$. This shows that $x \notin e\text{-}Cl_\theta(A)$. Thus $e\text{-}Cl_\theta(A) \subset Cl(A)$.

(h) $\Rightarrow$ (a): Let M be any closed set and $x \notin M$. Then $x \notin Cl(M)$ and $e\text{-}Cl_\theta(M) \subset Cl(M)$ and there exists $U \in E\Sigma(X, x)$, such that $e\text{-}Cl(U) \cap M = \emptyset$. Therefore, we have disjoint e-open sets U and $X \backslash e\text{-}Cl(U)$ such that $x \in U$ and $M \subset X \backslash e\text{-}Cl(U)$. This shows that X is e-regular.

Theorem 13. The following are equivalent for a space X:

a) X is e-regular;

b) For any closed set M and any point $x \notin M$, there exist disjoint e-regular sets U and V such that $x \in U$ and $M \subset V$;

c) For any closed set M and any point $x \notin M$, there exist disjoint e-θ-open sets U and V such that $x \in U$ and $M \subset V$.

Proof. The proof follows immediately from Lemma (2) [1].

Definition 7. A space X is called almost e-regular (resp. strongly e-regular) if for each regular closed (resp. e-closed) set $M \subset X$ and any point $x \in X \backslash M$, there exists disjoint e-open sets U and V such that $x \in U$ and $M \subset V$.

Theorem 14. The following are equivalent for a space X:

a) X is almost e-regular (resp. strongly e-regular);

b) For each $x \in X$ and for each $U \in RO(X, x)$ (resp. $U \in E\Sigma(X, x)$), there exists $V \in ER(X, T)$ such that $x \in V \subset U$;

c) For each $x \in X$ and for each $U \in RO(X, x)$ (resp. $U \in E\Sigma(X, x)$), there exists $V \in E\theta\Sigma(X, T)$, such that $x \in V \subset U$;

d)   Every regular open (resp. e-open) set is e-θ-open;
e)   Every regular closed (resp. e-closed) set is e-θ-closed;
f)   Every regular closed (resp. e-closed) set is the intersection of e-regular sets;
g)   Every regular open (resp. e-open) set is the union of e-regular sets;
h)   e-Cl_θ(A) $\subset Cl_\delta$(A)(resp. e-Cl_θ(A) $\subset$ e-Cl(A)) for every subset A of X.

Proof. We prove only the case of almost e-regularity. The proof of the other parts follows analogously.

(a) $\Rightarrow$ (b) $\Rightarrow$ (c) $\Rightarrow$ (d) $\Rightarrow$ (e) $\Rightarrow$ (f) $\Rightarrow$ (g) are clear.

(g) $\Rightarrow$ (h): Let A be any subset of X. Suppose that x $\notin$ Cl_δ(A). Then there exists a regular open set U containing x such that U $\cap$ A = Ø. By (g), there exists $F_x \in$ RC(X, T), such that x $\in$ $F_x \subset$ U. Therefore, we have $F_x \cap$ A = Ø. This shows that x $\notin$ e-Cl_θ(A). Thus e-Cl_θ(A) $\subset$ Cl_δ(A).

(h) $\Rightarrow$ (a): Let M be any regular closed set and x $\notin$ M. Then x $\notin$ Cl_δ(M) and e-Cl_θ(M) $\subset$ Cl_δ(M) and there exists U $\in$ EΣ(X, x), such that e-Cl(U) $\cap$ M = Ø. Therefore, we have disjoint e-open sets U and X\e-Cl(U) such that x $\in$ U and M $\subset$ X\e-Cl(U). This shows that X is almost e-regular.

Theorem 15. The following are equivalent for a space X:

a)   X is almost e-regular (resp. strongly e-regular);
b)   For any regular closed (resp. e-closed) set M and any point x $\notin$ M, there exists disjoint e-regular sets U and V such that x $\in$ U and M $\subset$ V;
c)   For any regular closed (resp. e-closed) set M and any point x $\notin$ M, there exists disjoint e-θ-open sets U and V such that x $\in$ U and M $\subset$ V.

Proof. The proof follows immediately from Lemma (2) [1].

Corollary 1. In a T_e-space e-regularity, strong e-regularity and regularity are equivalent.

A space X is called almost regular [14] if for any regular closed set M $\subset$ X and any point x $\in$ X\M there exists disjoint open sets U and V such that x $\in$ U and M $\subset$ V.

Theorem 16. Let X be a T_e-space, X is almost e-regular if and only if X is almost regular.

Proof. Clear.

Definition 8. A space X is said to be e-T_2 [1] if for each pair of distinct points x and y in X; there exists U $\in$ EΣ(X, x) and V $\in$ EΣ(X, y) such that U $\cap$ V = Ø.

Theorem 17. A space X is e-T_2 if and only if every singleton set is e-θ-closed.

Proof. Let x $\in$ X. For each x $\neq$ y, there exist disjoint e-open sets U_x and V_y containing x and y, respectively. Since e-Cl(V_y) is a e-regular set and does not contain x, we have {x} = $\cap_{x \neq y}$(X\e-Cl(V_y)) is e-θ-closed by Lemma 2.2 [3].

(Conversely) Let $x \neq y$. By hypothesis, there exist a e-regular set U containing x but not y. Clearly, U and X\U are the required disjoint e-open sets.

We recall that a space X is called Urysohn for each pair of points x_1, $x_2 \in X$ where $x_1 \neq x_2$, there exist open sets U_1 and U_2 containing x_1 and x_2, respectively, such that $Cl(U_1) \cap Cl(U_2) = \emptyset$.

Theorem 18. If f: $X \rightarrow Y$ is a θ-e-continuous, injection and Y is Urysohn, then X is e-T_2.

Proof. Let x_1, $x_2 \in X$ and $x_1 \neq x_2$. Since f is injective and since Y is Urysohn, then $f(x_1) \neq f(x_2)$ and there exist open sets V_1 and V_2 containing $f(x_1)$ and $f(x_2)$ respectively, such that $Cl(V_1) \cap Cl(V_2) = \emptyset$. Since f is θ-e-continuous, there exist e-open sets U_1 and U_2 such that $x_1 \in U_1$ and $x_2 \in U_2$ and $f(\text{e-}Cl(U_1)) \subset Cl(V_1)$ and $f(\text{e-}Cl(U_2)) \subset Cl(V_2)$. Therefore, we obtain disjoint e-open sets U_1 and U_2 containing x_1 and x_2, respectively. Thus X is e-T_2.

Lemma 5. ([15]) Let X be a space and A, B $\subset$ X. If A is o-open and B $\in$ EΣ(X, T), then A $\cap$ B$\in$ EΣ(X, T).

A function f: $X \rightarrow Y$ is called an R-map [16] if the preimage of every regular open subset of Y is a regular open subset of X.

Theorem 19. Let f, g: $X \rightarrow Y$ be functions and Y be a Hausdorff space. If f is θ-e-continuous and g is an R-map, then the set A={$x \in X$: $f(x) = g(x)$} is e-closed in X.

Proof. Let x $\notin$ A, so that $f(x) \neq g(x)$. Since Y is Hausdorff, there exist open sets V_1 and V_2 in Y such that $f(x) \in V_1$ and $g(x) \in V_2$ and $V_1 \cap V_2 = \emptyset$; hence $Cl(V_1) \cap Int(Cl(V_2)) = \emptyset$. Since f is θ-e-continuous, there exists H$\in$ EΣ(X, x) such that $f(\text{e-}Cl(H)) \subset Cl(V_1)$. Since g is an R-map, $g^{-1}(Int(Cl(V_2)))$ is a regular opens subset of X and x $\in$ $g^{-1}(Int(Cl(V_2)))$. Set U = H $\cap$ $g^{-1}(Int(Cl(V_2)))$. By Lemma (5), x $\in$ U $\in$ EΣ(X, x) and U $\cap$ A = $\emptyset$. Hence, x $\notin$ e-Cl(A). This completes the proof.

A space X is said to be, e-compact [15] if every cover of X by e-open sets has a finite subcover. It can be easily seen that every e-compact space is e-closed.

Theorem 20. If a topological space X is e-closed and strongly e-regular, then X is e-compact.

Proof. Let X be an e-closed and strongly e-regular space. Let {V_λ: $\lambda \in \Delta$} be any cover of X by e-open sets. For each x$\in$ X, there exists a $\lambda(x) \in \Delta$ such that x$\in$ $V_{\lambda(x)}$. Since X is strongly e-regular, there exists $U_x \in$ EΣ(X,T) such that x$\in$ $U_x \subset$ e-Cl(U_x) $\subset V_{\lambda(x)}$. Then {U_x: x$\in$ X} is a cover of the e-closed space X by e-open sets and hence there exist finitely many points say, x_1, x_2, ..., x_n such that X = $\bigcup_{i=1}^{n}$

e-Cl(U_{xi}) $\subset \bigcup_{i=1}^{n}$ $V_{\lambda(xi)}$ .This shows that X is e-compact.

7   Comparison of Some Functions

Definition 9. A function $f: X \rightarrow Y$ is called strongly θ-e-continuous [1] (resp. e-continuous[6]) if for each $x \in X$ and each open set V of Y containing $f(x)$, there is an e-open set U of X containing x such that $f(\text{e-Cl}(U)) \subset V$ (resp. $f(U) \subset V$).

Remark 3. The following implications hold:

$$\textbf{Strongly}~\theta\textbf{-e-continuity} \longrightarrow \theta\textbf{-e-continuity} \longrightarrow \textbf{e -continuity}$$

None of these implications are reversible; see [1] and following example.

Example 2. Let $X = Y = \{1, 2, 3, 4, 5\}$. Define a topology $T = \{\varnothing, X, \{1\}, \{3\}, \{1, 3\}, \{3, 4\}, \{1, 3, 4\}\}$ on X and a topology $T^* = \{\varnothing, X, \{1\}\}$ on Y. Then the identity function $f: (X, T) \rightarrow (X, T^*)$ is e-continuous but neither strongly θ-e-continuous nor θ-e-continuous.

Theorem 21. If $f: X \rightarrow Y$ is continuous, then f is θ-e-continuous.

Proof. Let V be an open subset of Y. Since $f^{-1}(V)$ is open, it follows by Lemma (1) that $\text{e-Cl}_\theta(f^{-1}(V)) \subset \text{Cl}(f^{-1}(V))$ and since is continuous $\text{Cl}(f^{-1}(V)) \subset f^{-1}(\text{Cl}(V))$. Therefore, we obtain that $\text{e-Cl}_\theta(f^{-1}(V)) \subset f^{-1}(\text{Cl}(V))$. This shows that f is θ-e-continuous by Theorem (6).

The following corollary follows from Theorem (12).

Corollary 2. ([1], Theorem (4.9)) A continuous function $f: X \rightarrow X$ is strongly θ-e-continuous if and only if X is e-regular.

Definition 10. A function $f: X \rightarrow Y$ is called weakly e-continuous if for each $x \in X$ and each open set V containing $f(x)$, there is a e-open set U containing x such that $f(U) \subset \text{Cl}(V)$.

Theorem 22. Let X be a strongly e-regular space. Then $f: X \rightarrow Y$ is θ-e-c. if and only if $f: X \rightarrow Y$ is weakly e-continuous.

Proof. ⇒: Obivious.

(Conversely) Let $x \in X$ and V be any open set in Y containing $f(x)$. Since f is weakly e-continuous, there exists $U \in E\Sigma(X, x)$ such that $f(U) \subset \text{Cl}(V)$. Since, X is strongly e-regular, there exists a e-open set H such that $x \in H \subset \text{e-Cl}(H) \subset U$. Therefore, we obtain that $f(\text{e-Cl}(H)) \subset \text{Cl}(V)$. This shows that f is θ-e-c. at x.

Theorem 23. [17] A function $f: X \to Y$ is weakly continuous if and only if $\mathrm{Cl}(f^{-1}(V)) \subset f^{-1}(\mathrm{Cl}(V))$ for every open subset V of Y.

Theorem 24. If X is an e-regular space and $f: X \to Y$ is weakly continuous, then f is θ-e-c.

Proof. Let V be any open set in Y. Since X is e-regular, by Theorem (12), e-$\mathrm{Cl}_\theta(f^{-1}(V)) \subset \mathrm{Cl}(f^{-1}(V))$. Since f is weakly continuous, $\mathrm{Cl}(f^{-1}(V)) \subset f^{-1}(\mathrm{Cl}(V))$. Thus e-$\mathrm{Cl}_\theta(f^{-1}(V)) \subset f^{-1}(\mathrm{Cl}(V))$ and by Theorem (6) f is θ-e-c.

Definition 11. A function $f: X \to Y$ is called contra e-θ-continuous if $f^{-1}(V)$ is e-θ-closed set in X for every open set V in Y.

Theorem 25. If $f: X \to Y$ is contra e-θ-continuous, then f is θ-e-c.

Proof. Let V be any open set in Y. Then, since $f^{-1}(V)$ is e-θ-closed, then e-$\mathrm{Cl}_\theta(f^{-1}(V))$ $= f^{-1}(V) \subset f^{-1}(\mathrm{Cl}(V))$. This shows that f is θ-e-c. by Theorem (6).

8   Graphs of θ-e-Continuous Functions

Definition 12. The graph G (f) of a function $f: X \to Y$ is said to be θ-e-closed if for each $(x, y) \in (X \times Y)\backslash G\,(f)$, there exists $U \in E\Sigma(X, x)$ and an open set V containing y such that $(\text{e-Cl}(U) \times \mathrm{Cl}(V)) \cap G\,(f) = \emptyset$.

Lemma 6. For the graph G (f) of a function $f: X \to Y$, the following properties are equivalent:
   a) G (f) is θ-e-closed in X×Y;
   b) For each point $(x, y) \in (X \times Y)\backslash G\,(f)$, there exist a e-open set U containing x and an open set V containing y such that $f(\text{e-Cl}(U)) \cap \mathrm{Cl}(V) = \emptyset$;
   c) For each point $(x, y) \in (X \times Y)\backslash G\,(f)$, there exist a e-regular set U containing x and an open set V containing y such that $f(U) \cap \mathrm{Cl}(V) = \emptyset$.

Proof. This proof follows immediately from the definition (12).

Theorem 26. If $f: X \to Y$ is θ-e-c. and Y is Urysohn, then the graph G (f) of f is θ-e-closed in X×Y.

Proof. Let $(x, y) \notin G\,(f)$, so that $y = f(x)$. Since Y is Urysohn, then there exist open sets V_1 and V_2 containing $f(x)$ and y, respectively, such that $\mathrm{Cl}(V_1) \cap \mathrm{Cl}(V_2) = \emptyset$. Since f is θ-e-c. there exists a e-open set U containing x such that $f(\text{e-Cl}(U)) \subset \mathrm{Cl}(V_1)$. Therefore, $f(\text{e-Cl}(U)) \cap \mathrm{Cl}(V_2) = \emptyset$ and G (f) is θ-e-closed in X×Y.

It is known that a space (X, T) is Hausdorff if and only if $\{x\} = \cap \{\mathrm{Cl}(V): x \in V \in T\}$ for each $x \in X$.

Theorem 27. If $f: X \rightarrow Y$ is a surjection with a θ-e-closed graph, then Y is Hausdorff.

Proof. Let y, y_1 be any distinct points of Y. Since f is surjective, there exists $x \in X$ such that $f(x) = y_1$. Then $(x, y) \notin G(f)$, and by Lemma (6) there exist a e-open set U containing x and an open set V containing y such that f (e-Cl(U)) $\cap$ Cl(V)= Ø. Then $y_1 \in f$ (e-Cl(U)), $y_1 \notin$ Cl(V). This shows that Y is Hausdorff.

Theorem 28. If $f: X \rightarrow Y$ has a θ-e-closed graph and f is injective, then X is e-T_2.

Proof. Let x and y be any two distinct points of X. Then, we have $f(x) \neq f(y)$ and so $(x, f(y)) \in (X \times Y) \backslash G(f)$. By θ-e-closedness of the graph G (f), there exist an e-open set U of X containing x and an open set V of Y containing $f(y)$ such that $(e\text{-}Cl(U) \times Cl(V)) \cap G(f) = Ø$. Then, we have f (e-Cl(U)) $\cap$ Cl(V) = Ø, hence e-Cl(U) $\cap f^{-1}$(Cl(V)) = Ø. Clearly, by Lemma (2), e-Cl(U) and X\e-Cl(U) are the required disjoint e-open sets containing x and y, respectively.

Definition 13. The graph G (f) of a function $f: X \rightarrow Y$ is said to be strongly e-closed [1] if for each $(x, y) \in (X \times Y) \backslash G(f)$, there exists $U \in e\text{-}\Sigma(X, x)$ and an open set V containing y such that $(e\text{-}Cl(U) \times V) \cap G(f) = Ø$.

It is easy to see that if the graph of a function is θ-e-closed, then it is strongly e-closed.

Theorem 29. If $f: X \rightarrow Y$ is θ-e-c. and Y is Hausdorff, then the graph G (f) of f is strongly e-closed in X×Y.

Proof. Let $(x, y) \notin G(f)$, so that $y \neq f(x)$. Since Y is Hausdorff, there exist disjoint open sets V_1 and V_2 containing $f(x)$ and y, respectively. Thus $f(x) \notin$ Cl(V_2) and hence $x \notin f^{-1}$(Cl(V_2)). Since f is θ-e-c. it follows by Theorem (6), that $x \notin$ e-Cl$_\theta(f^{-1}(V_2))$. So there exists a e-open set U containing x such that e-Cl(U)$\cap f^{-1}(V_2) = Ø$. Therefore, $(x, y) \in$ e-Cl(U) $\times V \subset (X \times Y) \backslash G(f)$, which proves that G (f) is strongly e-closed in X×Y.

The following corollary follows from Remark (3).

Corollary 3. ([1], Theorem (5.5)) If $f: X \rightarrow Y$ is strongly θ-e-continuous and Y is Hausdorff, then the graph G (f) of f is strongly e-closed in X×Y.

Recall that a space X is called submaximal if each dense subset of X is open. A space X is called extremally disconnected if the closure of each open subset of X is open. A function $f: X \rightarrow Y$ is called strongly θ-continuous [18] if for each $x \in X$ and each open set V containing $f(x)$, there is an open set U containing x such that f (Cl(U)) $\subset$ V.

Corollary 4. [1] Let X be a submaximal extremally disconnected space and Y be a compact Hausdorff space. Then the following properties are equivalent for a function $f: X \rightarrow Y$:

a) f is strongly θ-e-continuous;
b) G (f) is a strongly e-closed subset of X×Y;
c) f is strongly θ-continuous;
d) f is continuous;
e) f is e-continuous.

The following example shows that Corollary (4) is not true when we remove the condition "submaximal".

Example 3. Let X = {1, 2, 3, 4} and let T= {Ø, X, {1, 2}, {3, 4}} be a topology on X. Consider the set R of real numbers with the usual topology. Define the function f:(X,T) → ([0,1],T**) by $f(1)$ = 0, $f(2)$ =1/3, $f(3)$=2/3, $f(4)$=1 where T** is the subspace topology. Then (X, T) is extremally disconnected but it is not submaximal. Moreover, ([0, 1], T**) is a compact Hausdorff space and f is strongly θ-e-continuous on X, but it is not strongly θ-continuous.

The following corollary is a slight improvement of the above corollary (4) of Murad ÄOzkoc and GÄulhan Aslim [1].

Corollary 5. Let X be a submaximal extremally disconnected space and Y be a compact Hausdorff space. Then the following properties are equivalent for a function $f: X \rightarrow Y$:

a) f is strongly θ-e-continuous;
b) f is θ-e-continuous;
c) G (f) is a θ-e-closed subset of X×Y;
d) G (f) is a strongly e-closed subset of X×Y;
e) f is strongly θ-e-continuous;
f) f is continuous;
g) f is e-continuous.

Acknowledgments. The authors are thankful to the referees for giving the valuable comments to improve the paper.

References

1. Ozkoc, M., Aslim, G.: On Strongly θ-e-Continuous Functions. Bull. Korean Math. Soc. 47(5), 1025–1036 (2010)
2. Porter, J., Thomas, J.: On H-Closed and Minimal Hausdorff Spaces. Trans. Amer. Math. Soc. 138, 159–170 (1969)
3. Thompson, T.: s-Closed Spaces. Proc. Amer. Math. Soc. 60, 335–338 (1976)
4. Di Maio, G., Noiri, T.: On s-Closed Spaces. Indian J. Pure Appl. Math. 18(3), 226–233 (1987)

5. Abo-Khadra, A.: On Generalized Forms of Compactness. Master's Thesis, Faculty of Science, Tanta University, Egypt (1989)
6. Ekici, E.: On e-Open Sets, DP*-Sets and DPC*-Sets and Decompositions of Continuity. The Arabian J. for Sci. Eng. 33(2A), 269–282 (2008)
7. Joseph, J.E., Kwack, M.H.: On s-Closed Spaces. Proc. Amer. Math. Soc. 80(2), 341–348 (1980)
8. Dontchev, J., Ganster, M., Noiri, T.: On p-Closed Spaces. Int. J. Math. Sci. 24, 203–212 (2000)
9. Kucuk, M., Zorlutuna, I.: A Unification of Compactness and Closedness. Soochow J. Math. 29(3), 221–233 (2003)
10. Velicko, N.V.: H-Closed Spaces. Amer. Math. Soc. Trans. 78(2), 103–118 (1968)
11. Levine, N.: Semi-Open Sets and Semi-Continuity in Topological Spaces. Amer. Math. Monthly 70, 36–41 (1963)
12. Mashhour, S., Abd-ElMonsef, M.E., El-Deeb, S.N.: On Pre-Continuous and Weak Precontinuous Mappings. Proc. Math. Phys. Soc. 53, 47–53 (1982)
13. Caldas, M., Jafari, S.: On Strongly Faint e-Continuous Functions. Projécciones J. Math. 30(1), 29–41 (2011)
14. Singal, M.K., Arya, S.P.: On Almost Regular Spaces. Glas. Mat. Ser. 4(4), 89–99 (1969)
15. Ekici, E.: Some Generalizations of Almost Contra-Super-Continuity. Filomat. 21(2), 31–44 (2007)
16. Carnahan, D.: Some Properties Related to Compactness in Topological Spaces. Ph.D. Thesis, University of Arkansas (1973)
17. Rose, D.A.: Weak Continuity and Almost Continuity. Internat. J. Math. Math. Sci. 7, 311–318 (1984)
18. Noiri, T.: On δ-Continuous Functions. J. Korean Math. Soc. 16(2), 161–166 (1980)

RETRACTED CHAPTER: On θ-e-Irresolute Functions

Alaa. M.F. AL. Jumaili and Xiao-Song Yang

Department of Mathematics,
Huazhong University of Science and Technology
Wuhan, 430074, China
alaa_mf1970@yahoo.com

Abstract. In this paper we study and investigate a new class of functions called θ-e-Irresolute functions and obtain several characterizations of θ-e-Irresolute functions. The concepts of e-open sets were introduced by Erdal Ekici [1] and e-θ-open sets were introduced by ÄOzkoc and GÄulhan Aslim [2]. Also we investigate relationships between strongly e-irresolute functions and graphs.

Keywords: e-Open Set, e-θ-Closed Set, θ-e-Irresolute Functions, Weakly e-Irresolute.

1 Introduction

Is common viewpoint of many topologists that generalized open sets are important ingredients in General Topology and they are now the research topics of many topologists worldwide of which lots of important and interesting results emerged. Indeed a significant theme in General Topology and Real Analysis concerns the variously modified forms of continuity, separation axioms etc by using generalized open sets. As a generalization of open sets, e-open sets, e-θ-open sets were introduced and studied by E.Ekici [1], ÄOzkoc and GÄulhan Aslim [2]. In this paper, we will continue the study of related functions with e-open sets [1]; e-θ-open sets [2]. Also we introduce and characterize the concepts of θ-e-irresolute functions and relationships between strongly e-irresolute functions and graphs are investigated.

2 Preliminaries

Throughout this paper, (X, T) and (Y, T*) (or simply X and Y) mean topological spaces in which no separation axioms are assumed unless explicitly stated. For any subset A of X, the closure and interior of A are denoted by Cl(A) and Int(A), respectively. We recall the following definitions, which will be used often throughout this paper.

A point $x \in X$ is called a θ-cluster (resp. δ-cluster) point of A if Cl(V) $\cap$ A ≠ Ø (resp. Int(Cl(V)) $\cap$ A ≠ Ø) for every open subset V of X containing x. The set of all θ-cluster (resp. δ-cluster) points of A is called the θ-closure (resp. δ-closure) of A [3] and is denoted by $Cl_\theta(A)$ (resp. $Cl_\delta(A)$). If A = $Cl_\theta(A)$ (resp. A = $Cl_\delta(A)$), then A is

---

The paper starting on page 47 of this volume has been retracted for reasons of plagiarism. In addition, the name of the second author was added without his knowledge or consent. The erratum to this chapter is available at DOI: 10.1007/978-3-642-28926-2_63

P. Balasubramaniam and R. Uthayakumar (Eds.): ICMMSC 2012, CCIS 283, pp. 47–53, 2012.
© Springer-Verlag Berlin Heidelberg 2012

said to be θ-closed (resp. δ-closed). The complement of a θ-closed (resp. δ-closed) set is said to be θ-open (resp. δ-open) set. A subset A of a space X is called e-open [1] if $A \subset Cl(Int_\delta(A)) \cup Int(Cl_\delta(A))$; the complement of an e-open set is called e-closed. The intersection of all e-closed sets containing A is called the e-closure of A [1] and is denoted by e-Cl(A). A set A is e-closed [4] if and only if e-Cl(A) = A. The union of all e-open sets of X contained in A is called the e-interior [1] of A and is denoted by e-Int(A). A subset A of a topological space X is e-regular [2] if it is e-open and e-closed. A point x of X is called an e-θ-cluster point of A if e-Cl(V) $\cap$ A $\neq$ Ø for every e-open set V containing x. The set of all e-θ-cluster points of A is called e-θ-closure [2] of A and is denoted by e-$Cl_\theta(A)$. A subset A is said to be e-θ-closed if A = e-$Cl_\theta(A)$. The complement of an e-θ-closed set is said to be e-θ-open. The set of all e-θ-interior points of A is said to be the e-θ-interior of A and denoted by e-$Int_\theta(A)$. The family of all e-open (resp. e-closed, e-regular, e-θ-open) subsets of X containing a point x $\in$ X is denoted by EΣ(X, x) (resp. EC(X, x), ER(X, x), EθΣ(X, x)). The family of all e-open (resp. e-closed, e-regular, e-θ-open) sets in X are denoted by EΣ(X, T) (resp. EC(X, T), ER(X, T), EθΣ(X, T).

Theorem 1. [2] For a subset A of a topological space X; the following properties hold:

 a) A $\in$ EΣ(X, T) if and only if e- Cl(A) $\in$ ER(X, T).
 b) A $\in$ EC(X, T) if and only if e-Int(A)$\in$ ER(X, T).

Theorem 2. [2] For a subset A of a topological space X; the following properties hold:

 a) If A $\in$ EΣ(X, T), then e-Cl(A) = e-$Cl\theta(A)$,
 b) A $\in$ ER(X, T) if and only if A is e-θ-open and e-θ-closed.

Definition 1. [2] A space X is called e-regular if for each closed set M $\subset$ X and each point x$\in$ X\M, there exist disjoint e-open sets U and V such that x$\in$ U and M $\subset$ V.

Lemmas 1. [2] For a space X the following are equivalent:

 a) X is e-regular;
 b) For each point x $\in$ X and for each open set U of X containing x; there exists V $\in$ EΣ(X, T) such that x $\in$ V $\subset$ e- Cl(V) $\subset$ U;
 c) If each open set U and each x $\in$ U, there exists V $\in$ ER(X,T) such that x $\in$ V $\subset$ U.

Definition 2. [5] A function f: X $\rightarrow$ Y is said to be e-irresolute if f^{-1}(V) $\in$ EΣ(X, T) for every V $\in$ EΣ(Y, T).

3   Characterizations of θ-e- Irresolute Functions

Definition 3. A function f: X $\rightarrow$ Y is said to be θ-e-Irresolute if for each x $\in$ X and each V $\in$ EΣ(Y, f(x)) there exists a U $\in$ EΣ(X, x) such that f(e-Cl(U)) $\subset$ e-Cl(V).

Definition 4. A function $f: X \to Y$ is said to be weakly e-irresolute if for each $x \in X$ and each $V \in E\Sigma(Y, f(x))$ there exists a $U \in E\Sigma(X, x)$ such that $f(U) \subset$ e-Cl(V).

Clearly, every e-irresolute function is θ-e-Irresolute and every θ-e-Irresolute function is weakly e-irresolute. But the converses are not true as shown by the following example.

Example 1. Let $X = Y = \{1, 2, 3\}$. Define a topology $T = \{\emptyset, X, \{1\}\}$ on X and a topology $T^* = \{\emptyset, X, \{2\}\}$ on Y. Then the identity function $f: (X, T) \to (X, T^*)$ is θ-e- Irresolute but not e-irresolute.

Theorem 3. For a function $f: X \to Y$ the following properties are equivalent:

a) f is θ-e-irresolute;
b) e-Cl$_\theta(f^{-1}(B)) \subset f^{-1}(e$-Cl$_\theta(B))$ for every subset B of Y ;
c) $f(e$-Cl$_\theta(A)) \subset e$-Cl$_\theta(f(A))$ for every subset A of X.

Proof. (a) $\Rightarrow$ (b): Let B be any subset of Y. Suppose that $x \notin f^{-1}(e$-Cl$_\theta(B))$. Then $f(x) \notin e$-Cl$_\theta(B)$ and there exits $V \in E\Sigma(Y, f(x))$ such that e-Cl$(V) \cap B = \emptyset$. Since f is θ-e-irresolute, there exists $U \in E\Sigma(X, x)$ such that $f(e$-Cl$(U)) \subset e$-Cl(V). Therefore, we have $f(e$-Cl$(U)) \cap B = \emptyset$ and e-Cl$(U) \cap f^{-1}(B) = \emptyset$. This shows that $x \notin$ e-Cl$_\theta(f^{-1}(B))$. Hence, we obtain e-Cl$_\theta(f^{-1}(B)) \subset f^{-1}(e$-Cl$_\theta(B))$.

(b) $\Rightarrow$ (c): Let A be any subset of X. Then we have e-Cl$_\theta(A)) \subset e$-Cl$_\theta(f^{-1}(f(A)))$ $\subset f^{-1}(e$-Cl$_\theta(f(A)))$ and hence $f(e$-Cl$_\theta(A)) \subset e$-Cl$_\theta(f(A))$.

(c) $\Rightarrow$ (b): Let B be a subset of Y. We have $f(e$-Cl$_\theta(f^{-1}(B)) \subset e$-Cl$_\theta(f(f^{-1}(B)))$ $\subset e$-Cl$_\theta(B)$ and hence e-Cl$_\theta(f^{-1}(B)) \subset f^{-1}(e$-Cl$_\theta(B))$.

(b) $\Rightarrow$ (a): Let $x \in X$ and $V \in E\Sigma(Y, f(x))$. Then we have e-Cl$(V) \cap (Y - e$-Cl$(V)) = \emptyset$ and $f(x) \notin e$-Cl$_\theta(Y - e$-Cl$(V))$. Hence, $x \notin f^{-1}(e$-Cl$_\theta(Y - e$-Cl$(V)))$ and $x \notin e$-Cl$_\theta(f^{-1}(Y - e$-Cl$(V))$. There exists $U \in E\Sigma(X, x)$ such that e-Cl$(U) \cap f^{-1}(Y - e$-Cl$(V)) = \emptyset$ and hence $f(e$-Cl$(U)) \subset e$-Cl(V). This shows that f is θ-e- irresolute.

Theorem 4. For a function $f: X \to Y$, the following properties are equivalent:

a) f is θ-e-irresolute;
b) $f^{-1}(V) \subset e$-Int$_\theta(f^{-1}(e$-Cl$(V)))$ for every $V \in E\Sigma(Y, T)$;
c) e-Cl$_\theta(f^{-1}(V)) \subset f^{-1}(e$-Cl$(V))$ for every $V \in E\Sigma(Y, T)$.

Proof. (a) $\Rightarrow$ (b): Suppose that $V \in E\Sigma(Y, T)$ and $x \in f^{-1}(V)$. Then $f(x) \in V$ and there exists $U \in E\Sigma(X, x)$ such that $f(e$-Cl$(U)) \subset (e$-Cl$(U))$. Therefore, $x \in U \subset$ e-Cl$(U) \subset f^{-1}(e$-Cl$((V))$. This shows that $x \in$ e-Int$_\theta(f^{-1}(e$-Cl$(V)))$. There for we have $f^{-1}(V) \subset e$-Int$_\theta(f^{-1}(e$-Cl$(V)))$.

(b) $\Rightarrow$ (c): Suppose that $V \in E\Sigma(Y, T)$ and $x \notin f^{-1}(e$-Cl$(V))$. Then $f(x) \notin e$-Cl(V) and there exists $U \in E\Sigma(Y, f(x))$ such that $U \cap V = \emptyset$ and hence e-Cl$(U) \cap V = \emptyset$. Therefore, we have $f^{-1}(e$-Cl$(U)) \cap f^{-1}(V) = \emptyset$. Since $x \in f^{-1}(U)$, by (b), $x \in e$- Int$_\theta$ $(f^{-1}(e$-Cl$(U)))$. There exists $H \in E\Sigma(X, x)$ such that e-Cl$(H) \subset f^{-1}(e$-Cl$(U))$. Thus, we have e-Cl$(H) \cap f^{-1}(V) = \emptyset$ and hence $x \notin e$-Cl$_\theta(f^{-1}(V))$. This shows that e-Cl$_\theta$ $(f^{-1}(V)) \subset f^{-1}(e$-Cl$(V))$.

(c) $\Rightarrow$ (a): Suppose that $x \in X$ and $V \in E\Sigma(Y, f(x))$.Then (V) $\cap (Y - e\text{-Cl}(V)) =$ Ø and $f(x) \notin e\text{-Cl}(Y - e\text{-Cl}(V))$. Therefore, $x \notin f^{-1}(e\text{-Cl}(Y - e\text{-Cl}(V)))$. And by (c), $x \notin e\text{-Cl}_\theta(f^{-1}(Y - e\text{-Cl}(V)))$. There exists $U \in E\Sigma(X, x)$ such that $e\text{-Cl}(U) \cap f^{-1}(Y - e\text{-Cl}(V)) = \emptyset$. Therefore, we obtain $f(e\text{-Cl}(U)) \subset e\text{-Cl}(V)$. This shows that f is θ-e-irresolute.

Definition 5. A function $f: X \to Y$ is said to be strongly e-irresolute if for each point x $\in X$ and each $V \in E\Sigma(Y, f(x))$, there exists $U \in E\Sigma(X, x)$ such that $f(e\text{-Cl}(U)) \subset V$.

Theorem 5. Let Y be an e-regular space. Then for a function $f: X \to Y$ the following properties are equivalent:

   a) f is strongly e-irresolute;
   b) f is e-irresolute;
   c) f is θ-e-irresolute.

Proof. (a) $\Rightarrow$ (b): This is obvious.

   (b) $\Rightarrow$ (c): Suppose that $x \in X$ and $V \in E\Sigma(Y, f(x))$. Since f is e-irresolute, $f^{-1}(V)$ is e-open and $f^{-1}(e\text{-Cl}(V))$ is e-closed in X. Now, set $U = f^{-1}(V)$. Then we have $U \in E\Sigma(X, x)$ and $e\text{-Cl}(U) \subset f^{-1}(e\text{-Cl}(V))$. Therefore, we obtain $f(e\text{-Cl}(U)) \subset e\text{-Cl}(V)$. This shows that f is θ-e-irresolute.

   (c) $\Rightarrow$ (a): Suppose that $x \in X$ and $V \in E\Sigma(Y, f(x))$. Since Y is e-regular, there exists $H \in E\Sigma(Y, T)$ such that $f(x) \in H \subset e\text{-Cl}(H) \subset V$. Since f is θ-e-irresolute, there exists $U \in E\Sigma(X, x)$ such that $f(e\text{-Cl}(U)) \subset e\text{-Cl}(H) \subset V$. This shows that f is strongly e-irresolute.

Theorem 6. Let X be an e-regular space. Then a function $f: X \to Y$ is θ-e-irresolute if and only if it is weakly e-irresolute.

Proof. Suppose that f is weakly e-irresolute. Let $x \in X$ and $V \in E\Sigma(Y; f(x))$.Then, there exists $U \in E\Sigma(X, x)$ such that $f(U) \subset e\text{-Cl}(V)$. Since X is e-regular, there exists $U_o \in E\Sigma(X, x)$ such that $x \in U_o \subset e\text{-Cl}(U_o) \subset U$. Therefore, we obtain $f(e\text{-Cl}(U_o)) \subset e\text{-Cl}(V)$. This shows that f is θ-e-irresolute.

Theorem 7. Let $f: X \to Y$, $g: Y \to Z$ be functions and $g \circ f: X \to Z$ be the composition. Then the following properties hold:

   a) If f and g are θ-e-irresolute, then $g \circ f$ is θ-e-irresolute;
   b) If f is strongly e-irresolute and g is weakly e-irresolute, then $g \circ f$ is θ-e-Irresolute;
   c) If f is weakly e-irresolute and g is θ-e-irresolute, then $g \circ f$ is weakly e-Irresolute;
   d) If f is θ-e-irresolute and g is strongly e-irresolute, then $g \circ f$ is strongly e-Irresolute.

Proof. The proof follows directly from the definitions.

4   Graphs of θ-e-Irresolute Functions

Definition 6. [2] A space X is said to be e-T_2 if for each pair of distinct points x and y in X; there exist U ∈ EΣ(X, x) and V ∈ EΣ(X, y) such that e-Cl(U) ∩ e-Cl(V) = Ø.

Recall that for a function $f: X → Y$, the subset $\{(x, f(x)): x ∈ X\}$ of X × Y is called the graph of f and is denoted by G (f).

Definition 7. [2] The graph G (f) of a function $f : X → Y$ is said to be strongly e-closed (resp. e-θ-closed) if for each (x, y) ∈ (X × Y)\G(f), there exist U ∈ EΣ(X, x) and V ∈ EΣ(Y, y) such that (e-Cl(U) × V) ∩ G (f) = Ø (resp. (e-Cl(U) × e-Cl(V)) ∩ G (f) = Ø).

Lemma 2. The graph G (f) of a function $f : X → Y$ is said to be e-θ-closed in (X ×Y) if and only if for each point (x, y) ∈ (X × Y)\G(f), there exist U ∈ EΣ(X, x) and V ∈ EΣ(Y, y) such that f(e-Cl(U)) ∩ e-Cl(V) = Ø.

Proof. The proof follows directly from the definitions.

Theorem 8. Let $f: X → Y$ be a θ-e-irresolute and Y e-T_2, then G (f) is e-θ-closed in X × Y.

Proof. Let (x, y) ∈ (X ×Y)\G (f). It follows that f(x) ≠ y. Since Y is e-T_2, there exist e-open sets V and H in Y containing f(x) and y, respectively, such that e-Cl(V)) ∩ e-Cl(H) = Ø. Since f is θ-e-irresolute, there exists U ∈ EΣ(X; x) such that f(e-Cl(U)) ⊂ e-Cl(V). Therefore f(e-Cl(U)) ∩ e-Cl(H) = Ø and by Lemma (2), G (f) is e-θ-closed in X × Y.

Recall that A space X is said to be e-T_2 [2] if for each pair of distinct points x and y in X; there exist U ∈ EΣ(X, x) and V ∈ EΣ(X, y) such that U ∩ V = Ø.

Theorem 9. Let $f: X → Y$ be a strongly e-irresolute and Y is e-T_2, then G (f) is e-θ-closed in X × Y.

Proof. Let (x, y) ∈ (X ×Y)\G (f), It follows that f (x) ≠ y. Since Y is e-T_2, there exist e-open sets V and H in Y containing f(x) and y, respectively, such that V∩H = Ø and hence V ∩ e-Cl(H) = Ø. Since f is strongly e-irresolute, there exists U ∈ EΣ(X, x) such that f(e-Cl(U)) ⊂ V. Therefore, f(e-Cl(U)) ∩ e-Cl(H) = Ø and by Lemma (2) G (f) is e-θ-closed in X × Y.

Theorem 10. Let f, $g: X → Y$ be a function. If G (f) is e-θ-closed and g is θ-e-irresolute, then the set $\{(x_1, x_2): f(x_1) = g(x_2)\}$ is e-θ-closed in the product space X× X.

Proof. Let A = $\{(x_1, x_2): f(x_1) = g(x_2)\}$. Suppose $(x_1; x_2)$ ∈ A. Then $f(x_1) ≠ g(x_2)$ and hence $(x_1; g(x_2))$ ∈ G (f). Since G (f) is e-θ-closed, there exist U ∈ EΣ(X, x_1) and H ∈ EΣ(Y, $g(x_2)$) such that, f(e-Cl(U)) ∩ e-Cl(H) = Ø. Since g is θ-e-irresolute, Then there exists $U_o \in E\Sigma(X; x_2)$ such that g (e-Cl(U_o)) $\subset$ e-Cl(H) and hence f (e-Cl(U)) $\cap$ g (e-Cl(U_o)) = $\emptyset$. Therefore, we obtain (e-Cl(U) × e-Cl(U_o)) $\cap$ A = $\emptyset$ and hence A is e-θ-closed.

Theorem 11. Let f: X → Y be a θ-e-irresolute function and Y is e-T_2, then the subset A = {(x; y): f(x) = f(y)} is e-θ-closed in X × X.

Proof. Since f is θ-e-irresolute function and Y is e-T_2, by Theorem (8) G (f) is e-θ-closed. Therefore, by Theorem (10) A is e-θ-closed.

Definition 8. [2] A space X is said to be:

   a)   e-closed if every cover of X by e-open sets has a finite subcover whose e-closures cover X;
   b)   Countably e-closed if every countable cover of X by e-open sets has a finite subcover whose e-closures cover X.

A subset H of a space X is said to be e-closed relative to X [2] if for every cover {V_λ: $\lambda \in \Delta$} of H by e-open sets of X, there exists a finite subset Δ_o of Δ such that H $\subset \bigcup$ {e-Cl(V_λ): $\lambda \in \Delta_o$}.

Theorem 12. Let f: X → Y be a θ-e-irresolute function and H is e-closed relative to X. Then f (H) is e-closed relative to Y.

Proof. Suppose that f: X → Y is θ-e-irresolute and H is e-closed relative to X. Let {V_λ: $\lambda \in \Delta$} be a cover of f (H) by e-open sets of X. For each point x $\in$ H, there exists λ(x) $\in \Delta$ such that f (x) $\in V_{\lambda(x)}$. Since f is θ-e-irresolute, then there exists $U_x \in$ EΣ(X, x) such that f (e-Cl(U_x)) $\subset$ e-Cl($V_{\lambda(x)}$). The family {U_x: x $\in$ H} is a cover of H by e-open sets of X and hence there exists a finite subset H_1 of H such that H $\subset \bigcup_{x \in H_1} e - Cl(U_x)$ Therefore we obtain f (H) $\subset \bigcup_{x \in H_1} e - Cl(V_{\lambda(x)})$. This shows that f (H) is e-closed relative to Y.

Corollary 1. Let f: X → Y be a θ-e-irresolute surjection function. Then the following properties hold:

   a)   If X is e-closed, then Y is e-closed;
   b)   If X is countably e-closed, then Y is countably e-closed.

Acknowledgments. The authors are thankful to the referees for giving the valuable comments to improve the paper.

References

1. Ekici, E.: On e-Open Sets, DP*-Sets and DPC*-Sets and Decompositions of Continuity. The Arabian J. for Sci. Eng. 33(2A), 269–282 (2008)

2. Ozkoc, M., Aslim, G.: On Strongly θ-e-Continuous Functions. Bull. Korean Math. Soc. 47(5), 1025–1036 (2010)
3. Velicko, N.V.: H-Closed Spaces. Amer. Math. Soc. Trans. 78(2), 103–118 (1968)
4. Caldas, M., Jafari, S.: On Strongly Faint e-Continuous Functions. Proyecciones J. Math. 30(1), 29–41 (2011)
5. Ekici, E.: New Forms of Contra-Continuity. Carpathian J. Math. 24(1), 37–45 (2008)

On the Strictness of a Bound
for the Diameter of Cayley Graphs Generated
by Transposition Trees

Ashwin Ganesan

Department of Mathematics,
Amrita School of Engineering,
Amrita Vishwa Vidyapeetham,
Amritanagar, Coimbatore 641112, India
ashwin.ganesan@gmail.com, g_ashwin@cb.amrita.edu

Abstract. Cayley graphs have been well-studied as a model for inter-connection networks due to their low diameter, optimal fault tolerance, and algorithmic efficiency, among other properties. A problem of practical and theoretical interest is to determine or estimate the diameter of Cayley graphs. Let Γ be a Cayley graph on $n!$ vertices generated by a transposition tree on vertex set $\{1, 2, \ldots, n\}$. In an oft-cited paper [1], it was shown that the diameter of Γ is bounded as:

$$\mathrm{diam}(\Gamma) \leq \max_{\pi \in S_n} \left\{ c(\pi) - n + \sum_{i=1}^{n} \mathrm{dist}_T(i, \pi(i)) \right\},$$

where the maximization is over all permutations π in the symmetric group, $c(\pi)$ denotes the number of cycles in π, and dist_T is the distance function in T. It is of interest to determine how far away this upper bound can be from the true diameter value in the worst case and for which families of graphs this bound should be utilized or not utilized. In this work, we investigate the worst case performance of this upper bound. We show that for every n, there exists a transposition tree on n vertices such that the maximum possible difference Δ_n between the upper bound and the true diameter value is at least $n - 4$. The lower bound we provide for Δ_n is seen to be best possible, and an open problem is to determine an upper bound for Δ_n.

Keywords: Cayley graphs, Transposition trees, Permutation groups, Diameter, Interconnection networks.

1 Introduction

Cayley graphs generated by transposition trees were shown in the oft-cited paper by Akers and Krishnamurthy [1] to have diameter that is sublogarithmic in the number of vertices. This is one of the main reasons such Cayley graphs were considered to be a superior model to hypercubes for consideration as the topology of interconnection networks [13], [15]. It is now known that Cayley

P. Balasubramaniam and R. Uthayakumar (Eds.): ICMMSC 2012, CCIS 283, pp. 54–61, 2012.

graphs possess additional desirable properties such as optimal fault-tolerance [2], algorithmic efficiency [3], optimal gossiping protocols [5], and optimal routing algorithms [8], among others, and so have been widely studied in the field of interconnection networks and parallel and distributed computing [15].

The diameter of a network represents the maximum communication delay between two nodes in the network. The design and performance of bounds or algorithms that determine or estimate the diameter of various families of Cayley graphs of permutation groups is thus of much theoretical and practical interest. This diameter problem is difficult even for the simple case when the symmetric group is generated by cyclically adjacent transpositions (i.e. a set of transpositions whose transposition graph is a Hamilton cycle) [14]. When a bound is proposed in the literature for this problem, it is of interest to determine how far away this bound can be from the true diameter value in the worst case. The purpose of this present work is to investigate this strictness of a well-known upper bound on the diameter of Cayley graphs.

A transposition is a permutation of the elements of a set that interchanges just two elements of the set. Let S be a set of transpositions of $\{1, 2, \ldots, n\}$. The transposition graph $T(S)$ is the simple, undirected graph with vertex set $\{1, 2, \ldots, n\}$ and with two vertices i, j being adjacent whenever the transposition $(i, j) \in S$. Let Γ denote the Cayley graph (also known as the Cayley diagram) generated by S. Thus, the vertex set of Γ is the permutation group generated by S and there is an arc in Γ from π to τ if and only if $\tau = \pi(i, j)$ for some $(i, j) \in S$ [7] [6]. Since every transposition is its own inverse, we can assume that Γ is undirected. It is well known that a given set of transpositions S of $\{1, 2, \ldots, n\}$ generates the entire symmetric group S_n if and only if the transposition graph $T(S)$ contains a spanning tree [4],[9]. When $T(S)$ is a tree, it is called a transposition tree. Throughout this work, we focus on the case where $T(S)$ is a tree. We often use the symbol T to denote both the tree as well as the set of transpositions S, and we use the symbol (i, j) for both an edge of the tree as well as a transposition in S.

1.1 Notations and Prior Work

We let S_n denote the symmetric group on the n-element set $[n] := \{1, 2, \ldots, n\}$. We represent a permutation $\pi \in S_n$ as an arrangement of $[n]$, either in the form $[\pi(1), \pi(2), \ldots, \pi(n)]$ or in cycle notation. $c(\pi)$ denotes the number of cycles in π, including cycles of length 1. We let $\mathrm{inv}(\pi)$ denote the number of inversions of π (cf. [4]). Thus, if $\pi = [3, 5, 1, 4, 2] = (1, 3)(2, 5) \in S_5$, then $c(\pi) = 3$ and $\mathrm{inv}(\pi) = 6$. For $\pi, \tau \in S_n$, $\pi\tau$ is the permutation obtained by applying τ first and then π. If $\pi \in S_n$ and $\tau = (i, j)$ is a transposition, then $c(\tau\pi) = c(\pi) + 1$ if i and j are part of the same cycle of π, and $c(\tau\pi) = c(\pi) - 1$ if i and j are in different cycles of π; and similarly for $c(\pi\tau)$. We assume throughout that $n \geq 5$, since the problem is easily solved by using brute force for all smaller trees.

Let $\mathrm{dist}_G(u, v)$ denote the distance between vertices u and v in an undirected graph G, and let $\mathrm{diam}(G)$ denote the diameter of G. Note that $\mathrm{dist}_\Gamma(\pi, \sigma) = \mathrm{dist}_\Gamma(I, \pi^{-1}\sigma)$, where I denotes the identity permutation. Thus, the diameter of Γ is the maximum of $\mathrm{dist}_\Gamma(I, \pi)$ over $\pi \in S_n$.

Throughout this work, Γ denotes the Cayley graph generated by a transposition tree T. We now recall the following two bounds from the literature:

Theorem 1. *[1] Let T be a tree and let $\pi \in S_n$. Let Γ be the Cayley graph generated by T. Then*

$$\mathrm{dist}_\Gamma(I, \pi) \le c(\pi) - n + \sum_{i=1}^{n} \mathrm{dist}_T(i, \pi(i)).$$

By taking the maximum over both sides, it follows that

Corollary 2. *[12, p.188]*

$$\mathrm{diam}(\Gamma) \le \max_{\pi \in S_n} \left\{ c(\pi) - n + \sum_{i=1}^{n} \mathrm{dist}_T(i, \pi(i)) \right\} =: f(T).$$

In the sequel, we refer to the second upper bound $f(T)$ as the *diameter upper bound*. This bound was subsequently also derived in [17], and related work includes [18],[16].

1.2   Summary of Our Results

The diameter upper bound from the oft-cited paper [1] is investigated in this work. It is of interest to determine how far away this upper bound can be from the true diameter value in the worst case, and it is of interest to know for what families of trees this bound gives a good estimate of the diameter or for what families of trees this bound should not be utilized. To this end, in this work we show that for every n, there exists a transposition tree on n vertices such that the maximum possible difference Δ_n between the diameter upper bound and the true diameter value of the Cayley graph is at least $n - 4$. This result gives a lower bound on the strictness Δ_n of the diameter upper bound. This $n - 4$ lower bound is seen to be best possible in the sense that it is attained for some values of n. We leave it as an open problem to determine an upper bound for this difference.

2   Strictness of the Diameter Upper Bound

Define the worst case performance of the diameter upper bound by the quantity

$$\Delta_n := \max_{T \in \mathcal{T}_n} |f(T) - \mathrm{diam}(\Gamma)|,$$

where $\mathcal{T}_n$ denotes the set of all trees on n vertices. In our proof, we will use the following result:

Theorem 3. *[10] Let Γ be the Cayley graph generated by a transposition tree T. Then the diameter upper bound inequality*

$$\operatorname{diam}(\Gamma) \leq \max_{\pi \in S_n} \left\{ c(\pi) - n + \sum_{i=1}^{n} \operatorname{dist}_T(i, \pi(i)) \right\} = \max_{\pi \in S_n} f_T(\pi)$$

holds with equality if T is a path, and in this case equals $\binom{n}{2}$.

Our main result is the following:

Theorem 4. *For every $n \geq 5$, there exists a tree on n vertices such that the difference between the actual diameter of the Cayley graph and the diameter upper bound is at least $n - 4$; in other words, $\Delta_n \geq n - 4$.*

Proof. Throughout this proof, we let T denote the transposition tree defined by the edge set $\{(1,2), (2,3), \ldots, (n-3, n-2), (n-2, n-1), (n-2, n)\}$, which is shown in Figure 1. For conciseness, we let $d(i,j)$ denote the distance in T between vertices i and j. Also, for leaf vertices i, j of T, we let $T - \{i, j\}$ denote the tree on $n - 2$ vertices obtained by removing vertices i and j of T.

Our proof is in two parts. In the first part we establish that $f(T)$ is equal to $\binom{n}{2} - 2$. In the second part we show that the diameter of the Cayley graph generated by T is at most $\binom{n-1}{2} + 1$. Together, this yields the desired result. We now present the first part of the proof; we establish that $f(T)$, defined by

$$f(T) := \max_{\sigma \in S_n} \left\{ c(\sigma) - n + \sum_{i=1}^{n} \operatorname{dist}_T(i, \sigma(i)) \right\},$$

is equal to $\binom{n}{2} - 2$. We prove this result by examining several sub-cases. Define

$$f_T(\sigma) := c(\sigma) - n + S_T(\sigma), \quad S_T(\sigma) := \sum_{i=1}^{n} \operatorname{dist}_T(i, \sigma(i)).$$

We consider two cases, (1) and (2), depending on whether 1 and n are in the same or different cycle of σ; each of these cases will further involve subcases. In most of these subcases, we show that for a given σ, there is a σ' such that $f_T(\sigma) \leq f_T(\sigma')$ and $f_T(\sigma') \leq \binom{n}{2} - 2$.

(1) Assume 1 and n are in the same cycle of σ. So $\sigma = (1, k_1, \ldots, k_s, n, j_1, \ldots, j_\ell)\hat{\sigma}$. The different subcases consider the different possible values for s and ℓ.

(1.1) Suppose $s = 0, \ell = 0$. So $\sigma = (1, n)\hat{\sigma} = (1, n)\sigma_2 \ldots \sigma_r$. Then, $f_T(\sigma) = c(\sigma) - n + S_T(\sigma) = r - n + 2(n-2) + S_{T-\{1,n\}}(\hat{\sigma}) = 2n - 5 + (r-2) + (n-2) + S_{T-\{1,n\}}(\hat{\sigma}) = 2n - 5 + c(\hat{\sigma}) + (n-2) + S_{T-\{1,n\}}(\hat{\sigma}) = 2n - 5 + f_{T-\{1,n\}}(\hat{\sigma}) \leq 2n - 5 + \binom{n-2}{2} = \binom{n}{2} - 2$, where by Theorem 3 the inequality holds with equality for some $\hat{\sigma}$. Thus, the maximum of $f_T(\sigma)$ over all permutations that contain $(1, n)$ as a cycle is equal to $\binom{n}{2} - 2$. It remains to show that for all other kinds of permutations σ in the symmetric group S_n, $f_T(\sigma) \leq \binom{n}{2} - 2$.

(1.2) Suppose $s = 1, \ell = 0$. So $\sigma = (1, i, n)\sigma_2 \ldots \sigma_r = (1, i, n)\hat{\sigma}$. We consider some subcases.

(1.2.1) Suppose $i = n - 1$. Then, $f_T(\sigma) = r - n + (2n - 2) + S_{T-\{1, n-1, n\}}(\hat{\sigma}) = 2n - 4 + f_{T-\{1, n-1, n\}}(\hat{\sigma}) \leq 2n - 4 + \binom{n-3}{2} \leq \binom{n}{2} - 2$, where the inequality is by Theorem 3.

(1.2.2) Suppose $2 \leq i \leq n - 2$; so $\sigma = (1, i, n)\hat{\sigma}$. Let $\sigma' = (1, n)(i)\hat{\sigma}$. It is easily verified that $f_T(\sigma) \leq f_T(\sigma')$, and so the desired bound follows from applying subcase (1.1) to $f_T(\sigma')$.

(1.3) Suppose $s = 0, \ell = 1$, so $\sigma = (1, n, i)\hat{\sigma}$. Since $f_T(\sigma) = f_T(\sigma^{-1})$, this case also is settled by (1.2).

(1.4) Suppose $s = 0, \ell \geq 2$, so $\sigma = (1, n, j_1, \ldots, j_\ell)\hat{\sigma}$. Let $\sigma' = (1, n)(j_1, \ldots, j_\ell)\hat{\sigma}$. Observe that $f_T(\sigma) \leq f_T(\sigma')$ iff $d(n, j_1) + d(j_\ell, 1) \leq d(n, 1) + d(j_\ell, j_1) + 1$. We prove the latter inequality by considering 4 subcases:

(1.4.1) Suppose $j_1 < j_\ell \leq n - 2$. Then, an inspection of the tree in Figure 1 shows that $d(n, j_1) + d(j_\ell, 1) = d(n, 1) + d(j_\ell, j_1)$, and so the inequality holds.

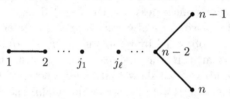

Fig. 1. Positions of j_1 and j_ℓ arising in subcase (1.4.1)

(1.4.2) Suppose $j_1 > j_\ell$ and $j_1, j_\ell \leq n - 2$. Then, $d(n, j_1) + d(j_\ell, 1) \leq d(1, n)$, and so the inequality holds.

(1.4.3) Suppose $j_1 = n - 1$. Then $d(n, j_1) = 2$. Also, $d(j_\ell, 1) \leq d(n, 1)$ and $d(j_\ell, j_1) \geq 1$, and so the inequality holds.

(1.4.4) Suppose $j_\ell = n - 1$. Then, $d(j_\ell, 1) = d(n, 1)$ and $d(n, j_1) = d(j_\ell, j_1)$, and so again the inequality holds.

(1.5) Suppose $s = 1, \ell = 1$, so $\sigma = (1, i, n, j)\hat{\sigma}$. Let $\sigma' = (1, n)(i, j)\hat{\sigma}$.

(1.5.1) If $i = n - 1$, by symmetry in T between vertices n and $n - 1$, this subcase is resolved by subcase (1.4).

(1.5.2) Let $2 \leq i \leq n - 2$. Then $d(1, i) + d(i, n) = d(1, n)$. So $f_T(\sigma) \leq f_T(\sigma')$ iff $d(n, j) + d(j, 1) \leq d(1, n) + d(i, j) + d(j, i) + 1$, which is true since $d(n, j) + d(j, 1) \leq d(1, n) + 2$.

(1.6) Suppose $s = 1, \ell \geq 2$. So $\sigma = (1, i, n, j_1, \ldots, j_\ell)\hat{\sigma}$. Let $\sigma' = (1, n)(i, j_1, \ldots, j_\ell)\hat{\sigma}$. It suffices to show that $f_T(\sigma) \leq f_T(\sigma')$, i.e., that $d(1, i) + d(i, n) + d(n, j_1) + d(j_\ell, 1) \leq d(1, n) + d(1, n) + d(i, j_1) + d(j_\ell, i) + 1$. We examine the terms of this latter inequality for various subcases:

(1.6.1) Suppose $2 \leq i \leq n - 2$. Then $d(1, i) + d(i, n) = d(1, n)$.

(1.6.1a) If $j_\ell = n - 1$, then $d(j_\ell, i) = n - i - 1$ and $d(i, j_1) = |i - j_1|$, and so the inequality holds iff $-1 \leq |j_1 - i| + j_1 - i$, which is clearly true.

(1.6.1b) Suppose $2 \leq j_\ell \leq n - 2$. Then, the inequality holds iff $d(n, j_1) + j_\ell - 1 \leq n - 2 + |i - j_1| + |i - j_\ell| + 1$, which can be verified separately for the cases $j_1 = n - 1$ and $2 \leq j_1 \leq n - 2$.

(1.6.2) Suppose $i = n - 1$. By symmetry in T of the vertices n and $n - 1$, this case is resolved by (1.4).

(1.7) Suppose $s \geq 2, \ell = 0$, so $\sigma = (1, k_1, \ldots, k_s, n)\hat{\sigma}$. Since $f_T(\sigma) = f_T(\sigma^{-1})$, this case is resolved by (1.4).

(1.8) Suppose $s \geq 2, \ell = 1$, so $\sigma = (1, k_1, \ldots, k_s, n, j_1)\hat{\sigma}$. Let $\sigma' = (1, n)(k_1, \ldots, k_s, j_1)\hat{\sigma}$. We can assume that $2 \leq k_1 \leq n - 2$ since the $k_1 = n - 1$ case is resolved by (1.4) due to the symmetry in T. To show $f_T(\sigma) \leq f_T(\sigma')$, it suffices to prove the inequality $d(1, k_1) + d(k_s, n) + d(n, j_1) + d(j_1, 1) \leq d(1, n) + d(n, 1) + d(k_s, j_1) + d(j_1, k_1) + 1$. We prove this inequality by separately considering whether $j_1 = n - 1$ or $k_s = n - 1$ or neither:

(1.8.1) Suppose $j_1 = n - 1$. Substituting $d(k_s, n) = n - k_s - 1, d(n, j_1) = 2, d(j_1, 1) = j_1 - 1$, etc, we get that the inequality holds iff $k_1 \leq |j_1 - k_1| + n - 2$, which is clearly true.

(1.8.2) Suppose $2 \leq j_1 \leq n-2$. Then $d(n, j_1) = n - j_1 - 1$, and so the inequality holds iff $k_1 + d(k_s, n) + n - 3 \leq 2n - 3 + d(k_s, j_1) + d(j_1, k_1)$. If $k_s = n - 1$, this reduces to $j_1 + k_1 \leq 2n - 3 + |j_1 - k_1|$, and is true, whereas if $2 \leq k_s \leq n - 2$, this reduces to $k_1 - k_s \leq 1 + |j_1 - k_s| + |j_1 - k_1|$, which is true due to the triangle inequality.

(1.9) Suppose $s, l \geq 2$, so $\sigma = (1, k_1, \ldots, k_s, n, j_1, \ldots, j_\ell)\hat{\sigma}$.

Let $\sigma' = (1, k_1, \ldots, k_s, n)(j_1, \ldots, j_\ell)\hat{\sigma}$. It suffices to show that $S_T(\sigma) \leq S_T(\sigma') + 1$, i.e., that $d(n, j_1) + j_\ell \leq n + d(j_1, j_\ell)$.

(1.9.1) If $j_1 < j_\ell$, then $j_1 \leq n - 2$, and so $d(n, j_1) = n - j_1 - 1$ and $d(j_1, j_\ell) = j_\ell - j_1$; the inequality thus holds.

(1.9.2) If $j_1 > j_\ell$, then $d(j_1, j_\ell) = j_1 - j_\ell$, and so it suffices to show that $d(n, j_1) \leq n + j_1 - 2j_\ell$. It can be verified that this holds if $j_1 = n - 1$ and also if $2 \leq j \leq n - 2$.

(2) Now suppose 1 and n are in different cycles of σ. So let $\sigma = (1, k_1, \ldots, k_s)(n, j_1, \ldots, j_\ell)\hat{\sigma}$.

(2.1) Suppose $s = 0$. Then $f_T(\sigma) \leq \binom{n-1}{2} - 2$, by induction on n.

(2.2) Suppose $s = 1$. So let $\sigma = (1, i)(n, j_1, \ldots, j_\ell)\hat{\sigma}$. By symmetry in T between vertices n and $n - 1$ and subcase (1.1), we may assume $i \neq n - 1$. Let $\sigma' = (1, n)(i, j_1, \ldots, j_\ell)\hat{\sigma}$. It suffices to show that $S_T(\sigma) \leq S_T(\sigma')$. If $\ell = 0$ this is clear since $d(1, i) \leq d(1, n)$. Suppose $\ell \geq 2$. Then, by the triangle inequality, $d(n, j_1) + d(j_\ell, n) \leq d(j_1, i) + d(i, n) + d(i, j_\ell) + d(i, n) = d(j_1, i) + d(i, j_\ell) + (n - i - 1)2$. Also, $d(1, n) = d(1, i) + d(i, n) = d(1, i) + n - i - 1$. Hence, $2d(1, i) + d(n, j_1) + d(j_\ell, n) \leq 2d(1, n) + d(i, j_1) + d(j_\ell, i)$. Hence, $S_T(\sigma) \leq S_T(\sigma')$. The case $\ell = 1$ can be similarly resolved by substituting j_1 for j_ℓ in the $l \geq 2$ case here.

(2.3) Suppose $s \geq 2, \ell = 0$. Then, by Theorem 3, $f_T(\sigma) \leq \binom{n-1}{2}$.

(2.4) Suppose $s \geq 2, \ell = 1$, so $\sigma = (1, k_1, \ldots, k_s)(n, j_1)\hat{\sigma}$. Let $\sigma' = (1, n)(k_1, \ldots, k_s, j_1)\hat{\sigma}$. It suffices to show that $d(1, k_1) + d(1, k_s) + 2d(n, j_1) \leq 2d(1, n) + d(k_s, j_1) + d(k_1, j_1)$. This inequality is established by considering the two subcases:

(2.4.1) Suppose $j_1 = n - 1$. Then the inequality holds iff $2k_s + k_1 \leq 3n - 7 + |k_1 - j_1|$, which is true since $k_1, k_2 \leq n - 2$ and $|k_1 - j_1| \geq 1$.

(2.4.2) Suppose $j_1 \neq n - 1$. Then the inequality holds iff $k_1 - j_1 + k_s - j_1 \leq |k_1 - j_1| + |k_s - j_1|$, which is clearly true.

(2.5) Suppose $s, \ell \geq 2$, so $\sigma = (1, k_1, \ldots, k_s)(n, j_1, \ldots, j_\ell)\hat{\sigma}$.

Let $\sigma' = (1, n)(k_1, \ldots, k_s, j_1, \ldots, j_\ell)\hat{\sigma}$. To show $f_T(\sigma) \leq f_T(\sigma')$, it suffices to show that $d(1, k_1) + d(k_s, 1) + d(n, j_1) + d(j_\ell, n) \leq 2d(n, 1) + d(k_s, j_1) + d(j_\ell, k_1)$. By symmetry in T between vertices n and $n - 1$, we may assume $k_1, \ldots, k_s \neq n - 1$, since these cases were covered in (1). We establish this inequality as follows:

(2.5.1) Suppose $j_1 = n - 1$. Then $d(n, j_1) = 2$ and $d(n, j_\ell) = n - j_\ell - 1$. So the inequality holds iff $2k_s \leq 2(n - 2) + |j_\ell - k_1| + j_\ell - k_1$, which is true since $k_s \leq n - 2$ and $|j_\ell - k_1| + j_\ell - k_1 \geq 0$.

(2.5.2) Suppose $j_1 \neq n - 1$. Then $d(n, j_1) = n - j_1 - 1$. If $j_\ell = n - 1$, the inequality holds iff $2k_1 \leq 2(n-2) + j_1 - k_s + |j_1 - k_s|$, which is true since $k_1 \leq n-2$. If $j_\ell \neq n - 1$, the inequality holds iff $k_s - j_1 + k_1 - j_\ell \leq |k_s - j_1| + |k_1 - j_\ell|$, which is true.

This concludes the first part of the proof.

We now provide the second part of the proof. Let Γ be the Cayley graph generated by T. We show that $\mathrm{diam}(\Gamma) \leq \binom{n-1}{2} + 1$. Let $\pi \in S_n$, and suppose each vertex i of T has marker $\pi(i)$. We show that all markers can be homed using at most the proposed number of transpositions. Since $\mathrm{diam}(T) = n - 2$, marker 1 can be moved to vertex 1 using at most $n - 2$ transpositions. Now remove vertex 1 from the tree T, and repeat this procedure for marker 2, and then for marker 3, and so on, removing each vertex from T after its marker is homed. Continuing in this manner, we eventually arrive at a star $K_{1,3}$, whose Cayley graph has diameter 4. Hence, the diameter of Γ is at most $[(n - 2) + (n - 3) + \ldots + 5 + 4 + 3] + 4 = \binom{n-1}{2} + 1$. This completes the proof.

3   Concluding Remarks

Cayley graphs have been studied as a suitable model for the topology of inter-connection networks, and a problem of both theoretical and practical interest is to obtain bounds for the diameter of Cayley graphs. In this work, we investigated an upper bound on the diameter of Cayley graphs generated by transposition trees. This bound was first proposed in the oft-cited paper [1]. We showed that for every n, there exists a transposition tree on n vertices such that the difference Δ_n between the diameter upper bound and the true diameter value is at least $n - 4$. Such results are of interest because they give us insight as to how far away these bounds can be from the true diameter value in the worst case and sometimes tell us for which families of graphs this bound can be utilized or not utilized.

The results in this paper along with the results in [10] imply that the $n - 4$ lower bound on Δ_n is best possible in the sense that it is attained for some values of n; for example, it is attained for $n = 5$. Now consider the tree on 9 vertices consisting of the edges $(1, 2), (2, 3), (3, 4), (4, 5), (5, 6), (6, 7), (6, 8), (6, 9)$. Then, it can be confirmed (with the help of a computer) that the diameter of the Cayley graph generated by this tree is 24, and the diameter upper bound $f(T)$ for this tree evaluates to 30. Hence, this $n - 4$ lower bound is not the exact value of Δ_n, and an open problem is to obtain an upper bound for Δ_n.

References

1. Akers, S.B., Krishnamurthy, B.: A group-theoretic model for symmetric intercon- nection networks. IEEE Trans. Comput. 38(4), 555–566 (1989)
2. Alspach, B.: Cayley graphs with optimal fault tolerance. IEEE Trans. Com- put. 41(10), 1337–1339 (1992)
3. Annexstein, F., Baumslag, M., Rosenberg, A.L.: Group action graphs and parallel architectures. SIAM J. Comput. 19(3), 544–569 (1990)
4. Berge, C.: Principles of Combinatorics. Academic Press, New York (1971)
5. Bermond, J.C., Kodate, T., Perennes, S.: Gossiping in Cayley Graphs by Pack- ets. In: Deza, M., Manoussakis, I., Euler, R. (eds.) CCS 1995. LNCS, vol. 1120, pp. 301–315. Springer, Heidelberg (1996)
6. Biggs, N.L.: Algebraic Graph Theory, 2nd edn. Cambridge University Press, Cam- bridge (1994)
7. Bollobás, B.: Modern Graph Theory. Graduate Texts in Mathematics, vol. 184. Springer, New York (1998)
8. Chen, B., Xiao, W., Parhami, B.: Internode distance and optimal routing in a class of alternating group networks. IEEE Trans. Computers 55(12), 1645–1648 (2006)
9. Godsil, C., Royle, G.: Algebraic Graph Theory. Graduate Texts in Mathematics, vol. 207. Springer, New York (2001)
10. Ganesan, A.: On a bound for the diameter of Cayley networks of symmetric groups generated by transposition trees (2011), http://arxiv.org/abs/1111.3114v1
11. The GAP Group. GAP-Groups, Algorithms, and Programming, Version 4.4.12 (2008), http://www.gap-system.org
12. Hahn, G., Sabidussi, G. (eds.): Graph Symmetry: Algebraic Methods and Appli- cations. Kluwer Academic Publishers, Dordrecht (1997)
13. Heydemann, M.C.: Cayley graphs and interconnection networks. In: Graph Sym- metry: Algebraic Methods and Applications, pp. 167–226. Kluwer Academic Pub- lishers, Dordrecht (1997)
14. Jerrum, M.: The complexity of finding minimum length generator sequences. The- oretical Computer Sci. 36, 265–289 (1985)
15. Lakshmivarahan, S., Jho, J.-S., Dhall, S.K.: Symmetry in interconnection networks based on Cayley graphs of permutation groups: A survey. Parallel Computing 19, 361–407 (1993)
16. Smith, J.H.: Factoring, into edge transpositions of a tree, permutations fixing a terminal vertex. J. Combinatorial Theory Series A 85, 92–95 (1999)
17. Vaughan, T.P.: Bounds for the rank of a permutation on a tree. J. Combinatorial Math. Combinatorial Computing 19, 65–81 (1991)
18. Vaughan, T.P., Portier, F.J.: An algorithm for the factorization of permutations on a tree. J. Combinatorial Math. Combinatorial Computing 18, 11–31 (1995)

Power Graphs of Finite Groups of Even Order

Sriparna Chattopadhyay and Pratima Panigrahi

Department of Mathematics,
Indian Institute of Technology Kharagpur,
Kharagpur - 721302,
West Bengal, India
sriparnamath@gmail.com, pratima@maths.iitkgp.ernet.in

Abstract. The power graph $\mathscr{G}(G)$ of a finite group G is the graph with vertex set G, having an edge joining x and y whenever one is a power of the other. In this paper we study some properties of $\mathscr{G}(S_n)$ and $\mathscr{G}(D_n)$, $(n \geq 3)$, where S_n and D_n are the symmetric group on n letters and dihedral group of degree n respectively. Finally we discuss the details about the power graphs of finite non-abelian groups of order up to 14.

Keywords: Group, Euler's ϕ function, Power graph, Planar graph, Eulerian graph, Chromatic number of a graph.

1 Introduction

Kelarav and Quinn [5] defined the directed power graph of a semigroup S as a directed graph in which the vertex set is S and for $x, y \in S$ there is an arc from x to y if and only if $x \neq y$ and $y = x^m$ for some positive integer m. Following this Chakrabarty et al. [3] defined the (undirected) power graph $\mathscr{G}(S)$ of a semigroup S with vertex set S and two vertices $x, y \in S$ are adjacent if and only if $x \neq y$ and $x^m = y$ or $y^m = x$ for some positive integer m and also they have proved that $\mathscr{G}(G)$ is connected for any finite group G. Moreover in [3] it was shown that for any finite group G, the power graph of a subgroup of G is an induced subgraph of $\mathscr{G}(G)$ and $\mathscr{G}(G)$ is complete if and only if G is a cyclic group of order 1 or p^m, for some prime p and for some $m \in \mathbb{N}$. In [2] Cameron and Ghosh proved that finite abelian groups with isomorphic power graphs must be isomorphic. Also in [2] it was shown that two finite groups with isomorphic directed power graphs have the same number of elements of each order. In [1] Cameron showed that the undirected power graph determines the directed power graph up to isomorphism and hence two finite groups which have isomorphic undirected power graphs have the same number of elements of each order.

In [3] the authors have studied properties of power graphs of finite commutative groups, in particular finite cyclic groups. However in this paper we study properties of power graphs of finite non-commutative groups. Particular attention is given to the power graphs of the dihedral group D_n of degree n and of the symmetric group S_n on n letters, $(n \geq 3)$. It is known [6] that the chromatic

P. Balasubramaniam and R. Uthayakumar (Eds.): ICMMSC 2012, CCIS 283, pp. 62–67, 2012.
© Springer-Verlag Berlin Heidelberg 2012

number $\chi(G)$ of an (undirected) graph G is the smallest positive integer k so that vertices of G can be colored using k colors satisfying no two adjacent vertices get the same color (proper k-coloring) and a graph G is said to be planar if it can be drawn on a plane without any crossing of its edges. Here we prove that if $n = p^m$, for some prime p and for some $m \in \mathbb{N}$, then $\chi(\mathscr{G}(D_n)) = n$ and we find the values of n for which $\mathscr{G}(D_n)$ is not planar.

2 Properties of $\mathscr{G}(D_n)$ and $\mathscr{G}(S_n)$

It is known that a connected graph G is Eulerian if and only if degree of every vertex of G is even.

Theorem 1. *The power graph of any finite group of even order is not Eulerian.*

Proof. Let G be a finite group with $\mid G \mid = 2m, m \in \mathbb{N}$. Then $\mathscr{G}(G)$ is connected and the identity element e of G is adjacent to every other vertices in $\mathscr{G}(G)$ [3]. Thus degree of e in $\mathscr{G}(G)$ is $2m - 1$, which is an odd number and so $\mathscr{G}(G)$ is not Eulerian.

We want to give particular attention to the dihedral group and the symmetric group. For each $n \geq 3$, the dihedral group [4] $D_n = <a, b>$ is a group of order $2n$ whose generators a and b satisfy:
(i) $o(a) = n, o(b) = 2$
(ii) $ba = a^{-1}b = a^{n-1}b$.
Since $o(a) = n$, $H = <a>$ is a cyclic subgroup of D_n with $\mid H \mid = n$. As dihedral group $D_n(n \geq 3)$ is a finite group of even order, the following is immediate.

Corollary 1. *The power graph of any dihedral group $D_n(n \geq 3)$ is not Eulerian.*

Let $K_n, K_{m,n}$ be the complete graph with n vertices and complete bipartite graph with a bipartition into two sets of m and n vertices respectively. Then we know that if a graph G contains K_n as a subgraph then $\chi(G) \geq n$ and if G is a supergraph of the non-planar graphs K_5 or $K_{3,3}$ then G is non-planar [6]. Then we have the following:

Theorem 2. *If $n = p^m$, for some prime p and for some $m \in \mathbb{N}$, then*
(i) $\chi(\mathscr{G}(D_n)) = n$ *and*
(ii) $\chi(\mathscr{G}(S_n)) \geq n$. *Moreover for all such n with $n \geq 5$, $\mathscr{G}(S_n)$ is non-planar.*

Proof. (i) If $n = p^m$, for some prime p and for some $m \in \mathbb{N}$, then $\mathscr{G}(H) = K_n$ [3]. As H is a subgroup of D_n, $\mathscr{G}(D_n)$ contains K_n as an induced subgraph [3]. So $\chi(\mathscr{G}(D_n)) \geq n$.

Now $(ab)^2 = a(ba)b = a(a^{n-1}b)b = a^n b^2 = e$ and $(a^{n-1}b)^2 = a^{n-1}(ba)a^{n-2}b = a^{n-1}(a^{n-1}b)a^{n-2}b = a^n(a^{n-2}b)^2 = (a^{n-2}b)^2$. Proceding in this way it can be easily verified that $(a^{n-1}b)^2 = (a^{n-2}b)^2 = \cdots = (a^2b)^2 = (ab)^2 = e$. So for $1 \leq k \leq n-1$ power of any of $\{b, a^k b\}$ is either itself or e. Again for $1 \leq k \leq n-1$ power of any a^k is some power of a, which is none of $\{b, a^k b\}$. Thus b and $a^k b$ are adjacent to only e, for all k satisfying $1 \leq k \leq n - 1$. Therefore the graph $\mathscr{G}(D_n)$ is of the form given in Fig. 1.

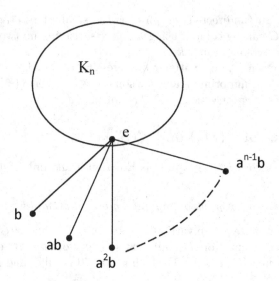

Fig. 1. Power Graph of $\mathscr{G}(D_n)$ where n is power of some prime

Assigning color 1 to e, color 2 to a, color 3 to $a^2, \cdots$, color n to a^{n-1}, that is n distinct colors to n distinct vertices of $\mathscr{G}(H) = K_n$ and color 2 to all of $\{b, ab, a^2b, \cdots, a^{n-1}b\}$ gives a proper n-coloring for $\mathscr{G}(D_n)$. So $\chi(\mathscr{G}(D_n)) \leq n$.
Hence $\chi(\mathscr{G}(D_n)) = n$.

(ii) It is known that for each $n \geq 3$, $T = < (12 \ldots n) >$ is a cyclic subgroup of order n of the non-commutative symmetric group S_n [4]. If $n = p^m$, for some prime p and for some $m \in \mathbb{N}$, then $\mathscr{G}(T) = K_n$ [3]. As T is a subgroup of S_n, $\mathscr{G}(S_n)$ contains K_n as an induced subgraph [3]. Hence $\chi(\mathscr{G}(S_n)) \geq n$. Moreover if all such n be ≥ 5 then $\mathscr{G}(T) = K_n$ is a supergraph of K_5. Thus $\mathscr{G}(S_n)$ is a supergraph of K_5 and so $\mathscr{G}(S_n)$ is non-planar.

Theorem 3. $\mathscr{G}(D_n)$ *is non-planar for the following values of n:*
(i) $n = p^m$, *for some prime p, $m \in \mathbb{N}$ and $n \geq 5$;*
(ii) $n = 2p$, *for some odd prime p.*

Proof. (i) For $n = p^m$ and $n \geq 5$, similar argument as in Theorem 2(i) shows that $\mathscr{G}(H)$ and hence $\mathscr{G}(D_n)$ is a supergraph of K_5. So $\mathscr{G}(D_n)$ is non-planar.

(ii) For $n = 2p$, $| H | = | < a > | = 2p$. The number of generators of H is $\phi(2p) = \phi(p) = p - 1$ and so there are exactly $[2p - (p - 1) - 1] = p$ elements of H which are neither identity nor generators. Since p is an odd prime, $p \geq 3$. Therefore there are at least two generators of H say a, a^i for $gcd(i, 2p) = 1$ and at least three elements of H say $a^{r_1}, a^{r_2}, a^{r_3}$ which are neither identity nor generators. Now $\{e, a, a^i\}$ be the set of vertices of $\mathscr{G}(H)$ which are adjacent to all other vertices of $\mathscr{G}(H)$ [1], in particular to all of $\{a^{r_1}, a^{r_2}, a^{r_3}\}$. Thus the subgraph of $\mathscr{G}(H)$ induced by $\{e, a, a^i, a^{r_1}, a^{r_2}, a^{r_3}\}$ is a supergraph of $K_{3,3}$. Hence $\mathscr{G}(D_n)$ being a supergraph of $\mathscr{G}(H)$ contains $K_{3,3}$ as a subgraph [3]. So $\mathscr{G}(D_n)$ is non-planar.

3   Power Graphs of Non-commutative Groups of Order Up to 14

It is known from the classification [4, p.98] of finite groups that all non-commutative groups of order up to 14 are of even order. So by Theorem 1 one gets that, the power graph of any non-commutative group of order up to 14 is not Eulerian. Up to isomorphism non-commutative groups of order up to 14 are $S_3(\simeq D_3), D_4, Q_8, D_5, A_4, D_6, Q_{12}, D_7$, where Q_8, A_4, Q_{12} are the quaternion group, the alternating group of degree 4 and the dicyclic group of order 12 respectively [4]. It is known that [3] for a finite group G the number of edges of $\mathscr{G}(G)$ is $\frac{1}{2}\sum_{a\in G}\{2o(a) - \phi(o(a)) - 1\}$. Here is a short description of all non-commutative groups of order up to 14 and of their power graphs.

(i) $S_3 = \{e, (12), (13), (23), (123), (132)\}$, where $(12)^2 = (13)^2 = (23)^2 = e$, $(123)^2 = (132), (132)^2 = (123), (123)^3 = (132)^3 = e$. So all of $\{(12), (13), (23)\}$ are adjacent to only e and $\{e, (123), (132)\}$ induces K_3. No other pair of vertices are adjacent as the number of edges in $\mathscr{G}(S_3)$ is 6. Since K_3 is a subgraph of $\mathscr{G}(S_3), \chi(\mathscr{G}(S_3)) \geq 3$. Assigning color 1 to e, color 2 to (123), color 3 to (132) and again color 2 to all of $\{(12), (13), (23)\}$ gives a proper 3-coloring for $\mathscr{G}(S_3)$ and so $\chi(\mathscr{G}(S_3)) \leq 3$. Hence $\chi(\mathscr{G}(S_3)) = 3$.

(ii) $D_4 = \{e, a, a^2, a^3, b, ab, a^2b, a^3b\}$, where $o(a) = 4, o(b) = 2, ba = a^3b$. An easy calculation shows that $o(a^3) = 4, o(a^2) = o(ab) = o(a^2b) = o(a^3b) = 2$. Arguing same as Theorem 2(i) one can conclude that all of $\{b, ab, a^2b, a^3b\}$ are adjacent to only e and $H = \{e, a, a^2, a^3\}$ induces K_4 and also $\chi(\mathscr{G}(D_4)) = 4$.

(iii) $Q_8 = \{e, a, a^2, a^3, b, ab, a^2b, a^3b\}$, where $o(a) = 4, a^2 = b^2, ba = a^3b$. It can be easily verified that $(b)^2 = (a^3)^2 = (ab)^2 = (a^2b)^2 = (a^3b)^2 = a^2, o(a^2) = 2$. These show that order of every element except e and a^2 is 4 and a^2 is adjacent to all the other vertices of $\mathscr{G}(Q_8)$. Each of $\{e, a, a^2, a^3\}, \{e, b, a^2, a^2b\}, \{e, ab, a^2, a^3b\}$, being cyclic subgroup of Q_8 of order $4 = 2^2$, induces K_4. No other pair of vertices are adjacent as the number of edges in $\mathscr{G}(Q_8)$ is 16. Since K_4 is a subgraph of $\mathscr{G}(Q_8)$ so $\chi(\mathscr{G}(Q_8)) \geq 4$. Now assigning color 1 to e, color 2 to a^2, color 3 to all of $\{a, b, ab\}$ and color 4 to all of $\{a^3, a^2b, a^3b\}$ gives a proper 4-coloring for $\mathscr{G}(Q_8)$ and so $\chi(\mathscr{G}(Q_8)) \leq 4$. Hence $\chi(\mathscr{G}(Q_8)) = 4$.

(iv) $D_5 = \{e, a, a^2, a^3, a^4, b, ab, a^2b, a^3b, a^4b\}$, where $o(a) = 5, o(b) = 2, ba = a^4b$. An easy calculation shows that $o(a^2) = o(a^3) = o(a^4) = 5, o(ab) = o(a^2b) = o(a^3b) = o(a^4b) = 2$. Arguing same as Theorem 2(i) we can conclude that all of $\{b, ab, a^2b, a^3b, a^4b\}$ are adjacent to only e and $H = \{e, a, a^2, a^3, a^4\}$ induces K_5 and also $\chi(\mathscr{G}(D_5)) = 5$.

(v) $A_4 = \{e, (123), (132), (124), (142), (134), (143), (234), (243), (12)(34), (13)(24), (14)(23)\}$, where order of each 3-cycle is 3 and $o((12)(34)) = o((13)(24)) = o((14)(23)) = 2$. Also for distinct a, b, c any 3-cycle (abc) with $1 \leq a, b, c \leq 4$ satisfies $(abc)^2 = (acb)$. So all of $\{(12)(34), (13)(24), (14)(23)\}$ are adjacent to only

e and each of $\{e, (123), (132)\}, \{e, (124), (142)\}, \{e, (134), (143)\}, \{e, (234), (243)\}$ being cyclic subgroup of A_4 of order 3 induces K_3. No other pair of vertices are adjacent as the number of edges in $\mathscr{G}(A_4)$ is 15. Since K_3 is a subgraph of $\mathscr{G}(A_4)$ so $\chi(\mathscr{G}(A_4)) \geq 3$. Assigning color 1 to e, color 2 to all of $\{(123), (124), (134), (234)\}$, color 3 to all of $\{(132), (142), (143), (243)\}$ and again color 2 to all of $\{(12)(34), (13)(24), (14)(23)\}$ gives a proper 3-coloring for $\mathscr{G}(A_4)$ and so $\chi(\mathscr{G}(A_4)) \leq 3$. Hence $\chi(\mathscr{G}(A_4)) = 3$.

(vi) $D_6 = \{e, a, a^2, a^3, a^4, a^5, b, ab, a^2b, a^3b, a^4b, a^5b\}$, where $o(a) = 6, o(b) = 2, ba = a^5b$. An easy calculation shows that $o(a^2) = o(a^4) = 3, o(a^3) = o(ab) = o(a^2b) = o(a^3b) = o(a^4b) = o(a^5b) = 2, o(a^5) = 6, (a^5)^3 = a^3$. So all of $\{b, ab, a^2b, a^3b, a^4b, a^5b\}$ are adjacent to only e. Also $\{e, a, a^2, a^4, a^5\}$ induces K_5 and a^3 is adjacent to all of $\{e, a, a^5\}$. No other pair of vertices are adjacent as the number of edges in $\mathscr{G}(D_6)$ is 19. Since K_5 is a subgraph of $\mathscr{G}(D_6)$ so $\chi(\mathscr{G}(D_6)) \geq 5$. Assigning color 1 to e, color 2 to a, color 3 to a^2, a^3, color 4 to a^4, color 5 to a^5 and again color 2 to all of $\{b, ab, a^2b, a^3b, a^4b, a^5b\}$ gives a proper 5-coloring for $\mathscr{G}(D_6)$ and so $\chi(\mathscr{G}(D_6)) \leq 5$. Hence $\chi(\mathscr{G}(D_6)) = 5$.

(vii) $Q_{12} = \{e, a, a^2, a^3, a^4, a^5, b, ab, a^2b, a^3b, a^4b, a^5b\}$, where $o(a) = 6, a^3 = b^2, ba = a^5b$. It can be easily verified that $o(b) = o(ab) = o(a^2b) = o(a^3b) = o(a^4b) = o(a^5b) = 4, o(a^2) = o(a^4) = 3, o(a^3) = 2, o(a^5) = 6, (a^5)^3 = a^3$. Each of $\{e, a^3, b, a^3b\}, \{e, a^3, ab, a^4b\}, \{e, a^2b, a^3, a^5b\}$, being cyclic subgroup of Q_{12} of order $4 = 2^2$, induces K_4 and $\{e, a, a^2, a^4, a^5\}$ induces K_5. Also a^3 is adjacent to a, a^5. No other pair of vertices are adjacent as the number of edges in $\mathscr{G}(Q_{12})$ is 28. Thus $\mathscr{G}(Q_{12})$, being a supergraph of K_5, is non-planar and $\chi(\mathscr{G}(Q_{12})) \geq 5$. Assigning color 1 to e, color 2 to all of $\{a^2, a^3\}$, color 3 to all of $\{a^5, b, ab, a^2b\}$, color 4 to all of $\{a, a^3b, a^4b, a^5b\}$ and color 5 to a^4 gives a proper 5-coloring for $\mathscr{G}(Q_{12})$ and so $\chi(\mathscr{G}(Q_{12})) \leq 5$. Hence $\chi(\mathscr{G}(Q_{12})) = 5$.

$(viii)$ $D_7 = \{e, a, a^2, a^3, a^4, a^5, a^6, b, ab, a^2b, a^3b, a^4b, a^5b, a^6b\}$, where $o(a) = 7, o(b) = 2, ba = a^6b$. An easy calculation shows that $o(a^2) = o(a^3) = o(a^4) = o(a^5) = o(a^6) = 7, o(ab) = o(a^2b) = o(a^3b) = o(a^4b) = o(a^5b) = o(a^6b) = 2$. Arguing same as Theorem $2(i)$ we can say that all of $\{b, ab, a^2b, a^3b, a^4b, a^5b, a^6b\}$ are adjacent to only e and $H = \{e, a, a^2, a^3, a^4, a^5, a^6\}$ induces K_7 and also $\chi(\mathscr{G}(D_5)) = 7$.

The other properties of the power graphs of non-commutative groups of order up to 14 given in Table 1 are obtained applying the results given in section 2.

Table 1. Power graphs of non-commutative groups of order up to 14

Power Graphs	Vertices	Vertex degree sequences	Number of edges	Properties
$\mathscr{G}(S_3)$	$\{e, (12), (13), (23), (123), (132)\}$	5,1,1,1,2,2	6	$\chi(\mathscr{G}(S_3)) = 3$ and $\mathscr{G}(S_3)$ is Planar
$\mathscr{G}(D_4)$	$\{e, a, a^2, a^3, b, ab, a^2b, a^3b\}$	7,3,3,3,1,1,1,1	10	$\chi(\mathscr{G}(D_4)) = 4$ and $\mathscr{G}(D_4)$ is Planar
$\mathscr{G}(Q_8)$	$\{e, a, a^2, a^3, b, ab, a^2b, a^3b\}$	7,3,7,3,3,3,3,3	16	$\chi(\mathscr{G}(Q_8)) = 4$ and $\mathscr{G}(Q_8)$ is planar
$\mathscr{G}(D_5)$	$\{e, a, a^2, a^3, a^4, b, ab, a^2b, a^3b, a^4b\}$	9,4,4,4,4,1,1,1,1,1	15	$\chi(\mathscr{G}(D_5)) = 5$ and by Theorem 3, $\mathscr{G}(D_5)$ is not planar
$\mathscr{G}(A_4)$	$\{e, (123), (132), (124), (142), (134), (143), (234), (243), (12)(34), (13)(24), (14)(23)\}$	11,2,2,2,2,2,2,2,1,1,1	15	$\chi(\mathscr{G}(A_4)) = 3$ and $\mathscr{G}(A_4)$ is planar
$\mathscr{G}(D_6)$	$\{e, a, a^2, a^3, a^4, a^5, b, ab, a^2b, a^3b, a^4b, a^5b\}$	11,5,4,3,4,5,1,1,1,1,1,1	19	$\chi(\mathscr{G}(D_6)) = 5$ and by Theorem 3, $\mathscr{G}(D_6)$ is not planar
$\mathscr{G}(Q_{12})$	$\{e, a, a^2, a^3, a^4, a^5, b, ab, a^2b, a^3b, a^4b, a^5b\}$	11,5,4,9,4,5,3,3,3,3,3,3	28	$\chi(\mathscr{G}(Q_{12})) = 5$ and $\mathscr{G}(Q_{12})$, being a supergraph of $K_{3,3}$, is not planar
$\mathscr{G}(D_7)$	$\{e, a, a^2, a^3, a^4, a^5, a^6, b, ab, a^2b, a^3b, a^4b, a^5b, a^6b\}$	13,6,6,6,6,6,6,1,1,1,1,1,1,1	28	$\chi(\mathscr{G}(D_7)) = 7$ and by Theorem 3, $\mathscr{G}(D_7)$ is not planar

References

1. Cameron, P.J.: The Power Graph of a Finite Group II. J. Group Theory 13, 779–783 (2010)
2. Cameron, P.J., Ghosh, S.: The Power Graph of a Finite Group. Discrete Mathematics 311, 1220–1222 (2011)
3. Chakrabarty, I., Ghosh, S., Sen, M.K.: Undirected Power Graphs of Semigroups. Semigroup Forum 78, 410–426 (2009)
4. Hungerford, T.W.: Algebra. Springer, New York (1974)
5. Kelarev, A.V., Quinn, S.J.: Directed Graph and Combinatorial Properties of Semigroups. J. Algebra 251, 16–26 (2002)
6. West, D.B.: Introduction to Graph Theory. Prentice-Hall, New Delhi (2003)

Global Vertex-Edge Domination Sets in Total Graph and Product Graph of Path P_n Cycle C_n

S. Chitra[1,2] and R. Sattanathan[3]

[1] Manonmaniam Sundaranar University,
[2] Valliammai Engineering College, Chennai
[3] Research and PG Dept., Dept. of Mathematics,
D.G. Vaishnav College, Chennai
chitrapremkumaar@yahoo.co.in, rsattanathan@gmail.com

Abstract. In a graph G, a subset S of V is Global vertex-edge dominating set if S is vertex-edge dominating set in both G and $\overline{G}$. In this paper, we have introduced the new concept, Global vertex-edge dominating set, and identified the Global vertex–edge domination number of Total graph of K_n, $K_{1,n-1}$, $K_{n,m}$, P_n, and C_n. An attempt is made to identify the Global vertex-edge domination number of Cartesian product of path P_n and cycle C_n.

Keywords: Global vertex-edge dominating set, domination number of total graph, domination number of Cartesian product of P_n and C_n, Concatenation of graphs.

1   Introduction

Let $G = (V, E)$ be a graph with a vertex set $V(G)$ and edge set $E(G)$. The set D is a *dominating set,* if every vertex $v \in V$ is either an element of D or is adjacent to an element of D [5].

The minimum cardinality dominating set of G is said to be *domination number* and is denoted by $\gamma(G)$. The maximum cardinality of a minimal dominating set of a graph G is *upper domination number* and is denoted by $\Gamma(G)$.

K.W Peters introduced two new graph theory concepts, vertex-edge domination and edge – vertex domination [1]. We can informally define vertex- edge domination by saying that a vertex v dominates the edges incident to v as well as the edges adjacent to those incident edges. Vertex- edge domination set is simply called as *ve-dominating set.* A set $S \subsetneq V(G)$ is a vertex-edge dominating set if for all edges $e \in E(G)$, there exist a vertex $v \in S$ such that v dominates e. For a graph $G = (V,E)$ a vertex $u \in V(G)$ ve-dominates an edge $vw \in E(G)$ if (i) $u = v$ or $u = w$ (u is incident to vw), or (ii) uv or uw is an edge in G (u is incident to an edge is adjacent to vw).

1.1   Global Vertex-Edge Domination

A subset S of V is *global vertex-edge dominating set* if S is vertex-edge dominating set in both G and its compliment $\overline{G}$. A subset S of V is said to be minimal global

P. Balasubramaniam and R. Uthayakumar (Eds.): ICMMSC 2012, CCIS 283, pp. 68–77, 2012.

vertex-edge dominating set if no proper subset of S is a global vertex-edge dominating set of G [2].The minimum cardinality of a global vertex-edge dominating set is called *global vertex-edge domination number* of a graph G and is denoted by $\gamma_{gve}(G)$.The maximum cardinality of a global vertex-edge dominating set is called as *upper global vertex-edge domination number* and is denoted by $\Gamma_{gve}(G)$.

Note. For any connected graph G, if the ve - domination number is $\gamma_{ve}(G)$ and for $\overline{G}$. ve – domination number is $\gamma_{ve}(\overline{G})$ then $\gamma_{gve}(G) = max\,\{\gamma_{ve}(G)\,,\gamma_{ve}(\overline{G})\}$ [4].

1.2   Cartesian Product

Let G and H be simple connected graph, the Cartesian product of these two graphs is denoted by $G \,\square\, H$ is a graph with vertex set $V(G \,\square\, H) = V(G) \circ V(H)$, that is the set $\{(g,h) \mid g \in G, h \in H\}$. The edge set of $G \,\square\, H$ consists of all pairs $[(g_1 h_1), (g_2 h_2)]$ of vertices with $(g_1 g_2) \in E(G)$ and $h_1 = h_2$ (or) $(h_1 h_2) \in E(H)$ and $g_1 = g_2$ [7].

1.3   Total Graph

Let G be a graph with vertex set V(G) and edge set E(G). The Total graph of G is denoted by T(G) is defined as follows. The vertex set of T(G) is $V(G) \cup E(G)$. Two vertices x, y in the vertex set of T(G) are adjacent in T(G) in case one of the following holds: (i) x,y are in V(G) and x is adjacent to y in G.(ii) x,y are in E(G) and x,y are adjacent in G (iii) x is in V(G), y is in E(G) and x,y are incident in G. [3]

1.4   Concatenation of Graphs

Concatenate means combine or join of two graphs my means of an edge, vertically, horizontally to get the required graph. In this paper, we have extended the edges to get required graph by using the concatenation concept as in [6].

2   Cartesian Product of Graphs

Theorem 1. The global vertex – edge domination number of $P_2 \,\square\, P_n$ is

$$\begin{cases} \dfrac{n+3}{3} & , \quad n+3 \equiv 0 \ (\bmod \ 3) \\[2mm] \dfrac{n+2}{3} & , \quad n+2 \equiv 0 \ (\bmod \ 3) \\[2mm] \dfrac{n+1}{3} & , \quad n+1 \equiv 0 \ (\bmod \ 3) \end{cases}$$

Proof

Let G be Cartesian product of Paths P_2 with P_n, to prove the result we make use of concatenation of blocks as in [6].

Case 1. $\dfrac{n+3}{3}$, $n+3 \equiv 0$ (mod 3)

We make use of the Blocks A, A^2, A^3, as given below

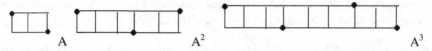

A A^2 A^3

We know that vertices of Cartesian product of P_2 and P_n is point of intersection of the lines P_2 and P_n, hence cardinality of $P_2 \square P_n$ is $2n$. The shaded vertex in each block represents the vertex edge dominating vertex. Let S be the vertex edge domination set, A^q means the concatenation of A with itself q times such that $\left| V(A^q) \right| = 2(2n+1)$.

Here $a_q = \left| V(A^q) \cap S \right|$, $a_1 = 2$, $a_2 = 3$, $a_3 = 4,\dots,a_n = n+1$, such that the vertex edge dominating set S in A^q contains a_q vertices.

There fore $\left| S \right| = a_q = \dfrac{n+3}{3}$, $n+3 \equiv 0$ (mod 3)

Case 2. $\dfrac{n+2}{3}$, $n+2 \equiv 0$ (mod 3)

We make use of the Blocks A, A^2, A^3, and B_1 as given below

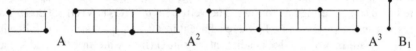

A A^2 A^3 B_1

Let us construct new blocks AB_1, $A^2 B_1$, $A^3 B_1$ as given below

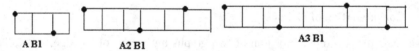

A Bl   A2 Bl   A3 Bl

Here $A^q B_r$ gives vertex edge dominating set S for $P_2 \square P_n$ where A^q is concatenation of A with itself q times such that $\left| V(A^q) \right| = 2(2n+1)$ and $\left| V(A^q B_r) \right| = 2(3n+1)$. Here $a_q = \left| V(A^q) \cap S \right|$, $b_r = \left| V(B_r) \cap S \right|$, $a_1 = 2$, $a_2 = 3$, $a_3 = 4,\dots,a_n = n+1$, $b_r = 0$ such that the vertex edge dominating set S in $A^q B_r$ contains $a_q + b_r$ vertices.

Therefore $\left| S \right| = a_q + b_r = \dfrac{n+2}{3}$, $n+2 \equiv 0$ (mod 3)

Case 3. $\dfrac{n+1}{3}$, $n+1 \equiv 0$ (mod 3)

We make use of the Blocks A, A^2, A^3, and B_2 as given below

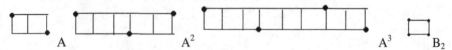

A A^2 A^3 B_2

Let us construct new blocks AB_2, $A^2 B_2$, $A^3 B_2$ as given below

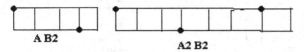

A B2   **A2 B2**

Here $A^q B_s$ gives vertex edge dominating set S for $P_2 \square P_n$ where A^q means the concatenation of A with itself q times such that $|V(A^q)| = 2(2n+1)$ and $|V(A^q B_s)| = 2(3n+2)$. Here $a_q = |V(A^q) \cap S|$, $b_s = |V(B_s) \cap S|$, $a_1 = 2$, $a_2 = 3$, $a_3 = 4,...,a_n = n+1$, $b_s = 0$ such that the vertex edge dominating set S in $A^q B_s$ contains $a_q + b_s$ vertices. Therefore

$$|S| = a_q + b_s = \frac{n+1}{3}, \quad n+1 \equiv 0 \ (\bmod \ 3)$$

In $(P_2 \square P_n)^c$ non adjacent vertices are adjacent and the upper bound of $\gamma_{gve}(G) \leq m/2$, the result is true for $(P_2 \square P_n)^c$.

Theorem 2. The global vertex – edge domination number of $P_3 \square P_n$ is

$$\begin{cases} \dfrac{n+1}{2}, & n+1 \equiv 0 \ (\bmod \ 2) \\ \dfrac{n}{2}, & n \equiv 0 \ (\bmod \ 2) \end{cases}$$

Proof

We can prove the result by method of mathematical induction.

Case i. $n+1 \equiv 0 \ (\bmod \ 2)$

When $n = 3$ the result is true, $\gamma_{gve} (P_3 \square P_3) = \dfrac{3+1}{2} = 2$. The result is trivial from the following graph.

P3 □ P3

Assume that the result is true for all $(2i-1) + 1 \equiv 0 \ (\bmod \ 2)$

The v-e domination number of $P_3 \square P_{2i-1}$ is $\dfrac{2i-1+1}{2} = i$ $(2i-1) + 1 \equiv 0 \ (\bmod \ 2)$

To prove the result is true for $n = 2(i+1) - 1$, the vertex edge domination number of $P_3 \square P_{2(i+1)-1}$ is $\dfrac{2(i+1) - 1+1}{2} = i+1$. From the graph, it is clear that $P_3 \square P_{2(i+1)-1}$ is concatenation of $P_3 \square P_{2i-1}$ and $P_3 \square P_2$.

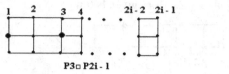

$$\gamma_{ve}(P_3 \square P_{2(i+1)-1}) = \gamma_{ve}(P_3 \square P_{2i-1}) + \gamma_{ve}(P_3 \square P_2)$$

$\dfrac{2(i+1)-1+1}{2} = \dfrac{2i-1+1}{2} + \dfrac{2}{2}$. Implies $i+1 = i+1$.

The result is true for $n = 2(i+1) -1$. By induction hypothesis the result is true for all values of n.

Case ii. $n \equiv 0 \pmod 2$

When n = 4 the result is true, $\gamma_{gve}(P_3 \square P_4) = \dfrac{4}{2} = 2$. The result is trivial from the graph.

Assume that the result is true for all $n = 2i$, $2i \equiv 0 \pmod 2$.

The vertex edge domination number of $P_3 \square P_{2i}$ is $\dfrac{2i}{2} = i$, $2i \equiv 0 \pmod 2$.

To prove the result is true for $n = 2(i+1)$, the vertex edge domination number of $P_3 \square P_{2(i+1)}$ is $\dfrac{2(i+1)}{2} = i+1$. From the graph it is clear that $P_3 \square P_{2(i+1)}$ is concatenation of $\quad P_3 \square P_{2i}$ and $P_3 \square P_2$.

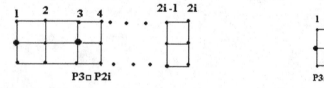

$$\gamma_{ve}(P_3 \square P_{2(i+1)}) = \gamma_{ve}(P_3 \square P_{2i}) + \gamma_{ve}(P_3 \square P_2)$$

$\dfrac{2(i+1)-1+1}{2} = \dfrac{2i-1+1}{2} + \dfrac{2}{2}$, implies $i+1 = i+1$. Result is true for $n = 2(i+1)$.

By induction hypothesis the result is true for all n.
Since the upper bound of $\gamma_{gve}(G) \le m/2$, the result is true for $(P_3 \square P_n)^c$.

Theorem 3. The global vertex – edge domination number of $C_3 \square C_n$ is

$$
\begin{cases}
\dfrac{n+1}{2}, & n+1 \equiv 0 \pmod 2 \\[2mm]
\dfrac{n}{2}, & n \equiv 0 \pmod 2
\end{cases}
\qquad n \geq 4.
$$

Proof

On following the similar lines of Theorem 2, we can prove the above result.

Theorem 4. The global vertex – edge domination number of $C_4 \square C_n$ is

$$
\begin{cases}
\dfrac{n+3}{2}, & n+3 \equiv 0 \pmod 2 \\[2mm]
\dfrac{n+2}{2}, & n+2 \equiv 0 \pmod 2
\end{cases}
$$

Proof

On following the similar lines of Theorem 1 we can prove the above result.

Theorem 5. The global vertex – edge domination number of $P_5 \square P_n = n$.

Proof

Let us consider the Cartesian product of $P_5 \square P_n$, where path P_5 is a path with $V(P_5) =$ { u_1, u_2, u_3, u_4, u_5} and P_n is path with $V(P_n) = $ { $v_1, v_2, v_3, ..., v_n$}. Thus $V(P_5 \square P_n) =$ { $R_1 R_2 R_3....R_n$, $S_1 S_2 S_3.....S_n$, $T_1 T_2 T_3...T_n$, $X_1 X_2 X_3...X_n$, $Y_1 Y_2 Y_3...Y_n$ }. Where $R_i = (v_i , u_1)$, $S_i = (v_i , u_2)$, $T_i = (v_i , u_3)$, $X_i = (v_i , u_4)$, $Y_i = (v_i , u_5)$.

Here the number of vertices in a Cartesian product of $P_5 \square P_n$ is 5n, also $S \subseteq V$ $(P_5 \square P_n)$ for which d $(y_1, y_2) \leq 4$, each path must have at least one dominating vertex, also all non adjacent vertices $x, y \in V(G) - S$ is dominated by n vertices of S. Since non adjacent vertices of $P_5 \square P_n$ are adjacent in $(P_5 \square P_n)^c$, the edge xy is dominated by n elements of S , which increases the domination number. Therefore $\gamma_{gve} (P_5 \square P_n)$ is n.

Note

The global vertex – edge domination number of $C_5 \square C_n = n$ can be proved by using the similar argument as in [7]. Thus $\gamma_{gve} (P_5 \square P_n) = \gamma_{gve} (C_5 \square C_n) = n$.

Theorem 6. The global vertex – edge domination number of $P_2 \square C_n$ is

$$
\begin{cases}
\left\lfloor \dfrac{n}{3} \right\rfloor + 1 & n \equiv 0 \pmod 3 \; n \neq 6, 12, 18, ... \\[2mm]
\left\lfloor \dfrac{n}{3} \right\rfloor, & n \equiv 0 \pmod 3 \; n = 6, 12, 18, ... \\[2mm]
\left\lfloor \dfrac{n-1}{3} \right\rfloor, & n-1 \equiv 0 \pmod 3 \\[2mm]
\left\lfloor \dfrac{n+1}{3} \right\rfloor, & n+1 \equiv 0 \pmod 3
\end{cases}
$$

Proof

Let G be Cartesian product of Path P_2 with cycle C_n, to prove the result we make use of concatenation of blocks. We make use of the Blocks A, B and C as given below

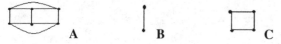

A   B   C

Case 1. $\left\lfloor \dfrac{n}{3} \right\rfloor + 1$, $n \equiv 0 \pmod 3$, n is not a multiple of 6.

Let us construct new blocks A^3, A^5, A^7 ....A^{2q+1}, Here A^q is concatenation of A with itself q times. A^{2q+1} is concatenation of A with itself $2q + 1$ times.

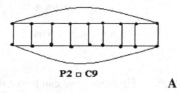

P2 □ C9

A^3

Let A^{2q+1} gives the vertex edge dominating set of $P_2 \square C_n$, $n \equiv 0 \pmod 3$, n is not multiple of 6, let it be a_{2q+1}, such that $a_{2q}+1 = \left|V(A^{2q+1}) \cap S\right|$. Here $\left|S\right| = a_{2q+1}$, for all $q = 1,2,3,\ldots$ and we get $a_3 = 4$, $a_5 = 6$, $a_7 = 8\ldots$ $a_m = m+1$ respectively. Thus $\left|S\right| = a_{2q+1} = \left\lfloor \dfrac{n}{3} \right\rfloor + 1$, $n \equiv 0 \pmod 3$ n is not multiple of 6.

Case 2. $\left\lfloor \dfrac{n}{3} \right\rfloor$, $n \equiv 0 \pmod 3$, n is multiple of 6.

Let us construct new blocks A^2, A^4, $A^6, \ldots, A^{2q}$. Here A^{2q} is concatenation of A with itself $2q$ times. Let A^{2q} gives the vertex edge dominating set of $P_2 \square C_n$, $n \equiv 0 \pmod 3$, n is multiple of 6, let it be a_{2q} such that $a_{2q} = \left|V(A^{2q}) \cap S\right|$. Here $\left|S\right| = a_{2q}$, for all $q = 1,2,3,\ldots$ and $a_2 = 2$, $a_4 = 4$, $a_6 = 6,\ldots a_{2q} = 2q$, respectively.

Thus $\left|S\right| = a_{2q} = \left\lfloor \dfrac{n}{3} \right\rfloor$, $n \equiv 0 \pmod 3$

Case 3. $\left\lfloor \dfrac{n-1}{3} \right\rfloor + 1$, $n-1 \equiv 0 \pmod 3$

Let us construct new blocks A, A^2B, $A^3B,\ldots..$, Here A^qB is concatenation of A^q with B and A^q is concatenation of A with itself q times.

Let A^qB gives the vertex edge dominating set of $P_2 \square C_n$ $n-1 \equiv 0 \pmod 3$, let it be a_qb, such that $a_qb = \left|V(A^qB) \subsetneq S\right|$. Here $\left|S\right| = a_q + b$, the vertex edge dominating set of A^qB contains $a_q + b$ vertices. Here $a_1 = 1$, $a_2 = 2$, $a_3 = 3,\ldots$ b = 1, $P_2 \square C_4 = a_1 + b = 2$, $P_2 \square C_7 = a_2 + b = 3$ and $P_2 \square C_{10} = a_3 + b = 4$.

Thus $|S| = a_q + b = \left\lfloor \dfrac{n-1}{3} \right\rfloor + 1$, $n - 1 \equiv 0 \ (\bmod\ 3)$

Case 4. $\left\lfloor \dfrac{n+1}{3} \right\rfloor$, $n + 1 \equiv 0 \ (\bmod\ 3)$

Let us construct new blocks AC, A^2C, A^3C,... Here $A^q C$ is concatenation of A^q with C and A^qC gives the vertex edge dominating set of $P_2 \Box C_n$, $n+1 \equiv 0 \ (\bmod\ 3)$, let it be a_qc, such that $a_qc = |V(A^qC) \cap S|$. Here $|S| = a_q + c$, the vertex edge dominating set of $A^q C$ contains $a_q + c$ vertices. Here $c = 1$, $a_1 = 1$, $a_2 = 2$, $a_3 = 3$,... and $P_2 \Box C_5 = a_1 + c = 2$, $P_2 \Box C_8 = a_2 + c = 3$, $P_2 \Box C_{11} = a_3 + c = 4$. Thus

$$|S| = a_q + c = \left\lfloor \dfrac{n+1}{3} \right\rfloor,   n + 1 \equiv 0 \ (\bmod\ 3)$$

Since the upper bound of $\gamma_{gve}(G) \leq m/2$, the result is true for $(P_2 \Box C_n)^c$.

Theorem 7. The global vertex – edge domination number of $P_3 \Box C_n$ is

$$\begin{cases} \left\lfloor \dfrac{n}{2} \right\rfloor, & n \equiv 0 \ (\bmod\ 2) \\[2mm] \left\lfloor \dfrac{n+1}{2} \right\rfloor, & n + 1 \equiv 0 \ (\bmod\ 2) \end{cases}$$

Proof. On the similar lines of Theorem 6, we can prove the above result.

3   Global v-e Domination of Total Graph

Proposition 1. Global vertex-edge domination number of Total graph of Star graph is

$$\gamma_{gve}(T K_{1,n-1}) = 2$$

Proof

Let $TK_{1,n-1}$ be a total graph of star graph has vertex set $V(K_{1,n-1}) \cup E(K_{1,n-1})$ that is $|V(TK_{1,n-1})| = n+n-1$, that is $2n$ -1 vertices. In a star graph, we know that one vertex 'u' has degree n-1, all other vertices are adjacent to that vertex, and all the edges are incident to the vertex u. Thus u is adjacent to all the remaining $2n$ - 2 vertices, such that u dominates the incident edges of u and adjacent to incident edges. $\gamma_{gve}(K_{1,n-1}) = 1$.

Consider the complement of $TK_{1, n-1}$, which has $2n$-1 vertices, u is an isolated vertex in the compliment of $TK_{1,n-1}$, so one more vertex is needed to dominate all the incident and adjacent to incident edge. Implies $\gamma_{gve}(T K_{1,n-1}) = 2$. From the note $\gamma_{gve}(K_{1,n-1}) = \max\{ \gamma_{ve}(K_{1,n-1}), \gamma_{ve}(\overline{K_{1,n-1}}) \}$. Thus $\gamma_{gve}(T K_{1,n-1}) = 2$.

Proposition 2. Global vertex-edge domination number of Total graph of Star graph is

$$\gamma_{gve}(T\,Kn) = \begin{cases} \dfrac{n}{2} & \text{when } n \text{ is even} \\ \dfrac{(2m+1)+(2m+3)}{4} & \text{when } n \text{ is odd}, \ n=2m+1 \ \text{for all } m=1,2,... \end{cases}$$

Proof

We will prove the result by method of induction.

Proposition 3. Global vertex-edge domination number of Total graph of Path is given by

$$\gamma_{gve}(T\,Pn) = \begin{cases} \dfrac{(m+p)k \ + \ [(m+p)k+1] \ + \ [(m+p)k+2]+p}{8} \\[2mm] \dfrac{(pk+m) \ +[((pk+m)+1] \ -2}{k} \end{cases}$$

For all $n \geq 5$, $m = 3$, $k = 5$ *and* $p = 0,1,2,3....$*Where*

$n = (m+p)k$, $n \equiv 0 \ (\text{mod}5)$ $n = (m+p)k+1$, $n-1 \equiv 0 \ (\text{mod}5)$

$n = (m+p)k+2$, $n-2 \equiv 0 \ (\text{mod}5)$ $n = (m+pk)$, $n-3 \equiv 0 \ (\text{mod}5)$

$n = (m+pk)+1$, $n-4 \equiv 0 \ (\text{mod}5)$

Proof. By method of induction we can prove the result.

Proposition 4. Global vertex-edge domination number of Total graph of cycle is given by

$$\gamma_{gve}(T\,Cn) = \begin{cases} \left\lceil \dfrac{3n+1}{3} \right\rceil + 1, & 3n+1 \equiv 1 \ (\text{mod } 3) \\[2mm] \left\lceil \dfrac{3n+2}{3} \right\rceil, & 3n+2 \equiv 2 \ (\text{mod } 3) \\[2mm] \left\lceil \dfrac{3(n+1)}{3} \right\rceil, & 3(n+1) \equiv 0 \ (\text{mod } 3) \end{cases}$$

Proof. By method of induction we can prove the result.

Proposition 5. Global vertex-edge domination number of Total graph of cycle is

$$\gamma_{gve}(T\,Kn,m) = n \ \text{ if } n = m \ \forall \ n \geq 2$$

Proof. By method of induction we can prove the result.

4   Conclusion

In this paper, we have introduced new concept *Global vertex-edge dominating set*, and identified *Global vertex-edge domination number* of Total graph of K_n, $K_{1,n-1}$, $K_{n,m}$, P_n, and C_n. An attempt is made to identify the *Global vertex – edge domination number* of Cartesian product of path P_n and cycle C_n.

References

1. Jason, R.L.: Vertex-Edge and Edge-Vertex Parameters in Graphs. A Dissertation Presented to the School of Clemson University (2007)
2. Kulli, V.R., Janakiram, B.: The total global domination number of a graph. Indian J. Pure Appl. Math. 27, 37–54 (2008)
3. Michalak, D.: On middle and total graphs with coarseness number equal 1. In: Graph Theory, Lagow Proceedings, pp. 139–150. Springer, New York (1981)
4. Sampathkumar, E.: The global Domination number of graph. J. Math. Phys. Sci. 23, 377–385 (1989)
5. Teresa, W.H., Stephen, T.H., Peter, J.S.: Domination in graph, Advanced Topics. Marcel Dekkar, New York (1998)
6. Tamizh Chelvam, T., Grace Prema, G.S.: Inverse Domination in Grid Graphs. Algebra, Graph Theory Appl., 93–103 (2010)
7. Wilfried, I., Sandi, K., Douglas, F.R.: Topics in Graph Theory Graphs and their Cartesian products. A.K Peters (2008)

Incomparability Graphs of Lattices

Meenakshi Wasadikar and Pradnya Survase*

Department of Mathematics, Dr.B.A.M. University, Aurangabad 431004, India
wasadikar@yahoo.com, survase.pradnya5@gmail.com

Abstract. Let L be a finite lattice with atleast two atoms and $W(L) = \{x \mid$ there exists $y \in L$ such that $x \parallel y \}$. The incomparability graph of L, denoted by $\Gamma'(L)$, is a graph with vertex set $W(L)$ and two distinct vertices $a, b \in W(L)$ are adjacent if and only if they are incomparable. In this paper, we study the incomparability graphs of lattices. We prove that, a disconnected graph is a graph of a lattice L if and only if L is of the form $L_1 \bigoplus L_2$. We prove that, $\Gamma'(L)$ cannot be an n-gon for any $n \geq 5$. Some properties of incomparability graphs are obtained.

Keywords: Incomparability graph, Disconnected graph, Tree, Bipartite graph, Ladder graph, Pendant vertex.

1 Introduction

There are many papers which interlink Lattice Theory and Graph Theory. In [4] Filipov discuses the comparability graphs of partially ordered sets by defining adjacency between two elements of a poset by the comparability relation, that is a, b are adjacent if either $a \leq b$ or $b \leq a$. In [3] Duffus and Rival discuss the covering graph of a poset. The papers of Gadenova [5], Bollobas and Rival [1] discuss the properties of covering graphs derived from lattices. In [9] Nimbhorkar, Wasadikar and Pawar defined graphs on a lattice L with 0, by defining adjacency of two elements $x, y \in L$ by $x \wedge y = 0$.

Recently, Bostjan Bresar et.al., [2] introduced the cover incomparability graphs of posets and called these graphs as $C - I$ graphs of P. They defined the graph in which the edge set is the union of the edge set of the corresponding covering graph and the corresponding incomparability graph.

Definition 1. *In a lattice L with 0, a nonzero element $a \in L$ is called an atom if there is no $x \in L$ such that $0 < x < a$.*

In a lattice L, if a, b are incomparable then we write $a \parallel b$.

Throughout we assume that L is a finite lattice with atleast two atoms.

We associate a graph with L and denote it by $\Gamma'(L)$. Let $W(L) = \{a \in L \mid$ there exists $b \in L$ such that $a \parallel b\}$. We associate a simple graph to a lattice L in which, (a) the vertices of $\Gamma'(L)$ are the elements of $W(L)$; (b) Two elements

* The second named author gratefully acknowledges the financial assistance in the form of Rajiv Gandhi National Senior Research Fellowship from UGC, New Delhi.

P. Balasubramaniam and R. Uthayakumar (Eds.): ICMMSC 2012, CCIS 283, pp. 78–85, 2012.
© Springer-Verlag Berlin Heidelberg 2012

$a, b \in \Gamma'(L)$ are adjacent if and only if $a \parallel b$ in L. We call it the *incomparability graph* of L.

The undefined terms and notations are from Gratzer [6], Harary [7] and West [11].

2   Properties of $\Gamma'(L)$

In this section we obtain some properties of $\Gamma'(L)$.

Remark 1. The incomparability graph of a chain is the empty graph.

Example 1. We note that the lattices shown in Figures 1 and 2 are not isomorphic but their graph is the same see, Figure 3.

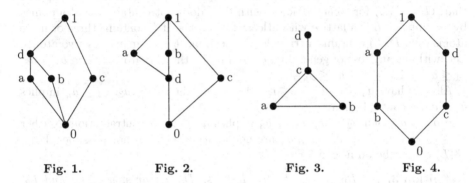

Fig. 1.   Fig. 2.   Fig. 3.   Fig. 4.

All connected graphs with less than or equal to four vertices can be realized as $\Gamma'(L)$ for some lattice L.

$\Gamma'(L)$ is complete if and only if $x \parallel y$ for all $x, y \in W(L)$.

For the lattice $L = M_n = \{0, a_1, \cdots, a_n, 1\}$ where $a_i \parallel a_j$ for all $i \neq j$, $\Gamma'(L)$ is the complete graph K_n.

Remark 2. Any two atoms in the incomparability graph of a lattice are adjacent.

Theorem 1. Let L be a lattice consisting of chains $C_1, C_2, \cdots, C_k$ between 0 and 1 such that $C_i \cap C_j = \{0, 1\}$ for $i \neq j$, $C_i \neq \{0, 1\}$ for each $1 \leq i \leq k$ and elements in two distinct chain are incomparable. Then $\Gamma'(L) = K_{n_1, n_2, \cdots, n_k}$, where $n_i = |C_i|$, $i = 1, 2, \cdots, k$.

Proof. Let $a \in C_i$, $b \in C_j$, $i \neq j$. Since $C_i \cap C_j = \{0, 1\}$, so a and b are incomparable and hence they are adjacent in $\Gamma'(L)$.

On the other hand all the elements of C_i are comparable so no two elements in C_i are adjacent with each other. This implies that $\Gamma'(L) = K_{n_1, n_2, \cdots, n_k}$.

Corollary 1. The incomparability graph of a lattice L with n atoms is an n - partite graph.

Definition 2. *A connected acyclic graph is called a tree.*

Remark 3. If the incomparability graph of a lattice L is a tree then L has exactly two atoms.

Proof. If L has three or more atoms then clearly $\Gamma'(L)$ contains a cycle and so $\Gamma'(L)$ cannot be a tree. Therefore L has exactly two atoms.

In general the converse of the Remark 3 is not true, see Fig. 4. This lattice contains two atoms and its incomparability graph is the cycle C_4.

Remark 4. If L has three atoms then $\Gamma'(L)$ contains a triangle. However in general the converse is not true, see Fig. 2 and Fig. 3.

Theorem 2. For any lattice L, $\Gamma'(L)$ cannot be an n - gon for $n \geq 5$.

Proof. Let the graph G be an n - gon $a_1, a_2, \cdots, a_n, a_1$ with $n \geq 5$. Suppose that, $G = \Gamma'(L)$ for some lattice L with 0. L does not contain one atom since by assumption L is a lattice with atleast two atoms. If L contains three or more atoms then $\Gamma'(L)$ contains a triangle. Hence L must contain exactly two atoms. Let, without any loss of generality, a_1 and a_2 be the two atoms. Then $a_1 \leq a_3$, $a_1 \leq a_4$ and $a_2 \leq a_n$, $a_2 \leq a_4$.

Also we have $a_3 \leq a_n$ or $a_n \leq a_3$. If $a_3 \leq a_n$ then $a_1 \leq a_3$, $a_3 \leq a_n$ implies $a_1 \leq a_n$, a contradiction.

If $a_n \leq a_3$ then $a_2 \leq a_n$, $a_n \leq a_3$ implies $a_2 \leq a_3$, a contradiction. Neither $a_3 \leq a_n$ nor $a_n \leq a_3$. Also a_3, a_n are nonadjacent which is not possible. Hence $\Gamma'(L)$ cannot be an n- gon for $n \geq 5$.

Definition 3. *Let L_1 and L_2 be two lattices. The linear sum of L_1 and L_2, denoted by $L_1 \oplus L_2$, is obtained by placing the diagram of L_1 directly below the diagram of L_2 and adding a line segment from the maximum element of L_1 to the minimum element of L_2.*

For example see Fig. 5.

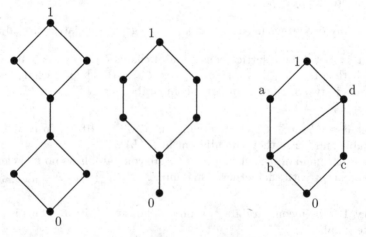

Fig. 5.   Fig. 6.   Fig. 7.

Theorem 3. If L is a linear sum $L_1 \oplus L_2$ of two lattices L_1 and L_2, each containing atleast two atoms then $\Gamma'(L)$ is disconnected.

Proof. Suppose L is of the form $L_1 \oplus L_2$.

Let x be the maximal element of L_1 and y be the minimal element of L_2. Then $x \leq y$. Hence for every element $t \in L_1$, $t \leq q$ for all $q \in L_2$ that is no element of L_1 is adjacent to any element of L_2. So we get two components in $\Gamma'(L)$ that is one from L_1 and the other from L_2. Hence $\Gamma'(L)$ is a disconnected graph.

Let G be a graph. For distinct vertices x and y of G let d(x,y) be the length of the shortest path from x and y.

The diameter of G is $diam G = sup\{d(x, y) \mid x$ and y are distinct vertices of G$\}$.

Nimbhorkar, Wasadikar and Pawar in [9] have associated a graph with a lattice L with 0 by saying that two elements are adjacent if and only if their meet is 0. Similarly we can define a graph with a lattice L with 0 by saying that two nonzero elements $a, b \in L$ are adjacent if and only if $a \wedge b = 0$ and call this graph as the zero divisor graph of L and denote it by $\Gamma(L)$.

The next theorem gives the relationship between $\Gamma(L)$ and $\Gamma'(L)$.

Theorem 4. $diam \Gamma(L) \leq diam \Gamma'(L)$.

Proof. Let $a_0 - a_1 - a_2 \cdots a_k$ be a path of length k in $\Gamma'(L)$. If $a_i \wedge a_{i+1} = 0$ for $i \in \{0, 1, \cdots, k\}$ then $a_0 - a_1 - a_2 \cdots a_k$ is a path of length k in $\Gamma(L)$. If $a_i \wedge a_{i+1} \neq 0$ for some $i \in \{0, 1, \cdots, k\}$ then $a_0 - a_1 - a_2 \cdots a_k$ is a path of length less than k in $\Gamma(L)$. Hence $diam \Gamma(L) \leq diam \Gamma'(L)$.

Definition 4. *A lattice L with 0 is called an integral lattice if for $a, b \in L$, $a \wedge b = 0$ implies $a = 0$ or $b = 0$.*

If $\Gamma'(L)$ is empty then $\Gamma(L)$ is empty. However converse does not hold.

For the converse, consider an integral lattice L. Then $\Gamma(L)$ is empty but $\Gamma'(L)$ need not be empty, see Fig. 6.

Remark 5. In Theorem 4, $diam \Gamma(L)$ is not necessarily equal to $diam \Gamma'(L)$. See Fig. 7, its $\Gamma(L)$ is Fig. 8 and $\Gamma'(L)$ is Fig. 9 and $diam \Gamma(L) = 2$ and $diam \Gamma'(L) = 3$.

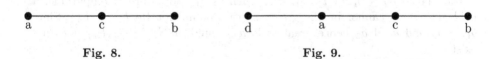

Fig. 8.   Fig. 9.

3   Properties of Some Special Graphs

Definition 5. *The ladder graph L_n is a planar undirected graph with $2n$ vertices and $n + 2(n-1)$ edges as shown in Fig. 10.*

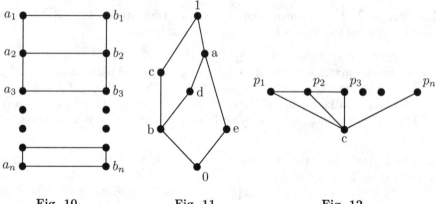

Fig. 10.   Fig. 11.   Fig. 12.

Theorem 5. The Ladder graph L_n, $n > 2$ cannot be realized as the incomparability graph of a lattice.

Proof. Consider, the ladder graph L_n, $n > 2$. Let the set of vertices of L_n be $A \bigcup B$ where $A = \{a_1, a_2, \cdots, a_n\}$ and $B = \{b_1, b_2, \cdots, b_n\}$ as shown in Figure 10. Suppose $L_n = \Gamma'(L)$ for some lattice L. Then L has exactly two atoms, since if L contains more than two atoms then $\Gamma'(L)$ must contain a triangle.

Let a_1 and a_2 be the two atoms in L from the set A. We have $a_1 \leq a_3$, $a_1 \leq b_3$ and $a_2 \leq b_3$, $a_2 \leq b_1$. Also we have $a_3 \leq b_1$ or $b_1 \leq a_3$.

If $a_3 \leq b_1$ then $a_1 \leq a_3$ and $a_3 \leq b_1$ implies $a_1 \leq b_1$, a contradiction, since a_1 and b_1 are adjacent.

If $b_1 \leq a_3$ then $a_2 \leq b_1$, $b_1 \leq a_3$ implies $a_2 \leq a_3$, a contradiction, since a_2 and a_3 are adjacent. Hence neither $a_3 \leq b_1$ nor $b_1 \leq a_3$ which is not possible.

The same proof can be given if L contains two atoms from the set B.

Now, let a_1 and b_1 be the two atoms in L from the two sets A and B respectively. We know that, in a lattice the greatest lower bound of any nonempty finite subset of L must exist.

We now show that the greatest lower bound of $b_3, b_4, \cdots, b_n$ does not exist. We note that the possible set of lower bounds of $\{b_3, b_4, \cdots, b_n\}$ is $\{0, a_1, b_1, a_2\}$, as b_i, b_{i+1} are incomparable with each other for $i = 3, 4, \cdots, n$. Also a_i, b_i are incomparable with each other for $i = 3, \cdots, n$.

Case (1) If $a_2 \leq b_i$ for $i = 3, \cdots, n$ then as a_1 and a_2 are incomparable, it is clear that a_2 cannot be the greatest lower bound for $\{b_3, b_4, \cdots, b_n\}$. Since $b_1 \leq a_2$ and $a_1 \parallel a_2$ hence greatest lower bound of $\{b_3, b_4, \cdots, b_n\}$ does not exists.

Case (2) If $a_2 \nleq b_i$ for some $i = 3, \cdots, n$. Then a_2 cannot be a lower bound for $\{b_3, b_4, \cdots, b_n\}$. In this case the only lower bounds of $\{b_3, b_4, \cdots, b_n\}$ are $\{0, a_1, b_1\}$. Since $a_1 \parallel b_1$ hence greatest lower bound of $\{b_3, b_4, \ldots, b_n\}$ does not exist.

Definition 6. *The bull graph is a graph with five vertices, consisting a triangle and two pendant vertices adjacent to two different vertices, as shown in the Figure 13.*

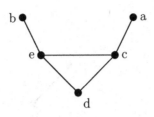

Fig. 13.

Remark 6. The bull graph is realizable as the incomparability graph of a lattice, see Fig. 11 and Fig. 13.

Definition 7. *The fan graph is denoted by $F_n = P_n \cup \{c\}$ where P_n is a path graph with n vertices and c is adjacent to all vertices of P_n, see Fig. 12.*

Theorem 6. The Fan graph $F_n = P_n \cup \{c\}$, $n \geq 5$ cannot be realized as the incomparability graph of a lattice.

Proof. Suppose F_n can be realized as $\Gamma'(L)$ for some lattice L. Suppose p_2 and p_3 are two atoms in L. Then $p_2 \leq p_4$, $p_2 \leq p_5$ and $p_3 \leq p_1$, $p_3 \leq p_5$. Also we have $p_1 \leq p_4$ or $p_4 \leq p_1$.

If $p_1 \leq p_4$ then $p_3 \leq p_1$, $p_1 \leq p_4$ implies $p_3 \leq p_4$, a contradiction, since p_3 and p_4 are adjacent.

If $p_4 \leq p_1$ then $p_2 \leq p_4$, $p_4 \leq p_1$ implies $p_2 \leq p_1$, a contradiction, since p_1 and p_2 are adjacent.

Hence neither $p_1 \leq p_4$ nor $p_4 \leq p_1$ holds which is not possible. We cannot take three atoms other than c because $p_i = 1, 2, \cdots, n$ are elements of the path P_n and do not form a triangle.

Theorem 7. If $\Gamma'(L)$ is a complete bipartite graph then L has exactly two atoms. However the converse need not hold.

Proof. Suppose that $\Gamma'(L)$ is a complete bipartite graph. Clearly L has atleast two atoms and a complete bipartite graph has no cycle of length three, L has no more than two atoms. Thus L has exactly two atoms.

For the converse see Fig. 2 and Fig. 3.

Remark 7. In $\Gamma(L)$, the lattice L has two atoms if and only if $\Gamma(L)$ is a complete bipartite graph.

It is shown in [10] that there is no lattice L whose zero divisor graph is a double star graph. However a double star graph can be realizable as $\Gamma'(L)$ for a lattice L, see Fig. 14 and Fig. 15.

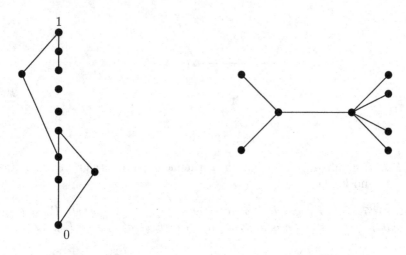

Fig. 14.   Fig. 15.

It is shown in [10] that diam$\Gamma(L) \leq 3$. However for $\Gamma'(L)$ we have the following remark.

Remark 8. There exists a lattice L with $diam\Gamma'(L) = 5$, see Fig. 16 and Fig. 17.

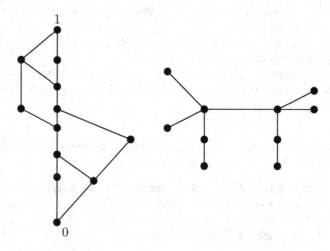

Fig. 16.   Fig. 17.

Theorem 8. Let $\Gamma'(L)$ be a bipartite but not complete bipartite graph.

i) If L has two atoms and an element which covers both the atoms then $\Gamma'(L)$ has exactly one pendant vertex.

ii) If L has two dual atoms and an element which is covered by both the dual atoms then $\Gamma'(L)$ has exactly one pendant vertex.

iii) If L satisfies both the conditions above then $\Gamma'(L)$ has two pendant vertices.

Proof. i) Suppose that L has two atoms and an element which covers both the atoms. Let a and b be the two atoms and c cover both a and b. So we have $a \leq c$ and $b \leq c$. Let V_1 and V_2 be the two partitions in $\Gamma'(L)$. Let $a \in V_1$ and $b \in V_2$. Suppose $c \in V_2$ then for all $d \in V_1$ we have $d \parallel b$. If not then we have $a, d \in V_1$, $b \leq d$ so $\{0, a, b\}$ is the set of lower bounds of c and d that is $c \wedge d$ does not exist. Hence $d \parallel b$.

Hence for all $d \in V_1$, $d \in N(b)$ that is $deg(b) > 1$. For any $e \in V_2$ ($e \neq c$) we have $a \leq e$. Hence a is comparable with every element from V_2 except b. So $deg(a) = 1$.

If $c \in V_1$ then in similar manner we have $deg(b) = 1$.

ii) Let x, y be the two dual atoms and z be covered by both x and y. We have $z \leq x$, $z \leq y$. Let V_1 and V_2 be two partitions in $\Gamma'(L)$. Let $x \in V_1$, $y \in V_2$ and suppose $z \in V_2$ then for all $u \in V_1$ we have $u \parallel y$. If not then $x, u \in V_1$, $u \leq y$. and hence, $\{1, x, y\}$ is the set of all upper bounds of z and u. Since, $x \parallel y$, $z \vee u$ does not exist. Hence $u \parallel y$ so for all $u \in V_1$, $u \in N(y)$ that is $deg(y) > 1$. For any $w \in V_2$ we have $w \leq x$. Hence x is comparable with every element from V_2 except y. Hence $deg(x) = 1$.

If $z \in V_1$ then in similar manner we have $deg(y) = 1$.

iii) By combining above two results we get $\Gamma'(L)$ is bipartite with two pendant vertices.

Remark 9. An incomparability graph $\Gamma'(L)$ is bipartite if and only if it contains no triangles.

Acknowledgments. The authors are thankful to Prof. S. K. Nimbhorkar, for his helpful suggestions.

References

1. Bollobas, B., Rival, I.: The maximal size of the covering graph of a lattice. Algebra Universalis 9, 371–373 (1979)
2. Bresar, B., Chhangat, M., Klavzar, S., Kovse, M., Mathews, J., Mathews, A.: Cover-incomparability graphs of posets. Order 25, 335–347 (2008)
3. Duffus, D., Rival, I.: Path lengths in the covering graph. Discrete Math. 19, 139–158 (1977)
4. Filipov, N.D.: Comparability graphs of partially ordered sets of different types. Collq. Maths. Soc. Janos Bolyai. 33, 373–380 (1980)
5. Gedenova, E.: Lattices, whose covering graphs are s-graphs. Colloq. Maths. Soc. Janos Bolyai. 33, 407–435 (1980)
6. Grätzer, G.: General Lattice Theory. Birkhäuser, Basel (1998)
7. Harary, F.: Graph Theory. Addison-Wesley, Reading (1972)
8. Nimbhorkar, S.K., Wasadikar, M.P., Demeyer, L.: Coloring of meet semilattices. Ars Combinatoria 84, 97–104 (2007)
9. Nimbhorkar, S.K., Wasadikar, M.P., Pawar, M.M.: Coloring of Lattices. Math. Slovaca. 60, 419–434 (2010)
10. Wasadikar, M.P., Survase, P.A.: Some Properties of Graphs derived from Lattices (preprint)
11. West, D.B.: Introduction to Graph theory. Prentice-Hall, New Delhi (1996)

A New Characterization
of Paired Domination Number of a Graph

G. Mahadevan[1], A. Nagarajan[2], A. Selvam[3], and A. Rajeswari[4]

[1] Department of Mathematics,
Anna University of Technology Tirunelveli,
Tirunelveli-627 002
gmaha2003@yahoo.co.in
[2] Department of Mathematics,
V. O. Chidambaram College, Tuticorin- 628 008
nagarajan.voc@gmail.com
[3] Department of Mathematics, V.H.N.S.N. College,
Virudhunagar-626 001
selvam_avadayappan@yahoo.co.in
[4] Aarupadai Veedu Institute of Technology,
Paiyanoor, Chennai -603 104
rajeswarivenket@yahoo.com

Abstract. Paired domination is a relatively interesting concept introduced by Teresa W. Haynes [9] recently with the following application in mind. If we think of each vertex s ∈ S, as the location of a guard capable of protecting each vertex dominated by S, then for a paired domination the guards location must be selected as adjacent pairs of vertices so that each guard is assigned one other and they are designated as a backup for each other. A set $S \subseteq V$ is a paired dominating set if S is a dominating set of G and the induced sub graph <S> has a perfect matching. The paired domination number $\gamma_{pr}(G)$ is the minimum cardinality taken over all paired dominating sets in G. The minimum number of colours required to colour all the vertices so that adjacent vertices do not receive the same colour and is denoted by $\chi(G)$. In this paper we characterize the class of all graphs whose sum of paired domination number and chromatic number equals to 2n – 7, for any n ≥ 4.

Keywords: Paired domination number, Chromatic number.

AMS subject Classification: 05C (primary).

1   Introduction

Throughout this paper, by a graph we mean a finite, simple, connected and undirected graph G(V, E). For notations and terminology, we follow [11]. The number of vertices in G is denoted by n. Degree of a vertex v is denoted by deg(v). We denote a cycle on n vertices by C_n, a path of n vertices by P_n, complete graph on n vertices by K_n. If S is a subset of V, then <S> denotes the vertex induced sub graph of G induced

P. Balasubramaniam and R. Uthayakumar (Eds.): ICMMSC 2012, CCIS 283, pp. 86–96, 2012.
© Springer-Verlag Berlin Heidelberg 2012

by S. A subset S of V is called a dominating set of G if every vertex in V-S is adjacent to at least one vertex in S. The domination number $\gamma(G)$ of G is the minimum cardinality of all such dominating sets in G. A dominating set S is called a total dominating set ,if the induced sub graph <S> has no isolated vertices. The minimum cardinality taken over all total dominating sets in G is called the total domination number and is denoted by $\gamma_t(G)$. One can get a comprehensive survey of results on various types of domination number of a graph in [10]. The chromatic number $\chi(G)$ is defined as the minimum number of colors required to color all the vertices such that adjacent vertices receive the same color.

Recently many authors have introduced different types of domination parameters by imposing conditions on the dominating set and/or its complement. Teresa W. Haynes [9] introduced the concept of paired domination number of a graph. If we think of each vertex s $\in$ S, as the location of a guard capable of protecting each vertex dominated by S, then for domination a guard protects itself, and for total domination each guard must be protected by another guard. For a paired domination the guards location must be selected as adjacent pairs of vertices so that each guard is assigned one other and they are designated as a backup for each other. Thus a paired dominating set S with matching M is a dominating set S = $\{v_1, v_2, v_3...v_{2t-1}, v_{2t}\}$ with independent edge set M = $\{e_1, e_2, e_3... e_t\}$ where each edge e_i is incident to two vertices of S, that is M is a perfect Matching in <S>. A set S $\subseteq$ V is a paired dominating set if S is a dominating set of G and the induced sub graph <S> has a perfect matching. The paired domination number $\gamma_{pr}(G)$ is the minimum cardinality taken over all paired dominating sets in G.

Several authors have studied the problem of obtaining an upper bound for the sum of a domination parameter and a graph theoretic parameter and characterized the corresponding extremal graphs. In [8], Paulraj Joseph J and Arumugam S proved that $\gamma + \kappa \le p$, where κ denotes the vertex connectivity of the graph. In [7], Paulraj Joseph J and Arumugam S proved that $\gamma_c + \chi \le p + 1$ and characterized the corresponding extremal graphs. They also proved similar results for γ and γ_t. In [6], Mahadevan G Selvam A, Iravithul Basira A characterized the extremal of graphs for which the sum of the complementary connected domination number and chromatic number. In [3], Paulraj Joseph J and Mahadevan G proved that $\gamma_{pr} + \chi \le 2n - 1$, and characterized the corresponding extremal graphs of order up to $2n - 6$. Motivated by the above results, in this paper we characterize all graphs for which $\gamma_{pr}(G) + \chi(G) = 2n - 7$ for any $n \ge 4$.

2   Main Result

We use the following preliminary results and notations for our consequent characterization:

Theorem 2.1[4] For any connected graph G of order $n \ge 3, \gamma_{pr}(G) \le n - 1$ and equality holds if and only if G = C_3, C_5 or subdivided star $S(K_{1,n})$.

Notation 2.2. $C_3(n_1 P_{m_1}, n_2 P_{m_2}, n_3 P_{m_3})$ is a graph obtained from C_3 by attaching n_1 times the pendent vertex of P_{m_1} (Path on m_1 vertices) to a vertex u_i of C_3 and attaching n_2 times the pendent vertex of P_{m_2} (Path on m_2 vertices) to a vertex u_j for $i \neq j$ of C_3 and attaching n_3 times the pendent vertex of P_{m_3} (Path on m_3 vertices) to a vertex u_k for $i \neq j \neq k$ of C_3.

Notation 2.3. $C_3(u(P_{m_1}, P_{m_2}))$ is a graph obtained from C_3 by attaching the pendent vertex of P_{m_1} (Path on m_1 vertices) and the pendent vertex of P_{m_2} (Paths on m_2 vertices) to any vertex u of C_3.

Notation 2.4. $K_5(n_1 P_{m_1}, n_2 P_{m_2}, n_3 P_{m_3}, n_4 P_{m_4}, n_5 P_{m_5})$ is a graph obtained from K_5 by attaching n_1 times the pendent vertex of P_{m_1} (Paths on m_1 vertices) to a vertex u_i of K_5 and attaching n_2 times the pendent vertex of P_{m_2} (Paths on m_2 vertices) to a vertex u_j for $i \neq j$ of K_5 and attaching n_3 times the pendent vertex of P_{m_3} (Paths on m_3 vertices) to a vertex u_k for $i \neq j \neq k$ of K_5 and attaching n_4 times the pendent vertex of P_{m_4} (Paths on m_4 vertices) to a vertex u_l for $i \neq j \neq k \neq l$ of K_5 and attaching n_5 times the pendent vertex of P_{m_5} (Paths on m_5 vertices) to a vertex u_m for $i \neq j \neq k \neq l \neq m$ of K_5.

Notation 2.5. $C_3(P_n)$ is the graph obtained from C_3 by attaching the pendant edge of P_n to any one vertices of C_3 and $K_n(P_m)$ is the graph obtained from K_n by attaching the pendant edge of P_m to any one vertices of K_n. For $n \leq p$, $K_p(n)$ is the graph obtained from K_p by adding a new vertex and join it with n vertices of K_p. $C_3(K_{1,n})$ is the graph obtained from C_3, by attaching the root vertex of $K_{1,n}$ to any one vertex of C_3.

Theorem 2.6. For any connected graph G of order n $(n \geq 3)$, $\gamma_{pr}(G) + \chi(G) = 2n - 7$ if and only if $G \cong C_3(P_5)$, $C_3(K_{1,3})$, $C_3(2P_2, P_2, P_2)$, $C_3(2P_2, P_2, 0)$, $C_3(2P_3, 0, 0)$, $C_3(P_3, P_2, P_2)$, $C_3(P_4, P_2, 0)$, $C_3(P_3, P_3, 0)$, $C_3(u(P_4, P_2))$, $C_3(u(P_3, P_2))$, $K_5(P_4)$, $K_5(P_3)$, $K_7(1)$, $K_7(2)$, $K_7(3)$, $K_7(4)$, $K_7(5)$, $K_7(6)$, $K_5(P_2, P_2, P_2, 0, 0)$, $K_5(P_3, P_2, 0, 0, 0)$, $K_5(P_2, P_2, 0, 0, 0)$, K_9 or any one of the graphs shown in Fig. 1.

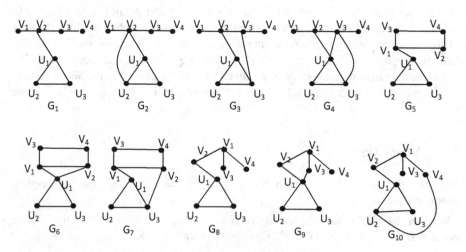

Fig. 1.

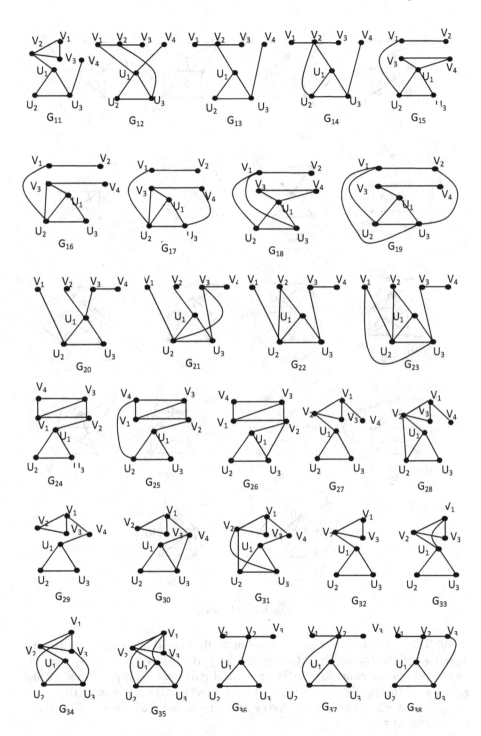

Fig. 1. (*Continued*)

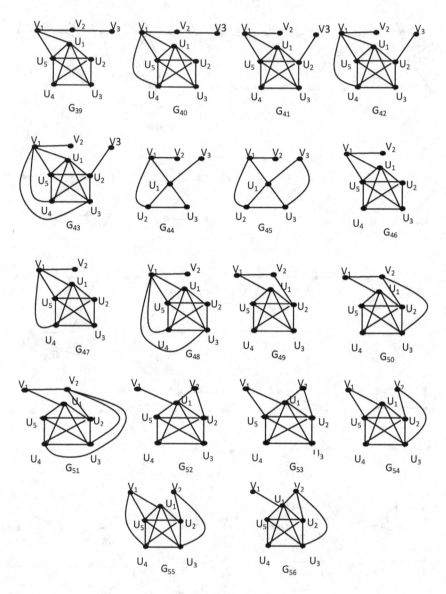

Fig. 1. (*Continued*)

Proof. If G is any one of the graphs stated in the theorem, then it can be verified that $\gamma_{pr}(G) + \chi(G) = 9 = 2n - 7$. Conversely, let $\gamma_{pr}(G) + \chi(G) = 2n - 7$. Then the various possible cases are (i) $\gamma_{pr}(G) = n - 1$ and $\chi(G) = n - 6$ (ii) $\gamma_{pr}(G) = n - 2$ and $\chi(G) = n - 5$ (iii) $\gamma_{pr}(G) = n - 3$ and $\chi(G) = n - 4$ (iv) $\gamma_{pr}(G) = n - 4$ and $\chi(G) = n - 3$ (v) $\gamma_{pr}(G) = n - 5$ and $\chi(G) = n - 2$ (vi) $\gamma_{pr}(G) = n - 6$ and $\chi(G) = n - 1$ (vii) $\gamma_{pr}(G) = n - 7$ and $\chi(G) = n$.

Case i. $\gamma_{pr} = n - 1$ and $\chi = n - 6$.

Since $\gamma_{pr} = n - 1$, by theorem, 2.1 $G \cong C_3$, C_5, or $S(K_{1,n})$. Then $\chi = 2$ or 3. If $\chi = 2$, $n = 8$, which is a contradiction. If $\chi = 3$, $n = 9$, which is a contradiction. Hence no graph exists.

Case ii. $\gamma_{pr} = n - 2$ and $\chi = n - 5$.

Since $\chi(G) = n - 5$, G contains a clique K on n−5 vertices (or) does not contains a clique K on n−5 vertices. By all the various possible cases, it can be verified that no graph exists satisfying the hypothesis.

Case iii. $\gamma_{pr} = n - 3$ and $\chi = n - 4$.

Since $\chi = n - 4$, G contains a clique K on $n - 4$ vertices or does not contain a clique K on $n - 4$ vertices. Let G contain a clique K on $n - 4$ vertices. Let $S = V(G) - V(K) = \{v_1, v_2, v_3, v_4\}$. Then the induced sub graph <S> has the following possible cases: $<S> = K_4$, $\overline{K_4}$, P_4, C_4, $K_{1,3}$, $P_3 \cup K_1$, $K_2 \cup K_2$, $K_3 \cup K_1$, $K_2 \cup \overline{K_2}$, $K_4 - \{e\}$, $C_3(1,0,0)$. If $<S> = K_4$, then it can be verified that no graph exists.

Sub case i. $<S> = \overline{K}_4$. Let $\{v_1, v_2, v_3, v_4\}$ be the vertices of $\overline{K}_4$. By the various possible cases, the only graph exists in this case is $C_3(2P_2, P_2, P_2)$.

Sub case ii. $<S> = P_4$.

Let $\{v_1, v_2, v_3, v_4\}$ be the vertices of P_4. Since G is connected, there exists a vertex u_i in K_{n-4} which is adjacent to v_1 (or v_4) (or) v_2 (or v_3). Let u_i be adjacent to v_1, then $\{u_i, v_1, v_2, v_3\}$ forms a γ_{pr} set of G so that $\gamma_{pr} = 4$ and $n = 7$. Hence $K = K_3 = <u_1, u_2, u_3>$. If u_1 is adjacent to v_1, then $G \cong C_3(P_5)$. If u_1 is adjacent to v_2, then $\{u_i, u_j, v_2, v_3\}$ forms a γ_{pr} - set of G so that $\gamma_{pr} = 4$ and $n = 7$. Hence $K = K_3 = <u_1, u_2, u_3>$. Let u_1 be adjacent to v_2, then $\deg(v_1) = \deg(v_4) = 1$, $\deg(v_2) = 3$, $\deg(v_3) = 2$, then $G \cong G_1$. Let u_1 be adjacent to v_2 and u_2 be adjacent to v_2. If $\deg(v_1) = \deg(v_4) = 1$, $\deg(v_2) = 4$, $\deg(v_3) = 2$, then $G \cong G_2$. Let u_1 be adjacent to v_2 and u_3 be adjacent to v_3. If $\deg(v_1) = \deg(v_4) = 1$, $\deg(v_2) = \deg(v_3) = 3$, then $G \cong G_3$. Let u_1 be adjacent to v_2 and v_3, and u_3 be adjacent to v_3. If $\deg(v_1) = \deg(v_4) = 1$, $\deg(v_2) = 3$, $\deg(v_3) = 4$, then $G \cong G_4$. All the remaining cases are not possible.

Sub case iii. $<S> = C_4$.

Let $\{v_1, v_2, v_3, v_4\}$ be the vertices of C_4. Since G is connected, there exists a vertex u_i in K_{n-4} which is adjacent to v_1. Let u_i for some i in K_{n-4}, be adjacent to v_1 and u_j for $i \neq j$, then $\{u_i, u_j, v_3, v_4\}$ forms a γ_{pr}-set of G so that $\gamma_{pr} = 4$ and $n = 7$. Hence $K = K_3 = <u_1, u_2, u_3>$. If u_1 is adjacent to v_1, then $\deg(v_1) = 3$, and so $G \cong G_5$. Let u_1 be adjacent to v_1 and v_2. If $\deg(v_1) = \deg(v_2) = 3$, then $G \cong G_6$. Let u_1 be adjacent to v_1 and u_3 be adjacent to v_2. If $\deg(v_1) = \deg(v_2) = 3$, then $G \cong G_7$.

Sub case iv. $<S> = K_{1,3}$.

Let v_1 be the root vertex and v_2, v_3,v_4 are adjacent to v_1. Since G is connected, there exists a vertex u_i in K_{n-4} which is adjacent to v_1 (or) any one of $\{v_2, v_3, v_4\}$ and v_4. Let u_i for some i in K_{n-4} be the vertex adjacent to v_1, then $\{u_i, v_1\}$ is a γ_{pr}-set of G so that $\gamma_{pr} = 2$ and n = 5, which is a contradiction. Hence no such graph exists. Since G is connected, there exists a vertex u_i in K_{n-4} which is adjacent to any one of $\{v_2, v_3, v_4\}$. Then u_i for some i, is adjacent to v_2. In this case, $\{u_i, u_j, v_1, v_2\}$ is an γ_{pr}-set of G so that $\gamma_{pr} = 4$ and n = 7, and hence $K = K_3 = <u_1, u_2, u_3>$. Let u_1 be adjacent to v_2. If deg(v_1) = 3, deg(v_3) = deg(v_4) = 1, deg(v_2) =2, then $G \cong G_8$. Let u_1 be adjacent to v_2 and v_3. If deg(v_1) = 3, deg(v_2) = deg(v_3) =2, deg(v_4) = 1, then $G \cong G_9$. Let u_1 be adjacent to v_2 and u_2 be adjacent to v_4. If deg(v_1) = 3, deg(v_2) = deg(v_4) = 2, deg(v_3) = 1, then $G \cong G_{10}$.

Sub case v. $<S> = K_3 \cup K_1$.

Let v_1 ,v_2 and v_3 be the vertices of K_3 and v_4 be the vertex of K_1. Since G is connected, there exists a vertex u_i in K_{n-4} which is adjacent to any one of $\{v_1, v_2, v_3\}$ and $\{v_4\}$. In this case $\{u_i, v_2\}$ is a γ_{pr}-set of G, so that $\gamma_{pr} = 2$ and n = 5, which is a contradiction. Hence no graph exists. Since G is connected, there exists a vertex u_i in K_{n-4} which is adjacent to v_2, and u_j for i ≠ j, is adjacent to v_4. In this case, $\{u_i, u_j, u_k, v_2\}$ is a γ_{pr}-set of G so that $\gamma_{pr} = 4$ and n = 7. Hence $K = K_3 = <u_1, u_2, u_3>$. Let u_1 be adjacent to v_2 and u_3 be adjacent to v_4. If deg(v_1) = deg(v_3) = 2, and deg(v_2) =3, deg(v_4) = 1, then $G \cong G_{11}$.

Sub case vi. $<S> = P_3 \cup K_1$.

Let v_1 ,v_2, v_3 be the vertices of P_3 and v_4 be the vertex of K_1. Since G is connected, there exists a vertex u_i in K_{n-4} which is adjacent to any one of $\{v_1, v_2, v_3\}$ and $\{v_4\}$. Then u_i be adjacent to v_1 and v_4. In this case $\{u_i, v_1, v_2, v_4\}$ is a γ_{pr}-set of G, so that $\gamma_{pr} = 4$ and n = 7. Hence $K = K_3 = <u_1, u_2, u_3>$. Let u_1 be adjacent to v_1 and v_4. If deg(v_1) = deg(v_2) =2, deg(v_3) = deg(v_4) = 1, then $G \cong C_3(u(P_4, P_2))$. Let u_1 be adjacent to v_1 and v_4, and let u_3 be adjacent to v_2. If deg(v_1) = 2 = deg(v_2) = 3, deg(v_3) = deg(v_4) = 1, then $G \cong G_{12}$. Since G is connected, there exists a vertex u_i in K_{n-4} which is adjacent to v_1 and there exists u_j for i ≠ j, is adjacent to v_4. In this case, $\{u_i, u_j, v_1, v_2\}$ is a γ_{pr}-set of G so that $\gamma_{pr} = 4$ and n = 7, and hence $K = K_3 = <u_1, u_2, u_3>$. Let u_1 be adjacent to v_1 and let u_3 be adjacent to v_4. If deg(v_1) = deg(v_2) =2, deg(v_3) = deg(v_4) = 1, then $G \cong C_3(P_4, P_2, 0)$. Since G is connected, there exists a vertex u_i in K_{n-4} which is adjacent to v_2 and v_4. In this case, $\{u_i, v_2\}$ is a γ_{pr}-set of G so that $\gamma_{pr} = 2$ and n = 5, which is a contradiction. Hence no graph exists. Since G is connected, there exists a vertex u_i in K_{n-4} which is adjacent to v_2 and u_j for i ≠ j is adjacent to v_4. In this case, $\{u_i, u_j, u_k, v_2\}$ is a γ_{pr}-set of G so that $\gamma_{pr} = 4$ and n = 7, and hence $K = K_3 = <u_1, u_2, u_3>$. Let u_1 be adjacent to v_2 and let u_3 be adjacent to v_4. If deg(v_1) = deg(v_3) = deg(v_4) = 1, deg(v_2) = 3, then $G \cong G_{13}$. Let u_1 be adjacent to v_2 and u_2 be adjacent to v_2 and u_3 be adjacent to v_4. If deg(v_1) = deg(v_3) = deg(v_4) = 1, deg(v_2) = 4, then $G \cong G_{14}$.

Sub case vii. $<S> = K_2 \cup K_2$.

Let v_1 and v_2 be the vertices of K_2 and v_3, v_4 be the vertices of K_2. Since G is connected, there exists a vertex u_i in K_{n-4} which is adjacent to any one of $\{v_1, v_2\}$ and any one of $\{v_3, v_4\}$. Let u_i be adjacent to v_1 and v_3. In this case $\{u_i, v_1, v_2, v_3\}$ forms a γ_{pr}-set of G, so that $\gamma_{pr} = 4$ and n = 7. Hence K = K_3 = $<u_1, u_2, u_3>$. Let u_1 be adjacent to v_1 and v_3. If $\deg(v_1) = \deg(v_3) = 2$, then G $\cong C_3(2P_3, 0, 0)$. Let u_1 be adjacent to v_3 and v_4, and let u_2 be adjacent to v_1, and u_3 be adjacent to v_2. If $\deg(v_1) = \deg(v_2) = \deg(v_3) = \deg(v_4) = 2$, then G $\cong G_6$. Since G is connected, there exists a vertex u_i in K_{n-4} which is adjacent to v_1 and there exists u_j for i $\neq$ j, is adjacent to v_3. In this case, $\{u_i, u_j, v_1, v_3\}$ is a γ_{pr}-set of G so that $\gamma_{pr} = 4$ and n = 7, and hence K = K_3 = $<u_1, u_2, u_3>$. Let u_1 be adjacent to v_3 and let u_2 be adjacent to v_1. If $\deg(v_1) = \deg(v_3) = 2$, $\deg(v_2) = \deg(v_4) = 1$, then G $\cong C_3(P_3, P_3, 0)$. Let u_1 be adjacent to v_3 and v_4, and let u_2 be adjacent to v_1. If $\deg(v_1) = \deg(v_3) = \deg(v_4) = 2$, $\deg(v_2) = 1$, then G $\cong G_{15}$. Let u_1 be adjacent to v_3 and let u_2 be adjacent to v_1 and v_3. If $\deg(v_1) = 2$, $\deg(v_2) = \deg(v_4) = 1$, $\deg(v_3) = 3$, then G $\cong G_{16}$. Let u_1 be adjacent to v_3, u_2 be adjacent to v_1 and v_3, and let u_3 is adjacent to v_4. If $\deg(v_1) = 2$, $\deg(v_2) = 1$, $\deg(v_4) = 2$, $\deg(v_3) = 3$, then G $\cong G_{17}$. Let u_1 be adjacent to v_3 and v_4, u_2 be adjacent to v_1 and let u_3 is adjacent to v_1. If $\deg(v_1) = 3$, $\deg(v_3) = 2$ $\deg(v_4) = 2$, $\deg(v_2) = 1$, then G $\cong G_{18}$. Let u_1 be adjacent to v_3 and u_2 be adjacent to v_1, and u_3 be adjacent to v_1 and v_2. If $\deg(v_1) = 3$, $\deg(v_2) = \deg(v_3) = 2$, $\deg(v_4) = 1$, then G $\cong G_{19}$.

Sub case viii. $<S> = K_2 \cup \bar{K}_2$.

Let v_1 and v_2 be the vertices of $\bar{K}_2$ and v_3, v_4 be the vertices of K_2. Since G is connected, there exists a vertex u_i in K_{n-4} which is adjacent to v_1 and v_2 and any one of $\{v_3, v_4\}$. Let u_i be adjacent to v_1, v_2, v_3. In this case $\{u_i, v_3\}$ forms a γ_{pr}- set of G, so that $\gamma_{pr} = 2$ and n = 5, which is a contradiction. Hence no graph exists. Since G is connected, there exists a vertex u_i in K_{n-4} which is adjacent to v_1 and there exists u_j for i $\neq$ j, is adjacent to v_2 and v_3. In this case, $\{u_i, u_j, u_k, v_3\}$ for i $\neq$ j $\neq$ k forms a γ_{pr}- set of G so that $\gamma_{pr} = 4$ and n = 7. Hence K = K_3 = $<u_1, u_2, u_3>$. Let u_1 be adjacent to v_2 and v_3 and let u_2 be adjacent to v_1. If $\deg(v_1) = \deg(v_2) = \deg(v_4) = 1$, $\deg(v_3) = 2$, then G $\cong G_{20}$. Since G is connected, there exists a vertex u_i in K_{n-4} which is adjacent to v_1 and there exists u_j for i $\neq$ j, is adjacent to v_2 and u_k for i $\neq$ j $\neq$ k is adjacent to v_3. In this case, $\{u_i, u_j, u_k, v_3\}$ for i $\neq$ j $\neq$ k forms a γ_{pr}-set of G so that $\gamma_{pr} = 4$ and n = 7. Hence K = K_3 = $<u_1, u_2, u_3>$. Let u_1 be adjacent to v_2 and let u_2 be adjacent to v_1 and u_3 be adjacent to v_3. If $\deg(v_1) = \deg(v_2) = \deg(v_4) = 1$, $\deg(v_3) = 2$, then G $\cong C_3(P_3, P_2, P_2)$. Let u_1 be adjacent to v_2, u_2 be adjacent to v_1 and v_3 and let u_3 is adjacent to v_3. If $\deg(v_1) = \deg(v_2) = \deg(v_4) = 1$, $\deg(v_3) = 3$, then G $\cong G_{21}$. Let u_1 be adjacent to v_2 and v_3, u_2 be adjacent to v_1 and let u_3 be adjacent to v_3. If $\deg(v_1) = \deg(v_2) = \deg(v_4) = 1$, $\deg(v_3) = 3$, then G $\cong G_{21}$. Let u_1 be adjacent to v_2, u_2 be adjacent to v_1 and let u_3 be adjacent to v_3 and v_1. If $\deg(v_1) = \deg(v_3) = 2$, $\deg(v_2) = \deg(v_4) = 1$, then G $\cong G_{18}$. Let u_1 be adjacent to v_2, u_2 is adjacent to v_1 and v_2 and let u_3 be adjacent to v_3. If $\deg(v_1) = \deg(v_4) = 1$, $\deg(v_2) = \deg(v_3) = 2$, then G $\cong G_{22}$. Let u_1 be adjacent to v_2, u_2 be adjacent to v_1 and v_2 and let u_3 is adjacent to v_1 and v_3. If $\deg(v_1) = 2$, $\deg(v_4) = 1$, $\deg(v_2) = 2$, $\deg(v_3) = 2$, then G $\cong G_{23}$.

Sub case ix. $<S> = K_4 - \{e\}$.

Let v_1, v_2, v_3, v_4 be the vertices of K_4. Let e be any one of the edges inside the cycle C_4. Since G is connected, there exists a vertex u_i in K_{n-4} which is adjacent to v_1. In this case $\{u_i, v_1\}$ is a γ_{pr}-set of G, so that $\gamma_{pr} = 2$ and n = 5, which is a contradiction. Hence no graph exists. Since G is connected, there exists a vertex u_i in K_{n-4} which is adjacent to v_2. In this case, $\{u_i, v_1\}$ is a γ_{pr}-set of G so that $\gamma_{pr} = 4$ and n = 7. Hence $K = K_3 = <u_1, u_2, u_3>$. Let u_1 be adjacent to v_2. If $\deg(v_1) = \deg(v_2) = \deg(v_3) = 3$, $\deg(v_4) = 2$, then $G \cong G_{24}$. Let u_1 be adjacent to v_2 and let u_2 adjacent to v_4. If $\deg(v_1) = 3$, $\deg(v_2) = 3$, $\deg(v_3) = 3$, $\deg(v_4) = 3$, then $G \cong G_{25}$. Let u_1 be adjacent to v_2 and let u_3 be adjacent to v_2. If $\deg(v_1) = \deg(v_3) = 3$, $\deg(v_2) = 4$, $\deg(v_4) = 2$, then $G \cong G_{26}$.

Sub case x. $<S> = C_3(1, 0, 0)$.

Let v_1, v_2, v_3 be the vertices of C_3 and let v_4 be adjacent to v_1. Since G is connected, there exists a vertex u_i in K_{n-4} which is adjacent to v_2 (or) there exists a vertex u_i in K_{n-4} which is adjacent to v_1 (or) there exists a vertex u_i in K_{n-4} which is adjacent to v_4. In all the cases, by various arguments, it can be verified that $G \cong G_{27}, G_{28}, G_{29}, G_{30}, G_{31}$. If G does not contain a clique K on n − 4 vertices, then it can be verified that no new graph exists.

Case iv. $\gamma_{pr} = n - 4$ and $\chi = n - 3$.

Since $\chi = n - 3$, G contains a clique K on n − 3 vertices or does not contains a clique K on n-3 vertices. Let $S = V(G) - V(K) = \{v_1, v_2, v_3\}$. Then the induced sub graph $<S>$ has the following possible cases. $<S> = K_3, \overline{K}_3, P_3, K_2 \cup K_1$.

Sub case i. $<S> = K_3$.

Let v_1, v_2, v_3 be the vertices of K_3. Since G is connected, there exists a vertex u_i in K_{n-3} which is adjacent to any one of $\{v_1, v_2, v_3\}$. Let u_i be adjacent to v_2, then $\{u_i, v_2\}$ is a γ_{pr}-set of G, so that $\gamma_{pr} = 2$ and n = 6. Hence $K = K_3 = <u_1, u_2, u_3>$. Let u_1 be adjacent to v_2. If $\deg(v_1) = \deg(v_3) = 2$, $\deg(v_2) = 3$, then $G \cong G_{32}$. Let u_1 be adjacent to v_1 and v_2. If $\deg(v_1) = \deg(v_2) = 3$, $\deg(v_3) = 2$, then $G \cong G_{33}$. Let u_1 be adjacent to v_2, u_2 be adjacent to v_1 and let u_3 be adjacent to v_3. If $\deg(v_1) = \deg(v_2), = \deg(v_3) = 3$, then $G \cong G_{34}$. Let u_1 be adjacent to both the vertices v_1, v_2, u_2 be adjacent to v_1 and let u_3 be adjacent to v_3. If $\deg(v_1) = 4$, $\deg(v_2) = \deg(v_3) = 3$, then $G \cong G_{35}$.

Sub case ii. $<S> = \overline{K}_3$.

Let v_1, v_2, v_3 be the vertices of $\overline{K}_3$. Since G is connected, one of the vertices of K_{n-3} say u_l is adjacent to all the vertices of S (or) u_i be adjacent to v_1, v_2 and u_j be adjacent to v_3 for $i \neq j$ (or) u_i be adjacent to v_1 and u_j be adjacent to v_2 and u_k be adjacent to v_3 for $i \neq j \neq k$. If u_i for some i, is adjacent to all the vertices of S, then $\{u_i, v\}$ for some v in K_{n-3}, is a γ_{pr}-set of G, so that $\gamma_{pr} = 2$ and n = 6. Hence $K = K_3 = <u_1, u_2, u_3>$. If u_1 is adjacent to all the vertices v_1, v_2, v_3, then $G \cong C_3(K_{1,3})$. Since G is connected, there exists a vertex u_i in K_{n-3} is adjacent to v_1 and u_j for $i \neq j$ is adjacent to v_2 and v_3, then $\{u_i, u_j\}$ is a γ_{pr}-set of G, so that $\gamma_{pr} = 2$ and $n = 6$. Hence $K = K_3 = <u_1, u_2, u_3>$. Let u_1 be adjacent to v_1 and v_2 and let u_3 be adjacent to v_3. If $\deg(v_1) = \deg(v_2) = \deg(v_3) = 1$ then $G \cong C_3(2P_2, P_2, 0)$. Since G is connected, there exists a vertex u_i in K_{n-3} is adjacent to v_1 and u_j for $i \neq j$ is adjacent to v_2 and u_k for $i \neq j \neq k$ in K_{n-3} is adjacent to v_3, then $\{u_i, u_j, u_k, v\}$ for some v in K_{n-3} is a γ_{pr}-set of G, so that $\gamma_{pr} = 4$ and $n = 8$. Hence $K = K_5 = <u_1, u_2, u_3, u_4, u_5>$. Let u_1 be adjacent to v_2, u_2 be adjacent to v_3 and let u_5 be adjacent to v_1. If $\deg(v_1) = \deg(v_2) = \deg(v_3) = 1$ then $G \cong K_5(P_2, P_2, P_2, 0, 0)$.

Sub case iii. $<S> = P_3$.

Let v_1, v_2, v_3 be the vertices of P_3. Since G is connected, there exists a vertex u_i in K_{n-3} which is adjacent to v_1 (or equivalently v_3) or v_2. If u_i is adjacent to v_2, then $\{u_i, v_2\}$ is a γ_{pr}-set of G, so that $\gamma_{pr} = 2$ and $n = 6$. Hence $K = K_3 = <u_1, u_2, u_3>$. Let u_1 be adjacent to v_2. If $\deg(v_2) = 3$, $\deg(v_1) = \deg(v_3) = 1$, then $G \cong G_{36}$. By increasing the degrees of the vertices, it can be verified that $G \cong G_{37}, G_{38}$. Since G is connected, there exists a vertex u_i in K_{n-3} is adjacent to v_1 then $\{u_i, u_j, v_1, v_2\}$ for $i \neq j$ is a γ_{pr}-set of G, so that $\gamma_{pr} = 4$ and $n = 8$. Hence $K = K_5 = <u_1, u_2, u_3, u_4, u_5>$. Let $u1$ be adjacent to v_1. If $\deg(v_1) = \deg(v_2) = 2$, $\deg(v_3) = 1$, then $G \cong K_5(P_4)$. By increasing the degrees of the vertices, it can be verified that $G \cong G_{39}, G_{40}$.

Sub case iv. $<S> = K_2 \cup K_1$.

Let v_1, v_2 be the vertices of K_2 and v_3 be the vertex of K_1. Since G is connected, there exists a vertex u_i in K_{n-3} which is adjacent to any one of $\{v_1, v_2\}$ and $\{v_3\}$ (or) u_i is adjacent to any one of $\{v_1, v_2\}$ and u_j for $i \neq j$ is adjacent to v_3.. In this case, $\{u_i, u_j, u_k, v_1\}$ for u_k in K_{n-3} for $i \neq j \neq k$ forms a γ_{pr}- set of G, so that $\gamma_{pr} = 4$ and $n = 8$. Hence $K = K_5 = <u_1, u_2, u_3, u_4, u_5>$. Let u_1 be adjacent to v_1 and let u_2 be adjacent to v_3. If $\deg(v_1) = 2$, $\deg(v_2) = \deg(v_3) = 1$, then $G \cong K_5(P_3, P_2, 0, 0, 0)$. By increasing the degrees of the vertices, it can be verified that $G \cong G_{41}, G_{42}, G_{43}$. Since G is connected, there exists a vertex u_i in K_{n-3} which is adjacent to v_1, v_3, so that $\{u_i, v\}$ is a γ_{pr}- set of G. Hence $\gamma_{pr} = 2$ and $n = 6$, so that $K = K_3 = <u_1, u_2, u_3>$. Let u_1 be adjacent to v_1. If $\deg(v_1) = 2$, $\deg(v_2) = \deg(v_3) = 1$, then $G \cong C_3(u(P_3, P_2))$. By increasing the degrees of the vertices, it can be verified that $G \cong G_{44}, G_{45}$. If G does not contain a clique K on $n - 3$ vertices, then it can be verified that no new graph exists.

Case v. $\gamma_{pr} = n - 5$ and $\chi = n - 2$.

Since $\chi = n - 2$, G contains a clique K on $n - 2$ vertices or does not contains a clique K on $n-2$ vertices. Let G contains a clique K on $n - 2$ vertices. Let $S = V(G) - V(K) = \{v_1, v_2\}$. Then the induced sub graph $<S>$ has the following possible cases. $<S> = K_2, \bar{K}_2$. In both cases, by the various possible arguments, it can be verified that $G \cong K_5(P_3)$, G_{46}, G_{47}, G_{48}, G_{49}, G_{50}, G_{51}, $K_5(P_2, P_2, 0, 0, 0)$, G_{52}, G_{53}, G_{54}, G_{55}, G_{56}. If G does not contain a clique K on $n - 2$ vertices, then it can be verified that no new graph exists.

Case vi. $\gamma_{pr} = n - 6$ and $\chi = n - 1$.

Since $\chi = n - 1$, G contains a clique K on $n - 1$ vertices. By various arguments it can be verified that $G \cong K_7(1)$, $K_7(2)$, $K_7(3)$, $K_7(4)$, $K_7(5)$, or $K_7(6)$.

Case vii. $\gamma_{pr} = n - 7$ and $\chi = n$.

Since $\chi = n$, $G \cong K_n$. But for K_n, $\gamma_{pr} = 2$ so that $n = 9$. Hence $G \cong K_9$.

References

1. Berge, C.: Theory of graphs and its applications. Methuen, London (1962)
2. Harary, F.: Graph Theory. Addison Wesley, Reading (1972)
3. Mahadevan, G.: On domination theory and related concepts in graphs. Ph.D. thesis, Manonmaniam Sundaranar University, Tirunelveli, India (2005)
4. Mahadevan, G., Avadayappan, S., Parveen, A.M.: Graphs whose sum of independent domination number and chromatic number equals to 2n-6 for any n > 3. Internat. J. Phys. Sci. (Ultra Science) 20(3)M, 757–762 (2008)
5. Mahadevan, G., Avadayappan, S., Hajmeral: On connected efficient domination number of a graph. Internat. J. Intelligent Inf. Process. 2(2), 313–319 (2008)
6. Mahadevan, G., Avadayappan, S., Basira, I.A.: Sum of complementary connected domination number and chromatic number of a graph. Internat. J. Comput. Math. Appl. 2(1-2), 159–169 (2008)
7. Joseph, J.P., Arumugam, S.: Domination in graphs. Internat. J. Management Syst. 11, 177–182 (1995)
8. Joseph, J.P., Arumugam, S.: Domination and connectivity in graphs. Internat. J. Management Syst. 8, 233–236 (1992)
9. Haynes, T.W.: Paired domination in graphs. Congr. Numer. 150 (2001)
10. Haynes, T.W., Hedetniemi, S.T., Slater, P.J.: Domination in graphs. Advanced Topics. Marcel Dekker, New York (1998)
11. Haynes, T.W., Hedetniemi, S.T., Slater, P.J.: Fundamentals of domination in graphs. Marcel Dekker, New York (1998)

Coding the Vertices of a Graph
with Rosenbloom-Tsfasman Metric

R. Rajkumar

Department of Mathematics
Gandhigram Rural Institute - Deemed University
Gandhigram - 624 302, Tamilnadu, India
rrajmaths@yahoo.co.in

Abstract. In this paper, we study the codability of graphs with respect to the Rosenbloom-Tsfasman metric. The possibilities of coding the vertices of connected graphs and disconnected graphs with this metric have been analyzed. Some comparisions between the codability of graphs with Rosenbloom-Tsfasman metric and Hamming metric have been made.

Keywords: Coding, Graphs, Rosenbloom-Tsfasman distance, Hamming distance.

1 Introduction

Let G be a finite, simple graph. G is connected if for any two vertices of G there exists a path between them; otherwise it is disconnected. A component of G is a maximal connected subgraph of G. If all vertices of G are adjacent, then G is said to be completely connected. Let $G_1(V_1, E_1)$ and $G_2(V_2, E_2)$ be two graphs. The union graph of G_1 and G_2 denoted by $G_1 \cup G_2$ has vertex set $V_1 \cup V_2$ and edge set $E_1 \cup E_2$. The number of vertices of a graph G is denoted by $|G|$.

A binary code C is a subset of $\{0,1\}^m$ for some positive integer m. The elements of C are called codewords. The concatenation of the code words c_i and c_j of C is denoted by $c_i \, c_j$. If r is a positive integer, the r-th order concatenation of the codeword c_i with itself is denoted by $c_i^{(r)}$. Let d be any metric defined on C. Associate with each vertex u_i of G a unique code word c_i from a binary code $C \subseteq \{0,1\}^m$. Let T be a threshold value. The coding problem for graphs can be described as follows. The graph G is said to be codable or in particular $[T, m]$ codable if there exist a unique code $C \subseteq \{0,1\}^m$ such that $d(c_i, c_j) \leq T$ whenever the vertices u_i and u_j are adjacent. T is chosen as smallest integer which satisfies this inequality. G is T codable means that there exist an m such that G is $[T, m]$ codable and G is m codable means that there exist a T such that G is $[T, m]$ codable.

Let C be a code in $\{0,1\}^m$. Let $x = (x_1, \ldots, x_m)$ and $y = (y_1, \ldots, y_m) \in C$. Then the Hamming distance d_H between x and y is defined as the number of places in which they are differ.

i.e., $\quad d_H(x, y) = \sum_{i=1}^{n} |x_i - y_i|.$

P. Balasubramaniam and R. Uthayakumar (Eds.): ICMMSC 2012, CCIS 283, pp. 97–103, 2012.

In [1, 2], Breuer etl., considered the problem of coding the vertices of a graph such that adjacency can be determined by the Hamming distance of the labels. These results have applications in the state assignment problem of digital automata. In particular, to that of error correction under component failure and that of the dynamic stability of digital kinetic system [1].

Recently, Rosenbloom and Tasfasman in [3] introduced a new metric called *m-metric* or *RT-metric* in the context of coding theory, which is a generalization of the classical Hamming metric. The RT distance is defined in the following way. For any two vectors $x = (x_1, \ldots, x_m)$ and $y = (y_1, \ldots, y_m) \in \{0,1\}^m$, the RT distance d_ρ between x and y is defined by

$$d_\rho(x, y) = max \ \{ \ i \mid x_i \neq y_i \ , 1 \leq i \leq n \}.$$

Then d_ρ is a metric on $\{0,1\}^m$. From the definitions of Hamming and RT metrics we observe that $d_H(x, y) \leq d_\rho(x, y)$, $\forall \ x, y \in \{0,1\}^m$.

The rest of the paper is arranged as follows: In Section 2, we study the various possibilities of codability of graphs with respect to the RT metric. In Section 3, the codability of graphs with RT metric has been compared with the Hamming metric case.

2   Codability of Graphs with RT Metric

Figure 1 indicates the codability of all connected graphs having the number of vertices $n = 2$, 3 and 4, where T and m are minimum. Figure 2 indicates the codability of disconnected graphs having the number of vertices $n = 2$, 3 and 4.

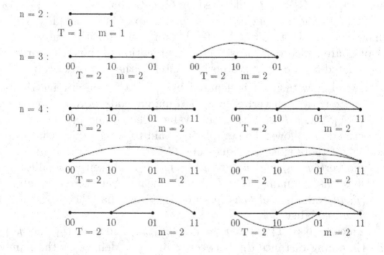

Fig. 1. Examples for codability of connected graphs with RT metric

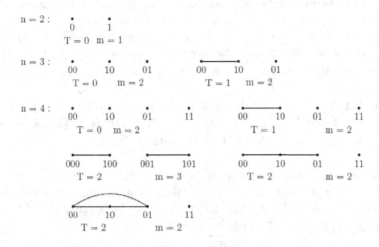

Fig. 2. Examples for codability of disconnected graphs with RT metric

The following theorem proves the codability of a connected graph with RT metric.

Theorem 2.1. *Let G be a connected graph with n vertices. Then G is T codable with RT metric, where T is the smallest positive integer such that $n \leq 2^T$ and a $[T, T]$ coding exist for G.*

Proof. For $n = 1$, the result is obvious.

Consider any code $C \subseteq \{0, 1\}^T$, where T in the smallest positive integer such that $n \leq 2^T$. Then, any one to one assignment of codewords of C to the vertices of G will satisfies $d_\rho(c_i, c_j) \leq T$. We prove that G is T codable with RT metric.

Assume $n \geq 2$. Suppose G is T' codable with RT metric for some $T' < T$. Then there exist a unique code $C' \subseteq \{0, 1\}^m$ (for some positive integer m) assigned to G. Since T in the smallest integer such that $n \leq 2^T$, we must have $2^{T'} < n \leq 2^T$.

Since $n \geq 2$, we can find atleast two vertices u_i and u_j in G such that they are adjacent. Then, there exist distinct codewords c_i and c_j in C corresponding to u_i and u_j respectively such that $d_\rho(c_i, c_j) \leq T'$. Then c_i and c_j are of the form, $c_i = (x_1, \ldots, x_{T'}, z_{T'+1}, \ldots, z_m)$ and $c_j = (y_1, \ldots, y_{T'}, z_{T'+1}, \ldots, z_m)$ with $(x_1, \ldots, x_{T'}) \neq (y_1, \ldots, y_{T'})$.

Thus, corresponding to each pair of adjacent vertices, we have a pair of distinct T' tuples. Since G is connected with n vertices, we must have correspondingly n distinct T' tuples from the total number $2^{T'}$ possible T' tuples. So $n \leq 2^{T'}$, which contradicts the fact that $2^{T'} < n$. Hence G is T codable with RT metric. For minimum value of m we choose $m = T$. So we have a $[T, T]$ coding exist for G. $\square$

To prove the codability of disconnected graph with RT metric we use the following notation. Let $\eta = (\eta_1, \ldots, \eta_k)$ be an ordered k tuple with $\eta_i \in \mathbb{R}^+, i = 1, \ldots, k$ and $\lambda \in \mathbb{R}^+$ with $\eta_i \leq \lambda, i = 1, \ldots, k$. The number denoted by (η, λ) and is defined to be the smallest positive integer r such that the set $I = \{1, \ldots, k\}$ can be partitioned into r sets $J_1, \ldots, J_r$ with $\sum_{i \in J_t} \eta_i \leq \lambda$, $t = 1, \ldots, r$.

Example 2.1. Let $\eta = (5, 2, 2, 1, 1, 3, 3, 4, 1, 2)$ and $\lambda = 6$. Here $I = \{1, \ldots, 10\}$ and $(\eta, \lambda) = 5$. Since $I = J_1 \cup J_2 \cup J_3 \cup J_4 \cup J_5$ where $J_1 = \{2, 3, 4, 5\}$, $J_2 = \{9, 10, 6\}$, $J_3 = \{1\}$, $J_4 = \{8\}$ and $J_5 = \{7\}$ with $\sum_{i \in J_t} \eta_i \leq \lambda$.

Note that with the above notation, if $(\eta, \lambda) = r$, then the existence of corresponding J_i, $(i = 1, \ldots, r)$ need not be unique. For example, in Example 2.1 we can write $I = J_1' \cup J_2' \cup J_3' \cup J_4' \cup J_5'$ where $J_1' = \{4, 5, 6, 9\}$, $J_2' = \{2, 3, 10\}$, $J_3' = \{1\}$, $J_4' = \{8\}$, $J_5' = \{7\}$ with $\sum_{i \in J_t'} \eta_i \leq \lambda$.

The next theorem shows the codability of disconnected graphs.

Theorem 2.2. *Let G be a disconnected graph with n vertices. Let $G_1, \ldots, G_k$ be the components of G with $n_1, \ldots, n_k$ vertices respectively. Then G is T codable with RT metric, where T is the smallest integer such that $N \leq 2^T$, $N = max\{n_1, \ldots, n_k\}$. If $n \leq 2^T$, then G is $[T, m]$ codable with $m = T$. Otherwise, G is $[T, m]$ codable with $m = ((n_1, \ldots, n_k), 2^T)$.*

Proof. Let T be the smallest integer such that $N \leq 2^T$.

Case 1: If $n \leq 2^T$, consider any code $C \subseteq \{0, 1\}^T$ with $|C| = n$. Then for any 1-1 assignment of codewords of C to the vertices of G, we have a T codability for G with RT metric. If we take $m = T$, then a $[T, m]$ coding exist for G.

Case 2: If $n > 2^T$, we will follow the procedure given below.

Let $\eta = (n_1, \ldots, n_k)$. Let $(\eta, 2^T) = r$. Then from the the definition of $(\eta, 2^T)$, the set $I = \{1, \ldots, k\}$ can be partitioned into r sets $J_1, \ldots, J_r$ such that $\sum_{i \in J_t} n_i \leq \lambda$, $t = 1, \ldots, r$.

Let $G_j' = \bigcup_{i \in J_j} G_i$. Then $G = G_1' \cup \ldots \cup G_r'$. Now corresponding to each G_j', consider a code $C_j' \subseteq \{0, 1\}^T$ with $|C_j'| = \sum_{i \in J} n_i$, $j = 1, \ldots, r$. Any 1-1 assignment of codewords in C_j' to the vertices of G_j' will satisfy $d_\rho(c_{it}, c_{is}) \leq T$, $\forall\ 1 \leq t, s \leq n_i, i \in J_j$ iff the vertices u_{it} and u_{is} are adjacent in G_i.

Let M be the smallest integer such that $r \leq 2^M$. Let $C'' \subseteq \{0, 1\}^M$ with $|C''| = r$. For each $i = 1, \ldots, r$, we fix an element in C''. Now for each G_i', we define a new code C_i by $C_i = \{c_{jt}c_i : c_{jt} \in C_j', c_i \in C'', 1 \leq t \leq n_i, j \in J_i\}, \forall i = 1, \ldots, r$. Then all the elements in C_i are distinct, since $c_{it} \neq c_{is}$, $\forall\ 1 \leq t, s \leq n_i, i \in J_j$. Also $C_i \cap C_j = \phi$, $\forall i, j \in \{1, \ldots, r\}, i \neq j$, since each c_i's are distinct. Moreover $d_\rho(c_{it}c_i, c_{is}c_i) = d_\rho(c_{it}, c_{is}) \leq T$, $\forall 1 \leq t, s \leq n_i, i \in J_j$.

Thus $C = C_1 \cup C_2 \cup \ldots \cup C_r$ is the required code assigned to the graph G with the property that G is T codable. Here the assignment of codewords to the vertices of G is made in the order as explained above. For the minimum value of m, we take $m = T + M$, then from the above construction we have a $[T, m]$ coding for G. □

Example 2.3. Let G be a disconnected graph with $|G| = 31$. Let G have 10 components, say G_i, $(i = 1, \ldots, 10)$ with $|G_1| = 5$, $|G_2| = 2$, $|G_3| = 3$, $|G_4| = 3$, $|G_5| = 7$, $|G_6| = 1$, $|G_7| = 1$, $|G_8| = 2$, $|G_9| = 3$ and $|G_{10}| = 4$.

Here $n = 30$ and $N = \max \{|G_i| : i = 1, \ldots, 10\} = 7$. So we have $T = 3$. In this problem, $\eta = (5, 2, 3, 3, 7, 1, 1, 2, 3, 4)$ and $I = \{1, \ldots, 10\}$. So for η and 2^T, $(\eta, 2^T) = 4$ with $I = J_1 \cup J_2 \cup J_3 \cup J_4$, where $J_1 = \{3, 4, 6, 7\}$, $J_2 = \{10, 2, 8\}$, $J_3 = \{1, 9\}$ and $J_4 = \{5\}$. So $M = 2$.

In this case $G'_1 = G_3 \cup G_4 \cup G_6 \cup G_7$, $G'_2 = G_{10} \cup G_2 \cup G_8$, $G'_3 = G_1 \cup G_9$ and $G'_4 = G_5$.

The code $C = C_1 \cup C_2 \cup C_3 \cup C_4$ is fully given below. This graph G is 3 codable with the code C with RT metric.

C_1		C_2		C_3		C_4	
$c_{31} =$	000 00	$c_{101} =$	000 10	$c_{11} =$	000 01	$c_{51} =$	000 11
$c_{32} =$	100 00	$c_{102} =$	100 10	$c_{12} =$	100 01	$c_{52} =$	100 11
$c_{33} =$	010 00	$c_{103} =$	010 10	$c_{13} =$	010 01	$c_{53} =$	010 11
$c_{41} =$	110 00	$c_{104} =$	110 10	$c_{14} =$	110 01	$c_{54} =$	110 11
$c_{42} =$	001 00	$c_{21} =$	001 10	$c_{15} =$	001 01	$c_{55} =$	001 11
$c_{43} =$	011 00	$c_{22} =$	011 10	$c_{91} =$	011 01	$c_{56} =$	011 11
$c_{61} =$	101 00	$c_{81} =$	101 10	$c_{92} =$	101 01	$c_{57} =$	101 11
$c_{71} =$	111 00	$c_{82} =$	111 10	$c_{93} =$	111 01		
	$C'_1 \quad c_1$		$C'_2 \quad c_2$		$C'_3 \quad c_3$		$C'_4 \quad c_4$

From Theorems 2.1 and 2.2 we have the following.

Corollary 2.1: Any graph G is codable with RT metric.

Theorem 2.3: If G is T' codable with RT metric then it is also T codable for $T = T' + p, p = 1, 2, \ldots$

Proof. G is T' codable with a code C having codewords c_i, $(i = 1, 2, \ldots, n)$. Then it is $T' + p$ codable with codewords cc_i, $(i = 1, 2, \ldots n)$, where $c \in \{0, 1\}^p$. □

The following result follows immediately from Theorem 2.3. Here we give another proof for this.

Corollary 2.2: If G is T' codable with RT metric, then it is also T codable for $T = pT'$, $p = 1, 2, \ldots$

Proof. If G is T' codable with a code C having codewords $c_i, (i = 1, 2, \ldots, n)$, then it is pT' codable with a code C' having codewords $c_i^{(p)}, \; i = 1, 2, \ldots, n$. $\square$

Theorem 2.4: *For every value of T and m, where $1 \leq T < m, m > 1$ and $n \leq 2^m$, there exist a graph with n vertices which is not $[T, m]$ codable with RT metric.*

Proof. For $T < m$, any connected graph with 2^m vertices is not $[T, m]$ codable with RT metric. $\square$

Theorem 2.5: *For each value of T and m, $(T < m)$, there exist a graph which is $[T, m]$ codable with RT metric but not $[T \pm 1, m]$ codable with RT metric.*

Proof. Let T and m be given, $(T < m)$. Let $M = m - T$. Now consider the graph G with 2^m vertices and having 2^M components $G_1, \ldots, G_{2^M}$ with $|G_i| = 2^T$, $i = 1, \ldots, 2^M$. By Theorem 2.2 , G is $[T, m]$ codable with RT metric but not $[T \pm 1, m]$ codable. $\square$

Theorem 2.6: *If G is T codable with RT metric, then any subgraph of G is T codable with RT metric.*

Proof. If G' is a subgraph of G, and f is a T coding of G, then the restriction of f to the vertices of G produces a T coding for G'. $\square$

The following result is direct.

Theorem 2.7: *If G is $[T, m]$ codable with RT metric, then it is also $[T, m']$ codable for all $m' \geq m$.*

3   Comparison with the Hamming Metric Case

In this section, we give some results which connects the codability of a graph with respect to Hamming metric and RT metric.

Theorem 3.1: *If G is $[T, m]$ codable with Hamming metric, then G need not be $[T, m]$ codable with RT metric.*

Proof. Consider any complete graph G with n vertices $(n \geq 5)$. Then G is $[2, n]$ codable with Hamming metric. Since $n \geq 5$, G is $[T, T]$ codable $(T > 3)$ with RT metric where T is the smallest positive integer such that $n \leq 2^T$. $\square$

Theorem 3.2: *If G is $[T, m]$ codable with RT metric, then G need not be $[T, m]$ codable with Hamming metric.*

Proof. Let $Q_n = \{0, 1\}^n$. Let f be a $1 - 1$ assignment of vertices to elements in Q_n. Now, join vertices u and v iff $d_H(f(u), f(v)) \leq n - 1$. The graph G so formed is $[n - 1, n]$ codable with Hamming metric but not $[n, n]$ codable.

G has 2^n vertices. Also G is connected. Since the codeword $(0, \ldots, 0)$ assigned to a vertex is adjacent to all other vertices, except the vertex which is assigned to the codeword $(1, \ldots, 1)$, as $d_H((0, \ldots, 0), (1, \ldots, 1)) = n$. The vertex assigned to $(1, 0, \ldots, 0)$ is adjacent to both the vertices which are assigned to the codewords $(0, \ldots, 0)$ and $(1, \ldots, 1)$ respectively, as $d_H((0, \ldots, 0), (1, 0, \ldots, 0)) = 1$ and $d_H((0, \ldots, 0), (1, 0, \ldots, 0)) = n - 1$.

Hence G is $[n, n]$ codable with RT metric. Thus, G is $[n, n]$ codable with RT metric but not $[n, n]$ codable with Hamming metric.   □

Theorem 3.3: *A graph G is $[T, m]$ codable with RT metric if and only if G is $[T', m]$ codable with Hamming metric, for some $T' \leq T$.*

Proof. Since G is T codable with RT metric, there exist a code C such that the vertices u_i, u_j in G are adjacent iff $d_\rho(c_i, c_j) \leq T$, $c_i, c_j \in C$.

Since $d_H(c_i, c_j) \leq d_\rho(c_i, c_j), \forall c_i, c_j \in C$, we have the vertices u_i, u_j are adjacent iff $d_H(c_i, c_j) \leq T, c_i, c_j \in C$.

Let $T' = \max \{d_H(c_i, c_j) : c_i, c_j \in C$ and u_i, u_j are adjacent $\}$

Then G is T' codable with the some code C with Hamming metric. with $T' \leq T$. By varying the code one can obtain T' still smaller.

By retracing the steps we can easily prove the converse part.   □

References

1. Breuer, M.A.: Coding the vertexes of a graph. IEEE Trans. Inform. Theory IT-12(2), 148–153 (1966)
2. Breuer, M.A., Folkman, J.: An unexpected result in coding the vertices of a graph. J. Math. Anal. Appl. 20, 583–600 (1967)
3. Rosenbloom, M.Y., Michael, A.: Tsfasman: Codes for the m-metric. Probl. Inf. Trans. 33, 55–63 (1997)

Application of Fuzzy Programming Method for Solving Nonlinear Fractional Programming Problems with Fuzzy Parameters

Animesh Biswas[*] and Koushik Bose

Department of Mathematics,
University of Kalyani
Kalyani – 741235, India
abiswaskln@rediffmail.com,
kobo.1980@gmail.com

Abstract. This paper presents a fuzzy programming procedure to solve nonlinear fractional programming problems in which the parameters involved in objective function are considered as fuzzy numbers. At first a multiobjective fuzzy nonlinear programming model is constructed by considering the numerator and denominator part, individually, of the fractional objective. Then by using tolerance ranges of the fuzzy parameters, the problem is decomposed and each decomposed objectives is solved independently within the feasible region to find the upper and lower tolerance values of the fuzzy objective goals in the decision making environment. Finally a Taylor's series linear approximation technique is applied to linearize the membership goals and thereby obtaining most satisfactory decision in the decision making arena. A numerical example is solved and the solution is compared with existing approach to establish the efficiency of the proposed methodology.

Keywords: Fractional programming, Quadratic programming, fuzzy programming, Triangular fuzzy number, Fuzzy goal programming, Taylor's series approximation.

1   Introduction

In real life decision making situations, the decision makers often face problems in making decision from linear/nonlinear fractional programming problems (FPPs) from the view point of its potential use in different real life planning problems, like engineering, finance, business, planning, economics etc. The generic form of an FPP is considered as a ratio of two functions in its objectives and/or constraints like profit/cost, inventory/sales, actual cost/standard cost, production/employee etc. In 1937 Neumann [1] first introduced the concept of FPP in the context of expanding economy. Thereafter plenty of work has been done in this field. Some good survey

---

[*] Corresponding author.

P. Balasubramaniam and R. Uthayakumar (Eds.): ICMMSC 2012, CCIS 283, pp. 104–113, 2012.
© Springer-Verlag Berlin Heidelberg 2012

work on the development and applications of FPPs are found in the study of active researchers [2–6]. Among the different fields of FPPs, nonlinear FPP (NLFPP) which maximize/minimize the ratio of two functions, at least one of which are in nonlinear form, is now playing an important role in the decision making context. Rodenas et. al. [7] studied algorithms for solving NLFPPs.

In most of the practical situations the possible value of the parameters involved in objective could not be defined precisely due to the lack of available data. The concept of fuzzy sets [8] is seemed to be most appropriate to deal with such imprecise data. The optimization problems involving these types of imprecise parameters in FPPs are called fuzzy FPPs (FFPPs). These types of FFPPs are further classified as fuzzy linear FPPs (FLFPPs), and fuzzy NLFPPs (FNLFPPs). Most of the FFPP models [9–11] developed so far are involved with only fuzzy goals of the objectives. But the imprecision involved with the parameters of the objective in FFPPs have not been studied extensively in the literature. In 2006, Zhang et. al. [12] proposed a branch and bound algorithm for solving FLFPP in a bilevel decision making environment. Further Biswas and Bose [13, 14] studied solution methodologies for solving linear and nonlinear programming problems involving fuzzy parameters in a hierarchical decision making context. But some strong methodologies for solving FNLFPPs are yet to appear in the literature.

This paper develops a solution procedure for solving fuzzy quadratic FPP (FQFPP) where the parameters involved in the objective function are fuzzy numbers and also the objective goal of the problem is fuzzily defined. Fuzzy goal programming (FGP) [15] procedure together with Taylor's Series Linear approximation Technique is applied to solve the problem.

2   Model Formulation

The generic form of a FQFPP is presented as:

Find $X(X_1, X_2,..., X_n)$ so as to

$$Max\tilde{F}(X) = \frac{\tilde{F}_1(X)}{\tilde{F}_2(X)}$$

subject to

$$AX \begin{pmatrix} \leq \\ = \\ \geq \end{pmatrix} b$$

$$X \geq 0. \tag{1}$$

where $\tilde{F}_i(X) = \tilde{C}_i X + \frac{1}{2} X^T \tilde{D}_i X$, $i = 1, 2$.

$\tilde{C}_i$ $(i = 1,2)$ are vectors of fuzzy numbers and $\tilde{D}_i$ $(i = 1,2)$ are symmetric matrices of fuzzy numbers. A represents a matrix of suitable dimension and b represents a vector. X^T denotes the transpose of decision vector X.

Now considering the fractional form of the objective, the model (1) can be converted into a multiobjective fuzzy quadratic problem with conflicting nature of objectives as:

Find X so as to

$$Max\tilde{F}_1(X) = \tilde{C}_1 X + \frac{1}{2} X^T \tilde{D}_1 X$$

$$Min\tilde{F}_2(X) = \tilde{C}_2 X + \frac{1}{2} X^T \tilde{D}_2 X$$

subject to the system constraints defined in (1)   (2)

Let the coefficients involved with the objective functions are taken as triangular fuzzy numbers.

The membership function of a triangular fuzzy number $\tilde{a} = \left(a^L, a, a^R\right)$ can be expressed as:

$$\mu_{\tilde{a}}(x) = \begin{cases} 0 & if \quad x < a^L \ and \ x > a^R \\ \dfrac{x - a^L}{a - a^L} & if \quad \quad a^L \le x \le a \\ \dfrac{a^R - x}{a^R - a} & if \quad \quad a \le x \le a^R \end{cases}$$ (3)

On the basis of the tolerance ranges of the triangular fuzzy numbers in (3), the model (2) can be decomposed as:

Find $X(X_1, X_2, ..., X_n)$ so as to

$$MaxF_1^L(X) = C_1^L X + \frac{1}{2} X^T D_1^L X$$

$$MaxF_1^R(X) = C_1^R X + \frac{1}{2} X^T D_1^R X$$

$$MinF_2^L(X) = C_2^L X + \frac{1}{2} X^T D_2^L X$$

$$MinF_2^R(X) = C_2^L X + \frac{1}{2} X^T D_2^R X$$

subject to the system constraints defined in (1)   (4)

where F_1^L, F_1^R are associated with the objective of F_1; and F_2^L, F_2^R are associated with the objective F_2.

2.1   Construction of Membership Functions

Now each decomposed objectives are solved individually to construct the membership functions of the objectives F_1 and F_2 in the decision making environment.

Let $\left(X^{iL} ; F_i^{LO} \right)$ and $\left(X^{iR} ; F_i^{RO} \right)$ for $i = 1,2$ be the individual optimal solutions of F_i^L and F_i^R (for $i = 1,2$), respectively.

Then the fuzzy goals of F_1 and F_2 are formed as $F_1 \gtrsim F_1^{RO}$ and $F_2 \lesssim F_2^{LO}$.

In this connection it is to be noted that, the lower tolerance limit of F_1 and F_2 would be F_1^{LO} and F_2^{RO}, respectively.

The membership functions for the corresponding fuzzy goals are presented as

$$\mu_{F_1}(X) = \begin{cases} 1 & if & F_1(X) \geq F_1^{RO} \\ \dfrac{F_1(X) - F_1^{LO}}{F_1^{RO} - F_1^{LO}} & if & F_1^{LO} < F_1(X) < F_1^{RO} \\ 0 & if & F_1(X) \leq F_1^{LO} \end{cases} \quad (5)$$

$$\mu_{F_2}(X) = \begin{cases} 0 & if & F_2(X) > F_2^{RO} \\ \dfrac{F_2^{RO} - F_2(X)}{F_2^{RO} - F_2^{LO}} & if & F_2^{LO} < F_2(X) \leq F_2^{RO} \\ 1 & if & F_2(X) \leq F_2^{LO} \end{cases} \quad (6)$$

Again the feasible region at which the decomposed part of numerator and denominator of the fractional objective attains their optimum values is found as

$$X^l = Min\left(X^{iL}, X^{iR} \right) \leq X \leq Max\left(X^{iL}, X^{iR} \right) = X^u , i = 1,2 . \quad (7)$$

The equivalent quadratic FGP (QFGP) model is presented in the following section.

3   QFGP Problem Formulation

In FGP, the membership functions are transformed into flexible goals by assigning unity as the aspiration level and introducing under- and over-deviational variables to each of them. Then the QFGP model for the defined membership functions can be presented as

Find $X(X_1, X_2, ..., X_n)$ so as to

Minimize $Z = w_1^- d_1^- + w_2^- d_2^-$

subject to

$$\frac{F_1(X) - F_1^{LO}}{F_1^{RO} - F_1^{LO}} + d_1^- - d_1^+ = 1$$

$$\frac{F_2^{RO} - F_2(X)}{F_2^{RO} - F_2^{LO}} + d_2^- - d_2^+ = 1$$

$$X^l \leq X_i \leq X^u$$

$$d_i^-, d_i^+ \geq 0 \text{ with } d_i^- . d_i^+ = 0 \text{ for } i = 1, 2. \tag{8}$$

where $d_i^-, d_i^+ (i = 1, 2)$ represent the under- and over-deviational variables, respectively, associated with the membership goals and the fuzzy weight $w_i^- (i = 1, 2)$ is determined as :

$$w_1^- = \frac{1}{F_1^{RO} - F_1^{LO}} \text{ and } w_2^- = \frac{1}{F_2^{RO} - F_2^{LO}} \tag{9}$$

Now a linear approximation procedure proposed by Ignigio [16] is used to linearize the modeled QFGP in the decision making context.

3.1   Linearization Technique for QFGP Model

It is difficult to find a satisfactory solution in a nonlinear decision making arena as most of the tools developed so far can solve only linear type models efficiently. In the context of nonlinear goal programming, J. P. Ignizio [16] proposed a methodology in 1976 by considering Taylor's series approximation technique. Here the concept of that technique is adopted to solve the QFGP model.

Let an approximate solution X^0 is determined by initial inspection from solution space defined in (7) and is declared as initial solution. Now, since the objectives of the model (4) are conflicting in nature, the initial solution may not provide optimal solution for both the objectives simultaneously. Under this situation, linear approximation technique is used to locate the satisfactory decision in the neighborhood of X^0.

Now using Taylor's series linear approximation technique, to the goals in neighborhood of X^0, the membership goals in (8), can be obtained as [14]:

$$G_i : \mu_{F_i}(X^0) + \left[\nabla \mu_{F_i}(X^0)\right]^T (V - W) + d_i^- - d_i^+ = I \qquad i = 1, 2. \tag{10}$$

where I denotes a column vector with two elements equal to1, $V - W = X - X^0$, $(V, W \geq 0)$ is used to represent that the value due to change of solution point from old point X^0 to the new point X is unrestricted in sign.

$$\nabla \mu_{F_i}(X^0) \text{ represents the gradient of } \mu_{F_i}(X^0).$$

Let v_k and w_k be the k-th element of the vector V and W, respectively,

$X_k^0 = k$ -th component of the present solution X^0,

$X_k^l = k$ -th component of the new solution X^l and

$X_k^u = k$ -th component of the new solution X^u.

Clearly, the resultant upper-bound restrictions on v_k and w_k are obtained as

$$0 \leq v_k \leq \min\{X_k^0 - X_k^l, X_k^u - X_k^0\} = p_k, 0 \leq w_k \leq \min\{X_k^0 - X_k^l, X_k^0\} = q_k.$$

$$k = 1, 2, ..., n. \tag{11}$$

Now, by applying the Taylor's series linear approximation technique, the QFGP model in (8) is converted into the following form as

Find X so as to

$$MinZ = \sum_i w_i^- d_i^- + \sum_k \left(a_k^+ + b_k^+\right)$$

and satisfy

$$\mu_{F_i}(X^0) + \left[\nabla \mu_{F_i}(X^0)\right]^T (V - W) + d_i^- - d_i^+ = I$$

$$v_k + a_k^- - a_k^+ = p_k$$

$$w_k + b_k^- - b_k^+ = q_k \tag{12}$$

Here $d_i^-, d_i^+, a_k^-, a_k^+, b_k^-, b_k^+ \geq 0$ with $d_i^- . d_i^+ = 0$, $a_k^- . a_k^+ = 0$ and $b_k^- . b_k^+ = 0$, for all $i = 1, 2$ and $k = 1, 2, ..., n$.

Now, the problem is solved to find the most satisfactory solution in the decision making environment.

4   A Numerical Example

To show the potential efficiency of the proposed approach, the fuzzified version of the problem recently studied by Khurana and Arora [17] is considered and solved by using the proposed methodology.

The FQFPP is considered as follows

Find $X(x_1, x_2, ..., x_6)$ so as to

$$Ma\tilde{F}(x_1, x_2) = \frac{\langle 1.5,2,3 \rangle x_1^2 + \langle 1,2,3 \rangle x_2^2 + \langle 3,4,6 \rangle x_1 x_2 + \langle 42,46,50 \rangle x_1 + \langle 40,46,48 \rangle x_2 + \langle 200,204,210 \rangle}{\langle -5,-4,-3 \rangle x_1^2 + \langle -2,-1,-0.5 \rangle x_2^2 + \langle 3,4,6 \rangle x_1 x_2 + \langle 7,8,10 \rangle x_1 + \langle -5,-4,-2 \rangle x_2 + \langle 160,165,170 \rangle}$$

subject to

$$x_1 + 2x_2 + x_3 = 2, \; 3x_1 + x_2 + x_4 = 4, \; -5x_1 + 3x_2 + x_5 = 0, \; -x_1 + x_2 + x_6 = 0,$$
$$x_1, x_2, x_3, x_4, x_5, x_6 \geq 0 \tag{13}$$

Now following the proposed technique, the FQFPP can be converted into the following form as

Find $X(x_1, x_2, ..., x_6)$ so as to

$$Ma\tilde{F}_1(x_1, x_2) = \langle 1.5, 2, 3 \rangle x_1^2 + \langle 1, 2, 3 \rangle x_2^2 + \langle 3, 4, 6 \rangle x_1 x_2 + \langle 42, 46, 50 \rangle x_1 + \langle 40, 46, 48 \rangle x_2$$
$$+ \langle 200, 204, 210 \rangle$$

$$Min\tilde{F}_2(x_1, x_2) = \langle -5, -4, -3 \rangle x_1^2 + \langle -2, -1, -0.5 \rangle x_2^2 + \langle 3, 4, 6 \rangle x_1 x_2 + \langle 7, 8, 10 \rangle x_1 + \langle -5, -4, -2 \rangle x_2$$
$$+ \langle 160, 165, 170 \rangle$$

subject to the system constrains defined in (13) $\tag{14}$

The above problem (14) is decomposed by considering the parameters as triangular fuzzy numbers in the following form as

Find $X(x_1, x_2, ..., x_6)$ so as to
$$MaxF_1^L(x_1, x_2) = 1.5x_1^2 + x_2^2 + 3x_1 x_2 + 42x_1 + 40x_2 + 200$$
$$MaxF_1^R(x_1, x_2) = 3x_1^2 + 3x_2^2 + 6x_1 x_2 + 50x_1 + 48x_2 + 210$$
$$MinF_2^L(x_1, x_2) = -5x_1^2 - 2x_2^2 + 3x_1 x_2 + 7x_1 - 5x_2 + 160$$
$$MinF_2^R(x_1, x_2) = -3x_1^2 - 0.5x_2^2 + 6x_1 x_2 + 10x_1 - 2x_2 + 170$$
subject to the system constrains in (13) $\tag{15}$

Now each of the decomposed objectives in (15) is solved independently under the system constraints defined in (13) to find the individual optimal solution and which are acting as the respective lower and upper tolerance values of F_1 and F_2 in the decision making environment. The achieved upper and lower tolerance value of F_1 and F_2 are presented in the following table.

Table 1. Upper and lower tolerance values of F_1 and F_2

Upper Tolerance values		Lower Tolerance Values	
$F_1^{RO} = 297$	$F_2^{LO} = 160$	$F_1^{LO} = 270$	$F_2^{RO} = 170$

Also the feasible region described by the system constraints are found by using (7) as

$$0 \le x_1 \le 1.2,\ 0 \le x_2 \le 0.675,\ 0 \le x_3 \le 2,\ 0 \le x_4 \le 4,\ 0 \le x_5 \le 4.8,\ 0 \le x_6 \le 0.8.$$

The membership functions are developed by the method described in (5) and (6) on the basis of achieved tolerance values of the objectives. Further by assigning unity as the aspiration level of the membership goals, the equivalent QFGP model is formulated as

Find $X(x_1, x_2, ..., x_6)$ so as to

$$Min = 0.037 d_1^- + 0.1 d_2^-$$

subject to

$$\frac{\left(2x_1^2 + 2x_2^2 + 4x_1 x_2 + 46x_1 + 46x_2 + 204\right) - 270}{27} + d_1^- - d_1^+ = 1$$

$$\frac{170 - \left(-4x_1^2 - 1x_2^2 + 4x_1 x_2 + 8x_1 - 4x_2 + 165\right)}{10} + d_2^- - d_2^+ = 1$$

$$0 \le x_1 \le 1.2,\ 0 \le x_2 \le 0.675,\ 0 \le x_3 \le 2,\ 0 \le x_4 \le 4,\ 0 \le x_5 \le 4.8,\ 0 \le x_6 \le 0.8.$$

$$x_1, x_2, x_3, x_4, x_5, x_6 \ge 0. \tag{16}$$

Now, to linearize the membership goals in (16), the initial approximation solution is considered as $\left(x_1^0, x_2^0, x_3^0, x_4^0, x_5^0, x_6^0\right) = (1, 0.5, 1, 2, 3, 0.5).$

Then applying the approximation methodology, the resultant FGP model can be presented as

Find $X(x_1, x_2, ..., x_6)$ so as to

$$Min = 0.037 d_1^- + 0.1 d_2^- + a_1^+ + a_2^+ + a_3^+ + a_4^+ + a_5^+ + a_6^+ + b_1^+ + b_2^+ + b_3^+ + b_4^+ + b_5^+ + b_6^+$$

subject to

$$1.92(v_1 - w_1) + 1.92(v_2 - w_2) + d_1^- - d_1^+ = 0.72$$

$$-0.2(v_1 - w_1) + 0.1(v_2 - w_2) + d_2^- - d_2^+ = 0.875$$

$$v_1 + a_1^- - a_1^+ = 0.2,\ v_2 + a_2^- - a_2^+ = 0.175,\ v_3 + a_3^- - a_3^+ = 1,$$

$$v_4 + a_4^- - a_4^+ = 2,\ v_5 + a_5^- - a_5^+ = 1.8,\ v_6 + a_6^- - a_6^+ = 0.3,$$

$$w_1 + b_1^- - b_1^+ = 1,\ w_2 + b_2^- - b_2^+ = 0.5,\ w_3 + b_3^- - b_3^+ = 1,$$

$$w_4 + b_4^- - b_4^+ = 2,\ w_5 + b_5^- - b_5^+ = 3,\ w_6 + b_6^- - b_6^+ = 0.5,$$

$$v_1, v_2, w_1, w_2 \ge 0, \tag{17}$$

$$d_i^-, d_i^+, a_j^-, a_j^+, b_j^-, b_j^+ \ge 0 \text{ with } d_i^- . d_i^+ = 0, a_j^- . a_j^+ = 0, b_j^- . b_j^+ = 0.\ i = 1,2.\ j = 1,2...,6.$$

The *software* **LINGO** (ver. **6.0**) is used to solve the above problem.

4.1   Result and Discussion

Now solving the model described in (17), the optimal solution is obtained as $(x_1, x_2, x_3, x_4, x_5, x_6) = (1.2, 0.675, 1, 2, 1.8, 0.3)$ with the optimal value of objective 1.76.

In this connection, it is to be noted that the solution obtained by Khurana and Arora [17] as $(x_1, x_2, x_3, x_4, x_5, x_6) = (1.2, 0.4, 0, 0, 4.8, 0.8)$ with the optimal value of objective 1.67. The comparison shows that a better solution is achieved by applying the proposed methodology. Further the developed model is flexible enough for assigning imprecise goal in the decision making context.

5   Conclusions and Scope for Future Studies

This paper presents a new efficient solution approach to FFPP in which the parameters involved in objective function are considered as fuzzy number. The developed methodology is free from any computational difficulty arising from the presence of nonlinearity in the decision making process. Still many areas need to be explored and developed in this direction. The proposed model can be extended to solve multiobjective FFPPs involving fuzzy goals and fuzzy parameters. Further the developed model can be applied to hierarchical decision making environment involving nonlinear fractional fuzzy goals and fuzzy parameters. However, it is hoped that the developed model may open up new vistas into the way of making decision from nonlinear fractional decision making problems under imprecise environment.

Acknowledgements. The authors are thankful to the anonymous referees for their constructive comments and helpful suggestions in improving the quality of the article.

References

1. Neumann, J.V.: Uber Ein Es Gleichungs System and Eine Verallgemeinerung Des Brouwerschen Fixpuntsatzes. Ergebnisse Eines Mathematicschen 8, 245–267 (1937)
2. Frenk, J.B.G., Schaible, S.: Fractional Programming. In: Floudas, C.A., Pardalos, P.M. (eds.) Encyclopedia of Optimization, vol. 11, pp. 162–172. Kluwer, Dordrecht (2001)
3. Frenk, J.B.G., Schaible, S.: Fractional Programming. In: Hadjisavvas, N., Komolosi, S., Schaible, S. (eds.) Handbook of Generalized Convexity and Generalized Monotonicity. Springer, Berlin (2004)
4. Schaible, S.: Fractional Programming. In: Horst, R., Pardalos, P.M. (eds.) Handbook of Global Optimization. Kluwer, Dordrecht (1995)
5. Schaible, S.: Fractional Programming-some Recent Developments. J. Infor. Opt. Sci. 10, 1–4 (1989)
6. Schaible, S.: Bibliography in Fractional Programming. Math. Meth. Ops. Res. 26, 211–241 (1982)
7. Rodenas, R.G., Lopez, M.L., Verastegui, D.: Extensions of Dinkelbach's Algorithm for Solving Non-linear Fractional Programming Problems. Sociedad de Estadistica e Investigacion Operativa Top 7, 33–70 (1999)
8. Zadeh, L.A.: Fuzzy Sets. Inform. Control 8, 333–353 (1965)

9. Chang, C.-T.: Fractional Programming with Absolute-value Functions: a Fuzzy Goal Programming Approach. Appl. Math. Comp. 167, 508–515 (2005)
10. Ahlatcioglu, M., Tiryaki, F.: Interactive Fuzzy Programming for Decentralized Two-level Linear Fractional Programming (DTLLFP) Problems. Omega 35, 432–450 (2007)
11. Mehra, A., Chandra, S., Bector, C.R.: Acceptable Optimality in Linear Fractional Programming with Fuzzy Coefficients. Fuzzy Opt. Dec. Making 6, 5–16 (2007)
12. Zhang, G., Lu, J., Dillon, T.: An Approximation Branch-and-bound Algorithm for Fuzzy Bilevel Decision Making Problems. In: Proceedings of the International Multiconference on Computer Science and Information Technology, pp. 223–231 (2006)
13. Biswas, A., Bose, K.: On Solving Bilevel Programming Problems with Fuzzy Parameters through Fuzzy Programming. In: Proceedings of the International Congress on Productivity, Quality, Reliability, Optimization and Modeling, pp. 229–242 (2011)
14. Biswas, A., Bose, K.: A Fuzzy Programming Approach for Solving Quadratic Bilevel Programming Problems with Fuzzy Resource Constraints. Int. J. Operational Res. 12, 142–156 (2011)
15. Pal, B.B., Moitra, B.N.: A Fuzzy Goal Programming Procedure for Solving Quadratic Bilevel Programming Problems. Int. J. Int. Syst. 18, 529–540 (2003)
16. Ignigio, J.P.: Goal Programming and Extensions, Massachusetts, Lexington (1976)
17. Khurana, A., Arora, S.R.: A Quadratic Fractional Program with Linear Homogeneous Constraints. African J. Math. Comp. Sci. Res. 4, 84–92 (2011)

Solving Bottleneck Bi-criteria Transportation Problems

D. Anuradha and P. Pandian

Department of Mathematics, School of Advanced Sciences,
VIT University, Vellore - 14, Tamilnadu, India
anuradhadhanapal1981@gmail.com, pandian61@rediffmail.com

Abstract. A new method namely, block-dripping method is proposed for finding a set of all efficient solutions to a bottleneck bi-criteria transportation problem which provides more efficient solutions than other existing methods [1, 14, 15] and the best compromise solution to the problem. Numerical example has been provided to illustrate the solution procedure. The proposed technique provides the necessary decision support to the decision makers, while they are handling time-oriented logistic problems having two objectives.

Mathematics Subject Classification: 90C08, 90C90.

Keywords: Bottleneck bi-criteria transportation problem, Block- Dripping method, Efficient solutions, Compromise solutions.

1 Introduction

In the classical transportation problem (TP), the traditional objective is of minimizing the total cost [3]. But in our daily life, TPs may be modeled more profitably with the simultaneous consideration of multiple objectives, because a decision maker is usually assumed to pursue multiple goals. For example, the objectives may be minimization of total cost, total deterioration of goods during transportation, consumption of certain scarce resources such as energy, minimization of delivery time of the commodities and so on [16]. Deterioration is relevant in the case of certain perishable or decaying items. The degree of deterioration may be dependent on the route, mode and time of transportation. The time minimizing TP is encountered in connection with transportation of perishable goods, with the delivery of emergency supplies, fire services or when military units are to be sent from their bases to the fronts. The classical TP focused on cost minimization as its objective function. Hammer [5], however, considered the time-minimizing version of this classical problem. His problem seeks to minimize the maximum time required to transport goods from sources of supply to different demand destination locations. By considering time minimization, Hammer's work represented a significant departure from the earlier studies of the TP. Subsequently, Garfinkel and Rao [2], Ilija Nikolic [6], Isserman [7], Szwarc [13], studied Hammer's problem further, seeking to provide better approaches. Hammer's problem was known as the time TP as well as the

P. Balasubramaniam and R. Uthayakumar (Eds.): ICMMSC 2012, CCIS 283, pp. 114–123, 2012.

bottleneck TP. A TP with two criteria and bottleneck bi-criteria was stated by Aneja and Nair [1]. They developed an algorithm to identify the set of non-dominated extreme points in the criteria space and also, in the bottleneck bi-criteria space. Rita Malhotra [12] provided an algorithm for finding the efficient points of a bi-criteria TP and also, the efficient points of a bottleneck bi-criteria TP. Tkachenko and Alhazov [14] discussed a method for solving a bi-criteria problem and they extended their work to a tri-criteria problem wherein the three criteria's were the minimization of transportation cost, the minimization of the total deterioration of goods during transportation and minimization of time. Tkachenko [15] proposed a generalized algorithm to find the extreme efficient points of a fractional multiple-objective TP. Pandian and Natarajan [10] have introduced the zero point method for finding an optimal solution to a classical TP without using an optimality checking method. Pandian and Natarajan [9] developed two algorithms, one for finding an optimal solution to bottleneck TP and the other for finding all efficient solutions of a bottleneck- cost TP. Pandian and Anuradha [11] have introduced the dripping method for finding a set of efficient solutions to a bi-objective TP.

In this paper, we propose a new method namely, block-dripping method for finding the set of all solutions to a bottleneck bi-criteria transportation problem (BBTP) which is a combination of two methods namely, blocking method [9] and dripping method [11]. It is a two step method in which first a possible solution is found and optimized by time and then, the time is sequentially traded for the cost, thus obtaining the set of all solutions. The obtained solutions set contain non-dominated and dominated solutions to the BBTP. The block-dripping method provides a set of transportation schedules to bottleneck bi-criteria transportation problems which helps the decision makers to select an appropriate transportation schedule, depending on his financial position and his time limit.

2   Preliminaries

In this paper, the following conventions for vectors in R^n will be followed. For any $x = (x_1, x_2, ..., x_n)$ and $y = (y_1, y_2, ..., y_n)$, we follow the notations of Mangasarian [8].

$x = y$ if and only if $x_i = y_i$, $i = 1, 2, ..., n$; $x \leq y$ if and only if $x_i \leq y_i$, $i = 1, 2, ..., n$;

$x \leq y$ if and only if $x_i \leq y_i$, $i = 1, 2, ..., n$ and $x_r < y_r$ for some $r \in \{1, 2, ..., n\}$ and

$x < y$ if and only if $x_i < y_i$, $i = 1, 2, ..., n$.

Consider the following bottleneck bi-criteria transportation problem (BBTP):

(P)   Minimize $Z_1 = \sum\limits_{i=1}^{m} \sum\limits_{j=1}^{n} c_{ij} x_{ij}$; Minimize $Z_2 = \sum\limits_{i=1}^{m} \sum\limits_{j=1}^{n} d_{ij} x_{ij}$;

$$\text{Minimize T} = [Maximize \ t_{ij} \ / \ x_{ij} > 0]$$
$$(i, j)$$

Subject to

$$\sum_{j=1}^{n} x_{ij} = a_i , \quad i = 1,2,...,m \tag{1}$$

$$\sum_{i=1}^{m} x_{ij} = b_j , \quad j = 1,2,...,n \tag{2}$$

$$x_{ij} \geq 0 , \quad \text{for all i and j and are integers} \tag{3}$$

where c_{ij} is the cost of transporting a unit from ith source to jth destination; d_{ij} is the deterioration of a unit while transporting from ith source to jth destination; t_{ij} is the time of transporting a unit from ith source to jth destination; a_i is the amount of the material available at ith source; b_j is the amount of the material required at jth destination and x_{ij} is the amount transported from ith source to jth destination.

The following three transportation problems can be constructed from the problem (P), namely first criteria TP (FCTP), second criteria TP (SCTP) and the bottleneck TP (BTP) of the given problem (P):

$$\text{(FCTP)} \quad \text{Minimize } Z_1 = \sum_{i=1}^{m} \sum_{j=1}^{n} c_{ij} x_{ij}$$

subject to (1), (2) and (3).

$$\text{(SCTP)} \quad \text{Minimize } Z_2 = \sum_{i=1}^{m} \sum_{j=1}^{n} d_{ij} x_{ij}$$

subject to (1), (2) and (3)

and (BTP) Minimize $T = [Maximize \ t_{ij} \ / \ x_{ij} > 0]$
$$(i, j)$$

subject to (1), (2) and (3).

Let $Z_1(X)$ and $Z_2(X)$ denote the values of Z_1 and Z_2 corresponding to $X = \{x_{ij}, i = 1,2,...,m; \ j = 1,2,...n\}$ respectively.

Definition 2.1. A set (X°, T°) where $X^{\circ} = \{x_{ij}^{\circ}, i = 1,2,...,m; \ j = 1,2,...n\}$ and T° is a time, is said to be feasible to the problem (P) if X° satisfies the conditions (1) to (3) within the total time transportation T°.

Definition 2.2. A feasible point (X°, T°) of the problem (P) is said to be an efficient (non-dominated) solution to the problem (P) if there exists no other feasible point (X, T) of (P) such that

$$(Z_1(X), Z_2(X), T) \leq (Z_1(X^\circ), Z_2(X^\circ), T^\circ).$$

Otherwise, it is called non-efficient (dominated) solution for (P).

For simplicity, a triplet $(Z_1(X^\circ), Z_2(X^\circ), T^\circ)$ is called a solution (an efficient / a non-efficient solution) to the problem (P) if (X°, T°) is a solution (efficient / non-efficient solution) to the problem (P).

3   Block-Dripping Method

We, now propose a new method namely, block-dripping method for finding all the solutions to the problem (P) which is a two step method in which first using the blocking algorithm [9], the given problem is reduced into a bi-criteria TP and then, the bi-criteria TP is solved by the dripping method [11].

The proposed method proceeds as follows:

Step 1: Construct the BTP of the given problem (P).

Step 2: Solve the BTP by the blocking method [9]. Let an optimal solution of BTP be T°.

Step 3: Solve the FCTP and SCTP of the problem (P) individually by the zero point method [10] and also, find their corresponding time transportation. Let them be T_{m1} and T_{m2} respectively. Let $T^m = \max\{T_{m1}, T_{m2}\}$.

Step 4: Construct the bi-criteria TP for each time M in $[T^\circ, T^m]$ which is obtained from the problem (P) after blocking the cells having more than the time M.

Step 5: Solve the bi-criteria TP obtained in step 5 using the dripping method [11].

Step 6: Each time M in $[T^\circ, T^m]$ together with the solution of the bi-criteria TP corresponding to it yields a solution to the problem (P).

Step 7: Combine all the solutions to the problem (P) obtained in the Step 7. This is the required set of solutions to the problem (P).

Remark 1: From Step 7, we can obtain a set of efficient solutions to the problem (P), a set of non-efficient solutions to the problem (P) and the best compromise solution to the problem (P) for each time in $[T^\circ, T^m]$ and also, a set of overall efficient solutions to the problem (P) and overall the best compromise solution to the problem (P).

4   Numerical Example

The block-dripping method for solving a BBTP is illustrated by the following example.

Example 4.1. Consider a transportation model of a company involving three factories, denoted by F_1, F_2 and F_3, and four warehouses, denoted by W_1, W_2, W_3 and W_4. A particular product is transported from the ith factory to the jth warehouse. Assume that there are three criteria's under consideration: (i) the minimization of total transportation cost consumed in transportation (ii) the minimization of total product deterioration during transportation and (iii) the minimization of maximum travel time during transportation. The cost of the transportation of a product, the deterioration cost of a product during transportation and the time of transportation are given in the following table:

		W_1		W_2		W_3		W_4		Supply
					Warehouses j					
	F_1	10	(1,4)	95	(2,4)	73	(7,3)	52	(7,4)	8
Factory i,	F_2	66	(1,5)	68	(9,8)	30	(3,9)	21	(4,10)	19
	F_3	97	(8,6)	63	(9,2)	19	(4,5)	23	(6,1)	17
Demand		11		3		14		16		

The goal is to locate the set of all solutions for the bottleneck bi-criteria TP.

Now, the bottleneck transportation problem of BBTP is given below:

	W_1	W_2	W_3	W_4	Supply
		Warehouses j			
F_1	10	95	73	52	8
Factory i, F_2	66	68	30	21	19
F_3	97	63	19	23	17
Demand	11	3	14	16	

Using blocking method [9], the optimal solution of the bottleneck TP of BBTP is 66.

Now, solving the FCTP and SCTP individually by zero point method [10], we obtain the following optimal solutions:

Optimal Solution							
FCTP				SCTP			
5	3	0	0	0	0	8	0
6	0	0	13	11	2	6	0
0	0	14	3	0	1	0	16

Bottleneck Bi-criteria Value (BBV)   (143, 265, 95)   (208, 167, 73)

Thus, we have $T_{m1} = 95$ and $T_{m2} = 73$. Therefore, $T^m = \max\{95,73\}$.

Now, since the overall time range is between 66 and 95, we have $[T^\circ, T^m] = \{66, 68, 73, 95\}$.

Now, the ideal solution to the problem is (143, 167, 66)

Case 1: M = 66

The bi-criteria TP for $M = 66$ is given below:

	Warehouses j				Supply
	W_1	W_2	W_3	W_4	
F_1	(1,4)	-	-	(7,4)	8
Factory i, F_2	(1,5)	-	(3,9)	(4,10)	19
F_3	-	(9,2)	(4,5)	(6,1)	17
Demand	11	3	14	16	

Now, by dripping method [11], the set of all solutions S of the BBTP for M=66 is :
S = {(158,283,66), (159,278,66), (160,273,66), (161,268,66), (162,263,66), (163,258,66), (164,253,66), (165,248,66), (166,243,66), (167, 238,66), (168,233,66), (169,228,66), (170, 223,66), (171,218,66), (172,213,66), (175,208,66), (178,203,66)}.

Thus, for the time 66, we obtain 17 efficient solutions to the problem and the best compromise solution to the problem is (172, 213, 66).

Case 2: M = 68

The bi-criteria TP for $M = 68$ is given below:

	Warehouses j				Supply
	W_1	W_2	W_3	W_4	
F_1	(1,4)	-	-	(7,4)	8
Factory i, F_2	(1,5)	(9,8)	(3,9)	(4,10)	19
F_3	-	(9,2)	(4,5)	(6,1)	17
Demand	11	3	14	16	

Now, by dripping method [11], the set of all solutions S of the BBTP for M=68 is:
S = {(178,203,68), (175,208,68), (172,213,68), (171,218,68), (170,223,68), (169,228,68), (168,233,68), (167,238,68), (166,243,68), (165,248,68), (164,253,68), (163,258,68), (162,263,68), (161,268, 68), (160,273,68), (159,278,68), (158,283,68)}.

Thus, for the time 68, we obtain 17 efficient solutions to the problem and the best compromise solution for the problem is (172, 213, 68).

Case 3: M=73

The bi-criteria TP for $M = 73$ is given below:

	Warehouses j				Supply
	W_1	W_2	W_3	W_4	
F_1	(1,4)	-	(7,3)	(7,4)	8
Factory i, F_2	(1,5)	(9,8)	(3,9)	(4,10)	19
F_3	-	(9,2)	(4,5)	(6,1)	17
Demand	11	3	14	16	

Now, by dripping method [11], the set of all solutions S of the BBTP for M= 73 is:
S = {(158,283,73), (159,278,73), (160,273,73), (161,268,73), (162,263,73), (163,258,73), (164,253,73), (165,248,73), (166,243,73), (167,238,73), (168,233,73), (169,228,73), (170,223,73), (171,218,73), (172,213,73), (175,208,73), (176,208,73), (178,203,73), (180,203,73), (182,198,73), (184,198,73), (186,193,73), (188,193,73), (190,188,73), (192, 188,73), (194,183,73), (196,183,73),(198,178,73), (200,178,73), (202,173,73), (203,173,73), (204,173,73), (205,170,73), (208,167,73) }.

Thus, for the time 73, we obtain 34 efficient solutions to the problem and the best compromise solution for the problem is (186, 193, 73).

Case 4: M=95
The bi-criteria TP for $M = 95$ is given below:

	Warehouses j				Supply
	W_1	W_2	W_3	W_4	
F_1	(1,4)	(2,4)	(7,3)	(7,4)	8
Factory i, F_2	(1,5)	(9,8)	(3,9)	(4,10)	19
F_3	-	(9,2)	(4,5)	(6,1)	17
Demand	11	3	14	16	

Now, by dripping method [11], the set of all solutions S of the BBTP for M=95 is :
S={(143,265,95),(144,260,95),(145,255,95),(146,250,95),(147,245,95),
(148,240,95),(149,235,95), (150,230,95), (151,225,95), (152,220,95), (153,215,95),
(154,210,95), (155,205,95), (156,200,95), (160,195,95), (164,190,95), (168,185,95),
(172,180,95), (176,175,95), (186,171,95), (187,173,95), (197,169,95), (198,171,95),
(209,169,95), (208,167,95)}.

Thus, for the time 95,   we obtain 22 efficient solutions and 3 non-efficient
solutions   and the best compromise solution to the given problem is (168, 185, 95).

Now, over all efficient solutions, P to the given problem is given below:
P   =   {158,283,66), (159,278,66), (160,273,66), (161,268,66), (162,263,66),
(163,258,66), (164,253,66),(165,248,66), (166,243,66), (167,238,66), (168,233,66),
(169,228,66), (170,223,66), (171,218,66), (172,213,66), (175,208,66), (178,203,66) ,
(182,198,73), (186,193,73), (190,188,73), (194,183,73),(198,178,73), (202,173,73),
(205,170,73), (208,167,73), (143,265,95), (144,260,95), (145,255,95), (146,250,95),
(147,245,95), (148,240,95), (149,235,95), (150,230,95), (151,225,95), (152,220,95),
(153,215,95), (154,210,95), (155,205,95), (156,200,95), (160,195,95), (164,190,95),
(168,185,95), (172,180,95), (176,175,95), (186,171,95)}

Thus, over all the best compromise solution to the given problem is (178,203,66)

where $Z_1(X^\circ) = 178$, $Z_2(X^\circ) = 203$, $X^\circ = \{x_{11} = 6, x_{14} = 2, x_{21} = 5, x_{23} = 14,$

$x_{32} = 3, x_{34} = 14\}$ and time T = 66.

Remark 2: In [14], **9** distinct efficient solutions and 11 distinct efficient solutions in
[15] to the problem (Example 4.1) were recorded but in this paper 45 distinct efficient
solutions to the same problem have been obtained.

Remark 3: In [14, 15], the best compromise solution to the problem (Example 4.1)
having least time is (178, 203, 68), but we have (178, 203, 66) in this paper.
Therefore, we provide the best compromise non-dominated solution to the problem
having least time which is better than all existing methods [14, 15].

Graphical representations

The solutions to the given problem for M =66, M=68, M=73 and M=95 in two
dimensional graphs are given below:

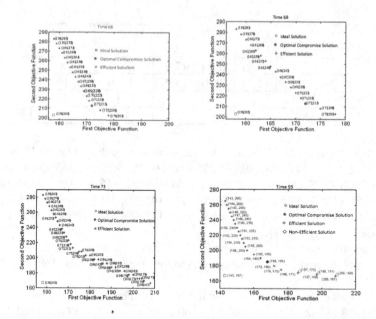

The overall efficient solutions to the given problem in three dimensional graph is given below:

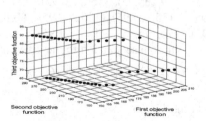

5   Conclusion

The time of transport might be a significant factor in several real life transportation problems. By the proposed method, a set of non-dominated solutions to a bottleneck bi-criteria problem is obtained and the best compromise solution to the problem for each time in the specified time interval and also, overall solutions are computed. The block-dripping method will provide the necessary decision support to the decision makers to evaluate the economical activities and make efficient appropriate managerial decisions while they are dealing with time related bi-criteria logistic problems.

References

1. Aneja, Y.P., Nair, K.P.K.: Bi-criteria transportation problem. Mgt. Sci. 21, 73–78 (1979)
2. Garfinkel, R.S., Rao, M.R.: The bottleneck transportation problem. Naval Research Logistics Quarterly 18, 465–472 (1971)
3. Glickman, T., Berger, P.D.: Cost-completion-date tradeoffs in the transportation problem. Op. Res. 25, 163–168 (1977)
4. Gupta, R.: Time-cost transportation problem. Econ-math Overview 4, 431–443 (1977)
5. Hammer, P.L.: Time minimizing transportation problems. Naval Res. Log. Qly. 16, 345–357 (1969)
6. Nikolić, I.: Total time minimizing transportation problem. Yugoslav J. Op. Res. 17, 125–133 (2007)
7. Issermann, H.: Linear bottleneck transportation problem. Asia Pacific J. Op. Res. 1, 38–52 (1984)
8. Mangasarian, O.L.: Nonlinear Programming. McGraw Hill, New York (1969)
9. Pandian, P., Natarajan, G.: A new method for solving bottleneck-cost transportation problems. Internat. Math. Forum. 6, 451–460 (2011)
10. Pandian, P., Natarajan, G.: A new method for finding an optimal solution for transportation problems. Internat. J. Math. Sci. Engg. Appl. 4, 59–65 (2010)
11. Pandian, P., Anuradha, D.: A new approach for solving bi-objective transportation problems. Australian J. Basic Appl. Sci. 5, 67–74 (2011)
12. Malhotra, R.: On bi-criteria transportation problem. In: Combinatorial Optimization: Some Aspects. Narosa Publishing House, New Delhi (2007)
13. Szwarc, W.: Some remarks on the transportation problem. Naval Res. Log. Qly. 18, 473–485 (1971)
14. Tkacenko, A., Alhazov, A.: The multiobjective bottleneck transportation problem. Comput. Sci. J. Moldova. Kishinev. 9, 321–335 (2001)
15. Tkacenko, A.: The generalized algorithm for solving the fractional multi-objective transportation problem. ROMAI J. 2, 197–202 (2006)
16. Zeleny, M.: Multiple criteria decision making. Mc Graw-Hill Book Company (1982)

A Fuzzy Goal Programming Approach
for Fuzzy Multiobjective Stochastic Programming
through Expectation Model

Animesh Biswas[*] and Nilkanta Modak

Department of Mathematics, University of Kalyani
Kalyani - 741235, India
abiswaskln@rediffmail.com, nmodak9@gmail.com

Abstract. In this paper a fuzzy goal programming method for modeling and solving multiobjective stochastic decision making problem involving fuzzy random variables associated with the parameters of the objectives as well as system constraints is developed. In the proposed approach, an expectation model is generated on the basis of the mean of the fuzzy random variables involved with the objectives of the problem. Then the problem is converted into an equivalent fuzzy programming model by considering the fuzzily defined chance constraints. Afterwards, the model is decomposed on the basis of the tolerance ranges of the fuzzy numbers associated with the fuzzy parameters of the problem. Now to construct the membership goals of the decomposed objectives under the extended feasible region defined by the decomposed system constraints, the individual optimal values of each objective is calculated in isolation. Then the membership functions are constructed to measure the degree of satisfaction of each decomposed objectives in the decision making environment. The membership functions are then converted into membership goals by assigning unity as the aspiration level of the membership goals. Then a fuzzy goal programming model is developed by minimizing the under deviational variables and thereby obtaining the optimal solution in the decision making environment.

Keywords: Stochastic programming, Chance constrained programming, Expectation model, Fuzzy random variables, Fuzzy numbers, Fuzzy programming, Fuzzy goal programming.

1   Introduction

With the development in computational aspects and scientific computing techniques, sophisticated optimization models [1,2] can now be solved efficiently. Yet many optimization models are affected by uncertainty in input data or in model relationship.

While research has progressed at a steady pace in the field of stochastic optimization [3,4] and fuzzy mathematical programming [5,6], each of which touches only on one type of uncertainty, the past decade, in particular, has witnessed in

---

[*] Corresponding author.

P. Balasubramaniam and R. Uthayakumar (Eds.): ICMMSC 2012, CCIS 283, pp. 124–135, 2012.
© Springer-Verlag Berlin Heidelberg 2012

developing interest with the situations where fuzziness and randomness are under one roof in an optimization frame work [7-9]. This interest has been motivated by the need for computing many human decision problems which is both fuzzily imprecise and probabilistically uncertain. The multidisciplinary research field that emerged as a result of this interest lies at the boundary of stochastic optimization and fuzzy mathematical programming is known as fuzzy stochastic programming. The general strategy of fuzzy stochastic optimization is to de-fuzzify and/or de-randomize fuzzy random variables to convert the problem into a deterministic problem [9,10].

Guo and Huang [11] developed a methodology in the context of solid waste management involving fuzzy and stochastic uncertainties which can not perform well for integrating the two types of uncertainties. Further this process fails to capture the technique of solving multiplicity of objectives involved in a decision making model. To deal with a solution having multiplicity of objectives Fuzzy goal programming (FGP) [12,13] as an extensions of goal programming, is appeared as a robust tool for making decision in a highly conflicting fuzzy environment. Again many real life problems are involved with conflicting goals under randomness and fuzziness simultaneously. This type of problem can be solved by using stochastic FGP technique [14-16]. In the recent past, Pal et al. [17] developed methodology for solving multiobjective stochastic programming by using interval valued goal programming.

However an efficient solution technique for solving fuzzy multiobjective stochastic programming in a highly conflicting fully fuzzified decision making environment is yet to appear in the literature.

In this work, FGP technique is used to solve the fuzzy multiobjective stochastic programming problem in which the parameters of the objectives as well as system constraints are considered as fuzzy random variables and fuzzy numbers. The objectives of the problem are first approximated by using Expectation-model (E-model) [18]. Then considering the tolerance ranges of the fuzzy numbers associated with the parameters of the objectives the model is decomposed in the decision making context. Further the fuzzily defined system constraints are converted into its fuzzy equivalent model by applying general chance constrained methodology. Afterwards the membership functions are constructed to measure the degree of satisfaction of the decomposed objectives within the extended feasible region defined by the decomposed system constraints by finding individual optimal values of them. Then an FGP model is developed to find out the most satisfactory solution in the decision making environment.

2   Problem Formulation

A fuzzy multiobjective stochastic programming model having K number of objectives involving fuzzy random variables with fuzzily defined chance constraints can be expressed in its general form as:

Find $X(x_1, x_2, ..., x_n)$ so as to

Maximize $Z_k = \sum_{j=1}^{n} \tilde{c}_{kj} x_j$; $k = 1, 2, ..., K$

$$\text{subject to } X \in S \left\{ X \in R^n / \Pr \left[\sum_{j=1}^{n} \tilde{a}_{ij} x_j \binom{\leq}{\geq} \tilde{b}_i \right] \geq p_i \right\}, \quad i = 1, 2, ..., m \tag{1}$$

where $X \geq 0$ denotes the vector of decision variables and Z_k represents the k-th objective function. The parameters $\tilde{c}_{kj}$, $\tilde{a}_{ij}$ and $\tilde{b}_i$ are normally distributed fuzzy random variables and p_i $(0 \leq p_i \leq 1)$ is the satisfying probability level defined for the randomness occurs in the i-th constraints. Then by considering the fuzzy random variable $\tilde{c}_{kj}$ associated with the objectives the following E-model is developed.

2.1   Fuzzy E-Model of Objective Function

Since $\tilde{c}_{kj}$ are normally distributed fuzzy random variables, let $E(\tilde{c}_{kj})$ be the mean values associated with $\tilde{c}_{kj}$ of the k-th objective Z_k. Then the E-Model of the objective can be presented as [18]

$$\text{Maximize } E(Z_k) = \sum_{j=1}^{n} E(\tilde{c}_{kj}) x_j \; ; \; k = 1, 2, ..., K \tag{2}$$

Since $E(\tilde{c}_{kj})$ is a fuzzy number, we introduce a fuzzy number $\tilde{M}_{kj}$ such that $\tilde{M}_{kj} = E(\tilde{c}_{kj})$

In a fuzzy decision making situation, it is to be assumed here that the mean associated with the fuzzy random variables $\tilde{c}_{kj}$ are triangular fuzzy numbers.

A triangular fuzzy number $\tilde{a}$ can be represented by a triple of three real numbers as $\tilde{a} = (a^L, a, a^R)$. The membership function of the triangular fuzzy number is of the form

$$\mu_{\tilde{a}}(x) = \begin{cases} 0 & if \quad x < a^L \, or \, x > a^R \\ \dfrac{x - a^L}{a - a^L} & if \quad a^L \leq x \leq a \\ \dfrac{a^R - x}{a^R - a} & if \quad a \leq x \leq a^R \end{cases}$$

where a^L and a^R denote, respectively, the left and right tolerance values of the fuzzy number $\tilde{a}$. Now considering the following triangular fuzzy number associated with the mean of the fuzzy random variable $\tilde{c}_{kj}$ as $\tilde{M}_{kj} = (M_{kj}^L, M_{kj}, M_{kj}^R)$, the objective function in (2) can be decomposed as

$$\text{Maximize } E(Z_k)^L = \sum_{j=1}^{n} \left\{ M_{kj}^L + (M_{kj} - M_{kj}^L)\alpha \right\} x_j \ ; \ k = 1, 2, ..., K \tag{3}$$

$$\text{Maximize } E(Z_k)^R = \sum_{j=1}^{n} \left\{ M_{kj}^R - (M_{kj}^R - M_{kj})\alpha \right\} x_j \ ; \ k = 1, 2, ..., K \ , \ 0 \le \alpha \le 1 \tag{4}$$

In the following subsection the method for converting this transformed model into an equivalent FP model is discussed.

2.2   FP Model Formulation

Here $\tilde{a}_{ij}$ and $\tilde{b}_i$ associated with the chance constraints are both random variables following normal distribution. Let $\tilde{y}_i = \left(\sum_{j=1}^{n} \tilde{a}_{ij} x_j - \tilde{b}_i \right)$, for $i = 1, 2, ..., m$. Then it is easy to realize that the newly introduced variable $\tilde{y}_i$ must also follow fuzzy normal distribution. Now considering $\ge$ type restriction involved with the probabilistic constraints described in (1), the chance constraints may take the form as $\Pr[\tilde{y}_i \ge 0] \ge p_i$ which expression can be generalized as

$$\Pr\left[\frac{\tilde{y}_i - E(\tilde{y}_i)}{\sqrt{\text{var}(\tilde{y}_i)}} \ge \frac{-E(\tilde{y}_i)}{\sqrt{\text{var}(\tilde{y}_i)}} \right] \ge p_i, \ \text{ i.e., } \ \Pr\left[\frac{\tilde{y}_i - E(\tilde{y}_i)}{\sqrt{\text{var}(\tilde{y}_i)}} \ge \frac{-E(\tilde{y}_i)}{\sqrt{\text{var}(\tilde{y}_i)}} \right] \le 1 - p_i \tag{5}$$

where $\dfrac{\tilde{y}_i - E(\tilde{y}_i)}{\sqrt{\text{var}(\tilde{y}_i)}}$ is a fuzzy standard normal variate. Therefore the equivalent FP

model of (5) can be expressed as $\dfrac{-E(\tilde{y}_i)}{\sqrt{\text{var}(\tilde{y}_i)}} \le \Phi^{-1}(p_i)$

i.e., $E(\tilde{y}_i) + \Phi^{-1}(p_i)\sqrt{\text{var}(\tilde{y}_i)} \ge 0 , i = 1, 2, ..., m.$ \tag{6}

Similarly considering $\le$ type restriction of the chance constraints in (1), the system constraints take the form

$$E(\tilde{y}_i) + \Phi^{-1}(p_i)\sqrt{\text{var}(\tilde{y}_i)} \le 0, \ i = 1, 2, ..., m. \tag{7}$$

Now the respective expression of (6) and (7) can be written as

$$\sum_{j=1}^{n} E(\tilde{a}_{ij}) x_j + \Phi^{-1}(p_i)\sqrt{\sum_{j=1}^{n} Var(\tilde{a}_{ij}) x_j^2 + Var(\tilde{b}_i)} \ge E(\tilde{b}_i), i = 1, 2, ..., m \tag{8}$$

$$\sum_{j=1}^{n} E(\tilde{a}_{ij})x_j + \Phi^{-1}(p_i)\sqrt{\sum_{j=1}^{n} Var(\tilde{a}_{ij})x_j^2 + Var(\tilde{b}_i)} \le E(\tilde{b}_i),\, i=1,2,...,m \tag{9}$$

Since $\tilde{a}_{ij}\,(i=1,2,...,m;\, j=1,2,...,n)$ are normally distributed fuzzy random variables and the decision variables $x_j \ge 0$, $j=1,2,...,n$ are unknowns, some fuzzy random variables, $\tilde{d}_i\,(i=1,2,...,m)$ can be introduced as $\tilde{d}_i = \sum_{j=1}^{n}\tilde{a}_{ij}\,x_j;\, i=1,2,...,m$. Then $\tilde{d}_i$ are normally distributed fuzzy random variables with the respective mean and variance given by $m_{\tilde{d}_i} = \sum_{j=1}^{n} E(\tilde{a}_{ij})x_j$, $\sigma_{\tilde{d}_i}^2 = \sum_{j=1}^{n} Var(\tilde{a}_{ij})x_j$, $i=1,2,...,m$.

Now applying mean and variance of the fuzzy random variables, the above respective expression in (8) & (9) are converted into the following form:

$$m_{\tilde{d}_i} + \Phi^{-1}(p_i)\sqrt{\sigma_{\tilde{d}_i}^2 + Var(\tilde{b}_i)} \ge E(\tilde{b}_i),\, i=1,2,...,m \tag{10}$$

$$m_{\tilde{d}_i} + \Phi^{-1}(p_i)\sqrt{\sigma_{\tilde{d}_i}^2 + Var(\tilde{b}_i)} \le E(\tilde{b}_i),\, i=1,2,...,m. \tag{11}$$

Characterization of Membership Functions. In a fuzzy decision making situation, it is to be assumed that the mean and variance associated with the fuzzy random variables $\tilde{d}_i$ are triangular fuzzy numbers, which are considered as follows:

$$m_{\tilde{d}_i} = \left(m_{\tilde{d}_i}^L, m_{\tilde{d}_i}, m_{\tilde{d}_i}^R\right)\quad \sigma_{\tilde{d}_i}^2 = \left(\sigma_{\tilde{d}_i}^{2^L}, \sigma_{\tilde{d}_i}^2, \sigma_{\tilde{d}_i}^{2^R}\right),\text{ for } i=1,2,...,m;\, j=1,2,...,n.$$

Also the expected values $E(\tilde{b}_i)\,(i=1,2,...,m)$ are considered as one sided fuzzy numbers which is represented by the following figure:

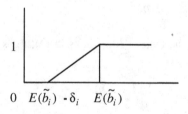

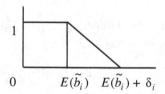

Fig. 1. Left sided fuzzy number   **Fig. 2.** Right sided fuzzy number

where δ_i be the tolerance limit of the fuzzy numbers $E(\tilde{b}_i)$. Again, let variances, $\sigma_{\tilde{b}_i}^2 = Var(\tilde{b}_i)$ are considered triangular fuzzy numbers as

$$\sigma_{\tilde{b}_i}^2 = \left(\sigma_{\tilde{b}_i}^{2^L}, \sigma_{\tilde{b}_i}^2, \sigma_{\tilde{b}_i}^{2^R} \right), \quad \text{for } i = 1, 2, ..., m.$$

In this connection it is to be noted that the left sided fuzzy numbers considered for the constraints described in (10) and right sided fuzzy numbers for the constraints described in (11) in the decision making context. Then, on the basis of above fuzzy numbers, (10) can be described as

$$\{m_{\tilde{d}_i}^L + (m_{\tilde{d}_i} - m_{\tilde{d}_i}^L)\alpha\} + \Phi^{-1}(p_i) \sqrt{\left\{ \sigma_{\tilde{d}_i}^{2^L} + (\sigma_{\tilde{d}_i}^2 - \sigma_{\tilde{d}_i}^{2^L})\alpha \right\} + \left\{ \sigma_{\tilde{b}_i}^{2^L} + (\sigma_{\tilde{b}_i}^2 - \sigma_{\tilde{b}_i}^{2^L})\alpha \right\}}$$
$$\geq \{E(b_i) - \delta_i + \alpha\delta_i\}$$

$$\{m_{\tilde{d}_i}^R - (m_{\tilde{d}_i}^R - m_{\tilde{d}_i})\alpha\} + \Phi^{-1}(p_i) \sqrt{\left\{ \sigma_{\tilde{d}_i}^{2^R} - (\sigma_{\tilde{d}_i}^{2^R} - \sigma_{\tilde{d}_i}^2)\alpha \right\} + \left\{ \sigma_{\tilde{b}_i}^{2^R} - (\sigma_{\tilde{b}_i}^{2^R} - \sigma_{\tilde{b}_i}^2)\alpha \right\}}$$
$$\geq \{E(b_i) - \delta_i + \alpha\delta_i\}$$

for $0 \leq \alpha \leq 1$, $i = 1, 2, ..., m$ (12)

Similarly the constraints in (11) can be written as

$$\{m_{\tilde{d}_i}^L + (m_{\tilde{d}_i} - m_{\tilde{d}_i}^L)\alpha\} + \Phi^{-1}(p_i) \sqrt{\left\{ \sigma_{\tilde{d}_i}^2 + (\sigma_{\tilde{d}_i}^2 - \sigma_{\tilde{d}_i}^{2^L})\alpha \right\} + \left\{ \sigma_{\tilde{b}_i}^{2^L} + (\sigma_{\tilde{b}_i}^2 - \sigma_{\tilde{b}_i}^{2^L})\alpha \right\}}$$
$$\leq \{E(b_i) + \delta_i - \alpha\delta_i\}$$

$$\{m_{\tilde{d}_i}^R - (m_{\tilde{d}_i}^R - m_{\tilde{d}_i})\alpha\} + \Phi^{-1}(p_i) \sqrt{\left\{ \sigma_{\tilde{d}_i}^{2^R} - (\sigma_{\tilde{d}_i}^{2^R} - \sigma_{\tilde{d}_i}^2)\alpha \right\} + \left\{ \sigma_{\tilde{b}_i}^{2^R} - (\sigma_{\tilde{b}_i}^{2^R} - \sigma_{\tilde{b}_i}^2)\alpha \right\}}$$
$$\leq \{E(b_i) + \delta_i - \alpha\delta_i\}$$

$0 \leq \alpha \leq 1$, $i = 1, 2, ..., m$, $j = 1, 2, ..., n$.   (13)

Now let $E(Z_k)_B^L$ and $E(Z_k)_W^L$ be the respective best and worst value of the objectives function obtained by solving (3) under the system constraints (12) and (13). Also let $E(Z_k)_B^R$ and $E(Z_k)_W^R$ be the best and worst values of the objectives respectively when solved (4) under the system constraints defined in (12) and (13).

Hence the fuzzy objective goals for each of the objectives can be expressed as
$E(Z_k)^L \gtrsim E(Z_k)^L_B$ and $E(Z_k)^R \gtrsim E(Z_k)^R_B$ for $k = 1, 2, ..., K$.

Thus the membership function for each of the decomposed objectives can be written as (for $k = 1, 2, ..., K$):

$$\mu_{E(Z_k)^L}(x) = \begin{cases} 0 & \text{if } E(Z_k)^L \leq E(Z_k)^L_W \\ \dfrac{E(Z_k)^L - E(Z_k)^L_W}{E(Z_k)^L_B - E(Z_k)^L_W} & \text{if } E(Z_k)^L_W \leq E(Z_k)^L \leq E(Z_k)^L_B \\ 1 & \text{if } E(Z_k)^L \geq E(Z_k)^L_B \end{cases} \qquad (14)$$

$$\mu_{E(Z_k)^R}(x) = \begin{cases} 0 & \text{if } E(Z_k)^R \leq E(Z_k)^R_W \\ \dfrac{E(Z_k)^R - E(Z_k)^R_W}{E(Z_k)^R_B - E(Z_k)^R_W} & \text{if } E(Z_k)^R_W \leq E(Z_k)^R \leq E(Z_k)^R_B \\ 1 & \text{if } E(Z_k)^R \geq E(Z_k)^R_B \end{cases} \qquad (15)$$

Now in the fuzzy decision making situation, the aim of the decision makers is to achieve the highest degree (unity) of each of the decomposed objectives to the extent possible. Considering this aspects FGP is seemed to be most appropriate to make the decision in the decision making context.

2.3   Construction of FGP Model

In FGP the membership functions are considered as flexible goals by assigning unity as the aspiration level and by introducing under and over deviational variables to each of them. Then under deviational variables are minimized to achieve the aspired levels of the goals to the extent possible in the decision making environment.

The minsum FGP model can be presented as

Find $X(x_1, x_2, ..., x_n)$ so as to

$$\text{Minimize D} = \sum_{k=1}^{K} w^L_k d^{L-}_k + \sum_{k=1}^{K} w^R_k d^{R-}_k$$

and satisfy $\dfrac{E(Z_k)^L - E(Z_k)^L_W}{E(Z_k)^L_B - E(Z_k)^L_W} + d^{L-}_k - d^{L+}_k = 1$

$$\dfrac{E(Z_k)^R - E(Z_k)^R_W}{E(Z_k)^R_B - E(Z_k)^R_W} + d^{R-}_k - d^{R+}_k = 1$$

subject to the system constraints defined in (12) and (13).   (16)

$$d_k^{L-} . d_k^{L+} = d_k^{R-} . d_k^{R+} = 0 \text{ and } d_k^{L-}, d_k^{L+}, d_k^{R-}, d_k^{R+} \geq 0$$

where w_k^L, $w_k^R \geq 0$ represent the numerical weights of importance of the fuzzy goals associated with the under deviational variables are determined as

$$w_k^L = \frac{1}{E(Z_k)_B^L - E(Z_k)_W^L}, \quad w_k^R = \frac{1}{E(Z_k)_B^R - E(Z_k)_W^R} .$$

The above model (16) is solved to get the most satisfactory solution in the decision making environment.

3   Numerical Example

To illustrate the proposed methodology, the problem considered by Pal et al. [17] is taken into account and is solved through the developed technique. The fuzzy multiobjective stochastic programming problem can be presented as

Maximize $Z_1 = \tilde{c}_{11}x_1 + \tilde{c}_{12}x_2 + \tilde{c}_{13}x_3$

Maximize $Z_2 = \tilde{c}_{21}x_1 + \tilde{c}_{22}x_2 + \tilde{c}_{23}x_3$

Subject to: $\Pr\left(\tilde{a}_{11}x_1 + \tilde{a}_{12}x_2 + \tilde{a}_{13}x_3 \geq \tilde{b}_1\right) \geq 0.90$; $\Pr\left(\tilde{a}_{21}x_1 + \tilde{a}_{22}x_2 + \tilde{a}_{23}x_3 \geq \tilde{b}_2\right) \geq 0.95$

(17)

where $\tilde{c}_{kj}$ $(k = 1, 2; \quad j = 1, 2, 3)$ are normally distributed fuzzy random variables with respective mean represented by the following triangular fuzzy numbers:

$$E(\tilde{c}_{11}) = (5.95, 6, 6.05), \ E(\tilde{c}_{12}) = (4.95, 5, 5.05), \ E(\tilde{c}_{13}) = (6.95, 7, 7.05)$$

$$E(\tilde{c}_{21}) = (6.95, 7, 7.05), \ E(\tilde{c}_{22}) = (7.95, 8, 8.05), \ E(\tilde{c}_{23}) = (5.95, 6, 6.05)$$

Also $\tilde{a}_{kj}, \tilde{b}_k$ $(k = 1, 2; \quad j = 1, 2, 3)$ are independent fuzzy normal variables. The mean and variance of the independent random variables are considered as:

$$E(\tilde{a}_{11}) = \tilde{5}, \text{var}(\tilde{a}_{11}) = \tilde{3} ; \ E(\tilde{a}_{12}) = \tilde{6}, \text{var}(\tilde{a}_{12}) = \tilde{4}; \ E(\tilde{a}_{13}) = \tilde{8}, \text{var}(\tilde{a}_{13}) = 5.\tilde{5};$$

$$E(\tilde{a}_{21}) = \tilde{8}, \text{var}(\tilde{a}_{21}) = \tilde{9}; \ E(\tilde{a}_{22}) = \tilde{9}, \text{var}(\tilde{a}_{22}) = \tilde{4}; \ E(\tilde{a}_{23}) = 1\tilde{0}, \text{var}(\tilde{a}_{23.}) = \tilde{6};$$

$$E(\tilde{b}_1) = \tilde{8}, \text{var}(\tilde{b}_1) = \tilde{5} ; \ E(\tilde{b}_2) = 1\tilde{6}, \text{var}(\tilde{b}_2) = 1\tilde{0} .$$

Now considering mean of the fuzzy random variables associated with the objectives and system constraints, the E-model of the given problem can be written as

Maximize $E(Z_1) = E(\tilde{c}_{11})x_1 + E(\tilde{c}_{12})x_2 + E(\tilde{c}_{13})x_3$

Maximize $E(Z_2) = E(\tilde{c}_{21})x_1 + E(\tilde{c}_{22})x_2 + E(\tilde{c}_{23})x_3$

subject to

$$E(\tilde{a}_{11})x_1 + E(\tilde{a}_{12})x_2 + E(\tilde{a}_{13})x_3 + 1.28\sqrt{\text{var}(\tilde{a}_{11})x_1^2 + \text{var}(\tilde{a}_{12})x_2^2 + \text{var}(\tilde{a}_{13})x_3^2 + \text{var}(\tilde{b}_1)} \geq E(\tilde{b}_1)$$
$$E(\tilde{a}_{21})x_1 + E(\tilde{a}_{22})x_2 + E(\tilde{a}_{23})x_3 + 1.645\sqrt{\text{var}(\tilde{a}_{21})x_1^2 + \text{var}(\tilde{a}_{22})x_2^2 + \text{var}(\tilde{a}_{23})x_3^2 + \text{var}(\tilde{b}_2)} \leq E(\tilde{b}_2)$$
$$x_j \geq 0, \quad j = 1, 2, 3 \tag{18}$$

As described in the proposed procedure, the fuzzy numbers and are considered as left and right sided fuzzy numbers, respectively which are given by

$$\mu_{E(b_1)}(x) = \begin{cases} 0 & \text{if} & E(b_1) \leq 7.5 \\ \dfrac{E(b_1) - 7.5}{0.5} & \text{if } 7.5 \leq E(b_1) \leq 8 \\ 1 & \text{if} & E(b_1) \geq 8 \end{cases}$$

$$\mu_{E(b_2)}(x) = \begin{cases} 1 & \text{if} & E(b_2) \leq 16 \\ \dfrac{16.5 - E(b_2)}{0.5} & \text{if } 16 \leq E(b_2) \leq 16.5 \\ 0 & \text{if} & E(b_2) \geq 16.5 \end{cases}$$

All other fuzzy numbers associated with the system constraints are taken as

$E(\tilde{a}_{11}) = \tilde{5} = (4.95, 5, 5.05) \text{ var}(\tilde{a}_{11}) = \tilde{3} = (2.95, 3, 3.05) \,; E(\tilde{a}_{12}) = \tilde{6} = (5.95, 6, 6.05),$

$\text{var}(\tilde{a}_{12}) = \tilde{4} = (3.95, 4, 4.05); \; E(\tilde{a}_{13}) = \tilde{8} = (7.95, 8, 8.05), \text{ var}(\tilde{a}_{13}) = 5.\tilde{5} = (5.45, 5.5, 5.55);$

$E(\tilde{a}_{21}) = \tilde{8} = (7.95, 8, 8.05), \text{ var}(\tilde{a}_{21}) = \tilde{9} = (8.95, 9, 9.05); \qquad E(\tilde{a}_{22}) = \tilde{9} = (8.95, 9, 9.05),$

$\text{var}(\tilde{a}_{22}) = \tilde{4} = (3.95, 4, 4.05); \; E(\tilde{a}_{23}) = 1\tilde{0} = (9.95, 10, 10.05), \text{ var}(\tilde{a}_{23}) = \tilde{6} = (5.95, 6, 6.05);$

$\text{var}(\tilde{b}_1) = \tilde{5} = (4.95, 5, 5.05) \,; \quad \text{var}(\tilde{b}_2) = 1\tilde{0} = (9.95, 10, 10.05).$

Now the FP model using the above defined fuzzy numbers takes the form as;

Maximize $E(Z_1)^L = (5.95 + 0.05\alpha)x_1 + (4.95 + 0.05\alpha)x_2 + (6.95 + 0.05\alpha)x_3$

Maximize $E(Z_1)^R = (6.05 - 0.05\alpha)x_1 + (5.05 - 0.05\alpha)x_2 + (7.05 - 0.05\alpha)x_3$

Maximize $E(Z_2)^L = (6.95 + 0.05\alpha)x_1 + (7.95 + 0.05\alpha)x_2 + (5.95 + 0.05\alpha)x_3$

Maximize $E(Z_1)^R = (7.05 - 0.05\alpha)x_1 + (8.05 - 0.05\alpha)x_2 + (6.05 - 0.05\alpha)x_3$

subject to

$$(4.95+0.05\alpha)x_1 + (5.95+0.05\alpha)x_2 + (7.95+0.05\alpha)x_3 +$$
$$1.28\sqrt{(2.95+0.05\alpha)x_1^2 + (3.95+0.05\alpha)x_2^2 + (5.45+0.05\alpha)x_3^2 + (4.95+0.05\alpha)} \geq (7.5+0.5\alpha);$$
$$(5.05-0.05\alpha)x_1 + (6.05-0.05\alpha)x_2 + (8.05-0.05\alpha)x_3 +$$
$$1.28\sqrt{(3.05-0.05\alpha)x_1^2 + (4.05-0.05\alpha)x_2^2 + (5.55-0.05\alpha)x_3^2 + (5.05-0.05\alpha)} \geq (7.5+0.5\alpha);$$
$$(7.95+0.05\alpha)x_1 + (8.95+0.05\alpha)x_2 + (9.95+0.05\alpha)x_3$$
$$+1.645\sqrt{(8.95+0.05\alpha)x_1^2 + (3.95+0.05\alpha)x_2^2 + (5.95+0.05\alpha)x_3^2 + (9.95+0.05\alpha)} \leq (16.5-0.5\alpha);$$
$$(8.05-0.05\alpha)x_1 + (9.05-0.05\alpha)x_2 + (10.05-0.05\alpha)x_3$$
$$+1.645\sqrt{(9.05-0.05\alpha)x_1^2 + (4.05-0.05\alpha)x_2^2 + (6.05-0.05\alpha)x_3^2 + (10.05-0.05\alpha)} \leq (16.5-0.5\alpha);$$
$$x_j \geq 0, \quad j = 1, 2, 3 \quad 0 \leq \alpha \leq 1. \tag{19}$$

Now solving each objective individually with respect to the system constraints the best and worst objective values are found as:

$$E(Z_1)_B^L = 7.254, \quad E(Z_1)_W^L = 3.449, \quad E(Z_1)_B^R = 7.366, \quad E(Z_1)_W^R = 3.519$$
$$E(Z_2)_B^L = 9.055, \quad E(Z_2)_W^L = 3.171, \quad E(Z_2)_B^R = 9.173, \quad E(Z_2)_W^R = 3.224$$

Hence the fuzzy objective goals for the respective objectives are expressed as

$$E(Z_1)^L \gtrsim 7.254, \ E(Z_1)^R \gtrsim 7.366, E(Z_2)^L \gtrsim 9.055, E(Z_2)^R \gtrsim 9.173.$$

So the FGP model for finding the most satisfactory decision of the given problem by constructing the membership functions as defined in (8) and (9) and converting them into membership goals by assigning unity as the aspiration level is appeared as:

Minimize D= $0.263d_1^{L-} + 0.260d_1^{R-} + 0.170d_2^{L-} + 0.168d_2^{R-}$

so as to

$$0.263[(5.95+0.05\alpha)x_1 + (4.95+0.05\alpha)x_2 + (6.95+0.05\alpha)x_3 - 3.449] + d_1^{L-} - d_1^{L+} = 1$$
$$0.260[(6.05-0.05\alpha)x_1 + (5.05-0.05\alpha)x_2 + (7.05-0.05\alpha)x_3 - 3.519] + d_1^{R-} - d_1^{R+} = 1$$
$$0.170[(6.95+0.05\alpha)x_1 + (7.95+0.05\alpha)x_2 + (5.95+0.05\alpha)x_3 - 3.171] + d_2^{L-} - d_2^{L+} = 1$$
$$0.168[(7.05-0.05\alpha)x_1 + (8.05-0.05\alpha)x_2 + (6.05-0.05\alpha)x_3 - 3.224] + d_2^{R-} - d_2^{R+} = 1$$
$$0 \leq \alpha \leq 1 \tag{20}$$

subject to the system constraints defined in (19).

The model is solved by using the software LINGO (ver6.0); and the solution is obtained as

$$x_1 = 0.550, \ x_2 = 0.298, \ x_3 = 0.325$$

with the objective values $Z_1 = [7.007, 7.124], Z_2 = [8.127, 8.244]$.

In the interval-valued approach of pal et al.[23] the solutions are given by

$$x_1 = 0, \ x_2 = 0.0304, \ x_3 = 0.7639 \text{ with } Z_1 = [2.999, 7.997], Z_2 = [2.558, 7.09].$$

The result shows that the proposed model provides more predictable decision than the method developed by Pal et al. [17] in terms of achieving objective values in the current decision making environment. Further, the model is flexible enough and may achieve satisfactory results by defining different tolerance ranges of the fuzzy numbers involved with the parameters of the objectives in the model formulation process. Again, in the developed methodology, the solution is not depending on the arbitrary target intervals as described by Pal et. al. [17], rather it is depending on the fuzzy goal values achieved in the decision making environment.

4   Conclusions

In this paper, an FGP model is developed for solving fuzzy multiobjective stochastic decision making problems efficiently, where achievement of each decomposed fuzzy goals to the highest membership value to the extent possible is taken into account. The proposed approach does not involve any computational load with evaluation of the problem in a repetitive manner unlike the previous FP and other approaches. Also, the proposed model is flexible enough to introduce the tolerance limits to the fuzzy goals initially to arriving at the satisfactory decisions on the basis of the needs and desires of the DM. Also this paper captures the fuzziness and randomness involved not only with the objectives, but also with the parameters of the resource constraints. An extension of the proposed approach to multiobjective fuzzy stochastic decision making problems in a hierarchical decision making arena may be introduced efficiently.

Acknowledgements. The authors are thankful to the anonymous reviewers for their constructive comments and valuable suggestions in improving the quality of the paper.

References

1. Zhou, K., Doyle, J., Glover, K.: Robust and Optimal Control. Prentice-Hall (1996)
2. Pachter, L., Sturmfels, B.: The mathematics of phylogenomics. SIAM Review 49, 3–31 (2007)
3. Kall, P.: Stochastic Linear Programming. Springer, Heidelberg (1976)
4. Wagner, M.R.: Stochastic 0-1 linear programming under limited distributional information. Operations Research Lett. 36(2), 150–156 (2008)
5. Lai, Y.J., Hwang, C.L.: Fuzzy mathematical programming. Springer, Berlin (1992)
6. Zimmerman, H.J.: Description and optimization of fuzzy systems. Internat. J. General Syst. 2(4), 209–215 (1976)
7. Liu, B.: Fuzzy random chance-constrained programming. IEEE Trans. Fuzzy Syst. 9(5), 713–720 (2001)

8. Nanda, S., Panda, G., Dash, J.: A new solution method for fuzzy chance constrained programming problem. Fuzzy Optim. Decision Making 5(4), 355–370 (2006)
9. Luhandjula, M.K.: Optimization under hybrid uncertainty. Fuzzy Sets Syst. 146(2), 187–203 (2004)
10. Luhandjula, M.K.: Fuzzy stochastic linear programming: Survey and future research directions. European J. Operational Research 174(3), 1353–1367 (2006)
11. Guo, P., Huang, G.H.: Inexact fuzzy-chance-constrained two stage mixed integer programming approach for long term planning of municipal solid waste management-part A methodology. J. Environmental Management 91, 461–470 (2009)
12. Biswas, A., Bose, K.: A fuzzy programming approach for solving quadratic bi-level programming problems with fuzzy resource constraints. Internat. J. Operational Research 12, 142–156 (2011)
13. Biswas, A., Pal, B.B.: Application of fuzzy goal programming technique to land use planning in agricultural system. Omega 33, 391–398 (2005)
14. Iskander, M.G.: A fuzzy weighted additive approach for stochastic fuzzy goal programming. Appl. Math. Comput. 154, 543–553 (2004)
15. Liang, T.F.: Applying fuzzy goal programming to project management decisions with multiple goals in uncertain environments. Expert Syst. Appl. 37, 8499–8507 (2010)
16. Biswas, A., Modak, N.: A fuzzy goal programming method for solving chance constrained programming with fuzzy parameters. Commun. Computer Inf. Sci. 140, 187–196 (2011)
17. Pal, B.B., Sen, S., Kumar, M.: An interval valued goal programming approach to stochastic multiobjective decision making problems. In: Proc. Internat. Conf. Operations Management Sci., pp. 415–423. Excel India Publishers, New Delhi (2010)
18. Charnes, A., Cooper, W.W.: Chance Constrained Programming. Mgmt. Sci. 6, 73–79 (1959)

Economic Order Quantity for Fuzzy Inventory Model without or with Shortage

P. Rajendran and P. Pandian

Department of Mathematics, School of Advanced Sciences,
VIT University, Vellore - 632 014, India
prajendranmaths@yahoo.com, pandian61@rediffmail.com

Abstract. A fuzzy inventory model without or with shortage is considered. A new procedure is proposed for finding optimal solutions to the fuzzy inventory model based on the fuzzy differentiation and the ranking of fuzzy numbers. Numerical examples are taken to illustrate the procedure of finding the optimal solutions. The proposed procedure can help the managers to take concrete decision regarding inventory, as the data available to them are not certain in business and industry.

Keywords: Fuzzy inventory models, Triangular fuzzy numbers, Economic order quantity, Optimal maximal stock quantity, Optimal total fuzzy inventory cost.

1   Introduction

Inventory referring to the physical stock of materials, connects the demand and supply of goods pertaining to the customers from the wholesale dealers or the retailers. Inventory control is a major consideration in many situations because of its practical and economic importance. In business and industry, the questions must be constantly answered as to when and how much raw material should be ordered, when a production order should be released to the plant, what level of safety stock should be maintained at a retail outlet, or how in-process inventory is to be maintained in a production process. These questions are amenable to quantitative analysis with the help of inventory theory. In conventional inventory models, uncertainties are considered as randomness and are handled by probability theory. But we cannot estimate the probability distribution due to lack of historical data. In business and industry, it becomes very difficult for a manager to take concrete decision regarding inventory, as the data available to him are not certain. Because uncertainty arises in demand, set-up resources   and capacity constraints of an inventory planning system, it could be more justified to consider these factors in an elastic form. Therefore, fuzziness is applied in the inventory problems. In recent years, the study of inventory problems using the fuzzy parameters has become more popular.

In [4,5], a group of computing schemes for the economic order quantity as fuzzy values of the inventory with/without backorder was investigated and the fuzzy order

P. Balasubramaniam and R. Uthayakumar (Eds.): ICMMSC 2012, CCIS 283, pp. 136–147, 2012.

quantity as a fuzzy number was expressed. Yao et al. [12] considered the inventory problems with and without backorder models in fuzzy environments. With the help of extension principle, they found the membership function of fuzzy total cost and its crisp value using centroid method. Robert et al. [8] applied basic algebraic skill to derive the optimal solution for inventory models with shortage and complete backlogging without derivatives. In Syed et al. [10], inventory model in fuzzy environment was prepared without shortage cost by using triangular fuzzy number and the total inventory cost was computed by signed-distance method. In [11], a fuzzy expected value model and a fuzzy dependent chance programming model for a fuzzy inventory problem are constructed and solved by an intelligent algorithm. Chou et al. [9] have studied the inventory model without fuzzy constraints, directly solved the model and derived an explicit expression for the minimum solution that will be useful to find the optimal solution under fuzzy environment using the convex property. In De and Apurva [2], an inventory model without shortage has been considered in a fuzzy environment having the ordering and holding costs as triangular fuzzy numbers and has been obtained for the optimal order quantity in fuzzy environment using signed-distance method. In [7], real life inventory problems faced by an employee in connection with his change of jobs have been studied and could be solved by applying a fuzzy inventory model.

In this paper, a fuzzy inventory model without or with shortage having the stock cost, order cost, the shortage cost, the period and the total demand quantity of plan period as triangular fuzzy numbers and the maximal stock quantity and the order quantity for each time as real quantity have been studied. A new procedure is proposed for finding optimal solutions to the fuzzy inventory model based on the fuzzy derivative of real functions and the magnitude of fuzzy numbers. The economic order quantity and optimal maximal stock are obtained as real quantity. The optimal total fuzzy inventory cost is obtained as a triangular fuzzy number. Illustrative examples are presented to clarify the idea of the proposed procedure. The proposed procedure can help the managers to take concrete decision regarding inventory, as the data available to them are not certain in business and industry.

2   Fuzzy Numbers

We need the following mathematical oriented definitions of fuzzy set, fuzzy number and membership function which can be found in [3, 13, 14].

Definition 2.1. Let A be a classical set and $\mu_A(x)$ be a function from A into [0, 1]. A fuzzy set $\tilde{A}$ with the member ship function $\mu_A(x)$ is defined by

$$\tilde{A} = \left\{ (x, \mu_A(x)) : x \in A \text{ and } \mu_A(x) \in [0,1] \right\}.$$

Definition 2.2. A real fuzzy number $\tilde{a} = (a_1, a_2, a_3)$ is a fuzzy subset from the real line R with the membership function $\mu_{\tilde{a}}(x)$ satisfying the following conditions:

(i) $\mu_{\tilde{a}}(x)$ is a continuous mapping from R to the closed interval [0, 1],

(ii) $\mu_{\tilde{a}}(x) = 0$ for every $a \in (-\infty, a_1]$,

(iii) $\mu_{\tilde{a}}(x)$ is strictly increasing and continuous on $[a_1, a_2]$,

(iv) $\mu_{\tilde{a}}(x)$ is strictly decreasing and continuous on $[a_2, a_3]$ and

(v) $\mu_{\tilde{a}}(x) = 0$ for every $a \in [a_3, +\infty)$.

Definition 2.3. A fuzzy number $\tilde{a}$ is a triangular fuzzy number denoted by (a_1, a_2, a_3) where a_1, a_2 and a_3 are real numbers and its membership function $\mu_{\tilde{a}}(x)$ is given below.

$$\mu_{\tilde{a}}(x) = \begin{cases} \dfrac{x - a_1}{a_2 - a_1} & : a_1 \leq x \leq a_2 \\ \dfrac{a_3 - x}{a_3 - a_2} & : a_2 \leq x \leq a_3 \\ 0 & : \text{otherwise} \end{cases}$$

We need the following definitions of the basic arithmetic operators on fuzzy triangular numbers based on the function principle which can be found in [3, 13].

Definition 2.4. Let (a_1, a_2, a_3) and (b_1, b_2, b_3) be two triangular fuzzy numbers (TFN's). Then, (i) $(a_1, a_2, a_3) \oplus (b_1, b_2, b_3) = (a_1 + b_1, a_2 + b_2, a_3 + b_3)$.

(ii) $(a_1, a_2, a_3) \ominus (b_1, b_2, b_3) = (a_1 - b_3, a_2 - b_2, a_3 - b_1)$.

(iii) $k(a_1, a_2, a_3) = (ka_1, ka_2, ka_3)$, for $k \geq 0$.

(iv) $k(a_1, a_2, a_3) = (ka_3, ka_2, ka_1)$, for $k < 0$.

(v) $(a_1, a_2, a_3) \otimes (b_1, b_2, b_3) = (t_1, t_2, t_3)$ where $t_1 = \text{minimum}\{a_1b_1, a_1b_3\}$;
$t_2 = a_2b_2$ and $t_3 = \text{maximum}\{a_3b_1, a_3b_3\}$.

(vi) If $a_i > 0, i = 1, 2, 3.$, $\dfrac{1}{(a_1, a_2, a_3)} = \left(\dfrac{1}{a_3}, \dfrac{1}{a_2}, \dfrac{1}{a_1}\right)$.

We need the following definitions of ordering on the set of the fuzzy numbers based on the magnitude of a fuzzy number which can be found in [1].

Definition 2.5. The magnitude of the TFN $\tilde{u} = (x_o - \sigma, x_o, x_o + \beta)$ with parametric form $\tilde{u} = (\underline{u}(r), \overline{u}(r))$ where $\underline{u}(r) = x_o - \sigma + \sigma r$ and $\overline{u}(r) = x_o + \beta - \beta r$ is defined as

$$Mag.(\tilde{u}) = \frac{1}{2}\left(\int_0^1 (\underline{u}(r) + \overline{u}(r) + 2x_o) r \, dr\right), \text{ where } r \in [0, 1].$$

Remark 2.1. The magnitude of the TFN $\tilde{u} = (a, b, c)$ is given by

$$Mag.(\tilde{u}) = \frac{a + 10b + c}{12} \, .$$

Definition 2.6. Let $\tilde{u}$ and $\tilde{v}$ be two TFN's. The ranking of $\tilde{u}$ and $\tilde{v}$ by the $Mag(.)$ on E, the set of TFN's, is defined as follows:

      (i) $Mag.(\tilde{u}) > Mag.(\tilde{v})$ if and only if $\tilde{u} \succ \tilde{v}$;

      (ii) $Mag.(\tilde{u}) < Mag.(\tilde{v})$ if and only if $\tilde{u} \prec \tilde{v}$ and

      (iii) $Mag.(\tilde{u}) = Mag.(\tilde{v})$ if and only if $\tilde{u} \approx \tilde{v}$.

Result 2.1. Let $\tilde{u}$ and $\tilde{v}$ be two TFN's. Then,

    (i) $Mag.(\tilde{u} \oplus \tilde{v}) = Mag.(\tilde{u}) + Mag.(\tilde{v})$;   (ii) $Mag.(k\tilde{u}) = kMag.(\tilde{u})$

3   Fuzzy Differentiation

Using the results and definitions in [6], we easily obtain the following definitions and results.

Let $\tilde{F}(q) \approx \sum\limits_{i=1}^{n} \tilde{c}_i f_i(q)$ be a fuzzy function of a real variable q where $f_i, i = 1,2,3,...,n$ are real functions of the real variable q . Then, the derivative of $\tilde{F}(q)$ with respect to q, $\dfrac{d}{dq}(\tilde{F}(q))$ is defined as $\dfrac{d}{dq}(\tilde{F}(q)) \approx \sum\limits_{i=1}^{n} \tilde{c}_i \dfrac{d}{dq}(f_i(q))$.

In general, $\dfrac{d^k}{dq^k}(\tilde{F}(q)) \approx \sum\limits_{i=1}^{n} \tilde{c}_i \dfrac{d^k}{dq^k}(f_i(q))$, where k is a positive integer.

Definition 3.1. A point q^* is said to be

    (i) a stationary point of $\tilde{F}(q)$ if $\dfrac{d}{dq}(\tilde{F}(q)) \approx \tilde{0}$ at $q = q^*$.

    (ii) a minima of $\tilde{F}(q)$ if q^* is stationary and $\dfrac{d^2}{dq^2}(\tilde{F}(q)) \succ \tilde{0}$ at $q = q^*$.

    (iii) a maxima of $\tilde{F}(q)$ if q^* is stationary and $\dfrac{d^2}{dq^2}(\tilde{F}(q)) \prec \tilde{0}$ at $q = q^*$.

Let $\tilde{F}(q,s) \approx \sum\limits_{i=1}^{n} \tilde{c}_i f_i(q,s)$ be a fuzzy function of two real variables q and s where $f_i, i = 1,2,3,...,n$ are real functions of the real variable q and s . Then, the first and second order partial derivatives of $\tilde{F}(q,s)$ are defined as

$$\frac{\partial}{\partial q}(\tilde{F}(q,s)) \approx \sum_{i=1}^{n} \tilde{c}_i \frac{\partial}{\partial q}(f_i(q,s)), \qquad \frac{\partial}{\partial s}(\tilde{F}(q,s)) \approx \sum_{i=1}^{n} \tilde{c}_i \frac{\partial}{\partial s}(f_i(q,s)),$$

$$\frac{\partial^2}{\partial q^2}(\tilde{F}(q,s)) \approx \sum_{i=1}^{n} \tilde{c}_i \frac{\partial^2}{\partial q^2}(f_i(q,s)), \qquad \frac{\partial^2}{\partial q \partial s}(\tilde{F}(q,s)) \approx \sum_{i=1}^{n} \tilde{c}_i \frac{\partial^2}{\partial q \partial s}(f_i(q,s)) \text{ and}$$

$$\frac{\partial^2}{\partial s^2}(\tilde{F}(q,s)) \approx \sum_{i=1}^{n} \tilde{c}_i \frac{\partial^2}{\partial s^2}(f_i(q,s)).$$

Definition 3.2. A point (q^*,s^*) is said to be

(i) a stationary point of $\tilde{F}(p,q)$ if $\dfrac{\partial \tilde{F}}{\partial q} \approx \tilde{0}$ and $\dfrac{\partial \tilde{F}}{\partial s} \approx \tilde{0}$ at (q^*,s^*),

ii) a minimum point of $\tilde{F}(p,q)$ if (q^*,s^*) is stationary ,

$$\frac{\partial^2 \tilde{F}}{\partial q^2} \otimes \frac{\partial^2 \tilde{F}}{\partial s^2} \ominus \left(\frac{\partial^2 \tilde{F}}{\partial q \partial s}\right)^2 \succ \tilde{0} \text{ at } (q^*,s^*) \text{ and } \frac{\partial^2 \tilde{F}}{\partial q^2} \succ \tilde{0} \text{ or } \frac{\partial^2 \tilde{F}}{\partial s^2} \succ \tilde{0} \text{ at } (q^*,s^*).$$

(iii) a maximum point of $\tilde{F}(p,q)$ if (q^*,s^*) is stationary ,

$$\frac{\partial^2 \tilde{F}}{\partial q^2} \otimes \frac{\partial^2 \tilde{F}}{\partial s^2} \ominus \left(\frac{\partial^2 \tilde{F}}{\partial q \partial s}\right)^2 \succ \tilde{0} \text{ at } (q^*,s^*) \text{ and } \frac{\partial^2 \tilde{F}}{\partial q^2} \prec \tilde{0} \text{ or } \frac{\partial^2 \tilde{F}}{\partial s^2} \prec \tilde{0} \text{ at } (q^*,s^*).$$

4   Inventory Models

An inventory model is a mathematical model that helps a firm in determining the economic order quantity and the frequency of ordering to keep goods or services flowing to the customer without interruption or delay. In inventory models, the major objective consists of minimizing the total inventory cost and to balance the economics of large orders or large production runs against the cost of holding inventory and the cost of going short.

Before proceeding for inventory models, we consider briefly the various costs involved in inventory models. They may be classified as follows.

Order cost : The cost associated with inventory for placing an order.
Shortage cost : The penalty cost that arises as a result of running out of stock
Stock cost : The cost associated with storing the goods in the stock.

Economic order quantity is the size of order quantity that minimizes the total cost of inventory. It is denoted by q^* and is minimum when the stock cost is equal to the order cost.

Symbols used in the proposed models

The following symbols are used in the inventory models in connection with this paper: $\tilde{a}$ - the fuzzy stock cost ; $\tilde{c}$ - the fuzzy order cost ; $\tilde{T}$ - the fuzzy plan for the whole period; $\tilde{R}$ -the fuzzy total demand quantity of plan period; $\tilde{b}$ -fuzzy shortage cost s - the maximal stock quantity and q - the order quantity for each time.

4.1   Fuzzy Inventory Model without Shortage

We consider an inventory model without shortage in fuzzy environment having the stock cost, order cost, the period and the total demand quantity of plan period as triangular fuzzy numbers and the order quantity for each time as a real positive parameter. In this model $\tilde{T}$ represents the fuzzy plan for the whole period, $\tilde{a}$ is the fuzzy stock cost, $\tilde{c}$ is the fuzzy order cost, $\tilde{R}$ represents the fuzzy total demand quantity of plan period and q represents the order quantity for each time .

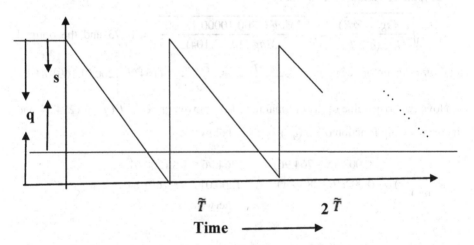

Fig. 1. Fuzzy inventory model without shortage

The total fuzzy inventory cost without shortage is $\tilde{F}(q) \approx \dfrac{(\tilde{a} \otimes \tilde{T})q}{2} \oplus \dfrac{\tilde{c} \otimes \tilde{R}}{q}$.

Now, for stationary condition, $\dfrac{d\tilde{F}}{dq} \approx \tilde{0} \ \Rightarrow \ \dfrac{\tilde{a} \otimes \tilde{T}}{2} \oplus (-1)\left(\dfrac{\tilde{c} \otimes \tilde{R}}{q^2} \right) \approx \tilde{0}.$

Since q is real and applying the magnitude concept, we have

$$Mag.\left(\dfrac{\tilde{a} \otimes \tilde{T}}{2} \oplus (-1)\left(\dfrac{\tilde{c} \otimes \tilde{R}}{q^2} \right) \right) = 0. \Rightarrow \ q = \sqrt{\dfrac{2Mag.(\tilde{c} \otimes \tilde{R})}{Mag.(\tilde{a} \otimes \tilde{T})}}.$$

Now, at $q = \sqrt{\dfrac{2Mag.(\tilde{c} \otimes \tilde{R})}{Mag.(\tilde{a} \otimes \tilde{T})}}$, we have $\dfrac{d^2\tilde{F}}{dq^2} \succ \tilde{0}$.

Therefore, the economic order quantity, $q^* = \sqrt{\dfrac{2Mag.(\tilde{c} \otimes \tilde{R})}{Mag.(\tilde{a} \otimes \tilde{T})}}$ and the optimal total

fuzzy inventory cost without shortage, $\tilde{F}(q^*) \approx \dfrac{(\tilde{a} \otimes \tilde{T})q^*}{2} \oplus \dfrac{\tilde{c} \otimes \tilde{R}}{q^*}$.

Remark 4.1. In this model, the economic order quantity   is a crisp value and the fuzzy total cost is a triangular fuzzy number.

Example 4.1. Let $\tilde{c} \approx (16, 20, 25)$, $\tilde{a} \approx (10, 12, 13)$, $\tilde{R} \approx (450, 500, 550)$ and $\tilde{T} \approx (4, 6, 8)$

Now, the economic order quantity

$$q^* = \sqrt{\frac{2Mag.(\tilde{c} \otimes \tilde{R})}{Mag.(\tilde{a} \otimes \tilde{T})}} = \sqrt{\frac{2Mag.(7200, 10000, 13750)}{Mag.(40, 72, 104)}} = 16.73 \text{ and the optimal}$$

total fuzzy inventory cost, $\tilde{F}(q^*) \approx \dfrac{(\tilde{a} \otimes \tilde{T})q^*}{2} \oplus \dfrac{\tilde{c} \otimes \tilde{R}}{q^*} \approx (764.96, 1200.01, 1691.84)$.

Now, the crisp value of the optimal total fuzzy inventory cost $\tilde{F}(q^*) = 1204.74$ and the membership function of $\tilde{F}(q^*)$ is given below:

$$\mu_{\tilde{F}(q^*)}(x) = \begin{cases} 0.0023(x - 764.96) & : 764.96 \le x \le 1200.01 \\ 0.002(1691.84 - x) & : 1200.01 \le x \le 1691.84 \ . \\ 0 & : \text{otherwise} \end{cases}$$

Example 4.2. Let $\tilde{c} \approx (16, 20, 25)$, $\tilde{a} \approx (10, 12, 13)$, $R = D$ and $T = 6$.

Now, the economic order quantity $q^* = \sqrt{\dfrac{2DMag.(\tilde{c})}{6Mag.(\tilde{a})}} = \sqrt{0.562D}$ and the crisp

value of the optimal total fuzzy inventory cost $\tilde{F}(q^*) = 35.76q^* + \dfrac{20.083D}{q^*}$.

Sensitivity Analysis

Demand (D)	The economic order quantity (q^*)	The crisp value of the optimal total inventory cost $\tilde{F}(q^*)$
450	15.90	1136.88
475	16.34	1168.04
500	16.76	1198.39
525	17.18	1227.98
550	17.58	1256.87

Remark 4.2. The values in the sensitivity analysis of Example 4.2 are better than values in [2].

4.2   Fuzzy Inventory Model with Shortage

We consider a fuzzy inventory model with shortage having the stock cost, order cost, the shortage cost, the period and the total demand quantity of plan period as triangular fuzzy numbers and maximal stock quantity and the order quantity for each time as real parameters.

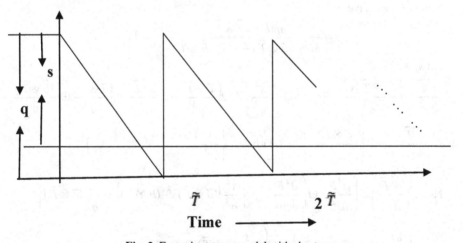

Fig. 2. Fuzzy inventory model with shortage

In this model, $\tilde{T}$ represents the fuzzy plan for the whole period, $\tilde{a}$ is the fuzzy stock cost, $\tilde{b}$ is the fuzzy shortage cost, $\tilde{c}$ is the fuzzy order cost, $\tilde{R}$ represents the fuzzy total demand quantity of plan period, s represents the maximal stock quantity and q represents the order quantity for each time.

The total fuzzy inventory cost with shortage is

$$\tilde{F}(q,s) \approx \frac{(\tilde{a} \otimes \tilde{T})s^2}{2q} \oplus \frac{(\tilde{b} \otimes \tilde{T})(q-s)^2}{2q} \oplus \frac{\tilde{c} \otimes \tilde{R}}{q}.$$

Now, for stationary condition, $\dfrac{\partial \tilde{F}}{\partial q} \approx \tilde{0}$ and $\dfrac{\partial \tilde{F}}{\partial s} \approx \tilde{0}$.

This implies that

$$(-1)\left(\frac{(\tilde{a}\otimes\tilde{T})s^2}{2q^2}\right) \oplus (\tilde{b}\otimes\tilde{T})\left(\frac{1}{2}-\frac{s^2}{2q^2}\right) \oplus (-1)\left(\frac{\tilde{c}\otimes\tilde{R}}{q^2}\right) \approx \tilde{0} \tag{1}$$

and

$$\frac{(\tilde{a}\otimes\tilde{T})s}{q} \oplus \frac{(\tilde{b}\otimes\tilde{T})(s-q)}{q} \approx \tilde{0} \tag{2}$$

Now, we have from (2), $\dfrac{s}{q} = \dfrac{Mag.(\tilde{b}\otimes\tilde{T})}{Mag.(\tilde{a}\otimes\tilde{T})+Mag.(\tilde{b}\otimes\tilde{T})}$. $\tag{3}$

Now, using (3) in (1), we have

$$q = \sqrt{\frac{2Mag.(\tilde{c}\otimes\tilde{R})\times\left(Mag.(\tilde{b}\otimes\tilde{T})+Mag.(\tilde{a}\otimes\tilde{T})\right)}{Mag.(\tilde{b}\otimes\tilde{T})\times Mag.(\tilde{a}\otimes\tilde{T})}}$$

From (3), we have

$$s = \frac{qMag.(\tilde{b}\otimes\tilde{T})}{Mag.(\tilde{a}\otimes\tilde{T})+Mag.(\tilde{b}\otimes\tilde{T})}$$

Now, we have

$$\frac{\partial^2 \tilde{F}}{\partial q^2} \approx \left(\frac{(\tilde{a}\otimes\tilde{T})s^2}{q^3}\right) \oplus (\tilde{b}\otimes\tilde{T})\left(\frac{s^2}{q^3}\right) \oplus 2\left(\frac{\tilde{c}\otimes\tilde{R}}{q^3}\right) \; ; \quad \frac{\partial^2 \tilde{F}}{\partial s^2} \approx \frac{(\tilde{a}\otimes\tilde{T})}{q} \oplus \frac{(\tilde{b}\otimes\tilde{T})}{q}$$

and $\dfrac{\partial^2 \tilde{F}}{\partial q\partial s} \approx (-1)\left(\dfrac{(\tilde{a}\otimes\tilde{T})s}{q^2}\right) \oplus (-1)(\tilde{b}\otimes\tilde{T})\left(\dfrac{s}{q^2}\right)$.

Now, $\left(\dfrac{\partial^2 \tilde{F}}{\partial q^2}\right) \otimes \left(\dfrac{\partial^2 \tilde{F}}{\partial s^2}\right) \ominus \left(\dfrac{\partial^2 F}{\partial q\partial s}\right)^2 \approx \left(\dfrac{s^2}{q^3}\left((\tilde{a}\otimes\tilde{T})\oplus(\tilde{b}\otimes\tilde{T})\right)\oplus \dfrac{2}{q^3}\left(\tilde{c}\otimes\tilde{R}\right)\right)\otimes$

$\left(\dfrac{1}{q}\left((\tilde{a}\otimes\tilde{T})\oplus(\tilde{b}\otimes\tilde{T})\right)\right) \ominus \left(\dfrac{s}{q^2}(\tilde{a}\otimes\tilde{T})\oplus(\tilde{b}\otimes\tilde{T})\right)^2$

$$\approx \left(\frac{2}{q^4}(\tilde{c}\otimes\tilde{R})\left((\tilde{a}\otimes\tilde{T})\oplus(\tilde{b}\otimes\tilde{T})\right)\right)$$

Now, at $q = \sqrt{\dfrac{2Mag.(\tilde{c}\otimes\tilde{R})\times\left(Mag.(\tilde{b}\otimes\tilde{T})+Mag.(\tilde{a}\otimes\tilde{T})\right)}{Mag.(\tilde{b}\otimes\tilde{T})\times Mag.(\tilde{a}\otimes\tilde{T})}}$ and

$$s = \frac{qMag.(\tilde{b} \otimes \tilde{T})}{Mag.(\tilde{a} \otimes \tilde{T}) + Mag.(\tilde{b} \otimes \tilde{T})},$$

We have $\left(\dfrac{\partial^2 \tilde{F}}{\partial q^2}\right) \otimes \left(\dfrac{\partial^2 \tilde{F}}{\partial s^2}\right) \ominus \left(\dfrac{\partial^2 F}{\partial q \partial s}\right)^2 \succ \tilde{0}$ and $\dfrac{\partial^2 \tilde{F}}{\partial s^2} \succ \tilde{0}$.

Therefore, the optimal order quantity q^* is given by

$$q^* = \sqrt{\frac{2Mag.(\tilde{c} \otimes \tilde{R}) \times \left(Mag.(\tilde{b} \otimes \tilde{T}) + Mag.(\tilde{a} \otimes \tilde{T})\right)}{Mag.(\tilde{b} \otimes \tilde{T}) \times Mag.(\tilde{a} \otimes \tilde{T})}}$$

and the optimal maximal stock quantity s^* is given by

$$s^* = \frac{q^* Mag.(\tilde{b} \otimes \tilde{T})}{Mag.(\tilde{a} \otimes \tilde{T}) + Mag.(\tilde{b} \otimes \tilde{T})}.$$

The optimal total fuzzy inventory cost with shortage is

$$\tilde{F}(q^*, s^*) \approx \frac{(\tilde{a} \otimes \tilde{T})s^{*2}}{2q^*} \oplus \frac{(\tilde{b} \otimes \tilde{T})(q^* - s^*)^2}{2q^*} \oplus \frac{\tilde{c} \otimes \tilde{R}}{q^*}$$

Remark 4.3: In this model, the economic order quantity and the optimal maximal stock quantity are crisp values and the fuzzy total cost is a triangular fuzzy number.

Example 4.3. Let $\tilde{c} \approx (16, 20, 25)$; $\tilde{a} \approx (10, 12, 13)$; $\tilde{R} \approx (450, 500, 550)$; $\tilde{T} \approx (4, 6, 8)$ and $\tilde{b} \approx (2, 3, 4)$.

Now, the economic order quantity,

$$q^* = \sqrt{\frac{2Mag.(\tilde{c} \otimes \tilde{R}) \times \left(Mag.(\tilde{b} \otimes \tilde{T}) + Mag.(\tilde{a} \otimes \tilde{T})\right)}{Mag.(\tilde{b} \otimes \tilde{T}) \times Mag.(\tilde{a} \otimes \tilde{T})}} = \sqrt{\frac{2 \times 10079.17 \times 90.33}{18.33 \times 72}}$$

$$= 37.14$$

and the optimal maximal stock quantity, $s^* = \dfrac{q^* Mag.(\tilde{b} \otimes \tilde{T})}{Mag.(\tilde{a} \otimes \tilde{T}) + Mag.(\tilde{b} \otimes \tilde{T})}$

$$= \frac{37.14 \times 18.33}{72 \times 18.33} = 7.54.$$

The optimal total fuzzy inventory cost with shortage,

$$\tilde{F}(q^*, s^*) \approx \frac{(\tilde{a} \otimes \tilde{T})s^{*2}}{2q^*} \oplus \frac{(\tilde{b} \otimes \tilde{T})(q^* - s^*)^2}{2q^*} \oplus \frac{\tilde{c} \otimes \tilde{R}}{q^*}$$

$$\approx 0.7654(40, 72, 104) \oplus 11.7945(8, 18, 32) \oplus 0.0269(7200, 10000, 13750)$$

$$\approx (318.659, \ 536.426, \ 826.93)$$

The crisp value of the optimal total fuzzy inventory cost $\tilde{F}(q^*, s^*)$ is 542.487 and the member ship function of $\tilde{F}(q^*, s^*)$ is given below:

$$\mu_{\tilde{F}(q^*, s^*)}(x) = \begin{cases} 0.005(x - 318.659) & : 318.659 \le x \le 536.426 \\ 0.003(826.93 - x) & : 536.426 \le x \le 826.93 \\ 0 & : \text{otherwise} \end{cases}$$

5　Conclusions

In this paper, a fuzzy inventory model without or with shortage is considered in which the stock cost, order cost, the shortage cost, the period and the total demand quantity of plan period are taken as triangular fuzzy numbers and the maximal stock quantity and the order quantity for each time are taken as real positive constants. Based on the fuzzy differentiation and the magnitude of fuzzy numbers, a new procedure for solving fuzzy inventory model is proposed. By the proposed procedure, the obtained optimal maximal stock quantity and the obtained economic order quantity are real and the obtained optimal total fuzzy inventory cost is a fuzzy quantity. In order to show, the application of the proposed models and procedures numerical examples are given. The proposed procedure can be served as an important tool for the decision makers to obtain optimal solutions when they are handling fuzzy inventory models.

References

1. Abbasbandy, S., Hajjari, T.: A new approach for ranking of trapezoidal fuzzy numbers. Comput. Math. Appl. 57, 413–419 (2009)
2. De, P.K., Rawat, A.: A Fuzzy inventory model without using shortages using Triangular Fuzzy Number. Fuzzy Inf. Engg. 1, 59–68 (2011)
3. Klir, G.J., Yuan, B.: Fuzzy sets and fuzzy logics: Theory and Applications. Prentice-Hall, NJ (1995)
4. Lee, H.M., Yao, J.S.: Economic order quantity in fuzzy sense for inventory without backorder level. Fuzzy Sets Syst. 105, 13–31 (1999)
5. Yao, J.S., Lee, H.M.: Fuzzy inventory with or without backorder for fuzzy order quantity with trapezoid fuzzy number. Fuzzy Sets Syst. 105, 311–337 (1999)
6. Mordeson, J.N., Wireman, M.J.: Differentiation of Fuzzy Functions. In: Annual Meeting of the North American Fuzzy Inf. Society - NAFIPS 1997, pp. 295–298 (1997)
7. Punniakrishnan, K., Kadambavanam, K.: Fuzzy Inventory Model with Shortages in Manpower Planning. Math. Theory Modeling 1, 19–25 (2011)
8. Ronalda, R., Yang, G.K., Chu, P.: The EOQ and EPQ models with shortages derived without derivatives. Int. J. Production Economics 92, 197–200 (2004)
9. Chou, S.-Y., Julian, P.C., Hung, K.-C.: A note on fuzzy inventory model with storage space and budget constraints. App. Math. Modeling 33, 4069–4077 (2009)
10. Syed, J.K., Aziz, L.A.: Fuzzy inventory models without using signed-distance method. Appl. Math. Inf. Sci. 1, 203–209 (2007)

11. Xiaobin, W., Wansheng, T., Ruiqing, Z.: Fuzzy economic order quantity inventory models without backordering. Tsinghua Science and Tech. 12, 91–96 (2007)
12. Yao, J.S., Chiang, J.: Inventory without backorder with fuzzy total cost and fuzzy storing cost defuzzified by centroid and signed-distance. Eur. J. of Op. Res. 148, 401–409 (2003)
13. Zadeh, L.A.: Fuzzy sets as a basis for a theory of possibility. Fuzzy Sets Syst. 100, 9–34 (1999)
14. Zimmermann, H.J.: Fuzzy set theory and its applications. Kluwer-Nijhoff, Boston (1996)

Improvement of Quality with the Reduction of Order Processing Cost by Just-In-Time Purchasing

R. Uthayakumar[1] and M. Rameswari[2]

[1] Department of Mathematics, Gandhigram Rural Institute - Deemed University,
Gandhigram - 624302, TamilNadu, India
[2] Department of Mathematics, SSM Institute of Engineering and Technology,
Dindigul, TamilNadu, India
{uthayagri,sivarameswari1977}@gmail.com

Abstract. This paper describes the characteristic features of the quality improvement investment in supply chain management for a single vendor - single buyer under conditions of order-processing time reduction for defective items. This article shows that the total annual variable cost function possesses some kinds of convexities. Finally, illustrative examples are provided to verify the theoretical results.

Keywords: Just-In-Time, Defective items, Order Processing time, Permissible delay in payment, Quality improvement.

1 Introduction

Just-In-Time (JIT) is a production and inventory control system in which materials are purchased and units are produced only as needed to meet actual customer demand. JIT approach can be used in both manufacturing and merchandising companies. It has the most profound effects, however, on the operations of manufacturing companies which maintain three class of inventories-raw material, Work in process, and finished goods. Traditionally, manufacturing companies have maintained large amounts of all three types of inventories to act as buffers so that operations can proceed smoothly even if there are unanticipated disruptions. Raw materials inventories provide insurance in case suppliers are late with deliveries. Work in process inventories are maintained in case a work station is unable to operate due to a breakdown or other reasons. Finished goods inventories are maintained to accommodate unanticipated fluctuations in demand. While these inventories provide buffers against unforeseen events, they have a cost. Although few companies have been able to reach this ideal, many companies have been able to reduce inventories only to a fraction of their previous level. The result has been a substantial reduction in ordering and warehousing costs, and much more efficient and effective operations.

Del Computer Corporation has finally tuned its Just-In-Time system so that an order for a customized personal computer that comes in over the internet at

P. Balasubramaniam and R. Uthayakumar (Eds.): ICMMSC 2012, CCIS 283, pp. 148–155, 2012.

9 AM. can be on a delivery truck to the customer by 9 P.M. In addition, Dell's low cost production system allows it to under price its rivals by 10% to 15%. This combination has made Dell the envy of the personal computer industry and has enabled the company to grow at five times the industry rate.

2   Literature Review

In a classical inventory model, it is implicitly assumed that the quality level is fixed at an optimal level, (i.e) all items are assumed to have perfect quality. However in the real production environment it can often be observed that there are defective items being produced due to imperfect production processes. The defective items must be rejected, repaired and reworked. In all cases, the customer substantial costs are incurred. Therefore, for the system with an imperfect production process, the manager may consider investing capital on quality improvement, so as to reduce the quality related costs.

Order Processing lead time is the length of time between the time when an order for an item is placed and when it actually becomes available to satisfy demand. Generally, it consists of the following components: order preparation, order transit to the supplier, supplier lead time, item transit time from the supplier, and time from order receipt until it is available on the shelf. This article develops an integrated inventory model to determine the optimal inventory policy under conditions of order-processing cost reduction. Hoque and Goyal [1] developed a heuristic solution procedure to minimize the total cost of setup, inventory holding and lead time crashing for an integrated inventory system under controllable lead time between a vendor and a buyer. Hill and Omar[2] relaxed an assumption regarding holding costs. They allowed for decreasing holding costs down the supply chain. This model has been modified by Zhou and wang [3]. Ouyang and Chang [4] investigated the impact of quality improvement on the modified lot size reorder point models involving variable lead time and partial backorders. Recently, Yang and Pan [5] have proposed an integrated inventory model involving variable lead time and quality improvement investment where the shortage cost is neglected and the reorder point is a predetermined parameter. Ouyang, Chen et al. [7] extended Moon and Choi's [6] model to include the possible relationship between quality and lot size and then investigate the joint effects of quality improvement and set up cost reduction in the model. Lee et al. [10] considered the continuous review inventory systems with variable lead time and back order discount, where the ordering cost can be reduced by capital investment. C.K. Huang [9] considered that the order-processing cost can be reduced at an extra crashing cost, which varies with the reduction in the order-processing time length. Huang, C-K [8] dealt with the order-processing cost reduction and permissible delay in payment problem in the single-vendor single-buyer integrated inventory model. In following, we extend Huang, C-K [8] model by considering defective items, the effect of investment on quality improvement and logarithmic function of the ordering cost.

3   Notations and Assumptions

For developing the proposed models, the following notations and assumptions are used throughout the paper.

3.1   Notations

To establish the proposed model, the following notations are used.

D - Annual demand rate of the buyer.
P - Annual production rate of the vendor $(P > D)$.
A_v - vendor setup cost.
h_v - holding cost for the vendor per item per unit time.
h_b - holding cost for the buyer per item per unit time.
F - Transportation cost per shipment.
Q - production lot size.
t - time period for permissible delay in settling accounts.
I - Carrying (interest) cost per year.
p - unit purchase price.
U - Order processing cost per unit time for the buyer.
K - Expenditure per year to operate the planned ordering system between the vendor and the buyer.
$L_i(K)$ - Planning order-processing time per shipment, which is a strictly decreasing function of K, with $L_i(K) = L_i e^{-rK}$, where $K = K_i$, i = 0,1,2,....m, $K_0 = 0$ and $L_0 = 0$
i_0 - the opportunity cost of capital per unit time.
n - number of shipments.
g - the cost of replacing a defective unit per unit time.
$JTC_i(n, Q, K)$ - The total annual cost per unit time of the integrated model.

3.2   Assumptions

The assumptions made in the paper are as follows:

1. There is only single vendor and single buyer for a single product.
2. The product is manufactured with a finite production rate and $P > D$.
3. Shortages are not allowed.
4. The transportation cost per unit from the vendor to the buyer is constant and independent of the ordering quantity.
5. The out of control probability θ is a continuous decision variable, and is described by a logarithmic investment function. The quality improvement and capital investment is represented by

$$q(\theta) = q_0 \ln(\frac{\theta_0}{\theta})$$

for $0 < \theta \leq \theta_0$, where θ_0 is the current probability that the production process can go out of control, and $q_0 = \frac{1}{\xi}$, and where ξ denoting the percentage decrease in θ per dollar increase in q(θ).

6. The integrated inventory model is designed for vendor production situations in which, once an order is placed, production begins and a constant number of units is added to the inventory each day until the production run has been completed. The expected number of defective items in a run of size nQ can be evaluated as $\frac{n^2Q^2\theta}{2}$.

7. The vendor allows the buyer a delay in payment.

8. The capital investment in reducing buyer's ordering cost is assumed a logarithmic function of the ordering cost.

4   Mathematical Model

4.1   Modeling Framework

The annual integrated total cost for both the vendor and buyer incorporating the defective cost and quality improvement can be represented by

$$JTC_i(Q,n,K_i) = \frac{D(A_v + n(F + UL_ie^{-rK}))}{Q} + \frac{Q}{2n}\left[(n-2)(1-\frac{D}{P})h_v + h_v + h_b\right.$$

$$\left. + \frac{pI}{1+It} + gn^2D\theta\right] + i_0q_0ln\left(\frac{\theta_0}{\theta}\right)$$

$for\ \ 0 < \theta \le \theta_0$

$$(1)$$

where i_0 is the opportunity cost of capital per unit time.

4.2   Solution Procedure

In order to find the minimum cost for this non-linear problem, we temporarily ignore the restriction subject to $0 < \theta \le \theta_0$ and minimize the total integrated cost function over (Q,n).

Theorem 1. *At a particular value of n and fixing K_i, let $JTC_i(Q) = JTC_i$ (Q,n,K_i), where $i = 0,1,2,....m$. Then $JTC_i(Q)$ is convex on $Q > 0$*

Let Q_i^* (i = 0,1,2......m) be the minimum point of $JTC_i(Q)$. Then we have

$$\frac{\partial JTC_i(Q)}{\partial Q} = \frac{-D(A_v + n(F + UL_ie^{-rK}))}{Q^2} + \frac{1}{2n}\left[(n-2)(1-\frac{D}{P})h_v + h_v + h_b\right.$$

$$\left. + gn^2D\theta + \frac{pI}{1+It}\right]$$

$$\frac{\partial^2 JTC_i(Q)}{\partial Q^2} = \frac{2D}{Q^3}[A_v + n(F + UL_ie^{-rK})] > 0$$

$$(2)$$

$JTC_i(Q)$ is convex on $Q > 0$. Therefore, we have completed the proof of Theorem 1.

$$\frac{\partial JTC_i(Q)}{\partial Q} = 0 \Rightarrow$$

$$\frac{D}{Q^2}[A_v + n(F + UL_ie^{-rK})] = \frac{1}{2n}\left[(n-2)(1-\frac{D}{P})h_v + h_v + h_b + \frac{pI}{1+It} + gn^2D\theta\right] \tag{3}$$

$$Q = \sqrt{\frac{2nD[A_v + n(F + UL_ie^{-rK})]}{\left[(n-2)(1-\frac{D}{P})h_v + h_v + h_b + \frac{pI}{1+It} + gn^2D\theta\right]}} \tag{4}$$

Therefore we can find the optimal order quantity Q^* when n and K_i are known. The total annual cost is given by

$$JTC_i(Q) = \sqrt{\frac{2D}{n}[A_v + n(F + UL_ie^{-rK})]\left[(n-2)(1-\frac{D}{P})h_v + h_v + h_b + \frac{pI}{1+It} + gn^2D\theta\right]} \tag{5}$$

After ignoring the constant and terms on the radical which are independent of n, minimizing JTC_i is equaivalent to minimizing

$$X_i(n) = n(1-\frac{D}{P})h_v(F + UL_ie^{-rK}) + A_vgD\theta + \frac{A_v}{n}\left[h_v + h_b + \frac{pI}{1+It} - 2h_v(1-\frac{D}{P})\right] \tag{6}$$

From Eqn. (6), $X_i(n)$ is convex on n. By using

$$X_i(n) \leq X_i(n_i - 1)$$
$$X_i(n) \leq X_i(n_i + 1) \tag{7}$$

we get

$$(n_i - 1)n_i(1-\frac{D}{P})h_v(F + UL_ie^{-rK}) \leq A_v\left[h_v + h_b + \frac{pI}{1+It} - 2h_v(1-\frac{D}{P})\right] \tag{8}$$

$$n_i(n_i + 1)(1-\frac{D}{P})h_v(F + UL_ie^{-rK}) \geq A_v\left[h_v + h_b + \frac{pI}{1+It} - 2h_v(1-\frac{D}{P})\right] \tag{9}$$

By combining Eqs. (8) & (9), we have

$$(n_i - 1)n_i \leq \frac{A_v\left[h_v + h_b + \frac{pI}{1+It} - 2h_v(1-\frac{D}{P})\right]}{(1-\frac{D}{P})h_v(F + UL_ie^{-rK})} \leq n_i(n_i + 1) \tag{10}$$

If $K = 0$, $g = 0$ and $\theta_0 = \theta$, then the total integrated cost function can be reduced to the Huang, C-K [8] model. That is, the Huang, C-K [8] model is a special case in our model, without considering the defective items. If $K = 0$, $g = 0$ and $\theta_0 = \theta$, then annual integrated total cost for both vendor and buyer (Huang, C-K [8]) is given by

$$JTC_{CK}(Q,n) = \frac{D(A_v + n(F + UL_i))}{Q} + \frac{Q}{2n}\left[(n-2)(1 - \frac{D}{P})h_v + h_v + h_b\right.$$
$$\left. + \frac{pI}{1+It}\right]$$

(11)

5 Algorithm

The following solution procedure determines the optimal $n_{opt}, Q_{opt} and K_{opt}$.

1. Find n_i^* using Eq.(8), i = 0,1,2....m.
2. Determine the value of Q_i^* using Equation (4), i = 0,1,2,.....m.
3. Compute the corresponding annual integrated cost $JTC_i(n_i^*, Q_i^*, K_i^*)$ by using Eq. (5), i = 0,1,2...m.
4. Set $JTC^*(n_{opt}, Q_{opt}, K_{opt}) = minJTC_i(n_i^*, Q_i^*, K_i^*)$ and the optimal solution $n_{opt}, Q_{opt}, K_{opt}$ can be obtained simultaneously.

Table 1. Order processing-time reduction data to illustrate the following example

Project	Investment	Order Processing time
i	$K_i(\$)$	L_i(years)
0	0	0.140
1	80	0.105
2	200	0.070
3	300	0.055

6 Numerical Example

To illustrate the results obtained in this paper, the proposed model is applied efficiently to solve the following numerical example. Consider an inventory system with the following characteristics: $D = 900$ units/year; $P = 1500$ units/year; $h_b = \$5.00$ /unit/year; $h_v = \$3.00$ /unit/year; $F = 15$ / shipment ; $A_v = \$15$ / order; $U = 100$; $g = \$10$ /defective unit; $\theta = 0.00009$; $\theta_0 = 0.000092$; $r = 0.0008$; $i_0 = \$0.05$ /year; $q_0 = 100$; $t = 0.25$; $I = 0.015$. The planning order processing time and investment expenditure per year have three situations shown in Table 1. Applying the proposed algorithm, the optimal solutions are $n_{opt} = 2$; $Q_{opt} = 123$; and $K_{opt} = 600$. The annual integrated total cost is **4234**. The solution results are summarized in Table 2.

Table 2. Solution results of the example by the solution procedure

i	K_i (years)	n_i^*	Q_i (years)	$JTC(n_i^*, Q_i^*, K_i^*)$ (years)
0	0	2	126	4456
1	150	2	125	4361
2	**600**	**2**	**123**	**4234**
3	2100	3	143	7352

Table 3. Summary of comparison for the illustrative example

	C-K Huang model	Proposed model
Optimal order lot size	255	123
Number of deliveries	5	2
Joint total annual cost	20255	4234

7 Conclusion

In this paper, we develop a single vendor supplies a single buyer with a product by presenting an integrated inventory model under conditions of order-processing time reduction and quality improvement for defective items. Using various methods to reduce costs has become the major focus for supply chain management. In order to decrease the total annual cost, the vendor and the buyer are willing to invest in reducing the ordering cost.

This model is also shown to provide a lower total cost, higher quality and smaller lot size. JIT purchasing has an enormous impact on a company's profitability, especially in a competitive environment characterized by small profit margins. Furthermore, the application of JIT technologies such as small lot size, order processing time reduction and quality improvement play a significant role in achieving JIT purchasing.

Acknowledgement. Authors thank the anonymous referees for their useful comments and suggestions for improving this paper. This research work is supported by UGC-SAP (Special Assistance Programme), Department of Mathematics, Gandhigram Rural Institute (Deemed University), Gandhigram, Tamil-Nadu, India.

References

1. Hoque, M.A., Goyal, S.K.: A heuristic solution procedure for an integrated inventory system under controllable lead-time with equal or unequal sized batch shipments between a vendor and a buyer. Int. J. Prod. Econ. 102, 217–225 (2006)

2. Hill, R.M., Omar, M.: Another look at the single-vendor single-buyer integrated production-inventory problem. Int. J. Prod. Res. 44, 791–800 (2006)
3. Zhou, Y., Wang, S.: Optimal production and shipment models for a single vendor-single-buyer integrated system. Eur. J. Prod. Econ. 180, 309–328 (2007)
4. Ouyang, L.Y., Chang, H.C.: Impact of investing in quality improvement on (Q,r,L) model involving imperfect production process. Prod. Plan. Cont. 11, 598–607 (2000)
5. Yang, J.S., Pan, J.C.-H.: Just-in-time purchasing: an integrated inventory model involving deterministic variable lead time and quality improvement investment. Int. J. Prod. Res. 42, 853–863 (2004)
6. Moon, I., Choi, S.: A note on lead time and distributional assumptions in continuous review inventory models. Comput. Oper. Res. 25, 1007–1012 (1998)
7. Ouyang, L.Y., Chen, C.K., Chang, H.C.: Quality improvement, set up cost and lead-time reductions in lot size reorder point models with an imperfect production process. Comput. Oper. Res. 29, 1701–1717 (2002)
8. Huang, C.K., Tsai, D.M., Wu, J.C., Chung, K.J.: An integrated vendor-buyer inventory model with order-processing cost reduction and permissible delay in payments. Eur. J. Operl. Res. 202, 473–478 (2010a)
9. Huang, C.K., Tsai, D.M., Wu, J.C., Chung, K.J.: An optimal integrated vendor-buyer inventory policy under conditions of order-processing time reduction and permissible delay in payments. Int. J. Prod. Econ. 128, 445–451 (2010b)
10. Lee, W.C., Wu, J.W., Lei, C.L.: Computational algorithmic procedure for optimal inventory policy involving ordering cost reduction and back-order discounts when lead time demand is controllable. App. Math. Comput. 189, 186–200 (2007)

Using Genetic Algorithm to Goal Programming Model of Solving Economic-Environmental Electric Power Generation Problem with Interval-Valued Target Goals

Bijay Baran Pal[1,*], Papun Biswas[2], and Anirban Mukhopadhyay[3]

[1]Department of Mathematics, University of Kalyani, Kalyani - 741235,
West Bengal, India
bbpal18@hotmail.com
[2]Department of Electrical Engineering, JIS College of Engineering,
Kalyani - 741235, West Bengal, India
papunbiswas@yahoo.com
[3]Department of Computer Science and Engineering, University of Kalyani, Kalyani - 741235,
West Bengal, India
anirban@klyuniv.ac.in

Abstract. This article presents how genetic algorithm (GA) method can be efficiently used to the goal programming (GP) formulation of Economic-Environmental Power Dispatch (EEPD) problem with target intervals in a power system operation and planning environment.

In the proposed approach, first the objectives of the problem, economic power generation and atmospheric emission reduction, are considered interval-valued goals in interval programming. Then, in the model formulation, the defined goals are converted into the standard goals in GP by using interval arithmetic technique [1].

In the solution process, an GA is introduced to reach the aspiration levels of the defined goals of the problem to the extent possible and thereby to arrive at a satisfactory decision in the decision making environment.

To illustrate the potential use of the approach, the problem is tested on IEEE 6-Generator 30-Bus System and the model solution is compared with the solutions obtained in the previous study.

Keywords: Economic-Environmental Power Dispatch, Goal Programming, Genetic Algorithm, Interval Programming.

1   Introduction

The rapid growth of living standards in society owing to the welfare of technology, demand of electric power is increasing in an alarming rate in the recent years. It is to be mentioned here that the main source of electricity supply is thermal power plants, where fossil-fuel is used as resource for power generation. The thermal power system

---

* Corresponding author.

P. Balasubramaniam and R. Uthayakumar (Eds.): ICMMSC 2012, CCIS 283, pp. 156–169, 2012.

operation and planning problems are actually optimization problems with various system constraints in the environment of power generation and dispatch.

The general mathematical programming model for optimal power generation was introduced by Dommel and Tinney [2] in 1968. The study on environmental power dispatch models developed from 1960s to 1970s was first surveyed by Happ [3] in 1977. Thereafter, different classical optimization models developed in the past century for EEPD problems were surveyed in [4], [5].

Now, in the context of thermal power plant operations, it is worthy to mention that use of coal as fossil-fuel to generate power produces atmospheric emissions, where Carbon oxides (CO_x), Sulfur oxides (SO_x) and the oxides of Nitrogen (NO_x) are the major and harmful gaseous pollutants. Pollution affects not only humans, but also other life-forms of our planet.

The constructive optimization model for emission minimization problem of thermal power plant was first proposed by Gent and Lament [6] in 1971, and thereafter various emission control models were studied by active researchers in [7], [8] among others.

Consideration of both the aspects of economic power generation and reduction of emissions in the framework of mathematical programming was introduced by Zahavi and Eisenberg [9] in 1975 and further study was made in [10], [11] in the past.

During 1990s, emissions control problems were seriously considered and different strategic optimization approaches were developed with the consideration of 1990's Clean Air Act Amendment [12] by the active researchers in the field and well documented in [13], [14].

Thereafter, different mathematical programming approaches have been studied [15], [16], [17], [18] for efficient management of EEPD problems. But in most of the previous approaches, decision deadlock often arises due to the use of single objective optimization techniques for solving such multiobjective decision problems.

The GP [19] based on the satisficing philosophy (coined by Simon[20]) as a robust and flexible tool for multiobjective decision analysis in crisp decision environment has been used [21] to obtain the goal-oriented solution for economic-emission power dispatch problems.

During the last twenty years, different multiobjective optimization methods for EEPD problems have also been studied [13], [22], [23] with the consideration of Clean Air Act Amendment.

Now, in most of the practical decision situations, it is to be observed that the parameters involved with the problems are often inexact in nature. The most prominent approaches for decision analysis in uncertain environments is stochastic programming (SP) [24]. The SP approaches to EEPD problems have been studied [11], [15] in the past.

In some decision situations, inexactness of decision parameters are not probabilistic in nature, but they are fuzzily described owing to the imprecise in nature of human judgments as well as inherent imprecision in model parameters. Fuzzy programming (FP) approach [25] based on Fuzzy Set Theory (FST) [26] has appeared as robust tool for solving multiobjective problems. FP approach to EEPD problems has also been discussed in [27]. But, the extensive study in this area is at an early stage.

Now, it is worthy to note that uses of the conventional multiobjective programming methods often lead to local optimal solution owing to competing in nature of the objectives of the proposed problem.

To overcome the above difficulty, GA [28] based on the natural selection and natural genetics in biological system and as a global search approach was first studied by Chen and Chang [29] in 1995. Then, several soft computing approaches to EEPD problems have been studied [17], [18], [22] by the active researchers in this field.

Now, in the context of solving multiobjective decision making (MODM) problems, it is worthy to note that although FP as well as Fuzzy Goal Programming (FGP) [30] as an extension of conventional GP have been successfully implemented to different real-life problems, the main difficulty of using such approaches is that it may not always be possible for the decision maker (DM) to assign fuzzy aspiration levels to the objectives and thereby defining of the tolerance ranges for goal achievement in actual decision situations.

To overcome the above difficulty, interval programming (IvP)) approach [31] has appeared as a prominent tool for solving multiobjective decision problems, where parameters are taken in interval form instead of assigning single parameter values in inexact environment. The problem with interval-valued parameters has been studied by Pal et al [32] in the recent past.

However, the methodological and modeling aspects of interval-valued MODM problems are still at an early stage. Further, interval programming approach to power system operation and planning problems is so far yet to be documented in the literature.

In this article, the objectives of the problems are considered interval-valued for achievement of objective values within certain specified ranges in the decision making context. In the model formulation, the interval-valued objectives are first transformed into the standard objective goals in conventional GP formulation by using interval arithmetic technique [1] and introducing under- and over-deviational variables to each of them. Then, both the aspects of GP, *minsum* GP [33] and *minmax* GP [34], are taken into consideration as convex combination of them to construct the regret function (achievement function) for achievement of goal values within the respective intervals specified in the decision making environment.

A case example of IEEE 6-Generator 30-Bus System is considered to illustrative the approach. The model solution is also compared with the approaches studied in [18] previously to expound the potential use of the approach.

Now, the formulation of general IvP model is presented in the Section 2.

2   Problem Formulation

In an inexact decision making environment, the objectives are generally described in the form of interval-valued goals whereas the system constraints as well as other structural constraints may be interval-valued or crisp and that depends on how inexact description of them is made in the decision making context. In the present decision situation, an inexact version of the problem for achieving the goal values to the extent possible within the respective specified ranges is taken into account in the decision making environment.

The formal descriptions of the goals are presented in the following Section 2.1.

2.1   Description of Interval-Valued Goals

It is to be noted that the objectives in EEPD problems are inherently quadratic in nature. The interval-valued goals in quadratic form with crisp coefficients can be expressed as:

$$\text{Optimize } Z_i : a_k X + \frac{1}{2} X^T B_k X + \alpha_k = [t_k^L, t_k^U], \quad k = 1, 2, ..., K \tag{1}$$

where X is the vector of decision variables, a_k is the vector of crisp coefficients and B_k is a symmetric coefficient matrix and α_k is constant. t_k^L and t_k^U (k=1, 2,..., K) represent the lower- and upper-limits of the target intervals of the kth objective and where L and U stand for lower- and upper-bounds, respectively.

2.2   Construction of Flexible Goals

Using interval arithmetic technique [1] and introducing under- and over-deviational variables, the goal expression in (1) can be explicitly presented as:

$$\sum_{j=1}^{n} a_{kj} x_j + \frac{1}{2} \sum_{j=1}^{n} \sum_{l=1}^{n} b_{kjl} x_j x_l + \alpha_k + \eta_{kL}^{-} - \eta_{kL}^{+} = t_k^L,$$

$$\sum_{j=1}^{n} a_{kj} x_j + \frac{1}{2} \sum_{j=1}^{n} \sum_{l=1}^{n} b_{kjl} x_j x_l + \alpha_k + \eta_{kU}^{-} - \eta_{kU}^{+} = t_k^U \qquad k=1, 2, ..., K \tag{2}$$

Now, construction of objective function (called the regret function) in the GP formulation of the problem for goal achievement is presented in the following Section 2.3.

2.3   Construction of Regret Function for Goal Achievement

In the decision situation, the DM's objective for achievement of goal values within the specified ranges means minimization of the associated deviational variables to the extent possible in the decision making environment.

In the present GP formulation, both the aspects of GP, *minsum* GP [33] for minimizing the sum of the weighted unwanted variables as well as *minmax* GP [34] for minimizing the maximum of the deviations, are simultaneously taken into account as a convex combination of them to reach a satisfactory solution within the specified target intervals of the objective goals.

Now, from the optimistic point of view of the DM, minimization of possible regrets associated with the regret intervals $(\eta_{kL}^-, \eta_{kU}^+)$ are taken into account.

The regret function appears as:

$$\text{Minimize } Z = \lambda\,(0 < \lambda < 1)\sum_{k=1}^{K}(w_{kL}^-.\eta_{kL}^- + w_{kU}^+.\eta_{kU}^+) + (1-\lambda)\ \max\ (\eta_{kL}^- + \eta_{kU}^+) \qquad (3)$$

where $w_{kL}^-, w_{kU}^+ \geq 0$ represent the relative numerical weights of importance of minimizing the respective deviational variables for goal achievement in the decision situation, λ is a parameter, and where

$$\sum_{k=1}^{K}(w_{kL}^- + w_{kU}^+) = 1.$$

Letting, $\max\,(\eta_{kL}^- + \eta_{kU}^+) = d$, the standard GP model of the problem can be presented as:

Find X so as to:

$$\text{Minimize } Z = \lambda\sum_{k=1}^{K}(w_{kL}^-.\eta_{kL}^- + w_{kU}^+.\eta_{kU}^+) + (1-\lambda)d,$$

$$\text{Subject to } (\eta_{kL}^- + \eta_{kU}^+) \leq d, \quad k=1, 2, \ldots, K \qquad (4)$$

and the environmental system constraints which are involved in the decision making horizon. Now, formulation of EEPD problem in the framework of the proposed model is described in the following Section 3.

3   Model Formulation of EEPD Problem

The different types of parameters and decision variables involved with the problem are defined as follows:

Definition of parameters

N   = Total Number of generators in the power supply system.
P_D = Total power demand (in per-unit (p.u.)).
P_L = Total transmission loss (in p.u) of power in the system.
FC = Total fuel-cost (in dollar/hour ($/h)) for operating all the generators of the system.
EM = Total emission (ton/h) discharged by all the generators during power generation.

Decision variable

P_{Gi} =   Generation of power from generator i, i= 1,2,…,N.

3.1   General EEPD Problems

A general EEPD problem involves two objective functions and a set of system constraints in the decision making context.

Definition of Objective Functions

Fuel-cost Minimization Function. The cost function for generation of power is generally quadratic in nature.

The total fuel- cost ($/h) can be expressed as:

$$FC = \sum_{i=1}^{N}(a_i + b_iP_{Gi} + c_iP_{Gi}^2) \; , \tag{5}$$

where a_i, b_i, c_i are the estimated cost coefficients associated with the i-th generator.

Environmental-emission Minimization Function. The major atmospheric pollutants created by fossil-fueled thermal units are the sulfur oxides (So_x), carbon oxides (CO_x) and the   oxides of nitrogen (NO_x).

The discharge of total emission (ton/h) can be expressed as:

$$EM = \sum_{i=1}^{N} 10^{-2}(\alpha_i + \beta_iP_{Gi} + \gamma_iP_{Gi}^2) + \zeta_i \exp(\lambda_iP_{Gi}), \tag{6}$$

where α_i, β_i, γ_i, ζ_i, λ_i are coefficients associated with the i-th generator.

Description of System Constraints

The system constraints are defined as follows:

Power Balance Constraint. The total power generation must cover the total demand P_D and total transmission loss P_L.

So, the power balance constraint takes the form:

$$\sum_{i=1}^{N}P_{Gi} - (P_D + P_L) = 0 \tag{7}$$

The transmission loss can be modeled as a function of the generator output and can be expressed as:

$$P_L = \sum_{i=1}^{N}\sum_{j=1}^{N}P_{Gi}B_{ij}P_{Gj} + \sum_{i=1}^{N}B_{0i}P_{Gi} + B_{00} \; , \tag{8}$$

where B_{ij}, B_{0i} and B_{00} are   called Kron's loss coefficients or B-coefficients [35] of the transmission network.

Generation Capacity constraint. Following the conventional power generation and dispatch system, the constraints on the generator outputs can be considered as:

$$P_{Gi}^{min} \leq P_{Gi} \leq P_{Gi}^{max} \; , \quad i = 1, 2, \ldots, N. \tag{9}$$

where min and max stands for minimum and maximum respectively.

3.2   Description of Interval-Valued Goals

Let, FC^L and FC^U be the upper- and lower-limits, respectively, of the total fuel-cost involved for generation of power. Again let, EM^L and EM^U be the permissible upper- and lower-limits, respectively, of total emission discharged to the environment during power generation.

Then, interval-valued goals appear as follows:

Fuel-cost Goal

$$\sum_{i=1}^{N}(a_i + b_i P_{Gi} + c_i P_{Gi}^2) = [FC^L, FC^U] \qquad (10)$$

Environmental-emission Goal

$$\sum_{i=1}^{N} 10^{-2}(\alpha_i + \beta_i P_{Gi} + \gamma_i P_{Gi}^2) + \zeta_i \exp(\lambda_i P_{Gi}) = [EM^L, EM^U] \qquad (11)$$

Now, the GA scheme used to determine the target intervals of the goals and thereby solving the formulated GP model of the problem is presented in the Section 4.

4   Design of GA Scheme

The genetic parameters and the core function adopted in the search process are presented in the following algorithmic steps:

Step 1: *Representation and Initialization*

Let V denote the binary coded representation of chromosome in a population, where $V = (x_1, x_2,...,x_n)$. The population size is defined by pop_size, and pop_size chromosomes are randomly initialized in its search domain.

Step2: *Fitness function*

The fitness [28] score of each individual is evaluated by the defined objective function. The fitness function can be presented as follows:

$$\text{eval }(V_\ell) = (Z)_\ell, \qquad \ell = 1,2,...,\text{pop-size} \qquad (12)$$

where Z represents the objective function to be evaluated.

Then, in the search process, the best chromosome with the highest fitness value at each generation is determined as:

$$V^* = \max \{\text{eval }(V_\ell) \mid \ell = 1, 2,..., \text{pop\_size}\},$$

and $V^* = \min\{\text{eval }(V_\ell) \mid \ell = 1, 2, ..., \text{pop\_size}\}$, respectively.

which depends on function searching of the best value of an objective.

Step 3: *Selection*

The selection process defines the choosing of individuals (the parent solutions) from the population for the mating purpose. The Roulette wheel scheme [28] is used for parent selection. In this scheme, the selection of two individuals is made on the basis of their successive highest fitness scores.

Step 4: *Crossover*

The parameter P_c is defined as the probability of crossover. The conventional single-point crossover [28] of a genetic system is applied here in the sense that the resulting offspring always satisfy the given system constraints set. In the crossover mechanism, a chromosome (solution individual) is selected as a parent for producing offspring, if for a defined random number $r \in [0, 1]$, $r < P_c$ is satisfied.

Here, for the two selected parents X_1 and $X_2 \in S$ (the feasible search space), the arithmetic crossover yields the offspring X_1^1 and X_2^1 as

$$X_1^1 = \alpha\, X_1 + (1 - \alpha)\, X_2, \quad X_2^1 = (1 - \alpha)\, X_1 + \alpha\, X_2, \qquad 0 < \alpha < 1,$$

where X_1^1 and $X_2^1 \in S$.

Step 5: *Mutation*

The parameter P_m, as in the conventional GA scheme, is defined as the probability of mutation. The mutation operation is performed on a bit-by-bit [28] basis, where for a random number $r \in [0, 1]$, a chromosome is selected for mutation provided that $r < P_m$.

Step 6: *Termination*

The execution of whole process terminates when the fitted chromosome is reported at certain generation number in the solution search process.

Now, conversion of interval-valued goals to the standard form of goals in the framework of GP and the model solution of the problem are presented via the case example in the Section 5.

5   A Demonstrative Case Example

The standard IEEE 30-bus 6-generator test system [22] is considered to illustrate the potential use of the approach.

The pictorial representation of single-line diagram of IEEE 30-bus test system is displayed in the Fig. 1.

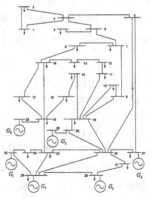

Fig. 1. Single-line diagram of IEEE 30-bus test system

The Fig. 1 shows that there are 6 generators and 41 lines and the total system demand for the 21 load buses is 2.834 p.u. The parameter values of the models are listed in Table-1, and the power generation limits are presented in the Table-2.

Table 1. Data Description of Power generation cost coefficients and emission coefficients

	G_1	G_2	G_3	G_4	G_5	G_6
Cost						
a	10	10	20	10	20	10
b	200	150	180	100	180	150
c	100	120	40	60	40	100
Emission						
α	4.091	2.543	4.258	5.326	4.258	6.131
β	-5.554	-6.047	-5.094	-3.550	-5.094	-5.555
γ	6.490	5.638	4.586	3.380	4.586	5.151
ζ	2.0E-4	5.0E-4	1.0E-6	2.0E-3	1.0E-6	1.0E-5
λ	2.857	3.333	8.000	2.000	8.000	6.667

Table 2. Data Description of Power Generation Limit (in p.u)

Limit	G_1	G_2	G_3	G_4	G_5	G_6
P_{Gi}^{min}	0.05	0.05	0.05	0.05	0.05	0.05
P_{Gi}^{max}	0.50	0.60	1.00	1.20	1.00	0.60

The B-coefficients [36] are as follows:

$$B = \begin{bmatrix} 0.1382 & -0.0299 & 0.0044 & -0.0022 & -0.0010 & -0.0008 \\ -0.0299 & 0.0487 & -0.0025 & 0.0004 & 0.0016 & 0.0041 \\ 0.0044 & -0.0025 & 0.0182 & -0.0070 & -0.0066 & -0.0066 \\ -0.0022 & 0.0004 & -0.0070 & 0.0137 & 0.0050 & 0.0033 \\ -0.0010 & 0.0016 & -0.0066 & 0.0050 & 0.0109 & 0.0005 \\ -0.0008 & 0.0041 & -0.0066 & 0.0033 & 0.0005 & 0.0244 \end{bmatrix},$$

$$B_0 = \begin{bmatrix} -0.0107 & 0.0060 & -0.0017 & 0.0009 & 0.0002 & 0.0030 \end{bmatrix},$$

$$B_{00} = 9.8573\ E-4.$$

Here, to employ the GA scheme defined in the Section 5, the following parameter values are introduced in the genetic search process.

- probability of crossover $P_c = 0.8$, probability of mutation $P_m = 0.08$
- population size = 100, chromosome length = 200.

Now, using the data Table-1, the individual best and worst solutions of the individual objectives are determined first by using the GA scheme, which are introduced as the upper- and lower-limits, respectively, of the respective target intervals of the objectives.

Then, following the procedure, the executable goal expressions of the problem are obtained as follows:

Fuel-cost Goals

$$(10+200P_{G1}+100P_{G1}^2+10+150P_{G2}+120P_{G2}^2+20+180P_{G3}+40P_{G3}^2+10+100P_{G4}$$
$$+60P_{G4}^2+20+180P_{G5}+40P_{G5}^2+10+150P_{G6}+100P_{G6}^2)+\eta_{1L}^--\eta_{1L}^+=590$$

$$(13)$$

$$(10+200P_{G1}+100P_{G1}^2+10+150P_{G2}+120P_{G2}^2+20+180P_{G3}+40P_{G3}^2+10+100P_{G4}$$
$$+60P_{G4}^2+20+180P_{G5}+40P_{G5}^2+10+150P_{G6}+100P_{G6}^2)+\eta_{1U}^--\eta_{1U}^+=632$$

$$(14)$$

Environmental-emission Goals

$$10^{-2}(4.091-5.554P_{G1}+6.490P_{G1}^2)+2.0E-4\exp(2.857P_{G1})+$$
$$10^{-2}(2.543-6.047P_{G2}+5.638P_{G2}^2)+5.0E-4\exp(3.333P_{G2})+$$
$$10^{-2}(4.258-5.094P_{G3}+4.586P_{G3}^2)+1.0E-6\exp8.000P_{G3}+$$
$$10^{-2}(5.326-3.550P_{G4}+3.380P_{G4}^2)+2.0E-3\exp(2.000P_{G4})+$$
$$10^{-2}(4.258-5.094P_{G5}+4.586P_{G5}^2)+1.0E-6\exp(8.000P_{G5})+$$
$$10^{-2}(6.131-5.555P_{G6}+5.151P_{G6}^2)+1.0E-5\exp(6.667P_{G6})+\eta_{2L}^--\eta_{2L}^+=0.19$$

$$(15)$$

$$10^{-2}(4.091-5.554P_{G1}+6.490P_{G1}^2)+2.0E-4\exp(2.857P_{G1})+$$
$$10^{-2}(2.543-6.047P_{G2}+5.638P_{G2}^2)+5.0E-4\exp(3.333P_{G2})+$$
$$10^{-2}(4.258-5.094P_{G3}+4.586P_{G3}^2)+1.0E-6\exp8.000P_{G3}+$$
$$10^{-2}(5.326-3.550P_{G4}+3.380P_{G4}^2)+2.0E-3\exp(2.000P_{G4})+$$
$$10^{-2}(4.258-5.094P_{G5}+4.586P_{G5}^2)+1.0E-6\exp(8.000P_{G5})+$$
$$10^{-2}(6.131-5.555P_{G6}+5.151P_{G6}^2)+1.0E-5\exp(6.667P_{G6})+\eta_{2U}^--\eta_{2U}^+=0.224$$

where $\eta_{iL}^-,\eta_{iL}^+,\eta_{iU}^-,\eta_{iU}^+\geq0$, $i=1,2$. $\qquad(16)$

In the sequel of formulating the model, the system constraints appear as follows:

Power balance constraint:

$$P_{G1}+P_{G2}+P_{G3}+P_{G4}+P_{G5}+P_{G6}-(2.834+P_L)=0 \qquad(17)$$

Generator output constraints:

$$0.05\leq P_{G1}\leq0.50, \qquad 0.05\leq P_{G2}\leq0.60, \qquad 0.05\leq P_{G3}\leq1.00,$$
$$0.05\leq P_{G4}\leq1.20, \qquad 0.05\leq P_{G5}\leq1.00, \qquad 0.05\leq P_{G6}\leq0.60$$

$$(18)$$

Now, in the decision making context, the deviational variables to be minimized in the regret function of a GP model depend on the needs and desires of the DM.

In the present decision situation, the executable GP model of the problem is constructed as follows:

Find $P_G(P_{G1},P_{G2},P_{G3},P_{G4},P_{G5},P_{G6})$ so as to:

Minimize $Z = \lambda \sum_{k=1}^{2} (w^-_{kL} \cdot \eta^-_{kL} + w^+_{kU} \cdot \eta^+_{kU}) + (1-\lambda) d$

subject to $(\eta^-_{kL} + \eta^+_{kU}) \leq d,$ k=1,2.

$$0 < \lambda < 1,$$

$$\sum_{k=1}^{K} (w^-_{kL} + w^+_{kU}) = 1 \qquad\qquad (19)$$

and the goal constraints in (13) – (16) and the system constraints in (17) and (18).

Now, for simplicity and without loss of generality, assigning equal weights to all the goals for their achievement and introducing $\lambda = 0.5$, the problem is solved by employing the proposed GA scheme.

In the solution search process, the fitness function appears as:

eval $(V_\ell) = (Z)_\ell$, $\ell = 1, 2, \ldots,$pop-size

where Z is the objective function expression defined in (19).

The best chromosome V^* with highest fitness score at a generation is determined as:

$$V^* = \min\{eval \ (V_\ell) \ | \quad \ell = 1, 2, \ldots, pop\_size\}$$

The GA based program is designed in Programming Language C^{++}. The execution is done in an Intel Pentium IV with 2.66 GHz. Clock-pulse and 1GB RAM.

The model solution is presented in the Table-3.

Table 3. Solution under the proposed approach

	P_{G1}	0.1394318
	P_{G2}	0.2753871
Generator Output (in p.u)	P_{G3}	0.4482069
	P_{G4}	0.9571830
	P_{G5}	0.5907677
	P_{G6}	0.4632497
Total Fuel-cost ($/hr)		600.8880
Total Environmental-emission (ton/hr)		0.2172952

The solutions obtained by using hybrid multi-objective optimization techniques in [18] are presented in the Table-4.

Table 4. Solutions under hybrid multiobjective optimization techniques

Optimization Scheme	Total generation cost ($/h)	Total emission (ton/h)
MO-DE/PSO	606.0073	0.220890
CMOPSO	606.0472	0.220468
SMOPSO	605.9909	0.220692
TV-MOPSO	606.4028	0.219770

Note: MO-DE/PSO= multi-objective optimization algorithm based on particle swarm optimization and differential evolution, CMOPSO= multi-objective particle swarm optimization algorithm proposed by Coello et al. [37], SMOPSO= multi-objective particle swarm optimization algorithm with sigma method, TV-MOPSO= time variant multi-objective particle swarm optimization.

The diagrammatic representations of total power generation and total emission obtained under the proposed model as well as for the use of different approaches are given in the Figs. 2 -3.

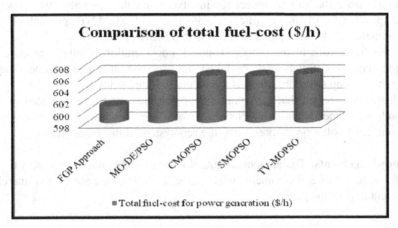

Fig. 2. Graphical representation of fuel-cost for power generation under different approaches

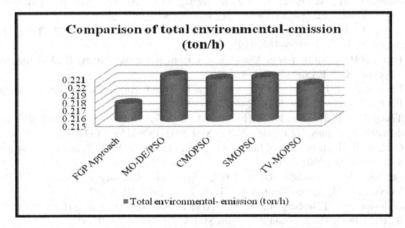

Fig. 3. Graphical representation of environmental-emission under different approaches

A comparison shows that the proposed GA based solution is a superior one over the solutions obtained in the previous study from the view point of achieving the environmental-economic electric power generation on the basis of needs in society.

6   Conclusions

In this paper, a GA based GP approach for solving the multiobjective electric power generation and dispatch problems with interval-valued target goal is presented.

The main advantage of the proposed approach is that the computational load and approximation error inherent to conventional linearization approaches can be avoided here with the use of the GA based solution method.

Again, since the various objectives involved with the optimal power generation and dispatch problem often conflict each other, the GA based IvP approach offers the most satisfactory decision in the decision making environment.

In the framework of the proposed model, consideration of other objectives and environmental constraints may be taken into account in a power generation and dispatch planning situation, which is a problem in future study.

Finally, it is hoped that the solution approach presented here may lead to future research for proper planning of electric power generation as well as controlling of environmental pollution for preserving the eco-system of this planet Earth.

Acknowledgements. The authors are grateful to the anonymous Reviewers for their helpful suggestions and comments which have led to improve the quality and clarity of presentation of the paper.

References

1. Inuiguchi, M., Kume, Y.: Goal Programming Problems with interval coefficients and target intervals. European J. Operational Res. 52, 345–360 (1991)
2. Dommel, H.W., Tinney, W.F.: Optimal Power Flow Solutions. IEEE Trans. Power Apparatus Syst. 87, 1866–1876 (1968)
3. Happ, H.H.: Optimal Power Dispatch - A Comprehensive Survey. IEEE Trans. Power Apparatus Syst. PAS-96(3), 841–854 (1977)
4. Chowdhury, B.H., Rahman, H.: A Review of Recent Advances in Economic Dispatch. IEEE Trans. Power Syst. 5(4), 1248–1258 (1990)
5. Talaq, J.H., El-Hawary, F., El-Hawary, M.E.: A Summary of Environmental/Economic dispatch algorithms. IEEE Trans. Power Syst. 9(3), 1508–1516 (1994)
6. Gent, M.R., Lament, J.W.: Minimum-Emission Dispatch. IEEE Trans. Power Apparatus Syst. 90, 2650–2660 (1971)
7. Sullivan, R.L., Hackett, D.F.: Air Quality Control Using a Minimum Pollution-Dispatching Algorithm. Environ. Sci. Tech. 7(11), 1019–1022 (1973)
8. Cadogan, J.B., Eisenberg, L.: Sulfur Oxide Emissions Management for Electric Power Systems. IEEE Trans. Power Apparatus Syst. 96(2), 393–401 (1977)
9. Zahavi, J., Eisenberg, L.: Economic- Environmental Power Dispatch. IEEE Trans. Syst. Man Cybernet. 5(5), 485–489 (1975)
10. Tsuji, A.: Optimal Fuel Mix and Load Dispatching under Environmental Constraints. IEEE Trans. Power Apparatus Syst. 100, 2357–2364 (1981)
11. Yokoyama, R., Bae, S.H., Morita, T., Sasaki, H.: Multiobjective generation dispatch based on probability security criteria. IEEE Trans. Power Syst. 3, 317–324 (1988)
12. Congressional Amendment to the Constitution, H.R. 3030/S. 1490 (1990)
13. El-Keib, A.A., Ma, H., Hart, J.L.: Economic Dispatch in view of the clean air act of 1990. IEEE Trans. Power Syst. 9(2), 972–978 (1994)
14. Srinivasan, D., Tettamanzi, A.: An evolutionary algorithm for evaluation of emission compliance options in view of the clean air act amendments. IEEE Trans. Power Syst. 12(1), 152–158 (1997)

15. Dhillon, J.S., Parti, S.C., Kothari, D.P.: Stochastic economic emission load dispatch. Electric Power Syst. Res. 26, 179–186 (1993)
16. Farag, A., Al-Baiyat, S., Cheng, T.C.: Economic load dispatch multiobjective optimization procedures using linear programming techniques. IEEE Trans. Power Syst. 10(2), 731–738 (1995)
17. Wang, L.F., Singh, C.: Environmental/economic power dispatch using a fuzzified multi-objective particle swarm optimization algorithm. Electric Power Syst. Res. 77(12), 1654–1664 (2007)
18. Gong, D., Zhang, Y., Qi, C.: Environmental/economic power dispatch using a hybrid multi-objective optimization algorithm. Electrical Power Energy Syst. 32, 607–614 (2010)
19. Ignizio, J.P.: Goal Programming and Extensions. D. C. Health, Lexington (1976)
20. Simon, H.A.: Administrative Behavior. The Free Press, New York (1945)
21. Nanda, J., Kothari, D.P., Lingamurthy, K.S.: Economic-emission load dispatch through goal programming techniques. IEEE Trans. Energy Conv. 3(1), 26–32 (1988)
22. Abido, M.A.: A novel multiobjective evolutionary algorithm for environmental/economic power dispatch. Electric Power Syst. Res. 65, 71–91 (2003)
23. AlRashidi, M.R., El-Hawary, M.E.: Pareto fronts of the emission-economic dispatch under different loading conditions. Internat. J. Electrical Power Energy Syst. Engg. 1(2), 68–71 (2008)
24. Kall, P., Wallace, S.W.: Stochastic Programming. John Wiley & Sons (1996)
25. Zimmermann, H.-J.: Fuzzy programming and linear programming with several objective functions. Fuzzy Sets Syst. 1, 45–55 (1978)
26. Zadeh, L.A.: Fuzzy Sets. Information and Control 8, 338–353 (1965)
27. Basu, M.: An interactive fuzzy satisfying-based simulated annealing technique for economic emission load dispatch with nonsmooth fuel cost and emission level functions. Electric Power Components Syst. 36, 163–173 (2004)
28. Goldberg, D.E.: Genetic Algorithms in Search, Optimization, and Machine Learning. Addison-Wesley, Reading (1989)
29. Chen, P.H., Chang, H.C.: Large-scale economic dispatch by genetic algorithm. IEEE Trans. Power Syst. 10(4), 1919–1926 (1995)
30. Tiwari, R.N., Dhamar, S., Rao, J.R.: Priority structure in fuzzy goal programming. Fuzzy Sets Syst. 19(3), 251–259 (1986)
31. Steuer, R.E.: Algorithm for linear programming problems with interval objectives function coefficients. Management Sci. 26, 333–348 (1981)
32. Pal, B.B., Gupta, S.: A goal programming approach for solving Interval valued multiobjective fractional programming problems using genetic algorithm. In: IEEE 10th Colloquium and International Conference on Industrial and Information Science 2008 (ICIIS 2008), vol. 440, pp. 1–6 (2008)
33. Biswas, A., Pal, B.B.: Application of fuzzy goal programming technique to land use planning in agricultural system. Omega 33, 391–398 (2005)
34. Romero, C.: Handbook of Critical Issues in Goal Programming. Pergamon Publ. Corp. (1991)
35. Saadat, H.: Power System Analysis. The McGraw-Hill Companies, New York (1999)
36. Wang, L.F., Singh, C.: Balancing risk and cost in fuzzy economic dispatch including wind power penetration based on particle swarm optimization. Electric Power Syst. Res. 78, 1361–1368 (2008)
37. Coello, C.A., Pulido, G.T., Lechuga, M.S.: Handling multiple objectives with particle swarm optimization. IEEE Trans. Evolutionary Comput. 8(3), 256–279 (2004)

Using Fuzzy Goal Programming for Long-Term Water Resource Allocation Planning in Agricultural System: A Case Study

Bijay Baran Pal[1,*], Subhendu Bikash Goswami[2], Shyamal Sen[3], and Durga Banerjee[1]

[1] Department of Mathematics, University of Kalyani
Kalyani - 741235, West Bengal, India
bbpal18@hotmail.com, agrud3@gmail.com
[2] Department of Agronomy, Bidhan Chandra Krishi Viswavidyalaya
Mohanpur - 741252, West Bengal, India
sbgoswami@yahoo.co.in
[3] Department of Mathematics, B.K.C. College
Kolkata - 700108, West Bengal, India
ssenk@yahoo.co.in

Abstract. This paper presents a fuzzy goal programming (FGP) method for modeling and solving farm planning problems for optimal production of several crops by proper allocation of irrigation water in different seasons in a planning year. In the model formulation, the priority based FGP is addressed for achievement of the highest value (unity) of the membership goals defined for the fuzzy goals of the problem to the extent possible on the basis of needs and desires of the decision maker (DM) in the decision making situation.

The case example of the Bardhaman District of West Bengal in India is considered to demonstrate the approach.

Keywords: Agricultural System, Fuzzy Goal Programming, Goal Programming, Membership Function, Water Resource Planning, Irrigation Water.

1   Introduction

Water is the unique substance for evolution of life on our planet. Among all the species, plants are the primitive species and the major constituent of any kind of plant is water. Again, the main source of foods for most of the moving species including human is plant species.

The history of human civilization has shown that mankind preferred to settle on the places where water and plants both are plenty, and originally human settled near river basin sides to meet the two basic needs food and water.

With the passing of time, due to the growth of human population and cultural evolution, cultivation started and water reservoir like pond, well and lake were prepared to meet the daily need as well as to farming for production of crops.

---

*Corresponding author.

P. Balasubramaniam and R. Uthayakumar (Eds.): ICMMSC 2012, CCIS 283, pp. 170–184, 2012.

Now, with the advancement of technologies and changes in lifestyle, although cropping and water supply systems were improved a lot prior to middle of the last century, the Green Revolution taken place during 1960s and planning models were then developed for water supply and farming systems to meet the needs in society.

Since, water resources are limiting day by day and availability of water mostly depends upon climatic conditions, water supply system highly depends on efficient management science models to meet the needs of multiple demand centers, namely domestics, industrial and commercial users.

The mathematical programming models for water resource planning problems with probabilistic as well as deterministic water supply constraints were deeply studied in [1, 2, 3, 4, 5] in the past.

Now, in the farm planning context, the constructive optimization models for optimal production of crops were deeply studied in [6, 7, 8].

Since irrigation water supply is a complicated issue, which involves socio-economic and environmental impacts, various uncertainties are associated with demand patterns and availability of water. Again, it is worthy to mention that industrial emissions have now become a great threat to the environment of the planet Earth. It has now become a great challenge to obtain fresh water for the users. Particularly, agricultural sectors are facing water scarcity challenges along with serious threat from water pollution and climate change issues.

Different management science models with multiplicity of objectives in deterministic and uncertain environments have been studied [9, 10, 11, 12] for optimal utilization of water resources in farm planning systems.

However, deep study in the area of irrigation water supply and farm management systems in uncertain decision environments is at an early stage. Further, the uses of fuzzy programming (FP) [13] as well as FGP [14] as an extension of conventional goal programming (GP) [15] for multiobjective decision analysis of real-life problems in inexact environment are yet to be widely circulated in the literature.

This paper presents how the priority based FGP method can be efficiently used for modeling and solving agricultural planning problems for achieving the aspiration levels of production of various seasonal crops cultivated in a plan period by allocating the arable land properly and utilizing the available productive resources efficiently throughout the planning year.

2   FP Problem Formulations

The generic form of a multiobjective FP problem can be presented as:

Find X $(x_1, x_2,..., x_n)$ so as to:

$$\text{satisfy } Z_k(X) \begin{pmatrix} \geq \\ \leq \end{pmatrix} g_k \text{ , } k = 1, 2,....,K. \tag{1}$$

$$\text{subject to } X \in S = \{X : X \in R^n \mid AX \geq b, X \geq 0, b \in R^m\}, \tag{2}$$

where X is the vector of decision variables, and where $\gtrsim$ and $\lesssim$ designate the fuzziness of $\geq$ and $\leq$ restrictions, respectively, in the sense of Zimmermann [13], and where g_k is the imprecise aspiration level of the k-th objective, b is a resource vector.

Now, in the field of FGP, the fuzzy goals are first characterized by their respective membership functions and then they are converted into conventional form of goals in GP (called membership goals) by assigning the highest membership value (unity) as aspiration level and introducing under-and over-deviational variables to each of them.

2.1   Construction of Membership Goals of Fuzzy Goals

The membership goal expression of the membership function, say $\mu_k(X)$, defined for the fuzzy goal $Z_k(X) \gtrsim g_k$ appears as [14]:

$$\mu_k(X): \frac{Z_k(X) - g_{lk}}{g_k - g_{lk}} + d_k^- - d_k^+ = 1, \ k \in K_1 \qquad (3)$$

where g_{lk} and $(g_k - g_{lk})$ represent the lower tolerance limit and tolerance range , respectively, for achievement of the associated k-th fuzzy goal. Also, $d_k^- \geq 0$ and $d_k^+ \geq 0$ are used to represent the under- and over- deviational variables, respectively, of the k-th membership goal $\mu_k(X)$.

Similarly, the membership goal expression of the fuzzy goal $Z_k(X) \lesssim g_k$ takes the form:

$$\mu_k(X): \frac{g_{uk} - Z_k(X)}{g_{uk} - g_k} + d_k^- - d_k^+ = 1, \ k \in K_2 \qquad (4)$$

where g_{uk} and $(g_{uk} - g_k)$ represent the upper tolerance limit and tolerance range, respectively, for achievement of the associated k-th fuzzy goal, and where $K_1 \cup K_2 = \{1, 2, ..., K\}$ with $K_1 \cap K_2 = \Phi$.

2.2   FGP Model Formulation

Now, in FGP formulation, the defined membership functions are transformed into membership goals by assigning the highest membership value (unity) as the aspiration level and introducing under- and over-deviational variables to each of them. Then, minimization of the under-deviational variables of the stated membership goals on the basis of their weights of importance of achieving the goal levels is taken into account.

In the proposed problem, since the objectives often conflict each other and they are not commensurable owing to different types of goals involved in the system, achievement of goals on the basis of priorities is taken into consideration.

In priority based FGP, the goals are rank ordered on the basis of the priorities of achieving the target levels of them. The goals which seem to be equally important from the view point of assigning a priority are included at the same priority level and numerical weights are given to them on the basis of their weights of importance of achieving their aspired levels at the same priority level.

The FGP model formulation under a pre-emptive priority structure can be presented as [14]:

Find $X(x_1, x_2, ..., x_n)$ so as to:

$$\text{Minimize } Z = [P_1(d^-), P_2(d^-), ... , P_r(d^-), ... , P_R(d^-)] \tag{5}$$

and satisfy the membership goals in (3) and (4) subject to the system constraints in (2), and where Z represents the vector of R priority achievement functions, $P_r(d^-)$ is a linear function of the weighted under-deviational variables, and where $P_r(d^-)$ is of the form

$$P_r(d^-) = \sum_{k \in I_1} w_{kr}^- d_{kr}^- \; , \; I_1 = \{1, 2, ..., K\}, \text{ where } w_{kr}^- (>0) \; (r = 1, 2, ... , R) \text{ associated}$$

with deviational variables represent the numerical weights of importance of achieving the goals at the r-th priority level, d_{kr}^- are renamed for d_k^- to represent them at the r-th priority level, and where values of w_{kr}^- are determined as:

$$w_{kr}^- = \begin{cases} \dfrac{1}{(g_k - g_{kl})_r}, & \text{for the defined } \mu_k(X) \text{ in } (3) \\[2mm] \dfrac{1}{(g_{ku} - g_k)_r}, & \text{for the defined } \mu_k(X) \text{ in } (4) \end{cases} \tag{6}$$

where $(g_k - g_{kl})_r$ and $(g_{ku} - g_k)_r$ are used to represent $(g_k - g_{kl})$ and $(g_{ku} - g_k)$ respectively, when k-th goal is included at the r-th priority level.

Here, it may be noted that the priority factors have the following relationship:

$P_1 >>> P_2 >>> ... >>> P_r >>> --- >>> P_R$, which means that the goal achievement under the priority factor P_r is preferred most to the next priority factor P_{r+1}, $r = 1,2,...,$ R-1, where >>>stands for much greater than.

3   Farm Planning Problem Formulation

3.1   Definition of Decision Variables and Resource Parameters

(a) Decision Variables

x_{cs}= Allocation of land for cultivating the crop c during season s, c = 1,2, ..., C;   s = 1, 2, . . . , S.

CW_s = Supply of canal-water during season s, s= 1, 2, ... , S.

GW_s = Abstraction of groundwater during season s, s= 1, 2, . . . , S.

(b) Productive Resource Utilization Parameters

Fuzzy Resources

MD_s = Estimated total man-days (in days) required during season s.

F_f = Estimated total amount of the fertilizer f (f = 1,2,..., F) (in quintals (qtls)) required during the plan period.

RS = Estimated total amount of cash (in Rupees (Rs.)) required in the plan period for the purchase of productive resources.

CW = Estimated total amount of canal-water (in million cubic meters (MCM)) supplied during the plan period.

GW = Estimated total amount of groundwater (in MCM) abstracted during plan period.

P_c = Annual production level (in qtls) of crop c.

MP = Estimated total market value (in Rs.) of all the crops yield during the plan period.

R_{ij} = Ratio of annual production of i-th and j-th crop (i, j = 1, 2, ... , C; i ≠ j).

Crisp Resources

LA_s = Total farmland (in hectares (ha)) available for cultivating the crops in season s.

ERW_s = Expected amount of rainwater (in MCM) precipitated during a cropping season s.

(c) Crisp Coefficients

MD_{cs} = Man days (in days) required per ha of land for cultivating crop c during season s.

F_{fcs} = Amount of the fertilizer f required per ha of land for cultivating crop c during season s.

P_{cs} = Estimated production of crop c per ha of land cultivated during season s.

A_{cs} = Average cost for the purchase of seeds and different farm assisting materials per ha of land cultivated for crop c during season s.

MP_{cs} = Market price (Rs / qtl) at the time of harvest of crop c cultivated during season s.

W_{cs} = Estimated amount of water consumption (in cubic metre (CM) / ha) of land for cultivating crop c during season s.

3.2   Description of Productive Resource Goals and Constraints

(a) Water Supply Goals and Constraints

Canal-water Goal

Canal-water supplied from River-barrage is imprecise in nature owing to capacity limitation of barrage, and after the certain level release of water is not permissible for preserving the eco-system of earth.

The fuzzy goal appears as:

$$\sum_{s=1}^{S} CW_s \lesssim CW$$

Groundwater Goal

Ground water is very scarce and after a certain limit, it cannot be abstracted to prevent from harmful mineral contamination.

The fuzzy goal appears as:

$$\sum_{s=1}^{S} GW_s \lesssim GW$$

Water Utilization Affinity Constraints

Since, both the canal-water and groundwater as reserved water are very limited, utilization of individual total capacity of each of them during major cropping seasons are to be considered with great care, and supply to other seasons are to be constrained to some extent depending on local climetic conditions.

The affinity constraints associated with reserved water utilization can be presented as:

$$CW_\ell \leq p_\ell \sum_{s=1}^{S} CW_s, \quad \ell \in \{1,2,...,S\}$$

$$GW_\ell \leq q_\ell \sum_{s=1}^{S} GW_s, \quad \ell \in \{1,2,...,S\}$$

where p_ℓ and q_ℓ designate the certain percentage of utilization of total canal-water and groundwater, respectively, during $\ell - $ th season.

Total Water Utilization Constraints

Since, all the three water sources are scarce, irrigation water demands in all the seasons are always constrained in the planning horizon.

The water utilization constraints appear as:

$$\sum_{c=1}^{C} w_{cs} \cdot x_{cs} - (CW_s + GW_s) \leq ERW_s, \quad s = 1,2,..., S$$

(b) Land Utilization Constraints. Availability of arable land generally differs from season to season due to duration of yielding the crops cultivated in a previous seasons as well as soil conditions.

The land utilization constraints appear as:

$$\sum_{c=1}^{C} x_{cs} \leq LA_s, \qquad s = 1, 2, \ldots, S.$$

(c) Farm Management Goals

Labourer Requirement Goals. A minimum number of labourers are to be employed throughout the planning period to avoid the trouble with hiring of extra labourers at the peak times.

The fuzzy goals take the form:

$$\sum_{c=1}^{C} MD'_{cs} \cdot x_{cs} \geq MD_s, \quad s = 1, 2, \ldots, S.$$

Fertilizer Requirement Goals. To achieve the optimal level of crop production, different types of fertilizers need be used in different seasons in the plan period.

The fuzzy goals take the form:

$$\sum_{s=1}^{S} \sum_{c=1}^{C} F_{fcs} \geq F_f, \quad f = 1, 2, \ldots, F.$$

Cash Expenditure Goal. An estimated amount of money (in Rs.) is involved for the purpose of purchasing seeds, fertilizers and various productive resources.

The fuzzy goals take the form:

$$\sum_{s=1}^{S} \sum_{c=1}^{C} A_{cs} \cdot x_{cs} \leq RS$$

Production Achievement Goals. To meet the demand of agricultural crops in society, a minimum achievement level of production of each type of the crops is needed.

The fuzzy goals appear as:

$$\sum_{s=1}^{S} P_{cs} \cdot x_{cs} \geq P_c, \quad c = 1, 2, \ldots, C.$$

Production-ratio Goals. To meet the demand of the main food products in society, certain ratios of total production of major crops should be maintained.

The production-ratio goals appear as:

$$\left(\sum_{s=1}^{S} P_{is} \cdot x_{is} \right) / \left(\sum_{s=1}^{S} P_{js} \cdot x_{js} \right) \geq R_{ij}, \quad i, j = 1, 2, \ldots, C, \text{ and } i \neq j.$$

where R_{ij} denotes ratios of i-th and j-th crops.

Net-benefit Goal. A certain level of net-benefit (in terms of Rs) from the farm is highly expected by the farm policy maker.

The fuzzy goal for net-benefit appears as:

$$\sum_{s=1}^{S} \sum_{c=1}^{C} (MP_{cs} \cdot P_{cs} - A_{cs}) \cdot x_{cs} \gtrsim MP$$

The modeling aspects of the proposed problem are demonstrated via the following case example in the Section 4.

4   An Illustrative Example: A Case Study

The land-use planning problem for production of the principal crops of the District Bardhaman of West Bengal, India is considered to illustrate the proposed FGP model. The three seasonal crop-cycles are: Pre-Kharif (Summer), Kharif (Rainy) and Rabi (Winter), which successively appear in West Bengal during a planning year. The data were collected from different farm related sources in [17, 18, 19].

The decision variables and different types of data involved with the problem are summarized in the Tables 1-5.

Table 1. Summary of decision variables for crop cultivation

Season (s)	Pre-Kharif (1)	Kharif (2)	Rabi (3)					
Crop (c)	Jute (1)	Aus-paddy (2)	Aman-paddy (3)	Boro-paddy (4)	Wheat (5)	Mustard (6)	Potato (7)	Pulses (8)
Variable (x_{cs})	x_{11}	x_{21}	x_{32}	x_{43}	x_{53}	x_{63}	x_{73}	x_{83}

Table 2. Summary of the supply of canal–water and groundwater

Water supply (in MCM)	Aspiration Level	Upper Tolerance Limit
Canal -water	1635.55	1839.99
Groundwater	1438.28	1618.06

Table 3. Data description of expected rainwater precipitation and land utilization

Season	1	2	3
Rainwater (in MCM)	975.767	5072.424	540.857
Land utilization (in '000 hectares)	457.80	457.80	457.80

Table 4. Data description of the aspired goal levels and tolerance limits

Goal	Aspiration Level	Tolerance Limit	
		Lower	Upper
1. a) Man-days (in days) :			
(i) Pre-Kharif season	2055	1849.5	----
(ii) Kharif season	24870	22134.43	----
(iii) Rabi season	16906.3	15384.46	----
b) Fertilizer requirement (in metric ton) :			
(i) Nitrogen	85.0	66.1	----
(ii) Phosphate	65.7	53.9	----
(iii) Potash	43.0	37.8	----
2. Production (in '000 metric ton) :			
a) Jute	50.83	28.27	----
b) Rice	1967.0	1467.87	----
c) Wheat	12.7	6.0	----
d) Mustard	41.4	30.0	----
e) Potato	1332.9	921.2	----
f) Rabi pulse	3.5	1.2	----
3. Cash expenditure (in Rs. Lac.)	128790.26	-----	154548.31
4. Profit (in Rs. Lac.)	177611.71	159850.54	----
5. Production ratio	2.32	1.6	

Table 5. Data description of productive resource utilization, cash expenditure and market price

Crops	W_{cs}	MD_s	F_f			PA	CE	MP
			N	P	K			
Jute	5030	90	40	20	20	2556	17297.00	1600
Aus	8636	60	40	20	20	2953	14331.80	1200
Aman	12700	60	40	20	20	2719	12849.20	1380
Boro	17780	60	100	50	50	3338	23721.60	1300
Wheat	3810	39	100	50	50	2718	11119.50	1179.00
Mustard	2540	30	80	40	40	850	8401.40	2715.50
Potato	4572	70	150	75	75	24520	37312.10	355.00
Pulses	2540	15	20	50	20	530	4942.00	4475.00

Note: W_{cs} = Water requirement (in CM/ha), MD_s = man-days (days/ha), F_f = fertilizer (kg/ha): N=Nitrogen, P = Phosphate, K = Potash; PA = production achievement (kg/ha), CE = cash expenditure (Rs / ha), MP = market price (Rs / qtl).

Now, using the data Tables 2-5, the membership goals of the defined fuzzy goals can be constructed by using the expressions in (3) and (4).

The fuzzy goals appear as follows:

4.1   Water Supply Goals and Constraints

Canal -water Goal

$$9.00 - 0.0049 \ (CW_1 + CW_2 + CW_3) + d_1^- - d_1^+ = 1 \qquad (7)$$

Groundwater Goal

$$4.239 - 0.0026 \ (GW_1 + GW_2 + GW_3) + d_2^- - d_2^+ = 1 \qquad (8)$$

Water Utilization Affinity Constraints

In west Bengal, the major crops are cultivated during rainy (Kharif) and winter (Rabi) seasons. Consequently, reserved water utilization constraints appear as:

$$CW_1 \leq 0.05 \sum_{s=1}^{3} CW_s$$

$$GW_1 \leq 0.06 \sum_{s=1}^{3} GW_s \qquad (9)$$

Total Water Utilization Constraints

Total water supply constraints in different seasons are presented as:

$$5.08 \, x_{11} + 8.636 \, x_{21} - (CW_1 + GW_1) \leq 975.767 \quad \text{(Pre-Kharif)}$$

$$12.7 \, x_{32} - (CW_2 + GW_2) \leq 5072.424 \qquad \text{(Kharif)}$$

$$17.78 \, x_{43} + 3.81 \, x_{53} + 2.54 \, x_{63} + 4.572 \, x_{73} + 2.54 \, x_{83} - (CW_3 + GW_3) \leq 540.857 \quad \text{(Rabi)} \quad (10)$$

4.2   Land Utilization Constraints

In Bardhaman district, the seasonal corps is cultivated as: Jute and Aus- paddy in Pri-Kharif season, Aman-paddy in Kharif season and Boro-paddy, Wheat, Mustard, Potato and Pulses in Rabi season.

The land utilization constraints in the three consecutive seasons appear as

$$x_{11} + x_{21} \leq 457.80 \qquad \text{(Pre-Kharif)}$$

$$x_{32} \leq 457.80 \qquad \text{(Kharif)}$$

$$x_{43} + x_{53} + x_{63} + x_{73} + x_{83} \leq 457.80 \qquad \text{(Rabi)} \qquad (11)$$

4.3   Farm Management Goals

Laborer Requirement Goal

$\mu_3 : 0..438x_{11} + 0.292x_{21} - 9.000 + d_3^- - d_3^+ = 1$ (Pre-Kharif)

$\mu_4 : 0.022x_{32} - 8.09 + d_4^- - d_4^+ = 1$ (Kharf)

$\mu_5 : 0.038\,x_{43} + 0.025x_{53} + 0.019\,x_{63} + 0.045\,x_{73} + 0.009\,x_{83} - 9.852 + d_5^- - d_5^+ = 1$ (Rabi) (12)

Fertilizer Requirement Goals

$\mu_6 : 0.002\ x_{11} + 0.002\ x_{21} + 0.002\ x_{32} + 0.005\ x_{43} + 0.0052\ x_{53} + 0.004\ x_{63}$
$\qquad + 0.008\ x_{73} + 0.001\ x_{83} - 3.497 + d_6^- - d_6^+ = 1$ (N)

$\mu_7 : 0.0017\ x_{11} + 0.0017\ x_{21} + 0.0017\ x_{32} + 0.004\ x_{43} + 0.004\ x_{53}$
$\qquad + 0.003\ x_{63} + 0.006\ x_{73} + 0.004\ x_{83} - 4.567 + d_7^- - d_7^+ = 1$ (P)

$\mu_8 : 0.007\ x_{11} + 0.004\ x_{21} + 0.004\ x_{32} + 0.009\ x_{43} + 0.009\ x_{53} + 0.007\ x_{63}$
$\qquad + 0.014\ x_{73} + 0.009\ x_{83} - 7.269 + d_8^- - d_8^+ = 1$ (K)

$$(13)$$

Cash Expenditure Goal

$\mu_9 : 6.00\ -\ (0.0067\ x_{11} + 0.0055\ x_{21} + 0.0049\ x_{32} + 0.0092\ x_{43}$
$\qquad + 0.0043\ x_{53} + 0.0033\ x_{63} + 0.0144\ x_{73} + 0.0019\ x_{83}) + d_9^- - d_9^+ = 1$

$$(14)$$

Production Achievement Goals

$\mu_{10} : 0.0059\ x_{21} + 0.0054\ x_{32} + 0.0067\ x_{43} + d_{10}^- - d_{10}^+ = 1$ (Rice)

$\mu_{11} : 0..0115\ x_{11} - 1.227 + d_{11}^- - d_{11}^+ = 1$ (Jute)

$\mu_{12} : 0.0456\ x_{53} - 0.8955 + d_{12}^- - d_{12}^+ = 1$ (Wheat)

$\mu_{13} : 0.0745\ x_{63} - 2.631 + d_{13}^- - d_{13}^+ = 1$ (Mustard)

$\mu_{14} : 0.0595\ x_{73} - 2.2375 + d_{14}^- - d_{14}^+ = 1$ (potato)

$\mu_{15} : 0.2304\ x_{83} - 2.2375 + d_{15}^- - d_{15}^+ = 1$ (Pulses)

$$(15)$$

Net-benefit Goal

$\mu_{16} : 0.0133x_{11} + 0.0119\ x_{21} + 0.0139\,x_{32} + 0.0111x_{43} + 0.0018\,x_{53} + 0.0082\ x_{63}$
$\qquad + 0.028\,x_{73} + 0.0105\,x_{83} - 8.9999 + d_{16}^- - d_{16}^+ = 1$

$$(16)$$

Production-Ratio Goal

The ratio of the two crops, rice and potato, are considered here as the major crops.
    The production ratio goal appears as:

$$\mu_{17} : 1.408 \ [(2.953 \ x_{21} + 2.719 \ x_{32} + 3.338 \ x_{43}) / (24.52 \ x_{73}) - 1.6] + d_{17}^{-} - d_{17}^{+} = 1 \qquad (17)$$

Now, the executable FGP model under the four assigned priorities appears as:
    Find $\{x_{cs} \mid c = 1,2,...,8; s = 1,2,3\}$ so as to:

Minimize $Z =$

$[P_1(0.0049 \ d_1^{-} + 0.0026 \ d_2^{-} + 0.0020 \ d_{10}^{-} + 0.1492 \ d_{11}^{-} + 0.0443 \ d_{12}^{-} + 0.0877 \ d_{13}^{-} + 0.0024 \ d_{14}^{-} + 0.00$

$20 \ d_{15}^{-}), P_2(0.00003 \ d_9^{-} + 0.00005 \ d_{16}^{-}), P_3(0.0048 \ d_3^{-} + 0.0003 \ d_4^{-} + 0.0006 \ d_5^{-} + 0.0529 \ d_6^{-} + 0.08$

$47 \ d_7^{-} + 0.1923 \ d_8^{-} + 1.408 \ d_{17}^{-})] \qquad (18)$

and satisfy the membership goals in (7), (8), (12)-(17), subject to the system
constraints in (9) – (11).
    The LINGO (ver. 12.0) solver (the permissible size of instance is 500 variables and
250 constraints) is used to solve the problem. The model (variable size 268, constraint
size 161) is executed in Pentium IV CPU with 2.66 GHz clock-pulse and 1GB RAM.
The required CPU time is 0.01 second.
    The model solutions for goal achievement are presented in the Table 6 and Table 7.

Table 6. Land allocation and crop production plan under the proposed model (2007-08)

Crop(c)	Jute	Rice	Wheat	Mustard	Potato	Pulses
Land allocation (in '000 ha)	19.88	687.91	4.67	116.31	46.90	6.60
Production (in '000 metric ton)	50.81	1966.95	12.69	98.86	1149.98	38.63

Table 7. Water utilization plan under the proposed model

Season	Pre-Kharif (1)	Kharif (2)	Rabi (3)
Water Utilization (in MCM) (CW, GW, RW)	(77.88, 77.27, 975.76)	(0.058, 741.57, 5072.42)	(1557.61, 417.58, 540.86)

The result shows that a satisfactory decision is achieved here in the decision
making environment.
    The existing farm management decisions are presented in the Table 8 and Table 9.

Table 8. Existing land allocation and crop production plan (2007-2008)

Crop(c)	Jute	Rice	Wheat	Mustard	Potato	Pulses
Land allocation (in '000 ha)	11.1	635.5	2.2	25.6	54.4	1.4
Production (in '000 metric ton)	28.37	1858.6	6.0	21.8	1332.9	0.8

Table 9. Existing water utilization plan

Season	1	2	3
Water Utilization (in MCM)	208.69	5264.15	3947.46

The graphical representations of the comparisons of model solutions with existing cropping plan are displayed in the Fig. 1 - 2.

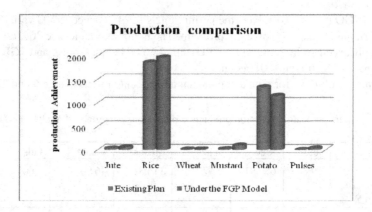

Fig. 1. Comparison of land allocation between the model solution and existing plan

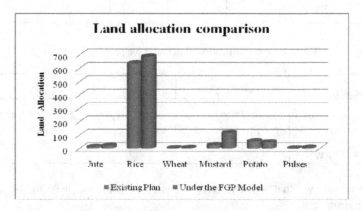

Fig. 2. Comparison of crop production between the model solution and existing plan

The comparisons reflect that the solution achieved under the proposed approach is superior over the existing cropping plan from the view point of achieving the aspired goal levels on the basis of needs and desires of the DM in the decision making environment.

5 Conclusion

The FGP approach to irrigation water supply and crop planning systems demonstrated in the paper provides a new look into the way of analyzing the different farm management activities in an imprecise decision-making environment.

Under the framework of the proposed model, different inexact constraints defined in the context of irrigation water supply management can easily be incorporated without involving any computational difficulty. Again, under the flexible nature of the priority based FGP model, priority structure can easily be changed to make proper crop production decision on the basis of the needs and desires of the DM in the planning situation.

However, it is hoped that the concept of solving the problems of irrigation water supply and cropping systems presented here can contribute to future research in the area of farm planning and management in uncertain decision environment

Acknowledgements. The authors are grateful to the anonymous Reviewers for their helpful suggestions and comments which have led to improve the quality and clarity of presentation of the paper.

References

1. Revelle, C., Joeres, E., Kirby, W.: The Linear Decision Rule in Reservoir Management and Design: Development of the Stochastic Model. Water Resources Res. 5, 767–777 (1969)
2. Askew, A.J.: Chance-Constrained Dynamic Programming and the Optimization of Water Resource Systems. Water Resources Research 10, 1099–1106 (1974)
3. Slowinski, R., Urbaniak, A., Weglarz, J.: Bicriterion Capacity Expansion Planning of a Water Supply System. Math. Operations Res. 46, 733–744 (1983)
4. Slowinski, R.: A Multicriteria Fuzzy Linear Programming Method for Water Supply System Development Planning. Fuzzy Sets Syst. 19, 217–237 (1986)
5. Yeh, W.W.G.: Reservoir Management and Operations Models: A State-of-the-Art Review. Water Resources Res. 21, 1797–1818 (1985)
6. Wheeler, B.M., Russell, J.R.M.: Goal Programming and Agricultural Planning. Operational Res. Qly. 28, 21–32 (1977)
7. Nix, J.S.: Farm management: The State-of-the-Art (or Science). Journal of Agricultural Economics 30, 277–292 (1979)
8. Glen, J.: Mathematical Models in Farm Planning: A Survey. Operations Research 35, 641–666 (1987)
9. Pal, B.B., Basu, I.: Selection of appropriate priority structure for optimal land allocation in agricultural planning through goal programming. Indian Journal of Agricultural Economics 51, 342–354 (1996)

10. Panda, S.N., Khepar, S.D., Kausal, M.P.: Interseasonal irrigation system planning for waterlogged sodic soils. J. Irrig. Drain. Engg. ASCE 122(3), 135–144 (1996)
11. Pal, B.B., Moitra, B.N.: Using Fuzzy Goal Programming for Long Range Production Planning in Agricultural Systems. Indian J. Agricultural Economics 59(1), 75–90 (2004)
12. Bravo, M., Ganzalez, I.: Applying Stochastic Goal Programming: A case study on water use planning. European J. Operational Res. 196, 1123–1129 (2009)
13. Zimmermann, H.-J.: Fuzzy Set Theory and Its Applications, 2nd Revised edn. Kluwer Academic Publishers, Boston (1991)
14. Pal, B.B., Moitra, B.N.: A Goal Programming Procedure for Solving Problems with Multiple Fuzzy Goals Using Dynamic Programming. European J. Operational Res. 144(3), 480–491 (2003)
15. Romero, C.: A Survey of Generalized Goal Programming. European J. Operational Res. 25(2), 183–191 (1986)
16. District Statistical Hand Book, Bardhaman. Department of Bureau of Applied Economics and Statistics. Govt. of West Bengal, India (2008)
17. Economic Review. Department of Bureau of Applied Economics and Statistics. Govt. of West Bengal, India
18. Basak, R.K.: Soil testing and fertilizer recommendation. Kalyani Publishers, New Delhi (2000)

Dynamics of Julia Sets
for Complex Exponential Functions

Bhagwati Prasad and Kuldip Katiyar

Department of Mathematics
Jaypee Institute of Information Technology
A-10, Sector-62, Noida, UP-201307, India
b_prasad10@yahoo.com, kuldipkatiyar.jiitn@gmail.com

Abstract. The Julia sets play an important role in the study of the complex analytic dynamics of functions. In this paper we study the patterns of Julia sets associated with complex exponential functions $E_\lambda(z) = \lambda e^z$ using Ishikawa iterative schemes. The bifurcation analysis of such maps is also discussed for the Mann iterates of the function.

Keywords: Julia sets, Complex exponential function, Mann iteration, Ishikawa iteration, Bifurcation.

1   Introduction

The field of complex analytic dynamics showed enthusiastic growth during the time of the French mathematician Gaston Julia. His strong inquisitiveness regarding the behaviour of complex function under iteration led him to attain a landmark in the field of fractal theory by obtaining the Julia set in 1918. The Julia set is of vital importance in the study of the complex dynamics of functions because it is the place where all the chaotic behaviour of a complex function occurs in [5]. A lot of work has been done on the structures of the Julia sets of the complex analytic function such as polynomial, rational and exponential functions. In the literature, the classes of maps most often studied are rational maps with quotients of two polynomials and transcendental functions without singularities in the finite plane besides polynomials, for instance [3], [4], [9], [18]. The importance of the transcendental functions lies in the fact that the Julia sets for such functions have an alternative characterization suitable for easier computations. Such functions are studied by authors of this paper in [14]. It is well known that a complex analytic function has a Julia set, for detailed discussion on Julia sets one may refer Barnsley [2], Devaney [5]-[8], Peitgen et. al [13] and several references thereof. Misiurewicz [12] was the first to explore the mathematical aspects of the complex exponential maps of the type $z_{n+1} = e^{z_n}$, Baker and Rippon [1] studied the complex exponential family $e^{\lambda z}$ and Devaney [6]-[7] extensively studied the family λe^z of maps. Thereafter, Romera et. al [17] studied the complex families

P. Balasubramaniam and R. Uthayakumar (Eds.): ICMMSC 2012, CCIS 283, pp. 185–192, 2012.

$e^{z^2 + \lambda}$ and $e^{\frac{z}{\lambda}}$ from graphical point of view and obtained interesting results. The Picard iteration scheme is used in all the above results. Recently Rani and Kumar [16] have explored the Julia sets using Mann iterative algorithm for logistic functions. Prasad and Katiyar [15] used the more general Ishikawa iteration scheme in their results.   In this paper, our aim is to study the complex function of the type $E_\lambda(z) = \lambda e^z$ by using Ishikawa iterative scheme. We also attempt to present the bifurcation analysis for the Mann iterates of this family of maps.

2   Preliminaries

In this section we present the basic definitions and concepts which are needed for our study.

Definition 2.1. [5] Let $F(z)$ be an entire transcendental function i.e., function like λe^z, $\lambda \sin z$, $\lambda \cos z$. The Julia set of $F(z)$, denoted by $J(F)$, is the set of points at which the family of iterates of F (i.e. F, $F o F = F^2$, F^3, ...) fails to be a normal family. Equivalently, $J(F)$ is the closure of the set of non attracting periodic points of F.

In the words of Peitgen et el. [13], we can say that the Julia set for such functions is the closure of the set of points whose orbits tend to infinity.

Notice that this definition of Julia set is quite different from the definition of the Julia set for polynomials. In the latter case, the Julia set formed the boundary of the set of escaping orbits. In the entire case, points whose orbits escape actually lie in the Julia set (Peitgen [13]). Peitgen (c.f. [13]) observed that there is nothing apparently chaotic about an orbit tending to infinity. However, in the entire case, infinity is an essential singularity, which is the main factor distinguishing these maps from polynomials. Also, the Julia set is the closure of the set of escaping orbits containing not only the escaping points, but also any limit point of a sequence of such points.

Definition 2.2. [2] Let X be a non empty set and $f: X \rightarrow X$. The orbit of a point z in X is defined as a sequence

$$\{ f^n(z): n = 0, 1, 2, \ldots \}.$$

If we generate the orbits by using Peano-Picard iteration, the generated orbit, represented by

$$PO(f, z_0) := \{z_n: z_n = f(z_{n-1}), n = 1, 2, \ldots\},$$

is called the Picard orbit. This iteration requires one number as input to return a new number as output and popularly called as one step feedback machine (Rani and Agarwal [16]). The Mann iteration [11] is defined in the following manner:

$$z_n = \alpha_n f(z_{n-1}) + (1 - \alpha_n) z_{n-1}, \tag{1}$$

for $n = 1, 2, 3, \ldots$, where $0 < \alpha_n \leq 1$ and $\{\alpha_n\}$ is convergent away from 0.

The sequence $\{z_n\}$ constructed as above is two step feedback system and its orbit is denoted by $MO(f, z_0, \alpha_n)$.

Now we define a three step feedback scheme essentially due to Ishikawa [10].

Definition 2.3. Let X be a non empty set and $f: X \rightarrow X$. For a Point z_0 in X, construct a sequence $\{z_n\}$ in the following manner.

$$z_n = \alpha_n f(v_{n-1}) + (1- \alpha_n) z_{n-1},$$

$$v_{n-1} = \beta_n f(z_{n-1}) + (1- \beta_n) z_{n-1}, \tag{2}$$

for $n = 1, 2, 3, \ldots$, where $0 < \alpha_n \leq 1$ and $0 \leq \beta_n \leq 1$ and $\{\alpha_n\}$ is convergent away from 0. Then sequence $\{z_n\}$ constructed above will be called the Ishikawa iterate of all iterates of a point z_0. We denote it by $IO(f, z_0, \alpha_n, \beta_n)$. In this paper, we shall study the Ishikawa orbit for $\alpha_n = \alpha$ and $\beta_n = \beta$.

It is remarked that (2) becomes (1) when we put $\beta_n = 0$ in it and (1) with $\alpha_n = 1$ is the Picard iteration.

Definition 2.4. [8] Define the horizontal strips

$$R_j = \{z \mid (2j-1)\pi < \operatorname{Im} z \leq (2j+1)\pi\}.$$

For each $z \in \mathbb{C}$ (the set of complex numbers), the itinerary of z is the sequence of integers $s = s_0 s_1 s_2 \ldots$ where $s_j \in \mathbb{Z}$ and $s_j = k$ iff $E_\lambda^j(z) \in R_k$. If s is a bounded sequence that consists of at most finitely many zeros, then the set of points that share this itinerary is a continuous curve homeomorphic to the half line $[0, \infty)$ and extended to ∞ in the right half the plane.

Definition 2.5. [7] Let $s = s_0 s_1 s_2 \ldots$. A continuous curve $H_s: [1, \infty) \rightarrow \mathbb{C}$ is called a hair with itinerary s if H_s satisfy:

1. If $\lambda = H_s(t)$ and $t > 1$, then $\operatorname{Re} E_\lambda^n(0) \rightarrow \infty$ and itinerary of λ under E_λ is s.
2. If $\lambda = H_s(1)$, then $E_\lambda(0) = \lambda = z_\lambda(s)$ where $z_\lambda(s)$ is the end point of hair with itinerary s in the dynamical plane. Hence the dynamics of λ under E_λ is bounded and has itinerary s.
3. $\displaystyle\lim_{t \rightarrow \infty} \operatorname{Re} H_s(t) = \infty$.

2.1   Escape Criterion for Exponentials

To produce computer graphical images of Julia sets of entire functions like λe^z, $\lambda \sin z$, $\lambda \cos z$ we need to search the points whose orbits go far from the origin. Generally, orbits tend to infinity in specific directions. We follow the given escape criterion stated by Peitgen [13] to generate graphical images of Julia sets for λe^z.

For the exponential family $E_\lambda(z) = \lambda e^z$ the orbits tend to infinity in the direction of the positive real axis. That is as

$$\lim_{n\to\infty} \mathrm{Re}(E_\lambda^n(z)) = \infty,$$

$$\lim_{n\to\infty} \left| E_\lambda^n(z) \right| = \infty \ \text{ also tends to infinity.}$$

Therefore if the real part of $E_\lambda^n(z)$ exceeds 50, then we say that the orbit of z escapes ([7], [13]).

2.2   Algorithm to Generate Graphical Images of Julia Set

The steps to generate Julia sets are given as:

- Compute the orbit of z up to N iterations using Ishikawa iteration.
- If the orbit of z enters the region $\mathrm{Re}\,z \geq 50$ at iteration $j \leq N$ then colour z with a colour corresponding to j. We follow the colour schemes of Pietgen et. al [13] along with the above defined escape criterion and algorithm for our study of the complex exponential functions. The colouring scheme of the graphics presented in the figures depends upon the rate of escape to infinity. A point z_0 is coloured black if the orbit of z_0 is not escaped within the first N iterates, red is used to denote points which escape to infinity fastest. Shades of orange, yellow and green are used to colour points which escaped less quickly and shades of blue and violet represent the points which escaped, but only after a significant number of iterations.
- If the orbit never enters this half plane, then colour z black and declare that $z \notin J(E_\lambda)$.

This colouring scheme is well depicted in the graphical patterns given in the Figs. 1-4 of Section 3.

3   Experimental Study Using Ishikawa Iteration

Let $z_n = x_n + iy_n$ and $\lambda = \lambda_1 + i\lambda_2$ then the real and imaginary part of $z_{n+1} = f(z_n) = \lambda e^{z_n}$ are given by

$$\mathrm{Re}\,z_{n+1} = \alpha\left\{ e^{x'_n} \left(\lambda_1 \cos y'_n - \lambda_2 \sin y'_n \right) \right\} + (1-\alpha)x_n,$$

$$\mathrm{Im}\,z_{n+1} = \alpha\left\{ e^{x'_n} \left(\lambda_1 \sin y'_n + \lambda_2 \cos y'_n \right) \right\} + (1-\alpha)y_n,$$

where

$$x'_n = \beta\left\{ e^{x_n} \left(\lambda_1 \cos y_n - \lambda_2 \sin y_n \right) \right\} + (1-\beta)x_n,$$

$$y'_n = \beta\left\{ e^{x_n} \left(\lambda_1 \sin y_n + \lambda_2 \cos y_n \right) \right\} + (1-\beta)y_n.$$

This gives us the formula to compute the main body of the loop in the program. Using the algorithm 2.2, we have generated the Julia sets for the Ishikawa iterates (2) of the function λe^z for the specific choices of parameters α and β by fixing $N = 25$, $\lambda = \dfrac{1}{e} + 0.1$ in Figs. 1-4.

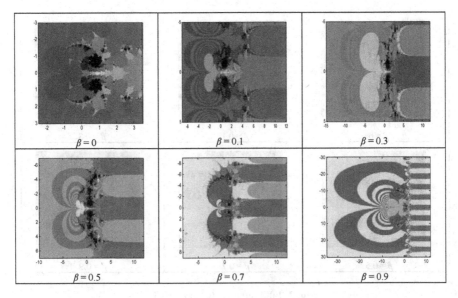

Fig. 1. Julia sets for λe^z at $\alpha = 1$

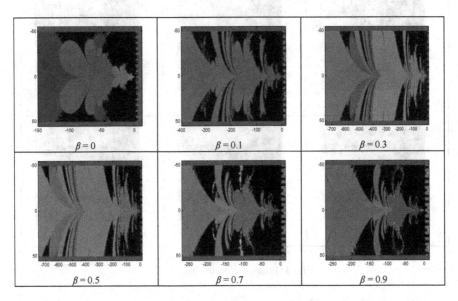

Fig. 2. Julia sets for λe^z at $\alpha = 0.7$

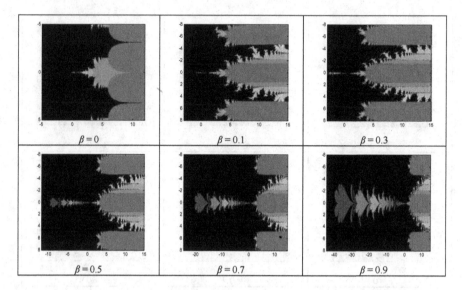

Fig. 3. Julia sets for λe^z at $\alpha = 0.3$

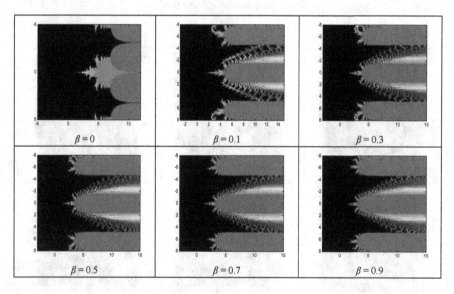

Fig. 4. Julia sets for λe^z at $\alpha = 0.1$

4   Bifurcation Analysis

Devaney [7] observed that the exponential family $E_\lambda(z) = \lambda e^z$ undergoes period doubling bifurcation at $\lambda = -e$ with Picard iteration scheme. On changing iteration scheme from Picard to Mann we observe a shift in the bifurcation point of the map.

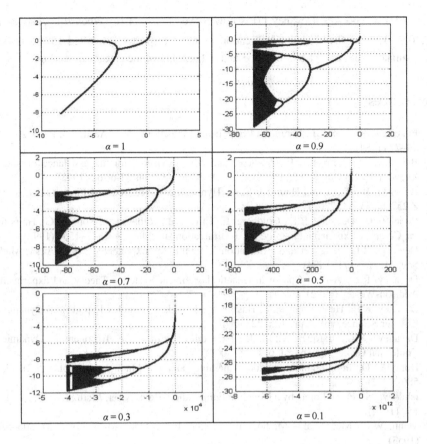

Fig. 5. Bifurcation diagram for λe^z at $\beta = 0$

It is noticed that on decreasing the values of the parameter α from 1 towards 0.1 the period doubling bifurcation points meet for higher values of λ as shown in Fig. 5.

5   Conclusions

We study the Julia sets obtained for different values of the parameter $\alpha_n = \alpha$, $\beta_n = \beta$. At $\alpha = 1$ and $\beta = 0$, the pattern is same as obtained by Peitgen [13] for Picard iteration (Fig. 1. for $\beta = 0$). On fixing α at 1 and varying β from 0.1 to 0.9 in the steps of 0.2, the hairs (in sense of Devaney [8]) in Julia sets get more closer and finally tied together as shown in Fig. 1. The attached hairs form flower like shape at $\alpha = 0.7$ and $\beta = 0$ which starts exploding as β is increased from 0.1 to 0.9 (Fig. 2). For $\alpha = 0.3$ and 0.1 temple like fractals are obtained (Fig. 3-4).

Further, when we take $\alpha = 0.9$ and $\beta = 0$ in our iteration scheme an appreciable shift in the bifurcation point of the maps is observed. The period doubling bifurcation which was observed at $\lambda = -e$ by Devaney [7] is now moved to $\lambda = -4.14$. When α is changed to 0.7, 0.5, 0.3 and 0.1 the first bifurcation point is respectively shifted to the higher approximate values of -11.87, -60, -163 and -3.36×10^9 of the parameter λ (Fig. 5.).

References

1. Baker, N., Rippon, P.J.: Iteration of Exponential Functions. Ann. Acad. Sci. Fenn. A 1(9), 49–77 (1984)
2. Barnsley, M.F.: Fractals Everywhere, 2nd edn. Revised with the assistance of and a foreword by Hawley Rising, III. Academic Press Professional, Boston (1993)
3. Barański, B.: Trees and Hairs for Some Hyperbolic Entire Maps of Finite Order. Math. Z 257(1), 33–59 (2007)
4. Bodelón, C., Devaney, R.L., Goldberg, L., Hayes, M., Hubbard, J., Roberts, G.: Hairs for the Complex Exponential Family. Int. J. Bifurcation and Chaos 9, 1517–1534 (1999)
5. Devaney, R.L.: Julia Set and Bifurcation Diagrams for Exponential Maps. Amer. Math. Soc. 11, 167–171 (1984)
6. Devaney, R.L.: A First Course in Chaotic Dynamical Systems: Theory and experiment. Addison-Wesley (1992)
7. Devaney, R.L., Henk Broer, F.T., Hasselblatt, B. (eds.): Complex Exponential Dynamics. Elsevier Science, vol. 3, pp. 125–223 (2010)
8. Devaney, R.L., Jarque, X.: Indecomposable Continua in Exponential Dynamics. Conformal Geom. Dynamical. 6, 1–12 (2002)
9. Dong, X.: On Iteration of a Function in the Sine Family. J. Math. Anal. Appl. 165(2), 575–586 (1992)
10. Ishikawa, S.: Fixed Points by a New Iteration Method. Proc. Amer. Math. Soc. 44(1), 147–150 (1974)
11. Mann, W.R.: Mean Value Methods in Iteration. Proc. Amer. Math. Soc. 4(3), 506–510 (1953)
12. Misiurewicz, M.: On Iterates of e^z: Ergod. Theor. Dyn. Syst. 1, 103–106 (1981)
13. Peitgen, H.O., Saupe, D.: The Science of Fractal Images. Springer, New York (1988)
14. Prasad, B., Katiyar, K.: Julia Sets for a Transcendental Function. In: Proc. of the IEEE International Conference on Computer Engineering and Technology, Jodhpur, India, pp. E59–E61 (2010)
15. Prasad, B., Katiyar, K.: Fractals via Ishikawa Iteration. In: Balasubramaniam, P. (ed.) ICLICC 2011. CCIS, vol. 140, pp. 197–203. Springer, Heidelberg (2011)
16. Rani, M., Kumar, V.: Superior Julia Set. J. Korea Soc. Math. Edu. Ser D: Res. Math. Edu. 84, 261–277 (2004)
17. Romera, M., Pastor, G., Alvarez, G., Montoya, F.: Growth in Complex Exponential Dynamics. Comput. Graph. 24(1), 115–131 (2000)
18. Schleicher, D., Zimmer, J.: Escaping Points of Exponential Maps. J. London Math. Soci. 67(2), 380–400 (2003)

Dynamics of Density-Dependent Closure Term in a Simple Plankton Model

A. Priyadarshi, S. Gakkhar, and Sandip Banerjee

Department of Mathematics,
Indian Institute of Technology Roorkee,
Roorkee - 247667, India
anupam240@gmail.com,
{sungkfma,sandofma}@iitr.ernet.in

Abstract. A three-component aquatic model, which consists of nutrient, phytoplankton and zooplankton has been investigated. To incorporate the effects of higher predation, the mortality of zooplankton is assumed to be density-dependent (sigmoidal form). The system has uniformly bounded and dissipative solutions in the non-negative octant. Some conditions on persistence of all three-species have been established. The parameter estimation technique "Marquardt-Levenberg (M-L) algorithm" has been used to estimate the values of some parameters, specially for density-dependent mortality. The bifurcation analysis reveals that model has periodic solutions for some parameter range. The amplitude of short term oscillations is varying with nutrient input present in the system. It is consistent with the field observational results. The nutrient input in the system may be one of the reason for short-term oscillations observed in the sea water.

Keywords: Nutrient input, periodic solutions, Density-dependent Closure term.

1 Introduction

Plankton (phyto- and zoo-plankton) form the basis of aquatic food webs throughout the worlds ocean, supporting a diverse range of marine and terrestrial life from shrimps and cod to blue whales and man. Therefore, to predict the future fish harvests and assess the possible consequences of global warming, it is essentials to understand the dynamics of plankton populations.

Three component simple plankton models are usually assembled with three components (i) source term, (ii) consumption term and (iii) mortality of top predator. The predations by higher-order predators are not included explicitly in the plankton model. It is supposed to be the part of the mortality of top predators (zooplankton), these last terms are called "closure terms", acknowledging their role in truncating the food webs [6,7].

The choice of functional form for zooplankton mortality is biologically controversial and potentially influential on the qualitative dynamics of the system [6].

P. Balasubramaniam and R. Uthayakumar (Eds.): ICMMSC 2012, CCIS 283, pp. 193–200, 2012.

Most food chains models include only density-independent mortality of the top predator, so that closure term itself is proportional to population density. If predation by higher order predators is to be incorporated in the model then the mortality of top predator may be density-dependent. Steel and Henderson [6] have suggested the closure terms

$$j(Z) = qZ^{\hat{\alpha}}Z, \ \hat{\alpha} \geq 0$$

(q is a constant), to incorporate the numerical and functional responses of higher-order predators to the density of their prey. The $\alpha = 0$ and 1 correspond to linear (qZ) and quadratic (qZ^2) closure terms respectively. The $(0 < \hat{\alpha} < 1)$, reflects the nonlinear closure terms particularly, the Hyperbolic ($j(Z) = \frac{\epsilon Z}{k_3 + Z} Z$) and Sigmoidal ($j(Z) = \frac{\epsilon Z^2}{k_3^2 Z^2} Z$) forms of closure terms.

The sigmoidal form behave somewhat similar to quadratic form (i.e. rate proportional to zooplankton biomass) at low zooplankton concentrations, whereas at high zooplankton concentration it is similar to the linear form with invariant specific rate. As Zooplankton density increases, the sigmoidal response leads to a more than linear increase in predation rate.

Consider a three-component aquatic model which consists of nutrient with density (N), phytoplankton with density (P) and zooplankton with density (Z). Now, using sigmoidal "closure term" and constant nutrient input (N_0) to the mixed layer in the aquatic model, the NPZ-model is given as:

$$\frac{dN}{dT} = -\frac{\mu NP}{k_1 + N} + rP + \frac{\beta \lambda P^2 Z}{k_2{}^2 + P^2} + \frac{\gamma \epsilon Z^2}{k_3^2 + Z^2} Z + k(N_0 - N)$$

$$\frac{dP}{dT} = \frac{\mu NP}{k_1 + N} - rP - \frac{\lambda P^2 Z}{k_2{}^2 + P^2} - (s + k)P \tag{1}$$

$$\frac{dZ}{dT} = \frac{\alpha \lambda P^2}{k_2{}^2 + P^2} Z - \frac{\epsilon Z^2}{k_3^2 + Z^2} Z$$

The system (1) is associated with the non-negative initial conditions $N(0) \geq 0$, $P(0) \geq 0$, $Z(0) \geq 0$ at time $t = 0$. All parameters are taken to be positive and their values are given in Table 1.

Obviously, the interaction functions of the system (1) are continuous and have continuous derivatives on the positive octant. Therefore, the system has an unique solution in non-negative octant $\Re^3_+$. The following theorem establishes the boundedness of the solution in $\Re^3_+$.

Theorem 1. *The system (1) has uniformly bounded solutions in $\Re^3_+$. The solutions are dissipative in the non-negative octant.*

Proof. From (1), we get,

$$\frac{d(N + P + Z)}{dT} = (\alpha + \beta - 1)\frac{\lambda P^2}{k_2{}^2 + P^2} Z + (\gamma - 1)\frac{\epsilon Z^2}{k_3^2 + Z^2} Z + k(N_0 - N) - (s + k)P$$

$$\leq kN_0 - kN - (s + k)P + (\gamma - 1)\epsilon Z \leq kN_0 - \hat{k}(N + P + Z); \ \hat{k} = \min(k, \epsilon(1 - \gamma))$$

The Standard theorem on differential inequality gives

$$0 \leq (N(t) + P(t) + Z(t)) \leq \frac{kN_0}{\hat{k}} + (N(0) + P(0) + Z(0)) \exp(-\hat{k}t)$$

As $t \to \infty$, $0 < (N(t) + P(t) + Z(t)) \leq (kN_0/\hat{k})$.

Thus, all the solutions of the system (1) enter into a topological region

$$B = \{(N, P, Z) : (N(t) + P(t) + Z(t)) \leq (kN_0/\hat{k}) + \xi, \, for \, any \, \xi > 0\}$$

such that B is a connected compact set and invariant attractor for the flow of the system (1). Hence, the system (1) is uniformly bounded.

Also, $\limsup_{t\to\infty}[N(t) + P(t) + Z(t)] \leq \frac{kN_0}{\hat{k}}$ holds in this case, which proves that it is dissipative in non-negative octant $\Re^3_+$.

2   Stability Analysis

In this section, the local behaviors of isolated singularities of system (1) are investigated and results are stated in the form of theorems.

Theorem 2.   (i) The axial singularity $E_1 = (N_0, 0, 0)$ of system (1) always exists. It has one-dimensional stable manifold along N-direction. It has center-manifold along Z-direction.
(ii) When $\frac{\mu N_0}{(k_1+N_0)} - (r + s + k) < 0$, the singularity E_1 is stable.
(iii) It is unstable when $\frac{\mu N_0}{(k_1+N_0)} - (r + s + k) > 0$.

The eigenvalues associated to singularity E_1 are $-k$, $\frac{\mu N_0}{(k_1+N_0)} - (r + s + k)$ and 0. Therefore, results are evident.

Note that system undergo transcritical bifurcation at

$$N_0 = \frac{k_1(r + s + k)}{\mu - r - s - k}, \quad or \quad \mu = \frac{(k_1 + N_0)(r + s + k)}{N_0}$$

The system has equilibrium point in the $(N - P) - plane$ if $(r + s + k) < \frac{\mu N_0}{k_1+N_0}$.

Theorem 3.   1. If $(r + s + k) < \frac{\mu N_0}{k_1+N_0}$ holds for the system. Then there exists a singularity $E_2 = (\hat{N} = \frac{k_1(r+s+k)}{\mu-(r+s+k)}, \hat{P} = \frac{k(N_0-\hat{N})}{(s+k)}, 0)$ of the system (1).
2. It is an attracting focus in the $(N - P)$ plane (when it exists).
3. It has one-dimensional repelling manifold in orthogonal direction to $(N - P)$ coordinate plane.

Proof. The Jacobian matrix associated with singularity E_2 is

$$J(E_2) = \begin{pmatrix} -\frac{k_1\mu\hat{P}}{(k_1+\hat{N})^2} - k & -\frac{\mu\hat{N}}{k_1+\hat{N}} + r & \frac{\beta\lambda\hat{P}^2}{k_2^2+P^2} \\ \frac{k_1\mu\hat{P}}{(k_1+\hat{N})^2} & \frac{\mu\hat{N}}{k_1+\hat{N}} - (r + s + k) & -\frac{\lambda\hat{P}^2}{k_2^2+P^2} \\ 0 & 0 & \frac{\alpha\lambda\hat{P}^2}{k_2^2+P^2} \end{pmatrix}$$

The associated eigenvalues $(\lambda_i; i = 1, 2, 3)$ are given by:

$$\lambda_3 = \frac{\alpha\lambda\hat{P}^2}{k_2^2 + P^2}, \; \lambda_1 + \lambda_2 = -\frac{k_1\mu\hat{P}}{(k_1 + \hat{N})^2} - k \; < 0$$

$$\lambda_1\lambda_2 = \frac{k_1\mu\hat{P}(s + k)}{(k_1 + \hat{N})^2} \; > 0$$

Since $\lambda_3 > 0$, system has repelling manifold in Z direction. Further, the singularity E_2 is attracting in $(N - P) - plane$.

Note that the existence of singularity E_2 is only possible when axial singularity E_1 becomes saddle. The persistence of the system is possible since singularity E_2 is always saddle in the interior.

Theorem 4. *The system is persistent in* $\Re_+^3$*, if*

$$(r + s + k) < \frac{\mu N_0}{k_1 + N_0} \tag{2}$$

Proof. By Theorems 2, the axial singularity E_1 is a saddle point having unstable manifold in orthogonal direction to $(N - Z)$ plane. Also, by Theorem 1, the singularity E_2 is an attracting focus when $(r + s + k) < \frac{\mu N_0}{k_1 + N_0}$ holds. However, it has repelling manifold in orthogonal direction to $N - P$ plane. Hence, by Butler-McGehee lemma [4], the inequality (2) implies the persistence of the system (1).

The system (1) has an interior equilibrium point $E_3 \equiv (N^{**}, P^{**}, Z^{**})$ when $(\alpha\lambda - \epsilon)P^2 < \epsilon k_2^2$. Due to intractable nature of the mathematical expressions, further calculations have been done numerically to investigate the behavior of the interior equilibrium points.

3 Parameter Estimation for the Parameters ϵ and k_3

As stated in the introduction that model (1) has equipped with two density dependent mortality of zooplankton. The parametric values of these parameters are not a good estimation [2,3]. The estimated values of these parameters are used in the aquatic modeling literatures. Different technique has been used for their estimation. The mortality term $\frac{\epsilon Z^2}{k_3 + Z}$ contains two parameters: (i) Zooplankton maximum loss rate ϵ and (ii) Half-saturation constant k_3.

In this paper, for estimation of these parameters "Marquardt-Levenberg algorithm" (M-L algorithm) Press et al. [5] have been used. The following procedure has been adoptable while using (M-L) algorithm. The behavior of density dependent mortality terms in somewhat lie between 'linear and quadratic' mortality terms $(0 < \hat{\alpha} < 1)$. Therefore, simulations have been done several times using linear / quadratic closure terms / between linear and quadratic (randomly chosen). Then sufficient numbers of data have been collected from the simulations.

Now, it is also well-known that at high zooplankton hyperbolic mortality term behaves like linear mortality term. Therefore, a sample has been developed by well-mixing of data obtained from the various simulations. The M-L algorithm applying on the data, which gives the estimated values of the parameters ϵ and k_3. Accordingly, the estimated values are obtained as $\epsilon = 0.21$ and $k_3 = 0.2$ respectively. The numerical estimation process have been shown in the figure 1. All further numerical calculations have been done using these parameters. Although, there are several literatures on plankton dynamics are available. The parametric values of other parameters given in Table 1 are mainly taken from [1,2,7].

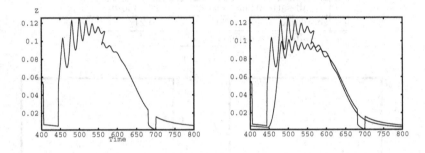

Fig. 1. The parameter estimation technique: M-L algorithm applying on mixed data to obtain the value of parameters ϵ and k_3

4   Numerical Investigation

Various numerical investigations have been carried out to know the global behavior of the system. The bifurcation analysis has been carried out for the parameters ϵ and k_3. The qualitative changes in the behavior have been observed due to change in the stability of the system over some parameter range.

4.1   Bifurcation over Parameter ϵ

With values of parameters in Table 1 the system (1) has a steady state $(N, P, Z) = (0.6135, 0.01024, 0.32536)$. The bifurcation analysis has been carried out with respect to parameter $\epsilon \in (0.2, 1.5)$. The bifurcation diagram of Z w. r. t. parameter ϵ is shown in Fig. 2. The system undergoes Andronov-Hopf bifurcations and consequently periodic solutions found in the system. The variation in the period is shown in Fig. 2. The variation is least and average period is $17 - 18$ days. At $\epsilon = 0.4128$, the first Hopf-point is observed. The periodic orbits start at this point. The maximum and minimum amplitude of these periodic orbits are shown by solid limit circles in the bifurcation diagrams. The limit orbits continue till second Hopf-point at $\epsilon = 0.8406$. All limit orbits are stable in this case which is shown by solid circles. Both Hopf-bifurcation points are supercritical.

Table 1. Symbol, Parameters and its Values

a	Maxim. growth rate P	$0.2m^{-1}day^{-1}$
b	Light-attenuation	$0.2m^{-1}$
c	P Self-shading coefficient	$0.4m^2g^{-1}$
k_1	Half-saturation constant	$0.09gm^{-3}$
k	Cross-thermocline exchange rate	$0.05day^{-1}$
r	P respiration rate	$0.15day^{-1}$
s	P sinking rate	$0.04day^{-1}$
N_0	Nutrient input	$0.6gm^{-3}$
α	Z assimilation efficiency	0.25
β	Z excretion fraction	0.33
γ	Regeneration of Z	0.5
λ	Maxim. Z grazing rate	$0.6day^{-1}$
k_2	Z half-saturation constant	$0.035gm^{-3}$
ϵ	zooplankton loss $rate^*$	0.21
k_3	Z half-saturation $constant^*$	0.2

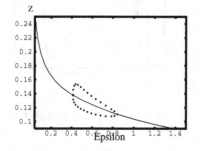 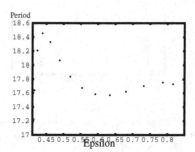

Fig. 2. Bifurcation diagram of Z and periods of periodic orbit w. r. t. parameter ϵ

4.2   Bifurcation over Parameter ϵ at Distinct Value of N_0

It is assumed that the concentration of input nutrient below to the mixed-layer is not always at one level (constant). Hence, due to wind-upwelling or other factors including temperature variations in sea or lakes, the concentration of nutrient input in the system is also vary time to time. To know the dynamics at different level of input nutrient, the bifurcation analysis have been done w. r. t. parameter ϵ. The behavior of the system change drastically on different values of parameter N_0. The bifurcation diagrams have been drawn at different values of $N_0 = 0.8,\ 1.5,\ 2.0$. At parameter $N_0 = 0.8$, the bifurcation diagrams have been shown in figure 2.

When parameter $N_0 = 1.5$, the stable periodic orbits emanating from the second Hopf point (at $\epsilon = 0.4234$). On decreasing parameter ϵ, these stable periodic orbits undergo fold bifurcations, which produce unstable periodic orbits and join the first Hopf-point at $\epsilon = 0.2571$. The homoclinic connection in which the unstable manifold of the saddle point joins up in a loop with its own stable

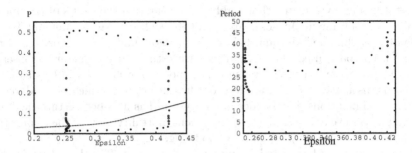

Fig. 3. Bifurcation diagram of Z and periods of periodic orbit w. r. t. parameter ϵ at input nutrient $N_0 = 1.5$

manifold, is observed in this case near first Hopf point at $\epsilon = 0.2574$. The first Hopf is subcritical while second is supercritical one. The bifurcation diagram and period w. r. t. parameter ϵ is shown in figure 3.

When parameter $N_0 = 2.0.$, then the limit orbits generated from Hopf-point $\epsilon = 0.31224$ and further, on decreasing the parameter value ϵ, limit orbits undergo fold bifurcations at $\epsilon = 0.2962404$ and $\epsilon = 0.2863783$. It created stable orbits which make homoclinic crossing near first Hopf point ($\epsilon = 0.2408804$). At this period doubling point, limit orbits has period 68.03. The complexity underlying in the dynamics has been shown in the bifurcation diagrams. The bifurcation diagram and period variation w. r. t. parameter ϵ is shown in figure 4. Here, the periods of limit cycles are higher than the previous results. It can be easily observed that the substantial increment in periods of the periodic solution is proportional to its nutrient input concentration to the mixed layer. Hence short periodic variation can be regarded as due to variations in input nutrient in the system.

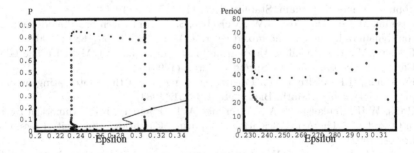

Fig. 4. Bifurcation diagram of Z and periods of periodic orbit w. r. t. parameter ϵ at input nutrient $N_0 = 2.0$

5   Discussions

The mortality of top predators an less (fewer) component models is always biologically controversial. In this paper, the nonlinear zooplankton mortality has been taken and investigated thoroughly. For estimating the values of parameters (i) Zooplankton maximum loss rate ϵ and (ii) Half-saturation constant k_3, the estimation technique "Marquardt-Levenberg algorithm" (M-L algorithm) Press et al. [5] has been used. The dynamics with estimated parameters is observed to be consistent with existing results. The model is well-posed in this case and system is dissipative. The analytic proof has been given which is validated by numerical simulations. Persistence of the system has also been established analytically in some cases. The density-dependent mortality may be the appropriate / suitable choice for simple plankton modeling.

It is observed and reported by many field ecologist [2,7] that there is short time oscillation in the open sea. The dynamics of system has also been investigated for different amount of the nutrient input, which is coming to mixed layer by temperature fluctuation or wind-waves. For this system, numerical analysis reveals that there is short time oscillations for some choice of parameters. The oscillations variation of the periodic orbits are according to amount of the nutrient input present in the system. The variation in periods is approximately proportional to increase in the nutrient input. The numerical results reveals that nutrient input may be regarded as one of the reason for short time oscillations in the sea-water dynamics.

Acknowledgement. We are thankful to anonymous reviewers for their valuable comments. The first author A. Priyadarshi wishes to thank *"Ministry of Human Resources and Development (MHRD), India"* for Senior Research Fellowship through Grant No. $MHR02 - 23 - 200 - 304$.

References

1. Edwards, A.M., Brindley, J.: Oscialltory behavior in a three-component plankton population model. Dynam. Stablity Syst. 11, 347–370 (1996)
2. Fasham, M.J.R., Ducklow, H.W., McKelvie, S.M.: A nitrogen-based model of plankton dynamics in the oceanic mixed layer. J. Mar. Res. 48, 591–639 (1990)
3. Fasham, M.J.R.: Modelling the marine biota. In: Heinmann, M. (ed.) The Global Carbon Cycle, pp. 457–504. Springer, Berlin (1993)
4. Freedman, H.I., Waltman, P.: Persistence in models of three intercating predator-prey populations. J. Math. Biosci. 68, 213–231 (1984)
5. Press, W.H., Teukolsky, S.A., Vetterling, W.T., Flannery, B.P.: Numerical Recipes in C, 2nd edn. The Art of Scientific Computing. Cambridge University Press, Cambridge (1992)
6. Steel, J.H., Henderson, E.W.: A simple plankton model. American Naturalist 117, 676–691 (1981)
7. Wroblewski, J.S.: An ocean basin scale model of phytoplankton dynamics in the North Atlantic. Solution for the climatological oceanographic condition in May. Global Biogeochem. 2, 199–218 (1988)

Crisis-Limited Chaotic Dynamics
in an Eco-epidemiological System of the Salton Sea

Sharada Nandan Raw[*], Ranjit Kumar Upadhyay, and Nilesh Kumar Thakur

Department of Applied Mathematics
Indian School of Mines, Dhanbad, Jharkhand - 826004, India
shardaraw@gmail.com

Abstract. In this paper, we have proposed a new eco-epidemiological model of the Salton Sea with Holling type II & IV functional responses. Numerical results are presented in the form of phase portraits, 2D scans, bifurcation analysis and basin boundary calculations. By these we have concluded that model system depicts the short-term recurrent chaos (STRC) and argued that crisis-limited chaotic dynamics can be commonly found in model eco-epidemiological systems.

Keywords: Eco-epidemiological model, Short-term recurrent chaos (STRC), 2D scan, Basin boundary structures, Bifurcation analysis.

1 Introduction

Eco-epidemiology is a young science but vitally important to help understand how chemical and physical factors interact to help shape the environment. Changes in the chemical or physical nature of a site can impact the fish, algae and invertebrates that live there. Ecological epidemiology, or eco-epidemiology, studies both the chemical and physical nature of the environment and how they both contribute to the health of the ecosystem. The first breakthrough in modern mathematical ecology was done by Lotka [1] and Volterra [2], for a predator-prey competing species. On the other hand, most models for the transmission of infectious diseases originated from the classic work of Kermack and McKendrick [3]. After these pioneering works in two different fields, lots of research works have been done both in theoretical ecology and epidemiology. Anderson and May [4] were the first who merged the above two fields and formulated a predator-prey model where prey species were infected by some disease. Chattopadhyay and Arino [5] coined the name eco-epidemiology for the study of such systems. In most theoretical studies of host-parasite-predation interactions, predator behavior is simplified and isolated from an ecosystem [6].

A crisis occurs when the environment of a species or a population changes in a way that destabilizes its continued survival. Sudden qualitative changes in properties of attractors are given the result from the collision of an unstable periodic orbit and a coexisting chaotic attractor may cause the crisis [7, 8, 9].

Upadhyay et al.[12] studied the work of Chattopadhyay et al. [10, 11] and presented in the respectively large range of parameters. Once again we are trying here

---

[*] Corresponding author.

P. Balasubramaniam and R. Uthayakumar (Eds.): ICMMSC 2012, CCIS 283, pp. 201–209, 2012.

to modified the eco-epidemiological model system of [11] and [10], both in a new developing eco-epidemiological model system of the Salton Sea and later on we will show that this model system displays the STRC and crisis limited chaotic dynamics. In Fig. 1. we present the sketch of eco-epidemiological model system of the Salton Sea. The paper is organized as follows: In Sect. 2 we discuss the assumptions and a developing eco-epidemiological model system. Sect. 3 depicts numerical simulation. Finally we summarize the results in last Sect. 4.

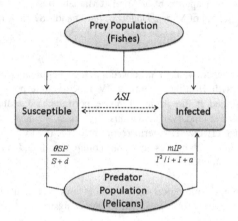

Fig. 1. Sketch of eco-epidemiological model system of the Salton Sea

2   Assumptions and Eco-epidemiological Model System

Consider an eco-epidemiological model system which consists the two parts i.e., fish population (Tilapia) and predator population (Pelicans). The prey (Tilapia) population density denoted by $N(t)$ at time t and the predator (Pelicans) population having population density $P(t)$ at time t. We impose the following assumption to formulating the basic differential equations.

Assumption 1. In the absence of bacterial infection, the fish population grows according to a logistic fashion with carrying capacity $K(K \in R_+)$ such that

$$\frac{dN}{dt} = rN\left(1 - \frac{N}{K}\right). \tag{1}$$

Assumption 2. In the presence of bacterial infection, we assume that the total fish population N is divided into two classes, namely, susceptible fish population, denoted by S, and infected fish population, denoted by I. Therefore, at any time t the total number of fish population is

$$N(t) = S(t) + I(t). \tag{2}$$

Assumption 3. We assume that only susceptible fish population, S, is capable of reproducing with logistic law (equation (1)) and the infective fish population, I, does not reproduce. However, the infective fish, I still contributes with S to population growth towards the carrying capacity.

Assumption 4. The mode of disease transmission follows the simple law of mass action. Therefore, the evolution equation for the susceptible fish population, S, according to equation (1) and assumptions (3) and (4), can be written as

$$\frac{dS}{dt} = rS\left(1 - \frac{S+I}{K}\right) - \lambda SI. \tag{3}$$

Where $\lambda(\lambda \in R_+)$ is the rate of transmission (or force of infection).

Assumption 5. The disease is spread among the prey population only and the disease is not genetically inherited. The infected population does not recover or become immune. Since prey population are infected by a (lethal) disease: infected preys are weakened and become easier to catch. Also they are present in the Salton Sea in considerable number, so we assume that predator's functional response to the infective prey follows Holling type IV functional response [13, 14] and included in the predator's growth equation with a positive sign. While susceptible preys easily escape and predation becomes difficult, so we assume that predator's functional response to the susceptible prey follows Holling type II predation [15] form which is included in the predator's growth equation with a positive sign. We write down following set of differential equations which describes the present model:

$$\frac{dS}{dt} = rS\left(1 - \frac{S+I}{K}\right) - \lambda SI - \frac{\theta_1 SP}{S+d}, \tag{4a}$$

$$\frac{dI}{dt} = \lambda SI - \frac{m_1 IP}{I^2/i+I+a} - \mu I, \tag{4b}$$

$$\frac{dP}{dt} = \frac{\theta_2 SP}{S+d} + \frac{m_2 IP}{I^2/i+I+a} - \delta P, \tag{4c}$$

$$\text{with} \quad S(0) > 0, \ I(0) > 0, \ P(0) > 0, \tag{4d}$$

where all the parameters are positive. θ_1, m_1 are the search rates of predator for susceptible and infected fishes respectively, θ_2 represents the conversion factor of susceptible Tilapia to Pelicans, m_2 represents the conversion factor of infected Tilapia to Pelicans, d measures the extent to which the environment provides protection to susceptible Tilapia, i a direct measure of pelican bird's immunity from or tolerance of infected Tilapia. a is the half saturation constant, total death rate of infected prey, $\mu = \mu_1 + \mu_2$; μ_1 is the natural death rates of infected prey population and μ_2 represents the rate of death of infected prey due to algal bloom. $\delta = \delta_1 + \delta_2$ is the total death of predator population. δ_1 ($\delta_1 \in R_+$) represents the natural death rate of predator population and δ_2 ($\delta_2 \in R_+$) represents the rate of death due to predation of infected prey.

3   Numerical Simulation Results

In this section, the global dynamical behavior of model system (4) in the interior of R_+^3 is investigated numerically. The objective is to detect the existence of complex dynamics of the model system including chaos. There are many way to detect the chaos in dynamical system, for investigating the deterministic behavior of the model system we are used the phase portraits, 2D scan concepts, basin boundary calculation and bifurcation theory. It is observed that for the following biologically feasible set of parameters values in equation (5), system (4) are displayed the chaotic attractor for the value $a = 16$ as shown in Fig. 2.

$$r = 22, K = 75, \lambda = 0.4, \theta_1 = 1, d = 60, m_1 = 15.5,$$
$$i = 40, \mu = 3.4, \theta_2 = 4, m_2 = 8.1, \delta = 3.6. \tag{5}$$

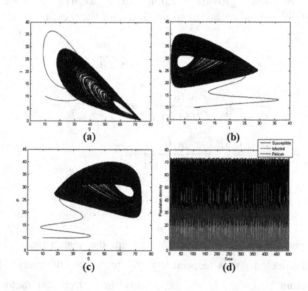

Fig. 2. Phase portraits of the chaotic attractor of the model system (4) for the parameter values given in equation (5) with $a = 16$ and initial condition [10, 10, 10] are (a) in SI -view. (b) in IP -view. (c) in SP -view. (d) Time series of chaotic attractor.

Under the bifurcation analysis of the model system (4), very rich and complex dynamics are observed, presenting various sequences of period-doubling bifurcation leading to chaotic dynamics or sequences of period-halving bifurcation leading to limit cycles [14]. For bifurcation diagram of model system (4) presented in Fig. 3(a), the value of the half saturation constant of prey population (infected fish) parameter a varies in the range [15.0, 24.5] and the values of other parameter is given in the equation (5) with respective of successive maxima of S in rage [12.0, 75.0]. Fig. 3(b) represents that magnified bifurcation diagram of the successive maxima of S in the range [25.0, 75.0] as a function of a is plotted in the range $18.07 \leq a \leq 18.13$ for the parameters given in equation (5).

The bifurcation diagrams are also generated for the successive maxima of the prey population S in the range [0.2, 75.0] as a function of total death rate of infected fish μ in the range $0.32 \leq \mu \leq 8.5$ (Fig. 4(a)) and the parameter values as given in equation (5) with $a = 18.3$. And Fig. 4(b) represents that magnified bifurcation diagram of the successive maxima of S in the range [30.0, 75.0] as a function of μ is plotted in the range $3.55 \leq \mu \leq 3.58$ for the parameters given in equation (5) with $a = 18.3$.

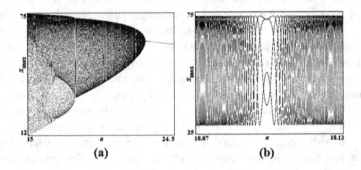

(a)   (b)

Fig. 3. Bifurcation diagram of the model system (4), (a) the successive maxima of S as a function of a is plotted in the range $15 \leq a \leq 24.5$ for the parameters given in equation (5). (b) magnified bifurcation diagram of the successive maxima of S as a function of a is plotted in the range $18.07 \leq a \leq 18.13$ for the parameters given in equation (5).

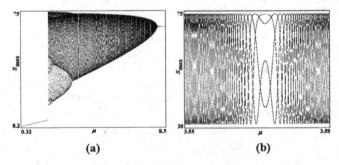

(a)   (b)

Fig. 4. Bifurcation diagram of the model system (4), (a) the successive maxima of S as a function of μ is plotted in the range $0.32 \leq \mu \leq 8.5$ for the parameters given in equation (5) with $a = 18.3$. (b) magnified bifurcation diagram of the successive maxima of S as a function of μ is plotted in the range $3.55 \leq \mu \leq 3.58$ for the parameters given in equation (5) with $a = 18.3$.

Based upon the concept of 2D scan we show the chaotic nature of the model system (4). Here we present 2D scan diagrams in various parameter spaces (Figs. 5 and 6). It is characterized by chaotic bursts repeated at unpredictable intervals. Chaotic dynamics is confined to a wedge-shaped region in the parameter spaces generated by λ with the parameters r, i, d, μ with step size of $r = 5$, $d = 5$, $i = 5$, $\lambda = 0.01$, $\mu = 0.25$ suggests that system (4) displays the STRC (Fig. 5).

We also present 2D scan diagrams in various parameter spaces (r, a) and (r, K) (Fig. 6) with step size of $r = 5$, $a = 0.5$, and $K = 5$, respectively and it suggests that system (4) displays also the STRC and the basin boundary calculations for chaotic attractors of the system (4) with respect to parameter spaces (Fig. 6) are presented in Fig. 7. The figures we have presented the SI -view of the basin boundary structure of chaotic attractor (shown in yellow color). The basin boundary calculations are performed using the basins and attractors structure (BAS) routine developed by *Maryland Chaos group*. We have used the dynamics software package of Nusse and Yorke [16], for all the basin boundary calculations. It is clear from these figures that basin boundaries of the chaotic attractor are fractal which show the dynamical complexities of the eco-epidemiological system (4). It is also seen that basin of attraction of different attractors are intermixed. The encroachment into the basin of chaotic attractor by basin of attractor at infinity (shown in green color) can be observed in Fig. 7. It appears between the first attractor (shown in green color) and its basin (shown in sky blue color). The interesting feature in the system (4) is that the riddled basin with fractal boundary lies in the basin of repeller which has many rectangular and square holes created by chaotic attractor. This complicated basin boundary structure suggests that the system dynamics may have loss of even qualitative predictability in the case of external disturbances.

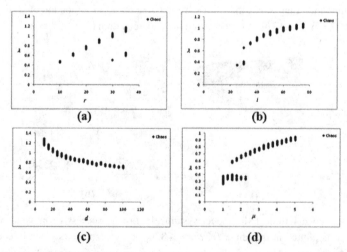

Fig. 5. 2D scan of the model system (5) in (a) (r, λ), (b) (i, λ), (c) (d, λ), (d) (μ, λ) parameter space. The parameter values of the other parameter are given in equation (5) except for $a = 18.3$.

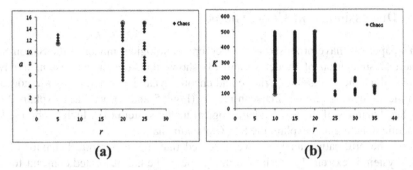

Fig. 6. 2D scan of the model system (5) in (a) (r, a), (b) (r, K) parameter space. The parameter values of the other parameter are given in equation (5) except for $a = 18.3$.

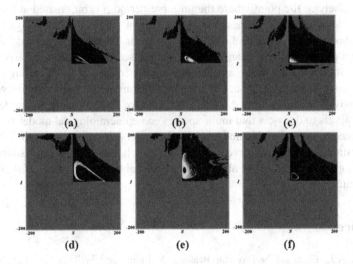

Fig. 7. Basin boundary structure for system (4) computed at different points in Fig. 6(a) for the chaotic attractor at (a) the top-right corner point $(r, a) = (25, 15)$; (b) the middle-top point $(r, a) = (20, 15)$; (c) the bottom-left corner point $(r, a) = (5, 11.5)$; and basin boundary structure for system (4) computed at different points in Fig. 6(b) for the chaotic attractor at (d) the top-right corner point $(r, K) = (35, 140)$; (e) the middle-top point $(r, K) = (20, 500)$; (f) the bottom-left corner point $(d, \mu) = (10, 95)$ all are in the domain $-200 \leq S, I \leq 200$. The meanings of the different colors are as follows: Green: color of first attractor, Sky Blue: basin of first attractor, Red: color of second attractor, Maroon: basin of second attractor, Brown: color of third attractor, White: basin of third attractor, Dark Blue: color of points that diverges from the screen area, Yellow: color of the chaotic attractor.

If we carefully examine the numerical results, one can find that there are two control parameters a and μ which affects the dynamics of the designed model for the Salton Sea.

4   Discussions and Conclusions

In this paper, we have proposed a new eco-epidemiological model of the Salton Sea. Numerical simulation of model system (4) shows that, the model system has rich complex dynamics including periodic and chaotic dynamics. From the phase portraits and time series (Fig. 2) and 2D scan results (Figs. 5 and 6), we have explained the complex dynamics of the eco-epidemiological model system (4) which is observed to be chaotic in nature and explain the STRC phenomena.

From the bifurcation analysis, we observed that the asymptotic behavior of the model system is extremely sensitive to the value of the half saturated constant for the infected prey a, and death rate of infected fish μ. Thus it is found that even small variation in parameters a and μ may cause a shift from chaos to limit cycles and vice-versa (Figs. 3(a) and 4(a)). From Figs. 3(b) and 4(b), we have seen that the crisis occurs precisely at the point where the unstable period-3 orbit created at the original saddle –node bifurcation intersects with the narrow chaotic region (Figs. 3(b) and 4(b)). It shows the evidence of the route to chaos through the cascade of period halving and other important change in the chaotic set include interior crisis in which a chaotic attractor undergoes a sudden increase in the size [7] along with the appearance or sudden enlargement of a fractal structure in the basin boundary.

This study gives crisis limited chaotic dynamics of the eco-epidemic model system (4) and supports to the view that multi-species eco-epidemiological model systems are able to generate unpredictable and complex behaviors. It points out to the difficulties in understanding the observed dynamical behavior of even simple eco-epidemiological systems in the absence of a reasonable a prior model of their growth and dynamics.

References

1. Lotka, A.J.: Elements of Physical Biology. Williams and Wilkins Co., Inc., Baltimore (1924)
2. Volterra, V.: Variazioni e Fluttuazioni del Numero d'individui in Specie Animali Conviventi. Mem. R. Accad. Naz. Dei Lincei. 2, 31–113 (1926)
3. Kermack, W.O., McKendrick, A.G.: A Contribution to the Mathematical Theory of Epidemics. Pro. Royal Soc. London, A 115, 700–721 (1727)
4. Anderson, R.M., May, R.M.: The Invasion, Persistence, and Spread of Infectious Diseases within Animal and Plant Communities. Philos. Trans. R. Soc. Lond. B 314, 533–570 (1986)
5. Chattopadhyay, J., Arino, O.: A Predator-Prey Model with Disease in the Prey. Nonlin. Analys. Theo. Meth. Appl. 36, 747–766 (1999)
6. Packer, C., Holt, R.D., Hudson, P.J., Lafferty, K.D., Dobson, A.P.: Keeping the Herds Healthy and Alert: Implications of Predator Control for Infectious Disease. Ecol. Lett. 6, 792–802 (2003)
7. Grebogi, C., Ott, E., Yorke, J.A.: Chaotic Attractor in Crisis. Phys. Rev. Lett. 42, 1507–1510 (1982)

8. Grebogi, C., Ott, E., Yorke, J.A.: Crises, Sudden Changes in Chaotic Attractors, and Transient Chaos. Physica 7D, 181–200 (1983)
9. Dangoisse, D., Glorieux, P., Hennequin, D.: Laser Chaotic in Crisis. Phys. Rev. Lett. 57, 2657–2660 (1986)
10. Chattopadhyay, J., Srinivasu, P.D.N., Bairagi, N.: Pelicans at Risk in Salton Sea - An Eco-epidemiological Model-II. Ecol. Model 167, 199–211 (2003)
11. Chattopadhyay, J., Bairagi, N.: Pelicans at Risk in Salton Sea - An Eco-epidemiological Model. Ecol. Model 136, 103–112 (2001)
12. Upadhyay, R.K., Bairagi, N., Kundu, K., Chattopadhyay, J.: Chaos in Eco-epidemiological Problem of the Salton Sea and its Possible Control. Appl. Math. Comp. 196, 392–401 (2008)
13. Andrews, J.F.: A Mathematical Model for the Continuous Culture of Microorganisms Utilizing Inhibitory Substrates. Biotech. Bioeng. 10, 707–723 (1968)
14. Upadhyay, R.K., Raw, S.N.: Complex Dynamics of a Three Species Food-Chain Model with Holling type IV Functional Response. Nonlin. Anal. Model. Cont. 16, 353–374 (2011)
15. Holling, C.S.: The Response of Predators to Prey Density and its Role in Mimicry and Population Regulation. Mem. Ent. Soc. Can. 45, 1–60 (1965)
16. Nusse, H.E., Yorke, J.A.: Dynamics: Numerical Exploration. Springer, NY (1994)

Delay-Dependent Stability Criteria for Uncertain Discrete-Time Lur'e Systems with Sector-Bounded Nonlinearity

Ramakrishnan Krishnan and Goshaidas Ray

Department of Electrical Engineering,
Indian Institute of Technology Kharagpur,
West Bengal 721302, India
{ramki,gray}@ee.iitkgp.ernet.in
http://www.iitkgp.ac.in

Abstract. In this paper, we consider the problem of delay-dependent stability of a class of discrete-time Lur'e systems with interval time-delay and sector-bounded nonlinearity using Lyapunov approach. By exploiting a candidate Lyapunov functional, and using slack matrix variables in the delay-dependent stability analysis, less conservative absolute and robust stability criteria are developed respectively for nominal and uncertain discrete-time Lur'e systems in terms of linear matrix inequalities (LMIs). For deriving robust stability conditions, time-varying norm-bounded uncertainties are considered in the system matrices. Finally, a numerical example is employed to demonstrate the effectiveness of the proposed results.

Keywords: Discrete-time Lur'e systems, Delay-dependent stability, Interval time-delay, Norm-bounded uncertainties, Sector-bounded nonlinearity.

1 Introduction

The problem of absolute stability of Lur'e systems has received considerable attention in control community, and many seminal contributions, such as Popov criterion, Circle criterion, and Kalman-Yakubovih-Popov lemma have been reported [1–3]. Recently reported results on master-slave synchronization of Lur'e system using time-delayed feedback control [4–6] has rekindled the interest in the stability studies of Lur'e system, and has given a fresh impetus to the problem. As time-delay is often encountered in physical systems like communication systems, air-craft stabilization, nuclear reactor, process systems, population dynamics etc., and is a source of poor performance and instability, the problem of absolute stability and robust stability of Lur'e systems with constant and time-varying delay have been investigated at length in [7–9]. Depending upon whether or not the stability criteria for Lur'e system contains the information of time-delay, the criteria can be classified respectively into two categories: namely, the delay-dependent stability criteria and delay-independent stability criteria. Since

P. Balasubramaniam and R. Uthayakumar (Eds.): ICMMSC 2012, CCIS 283, pp. 210–219, 2012.

delay-dependent criteria makes use of information on the length of the time-delay, they are less conservative than the delay-independent ones. In this context, most of the delay-dependent results available in literature are reported only for continuous-time Lur'e systems (see [10, 11], and the references cited therein). In the recently reported results [12] and [13], using Lyapunov function method, elegant robust stability and stabilization criteria are presented respectively for discrete-time Lur'e systems without time-delay. To the best of our knowledge, the delay-dependent absolute and robust stability criteria for discrete-time Lur'e systems with time-varying delay and sector-bounded nonlinearity have not been reported so far, and this motivates the present study.

In this paper, we investigate the delay-dependent absolute and robust stability of discrete-time Lur'e system with interval time-varying delay and sector-bounded nonlinearity using the Lyapunov approach. Subsequently, by exploiting a candidate Lyapunov functional, and using minimal number of slack matrix variables in the delay-dependent stability analysis, less conservative absolute and robust stability criteria are developed respectively for the nominal and uncertain discrete-time Lur'e systems in LMI framework. For deriving robust stability conditions, time-varying norm-bounded uncertainties are considered in the system matrices. The effectiveness of the proposed stability criteria is demonstrated using a numerical example.

2 System Description and Problem Formulation

Consider a class of discrete-time Lur'e system given by

$$\left.\begin{aligned}
x(k+1) &= Ax(k) + Bx(k-d(k)) + Du(k), \\
z(k) &= Mx(k) + Nx(k-d(k)), \\
u(k) &= -\varphi(k, z(k)), \\
x(k) &= \Phi(k), \ k = -d_2, \ -d_2+1, \ \ldots, \ 0
\end{aligned}\right\} \tag{1}$$

where $x(k) \in \mathbb{R}^n$, $u(k) \in \mathbb{R}^m$, and $z(k) \in \mathbb{R}^m$ are the state, input and output vectors of the system respectively; A, B, D, M and N are constant matrices of appropriate dimensions; the sequence $\Phi(k)$ is the initial condition; the time-varying delay $d(k)$ satisfies the following condition:

$$0 < d_1 \le d(k) \le d_2, \tag{2}$$

where d_1 and d_2 are non-negative integers representing the lower and upper bounds of the interval time-delay; it is assumed that $d_1 \ne d_2$. The nonlinear function $\varphi(k, x(k)) \in \mathbb{R}^m$ is discrete in k, and is globally Lipschitz in $z(k)$; they satisfy $\varphi(k, 0) = 0$, and the following sector condition:

$$(\varphi(k, z(k)) - K_1 z(k))^T (\varphi(k, z(k)) - K_2 z(k)) \le 0, \tag{3}$$

where K_1 and K_2 are real constant matrices of appropriate dimensions, and $K = K_2 - K_1$ is a symmetric positive-definite matrix. In other words, the nonlinear function $\varphi(k, z(k))$ is said to belong to the sector $[K_1, \ K_2]$. A schematic of $\varphi(k, z(k))$ versus $z(k)$ confined to the sector $[K_1, \ K_2]$ is shown in Fig. 1. We now state the definition of absolute stability of the system (1).

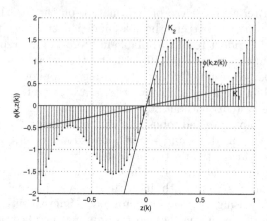

Fig. 1. Plot of nonlinear function $\varphi(k, z(k))$ versus $z(k)$

Definition 1. The discrete-time Lur'e system described by (1) and subject to (2) and (3) is said to be absolutely stable in the sector $[K_1, K_2]$, if the trivial solution $x(k) = 0$ is globally uniformly asymptotically stable for any nonlinear function $\varphi(k, z(k))$ satisfying (3).

For robust stability analysis, we consider the following uncertain Lur'e system:

$$\left.\begin{aligned}
x(k+1) &= (A + \Delta A(k))x(k) + (B + \Delta B(k))x(k - d(k)) + Du(k), \\
z(k) &= Mx(k) + Nx(k - d(k)), \\
u(k) &= -\varphi(k, z(k)), \\
x(k) &= \Phi(k), \ k = -d_2, \ -d_2 + 1, \ \ldots, \ 0
\end{aligned}\right\} \quad (4)$$

where the time-varying norm-bounded uncertainties are of the form:

$$\begin{bmatrix} \Delta A(k) & \Delta B(k) \end{bmatrix} = L\Delta(k)\begin{bmatrix} E_a & E_b \end{bmatrix}, \quad (5)$$

where L, E_a and E_b are known constant matrices of appropriate dimensions, and $\Delta(k)$ is an unknown real time-varying matrix satisfying

$$\Delta^T(k)\Delta(k) \le I. \quad (6)$$

This paper investigates the delay-dependent stability of the systems (1) and (4) satisfying the time-varying delay (2) and the sector-condition (3). Subsequently, less conservative stability criteria (sufficient conditions) are formulated in LMI framework for estimating the maximum allowable bound of the delay-range $[d_1, d_2]$, within which, the systems (1) and (4) remain absolutely stable and robustly stable respectively in the sense of Lyapunov. Following lemmas are indispensable in deriving the proposed stability criteria:

Lemma 1. [14] (Discrete Jenson inequality): For any constant matrix $W \in \mathbb{R}^{n \times n}$ with $W = W^T > 0$, integers $l_1 < l_2$, vector function $\omega : \{l_1, \, l_1 + 1, \, \ldots, \, l_2\} \mapsto \mathbb{R}^m$ such that the sums concerned are well defined, then

$$(l_2 - l_1 + 1) \sum_{i=l_1}^{l_2} \omega^T(i) W \omega(i) \geq \left(\sum_{i=l_1}^{l_2} \omega(i) \right)^T W \left(\sum_{i=l_1}^{l_2} \omega(i) \right). \tag{7}$$

Lemma 2. (Discrete Newton-Leibniz formula): If $u(i+1) - u(i) = t(i)$, then

$$\sum_{i=v}^{w} t(i) = u(w+1) - u(v). \tag{8}$$

Lemma 3. [14] Let U, $V(k)$, W and Q be real matrices of appropriate dimensions with Q satisfying $Q = Q^T$, then

$$Q + UV(k)W + W^T V^T(k) U^T < 0, \ \forall \ V^T(k) V(k) \leq I \tag{9}$$

if and only if there exists a scalar $\epsilon > 0$ such that

$$Q + \epsilon^{-1} U U^T + \epsilon W^T W < 0. \tag{10}$$

3   Main Result

In the following theorem, we state the absolute stability criterion for the system (1) with the nonlinear function $\varphi(k, z(k))$ belonging to the sector $[0, \, K]$, where $K > 0$ i.e. $\varphi(k, z(k))$ satisfies the following condition:

$$\varphi(k, z(k))^T (\varphi(k, z(k)) - K z(k)) \leq 0. \tag{11}$$

Theorem 1. The system (1) satisfying the time-varying delay (2) and sector condition (11) is absolutely stable for a given value of d_1 and d_2 with $d_{12} = d_2 - d_1$, if there exist real symmetric positive-definite matrices P, Q_i, $i = 1, 2, 3$, Z_j, $j = 1, 2$; matrices T_i, Y_i, $i = 1, 2, 3$ of appropriate dimensions such that the following LMIs hold:

$$\Upsilon_T = \begin{bmatrix} \Pi + \Pi_1 + \Pi_1^T & \bar{A}^T P & d_1 \hat{A}^T Z_1 & \hat{A}^T Z_2 & T_a \\ \star & -P & 0 & 0 & 0 \\ \star & \star & -Z_1 & 0 & 0 \\ \star & \star & \star & -\frac{Z_2}{d_{12}} & 0 \\ \star & \star & \star & \star & -\frac{Z_2}{d_{12}} \end{bmatrix} < 0, \tag{12}$$

$$\Upsilon_Y = \begin{bmatrix} \Pi + \Pi_1 + \Pi_1^T & \bar{A}^T P & d_1 \hat{A}^T Z_1 & \hat{A}^T Z_2 & Y_a \\ \star & -P & 0 & 0 & 0 \\ \star & \star & -Z_1 & 0 & 0 \\ \star & \star & \star & -\frac{Z_2}{d_{12}} & 0 \\ \star & \star & \star & \star & -\frac{Z_2}{d_{12}} \end{bmatrix} < 0, \tag{13}$$

where

$$
\Pi = \begin{bmatrix}
\Pi_{11} & 0 & Z_1 & 0 & -M^T K^T \\
\star & -Q_3 & 0 & 0 & -N^T K^T \\
\star & \star & -Q_1 - Z_1 & 0 & 0 \\
\star & \star & \star & -Q_2 & 0 \\
\star & \star & \star & \star & -2I
\end{bmatrix},
$$

$$\Pi_1 = \begin{bmatrix} 0 & -Y_a + T_a & Y_a & -T_a & 0 \end{bmatrix}, \quad Y_a = \begin{bmatrix} 0 & Y_1^T & Y_2^T & Y_3^T & 0 \end{bmatrix}^T,$$

$$T_a = \begin{bmatrix} 0 & T_1^T & T_2^T & T_3^T & 0 \end{bmatrix}^T, \quad \bar{A} = \begin{bmatrix} A & B & 0 & 0 & D \end{bmatrix}, \quad \hat{A} = \begin{bmatrix} A - I & B & 0 & 0 & D \end{bmatrix},$$

with $\Pi_{11} = -P + Q_1 + Q_2 + (d_{12} + 1)Q_3 - Z_1$.

Proof. Consider the Lyapunov functional $V(k) = \sum_{i=1}^{5} V_i(k)$ with

$$V_1(k) = x^T(k)Px(k), \tag{14}$$

$$V_2(k) = \sum_{i=k-d_1}^{k-1} x^T(i)Q_1x(i) + \sum_{i=k-d_2}^{k-1} x^T(i)Q_2x(i) + \sum_{i=k-d(k)}^{k-1} x^T(i)Q_3x(i), \tag{15}$$

$$V_3(k) = \sum_{i=-d_2+1}^{-d_1} \sum_{j=k+i}^{k-1} x^T(j)Q_3x(j), \tag{16}$$

$$V_4(k) = d_1 \sum_{i=-d_1}^{-1} \sum_{j=k+i}^{k-1} \eta^T(j)Z_1\eta(j), \tag{17}$$

$$V_5(k) = \sum_{i=-d_2}^{-d_1-1} \sum_{j=k+i}^{k-1} \eta^T(j)Z_2\eta(j), \tag{18}$$

where P, Q_i, $i = 1, 2, 3$, and Z_j, $j = 1, 2$ are real, symmetric, positive definite matrices and $\eta(j) = x(j + 1) - x(j)$. Define $\Delta V(k) = V(k + 1) - V(k)$. Then, along the solution of (1), we have,

$$
\begin{aligned}
\Delta V_1(k) = {} & x^T(k)(A^T PA - P)x(k) + 2x^T(k)A^T PBx(k - d(k)) \\
& + 2x^T(k)A^T PDu(k) + x^T(k - d(k))B^T PBx(k - d(k)) \\
& + 2x^T(k - d(k))B^T PDu(k) + u^T(k)D^T PDu(k),
\end{aligned}
\tag{19}
$$

$$
\begin{aligned}
\Delta V_2(k) \leq {} & x^T(k)(Q_1 + Q_2 + Q_3)x(k) \\
& - x^T(k - d_1)Q_1x(k - d_1) - x^T(k - d_2)Q_2x(k - d_2) \\
& - x^T(k - d(k))Q_3x(k - d(k)) + \sum_{j=k+1-d_2}^{k-d_1} x^T(j)Q_3x(j),
\end{aligned}
\tag{20}
$$

$$\Delta V_3(k) = d_{12}x^T(k)Q_3x(k) - \sum_{j=k+1-d_2}^{k-d_1} x^T(j)Q_3x(j), \tag{21}$$

$$\Delta V_4(k) = d_1^2\eta^T(k)Z_1\eta(k) - d_1 \sum_{j=k-d_1}^{k-1} \eta^T(j)Z_1\eta(j), \tag{22}$$

$$\Delta V_5(k) = d_{12}\eta^T(k)Z_2\eta(k) - \sum_{j=k-d_2}^{k-d_1-1} \eta^T(j)Z_2\eta(j). \tag{23}$$

The term $-d_1\sum_{j=k-d_1}^{k-1}\eta^T(j)Z_1\eta(j)$ in $\Delta V_4(k)$ is dealt using Lemma 1 and Lemma 2 as follows:

$$-d_1\sum_{j=k-d_1}^{k-1}\eta^T(j)Z_1\eta(j) \leq -\left(\sum_{j=k-d_1}^{k-1}\eta(j)\right)^T Z_1\left(\sum_{j=k-d_1}^{k-1}\eta(j)\right)$$

$$\leq -(x(k) - x(k-d_1))^T Z_1(x(k) - x(k-d_1)). \tag{24}$$

For handling the term $-\sum_{k-d_2}^{k-d_1-1}\eta^T(j)Z_2\eta(j)$ of $\Delta V_5(k)$, we make use of some free matrices $T = [T_1^T \ T_2^T \ T_3^T]^T$ and $Y = [Y_1^T \ Y_2^T \ Y_3^T]^T$ of appropriate dimensions. Define $\delta(k) = [x^T(k-d(k)) \ x^T(k-d_1) \ x^T(k-d_2)]^T$. Now, for any vectors z, y, and symmetric, positive definite matrix X, the following inequality holds:

$$-2z^Ty \leq z^TX^{-1}z + y^TXy. \tag{25}$$

Substituting $z = T^T\delta(k)$, $y = \sum_{j=k-d_2}^{k-d(k)-1}\eta(k)$ and $X = \frac{Z_2}{d_2-d(k)}$ in (25), we get

$$-2\delta^T(k)T\sum_{j=k-d_2}^{k-d(k)-1}\eta(k) \leq (d_2 - d(k))\delta^T(k)TZ_2^{-1}T^T\delta(k)$$

$$+\left(\sum_{j=k-d_2}^{k-d(k)-1}\eta(k)\right)^T\left(\frac{Z_2}{d_2-d(k)}\right)\left(\sum_{j=k-d_2}^{k-d(k)-1}\eta(k)\right) \tag{26}$$

which, using Lemma 2, can be expressed as follows:

$$-\sum_{j=k-d_2}^{k-d(k)-1}\eta^T(k)Z_2\eta(k) \leq (d_2 - d(k))\delta^T(k)TZ_2^{-1}T^T\delta(k)$$

$$+\delta^T(k)\left[[T \ 0 \ -T] + [T \ 0 \ -T]^T\right]\delta(k). \tag{27}$$

By analogy, we can derive the following inequality as well:

$$-\sum_{j=k-d(k)}^{k-d_1-1}\eta^T(k)Z_2\eta(k) \leq (d(k) - d_1)\delta^T(k)YZ_2^{-1}Y^T\delta(k)$$

$$+\delta^T(k)\left[[-Y \ Y \ 0] + [-Y \ Y \ 0]^T\right]\delta(k). \tag{28}$$

Summation of (27) and (28) yields the following bounding condition:

$$-\sum_{j=k-d_2}^{k-d_1-1}\eta^T(j)Z_2\eta(j) \leq \delta^T(k)\left[(d(k) - d_1)YZ_2^{-1}Y^T + (d_2 - d(k))TZ_2^{-1}T^T\right.$$

$$+\left.\left[[-Y+T \ Y \ -T] + [-Y+T \ Y \ -T]^T\right]\right]\delta(k). \tag{29}$$

Now, it follows from (11) that

$$- u^T(k)u(k) - u^T(k)K(Mx(k) + Nx(k - d(k))) \geq 0. \qquad (30)$$

By defining $\Sigma(k) = [x^T(k) \ \ \delta^T(k) \ \ u^T(k)]^T$, and substituting (24) and (29) respectively in (22) and (23), and adding the positive quantity in (30) to the resulting equations, we get

$$\Delta V(k) = \sum_{n=1}^{5} \Delta V_n(k) \leq \Sigma^T(k)(\Pi + \Pi_1 + \Pi_1^T + \bar{A}^T P\bar{A} + \hat{A}^T(d_1^2 Z_1)\hat{A}$$

$$+ \hat{A}^T(d_{12}Z_2)\hat{A} + (d(k) - d_1)Y_a Z_2^{-1} Y_a^T$$

$$+ (d_2 - d(k))T_a Z_2^{-1} T_a^T)\Sigma(k). \qquad (31)$$

If

$$\Pi + \Pi_1 + \Pi_1^T + \bar{A}^T P\bar{A} + \hat{A}^T(d_1^2 Z_1)\hat{A} + \hat{A}^T(d_{12}Z_2)\hat{A}$$

$$+ (d(k) - d_1)Y_a Z_2^{-1} Y_a^T + (d_2 - d(k))T_a Z_2^{-1} T_a^T < 0, \qquad (32)$$

then, there exists a sufficient small scalar $\alpha > 0$ such that $\Delta V(k) \leq -\alpha\|x(k)\|^2$, which, in turn, implies that the discrete-time system in (1) is absolutely stable in the sense of Lyapunov. By solving the convex LMI condition (32) non-conservatively at boundary conditions ($d(k) = d_1$ and $d(k) = d_2$), and by applying Schur complement to the resulting equations, we deduce the LMIs stated in Theorem 1. This completes the proof of Theorem 1.

Remark 1. In order to handle the nonlinearity $\varphi(k, z(k))$ belonging to the sector $[K_1, K_2]$, we apply loop transformation [15] and convert absolute stability of the system (1) in the sector $[K_1, K_2]$ to that of the following system:

$$\left.\begin{array}{l} x(k+1) = (A - DK_1M)x(k) + (B - DK_1N)x(k - d(k)) + D\bar{u}(k), \\ z(k) = Mx(k) + Nx(k - d(k)), \\ \bar{u}(k) = -\bar{\varphi}(k, z(k)), \\ x(k) = \Phi(k), \ k = -d_2, \ -d_2 + 1, \ \ldots, \ 0 \end{array}\right\} \qquad (33)$$

in the sector $[0, \ K_2 - K_1]$ i.e. $\bar{\varphi}(k, z(k))$ satisfies the following sector condition:

$$\bar{\varphi}(k, z(k))^T(\bar{\varphi}(k, z(k)) - (K_2 - K_1)z(k)) \leq 0. \qquad (34)$$

For this case, by Theorem 1, we have the following corollary for ascertaining delay-dependent absolute stability:

Corollary 1. The system (1) satisfying the time-varying delay (2) and sector condition (3) is absolutely stable for a given value of d_1 and d_2 with $d_{12} = d_2 - d_1$, if there exist real symmetric positive-definite matrices P, Q_i, $i = 1, 2, 3$, Z_j, $j = 1, 2$; matrices T_i, Y_i, $i = 1, 2, 3$ of appropriate dimensions such that the following LMIs hold:

$$
\Upsilon'_T =
\begin{bmatrix}
\Pi' + \Pi_1 + \Pi_1^T \bar{\bar{A}}^T P \, d_1 \hat{A}^T Z_1 \, \hat{A}^T Z_2 & T_a \\
\star & -P & 0 & 0 & 0 \\
\star & \star & -Z_1 & 0 & 0 \\
\star & \star & \star & -\frac{Z_2}{d_{12}} & 0 \\
\star & \star & \star & \star & -\frac{Z_2}{d_{12}}
\end{bmatrix} < 0,
\tag{35}
$$

$$
\Upsilon'_Y =
\begin{bmatrix}
\Pi' + \Pi_1 + \Pi_1^T \bar{\bar{A}}^T P \, d_1 \hat{A}^T Z_1 \, \hat{A}^T Z_2 & Y_a \\
\star & -P & 0 & 0 & 0 \\
\star & \star & -Z_1 & 0 & 0 \\
\star & \star & \star & -\frac{Z_2}{d_{12}} & 0 \\
\star & \star & \star & \star & -\frac{Z_2}{d_{12}}
\end{bmatrix} < 0,
\tag{36}
$$

where

$$
\Pi' =
\begin{bmatrix}
\Pi_{11} & 0 & Z_1 & 0 & -M^T(K_2 - K_1)^T \\
\star & -Q_3 & 0 & 0 & -N^T(K_2 - K_1)^T \\
\star & \star & -Q_1 - Z_1 & 0 & 0 \\
\star & \star & \star & -Q_2 & 0 \\
\star & \star & \star & \star & -2I
\end{bmatrix},
$$

$$
\bar{\bar{A}} = [A - DK_1M \; B - DK_1N \; 0 \; 0 \; D],
$$

$$
\hat{A} = [A - DK_1M - I \; B - DK_1N \; 0 \; 0 \; D].
$$

4 Robust Stability Criteria

The robust stability criteria for the uncertain system (4) satisfying the interval time-delay (2) and sector condition (11) is stated below:

Theorem 2. The uncertain system (4) satisfying the time-varying delay (2) and sector condition (11) is robustly stable for a given value of d_1 and d_2 with $d_{12} = d_2 - d_1$, if there exist real symmetric positive-definite matrices P, Q_i, $i = 1, 2, 3$, Z_j, $j = 1, 2$, scalar $\epsilon > 0$ and matrices T_i, Y_i, $i = 1, 2, 3$ of appropriate dimensions such that the following LMIs hold:

$$
\begin{bmatrix}
\Upsilon_T & \bar{P}L & \epsilon\bar{E}^T \\
\star & -\epsilon I & 0 \\
\star & \star & -\epsilon I
\end{bmatrix} < 0, \quad
\begin{bmatrix}
\Upsilon_Y & \bar{P}L & \epsilon\bar{E}^T \\
\star & -\epsilon I & 0 \\
\star & \star & -\epsilon I
\end{bmatrix} < 0,
$$

where $\bar{P} = [0 \; 0 \; 0 \; 0 \; 0 \; P \; d_1 Z_1 \; Z_2 \; 0]^T$ and $\bar{E} = [E_a \; E_b \; 0 \; 0 \; 0 \; 0 \; 0 \; 0]$.

Proof. For the uncertain system (4), from Theorem 1, following inequalities hold:

$$
\Upsilon_T + \bar{P}\Delta(k)\bar{E} + \bar{E}^T\Delta^T(k)\bar{P}^T < 0,
$$
$$
\Upsilon_Y + \bar{P}\Delta(k)\bar{E} + \bar{E}^T\Delta^T(k)\bar{P}^T < 0.
$$

Now, by successively applying Lemma 3 and Schur complement, we deduce the LMIs stated in Theorem 2. Similarly, following result can be readily deduced:

Corollary 2. The uncertain system (4) satisfying the time-varying delay (2) and sector condition (3) is robustly stable for a given value of d_1 and d_2 with $d_{12} = d_2 - d_1$, if there exist real symmetric positive-definite matrices P, Q_i, $i = 1, 2, 3$, Z_j, $j = 1, 2$, scalar $\epsilon > 0$ and matrices T_i, Y_i, $i = 1, 2, 3$ of appropriate dimensions such that the following LMIs hold:

$$\begin{bmatrix} \Upsilon_T' & \bar{P}L & \epsilon \bar{E}^T \\ \star & -\epsilon I & 0 \\ \star & \star & -\epsilon I \end{bmatrix} < 0, \qquad \begin{bmatrix} \Upsilon_Y' & \bar{P}L & \epsilon \bar{E}^T \\ \star & -\epsilon I & 0 \\ \star & \star & -\epsilon I \end{bmatrix} < 0.$$

5 Numerical Example

Example 1. Consider the nominal discrete-time Lur'e system in (1) with

$$A = \begin{bmatrix} 0.8 & 0 \\ 0.05 & 0.9 \end{bmatrix}, \quad B = \begin{bmatrix} -0.1 & 0 \\ -0.2 & -0.1 \end{bmatrix}, \quad D = \begin{bmatrix} -0.02 \\ -0.03 \end{bmatrix},$$

$$M = \begin{bmatrix} 0.3 & 0.1 \end{bmatrix}, \quad N = \begin{bmatrix} 0.1 & 0.2 \end{bmatrix}, \quad K_1 = 0.2, \quad K_2 = 0.5.$$

For a given d_1, the maximum allowable bound for d_2 provided by Corollary 1 that ensures the absolute stability of the system is given in Table 1. For demonstrating the effectiveness of the proposed robust stability criterion in Corollary 2, consider the uncertain discrete-time Lur'e system in (4) with the following matrices representing parametric uncertainties:

$$L = E_a = E_b = \begin{bmatrix} 0.1 & 0 \\ 0 & 0.1 \end{bmatrix}.$$

In this case, for a given d_1, the maximum allowable bound for d_2 that ensures the robust stability of the uncertain system is also given in Table 1. From the table,

Table 1. Upper Delay bound d_2 for different d_1

Method	d_1	4	6	8	10	12	15	17	18	20
Corollary 1	d_2	17	18	19	20	21	23	25	25	27
Corollary 2	d_2	12	13	14	15	16	18	20	21	22

we infer that in the presence of parametric uncertainties, the maximum allowable delay-range bound that ensures the delay-dependent stability of the uncertain system reduces compared to the nominal system without uncertainties.

6 Conclusion

In this paper, the problem of delay-dependent stability of a class of discrete-time Lur'e systems with interval time-delay and sector-bounded nonlinearity has been considered based on Lyapunov approach. By exploiting a candidate Lyapunov functional, and using slack matrix variables in the delay-dependent stability analysis, less conservative stability criteria are developed for discrete-time Lur'e systems in LMI framework. For deriving robust stability conditions, time-varying norm-bounded uncertainties are considered in the system matrices. Finally, to demonstrate the effectiveness of the proposed results for nominal and uncertain Lur'e systems, a numerical example is considered.

References

1. Popov, V.M.: Hyperstability of Control Systems. Springer, New York (1973)
2. Khalil, H.K.: Nonlinear Systems. Prentice-Hall, Upper Saddle River (1996)
3. Liao, X.X.: Absolute Stability of Nonlinear Control Systems. Science Press, Beijing (1993)
4. Yalcin, M.E., Suykens, J.A.K., Vandewalle, J.: Master-slave synchronization of Lur'e systems with time-delay. Internat. J. Bifurcation Chaos 11, 1707–1722 (2001)
5. Liao, X., Chen, G.: Chaos synchronization of general Lur'e systems via time-delay feedback control. Internat. J. Bifurcation Chaos 13, 207–213 (2003)
6. Cao, J., Li, H.X., Ho, D.W.C.: Synchronization criteria of Lur'e system with time-delay feedback control. Chaos Solitons Fractals 23, 1285–1298 (2005)
7. He, Y., Wu, M.: Absolute stability for multiple delay general Lur'e control systems with multiple nonlinearities. J. Comput. Appl. Math. 159, 241–248 (2003)
8. He, Y., Wu, M., She, J.H., Liu, G.P.: Robust stability for delay Lur'e systems with multiple nonlinearities. J. Comput. Appl. Math. 176, 371–380 (2005)
9. Han, Q.L., Yue, D.: Absolute stability of Lur'e systems with time-varying delay. IET Control Theory Appl. 1, 854–859 (2007)
10. Ramakrishnan, K., Ray, G.: Delay-range-dependent stability criteria for Lur'e system with interval time-varying delay. In: Proceedings of 49th IEEE Conf. Decision Cont., pp. 170–175 (2010)
11. Ramakrishnan, K., Ray, G.: Improved delay-range-dependent robust stability criteria for a class of Lur'e systems with sector-bounded nonlinearity. J. Franklin Inst. 348, 1769–1786 (2011)
12. Hao, F.: Absolute stability of uncertain discrete Lur'e systems and maximum admissible perturbed bounds. J. Franklin Inst. 347, 1511–1525 (2010)
13. Lee, S.M., Park, J.H.: Robust stabilization of discrete-time nonlinear Lur'e systems with sector and slope restricted nonlinearities. Appl. Math. Comput. 200, 429–436 (2008)
14. Huang, H., Feng, G.: Improved approach to delay-dependent stability analysis of discrete-time systems with time-varying delay. IET Cont. Theory Appl. 4, 2152–2159 (2010)
15. Yang, C., Zhang, Q., Zhou, L.: Strongly absolute stability problem of descriptor systems: Circle criterion. J. Franklin Inst. 345, 437–451 (2008)

Temporally Dependent Solute Dispersion in One-Dimensional Porous Media

Dilip Kumar Jaiswal[*]

Department of Mathematics & Astronomy, Lucknow University
Lucknow 226007, India
dilip3jais@gmail.com

Abstract. Analytical solutions are obtained for temporally dependent dispersion along a uniform flow velocity in a one-dimensional semi-infinite domain by using Laplace integral transform technique. Initially the domain is not solute free. It is combination of exponentially increasing function of space variable and ratio of zero order production and first decay which are inversely proportional to dispersion coefficient. Retardation factor is also considered. The solutions are obtained for two cases, first one for uniform and second for increasing input source. The variable coefficients in the advection–dispersion equation are reduced into constant coefficients by introducing new space and time variables. Illustrations are given with different graphs.

Keywords: Advection, Dispersion, Porous media, Laplace Transform.

1   Introduction

Earth's biosphere is endowed with extremely diverse kinds of environments which provide countless goods and services to human kind. Any component of the natural environment that can be utilized to promote for welfare is considered as a natural resource. Land, soil, water, forests, grasslands, etc. are examples of important natural resources. Some of the resources (e.g. soil, water, and atmosphere) are important components of the life-supporting system. Pollution is an undesirable change in physical, chemical or biological characteristics of our air, land or water caused by excessive accumulation of pollutants (substances causing pollution). Pollution affects biological species, humans and damages our industrial processes, living conditions.

Polluted environment where it occurs most (atmosphere, hydrosphere and lithosphere) it can be classified as air pollution, water pollution and soil pollution. Pollution may be natural (e.g. volcanic eruptions which add tons of toxic gases and particulate matter in the environment) or anthropogenic (man- made, such as industrial pollution, agricultural pollution, etc.). A list of previous works [1-3] related to pollution on natural resources, considered the uniform flow through a homogeneous and isotropic porous domain. Most of such works have been complied [4]. After that, in-homogeneity of the finite and the semi-infinite porous domains was defined and the dispersion problems have been solved numerically [5-6]. The transports of pollutants are subject to

---

[*] Corresponding author.

P. Balasubramaniam and R. Uthayakumar (Eds.): ICMMSC 2012, CCIS 283, pp. 220–228, 2012.

physical, chemical and biological activities, such as adsorption and desorption, retardation, degradation and chemical-biological reactions [7-10]. The works [11-12] included
a full, physically-based porous media transport model. But they simplified the solute
exchange between the soil and overland flow to a mixing-layer concept, which does
not explicitly account for solute diffusion from deeper soil.

The temporal moment solutions are obtained for one-dimensional advective-
dispersive solute transport with linear equilibrium sorption and first-order degradation
for time pulse sources to analyze soil column experimental data [13]. An analytical
solution has presented for the non-stationary two-dimensional advection-diffusion
equation to simulate the pollutant dispersion in the planetary boundary layer and
solved advection-diffusion equation by the application of the Laplace transform technique and the solution of the resulting stationary problem by the generalized integral
Laplace transform technique (GILTT) [14]. Analytical solutions are obtained for temporally and spatially dependent solute dispersion in semi-infinite porous domain by
using Laplace transform technique [15-19].

In the present study, analytical solutions are obtained for temporally dependent solute dispersion along a uniform flow in a semi-infinite medium for uniform and increasing input source condition.

2   Mathematical Model and Analytical Solutions

One-dimensional linear advection-dispersion equation with first order decay, zero
order production and retardation factor, may be written as

$$R\frac{\partial C}{\partial t} = \frac{\partial}{\partial x}\left[D(x,t)\frac{\partial C}{\partial x} - u(x,t)C\right] - \mu(x,t)C + \gamma(x,t) \qquad (1)$$

where C is the solute concentration at position x at time t, $D(x,t)$ represents the
solute dispersion and is called the dispersion coefficient if it is uniform and steady,
and $u(x,t)$ is the velocity of the medium transporting the solute particles.

Let us write $D(x,t)$ and $u(x,t)$ in Eq. (1) as

$$D(x,t) = D_0 f(mt) \quad \text{and} \quad u(x,t) = u_0, \qquad (2)$$

respectively, where m is a unsteady coefficient whose dimension is inverse of that of
the time variable t. $f(mt)$ is chosen such that for $m = 0$ or $t = 0$, $f(mt) = 1$. Thus
$f(mt)$ is an expression in the non-dimensional variable mt. $m = 0$ corresponds to
the temporally independent dispersion. Therefore $\mu(x,t)$ and $\gamma(x,t)$ become $\dfrac{\mu_0}{f(mt)}$

and $\dfrac{\gamma_0}{f(mt)}$ respectively, where D_0, u_0, μ_0 and γ_0 are constants. Eq. (1) is re-

written as

$$R\frac{\partial C}{\partial t} = \frac{\partial}{\partial x}\left[D_0 f(mt)\frac{\partial C}{\partial x} - u_0 C\right] - \frac{\mu_0}{f(mt)}C + \frac{\gamma_0}{f(mt)} \qquad (3)$$

Let us introduce a new independent variable, X by a transformation [15],

$$\frac{dX}{dx} = \frac{1}{f(mt)} \quad \text{or} \quad X = \int \frac{dx}{f(mt)} \tag{4}$$

Applying the transformation (4) on the partial differential equation, Eq. (3) becomes

$$Rf(mt)\frac{\partial C}{\partial t} = \frac{\partial}{\partial X}\left[D_0 \frac{\partial C}{\partial X} - u_0 C\right] - \mu_0 C + \gamma_0 \tag{5}$$

The partial differential equation (5) is solved analytically using with prescribed initial and boundary conditions for temporally dependent dispersion along a uniform flow. The domain is longitudinal semi-infinite but it is not solute free at the initial stage. Uniform and increasing input source concentrations are introduced at the origin of the domain.

Further another independent variable, T is introduced by the transformation, [20]

$$T = \int_0^t \frac{dt}{Rf(mt)} \tag{6}$$

Eq. (5) reduces to a partial differential equation with constant coefficients as,

$$\frac{\partial C}{\partial T} = D_0 \frac{\partial^2 C}{\partial X^2} - u_0 \frac{\partial C}{\partial X} - \mu_0 C + \gamma_0 \tag{7}$$

The dimension of variable X defined by Eq. (4) and variable T defined by Eq. (6) remain those of x and t, respectively and are referred to as the new space and new time variables, respectively. The initial solute concentration for one-dimensional advection-dispersion equation is assumed as,

$$C(x,t) \equiv A(x) = \frac{\gamma(x,t)}{\mu(x,t)} + \left(C_i - \frac{\gamma(x,t)}{\mu(x,t)}\right)\exp\left\{\frac{u(x,t) - v(x,t)}{2D(x,t)}x\right\}; \ x \geq 0, \ t = 0 \tag{8}$$

where $v(x,t) = u(x,t)\left\{1 + \frac{4\mu(x,t)D(x,t)}{u(x,t)^2}\right\}^{\frac{1}{2}}$,

Using expression (2), $\mu(x,t) = \frac{\mu_0}{f(mt)}$ and $\gamma(x,t) = \frac{\gamma_0}{f(mt)}$, initial condition becomes,

$$C(x,t) \equiv A(x) = \frac{\gamma_0}{\mu_0} + \left(C_i - \frac{\gamma_0}{\mu_0}\right)\exp\left\{\frac{u_0 - v_0}{2D_0}x\right\} \tag{9}$$

where $\quad v_0 = u_0\left\{1 + \frac{4\mu_0 D_0}{u_0^2}\right\}^{\frac{1}{2}} \tag{10}$

For $\mu_0 = 0$ in Eq. (10) and $\gamma_0 = 0$ in Eq. (9), the medium is supposed to possess an uniform solute concentration C_i, before an input concentration is introduced. Because in real scenario, the medium is already concentrated, so this type of initial condition is taken into account in the present study. The second boundary condition is assumed to be the flux type for both the input condition i.e. for uniform and increasing input conditions, which is

$$\frac{\partial C}{\partial x} = 0 \; ; \; x \to \infty \; , \; t \geq 0 \tag{11}$$

The input source can be broadly categorized into point sources and non-point sources. Point source, where the effluent discharge occurs at a specific site (sewage outlet of a municipal area or effluent outlet of a factory) while non-point source, occur over a large area (city storm water flow, agricultural runoff, constructional sediments, etc).

2.1   Uniform Input Source

An input concentration of the uniform nature is introduced at the origin. Thus the first boundary condition is as follows:

$$C(x,t) = C_0 \; ; \; x = 0 \; , \; t > 0, \tag{12}$$

The conditions (10) - (12) in terms of new space and time variables, X and T may be written as

$$C(X,T) = \frac{\gamma_0}{\mu_0} + \left(C_i - \frac{\gamma_0}{\mu_0} \right) \exp\left\{ \frac{u_0 - v_0}{2D_0} X \right\}; \; X \geq 0 \; , \; T = 0 \tag{13}$$

$$C(X,T) = C_0 \; ; \; X = 0 \; , \; T > 0 \tag{14}$$

$$\frac{\partial C}{\partial X} = 0 \; ; \; X \to \infty \; , T \geq 0 \tag{15}$$

Applying Laplace transformation technique (LTT) on Eq. (7) and Eqs. (13) – (15), one may get the desired analytical solution [21], as

$$C(x,T) = A(x) + (C_0 - C_i)F(x,T) \tag{16}$$

where $A(x)$ is the same as the initial condition,

$$F(x,T) = \frac{C_0}{2}\left[\exp\left\{ \frac{\beta - \left(\beta^2 + \mu_0\right)^{1/2} x}{f(mt)\sqrt{D_0}} \right\} erfc\left\{ \frac{\dfrac{Rx}{f(mt)} - \left(u_0^2 + 4\mu_0 D_0\right)^{1/2} T}{2\sqrt{D_0 RT}} \right\} \right]$$

$$+ \left[\exp\left\{ \frac{\beta + \left(\beta^2 + \mu_0\right)^{1/2} x}{f(mt)\sqrt{D_0}} \right\} erfc\left\{ \frac{\dfrac{Rx}{f(mt)} + \left(u_0^2 + 4\mu_0 D_0\right)^{1/2} T}{2\sqrt{D_0 RT}} \right\} \right],$$

$$\beta^2 = \frac{u_0^2}{4D_0} \quad \text{and} \quad T = \int_0^t \frac{dt}{Rf(mt)} \; .$$

2.2   Increasing Input Source

This type of situation may be defined by a mixed type input condition which is

$$-D(x,t)\frac{\partial C}{\partial x}+u(x,t)C=u_0C_0 \quad ; \ x=0 \ ,t>0 \tag{17}$$

Using condition (17) in the place of (12) and using the expressions from Eqs. (2), and applying Laplace transformation technique (LTT), we may get the desired analytical solution [21], as

$$C(x,T)=A(x)+(C_0-C_i)G(x,T) \tag{18}$$

where $A(x)$ is the same as the initial condition,

$$G(x,T)=\frac{u_0}{2\sqrt{D_0}\left\{\beta+\left(\beta^2+\mu_0\right)^{1/2}\right\}}$$

$$exp\left\{\frac{\beta-\left(\beta^2+\mu_0\right)^{1/2}x}{f(mt)\sqrt{D_0}}\right\}erfc\left\{\frac{\dfrac{Rx}{f(mt)}-\left(u_0^2+4\mu_0D_0\right)^{1/2}T}{2\sqrt{D_0RT}}\right\}$$

$$+\frac{u_0}{2\sqrt{D_0}\left\{\beta-\left(\beta^2+\mu_0\right)^{1/2}\right\}}exp\left\{\frac{\beta+\left(\beta^2+\mu_0\right)^{1/2}x}{f(mt)\sqrt{D_0}}\right\}erfc\left\{\frac{\dfrac{Rx}{f(mt)}+\left(u_0^2+4\mu_0D_0\right)^{1/2}T}{2\sqrt{D_0RT}}\right\}$$

$$+\frac{u_0^2}{2\mu_0D_0}exp\left\{\frac{u_0x}{D_0f(mt)}-\mu_0T\right\}erfc\left\{\frac{\dfrac{Rx}{f(mt)}+u_0T}{2\sqrt{D_0RT}}\right\},$$

$$\beta^2=\frac{u_0^2}{4D_0} \quad \text{and} \quad T=\int_0^t\frac{dt}{Rf(mt)} \ .$$

3   Numerical Examples and Discussion

The concentration values are evaluated from the analytical solutions described by Eqs. (16) and (18) in a finite domain $0\leq x\leq 10$ (m) of semi-infinite region. The input values are considered as: $C_0=1.0$, $C_i=0.1$, $D_0=1.05$ (m^2/day), $u_0=0.34$ (m/day). In addition to these, $\mu=0.1$ (day^{-1}), $\gamma=0.01$ (kg m^{-3}day^{-1}) and $R=1.0$ are considered. Graphs are drawn for $t=2.4$, 3.0 and 3.6 (day) in the figures (1) and (2). In Fig. 1, concentration values are started from one at position $x=0.0$, which is show the boundary condition of uniform nature and concentration values decreases at a particular position along the domain with increasing time. In Fig. 2, concentration values are increases when time increases at same position and constant far from the origin. This type of nature shows the increasing input condition.

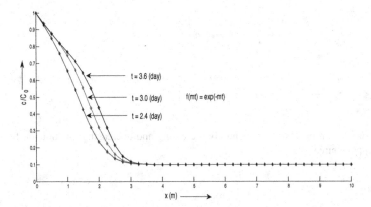

Fig. 1. Distribution of concentration at different time for uniform input source of exponentially decreasing function

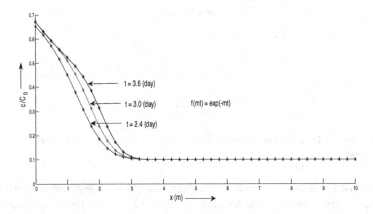

Fig. 2. Distribution of concentration at different time for increasing input source of exponentially decreasing function

Comparison between exponentially increasing and decreasing time dependent functions at a time $t = 3.0$ (day) are shown in Fig. 3 and Fig. 4, for uniform and increasing input source condition, respectively. In both the figures, the concentration values of increasing function is higher than decreasing function at particular position throughout the domain except some position. For other expression of $f(mt)$ these nature shows which cover our assumptions i.e for $m = 0$ or $t = 0$, $f(mt) = 1$.

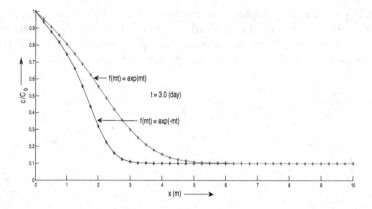

Fig. 3. Comparison between exponentially increasing and decreasing functions at fix time for uniform input source

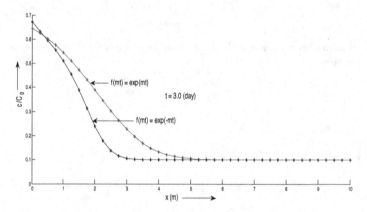

Fig. 4. Comparison between exponentially increasing and decreasing functions at fix time for increasing input source

The obtained solutions may help to determine the position and time to reach harmless concentration. The proposed solutions are applied to examine the applicability of numerical solutions of the time dependent dispersion model.

4   Conclusions

In the present study, solutions are obtained for two cases, first one for uniform and second for increasing input source for temporally dependent dispersion along a uniform flow velocity in a one-dimensional semi-infinite domain. Initially, the domain is not solute free. It is the combination of exponentially increasing function of space variable and ratio of zero order production and first decay which are inversely proportion to dispersion coefficient.

Acknowledgement. The author gratefully acknowledges for the financial assistance in the form of Dr. D. S. Kothari Post Doctoral Fellowship, University Grants Commission, India.

References

1. Banks, R.B., Ali, J.: Dispersion and adsorption in porous media flow. J. Hydraulics Div. 90, 13–31 (1964)
2. Ogata, A.: Theory of dispersion in a granular medium. U S Geological Survey, Pro. Pap. 411-I: 34 (1970)
3. Marino, M.: Flow against dispersion in non-adsorbing porous media. J. Hydrology 37, 149–158 (1978)
4. Lindstrom, F.T., Boersma, L.: Analytical solutions for convective dispersive transport in confined aquifers with different initial and boundary conditions. Water Resources Reseach 25, 241–256 (1989)
5. Lin, S.H.: Non-linear adsorbing in layered porous media flow. J. Hydraulics Div. 103, 951–958 (1977)
6. Kumar, N., Kumar, M.: Solute dispersion along unsteady groundwater flow in a semi-infinite aquifer. Hydrology Earth Syst. Sci. 2, 93–100 (1997)
7. Serrano, S.E.: Contaminant transport under non-linear sorption and decay. Water Research 35(6), 1525–1533 (2001)
8. Sen, T.K., Nalwaya, N., Khilar, K.C.: Colloid-associated contaminant transport in porous media: 2. Math. Modeling. AIChE Journal 48(10), 2375–2385 (2002)
9. Sandrin, S.K., Jordan, F.L., Maier, R.M., Brusseau, M.L.: Biodegradation during contaminant transport in porous media: 4. Impact of microbial lag and bacterial cell growth. J. Contanant Hydrology 50(3-4), 225–242 (2001)
10. Baverman, C., Moreno, L., Neretnieks, I.: A fast coupled geochemical and transport program and applications to waste leaching and contaminant transport. Internat. J. Rock Mechics Mining Sci. Geomechanics Abstracts 32(7), 317–318 (1995)
11. Wallach, R., van Genuchten, M.T.: A physically-based model for predicting solute transfer from soil solution to rainfall induced runoff water. Water Resources Research 26(9), 2119–2126 (1990)
12. Wallach, R., Grigorin, G., Rivlin, J.: A comprehensive mathematical model for transport of soil-dissolved chemicals by overland flow. J. Hydrol. 247(1-2), 85–99 (2001)
13. Pang, L., Goltz, M., Close, M.: Application of the method of temporal moments to interpret solute transport with sorption and degradation. J. Contaminant Hydrology 60, 123–134 (2003)
14. Moreira, D.M., Vilhena, M.T., Buske, D.E., Tirabassi, T.: The GILTT solution of the advection-diffusion equation for an inhomogeneous and non-stationary PBL. Atmos. Environ. 40, 3186–3194 (2006)
15. Jaiswal, D.K., Kumar, A., Kumar, N., Yadav, R.R.: Analytical solutions for temporally and spatially dependent solute dispersion of pulse type input concentration in one-dimensional semi-infinite media. J. Hydro-environment Research 2, 254–263 (2009)
16. Jaiswal, D.K., Kumar, A., Kumar, N., Singh, M.K.: Solute Transport along Temporally and Spatially Dependent flows through Horizontal Semi-infinite Media: Dispersion being proportional to square of velocity. J. Hydrologic Eng. (ASCE) 16(3), 228–238 (2011)

17. Kumar, A., Jaiswal, D.K., Kumar, N.: Analytical solutions to one-dimensional advection diffusion with variable coefficients in semi-infinite media. J. Hydrology 380(3-4), 330–337 (2010)
18. Yadav, R.R., Jaiswal, D.K., Yadav, H.K., Rana, G.: One-Dimensional temporally dependent advection-dispersion equation in porous media: Analytical solution. Natural Resources Modeling 23(4), 521–539 (2010)
19. Yadav, R.R., Jaiswal, D.K., Yadav, H.K., Rana, G.: Three-dimensional temporally dependent dispersion through porous media: Analytical solution. Environmental Earth Sciences (2011), doi:10.1007/s12665-011-1129-2
20. Crank, J.: The Mathematics of Diffusion. Oxford University Press, UK (1975)
21. van Genuchten, M.T., Alves, W.J.: Analytical solutions of the one dimensional convective-dispersive solute transport equation. US Department of Agriculture, Technical Bulletin No. 1661 (1982)

Instabilities and Patterns in Zooplankton-Phytoplankton Dynamics: Effect of Spatial Heterogeneity

Nilesh Kumar Thakur, Ranjit Kumar Upadhyay, and Sharada Nandan Raw[*]

Department of Applied Mathematics
Indian School of Mines, Dhanbad, Jharkhand - 826004, India
snr.ism@gmail.com

Abstract. In this paper, we have made an attempt to understand the instabilities and complex spatiotemporal patterns induced by spatial heterogeneity in a spatial Rosenzweig - McAurthur model for phytoplankton-zooplankton-fish interaction. We have examined the effect of heterogeneity, and fish predation on the stability of the model system. Based on these conditions and by performing a series of extensive simulations, we observed the irregular patterns in the presence of spatial heterogeneity and fish predation acts as a regularizing factor. Numerical simulation results reveal that the complex temporal dynamics in natural communities may arise through the spatial dimension. Spatially induced chaos may have an important role in spatial pattern generation in heterogeneous environments.

Keywords: Pattern formation, Spatial heterogeneity, Functional response, Diffusion, Chaotic dynamics.

1 Introduction

The fundamental importance of spatial and spatiotemporal pattern formation in aquatic ecology is self-evident. The issue of pattern formation in natural, social, ecological sciences focuses on how the nonlinearities conspire to form spatial patterns that are self-organized. Conceptual predator- prey models have successfully been used to model phytoplankton-zooplankton interactions and to elucidate mechanisms of spatiotemporal pattern formation like patchiness and blooming [1-5]. The spatiotemporal in prey-predator communities modeled by reaction-diffusion equations have always been an area of interest for researchers. The mechanisms of spatial pattern formation were suggested as a possible cause for the origin of planktonic patchiness in marine ecosystems [6]. Movement of phytoplankton and zooplankton population with different velocities can give rise to spatial patterns [7]. Malchow [8, 7] observed Turing patches in plankton community due to the effect of nutrients and planktivorous fish.

Recently, spatial heterogeneity of species has attracted much attention because it is closely related with the stability and coexistence of species in ecological systems.

---

[*] Corresponding author.

P. Balasubramaniam and R. Uthayakumar (Eds.): ICMMSC 2012, CCIS 283, pp. 229–236, 2012.
© Springer-Verlag Berlin Heidelberg 2012

Diffusion on a spatial gradient may drive a cyclic predator-prey system into chaotic state. Ecological models of diffusion driven instability with spatial heterogeneities have been studied for a variety of reasons [4, 9]. Upadhyay et al. [10, 11] had further extended these study by introducing environmental heterogeneity in the growth term (which varies linearly with x) of the model studied by Medvinsky et al. [9] in one space dimension. The transition between the regular and chaotic dynamics in reaction-diffusion systems was observed by Petrovskii and Malchow [12]. Sherratt *et al.* [13] have proposed a mechanism for generation of chaos. Nisbet *et al.* [14] observed how equilibrium states responded to spatial environmental heterogeneity in advective systems.

In this paper, we have considered a spatial Rosenzweig–McAurthur model with Holling type III functional responses for phytoplankton-zooplankton-fish interaction with spatial heterogeneity and self diffusion. The characteristic feature of Holling type III functional responses is that at low densities of the prey, the predator consumes it less proportionally than is available in the environment, relative to the predators' other prey [15]. We have investigated the effect of spatial heterogeneity on the stability of phytoplankton and zooplankton. Finally, we examine the effect of fish predation on the stability of the model system. Based on these conditions and by performing a series of simulations, we observed the complex spatiotemporal patterns. On fixed parameter space by increasing the fish predation rate we observed the stabilizing behavior of the model system.

2   The Mathematical Model

Consider a dimensionless form of phytoplankton-zooplankton-fish model where at any location (x, y) and time t, the phytoplankton $u(x, y, t)$ and zooplankton $v(x, y, t)$ populations satisfy the reaction- diffusion equation

$$\frac{\partial u}{\partial t} = r_x u (1-u) - \beta_1 \frac{u^2 v}{\left(u^2 + \alpha_1\right)} + d_1 \nabla^2 u, \tag{1a}$$

$$\frac{\partial v}{\partial t} = \beta_2 \frac{u^2 v}{\left(u^2 + \alpha_1\right)} - \gamma v - \frac{f v^2}{v^2 + \alpha_2} + d_2 \nabla^2 v, \tag{1b}$$

with initial condition

$$u(x, y, 0) > 0, \ v(x, y, 0) > 0 \text{ for } (x, y) \in \Omega = [0, L] \times [0, L], \tag{2}$$

$$\frac{\partial u}{\partial n} = \frac{\partial v}{\partial n} = 0, \ (x, y) \in \partial \Omega, \ t > 0 \tag{3}$$

where n is the outward normal to $\partial \Omega$.

The zero flux boundary condition (3) imply that no external input is imposed from outside. $u(x, y, t)$ and $v(x, y, t)$ are phytoplankton and zooplankton densities respectively, the prey growth rate $r_x (= s + lx)$, is a linear function of x. β_1 is the rate at which phytoplankton is grazed by zooplankton and it follows Holling type III functional response, α_1 and α_2 is the half-saturation constants for phytoplankton and zooplankton respectively. β_2 is the parameter measuring the ratio of product of conversion coefficient with grazing rate to the product of intensity of competition among individuals of phytoplankton with carrying capacity, γ be the per capita predator death rate. d_1, d_2 are self-diffusion for non-dimensionalized model, f is the maximum value of the total loss of zooplankton due to fish predation, which also follows Holling type-III functional response.

3   Spatiotemporal Model and Its Linear Stability Analysis

The linear stability analysis of the model (1a)-(1b) gives three stationary points of the model system. The equilibrium points of model system (1a)-(1b) can be obtained by solving $du/dt = 0$, $dv/dt = 0$. It can be seen that model system has three nonnegative equilibria, namely, $E_0(0, 0)$, $E_1(1, 0)$ and $E^*(u^*, v^*)$. In this section, we study the effect of spatial heterogeneity on the model system about the interior equilibrium point $E^*(u^*, v^*)$. Instability will occur due to diffusion when a parameter varies slowly in such a way that a stability condition is suddenly violated and it can bring about a situation wherein perturbation of a non-zero (finite) wavelength starts growing.

To study the effect of diffusion on the model system, we have considered the lineralized form of the model system (1a)-(1b) about $E^*(u^*, v^*)$ as follows:

$$\frac{\partial U}{\partial t} = a_{11}U + a_{12}V + d_1\frac{\partial^2 U}{\partial x^2}, \tag{4}$$

$$\frac{\partial V}{\partial t} = a_{21}U + a_{22}V + d_2\frac{\partial^2 V}{\partial x^2}, \tag{5}$$

where, $u = u^* + U$, $v = v^* + V$.

It may be noted that (U, V) are small perturbations of (u, v) about the equilibrium point (u^*, v^*). Let us assume the solutions of system of equations (4) and (5) of the form

$$\begin{pmatrix} U \\ V \end{pmatrix} = \begin{pmatrix} a \\ b \end{pmatrix} \exp(\lambda t)\cos(n\pi x/L),$$

where $L/n\pi$ is the critical wave length, L is the length of the system, $2\pi/n$ is the period of cosine term and λ is the frequency respectively.

Theorem 1. *The positive equilibrium* E^* *is locally asymptotically stable in the presence of diffusion if and only if*

$$\rho_1 > 0 \text{ and } \rho_2 > 0 \tag{6}$$

Proof: The Jacobian matrix of equations (4) - (5) corresponding to the positive equilibrium $E^*(u^*, v^*)$ is given by

$$J = \begin{bmatrix} a_{11} - d_1 \left(\dfrac{n\pi}{L} \right)^2 & a_{12} \\ a_{21} & a_{22} - d_2 \left(\dfrac{n\pi}{L} \right)^2 \end{bmatrix}$$

where,

$$a_{11} = -u^* \left\{ r_x - \frac{\beta_1 v^* \left(u^{*2} - \alpha_1 \right)}{\left(u^{*2} + \alpha_1 \right)^2} \right\}, \quad a_{12} = \frac{-\beta_1 u^{*2}}{\left(u^{*2} + \alpha_1 \right)}$$

$$a_{21} = \frac{2\alpha_1 \beta_2 u^* v^*}{\left(u^{*2} + \alpha_1 \right)^2}, \quad a_{22} = \frac{fv^* \left(v^{*2} - \alpha_2 \right)}{\left(v^{*2} + \alpha_2 \right)^2}. \tag{7}$$

The characteristic equation corresponding to the matrix J is given by

$$\lambda^2 + \rho_1 \lambda + \rho_2 = 0, \tag{8}$$

where,

$$\rho_1 = A + (d_1 + d_2)(n\pi / L)^2, \tag{9}$$

$$\rho_2 = d_1 d_2 (n\pi / L)^4 - (a_{22} d_1 + a_{11} d_2)(n\pi / L)^2 + B, \tag{10}$$

where A and B are defined below

$$A = \left[u^* \left\{ r_x - \frac{\beta_1 v^* \left(u^{*2} - \alpha_1 \right)}{\left(u^{*2} + \alpha_1 \right)^2} \right\} + \frac{fv^* \left(\alpha_2 - v^{*2} \right)}{\left(v^{*2} + \alpha_2 \right)^2} \right], \tag{11}$$

$$B = u^* v^* \left[\left\{ r_x - \frac{\beta_1 v^* \left(u^{*2} - \alpha_1 \right)}{\left(u^{*2} + \alpha_1 \right)^2} \right\} \left\{ \frac{f \left(\alpha_2 - v^{*2} \right)}{\left(v^{*2} + \alpha_2 \right)^2} \right\} + \frac{2\alpha_1 \beta_1 \beta_2 u^{*2} v^*}{\left(u^{*2} + \alpha_1 \right)^3} \right], \tag{12}$$

By Routh-Hurwitz criteria all eigenvalues of J will have negative real parts iff $\rho_1 > 0, \rho_2 > 0$. □

Diffusive instability sets in when at least one of the conditions in equation (6) is violated subject to the conditions $A > 0$ and $B > 0$ hold. But it is evident that the first condition $\rho_1 > 0$ is not violated when the condition $A > 0$ is met. Hence only the violation of condition $\rho_2 > 0$ gives rise to diffusive instability. Hence the condition for diffusive instability is given by

$$H\left((n\pi/L)^2\right) = d_1 d_2 (n\pi/L)^4 - (a_{22}d_1 + a_{11}d_2)(n\pi/L)^2 + B < 0, \qquad (13)$$

H is quadratic in $(n\pi/L)^2$ and the graph of $H\left((n\pi/L)^2\right) = 0$ is a parabola. The minimum of $H\left((n\pi/L)^2\right) = 0$ occurs at $(n\pi/L)^2 = (n\pi/L)_c^2$ where

$$(n\pi/L)_c^2 = \frac{1}{2d_1 d_2}\left[d_2 u^*\left\{\frac{\beta_1 v^*\left(u^{*2} - \alpha_1\right)}{\left(u^{*2} + \alpha_1\right)^2} - r_x\right\} + d_1 f v^*\left\{\frac{\left(v^{*2} - \alpha_2\right)}{\left(v^{*2} + \alpha_2\right)^2}\right\}\right] > 0, \qquad (14)$$

Consequently, the condition for diffusive instability is $H\left((n\pi/L)_c^2\right) < 0$. Therefore

$$\left[d_2 u^*\left\{r_x - \frac{\beta_1 v^*\left(u^{*2} - \alpha_1\right)}{\left(u^{*2} + \alpha_1\right)^2}\right\} + d_1 f v^*\left\{\frac{\left(\alpha_2 - v^{*2}\right)}{\left(v^{*2} + \alpha_2\right)^2}\right\}\right]^2 > 4d_1 d_2 B, \qquad (15)$$

where, B is defined in Eq. (12).

4   Numerical Simulation

In this section we perform simulation of model system (1a)-(1b) in one dimensional case. We choose the following set of parameters (other set of parameters may also exist) for the model system (1a)-(2b): $s = 2.0$, $\beta_1 = 0.95$, $\beta_2 = 0.90$, $\alpha_1 = 0.01$, $\alpha_2 = 0.02$, $\gamma = 0.6$, $f = 0.01$. With the above set of values of parameters, we note that the positive equilibrium point E^* exits and it is given by $(u^*, v^*) = (0.1499, 0.3877)$. For the numerical integration of the model system, we have used the Runge-Kutta fourth order procedure on the MATLAB 7.5 platform. The spatiotemporal dynamics of the model system (1) is studied with the help of numerical simulation. The spatiotemporal dynamics is studied by observing the effect of time and space in the density of phytoplankton and zooplankton populations. In order to avoid numerical artifacts we checked the sensitivity of the results to the choice of the time and space steps and their values were chosen sufficiently small. From a realistic biological point of view, we consider a non-monotonic form of initial condition which determines the initial spatial distribution of the species in the real community as

$$u(x, 0) = u^* + \varepsilon(x - x_1)(x - x_2)$$
$$v(x, 0) = v^* \qquad (16)$$

Where $(u^*, v^*) = (0.1499, 0.3877)$ is the non-trivial state for the co-existence of prey and predator and $\varepsilon = 10^{-8}$, $x_1 = 1200$, $x_2 = 2800$ is the parameter affecting the system dynamics. At the boundary, we assume the zero-flux boundary conditions as given in equation (3). This type of boundary conditions is specifically used for modeling the dynamics of spatially bounded aquatic ecosystems. The dynamics of the phytoplankton population is observed at the parameter values

$$s = 2.0, \, l = -1.4, \, \beta_1 = 0.95, \, \beta_2 = 0.90, \, \alpha_1 = 0.01,$$
$$\alpha_2 = 0.02, \, \gamma = 0.6, \, f = 0.01, \, d_1 = 10^{-4}, \, d_2 = 10^{-3}. \tag{17}$$

The simulation study of the model system (1) reflects the effect of spatial heterogeneity. In Fig.1 we observed the spatiotemporal chaos in the presence of heterogeneity. Similarly, in Fig.2 and 3 we have generated the time series and space series to see the effect spatial heterogeneity. Finally in fig.4 we observed the effect of fish predation as regularizing effect.

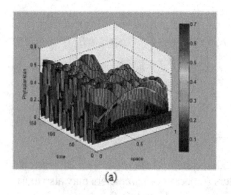 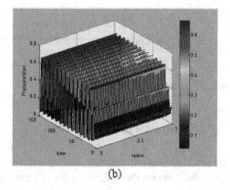

(a)   (b)

Fig. 1. Complex spatiotemporal patterns of phytoplankton density of the model system (1) for (a) with spatial heterogeneity (b) without spatial heterogeneity at the parameter values in equation (17)

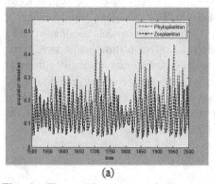 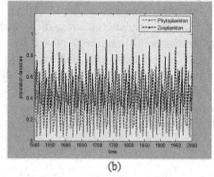

(a)   (b)

Fig. 2. Time series generated for (a) with spatial heterogeneity, (b) without spatial heterogeneity at fixed set of parameter values in equation (17)

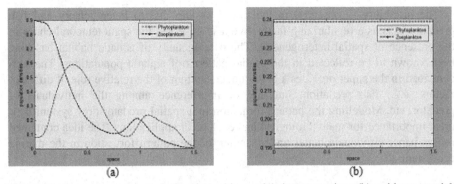

Fig. 3. Space series generated for (a) with spatial heterogeneity, (b) without spatial heterogeneity at fixed set of parameter values in equation (17)

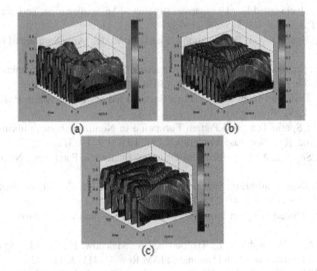

Fig. 4. Complex spatiotemporal patterns of phytoplankton density for model system (1) with spatial heterogeneity for different value (a) $f = 0.01$, (b) $f = 0.05$, (c) $f = 0.07$ at the fixed set of parameter values in equation (17)

5   Discussions and Conclusions

In this paper, we have considered a minimal model for phytoplankton-zooplankton-fish interaction. We have studied the reaction-diffusion model in one dimension and investigated its stability. The nontrivial equilibrium state E^* of zooplankton-phytoplankton coexistence is locally as well as globally asymptotically stable under a fixed region of attraction when certain conditions are satisfied. The important factors which govern spatiotemporal dynamics of aquatic systems are: spatial heterogeneity in the specific growth rate of the prey, intensity of fish predation and intensity of interference among individuals of predators.

The instability is observed in the model system due to heterogeneity and fish predation acts as a regularizing factor. We also observed the spatiotemporal chaos in the presence of spatial heterogeneity. The heterogeneity of aquatic habitat has long been known to be reflected in the spatial patterns of aquatic populations. The idea contained in the paper provides a better understanding of the relative role of different factors; e.g., fish predation, intensity of interference among the individuals of predator, etc. Modelling the pattern formation in a spatial predator-prey system is of great importance for many biological and ecological applications. The idea contained in this paper provides a better understanding of the pattern formation in the marine ecosystem.

References

1. Segel, L.A., Jackson, J.L.: Dissipative structure: An explanation and an ecological example. J. Theo. Biol. 37, 545–559 (1972)
2. Steele, J.H., Henderson, E.W.: A Simple Plankton Model. Am. Naturalist 117, 676–691 (1981)
3. Scheffer, M.: Should we expect Strange Attractors behind Plankton Dynamics - and if so, should we bother? J. Plank. Res. 13, 1291–1305 (1991)
4. Pascual, M.: Diffusion-Induced Chaos in a Spatial Predator-Prey System. Proc. Roy. Soc. London Ser. B 251, 1–7 (1993)
5. Malchow, H.: Spatio-Temporal Pattern Formation in Nonlinear nonequilibrium Plankton Dynamics. Proc. Roy. Soc. London Series B 251, 103–109 (1993)
6. Levin, S.A., Segel, L.A.: Hypothesis for Origin of Planktonic Patchiness. Nature 259, 659 (1976)
7. Malchow, H.: Non-equilibrium Spatio-Temporal Patterns in Models of non-linear Plankton Dynamics. Freshwater Biol. 45, 239–251 (2000)
8. Malchow, H.: Nonequilibrium Structures in Plankton Dynamics. Ecol. Mode. 76, 123–134 (1994)
9. Medvinsky, A.B., Petrovskii, S.V., Tikhonova, I.A., Malchow, H., Li, B.L.: Spatiotemporal Complexity of Plankton and Fish Dynamics. SIAM Rev. 44, 311–370 (2002)
10. Upadhyay, R.K., Nitu, K., Rai, V.: Wave of Chaos in a Diffusive System: Generating Realistic Patterns of Patchinesh in Plankton-Fish Dynamics. Chao. Solit. Fract. 40, 262–276 (2009)
11. Upadhyay, R.K., Nitu, K., Rai, V.: Exploring Dynamical Complexity in Diffusion Driven Predator-Prey Systems: Effect of Toxin Producing Phytoplankton and Spatial Heterogeneities. Chao. Solit. Fract. 42, 584–594 (2009)
12. Petrovskii, S.V., Malchow, H.: Wave of Chaos: New Mechanism of Pattern Formation in Spatio-Temporal Population Dynamics. Theo. Pop. Biol. 59, 157–174 (2001)
13. Sherratt, J.A., Lewis, M.A., Fowler, A.C.: Ecological Chaos in the Wake of Invasion. Proc. Natl. Acad. Sci. U S A 92, 2524–2528 (1995)
14. Nisbet, R.M., Anderson, K.E., McCauley, E., Lewis, M.A.: Response of Equilibrium States to Spatial Environmental Heterogeneity in Advective Systems. Math. Bios. Eng. 4, 1–13 (2007)
15. Kar, T.K., Matsuda, H.: Global Dynamics and Controllability of a Harvested Prey–Predator System with Holling type III Functional Response. Nonlinear Anal. Hybrid Systems 1, 59–67 (2007)

A Method for Estimating the Lyapunov Exponents of Chaotic Time Series Corrupted by Random Noise Using Extended Kalman Filter

Kamalanand Krishnamurthy

Department of Automobile Engineering, Madras Institute of Technology Campus,
Anna University, Chennai 600044, India
kkamalanandmit@gmail.com

Abstract. Identification of chaos in experimental data is essential for characterizing the system and for analyzing the predictability of the data under analysis. The Lyapunov exponents provide a quantitative measure of the sensitivity to initial conditions and are the most useful dynamical diagnostic for chaotic signals. However, it is difficult to accurately estimate the Lyapunov exponents of chaotic signals which are corrupted by a random noise. In this work, a method for estimation of Lyapunov exponents from noisy time series using Extended Kalman Filter (EKF) is proposed. The proposed methodology was validated using time series obtained from known chaotic maps. The proposed method seems to be advantageous since it can be used for online estimation of Lyapunov exponents. In this paper, the objective of the work, the proposed methodology and validation results are discussed in detail.

Keywords: Lyapunov exponents, Extended Kalman filter, Chaos theory, Neural networks.

1   Introduction

Deterministic chaos appears in variety of fields like engineering, biomedical and life sciences, social sciences, and physical sciences and recognizing the chaotic behavior of dynamical systems when only output data are available, is an important field of research [1]. Also, distinguishing deterministic chaos from noise has become an important problem in many diverse fields such as physiology and economics [2-4].

The Lyapunov exponents provides a quantitative measure of the sensitivity to initial conditions and is the most useful dynamical diagnostic for chaotic systems [5,6]. Positive Lyapunov exponents are considered evidence of chaos, Negative exponents of mean reverting behavior, and the value zero is characteristic of cyclic behavior [7]. Hence, the estimation of the Lyapunov exponents is an useful dynamical classifier for deterministic chaotic systems and is an important issue in nonlinear time series analysis [1].

There are several algorithms for estimation of Lyapunov exponents from experimental time series. However, most of them are usually unreliable except for

P. Balasubramaniam and R. Uthayakumar (Eds.): ICMMSC 2012, CCIS 283, pp. 237–244, 2012.

long and noise-free time series [2]. Wolf et al [5] presented the first algorithms for estimation of non-negative Lyapunov exponents from an experimental time series. Rosenstein et al [2] introduced a robust method for calculating the Lyapunov exponents of a time series using a simple measure of exponential divergence and by utilizing all the available data. Zeng et al [6] proposed a method for estimating the Lyapunov exponents from short time series of low precision. Eckmann et al [8] developed a method for obtaining the Lyapunov exponents corresponding to the large-time behavior of the chaotic system. Sano and Sawada [9] proposed a method to determine the spectrum of Lyapunov exponents from the observed time series of a single variable.

Also, several model based methods for estimating the Lyapunov exponents of a time series have emerged. Eckmann and Ruelle [10] and Eckmann et al. [8] proposed a method based on nonparametric regression which is known as the Jacobian method for obtaining the Lyapunov exponent from experimental time series. Ataei et al [1] estimated the Lyapunov exponents of a chaotic time series using a global polynomial model fitting to the given data followed by a Jacobian approach. McCaffrey et al [11] described procedures for estimating the Lyapunov exponents of time series data using four different modeling approaches namely, thin-plate splines, neural nets, radial basis functions and projection pursuit.

The estimation of Lyapunov exponents of a time series corrupted by a random noise is difficult. Most of the biological and economic systems are subjected to random perturbations and are observed over a limited period of time. In such time series, the dynamic information is limited by sample size and masked by noise [12]. Nychka et al [12] estimated the Lyapunov exponents of stochastic systems and have concluded that to certain extent it is possible to identify chaotic dynamics in short noisy systems using thin plate splines.

In this paper, a method based on extended Kalman filter is proposed for estimation of Lyapunov exponents of chaotic time series corrupted by a random noise. Further, the performance of the proposed method is analyzed using the time series obtained from well known chaotic maps [13-15].

2   Methodology

The proposed methodology for estimation of Lyapunov exponents from chaotic time series corrupted by random noise is described in Fig. 1. The first step of the proposed method is to develop the dynamic model from the measured time series. The generated model can be any nonlinear model but has to satisfy two conditions:

   a. The generated model must not have any discontinuities
   b. The generated model must be at least once differentiable

The next step is to apply the EKF algorithm to the measured time series using the developed model. Hence, a filtered estimate of the states is obtained. EKF is simultaneously used to develop a dynamic Artificial Neural Network (ANN) model from the estimated states. Further, the Jacobian matrix of the developed ANN model is computed and the Lyapunov exponent of the noisy time series is obtained from the

Jacobian matrix. The proposed method was validated using well known chaotic time series obtained using the logistic map, sine map, tent map and Ricker's map.

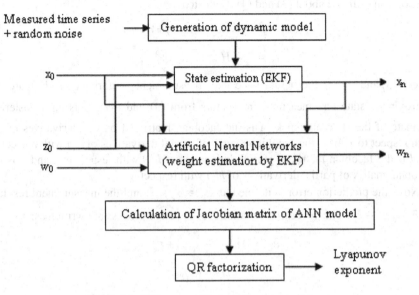

Fig. 1. Block diagram of the proposed methodology

2.1 The Extended Kalman Filter

A Kalman filter that linearize about the current mean and covariance is referred to as an extended Kalman filter or EKF. The EKF is used in that case where the dynamic model is nonlinear. In EKF, the estimation around the current estimate is linearized using the partial derivatives of the process and measurement functions to compute estimates [16, 17].

Let the state vector be $x \in \Re^n$ and the process is governed by the non-linear difference equation described by (1) and let the measurement be $z \in \Re^m$.

$$x_k = f\left(x_{k-1}, w_{k-1}\right) \tag{1}$$

$$z_k = h\left(x_k, v_k\right) \tag{2}$$

where the random variables w_k and v_k represent the process and measurement noise. The probability distributions of w and v are $p(w) \sim N(0,Q)$ and $p(v) \sim N(0,R)$. In practice, one does not know the individual values of the noise at each time step. However, one can approximate the state and measurement vector without them as:

$$\tilde{x}_k = f\left(\hat{x}_{k-1}, 0\right) \tag{3}$$

$$\tilde{z}_k = h\left(\tilde{x}_k, 0\right) \tag{4}$$

where, $\hat{x}_k$ is a posteriori estimate of the state (from a previous time step k). To estimate a process with non-linear relationships, new governing equations that linearize an estimate about (3) and (4), is written.

$$x_k \approx \tilde{x}_k + A\left(x_{k-1} - \hat{x}_{k-1}\right) + W w_{k-1} \tag{5}$$

$$z_k \approx \tilde{z}_k + H\left(x_k - \tilde{x}_k\right) + V v_k \tag{6}$$

where, x_k and z_k are the actual state and measurement vectors, $\tilde{x}_k$ and $\tilde{z}_k$ are the approximate state and measurement vectors from (3) and (4), $\hat{x}_k$ is an a posteriori estimate of the state at step k, A is the Jacobian matrix of partial derivatives of $f(.)$ with respect to x, W is the Jacobian matrix of partial derivatives of $f(.)$ with respect to w, H is the Jacobian matrix of partial derivatives of $h(.)$ with respect to x and V is the Jacobian matrix of partial derivatives of $h(.)$ with respect to v.

Now, the prediction error is defined as $\tilde{e}_{x_k} \equiv x_k - \tilde{x}_k$ and the measurement residual as $\tilde{e}_{z_k} \equiv z_k - \tilde{x}_k$. The governing equations for the error process are written as:

$$\tilde{e}_{x_k} \approx A\left(x_{k-1} - \hat{x}_{k-1}\right) + \varepsilon_k \tag{7}$$

$$\tilde{e}_{z_k} \approx H\tilde{e}_{x_k} + \eta_k \tag{8}$$

$$\hat{x}_k = \tilde{x}_k + \hat{e}_k \tag{9}$$

where, ε_k and η_k represent new independent random variables having zero mean and covariance matrices WQW^T and VRV^T. The Kalman filter equation used to estimate $\hat{e}_k$ is $\hat{e}_k = K_k \tilde{e}_{z_k}$. And hence,

$$\hat{x}_k = \tilde{x}_k + K_k \tilde{e}_{z_k} = \tilde{x}_k + K_k\left(z_k - \tilde{z}_k\right) \tag{10}$$

where, K_k is the Kalman gain [17].

2.2   Weight Estimation of Artificial Neural Network Using EKF

Let $y = f(x, \theta)$ be the transfer function of a single-layer feed forward neural network where y is the output, x is the input and θ is its parameter vector. Given a set of training data $\{x(i), y(i)\}_{i=1}^{N}$ the training of a neural network can be performed using the following set of equations [18, 19].

$$\theta(t) = \theta(t-1) + v(t) \tag{11}$$

$$y(t) = f\left(x(t), \theta(t)\right) + e(t) \tag{12}$$

where, $v(t)$ and $e(t)$ are zero mean Gaussian noise with variance $Q(t)$ and $R(t)$. A good estimation of the system parameter θ can thus be obtained by the extended Kalman filter method:

$$S(t) = H^T(t) \left[P(t-1) + Q(t) \right] H(t) + R(t) \tag{13}$$

$$L(t) = \left[P(t-1) + Q(t) \right] H(t) S^{-1}(t) \tag{14}$$

$$P(t) = \left(I_{n \times n} - L(t) H(t) \right) P(t-1) \tag{15}$$

$$\hat{\theta}(t) = \hat{\theta}(t-1) + L(t) \left(y(t) - f\left(x(t), \hat{\theta}(t-1) \right) \right) \tag{16}$$

where, $H(t) = \dfrac{\partial f}{\partial \theta} \bigg|_{\theta = \hat{\theta}(t-1)}$.

2.3   Calculation of Lyapunov Exponents

The Lyapunov exponent of the considered time series is calculated using the Jacobian method [1]. Consider the discrete dynamical system described in the following form:

$$x_k = F(x_{k-1}), k = 1, 2, 3 \ldots \tag{17}$$

where x_k is the state vector in the R^m space and $F(.)$ is the trained neural network. The linearized system for a small range around the operational trajectory in the phase space can be written as:

$$\delta x_k = J_{k-1} \delta x_{k-1} \quad \text{where} \quad J_{k-1} = \dfrac{\partial F}{\partial x} \bigg|_{x_{k-1}} \in R^{m \times m} \; k = 1, \ldots \tag{18}$$

where J_k is the Jacobian matrix in point k. Let $Y^k = J_{k-1}.J_{k-2}.\ldots.J_0$, then the following symmetric positive definite matrix exists:

$$A = \lim_{k \to x} \left(\left(Y^k \right)^T . Y^k \right)^{\frac{1}{2k}} \tag{19}$$

and the logarithms of its Eigen values are called the Lyapunov exponents. For large value of k, the fundamental solution Y^k may go to very large values and actually, the calculation of A is not possible. Hence, the QR factorization algorithm is used for approximation of Lyapunov exponents [1].

3   Results and Discussion

Fig. 2(a) shows the time series obtained from Ricker's population model. The obtained time series is shown in Fig. 2(b) along with an additive random noise. The states estimated using the Extended Kalman Filter for the considered time series is shown in Fig. 2(c). It is seen that the EKF algorithm is efficiently able to filter random noises from chaotic time series. Further, the states estimated using the neural network trained using Extended Kalman Filter, is shown in Fig. 2(d). It appears that the EKF method is efficient for developing dynamic ANN models of chaotic time series.

The actual values of Lyapunov exponents and the values obtained using the proposed method are compared in Table 1, for four different time series obtained from known chaotic maps such as logistic map, sine map, tent map and Ricker's map. The values estimated using the proposed method was found to be close to the actual values of the Lyapunov exponents. The deviation from the actual value was found to be high in the case of the sine map when compared to the other time series.

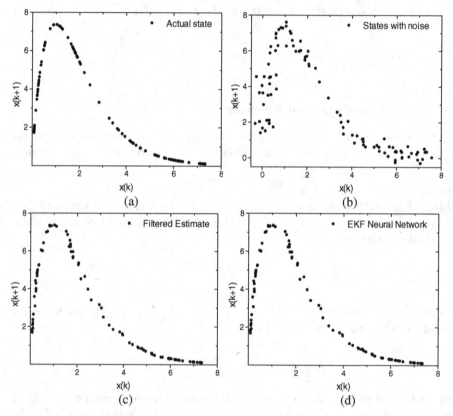

Fig. 2. (a) The actual states of Ricker's chaotic map, (b) the states of the Ricker's map with additive random noise (c) the states estimated using EKF, (d) the states estimated using EKF neural network

Table 1. The expected value of Lyapunov exponents and the values estimated using the proposed method

Time series	Lyapunov exponents		Deviation
	Expected value (Reference)	Proposed method	
Logistic map	0.67 (Rosenstein et al. 1993)	0.66	0.01
Tent map	0.69 (Devaney 1989)	0.64	0.05
Sine map	0.68 (Strogatz 1994)	0.56	0.12
Ricker's map	0.39 (Ricker 1954)	0.39	0

The error in estimation of Lyapunov exponents using the proposed method is shown as a function of the standard deviation of random noise for four different time series in Fig. 3. It is seen that the estimation error increases with increase in the standard deviation of the additive noise. The estimation error was found to be higher for the time series obtained from sine map when compared to the time series obtained from other chaotic maps. The error in estimation of Lyapunov exponents was found to be less and negligible for the chaotic time series obtained from Ricker's population model. Also, for the Ricker's time series, the estimation error was found to remain almost constant with increase in the standard deviation of the additive random noise.

The validation experiments demonstrate that the efficiency of the proposed method depends on the standard deviation of the random noise and also depends on the type of the chaotic time series.

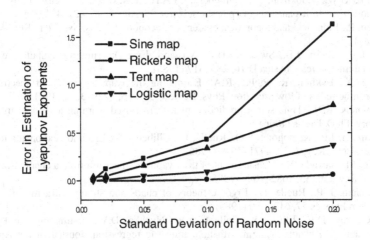

Fig. 3. The error in estimation of Lyapunov exponents shown as a function of the standard deviation of additive random noise for four different chaotic time series

4   Conclusions

Methods for analyzing experimental or observational data for evidence of chaos have been applied to data in such diverse fields as physics, geology, astronomy, neurobiology, ecology and economics [11]. The Lyapunov exponent is one of the basic quantities for characterizing the chaotic behavior of a system, [9, and 20]. However, the estimation of the Lyapunov exponents of chaotic time series corrupted by random noises is difficult [12].

In this work, a method using Extended Kalman Filter, is proposed for estimation of the Lyapunov exponents of chaotic time series corrupted by a random noise. The proposed method was validated using time series obtained from well known chaotic maps such as sine map, tent map, logistic map and Ricker's population model. The estimated exponents were compared with the actual values of the Lyapunov exponents for the considered time series. Results demonstrate that the error in estimation of Lyapunov exponents increases with increase in the standard deviation of the additive noise. It was also found that the estimation error depends on the nature of the chaotic time series. Further, the EKF neural network seems to be efficient for dynamic modeling of chaotic time series. It appears that the proposed methodology can be efficiently used for online estimation of Lyapunov exponents of chaotic time series measurements corrupted by random noises.

References

1. Ataei, M., Khaki-Sedigh, A., Lohmann, B., Lucas, C.: Estimating the Lyapunov exponents of chaotic time series: a model based method. In: European Control Conference (2003)
2. Rosenstein, M.T., Collins, J.J., De Luca, C.J.: A practical method for calculating largest Lyapunov exponents from small data sets. Physica D 65, 117–134 (1993)
3. Frank, G.W., Lookman, T., Nerenberg, M.A.H., Essex, C., Lemieux, J., Blume, W.: Chaotic time series analysis of epileptic seizures. Physica D 46, 427–438 (1990)
4. Chen, P.: Empirical and theoretical evidence of economic chaos. Sys. Dyn. Rev. 4, 81–108 (1988)
5. Wolf, A., Swift, J.B., Swinney, H.L., Vastano, J.A.: Determining Lyapunov exponents from a time series. Physica D 16, 285–317 (1985)
6. Zeng, X., Eykholt, R., Pielke, R.A.: Estimating the lyapunov-exponent spectrum from short time series of low precision. Phys. Review Lett. 66, 3229–3232 (1991)
7. Edmonds, A.N.: Time series prediction using supervised learning and tools from chaos theory. Ph.D Thesis (1996)
8. Eckmann, J.P., Kamphorst, S.O., Ruelle, D., Ciliberto, S.: Liapunov exponents from time series. Phys. Rev. A 34, 4971–4979 (1986)
9. Sano, M., Sawada, Y.: Measurement of the Lyapunov spectrum from a chaotic time series. Phys. Rev. Let. 55, 1082–1085 (1985)
10. Eckmann, J.-P., Ruelle, D.: Ergodic theory of chaos and strange attractors. Reviews of Modern Physics 57, 617–656 (1985)
11. McCaffrey, D.F., Ellner, S., Gallant, A.R., Nychka, D.W.: Estimating the Lyapunov exponent of a chaotic system with nonparametric regression. Journal of the Amer. Sta. Assoc. 87, 682–695 (1992)
12. Nychka, D., Ellner, S., Gallant, A.R., McCaHrey, D.: Finding chaos in noisy system. Journal of the Royal Statistical Society, Series B 54, 399–426 (1992)
13. Devaney, R.L.: An introduction to chaotic dynamical systems, 2nd edn. Addison Wesley, Redwood City (1989)
14. Strogatz, S.H.: Nonlinear dynamics and chaos. Perseus Publishing (1994)
15. Ricker, W.E.: Stock and recruitment. J. Fisheries Res. Board Can. 11, 559–623 (1954)
16. Welsh, G., Bishop, G.: An Introduction to the Kalman filter, University of North Carolina at Chapel Hill, Department of Computer Science, TR 95-041
17. Singhal, S., Wu, L.: Training multilayer perceptrons with the extended Kalman algorithm. In: Touretzky, D. (ed.) Advances in Neural Information Processing Systems, vol. 1, pp. 133–140. Morgan Kaufmann, San Mateo (1989)
18. Sum, J., Leung, C.-S., Young, G.H., Kan, W.-K.: On the Kalman filtering method in neural-network training and pruning. IEEE Trans. Neural Netw. 10, 161–166 (1999)
19. Iiguni, Y., Sakai, H., Tokumaru, H.: A real-time learning algorithm for a multilayered neural network based on the extended Kalman filter. IEEE Trans. Signal Process. 40, 959–966 (1992)
20. Skokos, C.: The Lyapunov characteristic exponents and their computation. Lect. Notes Phys. 790, 63–135 (2010)

Direct Delay Decomposition Approach to Robust Stability on Fuzzy Markov-Type BAM Neural Networks with Time-Varying Delays

R. Sathy[1],* and P. Balasubramaniam[2],**

[1] Horticultural College and Research Institute,
Tamil Nadu Agricultural University (TNAU),
Periyakulam - 625 604, India
maths_sathy@yahoo.co.in
[2] Department of Mathematics,
Gandhigram Rural Institute, Deemed University,
Gandhigram - 624 302, Tamilnadu, India
balugru@gmail.com

Abstract. This paper is concerned with the asymptotic stability analysis of fuzzy Markovian jumping bi-directional associative memory neural networks (FMJBAMNNs) with discrete time-varying delays. Direct delay decomposition method is employed for obtaining the maximum admissible upper bounds (MAUB) of discrete time-varying delays. By utilizing Lyapunov-Krasovskii functional, we show that the addressed BAMNNs is asymptotically stable with Markovian jumping parameter under T-S fuzzy model. A general stability condition is derived in the form of linear matrix inequality (LMI), and can be efficiently solved by LMI toolbox in MATLAB. A numerical example is given to illustrate the effectiveness of the proposed stability criterion.

Keywords: T-S fuzzy model, Markovian jumping, Bi-directional associative memory, Asymptotic stability, Direct delay decomposition.

1 Introduction

Kosko [1] proposed a new class of networks called bidirectional associative memory neural networks (BAMNNs), which have been used to obtain important advances in many fields such as pattern recognition and automatic control. Therefore, the BAMNNs have been one of the most interesting research topics and have attracted the attention of many researchers [2], [3]. However, in reality, time delay exhibits as a typical characteristic of signal transmission between neurons, and therefore, becomes one of the main sources for causing instability and poor performances of neural networks (NNs). Hence, it is important to investigate the stability of BAMNNs with time delays [4], [5]. In implementing and applications

---

* Assistant Professor, Corresponding author.
** Professor and Head.

P. Balasubramaniam and R. Uthayakumar (Eds.): ICMMSC 2012, CCIS 283, pp. 245–254, 2012.
© Springer-Verlag Berlin Heidelberg 2012

of BAMNNs, the neurons in one layer of BAMNNs are fully interconnected to the neurons in the other layer. This BAMNNs may also experiences an abrupt changes, so Markovian jumping parameters are introduced into BAMNNs. Authors in [6]-[8] have been studied the stability of delayed BAMNNs with Markovian jumping parameters. Moreover, if the time-varying parameters for the Markovian jumping BAMNNs (MJBAMNNs) are not accessible precisely, this kind of NNs can be modelled by the Takagi-Sugeno (T-S) fuzzy model [11] which is recognized as a popular and efficient tool in functional approximations see [12]-[14].

Based on the above discussions, our main aim is to derive a maximum admissible upper bounds (MAUB) of the time delays such that the T-S fuzzy Markov-type BAMNNs with discrete time-varying delays is robustly asymptotically stable. We use the direct delay decomposition approach proposed by [9] for determining the MAUB of discrete time-varying delays. As LKF defined in this approach is more general, the obtained delay-dependent stability result in the form of LMI leads less conservative results for the uncertain fuzzy Markov-type BAMNNs (FMJBAMNNs). Moreover, the Leibniz - Newton formula and many free-weighting matrix variables were not employed in this direct delay decomposition method, it is clear that the computational complexity of the obtained robust asymptotic stability criterion is lower than the existing literatures. Numerical examples are given to illustrate the effectiveness of the proposed method.

2   Problem Formulation and Preliminaries

Let $\rho(t) = \rho_t (t \geq 0)$, be a right-continuous homogeneous Markov chain on the probability space taking values in a finite state space $S = \{1, 2, \cdots, s\}$ with generator $\Pi = (\pi_{kk'})_{s \times s}, \forall k, k' \in S$ given by

$$Pr\{\rho_{t+\Delta t} = k' | \rho_t = k\} = \begin{cases} \pi_{kk'} \Delta t + o(\Delta t), & k \neq k', \\ 1 + \pi_{kk} \Delta t + o(\Delta t), & k = k' \end{cases}$$

$\Delta t > 0$ and $\lim_{\Delta t \to 0} \frac{o(\Delta t)}{\Delta t} = 0$. Here, $\pi_{kk'} \geq 0$ is the known transition rate from k to k' for $k \neq k'$, while $\pi_{kk} = -\sum_{k'=1, k' \neq k}^{s} \pi_{kk'}, \forall k, k' \in S$.

The continuous fuzzy system was proposed to represent a nonlinear system [11]. The system dynamics can be captured by a set of fuzzy rules which characterize local correlation in the state space. Each local dynamic described by the fuzzy IF-THEN rule has the property of linear input-output relation. Based on the T-S fuzzy model, i^{th} rule of T-S fuzzy model of delayed MJBAMNNs is given in the following form:

Plant Rule i : **IF** $\theta_1(t) = \{x_1(t)$ and $y_1(t)\}$ are $\eta_1^i, \ldots$, and $\theta_v(t) = \{x_v(t)$ and $y_v(t)\}$ are η_v^i, **THEN**

$$\dot{x}(t) = -\bar{A}_{(i,k)} x(t) + \bar{B}_{(i,k)} f(y(t)) + \bar{C}_{(i,k)} f(y(t - \sigma(t))), \tag{1}$$

$$\dot{y}(t) = -\bar{D}_{(i,k)} y(t) + \bar{E}_{(i,k)} g(x(t)) + \bar{F}_{(i,k)} g(x(t - \tau(t))), \tag{2}$$

$$x(t) = \varphi(t), \quad t \in [-\bar{\tau}, 0], \quad y(t) = \psi(t), \quad t \in [-\bar{\sigma}, 0],$$

where $\theta_j(t)$ are the premise variables; η_j^i, $(i = 1, 2, \cdots, m; j = 1, 2, \cdots, v)$ are the fuzzy sets with m is the number of **IF-THEN** rules. Also, where $x(t), y(t)$ are the state vectors associated with the n and m neurons, $f(\cdot)$ and $g(\cdot)$ are the neurons activation functions. $\tau(t)$, $\sigma(t)$ are discrete time-varying delays. $\bar{A}_{(i,k)} = A_{(i,k)} + \Delta A_{(i,k)}, \bar{D}_{(i,k)} = D_{(i,k)} + \Delta D_{(i,k)}$ are the matrices with positive diagonals, $\bar{B}_{(i,k)} = B_{(i,k)} + \Delta B_{(i,k)}, \bar{E}_{(i,k)} = E_{(i,k)} + \Delta E_{(i,k)}$ are the connection weight matrices and $\bar{C}_{(i,k)} = C_{(i,k)} + \Delta C_{(i,k)}, \bar{F}_{(i,k)} = F_{(i,k)} + \Delta F_{(i,k)}$ are the discretely delayed connection weight matrices. Also, $\Delta A_{(i,k)}, \Delta B_{(i,k)}, \Delta C_{(i,k)}, \Delta D_{(i,k)}, \Delta E_{(i,k)}, \Delta F_{(i,k)}$ are the parameter uncertainties.

We assume that the following assumptions are hold throughout this paper.

(A1) Discrete time-varying delays $\tau(t)$ and $\sigma(t)$ are continuous and differentiable functions satisfying

$$0 \le \tau(t) \le \bar{\tau}, \quad \dot{\tau}(t) \le \mu_1 < \infty, \quad 0 \le \sigma(t) \le \bar{\sigma}, \quad \dot{\sigma}(t) \le \mu_2 < \infty, \quad \forall t \ge 0,$$

where $\bar{\tau}, \bar{\sigma}, \mu_1$ and μ_2 are the constants.

(A2) Each neuron activation function $f(y(\cdot))$ and $g(x(\cdot))$ is bounded and satisfies Lipschitz condition

$$\mid g(x_1) - g(x_2) \mid \le W_1 \mid x_1 - x_2 \mid, \quad \forall x_1, x_2 \in \mathbb{R}, x_1 \ne x_2,$$

$$\mid f(y_1) - f(y_2) \mid \le W_2 \mid y_1 - y_2 \mid, \quad \forall y_1, y_2 \in \mathbb{R}, y_1 \ne y_2,$$

where $W_1 = diag(w_{11}, w_{12}, \cdots, w_{1n}) > 0$ and $W_2 = diag(w_{21}, w_{22}, \cdots, w_{2n}) > 0$ are the positive diagonal matrix.

(A3) Parameter uncertainties in (1)-(2) satisfy

$$[\Delta A_{(i,k)}, \Delta B_{(i,k)}, \Delta C_{(i,k)}, \Delta D_{(i,k)}, \Delta E_{(i,k)}, \Delta F_{(i,k)}] = $$
$$J_{(i,k)} H_{(i,k)}(t)[T_{1(i,k)}, T_{2(i,k)}, T_{3(i,k)}, T_{4(i,k)}, T_{5(i,k)}, T_{6(i,k)}],$$

where, the matrices in left hand side represents time-varying parameter uncertainties. The matrix $H_{(i,k)}(t)$ is the unknown time-varying matrix functions that satisfy $H_{(i,k)}^T(t)H_{(i,k)}(t) \le I$, and $J_{(i,k)}, T_{p(i,k)}(p = 1, 2, \cdots, 6)$ are the real known matrices with appropriate dimensions.

The defuzzified output of T-S fuzzy MJBAMNNs (FMJBAMNNs) (1)-(2) is inferred as:

$$\dot{x}(t) = \sum_{i=1}^{m} h_i(\theta(t))\Big\{ - \bar{A}_{(i,k)}x(t) + \bar{B}_{(i,k)}f(y(t)) + \bar{C}_{(i,k)}f(y(t - \sigma(t)))\Big\}$$

$$= -\bar{A}_\omega x(t) + \bar{B}_\omega f(y(t)) + \bar{C}_\omega f(y(t - \sigma(t))), \tag{3}$$

$$\dot{y}(t) = \sum_{i=1}^{m} h_i(\theta(t))\Big\{ - \bar{D}_{(i,k)}y(t) + \bar{E}_{(i,k)}g(x(t)) + \bar{F}_{(i,k)}g(x(t - \tau(t)))\Big\}$$

$$= -\bar{D}_\omega y(t) + \bar{E}_\omega g(x(t)) + \bar{F}_\omega g(x(t - \tau(t))) \tag{4}$$

where $\omega = (i,k), (\bar{A}_\omega, \bar{B}_\omega, \bar{C}_\omega, \bar{D}_\omega, \bar{E}_\omega, \bar{F}_\omega) = \sum_{i=1}^m h_i(\theta(t)) \{ \bar{A}_{(i,k)}, \bar{B}_{(i,k)}, \bar{C}_{(i,k)},$
$\bar{D}_{(i,k)}, \bar{E}_{(i,k)}, \bar{F}_{(i,k)} \}, h_i(\theta(t)) = \mathcal{M}_i(\theta(t)) / \sum_{i=1}^m \mathcal{M}_i(\theta(t)), \mathcal{M}_i(\theta(t)) = \prod_{j=1}^v \eta_j^i$
$(\theta_j(t))$, and $\eta_j^i(\theta_j(t))$ is the grade of membership of $\theta_j(t)$ in η_j^i. It is assumed that $\mathcal{M}_i(\theta(t)) \geq 0$, where $i = 1, 2, \cdots, m$ and $\sum_{i=1}^m \mathcal{M}_i(\theta(t)) > 0$ for all t. Therefore, $h_i(\theta(t)) \geq 0$ and $\sum_{i=1}^m h_i(\theta(t)) = 1$.

For the sake of convenience, following definition and lemma are introduced to prove the main results.

Definition 1. *The trivial solution or zero solution of FMJBAMNNs (3)-(4) is said to be asymptotically stable in the mean square, if for all finite $\varphi \in C_{\mathcal{F}_0}^2([-\delta_1, 0]; \mathbb{R}^n), \psi \in C_{\mathcal{F}_0}^2([-\delta_2, 0]; \mathbb{R}^n),$
$\rho(0) = k_0 \in S$, such that the following inequality holds:*

$$\lim_{t \to \infty} \left\{ \mathbb{E}\|x(t, \varphi, k_0)\|^2 + \mathbb{E}\|y(t, \psi, k_0)\|^2 \right\} = 0, \quad \forall t \geq 0.$$

Lemma 1. (*Yue et al* [10]). For any vectors $x, y \in \mathbb{R}^n$, matrices $A, P > 0$ are real with appropriate dimensions, $F^T(t)F(t) \leq I$ and scalar $\epsilon > 0$, the following inequalities hold

(a) $DF(t)N + N^T F^T(t)D^T \leq \epsilon^{-1}DD^T + \epsilon N^T N,$

(b) If $P^{-1} - \epsilon^{-1}HH^T > 0$, then

$$(A + HF(t)N)^T P(A + HF(t)N) \leq A^T(P^{-1} - \epsilon^{-1}HH^T)^{-1}A + \epsilon N^T N.$$

3   Stability Analysis

In this section, we consider a new LKF and derive the general robust asymptotic stability criterion for uncertain FMJBAMNNs (3)-(4) with discrete delays by direct delay decomposition approach.

Theorem 1. Let (A1) - (A3) hold and for the scalar $0 < \alpha < 1, 0 < \beta < 1$, the equilibrium solution of nonlinear uncertain FMJBAMNNs system (3)-(4) is robustly asymptotically stable in the mean square if there exist positive definite matrices $P_{1k} > 0$, $P_{2k} > 0$, $Q_\ell > 0$, $R_\ell > 0$, $(\ell = 1, 2, \cdots, 7)$, diagonal matrices $M \geq 0$ and $N \geq 0$ such that the following LMIs hold for $i = (1, 2, \cdots, m)$, $k = (1, 2, \cdots, s)$:

$$Q_5 + (1 - \mu_1)Q_7 > 0, \tag{5}$$
$$Q_6 + (1 - \mu_1)Q_7 > 0, \tag{6}$$
$$R_5 + (1 - \mu_2)R_7 > 0, \tag{7}$$
$$R_6 + (1 - \mu_2)R_7 > 0, \tag{8}$$

$$\Phi_{(i,k)} = \begin{bmatrix} \Omega_1 & \epsilon_1\Gamma_1 & \epsilon_2\Gamma_2 & \Gamma_3 & \Gamma_4 & \Gamma_5 & \Gamma_4 & \hat{P}_1 J_{(i,k)} & \hat{P}_2 J_{(i,k)} \\ -\epsilon_1 I & 0 & 0 & 0 & 0 & 0 & 0 & 0 \\ * & -\epsilon_2 I & 0 & 0 & 0 & 0 & 0 & 0 \\ * & * & -U_1 & U_1 J_{(i,k)} & 0 & 0 & 0 & 0 \\ * & * & * & -\epsilon_1 I & 0 & 0 & 0 & 0 \\ * & * & * & * & -V_1 & V_1 J_{(i,k)} & 0 & 0 \\ * & * & * & * & * & -\epsilon_2 I & 0 & 0 \\ * & * & * & * & * & * & -\epsilon_1 I & 0 \\ * & * & * & * & * & * & * & -\epsilon_2 I \end{bmatrix} < 0, \quad (9)$$

$$\Psi_{(i,k)} = \begin{bmatrix} \Omega_2 & \epsilon_1\Gamma_1 & \epsilon_2\Gamma_2 & \Gamma_6 & \Gamma_4 & \Gamma_7 & \Gamma_4 & \hat{P}_1 J_{(i,k)} & \hat{P}_2 J_{(i,k)} \\ -\epsilon_1 I & 0 & 0 & 0 & 0 & 0 & 0 & 0 \\ * & -\epsilon_2 I & 0 & 0 & 0 & 0 & 0 & 0 \\ * & * & -U_2 & U_2 J_{(i,k)} & 0 & 0 & 0 & 0 \\ * & * & * & -\epsilon_1 I & 0 & 0 & 0 & 0 \\ * & * & * & * & -V_2 & V_2 J_{(i,k)} & 0 & 0 \\ * & * & * & * & * & -\epsilon_2 I & 0 & 0 \\ * & * & * & * & * & * & -\epsilon_1 I & 0 \\ * & * & * & * & * & * & * & -\epsilon_2 I \end{bmatrix} < 0, \quad (10)$$

where $\Omega_1 = (\varphi_{ij})_{12\times12}$, $\Omega_2 = (\lambda_{ij})_{12\times12}$, with

$$\phi_{1,1} = -P_{1k}A_{(i,k)} - A_{(i,k)}^T P_{1k}^T + Q_1 + Q_3 + \sum_{k'=1}^{s} \pi_{kk'} P_{1k'} - (1/\alpha\bar{\tau})[Q_5$$

$$+(1-\mu_1)Q_7] + \epsilon_1 T_{1(i,k)}^T T_{1(i,k)}, \phi_{1,2} = (1/\alpha\bar{\tau})[Q_5 + (1-\mu_1)Q_7],$$

$$\phi_{1,5} = MW_1, \ \phi_{1,11} = P_{1k}B_{(i,k)}, \ \phi_{1,12} = P_{1k}C_{(i,k)}, \ \phi_{2,2} = -(1-\mu_1)Q_3$$

$$-(1/\alpha\bar{\tau})[Q_5 + (1-\mu_1)Q_7] - (1/\alpha\bar{\tau})Q_5, \ \phi_{2,3} = (1/\alpha\bar{\tau})Q_5,$$

$$\phi_{2,6} = MW_1, \ \phi_{3,3} = -Q_1 + Q_2 - (1/\alpha\bar{\tau})Q_5 - (1/(1-\alpha)\bar{\tau})Q_6,$$

$$\phi_{3,4} = (1/(1-\alpha)\bar{\tau})Q_6, \ \phi_{4,4} = -Q_2 - (1/(1-\alpha)\bar{\tau})Q_6, \ \phi_{5,5} = Q_4 - W_1$$

$$-W_1^T + \epsilon_2 T_{5(i,k)}^T T_{5(i,k)}, \ \phi_{5,7} = E_{(i,k)}^T P_{2k}, \ \phi_{6,6} = -(1-\mu_1)Q_4 - W_1$$

$$-W_1^T + \epsilon_2 T_{6(i,k)}^T T_{6(i,k)}, \ \phi_{6,7} = F_{(i,k)}^T P_{2k}, \ \phi_{7,7} = -P_{2k}D_{(i,k)} - D_{(i,k)}^T P_{2k}^T$$

$$+R_1 + R_3 + \sum_{k'=1}^{s} \pi_{kk'} P_{2k'} - (1/\beta\bar{\sigma})[R_5 + (1-\mu_2)R_7] + \epsilon_2 T_{4(i,k)}^T T_{4(i,k)},$$

$$\phi_{7,8} = (1/\beta\bar{\sigma})[R_5 + (1-\mu_2)R_7], \ \phi_{7,11} = NW_2, \ \phi_{7,17} = -D_{(i,k)}^T V_1,$$

$$\phi_{8,8} = -(1-\mu_2)R_3 - (1/\beta\bar{\sigma})[R_5 + (1-\mu_2)R_7] - (1/\beta\bar{\sigma})R_5,$$

$$\phi_{8,9} = (1/\beta\bar{\sigma})R_5, \ \phi_{8,12} = NW_2, \ \phi_{9,9} = -R_1 + R_2 - (1/\beta\bar{\sigma})R_5$$

$$-(1/(1-\beta)\bar{\sigma})R_6, \ \phi_{9,10} = (1/(1-\beta)\bar{\sigma})R_6,$$

$$\phi_{10,10} = -R_2 - (1/(1-\beta)\bar{\sigma})R_6, \ \phi_{11,11} = R_4 - W_2 - W_2^T + \epsilon_1 T_{2(i,k)}^T T_{2(i,k)},$$

$$\phi_{12,12} = -(1-\mu_2)R_4 - W_2 - W_2^T + \epsilon_1 T_{3(i,k)}^T T_{3(i,k)},$$

$$\lambda_{11} = \phi_{11}, \ \lambda_{13} = \phi_{12}, \ \lambda_{15} = \phi_{15}, \ \lambda_{1,11} = \phi_{1,11}, \ \lambda_{1,12} = \phi_{1,12},$$

$$\lambda_{22} = -(1-\mu_1)Q_3 - (1/(1-\alpha)\bar{\tau})[Q_6 + (1-\mu_1)Q_7] - (1/(1-\alpha)\bar{\tau})Q_6,$$

$$\lambda_{23} = (1/(1-\alpha)\bar{\tau})[Q_6 + (1-\mu_1)Q_7], \ \lambda_{24} = \phi_{34}, \ \lambda_{26} = \phi_{26}, \ \lambda_{33} = -Q_1 + Q_2$$
$$-\phi_{12} - \lambda_{23}, \ \lambda_{44} = \phi_{44}, \ \lambda_{55} = \phi_{55}, \ \lambda_{57} = \phi_{57}, \ \lambda_{66} = \phi_{66}, \ \lambda_{67} = \phi_{67},$$

$$\lambda_{77} = \phi_{77}, \ \lambda_{79} = \phi_{78}, \ \lambda_{7,11} = \phi_{7,11}, \ \lambda_{88} = -(1-\mu_2)R_3 - (1/(1-\beta)\bar{\sigma})[R_6$$
$$+ (1-\mu_2)R_7] - (1/(1-\beta)\bar{\sigma})R_6, \ \lambda_{89} = (1/(1-\beta)\bar{\sigma})[R_6 + (1-\mu_2)R_7],$$

$$\lambda_{8,10} = (1/(1-\beta)\bar{\sigma})R_6, \ \lambda_{8,12} = \phi_{8,12}, \ \lambda_{99} = -R_1 + R_2 - \phi_{78} - \lambda_{89},$$

$$\lambda_{10,10} = \phi_{10,10}, \ \lambda_{11,11} = \phi_{11,11}, \ \lambda_{12,12} = \phi_{12,12}, \ \hat{P}_1 = [P_{1k} \ 0 \cdots \ 0 \ 0]^T,$$

$$\hat{P}_2 = [0 \ 0 \ 0 \ 0 \ 0 \ 0 \ P_{2k} \ 0 \ 0 \ 0 \ 0]^T, \ \Gamma_1 = [-T_1 \ 0 \cdots \ 0 \ T_2 \ T_3]^T,$$

$$\Gamma_2 = [0 \ 0 \ 0 \ 0 \ T_5 \ T_6 \ -T_4 \ 0 \ 0 \ 0 \ 0 \ 0]^T, \ \Gamma_4 = [0_{(1\times 12)}]^T,$$

$$\Gamma_3 = [-U_1 A_{(i,k)} \ 0 \ 0 \ 0 \ 0 \ 0 \ 0 \ 0 \ 0 \ 0 \ U_1 B_{(i,k)} \ U_1 C_{(i,k)}]^T,$$

$$\Gamma_5 = [0 \ 0 \ 0 \ 0 \ V_1 E_{(i,k)} \ V_1 F_{(i,k)} \ -V_1 D_{(i,k)} \ 0 \ 0 \ 0 \ 0 \ 0]^T,$$

$$\Gamma_6 = [-U_2 A_{(i,k)} \ 0 \ 0 \ 0 \ 0 \ 0 \ 0 \ 0 \ 0 \ 0 \ U_2 B_{(i,k)} \ U_2 C_{(i,k)}]^T,$$

$$\Gamma_7 = [0 \ 0 \ 0 \ 0 \ V_2 E_{(i,k)} \ V_2 F_{(i,k)} \ -V_2 D_{(i,k)} \ 0 \ 0 \ 0 \ 0 \ 0]^T,$$

$$U_1 = \alpha \bar{\tau} Q_5 + (1-\alpha)\bar{\tau} Q_6 + \alpha \bar{\tau} Q_7, \ V_1 = \beta \bar{\sigma} R_5 + (1-\beta)\bar{\sigma} R_6 + \beta \bar{\sigma} R_7,$$

$$U_2 = \alpha \bar{\tau} Q_5 + (1-\alpha)\bar{\tau} Q_6 + \bar{\tau} Q_7, \ V_2 = \beta \bar{\sigma} R_5 + (1-\beta)\bar{\sigma} R_6 + \bar{\sigma} R_7.$$

Proof. Consider the following Lyapunov-Krasovskii functional:

$$V(i, k, x(t), y(t)) = \sum_{r=1}^{5} V_r(i, k, x(t), y(t)), \tag{11}$$

where

$$V_1(\cdot) = x^T(t)P_{1k}x(t) + y^T(t)P_{2k}y(t),$$

$$V_2(\cdot) = \int_{t-\alpha\bar{\tau}}^{t} x^T(s)Q_1 x(s)ds + \int_{t-\bar{\tau}}^{t-\alpha\bar{\tau}} x^T(s)Q_2 x(s)ds + \int_{t-\tau(t)}^{t} x^T(s)Q_3 x(s)ds,$$

$$V_3(\cdot) = \int_{t-\tau(t)}^{t} g^T(x(s))Q_4 g(x(s))ds + \int_{-\alpha\bar{\tau}}^{0}\int_{t+\theta}^{t} \dot{x}^T(s)Q_5 \dot{x}(s)dsd\theta$$
$$+ \int_{-\bar{\tau}}^{-\alpha\bar{\tau}}\int_{t+\theta}^{t} \dot{x}^T(s)Q_6 \dot{x}(s)dsd\theta + \int_{-\tau(t)}^{0}\int_{t+\theta}^{t} \dot{x}^T(s)Q_7 \dot{x}(s)dsd\theta,$$

$$V_4(\cdot) = \int_{t-\beta\bar{\sigma}}^{t} y^T(s)R_1 y(s)ds + \int_{t-\bar{\sigma}}^{t-\beta\bar{\sigma}} y^T(s)R_2 y(s)ds + \int_{t-\sigma(t)}^{t} y^T(s)R_3 y(s)ds,$$

$$V_5(\cdot) = \int_{t-\sigma(t)}^t f^T(y(s))R_4 f(y(s))ds + \int_{-\beta\bar\sigma}^0 \int_{t+\theta}^t \dot y^T(s)R_5\dot y(s)dsd\theta$$

$$+ \int_{-\bar\sigma}^{-\beta\bar\sigma} \int_{t+\theta}^t \dot y^T(s)R_6\dot y(s)dsd\theta + \int_{-\sigma(t)}^0 \int_{t+\theta}^t \dot y^T(s)R_7\dot y(s)dsd\theta.$$

Let $\mathcal{L}$ be the weak infinitesimal generator of random process $\{x(t), y(t), \rho_t, t \geq 0\}$ along the trajectory of (3)-(4). Then, we get

$$\mathcal{L}V = 2x^T(t)P_{1k}\dot x(t) + x^T(t)\sum_{k'=1}^s [\pi_{kk'}P_{1k'}]x(t) + 2y^T(t)P_{2k}\dot y(t) + y^T(t)$$

$$\times \sum_{k'=1}^s [\pi_{kk'}P_{2k'}]y(t) + x^T(t)Q_1 x(t) + x^T(t-\alpha\bar\tau)(-Q_1 + Q_2)x(t-\alpha\bar\tau)$$

$$- x^T(t-\bar\tau)Q_2 x(t-\bar\tau) + x^T(t)Q_3 x(t) - (1-\mu_1)x^T(t-\tau(t))Q_3 x(t-\tau(t))$$

$$+ g^T(x(t))Q_4 g(x(t)) - (1-\mu_1)g^T(x(t-\tau(t)))Q_4 g(x(t-\tau(t))) + \dot x^T(t)$$

$$\times [\alpha\bar\tau Q_5 + (1-\alpha)\bar\tau Q_6 + \tau(t)Q_7]\dot x(t) - \int_{t-\alpha\bar\tau}^t \dot x^T(s)Q_5\dot x(s)ds$$

$$- \int_{t-\bar\tau}^{t-\alpha\bar\tau} \dot x^T(s)Q_6\dot x(s)ds - (1-\mu_1)\int_{t-\tau(t)}^t \dot x^T(s)Q_7\dot x(s)ds$$

$$+ y^T(t)R_1 y(t) + y^T(t-\beta\bar\sigma)(-R_1 + R_2)y(t-\beta\bar\sigma) - y^T(t-\bar\sigma)R_2$$

$$\times y(t-\bar\sigma) + y^T(t)R_3 y(t) - (1-\mu_2)y^T(t-\tau(t))R_3 y(t-\tau(t)) + f^T(y(t))$$

$$\times R_4 f(y(t)) - (1-\mu_2)f^T(y(t-\sigma(t)))R_4 f(y(t-\sigma(t)))$$

$$+ \dot y^T(t)[\beta\bar\sigma R_5 + (1-\beta)\bar\sigma R_6 + \sigma(t)R_7]\dot y(t) - \int_{t-\beta\bar\sigma}^t \dot y^T(s)R_5\dot y(s)ds$$

$$- \int_{t-\bar\sigma}^{t-\beta\bar\sigma} \dot y^T(s)R_6\dot y(s)ds - (1-\mu_2)\int_{t-\sigma(t)}^t \dot y^T(s)R_7\dot y(s)ds. \tag{12}$$

For any diagonal matrices $M \geq 0$ and $N \geq 0$, it is obvious from the assumption (A2) that

$$-2g^T(x(t))W_1 g(x(t)) + 2x^T(t)MW_1 g(x(t)) \geq 0, \tag{13}$$

$$-2g^T(x(t-\tau(t)))W_1 g(x(t-\tau(t))) + 2x^T(t-\tau(t))MW_1 g(x(t-\tau(t))) \geq 0, \tag{14}$$

$$-2f^T(y(t))W_2 f(y(t)) + 2y^T(t)NW_2 f(y(t)) \geq 0, \tag{15}$$

$$-2f^T(y(t-\sigma(t)))W_2 f(y(t-\sigma(t))) + 2y^T(t-\sigma(t))NW_2 f(y(t-\sigma(t))) \geq 0. \tag{16}$$

Applying the method discussed in [9], by the Jensen's inequality, we estimate the upper bounds of integral terms in (12). Also, we introduce the following new vector as follows:

$$\xi(t) = \begin{bmatrix} x^T(t) & x^T(t-\tau(t)) & x^T(t-\alpha\bar\tau) & x^T(t-\bar\tau) & g^T(x(t)) & g^T(x(t-\tau(t))) \end{bmatrix}$$

$$\left. y^T(t)\ y^T(t-\sigma(t))\ y^T(t-\beta\bar{\sigma})\ y^T(t-\bar{\sigma})\ f^T(y(t))\ f^T(y(t-\sigma(t))) \right]^T \neq 0.$$

Using (3),(4),(A3), Lemma 1 in (12) and then combining (12)-(16), we get the following by using Schur Complement Lemma

$$\mathcal{LV}(i,k,x(t),y(t)) \leq \sum_{i=1}^{m} h_i(\theta(t)) \left\{ \xi^T(t)\Phi_{(i,k)}\xi(t) \right\}.$$

$$\mathcal{LV}(i,k,t,x(t),y(t)) \leq \sum_{i=1}^{m} h_i(\theta(t)) \left\{ \xi^T(t)\Psi_{(i,k)}\xi(t) \right\}.$$

Since $h_i(\theta(t)) \geq 0$, $\sum_{i=1}^{m} h_i(\theta(t)) = 1$, and $\xi(t) \neq 0$, it is easy to see that $\mathcal{LV}(i,k,t,x(t),y(t)) < 0$ holds if (5)-(10) are true. Then by Definition 1, we conclude that the uncertain FMJBAMNNs system (3)-(4) is robustly asymptotically stable for $i = 1,2,\cdots,m$ and $k \in (1,2,\cdots,s)$. This completes the proof.

Remark 1. Our main purpose is to determine the upper bounds $\bar{\tau}$ and $\bar{\sigma}$ such that the system (3)-(4) is robustly asymptotically stable with Markovian jumping parameters by using T-S fuzzy logic. It is worth noting that the stability criterion derived in Theorem 1 is possibly extended to the case of considering uncertainty in a $\pi_{kk'}$ (uncertain switching probabilities) for the dynamical systems such as Hopfield neural networks, Cohen-Grossberg neural networks, Cohen-Grossberg BAMNNs, etc..

4   Numerical Example

In this section, a numerical example is provided to show the effectiveness of the proposed method for the FMJBAMNNs (3) and (4) with two modes $k = 1,2$ and $m = 2$ and $g(x) = f(y) = tanh(y), h_1(\theta(t)) = e^{-2x_1^2(t)}$ and $h_2(\theta(t)) = 1 - h_1(\theta(t))$. Consider the following matrices:

$$A_{(1,1)} = \begin{bmatrix} 1.5 & 0 \\ 0 & 2.6 \end{bmatrix}, \ A_{(1,2)} = \begin{bmatrix} 2.8 & 0 \\ 0 & 1.6 \end{bmatrix}, \ A_{(2,1)} = \begin{bmatrix} 2.3 & 0 \\ 0 & 2.8 \end{bmatrix},$$

$$A_{(2,2)} = \begin{bmatrix} 3.1 & 0 \\ 0 & 1.7 \end{bmatrix}, \ B_{(1,1)} = \begin{bmatrix} 0.5 & -0.3 \\ -0.5 & 0.4 \end{bmatrix}, \ B_{(1,2)} = \begin{bmatrix} 0.7 & 0.2 \\ -0.9 & -1.1 \end{bmatrix},$$

$$B_{(2,1)} = \begin{bmatrix} 0.2 & 0.5 \\ 0 & 0.2 \end{bmatrix}, \ B_{(2,2)} = \begin{bmatrix} 0.5 & 0 \\ 0.1 & -0.7 \end{bmatrix}, \ C_{(1,1)} = \begin{bmatrix} 0.6 & 0.3 \\ 1.1 & -1.5 \end{bmatrix},$$

$$C_{(1,2)} = \begin{bmatrix} 0.8 & 0.6 \\ -0.3 & -0.9 \end{bmatrix}, \ C_{(2,1)} = \begin{bmatrix} 0.8 & 0.1 \\ 0.5 & 1.5 \end{bmatrix}, \ C_{(2,2)} = \begin{bmatrix} 0.3 & 0.2 \\ -0.5 & -0.9 \end{bmatrix},$$

$$D_{(1,1)} = \begin{bmatrix} 1.5 & 0 \\ 0 & 2.6 \end{bmatrix}, \ D_{(1,2)} = \begin{bmatrix} 1.9 & 0 \\ 0 & 3.5 \end{bmatrix}, \ D_{(2,1)} = \begin{bmatrix} 1.6 & 0 \\ 0 & 2.2 \end{bmatrix},$$

$$D_{(2,2)} = \begin{bmatrix} 1.9 & 0 \\ 0 & 2.4 \end{bmatrix}, \ E_{(1,1)} = \begin{bmatrix} 0.1 & 0 \\ 0 & -0.1 \end{bmatrix}, \ E_{(1,2)} = \begin{bmatrix} -0.6 & -0.8 \\ 0 & 0.1 \end{bmatrix},$$

$$E_{(2,1)} = \begin{bmatrix} -0.1 & -0.5 \\ 0 & -0.1 \end{bmatrix}, E_{(2,2)} = \begin{bmatrix} -0.7 & 0 \\ -0.3 & -0.5 \end{bmatrix}, F_{(1,1)} = \begin{bmatrix} 0.3 & 0.1 \\ 0.1 & 0.4 \end{bmatrix},$$

$$F_{(1,2)} = \begin{bmatrix} -0.4 & 0.1 \\ 0.1 & -0.7 \end{bmatrix}, F_{(2,1)} = \begin{bmatrix} 0.2 & 0.3 \\ 1 & 1.8 \end{bmatrix}, \quad F_{(2,2)} = \begin{bmatrix} 0.7 & 0.3 \\ 1.8 & 1.3 \end{bmatrix},$$

$$W_1 = W_2 = \begin{bmatrix} 0.5 & 0 \\ 0 & 0.5 \end{bmatrix}, \quad J_{1(i,k)} = J_{2(i,k)} = \begin{bmatrix} 0.2 & 0 \\ 0 & 0.2 \end{bmatrix}, \quad \Pi = \begin{bmatrix} -7 & 7 \\ 6 & -6 \end{bmatrix}$$

$$T_{1(i,k)} = T_{2(i,k)} = T_{3(i,k)} = T_{4(i,k)} = T_{5(i,k)} = T_{6(i,k)} = \begin{bmatrix} 1 & 0 \\ 0 & 1 \end{bmatrix}, i = (1,2), k = (1,2).$$

Using Theorem 1, it is found that the system (3) and (4) is robustly asymptotically stable in the mean square for all values of μ_1 and μ_2. The complete information about MAUB of $\sigma(t)$ and $\tau(t)$ are given in Table 1. State trajectories and the Markovian switching signals of (3) and (4) are given in Fig.1. with initial conditions $\varphi(t) = \psi(t) = [-0.1, 0.2]$ and $\tau(t) = \sigma(t) = 1.5 + 0.1sint.$

Table 1. Allowable upper bounds (σ, τ) for different $\mu(= \mu_1 = \mu_2)$ in Theorem 1

$(\alpha = \beta = 0.5)$	$\mu_1 = 0.8$	$\mu_1 \geq 1$
$\mu_2 = 0.8$	(1.5946, 1.5327)	(1.4536, 1.4267)

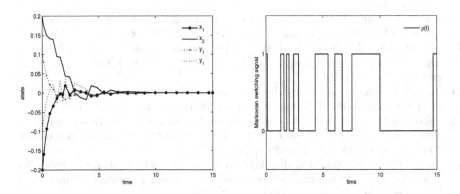

Fig. 1. First figure shows the state curves with uncertainties and the second shows the Markovian switching signal of system (3) and (4)

5   Conclusion

This paper investigates a robust asymptotic stability for the delayed uncertain FMJBAMNNs with discrete time-varying delays. Based on the Lyapunov-Krasovskii function method, a new and general delay-dependent robust asymptotic stability criteria has been derived for delayed uncertain BAMNNs with Markovian jumping parameters by T-S fuzzy model. Numerical example is given to illustrate the effectiveness of the proposed method.

References

1. Kosko, B.: Adaptive bi-directional associative memories. Appl. Optics 26, 4947–4960 (1987)
2. Hansan, R., Siong, K.: A parallel processing VLSI BAM engine. IEEE Trans. Neural Netw. 8, 424–436 (1997)
3. Rao, V.S.H., Phaneendra, R.M.: Global dynamics of bidirectional associative memory neural networks involving transmission delays and dead zones. Neural Netw. 12, 455–465 (1999)
4. Chen, A., Huang, L., Liu, Z., Cao, J.: Periodic bidirectional associative memory neural networks with distributed delays. J. Math. Anal. Appl. 317, 80–102 (2006)
5. Feng, C., Plamondon, R.: Stability anlysis of bi-directional associative memory networks with time delays. IEEE Trans. Neural Netw. 14, 1560–1565 (2003)
6. Liu, H., Ou, Y., Hu, J., Liu, T.: Delay-dependent stability analysis for continuous-time BAM neural networks with Markovian jumping parameters. Neural Netw. 23, 315–321 (2010)
7. Liu, Y., Wang, Z., Liu, X.: On global stability of delayed BAM neural networks with Markovian switching. Neural Process. Lett. 30, 19–35 (2009)
8. Lou, X.Y., Cui, B.T.: Stochastic exponential stability for Markovian jumping BAM neural networks with time-varying delays. IEEE Trans. Syst. Man Cybern. B 37, 713–719 (2007)
9. Zhu, X.L., Yang, G.H.: New results of stability analysis for systems with time-varying delay. Internat. J. Rob. Nonlin. Control 20, 596–606 (2010)
10. Yue, D., Tian, E., Zhang, Y., Peng, C.: Delay-distribution-dependent robust stability of uncertain systems with time-varying delay. Internat. J. Rob. Nonlin. Control 19, 377–393 (2009)
11. Takagi, T., Sugeno, M.: Fuzzy identification of systems and its application to modeling and control. IEEE Trans. Syst. Man Cybernet. B 15, 116–132 (1985)
12. Sakthivel, R., Arunkumar, A., Mathiyalagan, K., Anthoni, S.M.: Robust passivity analysis of fuzzy Cohen-Grossberg BAM neural networks with time-varying delays. Appl. Math. Comp. 218, 3799–3809 (2011)
13. Mathiyalagan, K., Sakthivel, R., Anthoni, S.M.: New robust passivity criteria for stochastic fuzzy BAM neural networks with time-varying delays. Commun. Nonlinear Sci. Numer. Simul. 17, 1392–1407 (2012)
14. Sakthivel, R., Samithurai, R., Anthoni, S.M.: New exponential stability criteria for stochastic BAM neural networks with impulses. Physica Scripta. 82, 458–462 (2010)

Robust Stabilization Results for TS Fuzzy Nonlinear Stochastic Systems with Discrete and Distributed Time Delays

P. Vadivel[1], K. Mathiyalagan[2], R. Sakthivel[3,*], and P. Thangaraj[4]

[1] Department of Mathematics, Kongu Engineering College,
Erode - 638 052, India
[2] Department of Mathematics, Anna University of Technology,
Coimbatore - 641 047, India
kmathimath@gmail.com
[3] Department of Mathematics, Sungkyunkwan University,
Suwon 440-746, South Korea
krsakthivel@yahoo.com
[4] Department of CSE, Bannari Amman Institute of Technoloy,
Sathiamangalam - 638 401, India

Abstract. This paper addresses the problem of stabilization issue for Takagi-Sugeno (TS) fuzzy nonlinear uncertain stochastic systems with discrete and distributed time delays. By utilizing a Lyapunov functional and employing the linear matrix inequality technique we derived delay-dependent sufficient conditions for the existence of state feedback controller in the terms of LMIs, which ensure the robust stabilization of the considered fuzzy system. Further a numerical example is provided to demonstrate the effectiveness of the obtained results.

Keywords: Nonlinear TS fuzzy systems; Linear matrix inequality; Robust Stabilization; Stochastic systems; Robust control.

1 Introduction

The time delays are unavoidable in hardware implementations. The existence of time delays may lead to oscillation, divergence, and even instability in dynamical systems. Also it is well known that the uncertainties are inevitable in dynamical systems because of the existence of modeling errors and external disturbance. Therefore, it is important to study the stability result in the presence of time delays and parameter uncertainties has become a topic of great theoretical and practical importance [3]. Moreover, stochastic models are widely involved in engineering and scientific systems [8–10]. On the other hand, fuzzy theory is considered as a more suitable setting for taking vagueness into consideration. Takagi-Sugeno (TS) model [2] in fuzzy system has attracted researchers in the recent years due to its simpler form in the defuzzification[5–7].

---

* Corresponding author.

P. Balasubramaniam and R. Uthayakumar (Eds.): ICMMSC 2012, CCIS 283, pp. 255–262, 2012.
© Springer-Verlag Berlin Heidelberg 2012

Further in practical systems, the analysis of mathematical model is usually an important work for a control engineer to control a system [4]. Recently, Senthilkumar et al. [1] discussed Robust $H\infty$ control for nonlinear uncertain stochastic TS fuzzy systems with time-delays. To the best of authors knowledge, so far the results on stabilization of stochastic fuzzy system with nonlinear perturbations, discrete and distributed time delays is not yet fully investigated this motivates our study. By constructing a new type Lyapunov-Krasovskii functional and utilizing some advanced techniques, new set of sufficient conditions are derived to ensure the robust stabilization of the fuzzy system. The criterian is expressed in the terms of LMIs, which can be efficiently solved via LMI toolbox. Finally we provide a numerical example to demonstrate the effectiveness of the proposed results.

2   Problem Formulation and Preliminaries

In this section, we start by introducing some notations, definitions and basic results that will be used throughout the paper. The superscripts "T" and "(-1)" stands for matrix transposition and matrix inverse respectively. $\mathbb{R}^n$ denotes the n-dimensional Euclidean space. $P > 0$ means that P is real symmetric and positive definite. In symmetric block matrices or long matrix expressions, we use an asterisk $(*)$ to represent a term that is induced by symmetry and diag$\{ \cdot \}$ stands for a block-diagonal matrix. $(\Omega, \mathcal{F}, \{\mathcal{F}_t\}_{t\geq0}, \mathcal{P})$ be a complete probability space with a filtration $\{\mathcal{F}_t\}_{t\geq0}$ satisfying all usual conditions, $|\cdot|$ denotes the Euclidean norm and $\|\cdot\|$ the $L_2[0,\infty)$ norm and $\mathbb{E}(\cdot)$ stands for the mathematical expectation.

Consider the following nonlinear uncertain stochastic TS fuzzy system with discrete and distributed time delays and nonlinear perturbations together with the ith rule as follows:

Plant Rule i: IF $\{\theta_1(t)$ is $M_1^i\}$ and $\ldots$ and $\{\theta_m(t)$ is $M_m^i\}$ THEN

$$dx(t) = \left[\bar{A}_i x(t) + \bar{A}_{di} x(t-\tau) + \bar{A}_{hi} \int_{t-\tau}^{t} x(s)ds + \bar{B}_{1i}u(t) + A_{fi}f(t,x(t),x(t-\tau))\right]dt$$

$$+ \left[\bar{D}_i x(t) + \bar{D}_{di} x(t-\tau) + \bar{D}_{hi} \int_{t-\tau}^{t} x(s)ds + \bar{B}_{2i}u(t)\right]dw(t), \qquad (1)$$

$$x(t) = \phi(t), \quad t \in [-\tau, 0],$$

where $\theta_1(t),\ldots,\theta_m(t)$ is the premise variable vector; $M_l^i(i = 1,2,\ldots,r,\ l = 1,2,\ldots,m)$ are fuzzy sets and r is the number of IF-THEN rules; $x(t)$ is the state vector; u(t) is the control input vector; τ is discrete time delay; $\phi(t)$ is the initial condition, $\bar{A}_i = A_i + \Delta A_i(t)$, $\bar{A}_{di} = A_{di} + \Delta A_{di}(t)$, $\bar{A}_{hi} = A_{hi} + \Delta A_{hi}(t)$, $\bar{B}_{1i} = B_{1i} + \Delta B_{1i}(t)$, $\bar{D}_i = D_i + \Delta D_i(t)$, $\bar{D}_{di} = D_{di} + \Delta D_{di}(t)$, $\bar{D}_{hi} = D_{hi} + \Delta D_{hi}(t)$, $\bar{B}_{2i} = B_{2i} + \Delta B_{2i}(t)$ in which A_i, A_{di}, A_{hi}, B_{1i}, D_i, D_{di}, D_{hi}, B_{2i} are known constant matrices with appropriate dimensions. $\Delta A_i(t)$, $\Delta A_{di}(t)$, $\Delta A_{hi}(t)$, $\Delta B_{1i}(t)$ $\Delta D_i(t)$, $\Delta D_{di}(t)$, $\Delta D_{hi}(t)$, $\Delta B_{2i}(t)$ denote the parameter uncertainties and are assumed to be the form: $[\Delta A_i\ \Delta A_{di}\ \Delta A_{hi}\ \Delta B_{1i}\ \Delta D_i\ \Delta D_{di}\ \Delta D_{hi}\ \Delta B_{2i}] = H_i F_i(t) [E_{1i}\ E_{2i}\ E_{3i}\ E_{4i}\ E_{5i}\ E_{6i}\ E_{7i}\ E_{8i}]$ where H_i, E_{1i}, E_{2i}, E_{3i}, E_{4i}, E_{5i}, E_{6i},

E_{7i} and E_{8i} are known real constant matrices and $F_i(t)$ denotes unknown time-varying matrix function satisfying $F_i^T(t)F_i(t) \leq I$, where I is the identity matrix. In the stochastic system (1), $w(t)$ is a scalar Wiener process (Brownian Motion) on $(\Omega, \mathcal{F}, \{\mathcal{F}_t\}_{t \geq 0}, \mathcal{P})$ with $\mathbb{E}[w(t)] = 0$, $\mathbb{E}[w^2(t)] = t$, $\mathbb{E}[w(i)w(j)] = 0$, $(i \neq j)$.

Assumption A1. There exist known real constant matrices G_{1i}, $G_{2i} \in \mathbb{R}^{n \times n}$, such that the unknown nonlinear function $f_i(\cdot, \cdot, \cdot) \in \mathbb{R}^n$ satisfies the following bounded condition:

$$|f_i(x(t), x(t - \tau), t)| \leq |G_{1i}x(t)| + |G_{2i}x(t - \tau)|, \quad i = 1, 2 \ldots, r$$

Based on the parallel distributed compensation strategy and the fuzzy state-feedback controller obeys the following rule.
Control Rule i: IF $\{\theta_1(t)$ is $M_1^i\}$ and ... and $\{\theta_r(t)$ is $M_r^i\}$ THEN

$$u(t) = K_i x(t), \quad i = 1, 2 \ldots, r,$$

where $x(t) \in \mathbb{R}^n$ is the input of the controller, $u(t) \in \mathbb{R}^m$ is the output of the controller, and K_i is the gain matrix of the state feedback controller. Thus, the overall fuzzy state feedback controller can be represented by the following input-output form:

$$u(t) = \sum_{i=1}^{r} \eta_i(\theta(t)) K_i x(t) \tag{2}$$

By inferring from the fuzzy models, the final output of stochastic fuzzy system is inferred as

$$dx(t) = \sum_{i=1}^{r} \eta_i(\theta(t)) \sum_{j=1}^{r} \eta_j(\theta(t)) \left\{ \left[(\bar{A}_i + \bar{B}_{1i}K_j)x(t) + \bar{A}_{di}x(t - \tau) + \bar{A}_{hi} \int_{t-\tau}^{t} x(s)ds \right. \right.$$

$$+ A_{fi}f(t, x(t), x(t - \tau)) \Big] dt + \Big[(\bar{D}_i + \bar{B}_{2i}K_j)x(t) + \bar{D}_{di}x(t - \tau)$$

$$+ \bar{D}_{hi} \int_{t-\tau}^{t} x(s)ds \Big] dw(t) \Bigg\}, \tag{3}$$

$\eta_i(\theta(t))$ be the normalized membership function of the inferred fuzzy set $w_i(\theta(t))$. i.e., $\eta_i(\theta(t)) = \frac{w_i(\theta(t))}{\sum\limits_{i=1}^{r} w_i(\theta(t))}$, where $w_i(\theta(t)) = \prod\limits_{l=1}^{m} M_l^i(\theta_l(t))$, here $M_l^i(\theta_l(t))$ is the grade of the membership function of $\theta_l(t)$ in M_l^i. According to the theory of fuzzy sets, it is obvious that $w_i(\theta(t)) \geq 0$, $i = 1, 2, \ldots, r$, $\sum\limits_{i=1}^{r} w_i(\theta(t)) > 0$ for any $\theta(t)$. Therefore, it implies that $\eta_i(\theta(t)) \geq 0$, $\sum\limits_{i=1}^{r} \eta_i(\theta(t)) = 1$, for any $\theta(t)$.

The following definition and lemmas will be used in the main proof.

Definition 1. *The trivial solution of (3) is robustly asymptotically stable in mean square if there exists a $\delta > 0$ and for any $\eta > 0$, satisfying that $\mathbb{E}||x(t)||^2 < \eta$ and $\lim\limits_{t \to \infty} \mathbb{E}||x(t)||^2 = 0$ when $t > t_0$ and $\mathbb{E}||x(t_0)||^2 < \delta$ holds for any initial conditions.*

Lemma 1. *[1] For any matrix $Z > 0$ and scalar $\tau > 0$, if there exists a Lebesque vector function $\omega : [t - \tau, t] \to \mathbb{R}^n$ then the following inequalities hold:*

$$-\int_{t-\tau}^{t} \omega^T(s)Z\omega(s)ds \leq -\frac{1}{\tau}\left(\int_{t-\tau}^{t} \omega^T(s)ds\right)Z\left(\int_{t-\tau}^{t} \omega(s)ds\right)$$

Lemma 2. *[1] Assume that Ω, M_i and E are real matrices with appropriate dimensions and $F_i(t)$ is a matrix function satisfying $F_i^T(t)F_i(t) \leq I$. Then, $\Omega + M_iF_i(t)E_i + [M_iF_i(t)E_i]^T < 0$ holds if and only if there exists a scalar $\epsilon > 0$ satisfying $\Omega + \epsilon^{-1}M_iM_i^T + \epsilon E_i^T E_i < 0$.*

Lemma 3. *[1] For any matrices A, P, D, E and F with appropriate dimensions such that $P > 0$, $F^T F \leq I$, and a scalar $\epsilon > 0$ if $P - \epsilon DD^T > 0$, then $(A + DFE)^T P^{-1}(A + DFE) \leq A^T(P - \epsilon DD^T)^{-1}A + \epsilon^{-1}E^T E$.*

3   State Feed Back

In this section, we derive the sufficient condition for Robust Stabilisation of T-S fuzzy nonlinear uncertain stochastic systems with discrete and distributed time delays given in equations (3).

Theorem 1. *Under assumption (A1), for given scalar $\tau > 0$ the closed-loop system (3) is robustly asymptotically stabilizable through the feedback controller (2), if there exist symmetric positive definite matrices $X > 0, \tilde{Q}_j > 0$, $j = 1, 2$, $\tilde{R}_1 > 0$ and scalars ϵ_i, ϵ_{1ij}, $\epsilon_{2ij} > 0$, such that the following LMI holds for $i = 1, 2, \ldots, r$*

$$\hat{\Pi}_{9\times 9} < 0 \tag{4}$$

where,

$\hat{\Pi}_{11} = 2(A_iX^T + B_{1i}Y_j) + \epsilon_{1ij}HH^T + \epsilon_i A_{fi}A_{fi}^T + \tilde{Q}_1 + \tau\tilde{R}_1$, $\hat{\Pi}_{12} = A_{di}X^T$,

$\hat{\Pi}_{13} = \tau A_{hi}X^T$, $\hat{\Pi}_{15} = \sqrt{2}XG_{1i}^T$, $\hat{\Pi}_{17} = XE_{1i}^T + Y_j^T E_{4i}^T$, $\hat{\Pi}_{18} = XD_i^T + Y_j^T B_{2i}^T$,

$\hat{\Pi}_{19} = XE_{5i}^T + Y_j^T E_{8i}^T$, $\hat{\Pi}_{22} = -\tilde{Q}_1$, $\hat{\Pi}_{26} = \sqrt{2}XG_{2i}^T$, $\hat{\Pi}_{27} = XE_{2i}^T$, $\hat{\Pi}_{28} = XD_{di}^T$,

$\hat{\Pi}_{29} = XE_{6i}^T$, $\hat{\Pi}_{33} = \tau^2\tilde{Q}_2 - \tau\tilde{R}_1$, $\hat{\Pi}_{37} = \tau XE_{3i}^T$, $\hat{\Pi}_{38} = \tau XD_{hi}^T$, $\hat{\Pi}_{39} = \tau XE_{7i}^T$,

$\hat{\Pi}_{44} = -\tau^2\tilde{Q}_2$, $\hat{\Pi}_{55} = -\epsilon_i I$; $\hat{\Pi}_{66} = -\epsilon_i I$, $\hat{\Pi}_{77} = -\epsilon_{1ij}I$, $\hat{\Pi}_{88} = -X + \epsilon_{2ij}HH^T$,

$\hat{\Pi}_{99} = -\epsilon_{2ij}I$ *and other parameters are zero.*

Proof. In order to prove the stabilization result, we consider the following Lyapunov-Krasovskii functional for system (3):

$$V(t, x(t)) = x^T(t)Px(t) + \int_{t-\tau}^{t} x^T(s)Q_1x(s)ds + + \int_{-\tau}^{0}\int_{t+\theta}^{t} x^T(s)R_1x(s)dsd\theta$$

$$+ \int_{t-\tau}^{t}\left(\int_{s-\tau}^{s} x^T(y)dy\right)Q_2\left(\int_{s-\tau}^{s} x(y)dy\right)ds \tag{5}$$

It can be derived by Ito's formula that,

$$dV(t, x(t)) = \mathcal{L}V(t, x(t)) + \sum_{i=1}^{r}\sum_{j=1}^{r} \eta_i \eta_j 2x^T(t)P\Big[(D_i(t) + B_{2i}(t)K_j)x(t)$$

$$+ D_{di}(t)x(t - \tau) + D_{hi}(t)\int_{t-\tau}^{t} x(s)ds\Big]dw(t) \tag{6}$$

and

$$\mathcal{L}V(t, x(t)) = \sum_{i=1}^{r}\sum_{j=1}^{r} \eta_i \eta_j \Big\{ 2x^T(t)P\Big[(A_i(t) + B_{1i}(t)K_j)x(t) + A_{di}(t)x(t - \tau)$$

$$+ A_{hi}(t)\int_{t-\tau}^{t} x(s)ds\Big] + 2x^T(t)P\Big[(\Delta A_i(t) + \Delta B_{1i}(t)K_j)x(t)$$

$$+ \Delta A_{di}(t)x(t - \tau) + \Delta A_{hi}(t)\int_{t-\tau}^{t} x(s)ds + A_{f_i}f(x(t), x(t - \tau), t)\Big]$$

$$+ \Big[(D_i(t) + B_{2i}(t)K_j)x(t) + D_{di}(t)x(t - \tau) + D_{hi}(t)\int_{t-\tau}^{t} x(s)ds\Big]^T$$

$$\times P\Big[(D_i(t) + B_{2i}(t)K_j)x(t) + D_{di}(t)x(t - \tau) + D_{hi}(t)\int_{t-\tau}^{t} x(s)ds\Big]$$

$$+ x^T(t)Q_1 x(t) - x^T(t - \tau)Q_1 x(t - \tau) + \int_{t-\tau}^{t} x^T(s)ds Q_2 \int_{t-\tau}^{t} x(s)ds$$

$$- \int_{t-2\tau}^{t-\tau} x^T(s)ds Q_2 \int_{t-2\tau}^{t-\tau} x(s)ds + \tau x^T(t)R_1 x(t) - \int_{t-\tau}^{t} x^T(s)R_1 x(s)ds \Big\}$$

$$\tag{7}$$

consider

$$- \int_{t-2\tau}^{t-\tau} x^T(s)ds Q_2 \int_{t-2\tau}^{t-\tau} x(s)ds = - \int_{t-\tau}^{t} x^T(s - \tau)ds Q_2 \int_{t-\tau}^{t} x(s - \tau)ds \tag{8}$$

then by using Lemma 1, it is easy to obtain the following inequality

$$- \int_{t-\tau}^{t} x^T(s)R_1 x(s)ds \le -\frac{1}{\tau}\int_{t-\tau}^{t} x^T(s)ds R_1 \int_{t-\tau}^{t} x(s)ds. \tag{9}$$

On the other hand, for positive scalars $\epsilon_{1ij} > 0$ and Lemma 2 we have

$$x^T(t)2P\Big[(\Delta A_i(t) + \Delta B_{1i}(t)K_j)x(t) + \Delta A_{di}(t)x(t - \tau) + \Delta A_{hi}(t)\int_{t-\tau}^{t} x(s)ds\Big]$$

$$= 2x^T(t)PH_i F_i(t)\Big[(E_{1i} + E_{4i}K_j)x(t) + E_{2i}x(t - \tau) + E_{3i}\int_{t-\tau}^{t} x(s)ds\Big]$$

$$\le \epsilon_{1ij}x^T(t)PH_i H_i^T P^T x(t) + \epsilon_{1ij}^{-1}\Big[(E_{1i} + E_{4i}K_j)x(t) + E_{2i}x(t - \tau)$$

$$+ E_{3i}\int_{t-\tau}^{t} x(s)ds\Big]^T\Big[(E_{1i} + E_{4i}K_j)x(t) + E_{2i}x(t - \tau) + E_{3i}\int_{t-\tau}^{t} x(s)ds\Big]. \tag{10}$$

Also, for positive scalar $\epsilon_{2ij} > 0$ and $P^{-1} - \epsilon_{2ij}H_iH_i^T > 0$ and Lemma 3, we get

$$\left[(D_i(t) + B_{2i}(t)K_j)x(t) + D_{di}(t)x(t-\tau) + D_{hi}(t)\int_{t-\tau}^t x(s)ds\right]^T$$

$$\times P\left[(D_i(t) + B_{2i}(t)K_j)x(t) + D_{di}(t)x(t-\tau) + D_{hi}(t)\int_{t-\tau}^t x(s)ds\right]$$

$$= \frac{1}{\tau^2}\int_{t-\tau}^t\int_{t-\tau}^t \xi^T(t,s)\left[\hat{D}_{ij} + H_iF_i(t)\hat{E}_{ij}\right]^T P\left[\hat{D}_{ij} + H_iF_i(t)\hat{E}_{ij}\right]\xi(t,\vartheta)dsd\vartheta$$

$$\leq \frac{1}{\tau^2}\int_{t-\tau}^t\int_{t-\tau}^t \xi^T(t,s)\left[\hat{D}_{ij}^T[P^{-1} - \epsilon_{2ij}H_iH_i^T]^{-1}\hat{D}_{ij} + \epsilon_{2ij}^{-1}\hat{E}_{ij}^T\hat{E}_{ij}\right]\xi(t,\vartheta)dsd\vartheta \quad (11)$$

where, $\hat{D}_{ij} = \begin{bmatrix} D_i + B_{2i}K_j & D_{di} & \tau D_{hi} & 0 \end{bmatrix}$, $\hat{E}_{ij} = \begin{bmatrix} E_{5i} + E_{8i}K_j & E_{6i} & \tau E_{7i} & 0 \end{bmatrix}$,
$\xi(t,s) = \begin{bmatrix} x(t) & x(t-\tau) & x(s) & x(s-\tau) \end{bmatrix}$.
fro any $\epsilon_i > 0$, consider the following

$$2x(t)^T PA_{f_i}f(x(t),x(t-\tau),t)$$

$$\leq \epsilon_i x(t)^T PA_{f_i}A_{f_i}^T Px(t) + \epsilon_i^{-1}f(x(t),x(t-\tau),t)^T f(x(t),x(t-\tau),t)$$

$$\leq \epsilon_i x(t)^T PA_{f_i}A_{f_i}^T Px(t) + 2\epsilon_i^{-1}\left[x(t)^T G_{1i}^T G_{1i}x(t) + x(t-\tau)^T G_{1i}^T G_{1i}x(t-\tau)\right] \quad (12)$$

Combining (7)-(12) we get

$$\mathcal{L}V(t,x(t)) \leq \frac{1}{\tau^2}\int_{t-\tau}^t\int_{t-\tau}^t\left[\sum_{i=1}^r\sum_{j=1}^r \eta_i\eta_j\xi^T(t,s)\Pi\xi(t,\vartheta)\right]dsd\vartheta, \quad (13)$$

where

$$\Pi = \tilde{\Pi} + \epsilon_{1ij}^{-1}\begin{bmatrix} E_{1i} + E_{4i}K_j & E_{2i} & \tau E_{3i} & 0 & 0 \end{bmatrix}^T\begin{bmatrix} E_{1i} + E_{4i}K_j & E_{2i} & \tau E_{3i} & 0 & 0 \end{bmatrix}$$

$$+ \left[\hat{D}_{ij}^T[P^{-1} - \epsilon_{2ij}H_iH_i^T]^{-1}\hat{D}_{ij} + \epsilon_{2ij}^{-1}\hat{E}_{ij}^T\hat{E}_{ij}\right],$$

and

$$\tilde{\Pi} = \begin{bmatrix} \tilde{\Pi}_{11} & PA_{di} & \tau PA_{hi} & 0 \\ * & 2\epsilon_i^{-1}G_{2i}^T G_{2i} - Q_1 & 0 & 0 \\ * & * & \tau^2 Q_2 - \tau R_1 & 0 \\ * & * & * & -\tau^2 Q_2 \end{bmatrix},$$

$$\tilde{\Pi}_{11} = 2P(A_i + B_{1i}K_j) + \epsilon_{1ij}PHH^T P^T + \epsilon_i PA_{fi}A_{fi}^T P^T + 2\epsilon_i^{-1}G_{1i}^T G_{1i} + Q_1 + \tau R_1$$

Letting $X = P^{-1}$, pre and post multiply Π by $N = \text{diag}\{X,X,X,X,X\}$ and N^T respectively. Then, by using Schur complement, from (13) it is easy to obtain $\hat{\Pi} = N\Pi N^T$ taking $Y_i = K_iX^T$ and $\hat{Q}_1 = XQ_1X^T$ similarly $\hat{Q}_2, \hat{R}_1, \hat{R}_2$. Hence, if $\hat{\Pi} < 0$ then $\mathcal{L}V(t,x(t)) < 0$. Then by taking expectaion on both sides implies that $\lim_{t\to\infty}\mathbb{E}\|x\|^2 = 0$. Therefore from Definition 1, the closed-loop fuzzy system (3) is robustly asymptotically stabilizable with the feedback control (2). This completes the proof.

4 Numerical Example

In this section, we present a numerical example to demonstrate the effectiveness of the proposed method.

Example 4.1. *Consider the following TS fuzzy nonlinear uncertain stochastic system:*

Plant Rule 1. IF $\{x_1(t) \ is \ \eta_1^1\}$ **THEN**

$$dx(t) = \left[(A_1(t) + B_{11}(t)K_1)x(t) + A_{d1}(t)x(t-\tau) + A_{h1}(t)\int_{t-\tau}^{t} x(s)ds \right.$$

$$\left. + A_{f_1}f(x(t), x(t-\tau), t)\right]dt + \left[(D_1(t) + B_{21}(t)K_1)x(t) + D_{d1}(t)x(t-\tau)\right.$$

$$\left. + D_{h1}(t)\int_{t-\tau}^{t} x(s)ds\right]dw(t),$$

Plant Rule 2. IF $\{x_2(t) \ is \ \eta_2^2\}$ **THEN**

$$dx(t) = \left[(A_2(t) + B_{12}(t)K_1)x(t) + A_{d2}(t)x(t-\tau) + A_{h2}(t)\int_{t-\tau}^{t} x(s)ds \right.$$

$$\left. + A_{f_2}f(x(t), x(t-\tau), t)\right]dt + \left[(D_2(t) + B_{22}(t)K_1)x(t) + D_{d2}(t)x(t-\tau)\right.$$

$$\left. + D_{h2}(t)\int_{t-\tau}^{t} x(s)ds\right]dw(t)$$

with the following parameters

$$A_1 = \begin{bmatrix} -7.4 & 1.2 \\ 1.2 & -5.3 \end{bmatrix}, \ A_{d1} = \begin{bmatrix} 0.1 & 0.6 \\ 0.1 & 1.6 \end{bmatrix}, \ A_{h1} = \begin{bmatrix} 0.2 & 0 \\ 0 & 0.2 \end{bmatrix}, \ A_{f1} = \begin{bmatrix} 0.1 & 0.6 \\ 0.1 & 1.6 \end{bmatrix},$$

$$D_1 = \begin{bmatrix} 2 & 0 \\ 0 & 1 \end{bmatrix}, \ D_{d1} = \begin{bmatrix} -0.1 & 0.1 \\ -0.3 & 0.6 \end{bmatrix}, \ D_{h1} = \begin{bmatrix} 0.3 & 0.2 \\ 0.2 & 0.5 \end{bmatrix}, \ B_{11} = \begin{bmatrix} -0.01 \\ 0.6 \end{bmatrix}, \ B_{21} = \begin{bmatrix} 2 \\ 0.4 \end{bmatrix},$$

$$E_{11} = \begin{bmatrix} 0.2 & -0.1 \\ 0.3 & 0.2 \end{bmatrix}, \ E_{21} = \begin{bmatrix} 0.4 & 0.2 \\ 0.1 & 0.1 \end{bmatrix}, \ E_{31} = \begin{bmatrix} 0.1 & -0.3 \\ 0.2 & 0.1 \end{bmatrix}, \ E_{41} = \begin{bmatrix} 0.1 \\ 0.2 \end{bmatrix}, \ E_{51} = \begin{bmatrix} 0.2 & 0 \\ 0 & 0.1 \end{bmatrix},$$

$$E_{61} = \begin{bmatrix} 0.1 & 0 \\ 0 & 0.3 \end{bmatrix}, \ E_{71} = \begin{bmatrix} 0.3 & 0 \\ 0 & 0.1 \end{bmatrix}, \ E_{81} = \begin{bmatrix} 0.3 \\ 0 \end{bmatrix}, \ G_{11} = \begin{bmatrix} 0.1 & 0 \\ 0 & 0.1 \end{bmatrix}, \ G_{21} = \begin{bmatrix} 0.1 & 0 \\ 0 & 0.1 \end{bmatrix},$$

$$H_1 = \begin{bmatrix} 1 & 0.2 \\ 0.3 & 1 \end{bmatrix}, \ A_2 = \begin{bmatrix} -6 & 3 \\ 2 & -4 \end{bmatrix}, \ A_{d2} = \begin{bmatrix} 0.2 & 0.5 \\ 0.4 & 0.5 \end{bmatrix}, \ A_{h2} = \begin{bmatrix} 0.2 & 0 \\ 0 & 0.2 \end{bmatrix}, \ A_{f2} = \begin{bmatrix} 0.1 & 0.6 \\ 0.1 & 1.6 \end{bmatrix},$$

$$D_2 = \begin{bmatrix} 0.8 & 0.5 \\ 0.6 & 0.5 \end{bmatrix}, \ D_{d2} = \begin{bmatrix} 0.2 & 0.2 \\ 0.2 & 0.1 \end{bmatrix}, \ D_{h2} = \begin{bmatrix} 0.8 & -0.4 \\ 0.9 & -0.5 \end{bmatrix}, \ B_{12} = \begin{bmatrix} 2 \\ 0.5 \end{bmatrix}, \ B_{22} = \begin{bmatrix} 0.6 \\ 0.3 \end{bmatrix},$$

$$E_{12} = \begin{bmatrix} 0.5 & 0.3 \\ 0.1 & 0.4 \end{bmatrix}, \ E_{22} = \begin{bmatrix} 0.2 & -0.1 \\ 0.1 & 0.3 \end{bmatrix}, \ E_{32} = \begin{bmatrix} 0.2 & 0.1 \\ -0.2 & -0.4 \end{bmatrix}, \ E_{42} = \begin{bmatrix} 0.2 \\ -0.2 \end{bmatrix},$$

$$E_{52} = \begin{bmatrix} 0.2 & 0.1 \\ 0.1 & -0.1 \end{bmatrix}, \ E_{62} = \begin{bmatrix} 0.9 & 0 \\ 0 & 0.2 \end{bmatrix}, \ E_{72} = \begin{bmatrix} 0.01 & 0.02 \\ 0.01 & 0.02 \end{bmatrix}, \ E_{82} = \begin{bmatrix} 0.1 \\ 0.2 \end{bmatrix},$$

$$G_{12} = \begin{bmatrix} 0.1 & 0 \\ 0 & 0.1 \end{bmatrix}, \ G_{22} = \begin{bmatrix} 0.1 & 0 \\ 0 & 0.1 \end{bmatrix}, \ H_2 = \begin{bmatrix} 1 & -0.1 \\ -0.2 & 1 \end{bmatrix}.$$

The membership functions for Rule 1 and Rule 2 are $\eta_1^1 = \frac{1}{e^{-2\eta_1(t)}}$ and $\eta_2^2 = 1 - \eta_1^1$. By solving the LMIs in Theorem 3.1, we can obtain the feasible solutions which are not fully given here due to the page limit, the resulting closed-loop system is robustly asymptotically stable and the maximum allowable time delay upper bound is $\tau = 1.1138$ and the corresponding stabilizing state feedback gain matrices are $K_1 = [-0.8392 \ -0.8896]$, $K_2 = [-2.0438 \ -1.6262]$.

5   Conclusion

In this paper, we have investigated the problem stabilization of TS fuzzy nonlinear stochastic system with discrete and distributed time delays. The criteria has been developed in the frame of LMIs, which can be easily solved via standard numerical software. Finally, an example is given to demonstrate applicability and usefulness of the developed theoretical results.

References

1. Senthilkumar, T., Balasubramaniam, P.: Robust $H\infty$ control for nonlinear uncertain stochastic T-S fuzzy systems with time-delays. Appl. Math. Lett. 24, 1986–1994 (2011)
2. Takagi, T., Sugeno, M.: Fuzzy identification of systems and its applications to modeling and control. IEEE Trans. Systems Man Cybernet. A 15, 116–132 (1985)
3. Fridman, E.: H_∞ control of distributed and discrete delay systems via discretized Lyapunov functional. European J. Cont. 1, 1–11 (2009)
4. Cao, S.G., Rees, N.W., Feng, G.: Analysis and design of a class of continuous time fuzzy control systems. Int. J. Control 64, 1069–1087 (1996)
5. Liu, X., Zhang, Q.L.: New approaches to controller designs based on fuzzy observers for TS fuzzy systems via LMI. Automatica 39, 1571–1582 (2003)
6. Yoneyama, J.: Robust stability and stabilizing controller design of fuzzy systems with discrete and distributed delays. Inf. Sci. 178, 1935–1947 (2008)
7. Guan, X.P., Chen, C.L.: Delay-dependent guaranteed cost control for TS fuzzy systems with time-delay. IEEE Trans. Fuzzy Syst. 12, 236–249 (2004)
8. Zhang, B., Xu, S., Zong, G., Zou, Y.: Delay-dependent stabilization for stochastic fuzzy systems with time delays. Fuzzy Sets Syst. 158, 2238–2250 (2007)
9. Hu, L., Yang, A.: Fuzzy model-based control of nonlinear stochastic systems with time-delay. Nonlinear Anal.: TMA 71, 2855–2865 (2009)
10. Wang, Z., Ho, D.W.C., Liu, X.: A note on the robust stability of uncertain stochastic fuzzy systems with time-delays. IEEE Trans. Systems Man Cybernet. A 34, 570–576 (2004)

Robust Passivity of Fuzzy Cohen-Grossberg Neural Networks with Time-Varying Delays

A. Arunkumar[1], K. Mathiyalagan[1], R. Sakthivel[2,*], and S. Marshal Anthoni[1]

[1] Department of Mathematics, Anna University of Technology,
Coimbatore - 641 047, India
kmathimath@gmail.com
[2] Department of Mathematics, Sungkyunkwan University,
Suwon 440-746, South Korea
krsakthivel@yahoo.com

Abstract. In this paper the problem of robust passivity analysis for a class of fuzzy Cohen-Grossberg neural networks with time varying delay is considered. By employing the Lyapunov technique and linear matrix inequality approach, delay independent criterion's are established for the robust passivity of fuzzy Cohen-Grossberg neural networks. The results are expressed in terms of LMIs, which can be easily solved by the MATLAB LMI toolbox. Finally, a numerical example is given to illustrate the effectiveness of the proposed results.

Keywords: Fuzzy Cohen-Grossberg neural networks, Passivity analysis, Linear matrix inequality, Delay fractioning technique.

1 Introduction

Neural networks have found much attention of researchers due to its large number of successful applications in various fields of science and engineering. In practice, time delays often occur due to finite switching speeds of the amplifiers and communication time. Moreover, it is observed both experimentally and numerically that time delay in neural networks may leads to poor performance. On the other hand, due to the presence of modelling error, external disturbance or parameter fluctuation during the physical implementation, uncertainty is unavoidable and which may affect the dynamical behaviour of the system. Hence, the neural networks with time varying delays and uncertainties are widely investigated [5, 6]. Fuzzy theory is considered as a more suitable setting for the sake of taking vagueness into consideration. Fuzzy systems in the form of the Takagi-Sugeno (TS) model [8] have attracted rapidly growing interest in recent years. TS fuzzy systems are nonlinear systems described by a set of IF-THEN rules [1].

The passivity framework is a promising approach to the stability analysis of delayed neural networks, because it can lead to general conclusions on stability. Recently, passivity analysis problem has been investigated for neural networks with

---

* Corresponding author.

P. Balasubramaniam and R. Uthayakumar (Eds.): ICMMSC 2012, CCIS 283, pp. 263–270, 2012.

time varying delays [2–4]. Further, while studying the passivity or stability issues the main criteria is, how to reduce the possible conservatism induced by the introduction of the new Lyapunov-Krasovskii functional when dealing with time delays. Moreover, the idea of delay partitioning, fractioning or decomposition techniques becomes an increasing interest of many researchers due to much less conservative results while studying the dynamical behaviors via LMI approach [4].

Motivated by the above discussions, in this paper study the passivity analysis of a class of fuzzy Cohen-Grossberg neural networks with time varying delays and parameter uncertainties is considered.

2   Problem Formulation and Preliminaries

In this section, we start by introducing some definitions, notations and basic results that will be used in this paper. The superscripts T and (-1) stands for matrix transposition and matrix inverse respectively; $\Re^{n \times n}$ denotes the $n \times n$-dimensional Euclidean space; the notation $P > 0$ means that P is real, symmetric and positive definite; I denote the identity matrix with compatible dimensions; diag$\{\cdot\}$ stands for a block-diagonal matrix; and sym(A) is defined as $A + A^T$. Matrices which are not explicitly stated are assumed to be compatible for matrix multiplications.

Consider the following uncertain Cohen-Grossberg neural networks with time-varying delays

$$\dot{u}_i(t) = -a_i(u_i(t))\big[b_i(u_i(t)) - \sum_{j=1}^{n} \bar{c}_{ji} f_j(u_j(t)) - \sum_{j=1}^{n} \bar{d}_{ji} f_j(u_j(t - \tau_j(t))) - J_i(t)\big] \quad (1)$$

where u_i are the activations of the ith neuron; $f_j(\cdot)$ is called the activation function; $a_i(u_i(t))$ is the positive constants, $b_i(u_i(t))$ is the behaved functions; $\bar{c}_{ji} = (c_{ji} + \Delta c_{ji})$, $\bar{d}_{ji} = (d_{ji} + \Delta d_{ji})$, in which c_{ji} and d_{ij} are the synaptic connection weights; $J_j(t)$ represent the external inputs; $\tau(t)$ represent time varying delays and satisfy, $0 \le \tau_1 \le \tau(t) \le \tau_2$, $\dot{\tau}(t) \le \mu$ here the time varying delay is represented into two parts: constant part and time-varying part as in [4], that is, $\tau(t) = \tau_1 + \tau^*(t)$, where $\tau^*(t)$ satisfies $0 \le \tau^*(t) \le \tau_2 - \tau_1$, and $\tau_2 > \tau_1 > 0$, $\mu > 0$ are constants, and we let $f_j(.)$ as the output of the neural networks.

Throughout this paper, we make the following assumptions

(H1) $a_i(u_i(t)) > 0, a_i$ are bounded, that is there exist $\underline{a_i}, \overline{a_i} > 0$ such that $\underline{a_i} \le a_i \le \overline{a_i}, i = 1, 2, \ldots, n = 1, 2, \bar{a} = \max\{\overline{a_i}\}, \underline{a} = \min\{\underline{a_i}\}.$

(H2) $\dfrac{b_i(x) - b_i(y)}{x - y} \ge \gamma_i > 0$ for any $x, y \in R$ and $x \ne y$.

For notational convenience, we shift the equilibrium point $(u_1^*, u_2^*, \ldots, u_n^*)^T$ to the origin by the transformations $x_i(t) = u_i - u_i^*$, $\alpha_i(x_i(t)) = a_i(x_i(t) + u_i^*) - a_i(u_i^*)$, $\beta_i(x_i(t)) = b_i(x_i(t) + u_i^*) - b_i(u_i^*)$, $f_j(x_i(t)) = f_j(x_i(t) + u_i^*) - f_j(u_i^*)$ $u_i(t) = J_i - J_i^*$ which yields the following system

$$\dot{x}_i(t) = -\alpha_i(x_i(t))\big[\beta_i(x_i(t)) - \sum_{j=1}^{n} \bar{c}_{ji} f_j(x_j(t)) - \sum_{j=1}^{n} \bar{d}_{ji} f_j(x_j(t - \tau_j(t))) - u_i(t)\big] \quad (2)$$

Then the system (2) is transformed to

$$\dot{x}(t) = -\alpha_1(x(t))\left\{\beta_1(x(t)) - \overline{C}f\left(x(t)\right) - \overline{D}f\left(x(t - \tau(t))\right) - u(t)\right\}, \qquad (3)$$

where $x(t) = [x_1(t), \ldots, x_n(t)]^T$, $f(x(t)) = [f(x_1(t)), \ldots, f(x_n(t))]^T$,

$\tau(t) = [\tau_1(t), \ldots, \tau_n(t)]^T$, $\alpha_1(x(t)) = \text{diag}\{\alpha_1(x_1(t)), \ldots, \alpha_1(y_n(t))\}^T$,

$\beta_1(x(t)) = \text{diag}\{\beta_1(x_1(t)), \ldots, \beta_1(x_n(t))\}^T$, $\overline{C} = (C + \Delta C)$, $\overline{D} = (D + \Delta D)$.

Further, the activation function satisfies the following assumption

(H3) For any $i = 1, 2, \ldots, n$ there exist constants F_i^- and F_i^+ such that

$$F_i^- \leq \frac{F_i(x_1) - F_i(x_2)}{x_1 - x_2} \leq G_i^+, \quad \text{for all } x_1, x_2 \in R, x_1 \neq x_2,$$

Next, we will consider a uncertain fuzzy neural networks with time varying delays. The k-th rule of this TS fuzzy model is of the following form :
Plant Rule η: IF $\{u_1(t)$ are $\theta_1^\eta\}$, ..., $\{u_p(t)$ are $\theta_p^\eta\}$ THEN

$$\dot{x}(t) = -\alpha_\eta(x(t))\left[\beta_\eta(x(t)) - \overline{C}_\eta f(x(t)) - \overline{D}_\eta f(x(t - \tau(t))) - u(t)\right] \qquad (4)$$

where θ_i^η, $(i=1, 2, \ldots, p, \eta=1, 2, \ldots, r)$ are fuzzy sets $(u_1(t), u_2(t), \ldots, u_p(t))^T$ are the premise variable vectors; $x(t)$ are the state variable and r is the number of IF-THEN rules. Further we assume that the parameter uncertainties $\Delta C_\eta, \Delta D_\eta$ are time varying and described by,

(A1) Structured perturbations: $[\Delta C_\eta \ \Delta D_\eta] = M_\eta F_\eta(t)[N_{1\eta} \ N_{2\eta}]$,

where $N_{1\eta}, N_{2\eta}$ and M_η are known constant matrices of appropriate dimensions and $F_\eta(t)$ is an known time varying matrix with Lebegue measurable elements bounded by, $F_\eta^T(t)F_\eta(t) \leq I$.

Let $\lambda_\eta(\xi(t))$ be the normalized membership function of the inferred fuzzy set $\beta_\eta(\xi(t))$. The defuzzifield system is as follows:

$$\dot{x}(t) = \sum_{\eta=1}^r \lambda_\eta(\xi(t))\left\{-\alpha_\eta(x(t))\left[\beta_\eta(x(t)) - \overline{C}_\eta f(x(t)) - \overline{D}_\eta f(x(t - \tau(t))) - u(t)\right]\right\} (5)$$

where $\lambda_\eta(\xi(t)) = \frac{\beta_\eta(\xi(t))}{\sum_{\eta=1}^r \beta_\eta(\xi(t))}$, $\beta_\eta(\xi(t)) = \prod_{i=1}^p \theta_i^\eta(z_i(t))$, and $\theta_i^\eta(z_i(t))$ is the grade of the membership function of $z_i(t)$ in θ_i^η. We assume $\beta_\eta(\xi(t)) \geq 0, \eta = 1, 2, \ldots, r$, $\sum_{\eta=1}^r \beta_\eta(\xi(t)) > 0$, and $\lambda_\eta(\xi(t))$ satisfy, $\lambda_\eta(\xi(t)) \geq 0, \eta = 1, 2, \ldots, r$, $\sum_{\eta=1}^r \lambda_\eta(\xi(t)) = 1$ for any $\xi(t)$.

Before giving main results, we will present the definitions and lemmas.

Definition 1. *The neural networks (5) is said to be passive if there exists a scalar ϑ such that,*

$$2\int_0^{t_p} f^T(x(s))u(s)ds \geq -\vartheta \int_0^{t_p} u^T(s)u(s)ds$$

for all $t_p \geq 0$ and for all solution of (5) with $x_0 = 0$.

Lemma 1. *[7]. Given a positive definite matrix $S \in \Re^{n \times n}$, $S = S^T > 0$ and scalars $\tau_1 < \tau(t) < \tau_2$, for vector function $x = [x_1(t), x_2(t), \ldots, x_n(t)]^T$, we have*

$$(\tau_2 - \tau_1) \int_{t-\tau_2}^{t-\tau_1} \dot{x}^T(s) S \dot{x}(s) ds \geq \left(\int_{t-\tau_2}^{t-\tau_1} \dot{x}(s) ds \right)^T S \left(\int_{t-\tau_2}^{t-\tau_1} \dot{x}(s) ds \right).$$

Lemma 2. *[7] Given matrices $\phi = \phi^T$, M, N and $F(t)$ with appropriate dimensions $\phi + MF(t)N + N^T F(t)^T M^T < 0$ for all $F(t)$ satisfying $F(t)^T F(t) \leq I$, if and only if there exists a scalar $\epsilon > 0$ such that $\phi + \epsilon M M^T + \epsilon^{-1} N^T N < 0$.*

3   Main Results

In this section, we consider the passivity criteria for neural networks (5) when $\Delta C_\eta = 0$, and $\Delta D_\eta = 0$. For presentation convenience, we denote

$$F_1 = \mathrm{diag}\left\{ F_1^- F_1^+, F_2^- F_2^+, \ldots, F_n^- F_n^+ \right\}, \quad F_2 = \mathrm{diag}\left\{ \frac{F_1^- + F_1^+}{2}, \frac{F_2^- + F_2^+}{2}, \ldots, \frac{F_n^- + F_n^+}{2} \right\}.$$

Theorem 1. *Given integers $l \geq 1$, the fuzzy Cohen-Grossberg neural networks (5) without uncertainty is passive, if there exist symmetric positive definite matrices $P_1 > 0$ and $Q_k > 0$, $k = 1, 2, 3, 4$ and $S_k > 0$, $k = 1, 2$, there exit a positive diagonal matrices $R_i > 0$, $i = 1, 2$, a scalar $\vartheta > 0$ and any appropriately dimensioned matrices L, X such that following LMI hold for $\eta = 1, 2, \ldots, r$:*

$$\Xi < 0, \tag{6}$$

where

$$\Xi = W_{P_1}^T \overline{P_1} W_{P_1} + W_{Q_1}^T \overline{Q_1} W_{Q_1} + W_{Q_2}^T \overline{Q_2} W_{Q_2} + W_{S_{11}}^T \overline{S}_{11} W_{S_{11}} + W_{S_1}^T S_1 W_{S_1} - W_{\vartheta_1}^T W_\vartheta$$
$$+ W_{S_2}^T \overline{S}_2 W_{S_2} + \mathrm{sym}\left(W_{\xi_3}^T L W_L \right) + W_{R_{11}}^T \overline{R}_1 W_{R_{12}} + W_{R_{21}}^T \overline{R}_2 W_{R_{22}} - \vartheta W_\vartheta^T W_\vartheta$$

and $\overline{Q_1} = \mathrm{diag}\left\{ Q_1, -Q_1, Q_2, -Q_2 \right\}$, $\overline{Q_2} = \mathrm{diag}\left\{ Q_3, -Q_3, Q_4, -Q_4 \right\}$,

$$\overline{P_1} = \begin{bmatrix} 0 & P_1 \\ P_1 & 0 \end{bmatrix}, \quad \overline{R}_1 = \begin{bmatrix} -R_1 & 0 \\ 0 & -R_1 \end{bmatrix}, \quad \overline{R}_2 = \begin{bmatrix} -R_2 & 0 \\ 0 & -R_2 \end{bmatrix}, \quad \overline{S}_{11} = \mathrm{diag}\left\{ S_1, S_2 \right\},$$

$$W_{Q_1} = \begin{bmatrix} I_{ln} & 0_{ln,7n} \\ 0_{ln,n} & I_{ln} & 0_{ln,6n} \\ I_n & 0_{n,(l+6)n} \\ 0_{n,(l+1)n} & \sqrt{1 - \mu_1} I_n & 0_{n,5n} \end{bmatrix}, \quad W_{Q_2} = \begin{bmatrix} I_n & 0_{n,(l+6)n} \\ 0_{n,(l+2)n} & I_n & 0_{n,4n} \\ 0_{n,(l+4)n} & I_n & 0_{n,2n} \\ 0_{n,(l+5)n} & \sqrt{1 - \mu_1} I_n & 0_{n,n} \end{bmatrix},$$

$$W_{P_1} = \begin{bmatrix} I_n & 0_{n,(l+6)n} \\ 0_{n,(l+3)n} & I_n & 0_{n,3n} \end{bmatrix}, \quad W_{S_1} = \begin{bmatrix} \sqrt{\frac{1}{\tau_1}} I_n & -\sqrt{\frac{1}{\tau_1}} I_n & 0_{n,(l+5)n} \end{bmatrix},$$

$$W_{S_2} = \begin{bmatrix} 0_{n,(l+1)n} & \frac{1}{\sqrt{\tau_2 - \tau_1}} I_n & -\frac{1}{\sqrt{\tau_2 - \tau_1}} I_n & 0_{n,4n} \\ 0_{n,ln} & \frac{1}{\sqrt{\tau_2 - \tau_1}} I_n & -\frac{1}{\sqrt{\tau_2 - \tau_1}} I_n & 0_{n,5n} \end{bmatrix},$$

$$W_{\xi_3} = \begin{bmatrix} I_{ln} & 0_{n,7n} \\ 0_{n,(l+1)n} & I_n & 0_{5n} \\ 0_{n,(l+3)n} & I_n & 0_{3n} \end{bmatrix}, \quad W_L = \begin{bmatrix} -\Gamma_\eta & 0_{n,(l+2)n} & -I_n & C_\eta & D_\eta & I_n \end{bmatrix},$$

$$W_{R_{12}} = \begin{bmatrix} I_n & 0_{n,(l+6)n} \\ 0_{n,(l+4)n} & I_n & 0_{n,2n} \end{bmatrix}, \quad W_{R_{22}} = \begin{bmatrix} 0_{n,(l+1)n} & I_n & 0_{n,5n} \\ 0_{n,(l+5)n} & I_n & 0_{n,n} \end{bmatrix},$$

$$W_\vartheta = \begin{bmatrix} 0_{n,(l+6)n} & I_n \end{bmatrix}, \quad W_{\vartheta_1} = \begin{bmatrix} 0_{n,(l+4)n} & I_n & 0_{n,2n} \end{bmatrix}, \quad \overline{S}_2 = \mathrm{diag}\left\{ -S_2, -S_2 \right\},$$

$$W_{R11} = \begin{bmatrix} F_1 & -F_2 \\ 0_{(l+3)n,n} & 0_{(l+3)n,n} \\ -F_2 & I_n \\ 0_{2n,n} & 0_{2n,n} \end{bmatrix}, \quad W_{R21} = \begin{bmatrix} 0_{(l+1)n,n} & 0_{(l+1)n,n} \\ F_1 & -F_2 \\ 0_{3n,n} & 0_{3n,n} \\ -F_2 & I_n \\ 0_{n,n} & 0_{n,n} \end{bmatrix}.$$

Proof. In order to prove the passivity criteria, we consider the following Lyapunov-Krasovskii functional based on the idea of delay fractioning,

$$V(t, x(t)) = x^T(t)P_1 x(t) + \int_{t-\frac{\tau_1}{l}}^{t} \gamma_1^T(s)Q_1\gamma_1(s)ds + \int_{t-\tau(t)}^{t} x^T(s)Q_2 x(s)ds$$

$$+ \int_{t-\tau_2}^{t} x^T(s)Q_3 x(s)ds + \int_{t-\tau(t)}^{t} f^T(x(s))Q_4 f(x(s))ds$$

$$+ \int_{-\frac{\tau_1}{l}}^{0} \int_{t+\theta}^{t} \dot{x}^T(s)S_1\dot{x}(s)ds d\theta + \int_{-\tau_2}^{-\tau_1} \int_{t+\theta}^{t} \dot{x}^T(s)S_2\dot{x}(s)ds d\theta \quad (7)$$

with $\gamma_1^T(s) = \left[x^T(s), x^T(s - \frac{\tau_1}{l}), \ldots, x^T(s - \frac{(l-1)}{l}\tau_1) \right]$, $l \geq 1$ is the fractioning number.

Taking the time derivative of $V(t, x(t))$ along the trajectories of (5), we have

$$\dot{V}(t, x(t)) = 2x^T(t)P_1\dot{x}(t) + \gamma_1^T(t)Q_1\gamma_1(t) - \gamma_1^T(t - \frac{\tau_1}{l})Q_1\gamma_1(t - \frac{\tau_1}{l})$$

$$+ x^T(t)Q_2 x(t) - (1 - \dot{\tau}(t))x^T(t - \tau(t))Q_2 x(t - \tau(t)) + x^T(t)Q_3 x(t)$$

$$- x^T(t - \tau_2)Q_3 x(t - \tau_2) + f^T(x(t))Q_4 f(x(t)) + \frac{\tau_1}{l}\dot{x}^T(t)S_1\dot{x}(t)$$

$$- (1 - \dot{\tau}(t))f^T(x(t - \tau(t)))Q_4 f(x(t - \tau(t))) + (\tau_2 - \tau_1)\dot{x}^T(t)S_2\dot{x}(t)$$

$$- \int_{t-\frac{\tau_1}{l}}^{t} \dot{x}^T(s)S_1\dot{x}(s)ds - \int_{t-\tau_2}^{t-\tau_1} \dot{x}^T(s)S_2\dot{x}(s)ds \quad (8)$$

By applying Lemma 2.2, we can get the following inequalities

$$- \int_{t-\frac{\tau_1}{l}}^{t} \dot{x}^T(s)S_1\dot{x}(s)ds \leq -\frac{l}{\tau_1}\left[\int_{t-\frac{\tau_1}{l}}^{t} \dot{x}(s)ds\right]^T S_1 \left[\int_{t-\frac{\tau_1}{l}}^{t} \dot{x}(s)ds\right], \quad (9)$$

$$- \int_{t-\tau_2}^{t-\tau_1} \dot{x}^T(s)S_2\dot{x}(s)ds = -\int_{t-\tau_2}^{t-\tau(t)} \dot{x}^T(s)S_2\dot{x}(s)ds - \int_{t-\tau(t)}^{t-\tau_1} \dot{x}^T(s)S_2\dot{x}(s)ds, (10)$$

$$- \int_{t-\tau_2}^{t-\tau(t)} \dot{x}^T(s)S_2\dot{x}(s)ds \leq -\frac{1}{\tau_2 - \tau_1}\left[\int_{t-\tau_2}^{t-\tau(t)} \dot{x}(s)ds\right]^T S_2 \left[\int_{t-\tau_2}^{t-\tau(t)} \dot{x}(s)ds\right], (11)$$

$$- \int_{t-\tau(t)}^{t-\tau_1} \dot{x}^T(s)S_2\dot{x}(s)ds \leq -\frac{1}{\tau_2 - \tau_1}\left[\int_{t-\tau(t)}^{t-\tau_1} \dot{x}(s)ds\right]^T S_2 \left[\int_{t-\tau(t)}^{t-\tau_1} \dot{x}(s)ds\right], \quad (12)$$

On the other hand, for any matrix L the following inequalities hold,

$$2\xi_3^T(t)L\left[-\Gamma_\eta x(t) + C_\eta f(x(t)) + D_\eta f(x(t - \tau(t))) + u(t) - \dot{x}(t)\right] = 0 \quad (13)$$

where $\xi_3^T(t) = \left[\gamma_1^T(t), x^T(t - \tau(t)), \dot{x}^T(t)\right]$.

From assumption $(H3)$, for any $i = 1, 2, \ldots, n$, we have

$$(f_i(x_i(t)) - F_i^- x_i(t))(f_i(x_i(t)) - F_i^+ x_i(t)) \le 0$$

which is equivalent to $\begin{bmatrix} x(t) \\ f(x(t)) \end{bmatrix}^T \begin{bmatrix} F_i^- F_i^+ e_i e_i^T & -\frac{F_i^- + F_i^+}{2} e_i e_i^T \\ -\frac{F_i^- + F_i^+}{2} e_i e_i^T & e_i e_i^T \end{bmatrix} \begin{bmatrix} x(t) \\ f(x(t)) \end{bmatrix} \le 0,$

where e_i denotes the units column vector having element 1 on its ith row and zeros elsewhere. Let $R_1 = \text{diag} \{r_{11}, r_{12}, \ldots, r_{1n}\}, \quad R_2 = \text{diag} \{r_{21}, r_{22}, \ldots, r_{2n}\},$

$$\sum_{i=1}^{n} r_i \begin{bmatrix} x(t) \\ f(x(t)) \end{bmatrix}^T \begin{bmatrix} F_i^- F_i^+ e_i e_i^T & -\frac{F_i^- + F_i^+}{2} e_i e_i^T \\ -\frac{F_i^- + F_i^+}{2} e_i e_i^T & e_i e_i^T \end{bmatrix} \begin{bmatrix} x(t) \\ f(x(t)) \end{bmatrix} \le 0.$$

That is,

$$\begin{bmatrix} x(t) \\ f(x(t)) \end{bmatrix}^T \begin{bmatrix} F_1 R_1 & -F_2 R_1 \\ -F_2 R_1 & R_1 \end{bmatrix} \begin{bmatrix} x(t) \\ f(x(t)) \end{bmatrix} \le 0. \tag{14}$$

Similar to this, one can get

$$\begin{bmatrix} x(t - \tau(t)) \\ f(x(t - \tau(t))) \end{bmatrix}^T \begin{bmatrix} F_1 R_2 & -F_2 R_2 \\ -F_2 R_2 & R_2 \end{bmatrix} \begin{bmatrix} x(t - \tau(t)) \\ f(x(t - \tau(t))) \end{bmatrix} \le 0 \tag{15}$$

From (8)-(15) and Definition 2.1, we have

$$\dot{V}(t, x(t)) - 2f^T(x(t))u(t) - \vartheta u^T(t)u(t) \le \xi^T(t)\Xi\xi(t), \tag{16}$$

where $\xi^T(t) = \begin{bmatrix} \gamma_1^T(t) & x^T(t - \tau_1) & x^T(t - \tau(t)) & x^T(t - \tau_2) & \dot{x}^T(t) & f^T(x(t)) & f^T(x(t - \tau(t))) & u^T(t) \end{bmatrix}.$

Thus we concluded that, if the LMI (6) holds then,

$$\dot{V}(t, x(t)) - 2f^T(x(t))u(t) - \vartheta u^T(t)u(t) \le 0. \tag{17}$$

By integrating the equation (17), with respect to t over the time period $[0, t_p]$, using $V(x(0)) = 0$ we have,

$$2 \int_0^{t_p} f^T(x(s))u(s)ds \ge -\vartheta \int_0^{t_p} u^T(s)u(s)ds.$$

Hence the neural network (5) is passive in the sense of Definition 1. This completes the proof.

Next, The results of the previous section will be extended to obtain the robust passivity analysis of fuzzy Cohen-Grossbergneural networks (5).

Theorem 2. *Under the assumption* $(A1)$*, given integers* $l \ge 1$*, the neural networks (5) is robustly passive, if there exist symmetric positive definite matrices* $P_1 > 0$ *and* $Q_k > 0$*,* $k = 1, 2, 3, 4$ *and* $S_k > 0$*,* $k = 1, 2$*, there exit a positive diagonal matrices* $R_i > 0$*,* $i = 1, 2$*, a scalars* $\vartheta > 0$*,* $\epsilon_1 > 0$ *and any appropriately dimensioned matrices* L, X *such that following LMI hold for* $\eta = 1, 2, \ldots, r$:

$$\begin{bmatrix} \Theta & W_{\xi_3}^T L^T M_\eta \\ * & -\epsilon_1 I \end{bmatrix} < 0, \tag{18}$$

where $\Theta = \Xi + \epsilon_1 N_1^T N_1$, $N_1^T = \begin{bmatrix} 0_{n,(l+4)n} & N_{11\eta} & N_{12\eta} & 0_n \end{bmatrix}$ *and the remaining parameters are same as defined in Theorem 3.1.*

Proof. By considering the uncertainties described in (A1), replacing the matrices $\overline{C}_\eta$ and $\overline{D}_\eta$ in (5) by $C_\eta + M_\eta F_\eta(t)N_{11\eta}$ and $D_\eta + M_\eta F_\eta(t)N_{12\eta}$ and following the similar steps in Theorem 3.1 we can get,

$$\varXi + W_{\xi_3}^T LM_\eta F_\eta(t)N_1 + N_1^T F_\eta^T(t)M_\eta^T L^T W_{\xi_3} < 0.$$

By using Lemma 2.4 in the above inequality, we obtain

$$\varXi + \epsilon_1^{-1} W_{\xi_3}^T LM_\eta M_\eta^T L^T W_{\xi_3} + \epsilon_1 N_1^T N_1 < 0. \tag{19}$$

Applying the schur complement (19) is equivalent to (18). This completes the proof.

4   Numerical Simulation

In this section, we present an example to demonstrate the effectiveness of the proposed results.

Example 1. Consider the fuzzy Cohen-Grossberg neural networks (5) in the absence of parameter uncertainties with the following parameters

$$\varGamma_1 = \begin{bmatrix} 1.6 & 0 \\ 0 & 1.4 \end{bmatrix}, \quad C_1 = \begin{bmatrix} 0.1 & 0.2 \\ 0.1 & 0.2 \end{bmatrix}, \quad D_1 = \begin{bmatrix} 0.3 & 0.2 \\ 0.2 & 0.1 \end{bmatrix},$$

$$\varGamma_2 = \begin{bmatrix} 1.6 & 0 \\ 0 & 1.8 \end{bmatrix}, \quad C_2 = \begin{bmatrix} 0.12 & -0.13 \\ -0.12 & 0.16 \end{bmatrix}, \quad D_2 = \begin{bmatrix} 0.15 & -0.02 \\ -0.12 & 0.17 \end{bmatrix},$$

The activation functions are described by $f_1(x) = \tanh(-2x)$, $f_2(x) = \tanh(-8x)$, the membership functions for Rule 1 and Rule 2 are $\theta_1^1 = \frac{1}{e^{-2\theta_1(t)}}$ and $\theta_2^2 = 1 - \eta_1^1$. Clearly, it can be seen that the assumption (H3) is satisfied with $F_1^- = -2$, $F_1^+ = 0$, $F_2^- = -8$, $F_2^+ = 0$. Thus $F_1 = \begin{bmatrix} 0 & 0 \\ 0 & 0 \end{bmatrix}$, $F_2 = \begin{bmatrix} -1 & 0 \\ 0 & -4 \end{bmatrix}$.

If the delay-fractioning number l, time delay lower bounds τ_1 and time derivative limit μ are given, we can obtain the feasible solutions by solving the LMI's in Theorem 3.1 via Matlab LMI toolbox which is not given here due to the page limit, it clear that the model (5) with above given parameters is passive. If we set $l = 1,2$ and the lower bound τ_1 then the obtained time delay upper bounds for different values of μ are given in Table.1. It is clear that the calculated upper bound τ_2 increases as the fractioning time l increases. It is noted that as l increases, the computational complexity also increases correspondingly, so the optimum result should be less conservative. If the external inputs $u(t) = 0$ and we choose our initial values of the state variables as $[x_1(t), x_2(t)] = [0.5, -1]$, then the trajectories of the state variables are shown in Fig.1. The simulation results reveals that $x(t)$ is converging to the equilibrium point zero, so we conclude that the considered fuzzy Cohen-Grossberg neural networks is internally stable.

Table 1. Calculated upper bound of τ_2 for various values of μ

μ	0.74	0.78	0.8	0.84	0.88	$0.9 \le \mu \le 2.0$
$l = 1$	$\tau_2 = 7.59$	4.71	4.40	4.099	3.98	3.96
$l = 2$	$\tau_2 = 10.41$	5.25	4.67	4.099	4.00	3.96

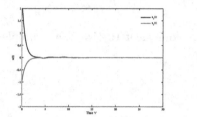 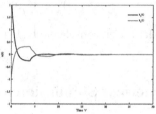

Fig. 1. State trajectories of the fuzzy neural networks when $\eta = 1, \eta = 2$ and $\tau_2 = 3.96$

References

1. Balasubramaniam, P., Syed Ali, M.: Stability analysis of Takagi-Sugeno fuzzy Cohen Grossberg BAM neural networks with discrete and distributed time-varying delays. Math. Comput. Model. 53, 151–160 (2011)
2. Chen, Y., Li, W., Bi, W.: Improved results on passivity analysis of uncertain neural networks with time-varying discrete and distributed delays. Neural Process. Lett. 30, 155–169 (2009)
3. Fu, J., Zhang, H., Ma, T., Zhang, Q.: On passivity analysis for stochastic neural networks with interval time-varying delay. Neurocomputing 73, 795–801 (2010)
4. Li, H., Gao, H., Shi, P.: New passivity analysis for neural networks with discrete and distributed delays. IEEE Trans. Neural Netw. 21, 1842–1847 (2010)
5. Lu, C.Y., Tsai, H.H., Su, T.J.: Delay-dependent approach to passivity analysis for uncertain neural networks with time-varying delay. Neural Process. Lett. 27, 237–246 (2008)
6. Kwon, O.M., Park, J.H.: Exponential stability analysis for uncertain neural networks with interval time-varying delays. Appl. Math. Comput. 212, 530–541 (2009)
7. Sakthivel, R., Arunkumar, A., Mathiyalagan, K., Anthoni, S.M.: Robust passivity analysis of fuzzy Cohen–Grossberg BAM neural networks with time-varying delays. Appl. Math. Comput. 218, 3799–3809 (2011)
8. Takagi, T., Sugeno, M.: Fuzzy identification of systems and its applications to modeling and control. IEEE Trans. Syst. Man Cybernet. 15, 116–132 (1985)

Asymptotic Stability Criterion
for Fuzzy Recurrent Neural Networks
with Interval Time-Varying Delay

R. Chandran[1,2] and P. Balasubramaniam[2]

[1] Department of Computer Science,
Government Arts College, Melur - 625 106,
Madurai(Dt), Tamilnadu, India
rchandran62@gmail.com
[2] Department of Mathematics,
Gandhigram Rural Institute - Deemed University,
Gandhigram - 624 302, Tamilnadu, India
balugru@gmail.com

Abstract. This paper focuses on the delay-dependent asymptotic stability for fuzzy recurrent neural networks (FRNNs) with interval time-varying delay. The delay interval is decomposed into multiple uniform subintervals, Lyapunov-Krasovskii functionals (LKFs) are constructed on these intervals. By employing these LKFs, new delay-dependent asymptotic stability criterion is proposed in terms of Linear Matrix Inequalities (LMIs), which can be easily solved by MATLAB LMI toolbox. Numerical example is given to illustrate the effectiveness of the proposed method.

Keywords: Fuzzy recurrent neural networks, Interval time-varying delay, Delay decomposition approach, Linear Matrix Inequalities, Maximum Admissible Upper Bound (MAUB).

1 Introduction

Recurrent neural networks including Hopfield neural networks (HNNs) and Cellular neural networks (CNNs) have been studied extensively over the recent decades ([1]-[13]), because of its potential applications in pattern recognition, image processing, associative memory, optimization problems etc. Some of the applications require the knowledge of dynamical behaviors of neural networks, such as uniqueness and stability of equilibrium point of a designed neural networks. Time delay, which will occur in the interaction among the neurons, may affect the stability of the neural networks. Thus, it is very important to consider influences of time delay, when we analyze the stability of neural networks. In literature, the time delay $\tau(t)$, often used in stability analysis of recurrent neural networks (RNNs), belongs to the interval $0 \leq \tau(t) \leq h$. Moreover, there exists a special type of time delay in practical engineering systems, namely, interval time-varying delay, in which the lower bound is not restricted to be zero, that is, $0 < \varphi \leq \tau(t) \leq h$. Interval time-varying delays are used to indicate that the propagated speed of signals is finite and uncertain in networks such as

P. Balasubramaniam and R. Uthayakumar (Eds.): ICMMSC 2012, CCIS 283, pp. 271–282, 2012.
© Springer-Verlag Berlin Heidelberg 2012

network control systems. This leads to the initial study of RNNs with interval time-varying delay. Authors in ([8]-[13]) have derived some sufficient conditions for the stability of the neural networks with interval time varying-delay.

In application, there will be some parameter variations in the structures of neural networks. These changes may be abrupt or continuous variation. If the time varying parameters are not accessible precisely, then this kind of neural networks can be modeled by Takagi-Sugeno (T-S) fuzzy model which is recognized as a popular and efficient tool in functional approximation. Therefore, it is of practical importance to study the fuzzy effects on the stability of delayed neural networks. Takagi-Sugeno (T-S) fuzzy systems [14] are nonlinear system described by a set of IF-THEN rules. The stability of fuzzy neural networks were studied by few authors see [4]-[7] and references therein.

In this paper, as continuation of our previous published results [7], the delay decomposition approach, proposed by Zhang and Han ([16]), for delay-dependent asymptotic stability analysis of FRNNs with interval time-varying delay is further exploited. The main aim of this paper is to derive MAUB of $\tau(t)$ such that FRNN system is asymptotically stable. The larger MAUB of time delay implies less conservatism of delay-dependent stability criterion. Numerical example is illustrated to show the applicability of the proposed method.

2 Problem Description and Preliminaries

In this paper, consider the RNNs with interval time-varying delay described by the following delay differential equation

$$\dot{x}_i(t) = -a_i x_i(t) + \sum_{j=1}^{n} b_{ij} g_j(x_j(t)) + \sum_{j=1}^{n} c_{ij} g_j(x_j(t - \tau(t))) + J_i,$$

$$x_i(s) = \phi_i(s), \quad \text{for} \quad s \in [-h, 0], \quad i = 1, 2, \cdots, n, \tag{1}$$

where $x_i(t)$ corresponds to the state of the ith unit at time t, $a_i > 0$ is real constant, $g_j(\cdot)$ is the activation function, b_{ij} and c_{ij} are normal and delayed connection weight coefficients respectively. J_i denotes the constant external input. The time varying-delay $\tau(t)$ is bounded and satisfying

$$0 < \varphi \le \tau(t) \le h, \quad 0 \le \dot{\tau}(t) \le \mu,$$

where φ, h and μ are real scalars.

Assumption (A1). The neuron activation functions $g_i(\cdot)$ $(i = 1, 2, \ldots, n)$ are assumed to be bounded and satisfy the Lipschitz condition

$$0 \le \frac{g_i(x) - g_i(y)}{x - y} \le l_i, \ \forall \ x, \ y \in \mathbb{R}, \ x \ne y, \quad \text{where } l_i > 0 \text{ are known real scalars}$$

for $i = 1, 2, \ldots, n$.

Then by (A1), we can have

$$g_i(x_i) \le l_i x_i \quad \forall \ x_i \in \mathbb{R} \tag{2}$$

By the fixed point theorem [1], the existence of an equilibrium point of the system (1) is guaranteed. Now let $x^* = (x_1^*, x_2^*, \cdots, x_n^*)^T \in \mathbb{R}^n$ be an equilibrium point of the system (1). The transformations

$$y_i(t) = x_i(t) - x_i^* \text{ and } f_i(y_i(t)) = g_i(y_i(t) + x_i^*) - g_i(x_i^*)$$

yield the following system for equation (1)

$$\dot{y}_i(t) = -a_i y_i(t) + \sum_{j=1}^{n} b_{ij} f_j(y_j(t)) + \sum_{j=1}^{n} c_{ij} f_j(y_j(t - \tau(t))). \tag{3}$$

By definition of f_i and by (A1), we get

$$f_i(y_i(t)) \leq l_i y_i(t). \tag{4}$$

The compact form of the system (3) is equivalent to

$$\dot{y}(t) = -Ay(t) + Bf(y(t)) + Cf(y(t - \tau(t))) \tag{5}$$

where $y(t) = [y_1(t), y_2(t), \cdots, y_n(t)]^T$, $f(y(\cdot)) = [f_1(y_1(\cdot)), f_2(y_2(\cdot)), \cdots, f_n(y_n(\cdot))]^T$, $A = \text{diag}\{a_1, a_2, \cdots, a_n\}$, $B = (b_{ij})_{n \times n}$ and $C = (c_{ij})_{n \times n}$.

T-S fuzzy model is composed of a set of fuzzy implication and each implication is expressed by a linear model. The RNNs with interval time-varying delay (5) can be represented by T-S fuzzy model. The kth rule of this fuzzy model is described by the following IF-THEN form:

Plant Rule k: **IF** $\{\theta_1(t) \text{ is } M_{k1}\}$ and $\cdots$ and $\{\theta_r(t) \text{ is } M_{kr}\}$ **THEN**

$$\dot{y}(t) = -A_k y(t) + B_k f(y(t)) + C_k f(y(t - \tau(t))), \tag{6}$$

where $\theta_i(t)$ $(i = 1, 2, \ldots, r)$ is known variable, M_{kl} $(k \in \{1, 2, \ldots, m\}, l \in \{1, 2, \ldots, r\})$ is the fuzzy set and m is number of model rules. By inferring from the fuzzy models, the final output of delayed fuzzy recurrent neural networks (FRNNs) is obtained by

$$\dot{y}(t) = \sum_{k=1}^{m} \omega_k(\theta(t))\{-A_k y(t) + B_k f(y(t)) + C_k f(y(t - \tau(t)))\} \Big/ \sum_{k=1}^{m} \omega_k(\theta(t)) \tag{7}$$

The weight and average weight of each fuzzy rule are denoted by

$$\omega_k(\theta(t)) = \Pi_{l=1}^{r} M_{kl}(\theta(t)) \text{ and } \eta_k(\theta(t)) = \omega_k(\theta(t)) \Big/ \sum_{k=1}^{m} \omega_k(\theta(t)).$$

The term $M_{kl}(\theta(t))$ is the grade membership of $\theta(t)$ in M_{kl}. We assume that $\omega_k(\theta(t)) > 0$, $\forall\, k \in \{1, 2, \ldots, m\}$ and $\sum_{k=1}^{m} \eta_k(\theta(t)) = 1$, $\forall\, t \geq 0$.

So the fuzzy system (7) can be rewritten as

$$\dot{y}(t) = \sum_{k=1}^{m} \eta_k(\theta(t))\{-A_k y(t) + B_k f(y(t)) + C_k f(y(t - \tau(t)))\}. \tag{8}$$

Now we are stating the following lemmas which will be more useful in the sequel.

Lemma 2.1.   (Han [18]) For any constant matrix $R \in \mathbb{R}^{n \times n}$, $R = R^T > 0$, scalar h with $0 \leq \tau(t) \leq h < \infty$ and a vector-valued function $\dot{x} : [t - h, t] \rightarrow \mathbb{R}^n$, the following integration is well defined

$$-\tau(t) \int_{t-\tau(t)}^{t} \dot{x}^T(s) R \dot{x}(s) ds \leq \begin{bmatrix} x(t) \\ x(t - \tau(t)) \end{bmatrix}^T \begin{bmatrix} -R & R \\ * & -R \end{bmatrix} \begin{bmatrix} x(t) \\ x(t - \tau(t)) \end{bmatrix}.$$

For deriving stability criterion for the system (8), consider the following two cases for FRNNs.

Case I: $\tau(t)$ is a continuous function satisfying
$$0 < \varphi \leq \tau(t) \leq h < \infty, \quad \forall \ t \geq 0,$$
Case II: $\tau(t)$ is a differentiable function satisfying
$$0 < \varphi \leq \tau(t) \leq h < \infty, \quad \dot{\tau}(t) \leq \mu < \infty, \quad \forall \ t \geq 0,$$
where φ, $\bar{\tau}$ and μ are real scalars.

3   Main Results

In this section, introduce LKFs to derive some new delay-dependent asymptotic stability criterion for FRNNs with interval time-varying delay described by (8).

Theorem 3.1. Under case I and assumption (A1), for given scalars $\varphi > 0$ and $h > 0$, the equilibrium point of FRNNs (8) is asymptotically stable if there exist positive definite symmetric matrices P and $R_i > 0$, symmetry matrices Q_i, U_i and V_i satisfying

$$\tilde{Q}_i = \begin{bmatrix} Q_i & U_i \\ * & V_i \end{bmatrix} > 0 \quad (i = 0, 1, 2, \cdots, N),$$

$T_j = diag\{t_{j1}, t_{j2}, \cdots, t_{jn}\} \geq 0 \ (j = 0, 1, 2, \cdots, N+2)$ and $\Lambda = diag\{\lambda_1, \lambda_2, \cdots, \lambda_n\} \geq 0$ such that the following LMI holds for all $k \in \{1, 2,, \ldots, m\}$ and $l \in \{1, 2, \ldots, N\}$,

$$\Omega^k = \begin{bmatrix} \Omega_{11}^k & \Omega_{12}^k & \varphi\Gamma_1^T R_0 & \delta\Gamma_1^T \mathcal{R} \\ * & \Omega_{22}^k & \varphi\Gamma_2^T R_0 & \delta\Gamma_2^T \mathcal{R} \\ * & * & -R_0 & 0 \\ * & * & * & -\mathcal{R} \end{bmatrix} < 0, \tag{9}$$

where

$$\Omega_{11}^k = \begin{bmatrix} \gamma_0 & R_0 & 0 & 0 & \ldots & 0 & 0 & \ldots & 0 & 0 & 0 \\ * & \gamma_1 & R_1 & 0 & \ldots & 0 & 0 & \ldots & 0 & 0 & 0 \\ * & * & \gamma_2 & R_2 & \ldots & 0 & 0 & \ldots & 0 & 0 & 0 \\ \vdots & \vdots & \vdots & \vdots & \ddots & \vdots & \vdots & \ddots & \vdots & \vdots & \vdots \\ * & * & * & * & \ldots & \gamma_l & 0 & \ldots & 0 & 0 & R_l \\ * & * & * & * & \ldots & * & \gamma_{l+1} & \ldots & 0 & 0 & R_l \\ \vdots & \vdots & \vdots & \vdots & \ddots & \vdots & \vdots & \ddots & \vdots & \vdots & \vdots \\ * & * & * & * & \ldots & * & * & \ldots & \gamma_N & R_N & 0 \\ * & * & * & * & \ldots & * & * & \ldots & * & \gamma_{N+1} & 0 \\ * & * & * & * & \ldots & * & * & \ldots & * & * & -2R_l \end{bmatrix}_{(N+3)\times(N+3)}, \tag{10}$$

$$\Omega_{12}^k = \begin{bmatrix} \alpha_0 & 0 & 0 & 0 & \ldots & 0 & 0 & \ldots & 0 & 0 & PC_k \\ 0 & \alpha_1 & 0 & 0 & \ldots & 0 & 0 & \ldots & 0 & 0 & 0 \\ 0 & 0 & \alpha_2 & 0 & \ldots & 0 & 0 & \ldots & 0 & 0 & 0 \\ \vdots & \vdots & \vdots & \vdots & \ddots & \vdots & \vdots & \ddots & \vdots & \vdots & \vdots \\ 0 & 0 & 0 & 0 & \ldots & \alpha_l & 0 & \ldots & 0 & 0 & 0 \\ 0 & 0 & 0 & 0 & \ldots & 0 & \alpha_{l+1} & \ldots & 0 & 0 & 0 \\ \vdots & \vdots & \vdots & \vdots & \ddots & \vdots & \vdots & \ddots & \vdots & \vdots & \vdots \\ 0 & 0 & 0 & 0 & \ldots & 0 & 0 & \ldots & \alpha_N & 0 & 0 \\ 0 & 0 & 0 & 0 & \ldots & 0 & 0 & \ldots & 0 & \alpha_{N+1} & 0 \\ 0 & 0 & 0 & 0 & \ldots & 0 & 0 & \ldots & 0 & 0 & LT_{N+2}^T \end{bmatrix}_{(N+3)\times(N+3)}, \tag{11}$$

$$\Omega_{22}^k = \begin{bmatrix} \beta_0 & 0 & 0 & 0 & \dots & 0 & 0 & \dots & 0 & 0 & \Lambda C_k \\ * & \beta_1 & 0 & 0 & \dots & 0 & 0 & \dots & 0 & 0 & 0 \\ * & * & \beta_2 & 0 & \dots & 0 & 0 & \dots & 0 & 0 & 0 \\ \vdots & \vdots & \vdots & \vdots & \ddots & \vdots & \vdots & \ddots & \vdots & \vdots & \vdots \\ * & * & * & * & \dots & \beta_l & 0 & \dots & 0 & 0 & 0 \\ * & * & * & * & \dots & * & \beta_{l+1} & \dots & 0 & 0 & 0 \\ \vdots & \vdots & \vdots & \vdots & \ddots & \vdots & \vdots & \ddots & \vdots & \vdots & \vdots \\ * & * & * & * & \dots & * & * & \dots & \beta_N & 0 & 0 \\ * & * & * & * & \dots & * & * & \dots & * & \beta_{N+1} & 0 \\ * & * & * & * & \dots & * & * & \dots & * & * & -2T_{N+2} \end{bmatrix}_{(N+3)\times(N+3)}, \tag{12}$$

$$\gamma_i = \begin{cases} -2PA_k + Q_0 - R_0, & i = 0 \\ Q_i - Q_{i-1} - R_i - R_{i-1}, & i = 1, 2 \dots N \\ -Q_N - R_N, & i = N+1 \end{cases} \tag{13}$$

$$\alpha_i = \begin{cases} PB_k + U_0 + LT_0^T - A_k^T \Lambda, & i = 0 \\ U_i - U_{i-1} + LT_i^T, & i = 1, 2 \dots N \\ -U_N + LT_{N+1}^T, & i = N+1 \end{cases} \tag{14}$$

$$\beta_i = \begin{cases} 2\Lambda B_k + V_0 - 2T_0, & i = 0 \\ V_i - V_{i-1} - 2T_i, & i = 1, 2 \dots N \\ -V_N - 2T_{N+1}, & i = N+1 \end{cases} \tag{15}$$

$$\mathcal{R} = R_1 + R_2 + \cdots + R_N, \tag{16}$$

$$\Gamma_1 = [-A_k \ \ 0 \ \ 0 \ \ \dots \ \ 0 \ \ 0 \ \ 0]_{1\times(N+3)}, \tag{17}$$

$$\Gamma_2 = [B_k \ \ 0 \ \ 0 \ \ \dots \ \ 0 \ \ 0 \ \ C_k]_{1\times(N+3)}, \tag{18}$$

$L = diag\{l_1, l_2, \cdots, l_n\}$ and '0' represents zero matrix with appropriate dimension.

Proof. The delay interval $[-h, 0]$ is decomposed into $[-h, -\varphi] \cup [-\varphi, 0]$. Let $N > 0$ be a positive integer and τ_i $(i = 0, 1, 2, \cdots, N)$ be some scalars satisfying $\varphi = \tau_0 < \tau_1 < \tau_2 < \cdots < \tau_N = h$. Then the delay interval $[-h, -\varphi]$ is uniformly decomposed into N segments, that is,
$[-h, -\varphi] = \cup_{i=1}^N [-\tau_i, \tau_{i-1}].$

Let $\delta = (h - \varphi)/N$ be the length of the interval $[-\tau_i, -\tau_{i-1}]$ for $i = 1, 2, \cdots, N$ respectively. The following LKFs are introduced for analyzing the asymptotic stability for the system (8)

$$V(y_t) = V_1(y_t) + V_2(y_t) + V_3(y_t), \tag{19}$$

where

$$V_1(y_t) = y^T(t)Py(t) + 2\sum_{i=1}^{n} \lambda_i \int_0^{y_i(t)} f_i(s)ds,$$

$$V_2(y_t) = \int_{-\varphi}^{0} \begin{bmatrix} y(t+s) \\ f(y(t+s)) \end{bmatrix}^T \tilde{Q}_0 \begin{bmatrix} y(t+s) \\ f(y(t+s)) \end{bmatrix} ds + \sum_{j=1}^{N} \int_{-\tau_j}^{-\tau_{j-1}} \begin{bmatrix} y(t+s) \\ f(y(t+s)) \end{bmatrix}^T \tilde{Q}_j \begin{bmatrix} y(t+s) \\ f(y(t+s)) \end{bmatrix} ds,$$

$$V_3(y_t) = \varphi \int_{-\varphi}^{0} \int_{t+\theta}^{t} \dot{y}^T(s)R_0\dot{y}(s)dsd\theta + \sum_{j=1}^{N} \delta \int_{-\tau_j}^{-\tau_{j-1}} \int_{t+\theta}^{t} \dot{y}^T(s)R_j\dot{y}(s)dsd\theta.$$

Taking the derivative of $V(y_t)$ in (18) with respect to t along the trajectory of (8) yields

$$\dot{V}(y_t) = \dot{V}_1(y_t) + \dot{V}_2(y_t) + \dot{V}_3(y_t), \tag{20}$$

where

$$\dot{V}_1(y_t) = y^T(t)2P\dot{y}(t) + f^T(y(t))2\Lambda\dot{y}(t) \tag{21}$$

$$= \sum_{k=1}^{m} \eta_k(\theta(t))[y^T(t)(-2PA_k)y(t) + y^T(t)(2PB_k)f(y(t))$$
$$+ y^T(t)(2PC_k)f(y(t-\tau(t))) + f^T(y(t))(-2\Lambda A_k)y(t)$$
$$+ f^T(y(t))(2\Lambda B_k)f(y(t)) + f^T(y(t))(2\Lambda C_k)f(y(t-\tau(t)))],$$

$$\dot{V}_2(y_t) = [y^T(t)Q_0 y(t) + y^T(t-\varphi)(Q_1 - Q_0)y(t-\varphi) + y^T(t-\tau_1)(Q_2 - Q_1)$$
$$\times y(t-\tau_1) + \cdots + y^T(t-\tau_N)(-Q_N)y(t-\tau_N) + y^T(t)2U_0 f(y(t))$$
$$+ y^T(t-\varphi)2(U_1 - U_0)f(y(t-\varphi)) + y^T(t-\tau_1)2(U_2 - U_1)f(y(t-\tau_1))$$
$$+ \cdots + y^T(t-\tau_N)(-2U_N)f(y(t-\tau_N)) + f^T(y(t))V_0 f(y(t))$$
$$+ f^T(y(t-\varphi))(V_1 - V_0)f(y(t-\varphi)) + f^T(y(t-\tau_1))(V_2 - V_1)$$
$$\times f(y(t-\tau_1)) + \cdots + f^T(y(t-\tau_N))(-V_N)f(y(t-\tau_N))], \tag{22}$$

$$\dot{V}_3(y_t) = \dot{y}^T(t)(\varphi^2 R_0)\dot{y}(t) - \varphi \int_{t-\varphi}^{t} \dot{y}^T(s)R_0\dot{y}(s)ds$$

$$+ \dot{y}^T(t)(\delta^2 \mathcal{R})\dot{y}(t) - \sum_{j=1}^{N} \delta \int_{t-\tau_j}^{t-\tau_{j-1}} \dot{y}^T(s)R_j\dot{y}(s)ds. \tag{23}$$

From the sector condition (4), the following inequalities hold

$$f^T(y(t))2LT_0 y(t) - f^T(y(t))2T_0 f(y(t) \geq 0, \tag{24}$$

$$\sum_{j=0}^{N} \left[f^T(y(t-\tau_j))2LT_{j+1}y(t-\tau_j) - f^T(y(t-\tau_j))2T_{j+1}f(y(t-\tau_j)) \right] \geq 0, \tag{25}$$

$$f^T(y(t-\tau(t)))2LT_{N+2}y(t-\tau(t))) - f^T(y(t-\tau(t)))2T_{N+2}f(y(t-\tau(t))) \geq 0. \tag{26}$$

The following inequalities are obtained from the the system (8)

$$\dot{y}^T(t)(\varphi^2 R_0 + \delta^2 \mathcal{R})\dot{y}(t) = \sum_{k=1}^{m} \eta_k(\theta(t))\left(\xi^T(t)[\Gamma^T(\varphi^2 R_0 + \delta^2 \mathcal{R})\Gamma]\xi(t)\right) \tag{27}$$

where

$$\xi(t) = [\xi_1^T(t)\ \xi_2^T(t)]^T, \quad \Gamma = [\Gamma_1 \Gamma_2],$$
$$\xi_1(t) = [y^T(t)\ y^T(t-\varphi)\ y^T(t-\tau_1)\ y^T(t-\tau_2) \ldots\ y^T(t-\tau_N)\ y^T(t-\tau(t))]^T, \tag{28}$$
$$\xi_2(t) = [f^T(y(t))\ f^T(y(t-\varphi))\ f^T(y(t-\tau_1)) \ldots\ f^T(y(t-\tau_N))\ f^T(y(t-\tau(t)))]^T, \tag{29}$$

and Γ_1, Γ_2 are defined in (17) and (18) respectively.

Using (21)-(23) and (27) in (20) and adding (24)-(26) to (20), we obtain the following

$$\dot{V}(y_t) = \sum_{k=1}^{m} \eta_k(\theta(t))\left[\xi^T(t)\left(\Phi^k + \Gamma^T(\varphi^2 R_0 + \delta_2 \mathcal{R})\Gamma\right)\xi(t)\right] -$$

$$\varphi \int_{t-\varphi}^{t} \dot{y}^T(s)R_0\dot{y}(s)ds - \sum_{j=1}^{N}\delta \int_{t-\tau_j}^{t-\tau_{j-1}} \dot{y}^T(s)R_j\dot{y}(s)ds, \qquad (30)$$

where

$$\Phi^k = \begin{bmatrix} \Phi_{11}^k & \Omega_{12}^k \\ * & \Omega_{22}^k \end{bmatrix}, \qquad (31)$$

and

$$\Phi_{11}^k = \begin{bmatrix} -2PA_k+Q_0 & 0 & 0 & 0 & \cdots & 0 & 0 & \cdots & 0 & 0 & 0 \\ * & Q_1-Q_0 & 0 & 0 & \cdots & 0 & 0 & \cdots & 0 & 0 & 0 \\ * & * & Q_2-Q_1 & 0 & \cdots & 0 & 0 & \cdots & 0 & 0 & 0 \\ \vdots & \vdots & \vdots & \ddots & & \vdots & & \ddots & \vdots & \vdots & \vdots \\ * & * & * & * & \cdots & Q_l-Q_{l-1} & 0 & \cdots & 0 & 0 & 0 \\ * & * & * & * & \cdots & * & Q_{l+1}-Q_l & \cdots & 0 & 0 & 0 \\ \vdots & \vdots & \vdots & \vdots & & \vdots & & \ddots & \vdots & \vdots & \vdots \\ * & * & * & * & \cdots & * & * & \cdots & Q_N-Q_{N-1} & 0 & 0 \\ * & * & * & * & \cdots & * & * & \cdots & * & -Q_N & 0 \\ * & * & * & * & \cdots & * & * & \cdots & * & * & 0 \end{bmatrix}, \quad (32)$$

Ω_{12}^k, Ω_{22}^k are defined in (11) and (12) respectively.

As in [7,16] and using the Lemma 2.1, the following inequality is obtained

$$-\varphi \int_{t-\varphi}^{t} \dot{y}^T(s)R_0\dot{y}(s)ds - \sum_{j=1}^{N}\delta \int_{t-\tau_j}^{t-\tau_{j-1}} \dot{y}^T(s)R_j\dot{y}(s)ds \le \xi_1^T(t)\Psi\xi_1(t), (33)$$

where $\xi_1(t)$ is defined in (28) and

$$\Psi = \begin{bmatrix} -R_0 & R_0 & 0 & 0 & \cdots & 0 & 0 & \cdots & 0 & 0 & 0 \\ * & -R_1-R_0 & R_1 & 0 & \cdots & 0 & 0 & \cdots & 0 & 0 & 0 \\ * & * & -R_2-R_1 & R_2 & \cdots & 0 & 0 & \cdots & 0 & 0 & 0 \\ \vdots & \vdots & \vdots & \vdots & \ddots & \vdots & & \ddots & \vdots & \vdots & \vdots \\ * & * & * & * & \cdots & -R_l-R_{l-1} & 0 & \cdots & 0 & 0 & R_l \\ * & * & * & * & \cdots & * & -R_{l+1}-R_l & \cdots & 0 & 0 & R_l \\ \vdots & \vdots & \vdots & \vdots & \ddots & \vdots & & \ddots & \vdots & \vdots & \vdots \\ * & * & * & * & \cdots & * & * & \cdots & -R_N-R_{N-1} & R_N & 0 \\ * & * & * & * & \cdots & * & * & \cdots & * & -R_N & 0 \\ * & * & * & * & \cdots & * & * & \cdots & * & * & -2R_l \end{bmatrix}. \quad (34)$$

Substituting (33) into (30) yields

$$\dot{V}(y_t) \le \sum_{k=1}^{m} \eta_k(\theta(t))\left[\xi^T(t)(\Phi^k + \Psi + \Gamma^T(\varphi^2 R_0 + \delta^2 \mathcal{R})\Gamma)\xi(t)\right]$$

$$\le \sum_{k=1}^{m} \eta_k(\theta(t))\left[\xi^T(t)(\Delta^k + \Gamma^T(\varphi^2 R_0 + \delta^2 \mathcal{R})\Gamma)\xi(t)\right], \qquad (35)$$

where

$$\Delta^k = \begin{bmatrix} \Omega_{11}^k & \Omega_{12}^k \\ * & \Omega_{22}^k \end{bmatrix}, \quad \text{where } \Omega_{11}^k \text{ is defined in (10).}$$

A sufficient condition for asymptotic stability of the system described by (8) is that there exist positive definite symmetric matrices P and R_i, symmetric matrices Q_i, U_i and V_i satisfying

$$\tilde{Q}_i = \begin{bmatrix} Q_i & U_i \\ * & V_i \end{bmatrix} > 0 \quad (i = 0, 1, 2, \cdots, N)$$

and diagonal matrices $T_j \geq 0$ $(j = 0, 1, 2, \cdots, N + 2)$, and $\Lambda \geq 0$ such that

$$\dot{V}(y_t)) \leq \sum_{k=1}^m \eta_k(\theta(t)) \left[\xi^T(t)(\Delta^k + \Gamma^T(\varphi^2 R_0 + \delta^2 \mathcal{R})\Gamma)\xi(t) \right]$$

$$\leq -(\lambda)y^T(t)y(t) < 0 \quad \forall \, y(t) \neq 0 \quad \text{where } \lambda > 0. \tag{36}$$

In order to guarantee (36), we require the following condition

$$\Delta^k + \Gamma^T(\varphi^2 R_0 + \delta^2 \mathcal{R})\Gamma < 0. \tag{37}$$

Notice that by applying Schur complement lemma, (37) is equivalent to $\Omega^k < 0$, which is defined in (9). Considering all possibilities of l in the set $\{1, 2, \ldots, N\}$, we arrive at the conclusion that that (9) holds for any $l \in \{1, 2, \ldots, N\}$ and $k \in \{1, 2, \ldots, m\}$. This completes the proof.

We now consider the Case II, in which $\tau(t)$ is a differentiable function and the derivative of the time-varying delay is available. We will use this additional information to provide a less conservative result. For this goal, we will modify $V(y_t)$ as

$$\hat{V}(y_t) = V(y_t) + \int_{t-\tau(t)}^t \begin{bmatrix} y(s) \\ f(y(s)) \end{bmatrix}^T W \begin{bmatrix} y(s) \\ f(y(s)) \end{bmatrix} ds, \tag{38}$$

where W is positive definite symmetric matrix. Now, by employing the LKF candidate (38) and using the similar proof of Theorem 3.1, we have the following.

Theorem 3.3. Under case II and Assumption 1, for a given scalars $\varphi > 0$ and $h > 0$, the equilibrium point of FRNNs (8) is asymptotically stable, if there exist positive definite symmetric matrices P, W and R_i, symmetric matrices Q_i, U_i and V_i satisfying $\tilde{Q}_i = \begin{bmatrix} Q_i & U_i \\ * & V_i \end{bmatrix} > 0$ $(i = 0, 1, 2, \cdots, N)$, diagonal matrices $T_j = diag\{t_{j1}, t_{j2}, \cdots, t_{jn}\} \geq 0$ $(j = 0, 1, 2, \cdots, N + 2)$ and $\Lambda = diag\{\lambda_1, \lambda_2, \cdots, \lambda_n\} \geq 0$ such that the following LMI holds for all $k \in \{1, 2, \ldots, m\}$ and $l \in \{1, 2, \ldots, N\}$,

$$\tilde{\Omega}^k = \begin{bmatrix} \tilde{\Omega}_{11}^k & \Omega_{12}^k & \varphi \Gamma_1^T R_0 & \delta \Gamma_1^T \mathcal{R} \\ * & \tilde{\Omega}_{22}^k & \varphi \Gamma_2^T R_0 & \delta \Gamma_2^T \mathcal{R} \\ * & * & -R_0 & 0 \\ * & * & * & -\mathcal{R} \end{bmatrix} < 0, \tag{39}$$

where

$$\tilde{\Omega}_{11}^k = \Omega_{11}^k + diag\{W, 0, \ldots, 0, -(1-\mu)W\},$$
$$\tilde{\Omega}_{22}^k = \Omega_{22}^k + diag\{W, 0, \ldots, 0, -(1-\mu)W\},$$

Ω_{11}^k, Ω_{12}^k, Ω_{22}^k and Γ_i $(1 = 1, 2)$ are defined in (10-(12) and (17), (18) respectively.

Remark 3.4. In the above cases, if we put $\varphi = 0$, we will get the system (8) with delay $\tau(t)$ satisfying $0 \le \tau(t) \le h$. Then we will get another result as described in the following corollary.

Corollary 3.5. Under case I and Assumption A1, for a given scalar $h > 0$, the equilibrium point of FRNNs (8) is asymptotically stable, if there exist positive definite symmetric matrices P and R_i, symmetric matrices Q_i, U_i and V_i satisfying $\tilde{Q}_i = \begin{bmatrix} Q_i & U_i \\ * & V_i \end{bmatrix} > 0$, $(i = 1, 2, \cdots, N)$, diagonal matrices $T_j = diag\{t_{j1}, t_{j2}, \cdots, t_{jn}\} \ge 0$ $(j = 0, 1, 2, \cdots, N+1)$ and $\Lambda = diag\{\lambda_1, \lambda_2, \cdots, \lambda_n \ge 0\}$ such that the following LMI holds for all $k \in \{1, 2, , \ldots, m\}$ and $l \in \{1, 2, \ldots, N\}$,

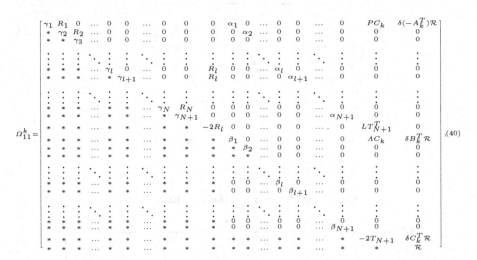

$\gamma_1 = -2PA_k + Q_1 - R_1$, $\alpha_1 = PB_k + U_1 + LT_1^T - A_k^T \Lambda$, and $\beta_1 = 2\Lambda B_k + V_1 - 2T_1$, γ_i, α_i and β_i for $(i = 2, 3, \cdots, N+1)$ are already defined in (13)-(15). It should be pointed out here is the dimension of Ω_{11}^k is $(2N + 5) \times (2N + 5)$.

4 Numerical Example

In this section, the following example is provided to show the effectiveness of the proposed methods.

Example 4.1. Consider the fuzzy Hopfield neural networks (FHNNs) with interval time varying delay together with i^{th} rule defined by

Plant Rule 1. IF $\theta(t) = z_2(t)/0.5$ is about 0 **THEN**
$$dz(t) = \Big[-A_1 z(t) + B_1 f(z(t)) + C_1 f(z(t - \tau(t)))\Big]dt$$

Plant Rule 2. IF $\theta(t) = z_2(t)/0.5$ is about π or $-\pi$ **THEN**
$$dz(t) = \Big[-A_2 z(t) + B_2 f(z(t)) + C_2 f(z(t - \tau(t)))\Big]dt, \text{ where}$$

$$A_1 = \begin{bmatrix} 2 & 0 \\ 0 & 2 \end{bmatrix}, \quad B_1 = \begin{bmatrix} 0 & 0.2 \\ 0.2 & 0 \end{bmatrix}, \quad C_1 = \begin{bmatrix} 1.8 & 1 \\ 2 & 1.8 \end{bmatrix},$$

$$A_2 = \begin{bmatrix} 3 & 0 \\ 0 & 3 \end{bmatrix}, \quad B_2 = \begin{bmatrix} 1 & 0 \\ 1 & 0 \end{bmatrix}, \quad C_2 = \begin{bmatrix} 2.6 & 0 \\ 0 & 2.5 \end{bmatrix},$$

Table 1. MAUB of $\tau(t)$

φ	N=2	3	4	5
0	2.7780	3.2913	3.6948	4.0260
0.512	3.7140	3.7749	3.9664	4.1560

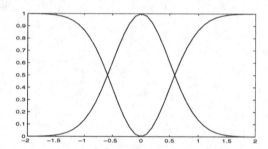

Fig. 1. Membership functions for Example.1

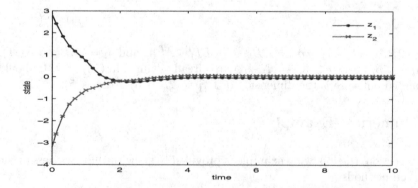

Fig. 2. State Trajectories for Example.1

with $f(z) = (1/5) \tanh(z)$. The time varying delay is assumed as $\tau(t) = 0.5 + 0.5|\sin(t)|$. Assumption (A1) satisfies $L = \text{diag}\{0.2, 0.2\}$. Then the final outputs of the FHNNs are inferred as follows

$$dz(t) = \sum_{k=1}^{2} \eta_k \{-A_k z(t) + B_k f(z(t)) + C_K f(z(t - \tau t))\},$$

where $\eta_1(z_1(t)) = e^{\frac{-z_1^2}{2}}$, $\eta_2(z_2(t)) = 1 - \eta_1(z_1(t))$.

The membership functions are shown in Fig.1. The state trajectories of y_1 and y_2 are shown in Fig. 2.

Our purpose is to determine the various upper bounds h of delay $\tau(t)$, for which the system is asymptotically stable. For $\varphi = 0.512$, Using Theorem 3.1 with $\varphi = 0.512$ and Theorem 3.5, we get the following Table 1.

5 Conclusion

In this paper, the problem of asymptotic stability analysis for FRNNs with interval time-varying delay has been studied. Applying the delay decomposition approach, a novel asymptotic stability criterion has been given in terms of LMIs, which is easily solved by Matlab LMI Toolbox. Finally, numerical example has been presented to illustrate the effectiveness and usefulness of the proposed results.

Acknowledgement. The research is supported by UGC-SAP(DRS-II) and University Grant Commission, Government of India, under Faculty Development Programme, XI plan grant.

References

1. Arik, S.: An analysis of exponential stability of delayed neural networks with time varying delays. Neural Netw. 17, 1027–1031 (2004)
2. Mathiyalagan, K., Sakthivel, R., Anthoni, S.M.: New stability and stabilization criteria for fuzzy neural networks with various activation functions. Phys. Scr. 84, 015007 (2011)
3. Fang, Q., Tong, C.B., Yan, J.: A delay decomposition approach for stability of neural networks with time-varing delay. Chinese Phys. B 18, 5203 (2009)
4. Song, Q., Cao, J.: Impulsive effects on stability of fuzzy Cohen-Grossberg neural networks with time varying delays. IEEE Trans. Syst. Man Cybern. 37, 733–741 (2007)
5. Ali, M.S., Balasubramaniam, P.: Stability analysis of uncertain fuzzy Hopfield neural networks with time delays. Commun. Nonlinear Sci. Numer. Simulat. 14, 2776–2783 (2009)
6. Li, S., Goa, M., Yang, H.: Delay dependent robust stability for uncertain stochastic fuzzy Hopfield neural networks with time varying delays. Fuzzy Sets Syst. 160, 3503–3517 (2009)

7. Balasubramaniam, P., Chandran, R.: Delay decomposition approach to stability analysis for uncertain fuzzy Hopfield neural networks with time varying delay. Commun. Nonlinear Sci. Numer. Simulat. 16, 2098–2108 (2011)
8. Mahmoud, M.S., Xia, Y.: Improved exponential stability analysis for delayed recurrent neural networks. J. Franklin Inst. 348, 201–211 (2010)
9. Tian, J., Zhou, X.: Improved asymptotic stability criteria for neural networks with interval time varying delay. Expert Syst. Appl. 37, 7521–7525 (2010)
10. Li, C., Feng, G.: Delay-interval-dependent stability of recurrent neutral networks with e time varying delay. Neurocomputing 72, 1179–1183 (2009)
11. He, Y., Liu, G., Rees, D., Wu, M.: Stability analysis for neural networks with time-varying interval delay. IEEE Trans. Neural Netw. 18, 1850–1854 (2007)
12. Chen, J., Sun, J., Liu, G.P., Rees, D.: New delay-dependent stability criteria for neural networks with time-varying interval delay. Physics Lett. A 374, 4397–4405 (2010)
13. Souza, F.O., Palhares, R.M.: Interval time-varying delay stability for neural networks. Neurocomputing 73, 2789–2792 (2010)
14. Takagi, T., Sugeno, M.: Fuzzy identification of system and its applications to modelling and control. IEEE. Trans. Syst. Man Cybern. 15, 116–132 (1985)
15. Takagi, T., Sugeno, M.: Stability analysis and design of fuzzy control system. Fuzzy Sets Syst. 45, 135–156 (1992)
16. Zhang, X.M., Han, Q.-L.: A delay decomposition approach to delay dependent stability for linear systems with time varying delays. Int. J. Robust Nonlinear Control 19, 1922–1930 (2009)
17. Xie, L.: Output feedback H_∞ control of systems with parameter uncertainty. Int. J. of Control 63, 741–750 (1996)
18. Han, Q.-L.: A new delay dependent stability creterion for linear neutral systems with norm-bounded uncertainities in all system matrices. Int. J. Syst. Sci. 36, 469–475 (2005)

Stability of Uncertain Reaction-Diffusion Stochastic BAM Neural Networks with Mixed Delays and Markovian Jumping Parameters

C. Vidhya[1] and P. Balasubramaniam[2,*]

[1] Department of Mathematics,
School of Advanced Sciences,
VIT University - Chennai Campus,
Chennai - 600 048
vidhya.c@vit.ac.in
[2] Department of Mathematics,
Gandhigram Rural Institute - Deemed University,
Gandhigram - 624 302, Tamil Nadu, India
balugru@gmail.com

Abstract. In this paper, sufficient conditions are proposed to study robust asymptotic stability of uncertain reaction-diffusion stochastic Bi-directional Associative Memory (BAM) neural network with time-varying delays and Markovian jumping parameters by using suitable Lyapunov-Krasovokii functional, inequality techniques and Linear Matrix Inequality (LMI). A numerical example is given to illustrate the theory.

Keywords: Global asymptotic stability, Reaction-diffusion, Stochastic BAM neural networks.

1 Introduction

The stability theories of recurrent neural networks (RNNs) including cellular neural networks and Hopfield neural networks are very important in the study of associative content - addressable memories, pattern - recognition and optimization. A special class of RNNs that can store bipolar vector pairs is called BAM. The BAM neural networks model was first introduced by Kosko (see, e.g., [1–3]). In BAM neurons in one layer are fully interconnected to the neurons in the other layer, while there are no interconnections among the neurons in the same layer. Recently, some sufficient conditions have been obtained for global asymptotic stability of delayed BAM neural networks. In 2003, Cao [4] had considered the global asymptotic stability and existence of the equilibrium for BAM neural networks with axonal signal transmission delay by using Lyapunov functional method and derived new sufficient conditions.

Diffusion effect cannot be avoided in neural networks when electrons are moving in asymmetric electromagnetic fields. So it is more important to consider that

---

* Corresponding author.

P. Balasubramaniam and R. Uthayakumar (Eds.): ICMMSC 2012, CCIS 283, pp. 283–292, 2012.

the activation function vary in space as well as in time. Recently several authors [5–7] have considered the stability of neural networks with reaction-diffusion terms, which are expressed by partial differential equations. The function of actual delayed systems are influenced by unknown disturbances, which may be regarded as stochastic. In order to fix these problems, the system dynamics are suitably approximated by a stochastic linear or non-linear delayed system. Thus, stochastic delay neural networks have their own characteristic and it is desirable to obtain stability criteria that make full use of these characteristics.

Motivated by the above discussion, in this paper, we aim to challenge the robust asymptotic stability of uncertain reaction-diffusion stochastic BAM neural network with time-varying delays and Markovian jumping parameters. We have derived some easy-to-test sufficient conditions for the robust stability for uncertain reaction-diffusion stochastic BAM neural networks with mixed delays and Markovian jumping parameter. The result of this paper is new and it complements previously known results. An example is employed to demonstrate the usefulness of the proposed result.

2   Model Descriptions

We consider the following reaction diffusion uncertain stochastic BAM neural networks with mixed delays described by

$$
\left.\begin{aligned}
dz_1(t,x) &= \Big[\nabla \cdot (D(t,x,z_1) \circ \nabla z_1) - [C(r(t)) + \Delta C(r(t))]z_1(t,x) \\
&\quad + [W1(r(t)) + \Delta W1(r(t))]g(z_2((t-\tau(t)),x)) \\
&\quad + [W2(r(t)) + \Delta W2(r(t))] \int_{-\infty}^{t} K(t-s)g(z_2(s,x))ds\Big]dt \\
&\quad + \sigma_1(r(t), z_1(t,x), z_2(t-\tau(t),x))d\omega(t) \\
dz_2(t,x) &= \Big[\nabla \cdot (D^*(t,x,z_2) \circ \nabla z_2) - [E(r(t)) + \Delta E(r(t))]z_2(t,x) \\
&\quad + [H1(r(t)) + \Delta H1(r(t))]f(z_1((t-\rho(t)),x)) \\
&\quad + [H2(r(t)) + \Delta H2(r(t))] \int_{-\infty}^{t} N(t-s)f(z_1(s,x))ds\Big]dt \\
&\quad + \sigma_2(r(t), z_2(t,x), z_1(t-\rho(t),x))d\omega(t)
\end{aligned}\right\} \quad (1)
$$

where $z_1(t,x)$ and $z_2(t,x)$ are the state neurons at time t and in space x, f and g denote the signal functions of the neuron at time t and in space x, $C > 0$ and $E > 0$ denote the rate with which the neuron will reset their potential to the resting state in isolation when disconnected from the networks and the external inputs, $W1, W2, H1, H2$ denote the connection weights, smooth functions $D(t,x,z_1) \geq 0$ and $D^*(t,x,z_2) \geq 0$ denote the transmission diffusion operators along the neuron, $\tau(t)$ and $\rho(t)$ denote bounded time - varying delays in the state and respectively satisfying $0 \leq \tau(t) < \tau$ and $0 \leq \rho(t) < \rho$. Further $\dot{\tau}(t)$ and $\dot{\rho}(t)$ are the derivatives of $\tau(t)$ and $\rho(t)$ respectively satisfying $\dot{\tau}(t) \leq \eta < 1$ and $\dot{\rho}(t) \leq \gamma < 1$. $x = [x_1 \ x_2 \ x_3 \ \ldots \ x_l] \in S \subset R^l$, S is a bounded compact set with smooth boundary ∂S and mes $S > 0$ in space R^l, $\|z\|_1$ denotes 1-norm of x and define

$$\|z_1(t,x)\|_2 = \left[\int_S |z_1(t,x)dx|^2\right]^{\frac{1}{2}}.$$

$\circ$ denotes Hadmard product of matrix D and ∇u, $\omega(t)$ is an m-dimensional Brownian motion defined on a complete probability space $(\Omega, \mathcal{F}, \{\mathcal{F}_t\}_{t\geq 0}, \wp)$, $\sigma_1(\cdot) : R_+ \times R^n \times R^m \to R^{n\times m}$ and $\sigma_2(\cdot) : R_+ \times R^m \times R^n \to R^{n\times m}$ be locally Lipschitz continuous and satisfy the linear growth condition as well, $\{r(t), t \geq 0\}$ is a right-continuous Markov process on the probability space which takes values in the finite space $\mathcal{H} = \{1, 2, \cdots, N\}$ with generator $\Gamma = \{\gamma_{ij}\}$ $(i, j \in \mathcal{H})$ (also called jumping transfer matrix) given by

$$P\{r(t+\Delta) = j|r(t) = i\} = \begin{cases} \gamma_{ij}\Delta + o(\Delta) & if \quad i \neq j \\ 1 + \gamma_{ii}\Delta + o(\Delta) & if \quad i = j \end{cases}$$

$\Delta > 0$ and $\lim_{\Delta \to 0} o(\Delta)/\Delta = 0$, $\gamma_{ij} \geq 0$ is the transition rate from i to j if $i \neq j$ and $\gamma_{ii} = -\sum_{j\neq i} \gamma_{ij}$.

For a fixed network mode, $C(r(t))$, $E(r(t))$, $W1(r(t))$, $W2(r(t))$, $H1(r(t))$, $H2(r(t))$ are known constant matrices with appropriate dimensions.

3 Preliminaries

In this section, we state some results and definitions that are needed to prove the main theorem.

Lemma 1. *[8]   If $u(t,x)$ and $v(t,x)$ are solutions of (2), then*

$$\int_S u^T(t,x)\nabla\cdot(D(t,x,u)\circ\nabla u(t,x))dx = -\int_S (D(t,x,u)\cdot(\nabla u(t,x)\circ\nabla u(t,x)))Edx$$

where

$$\nabla u \circ \nabla u = \left[(\nabla u_1 \circ \nabla u_1) \ \ldots \ (\nabla u_n \circ \nabla u_n)\right]^T, \ \nabla u_i \circ \nabla u_i = \left[(\frac{\partial u_i}{\partial x_1})^2, \ldots, (\frac{\partial u_i}{\partial x_l})^2\right]^T,$$

$$D \cdot (\nabla u \circ \nabla u) = \left[(D_1 \cdot (\nabla u_1 \circ \nabla u_1))\ldots(D_n \cdot (\nabla u_n \circ \nabla u_n))\right], \ D = \left[D_1 \ldots D_n\right]^T,$$

$$D_i = \left[D_{i1} \ \ldots \ D_{il}\right]^T \ and \ E = \left[1 \ldots 1\right]^T for \ (i = 1, 2, \ldots, n).$$

Further,

$$\int_S v^T(t,x)\nabla \cdot (D^*(t,x,v) \circ \nabla v(t,x))dx = -\int_S (D^*(t,x,v) \cdot (\nabla v(t,x) \circ \nabla v(t,x)))Edx.$$

Lemma 2. *[9]   For any positive definite matrix $M > 0$, any scalars a and b with $a < b$ and the vector function $x(t) : [a,b] \to \mathbb{R}^n$ such that the integrations concerned are well defined, the following inequality holds*

$$\left[\int_a^b x(s)ds\right]^T M\left[\int_a^b x(s)ds\right] \leq (b-a)\left[\int_a^b x^T(s)Mx(s)ds\right].$$

Lemma 3. *[10]   Let $x \in \mathbb{R}^n, y \in \mathbb{R}^n$ and any scalar $\varepsilon > 0$. Then we have*

$$2x^T y \le \varepsilon^{-1} x^T x + \varepsilon y^T y.$$

Lemma 4. *[7] Given real matrix Q_1, Q_2 with compatible dimension, the following inequality holds,*

$$g^T(z_2(t,x))Q_2g(z_2(t,x)) \le z_2^T(t,x)L_0^T Q_2 L_0 z_2(t,x),$$

$$f^T(z_1(t,x))Q_1f(z_1(t,x)) \le z_1^T(t,x)L_1^T Q_1 L_1 z_1(t,x).$$

Lemma 5. *[11] Given constant matrices M, P, Q where $P = P^T$ and $Q = Q^T > 0$, then $P + M^T Q^{-1} M < 0$ if and only if*

$$\begin{bmatrix} P & M^T \\ M & -Q \end{bmatrix} < 0, \ or \quad \begin{bmatrix} -Q & M \\ M^T & P \end{bmatrix} < 0.$$

We assume that the activation functions and delay kernel functions satisfy the following properties

(A1) The neuron activation functions f_i and g_j are bounded and Lipschitz continuous, that is, there exist constants $L_{0j} > 0$ and $L_{1i} > 0$ such that

$$|g_j(\xi_1) - g_j(\xi_2)| \le |L_{0j}(\xi_1 - \xi_2)|$$

$$|f_i(\xi_1) - f_i(\xi_2)| \le |L_{1i}(\xi_1 - \xi_2)|$$

for all $\xi_1, \xi_2 \in R$.

(A2) $tr[\sigma_{1i}^T P_{1i} \sigma_{1i}] \le z_1^T(t,x)X_{0i}z_1(t,x) + z_2^T(t - \tau(t), x)X_{1i}z_2(t - \tau(t), x)$

$tr[\sigma_{2i}^T P_{2i} \sigma_{2i}] \le z_2^T(t,x)\overline{X}_{0i}z_2(t,x) + z_1^T((t - \rho(t), x)\overline{X}_{1i}z_2(t - \rho(t), x)$.

(A3) The delay kernels K_{ji}, N_{ij} is a real valued non-negative continuous function defined on $[0, \infty)$ and satisfies

(i) $\int_0^\infty K_{ji}(s)ds = \int_0^\infty N_{ij}(s)ds = 1$,

(ii) $\int_0^\infty sK_{ji}(s)ds < \infty$, $\int_0^\infty sN_{ij}(s)ds < \infty$ and

(iii) there exists a positive number μ such that
$\int_0^\infty se^{\mu s}K_{ji}(s)ds < \infty$,
$\int_0^\infty se^{\mu s}N_{ij}(s)ds < \infty$, for each $(i = 1, 2, \dots n$ and $j = 1, 2, \dots m)$.

4   Main Result

Lemma 6. *The general stochastic system,*

$$dz = F(t, z_1(t), z_2(t))dt + G(t, z_1(t), z_2(t))d\omega(t),$$

$$z_1(t) = \phi_1(t), \ for \ every \ t \in [-\tau, 0],$$

$$z_2(t) = \phi_2(t), \ for \ every \ t \in [-\rho, 0],$$

on $t \in [0, t_1]$ where $F : R_+ \times R^n \times R^m \to R^n$ and $G : R_+ \times R^n \times R^m \to R^{n \times m}$, is robustly globally asymptotically stable in the mean square if there exists a function $V(t, z_1, z_2) \in R_+ \times R^n \times R^m$ which is positive definite in the Lyapunov sense and satisfies

$$\mathcal{L}V(t, z_1(t), z_2(t)) = \frac{\partial V}{\partial t} + grad(V)F + \frac{1}{2}tr(G^T G)Hess(V) < 0.$$

The matrix $Hess(V)$ is the Hessian matrix of second-order partial derivatives.

For our convenience, the equation (1) can be rewritten in the following from

$$\left.\begin{aligned}
dz_1(t, x) &= \Big[\nabla \cdot (D(t, x, z_1) \circ \nabla z_1) - [C_i + \Delta C_i]z_1(t, x) \\
&\quad + [W1_i + \Delta W1_i]g(z_2((t - \tau(t)), x)) \\
&\quad + [W2_i + \Delta W2_i] \int_{-\infty}^{t} K(t - s)g(z_2(s, x))ds \Big] dt \\
&\quad + \sigma_{1i}(z_1(t, x), z_2(t - \tau(t), x))d\omega(t), \\
dz_2(t, x) &= \Big[\nabla \cdot (D^*(t, x, z_2) \circ \nabla z_2) - [E_i + \Delta E_i]z_2(t, x) \\
&\quad + [H1_i + \Delta H1_i]f(z_1((t - \rho(t)), x)) \\
&\quad + [H2_i + \Delta H2_i] \int_{-\infty}^{t} N(t - s)f(z_1(s, x))ds \Big] dt \\
&\quad + \sigma_{2i}(z_2(t, x), z_1(t - \rho(t), x))d\omega(t),
\end{aligned}\right\} \tag{2}$$

where the Markovian process $\{r(t), t \geq 0\}$ takes values in the finite space $\mathcal{H} = 1, 2, \ldots, N$ and $r(t) = i, \forall i \in \mathcal{H}$,
$C(i) = C_i, W1(i) = W1_i, W2(i) = W2_i, E(i) = E_i, H1(i) = H1_i, H2(i) = H2_i$.

Assume that there exist constant matrices with appropriate dimensions $M_i, N1_i, N2_i, N3_i, N4_i, N5_i, N6_i$ such that

$$\Big[\Delta C_i \; \Delta W1_i \; \Delta W2_i \; \Delta E_i \; \Delta H1_i \; \Delta H2_i\Big] = M_i \, \theta(t) \Big[N1_i \; N2_i \; N3_i \; N4_i \; N5_i \; N6_i\Big]$$

$\theta^T(t)\theta(t) \leq I$, where $\theta(t)$ is an unknown matrix representing the parameter uncertainty.

Theorem 1. The equilibrium solution to the model (2) is robustly globally asymptotically stable in the mean square, if there exists symmetric positive definite matrices $P_{1i}, P_{2i}, Q_1, Q_2, R_1, R_2, G_1, G_2$ and positive scalars ε_l, $l = 1, 2, 3, \ldots, 10$ such that the following LMI holds

$$\Sigma = \Phi^{(1)} + \Phi^{(2)} < 0, \tag{3}$$

where

$$\Phi^{(1)} = diag\Big[\Lambda_{11}, \; \Lambda_{22}, \; \Omega_{33}, \; \Omega_{44}, \; \Omega_{55}, \; \Omega_{66}, \; \Omega_{77}, \; \Omega_{88}, \; -\varepsilon_1 I, \; -\varepsilon_2 I, \; -\varepsilon_3 I, \; -\varepsilon_4 I,$$

$$- \varepsilon_5 I, \; -\varepsilon_6 I, -\varepsilon_7 I, -\varepsilon_8 I, \; -\varepsilon_9 I, \; -\varepsilon_{10} I\Big],$$

$$\Phi^{(2)} = (\phi_{i,j})_{(18 \times 18)}$$

with

$$\Lambda_{11} = -P_{1_l}C_l - C_l^T P_{1_l} + \varepsilon_1 N1_l^T N1_l + L_1^T Q_1 L_1 + \rho R_1 + L_1^T Q_2 L_1 + X_{0_l} + \sum_{j=1}^{n} \gamma_{lj} P_{1j},$$

$$\Lambda_{22} = -P_{2_l}E_l - E_l^T P_{2_l}^T + \varepsilon_6 N4_l^T N4_l + L_0^T G_1 L_0 + \tau R_2 + L_0^T G_2 L_0 + \overline{X}_{0_l} + \sum_{j=1}^{n} \gamma_{lj} P_{2j},$$

$$\Omega_{33} = \varepsilon_7 L_1^T L_1 + \varepsilon_8 L_1^T N5_l^T N5_l L_1 + \overline{X}_{1_l} - (1-\gamma)L_1^T Q_1 L_1,$$

$$\Omega_{44} = \varepsilon_2 L_0^T L_0 + \varepsilon_3 L_0^T N2_l^T N2_l L_0 + X_{1_l} - (1-\eta)L_0^T Q_2 L_0, \quad \Omega_{55} = -\frac{R_1}{\rho},$$

$$\Omega_{66} = -\frac{R_2}{\tau}, \quad \Omega_{77} = \varepsilon_9 + \varepsilon_{10} N6_l^T N6_l - Q_2, \quad \Omega_{88} = \varepsilon_4 + \varepsilon_5 N3_l^T N3_l - G_2,$$

$$\phi_{1,9} = P_{1_l}M_l, \ \phi_{1,10} = P_{1_l}W1_l, \ \phi_{1,11} = P_{1_l}M_l, \ \phi_{1,12} = P_{1_l}W2_l, \ \phi_{1,13} = P_{1_l}M_l,$$

$$\phi_{2,14} = P_{2_l}M_l, \ \phi_{2,15} = P_{2_l}H1_l, \ \phi_{2,16} = P_{2_l}M_l, \ \phi_{2,17} = P_{1_l}H2_l, \ \phi_{2,18} = P_{2_l}M_l.$$

and remaining other terms are zero.

Proof. Choose a Lyapunov-Krasovskii functional as

$$V(t, z_1(t), z_2(t)) = \int_S \Big\{ V_1(t, z_1(t), z_2(t)) + V_2(t, z_1(t), z_2(t)) \Big\} dx,$$

where

$$
\begin{aligned}
V_1(t, z_1(t), z_2(t)) &= z_1^T(t) P_{1_l} z_1(t) + \int_{t-\rho(t)}^{t} f^T(z_1(s)) Q_1 f(z_1(s)) ds \\
&+ \int_{-\rho}^{0} \int_{t+s}^{t} z_1^T(\theta) R_1 z_1(\theta) d\theta ds + Q_2 \Big[\int_0^{\infty} N(s) \int_{t-s}^{t} f^2(z_1(\theta)) d\theta ds \Big],
\end{aligned}
$$

$$
\begin{aligned}
V_2(t, z_1(t), z_2(t)) &= z_2^T(t) P_{2_l} z_2(t) + \int_{t-\tau(t)}^{t} g^T(z_2(s)) G_1 g(z_2(s)) ds \\
&+ \int_{-\tau}^{0} \int_{t+s}^{t} z_2^T(\theta) R_2 z_2(\theta) d\theta ds + G_2 \Big[\int_0^{\infty} K(s) \int_{t-s}^{t} g^2(z_2(\theta)) d\theta ds \Big].
\end{aligned}
$$

Now calculating the stochastic derivatives of $V(t)$ along the trajectories (2), we obtain the following equation

$$\mathcal{L}V(t, z_1(t), z_2(t)) = \int_S \Big\{ \mathcal{L}V_1(t, z_1(t), z_2(t)) + \mathcal{L}V_2(t, z_1(t), z_2(t)) \Big\} dx.$$

By using Lemma 2, 3, 4 and Assumptions (A2), (A3) after some manipulations and using the Lemma 1 we get

$$\mathcal{L}V(t, z_1(t), z_2(t)) \le \int_S N^T(t, x) \Xi N(t, x) dx,$$

where

$$N^T(t) = \Big[z_1^T(t, x) \ z_2^T(t, x) \ z_1^T(t-\rho(t)x), \ z_2^T(t-\tau(t), x) \ \Big(\int_{t-\rho}^{t} z_1(s, x) ds \Big)^T$$

$$\left(\int_{t-\tau}^{t} z_2(s,x)ds \right)^T \left(\int_{-\infty}^{t} N(t-s)f(z_1(s,x))ds \right)^T$$

$$\left(\int_{-\infty}^{t} K(t-s)g(z_1(s,x))ds \right)^T \right]$$

$$\Xi = diag\{\Omega_{11},\ \Omega_{22},\ \Omega_{33},\ \Omega_{44},\ \Omega_{55},\ \Omega_{66},\ \Omega_{77},\ \Omega_{88}\}$$

with

$$\Omega_{11} = -P_{1\imath}C_{\imath} - C_{\imath}^T P_{1\imath}^T + \varepsilon_1 N1_{\imath}^T N1_{\imath} + L_1^T Q_1 L_1 + \rho R_1 + L_1^T Q_2 L_1 + X_{0\imath} + \sum_{j=1}^{n} \gamma_{\imath j} P_{1j}$$

$$+ \varepsilon_1^{-1} P_{1\imath} M_{\imath} M_{\imath}^T P_{1\imath} + \varepsilon_2^{-1} P_{1\imath} W1_{\imath} W1_{\imath}^T P_{1\imath} + \varepsilon_3^{-1} P_{1\imath} M_{\imath} M_{\imath}^T P_{1\imath} + \varepsilon_4^{-1} P_{1\imath} W2_{\imath} W2_{\imath}^T P_{1\imath}$$

$$+ \varepsilon_5^{-1} P_{1\imath} M_{\imath} M_{\imath}^T P_{1\imath},$$

$$\Omega_{22} = -P_{2\imath}E_{\imath} - E_{\imath}^T P_{2\imath}^T + \varepsilon_6 N4_{\imath}^T N4_{\imath} + L_0^T G_1 L_0 + \tau R_2 + L_0^T G_2 L_0 + \overline{X}_{0\imath} + \sum_{j=1}^{n} \gamma_{\imath j} P_{2j}$$

$$+ \varepsilon_6^{-1} P_{2\imath} M_{\imath} M_{\imath}^T P_{2\imath} + \varepsilon_7^{-1} P_{2\imath} H1_{\imath} H1_{\imath}^T P_{2\imath} + \varepsilon_8^{-1} P_{2\imath} M_{\imath} M_{2\imath}^T P_{2\imath} + \varepsilon_9^{-1} P_{2\imath} H2_{\imath} H2_{\imath}^T P_{2\imath}$$

$$+ \varepsilon_{10}^{-1} P_{2\imath} M_{\imath} M_{\imath}^T P_{2\imath},$$

$$\Omega_{33} = \varepsilon_7 L_1^T L_1 + \varepsilon_8 L_1^T N5_{\imath}^T N5_{\imath} L_1 + \overline{X}_{1\imath} - (1-\gamma)L_1^T Q_1 L_1,$$

$$\Omega_{44} = \varepsilon_2 L_0^T L_0 + \varepsilon_3 L_0^T N2_{\imath}^T N2_{\imath} L_0 + X_{1\imath} - (1-\eta)L_0^T Q_2 L_0, \quad \Omega_{55} = -\frac{R_1}{\rho},$$

$$\Omega_{66} = -\frac{R_2}{\tau}, \quad \Omega_{77} = \varepsilon_9 + \varepsilon_{10} N6_{\imath}^T N6_{\imath} - Q_2, \quad \Omega_{88} = \varepsilon_4 + \varepsilon_5 N3_{\imath}^T N3_{\imath} - G_2$$

By using the Schur complement Lemma 5 and LMI (3), we can rewrite the above inequality as follows

$$\mathcal{L}V(t, z_1(t), z_2(t)) < 0.$$

By using the Lyapunov-Krasovskii stability theorem, we can conclude that the uncertain reaction-diffusion stochastic BAM system (2) is robustly globally asymptotically stable. This completes the proof.

When Markovian jumping parameter is not present in the system (2) then the system becomes

$$\left.\begin{aligned}
dz_1(t,x) &= \Big[\nabla \cdot (D(t,x,z_1) \circ \nabla z_1) - [C + \Delta C]z_1(t,x) \\
&\quad + [W1 + \Delta W1]g(z_2((t-\tau(t)),x)) \\
&\quad + [W2 + \Delta W2]\int_{-\infty}^{t} K(t-s)g(z_2(s,x))ds \Big]dt \\
&\quad + \sigma_1(t, z_1(t,x), z_2(t-\tau(t),x))d\omega(t), \\
dz_2(t,x) &= \Big[\nabla \cdot (D^*(t,x,z_2) \circ \nabla z_2) - [E + \Delta E]z_2(t,x) \\
&\quad + [H1 + \Delta H1]f(z_1((t-\rho(t)),x)) \\
&\quad + [H2 + \Delta H2]\int_{-\infty}^{t} N(t-s)f(z_1(s,x))ds \Big]dt \\
&\quad + \sigma_2(t, z_2(t,x), z_1(t-\rho(t),x))d\omega(t).
\end{aligned}\right\} \quad (4)$$

Theorem 2. *The equilibrium solution to the model (4) is robustly globally asymptotically stable in the mean square, if there exist symmetric positive definite matrices $P_1, P_2, Q_1, Q_2, R_1, R_2, G_1, G_2$ and positive scalars ε_l, $l = 1, 2, 3, \cdots, 10$ such that the following LMI holds*

$$\Sigma = \Phi^{(1)} + \Phi^{(2)} < 0, \tag{5}$$

where

$$\Phi^{(1)} = diag\Big[\Lambda_{11}, \Lambda_{22}, \Omega_{33}, \Omega_{44}, \Omega_{55}, \Omega_{66}, \Omega_{77}, \Omega_{88}, -\varepsilon_1 I, -\varepsilon_2 I, -\varepsilon_3 I, -\varepsilon_4 I,$$

$$-\varepsilon_5 I, -\varepsilon_6 I, -\varepsilon_7 I, -\varepsilon_8 I, -\varepsilon_9 I, -\varepsilon_{10} I\Big],$$

$$\Phi^{(2)} = (\phi_{i,j})_{(18\times18)}$$

with

$$\Lambda_{11} = -P_1 C - C^T P_1 + \varepsilon_1 N1^T N1 + L_1^T Q_1 L_1 + \rho R_1 + L_1^T Q_2 L_1 + X_0,$$

$$\Lambda_{22} = -P_2 E - E^T P_2 + \varepsilon_6 N4^T N4 + L_0^T G_1 L_0 + \tau R_2 + L_0^T G_2 L_0 + \overline{X}_0,$$

$$\Omega_{33} = \varepsilon_7 L_1^T L_1 + \varepsilon_8 L_1^T N5^T N5 L_1 + \overline{X}_1 - (1 - \gamma) L_1^T Q_1 L_1,$$

$$\Omega_{44} = \varepsilon_2 L_0^T L_0 + \varepsilon_3 L_0^T N2^T N2 L_0 + X_1 - (1 - \eta) L_0^T Q_2 L_0, \quad \Omega_{55} = -\frac{R_1}{\rho},$$

$$\Omega_{66} = -\frac{R_2}{\tau}, \quad \Omega_{77} = \varepsilon_9 + \varepsilon_{10} N6^T N6 - Q_2, \quad \Omega_{88} = \varepsilon_4 + \varepsilon_5 N3^T N3 - G_2,$$

$$\phi_{1,9} = P_1 M, \quad \phi_{1,10} = P_1 W1, \quad \phi_{1,11} = P_1 M, \quad \phi_{1,12} = P_1 W2, \quad \phi_{1,13} = P_1 M,$$

$$\phi_{2,14} = P_2 M, \quad \phi_{2,15} = P_2 H1, \quad \phi_{2,16} = P_2 M, \quad \phi_{2,17} = P_1 H2, \quad \phi_{2,18} = P_2 M.$$

and remaining other terms are zero.

5   Example

Consider the model uncertain reaction-diffusion stochastic BAM neural networks (2) with the following parameter.

$$\theta(t) = diag\{-0.15 sin(t), -0.15 cos(t)\}, \quad L_0 = L_1 = diag\{0.1, 0.1\}, \quad \rho = \tau = 0.9,$$

$$\eta = \gamma = 0.5, \quad C_{11} = diag\{2.41, 0.1\}, \quad C_{12} = diag\{2.05, 0.7\}, \quad E_1 = diag\{1.41, 1.1\},$$

$$E_2 = diag\{1.15, 1\}, \quad W1_{11} = diag\{1.3333, 1.6667\}, \quad W1_{12} = diag\{1.3, 0.4\},$$

$$W2_{11} = diag\{1.6667, 1.3333\}, \quad W2_{12} = diag\{1.18, 1.1\}, \quad H1_{12} = diag\{1.01, 1.201\},$$

$$H2_{12} = diag\{1.08, 1.1\}, \quad X_{01} = diag\{0.21, 0.12\}, \quad X_{02} = diag\{0.11, 0.12\},$$

$$\overline{X}_{01} = diag\{0.20, 0.10\}, \quad \overline{X}_{02} = diag\{0.22, 0.12\}, \quad X_{11} = diag\{0.21, 0.22\},$$

$$X_{12} = diag\{0.11, 0.11\}, \quad \overline{X}_{11} = diag\{0.11, 0.12\}, \quad \overline{X}_{12} = diag\{0.10, 0.12\},$$

$$H1_{11} = \begin{bmatrix} 0.3333 & 0.6667 \\ 1.3333 & 0.6667 \end{bmatrix}, \quad H2_{11} = \begin{bmatrix} 1.6667 & 0.3333 \\ 0.6667 & 1.3333 \end{bmatrix},$$

$$f_1(\xi) = f_2(\xi) = g_1(\xi) = g_2(\xi) = 0.5(|\xi + 1| - |\xi - 1|),$$

$$N_{ij}(t) = K_{ji}(t) = te^{-t}, \ i,j = 1,2.$$

By Theorem 4.2 and using MATLAB LMI toolbox, we obtained the following feasible solutions

$$P_{11} = \begin{bmatrix} 111.0156 & -1.7643 \\ -1.7643 & 193.0625 \end{bmatrix}, \quad P_{12} = \begin{bmatrix} 20.4857 & -0.2203 \\ -0.2203 & 24.2344 \end{bmatrix},$$

$$P_{21} = \begin{bmatrix} 90.2718 & -13.8740 \\ -13.8740 & 152.1060 \end{bmatrix}, \quad P_{22} = \begin{bmatrix} 2.4820 & -5.1896 \\ -5.1896 & 42.5749 \end{bmatrix},$$

$$R_1 = \begin{bmatrix} 215.1160 & 0.0716 \\ 0.0716 & 233.0988 \end{bmatrix}, \quad R_2 = \begin{bmatrix} 40.7664 & -36.4945 \\ -36.4945 & 326.2113 \end{bmatrix},$$

$$Q_1 = 10^3 \times \begin{bmatrix} 6.8426 & 0.3880 \\ 0.3880 & 2.6222 \end{bmatrix}, \quad Q_2 = 10^3 \times \begin{bmatrix} 2.2701 & 0.1287 \\ 0.1287 & 2.2975 \end{bmatrix},$$

$$G_1 = \begin{bmatrix} 291.5248 & -11.1768 \\ -11.1768 & 378.9172 \end{bmatrix}, \quad G_2 = 10^3 \times \begin{bmatrix} 1.6889 & 0.1262 \\ 0.1262 & 0.9150 \end{bmatrix}.$$

The above result shows that all conditions stated in Theorem 1 have been satisfied. Hence the uncertain reaction-diffusion stochastic BAM neural networks with mixed delays and Markovian parameters is robustly globally asymptotically stable in the mean square.

6   Conclusion

In this paper, sufficient delay-dependent conditions have been derived for checking robust asymptotic stability of uncertain reaction-diffusion stochastic BAM neural network with time-varying delays and Markovian jumping parameters by using suitable Lyapunov-Krasovokii functional, inequality techniques and LMI. The designed reaction-diffusion stochastic BAM neural networks play an important role in the design and applications.

References

1. Kosko, B.: Adaptive bi-directional associative memories. Appl. Opt. 26, 4947–4960 (1987)
2. Kosko, B.: Bi-directional associative memoriees. IEEE Trans. Syst. Man Cybern. 18, 49–60 (1988)
3. Kosko, B.: Neural Networks and Fuzzy System - A Dynamical Systems Approach to Machine Intelligence. Prentice-Hall, NJ (1992)

4. Cao, J.: Global asymptotic stability of delayed bi-directional associative memory neural networks. Appl. Math. Comput. 142, 333–339 (2003)
5. Balasubramaniam, P., Rakkiyappan, R., Sathy, R.: Delay dependent stability results for fuzzy BAM neural networks with Markovian jumping parameters. Expert Syst. Appl. 38, 121–130 (2011)
6. Balasubramaniam, P., Vidhya, C.: Global asymptotic stability of stochastic BAM neural networks with distributed delays and reaction diffusion terms. J. Comput. Appl. Math. 234, 3458–3466 (2010)
7. Wang, L.S., Xu, D.Y.: Global exponential stability of Hopfield reaction-diffusion neural networks with time-varying delays. Science China (Series F) 46, 466–474 (2005)
8. Wang, L., Zhang, Z., Wang, Y.: Stochastic exponential stability of the delayed reaction-diffusion recurrent neural networks with markovian jumping parameters. Phys. Lett. A 372, 3201–3209 (2008)
9. Gu, K.: An integral inequality in the stability problems of time delay system. In: Proceeding of 39th IEEE Conference on Decision and Control, Sydney, Australia, pp. 2805–2810 (2000)
10. Wang, Z., Liu, Y., Fraser, K., Liu, X.: Stochastic stability of uncertain Hopfield neural networks with discrete and distributed delays. Phys. Lett. A 354, 288–297 (2006)
11. Boyd, S., Ghoui, L., Feron, E., Balakrishnan, V.: Linear Matrix Inequlities in system and control theory. SIAM, Philadephia (1994)

Controllability of Semilinear Evolution Differential Systems in a Separable Banach Space

Bheeman Radhakrishnan*

Department of Mathematics, Bharathiar University,
Coimbatore - 641046, Tamil Nadu, India
radhakrishnanb1985@gmail.com

Abstract. In this paper, we study the controllability of semilinear evolution differential systems with nonlocal initial conditions in a separable Banach space. The results are obtained by using Hausdorff measure of noncompactness and a new calculation method.

Keywords: Controllability, Semilinear differential system, Evolution system, Measure of noncompactness.

1 Introduction

There have appeared a lot of papers concerned with the existence of semilinear evolution equation with nonlocal conditions [4,9,11,12]. Controllability means that it is possible to steer a dynamic system from an arbitrary initial state to an arbitrary final state using the set of admissible controls. Controllability of linear and nonlinear systems represented by ordinary differential equations in finite-dimensional spaces has been extensively investigated. Several papers have appeared on finite dimensional controllability of linear systems [8] and infinite dimensional systems in abstract spaces [6,7]. Of late, the controllability of nonlinear systems in finite-dimensional space by means of fixed point principles [1]. Several authors have extended the concept of controllability to infinite-dimensional spaces by applying semigroup theory [5,10,13]. Controllability of nonlinear systems with different types of nonlinearity has been studied by many authors with the help of fixed point principles [2].

2 Preliminaries

Consider the semilinear evolution differential system with nonlocal conditions of the form

$$x'(t) = A(t)x(t) + Bu(t) + f(t, x(t)), \quad t \in J, \tag{1}$$
$$x(0) = g(x), \tag{2}$$

---

* Corresponding author.

P. Balasubramaniam and R. Uthayakumar (Eds.): ICMMSC 2012, CCIS 283, pp. 293–301, 2012.

where the state variable $x(\cdot)$ takes values in the separable Banach space X with norm $\| \cdot \|$, $A(t) : D_t \subset X \to X$ generates an evolution system $\{U(t,s)\}_{0 \le s \le t \le b}$ on a separable Banach space X, $g : C(J,X) \to X$ and $f : J \times X \to X$ are given mappings. And the control function $u(\cdot)$ is given in $\mathcal{L}^2(J,U)$, a Banach space of admissible control functions with U as a Banach space and $J = [0,b]$. In this paper, we give conditions guaranteeing the controllability results for differential evolution system $(1) - (2)$ without assumptions on the compactness of f, g and the evolution system $\{U(t,s)\}$ is strongly continuous. The results obtained in this paper are based on the new calculation method which employs the technique of measure of noncompactness.

Denote by $\mathbb{B}(x,r)$ the closed ball in X centered at x and with radius r. The collections of all linear and bounded operators from X into itself will be denoted by $\mathcal{B}(X)$. If Y is a subset of X we write $\overline{Y}, ConvY$ in order to denote the closure and convex closure of Y, respectively. Moreover, we denote by $\mathcal{F}_X$ the family of all nonempty and bounded subsets of X and by $\mathcal{G}_X$ its subfamily consisting of relatively compact sets.

Definition 1. *[3] A function $\chi : \mathcal{F}_X \to \mathbb{R}_+$ is said to be a regular measure of noncompactness if it satisfies the following conditions:*

(i) The family $ker\chi = \{Y \in \mathcal{F}_X : \chi(Y) = 0\}$ is nonempty and $ker\chi \subset \mathcal{G}_X$.
(ii) $Y \subset Z \Rightarrow \chi(Y) \le \chi(Z)$.
(iii) $\chi(ConvY) = \chi(Y)$.
(iv) $\chi(\lambda Y + (1-\lambda)Z) \le \lambda\chi(Y) + (1-\lambda)\chi(Z)$, for $\lambda \in [0,1]$.
(v) If $\{Y_n\}_{n=1}^{\infty}$ is a decreasing sequence of nonempty, bounded and closed subset of X such that $Y_{n+1} \subset Y_n$ $(n = 1,2,...)$ and if $\lim_{n\to\infty} \chi(Y_n) = 0$, then the intersection $Y_\infty = \bigcap_{n=1}^{\infty} Y_n$ is nonempty and compact in X.

Remark. Let us notice that the intersection set Y_∞ described in axiom (v) is the kernel of measure of noncompactness χ. In fact, the inequality $\chi(Y_\infty) \le \chi(Y_n)$, for $n = 1,2,...$ implies that $\chi(Y_\infty) = 0$. Hence $Y_\infty \in ker\chi$. This property of the set Y_∞ will be important in the investigations.

Definition 2. *A family of operators $\{U(t,s) : 0 \le s \le t \le b\} \subset \mathcal{L}(X)$ is called a evolution family of operators, if the following properties hold:*

(a) $U(t,s)U(s,\tau) = U(t,\tau)$ and $U(t,t)x = x$, for every $s \le \tau \le t$ and all $x \in X$;
(b) For each $x \in X$, the functions for $(t,s) \to U(t,s)x$ is continuous and $U(t,s) \in \mathcal{L}(X)$ for every $t \ge s$ and
(c) For $0 \le s \le t \le b$, the function $t \to U(t,s)$, for $(s,t] \in \mathcal{L}(X)$, is differentiable with $\frac{\partial}{\partial t}U(t,s) = A(t)U(t,s)$.

The most frequently applied measure of noncompactness is that introduced by Hausdorff and defined in the following way

$$\beta(Y) = \inf\{r > 0 : Y \text{ can be covered by a finite number of balls with radii } r\}.$$

The measure β is called the Hausdorff measure of noncompactness.

The space $\mathcal{C}(J, X)$ is furnished with standard norm $\|x\|_{\mathcal{C}} = \sup\{\|x(t)\| : t \in J\}$. In order to define the measure let us fix a nonempty bounded subset Y of the space $\mathcal{C}(J, X)$ and a positive number $t \in J$. For $y \in Y$ and $\epsilon \geq 0$ denoted by $\omega^t(y, \epsilon)$ the *modulus of continuity* of the function y on the interval $[0, t]$, that is,

$$\omega^t(y, \epsilon) = \sup\{\|y(t_2) - y(t_1)\| : t_1, t_2 \in [0, t], \ |t_2 - t_1| \leq \epsilon\}.$$

Further, let us put

$$\omega^t(Y, \epsilon) = \sup\{\omega^t(y, \epsilon) : y \in Y\},$$

$$\omega_0^t(Y) = \lim_{\epsilon \to 0+} \omega^t(Y, \epsilon).$$

Apart from this put

$$\overline{\beta}(Y) = \sup\{\beta(Y(t)) : t \in J\},$$

where β denotes Hausdroff measure of noncompactnesss in X. Finally, we define the function χ on the family $\mathcal{F}_{\mathcal{C}(J,X)}$ by putting

$$\chi(Y) = \omega_0^t(Y) + \overline{\beta}(Y).$$

It may be shown that the function χ is the measure of noncompactness in the space $\mathcal{C}(J, X)$. The kernel ker χ is the family of all nonempty and bounded subsets Y such that functions belonging to Y are equicontinuous on J and the set $Y(t)$ is relatively compact in X, $t \in J$. Next, for a given set $Y \in \mathcal{F}_{\mathcal{C}(J,X)}$, let us denote $\displaystyle\int_0^t Y(s)ds = \left\{\int_0^t y(s)ds : y \in Y\right\}$, $Y([0,t]) = \{y(s) : y \in Y, \ s \in [0, t]\}$.

Lemma 1. If the Banach space X is *separable* and a set $Y \subset \mathcal{C}(J, X)$ is bounded, then the function $t \to \beta(Y(t))$ is measurable and

$$\beta\left(\int_0^t Y(s)ds\right) \leq \int_0^t \beta(Y(s))ds, \quad \text{for each } t \in J.$$

Lemma 2. Assume that a set $Y \subset \mathcal{C}(J, X)$ is bounded. Then

$$\beta(Y([0,t])) \leq \omega_0^t(Y) + \sup_{s \leq t} \beta(Y(s)), \quad \text{for } t \in J. \tag{3}$$

Proof. For arbitrary $\delta > 0$. Then there exists $\epsilon > 0$ such that

$$\omega^t(Y, \epsilon) \leq \omega_0^t(Y) + \delta/2. \tag{4}$$

Let us take a partition $0 = t_0 < t_1 < ... < t_k = t$ such that $t_i - t_{i-1} \leq \epsilon$, for $i = 1, 2, ..., k$. Then for each $t' \in [t_{i-1}, t_i]$ and $y \in Y$ the following inequality is fulfilled

$$\|y(t') - y(t_i)\| \leq \omega_0^t(Y) + \delta/2. \tag{5}$$

Let us notice that, for each $i = 1, 2, ..., k$ there are points $z_{ij} \in X (j = 1, 2, ..., n_i)$ such that

$$Y(t_i) \subset \bigcup_{j=1}^{n_i} B(z_{ij}, \sup_{s \le t} \beta(Y(s)) + \delta/2). \tag{6}$$

We show that $Y([0, t]) = \bigcup_{i=1}^{k} \bigcup_{j=1}^{n_i} B(z_{ij}, \sup_{s \le t} \beta(Y(s)) + \omega_0^t(Y) + \delta/2). \tag{7}$

Let us choose an arbitrary element $q \in Y([0, t])$. Then, we can find $t' \in [0, t]$ and $y \in Y$, such that $q = y(t')$. Choosing i such that $t' \in [t_{i-1}, t_i]$ and j such that $B(z_{ij}, \sup_{s \le t} \beta(Y(s)) + \delta/2)$ we obtain from (5) and (6)

$$\|q - z_{ij}\| = \|y(t') - z_{ij}\| \le \|y(t') - y(t_i)\| + \|y(t_i) - z_{ij}\|$$
$$\le \omega_0^t(Y) + \sup_{s \le t} \beta(Y(s)) + \delta,$$

and this verifies (7). Condition (7) yields that

$$\beta(Y([0, t])) \le \omega_0^t(Y) + \sup_{s \le t} \beta(Y(s)) + \delta.$$

Letting $\delta \to 0+$ we get (4). □

Definition 3. *A solution* $x(\cdot) \in C([0, b], X)$ *is said to be a mild solution of* $(1) - (2)$ *if* $x(s) = g(x), s \in [0, b]$, *then for each* $0 \le s \le t \le b$, *and the following integral equation is satisfied.*

$$x(t) = U(t, 0)g(x) + \int_0^t U(t, s)Bu(s)ds + \int_0^t U(t, s)f(s, x(s))ds, \quad t \in J.$$

To study the controllability problem we assume the following hypotheses:

(H1) $A(t)$ generates strongly continuous semigroup of a family of evolution operators $U(t, s)$ and there exist constants $N_1 > 0$, $N_0 > 0$ such that

$$\|U(t, s)\| \le N_1, \quad \text{for } 0 \le s \le t \le b, \text{ and } N_0 = \sup\{\|U(s, 0)\| : 0 \le s \le t\}.$$

(H2) The linear operator $W : \mathcal{L}^2(J, U) \to X$ defined by $Wu = \int_0^b U(b, s)Bu(s)ds$ has an inverse operator W^{-1}, which takes values in $\mathcal{L}^2(J, U)/kerW$ and there exists a positive constant K_1 such that $\|BW^{-1}\| \le K_1$.

(H3) (i) The mapping $f : J \times X \to X$ satisfies the Caratheodory condition, i.e. $f(\cdot, x)$ is measurable for $x \in X$ and $f(t, \cdot)$ is continuous for a.e. $t \in J$.

    (ii) The mapping f is bounded on bounded subsets of $C(J, X)$.

    (iii) There exists a constant $m_f > 0$ such that for any bounded set $Y \subset C(J, X)$, the inequality $\beta(f([0, t] \times Y)) \le m_f \beta(Y([0, t]))$ holds for $t \in J$, where $f([0, t] \times Y) = \{f(s, x(s)) : 0 \le s \le t, x \in Y\}$.

(H4) The function $g : C(J, X) \to X$ is continuous and there exists a constant $m_g \ge 0$ such taht $\beta(g(Y)) \le m_g \beta(Y(J))$, for each bounded set $Y \subset C(J, X)$.

(H5) There exists a constant $r > 0$ such that

$$(1+bN_1K_1)\left[N_0 \sup_{x\in\mathbb{B}(\theta,r)} \|g(x)\|+N_1 \sup_{x\in\mathbb{B}(\theta,r)} \int_0^b \|f(\tau,x(\tau))\|d\tau\right]+bN_1K_1\|x_1\| \leq r,$$

for $t \in J$, where $\mathbb{B}(\theta,r)$ in $\mathcal{C}(J,X)$ centered at θ and with radius r.

Definition 4. *Systems (1) and (2) is said to be controllable on the interval J, if for every initial function $x_0 \in X$ and $x_1 \in X$, there exists a control $u \in \mathcal{L}^2(J,U)$ such that the solution $x(\cdot)$ of (1) and (2) satisfies $x(0) = x_0$ and $x(b) = x_1$.*

3   Controllability Result

Using (H2) for an arbitrary function $x(\cdot) \in \mathcal{C}(J,X)$, we define the control

$$u(t) = W^{-1}\left[x_1 - U(b,0)g(x) - \int_0^b U(b,s)f(s,x(s))ds\right](t). \tag{8}$$

Consider $\mathcal{Z} = \mathcal{C}(J,X)$ with norm $\|x\| = \sup\{|x(t)| : t \in J\}$. We shall show that when using the control $u(t)$, the operator $\Psi : \mathcal{Z} \to \mathcal{Z}$ defined by

$$(\Psi x)(t) = U(t,0)g(x) + \int_0^t U(t,s)f(s,x(s))ds$$

$$+ \int_0^t U(t,s)BW^{-1}\left[x_1 - U(b,0)g(x) - \int_0^b U(b,s)f(s,x(\tau))d\tau\right](s)ds$$

has a fixed point $x(\cdot)$. This fixed point is the mild solution of the systems (1) and (2), which implies that the system is controllable on J.

Next, consider the operators v_1, v_2, $v_3 : \mathcal{C}(J,X) \to \mathcal{C}(J,X)$ defined by

$$(v_1 x)(t) = U(t,0)g(x), \quad (v_2 x)(t) = \int_0^t U(t,s)f(s,x(s))ds$$

$$(v_3 x)(t) = \int_0^t U(t,s)BW^{-1}\left[x_1 - U(b,0)g(x) - \int_0^b U(b,\tau)f(\tau,x(\tau))d\tau\right](s)ds.$$

Lemma 3. *Assume that (H1) and (H3) are satisfied and a set $Y \subset \mathcal{C}(J,X)$ is bounded. Then*

$$\omega_0^t(v_2 Y) \leq 2bN_1\beta(f([0,b] \times Y)), \text{ for } t \in J.$$

Proof. Fix $t \in J$ and denote $Q = f([0,t] \times Y)$,

$$q^t(\epsilon) = \sup\left\{\|(U(t_2,s) - U(t_1,s))q\| : 0 \leq s \leq t_1 \leq t_2 \leq t, \ t_2 - t_1 \leq \epsilon, \ q \in Q\right\}.$$

At the beginning we show that

$$\lim_{\epsilon\to 0+} q^t(\epsilon) \leq 2N_1\beta(Q). \tag{9}$$

Suppose contrary. Then there exists a number d such that

$$\lim_{\epsilon \to 0+} q^t(\epsilon) > d > 2N_1\beta(Q). \tag{10}$$

Fix $\delta > 0$ such that

$$\lim_{\epsilon \to 0+} q^t(\epsilon) > d + \delta > d > 2N_1(\beta(Q) + \delta). \tag{11}$$

Condition (10) yields that there exist sequences $\{t_{2,n}\}, \{t_{1,n}\}, \{s_n\} \subset J$ and $\{q_n\} \subset Q$, such that $t_{2,n} \to t'$, $t_{1,n} \to t'$, $s_n \to s$ and

$$\|(U(t_{2,n}, s_n) - U(t_{1,n}, s_n))q_n\| > d. \tag{12}$$

Let the points $z_1, z_2, ..., z_k \in X$ be such that $Q \subset \bigcup_{i=1}^{k} B(z_i, \beta(Q) + \delta)$. Then there exists a point z_i and a subsequence $\{q_n\}$, such that $\{q_n\} \in B(z_i, \beta(Q)+\delta)$, that is, $\|z_j - q_n\| \leq \beta(Q) + \delta$, for $n = 1, 2, ...$ Further we obtain

$$\|U(t_{2,n}, s_n)q_n - U(t_{1,n}, s_n)q_n\| \leq 2N_1(\beta(Q) + \delta) + \|U(t_{2,n}, s_n)z_j - U(t_{1,n}, s_n)z_j\|.$$

Letting $n \to \infty$, and using the properties of the evolution system $\{U(t, s)\}$, from the above estimate we get

$$\limsup_{n\to\infty} \|U(t_{2,n}, s_n)q_n - U(t_{1,n}, s_n)q_n\| \leq 2N_1(\beta(Q) + \delta).$$

This contradicts (9) and (10).

Now, fix $\epsilon > 0$ and t_1, $t_2 \in [0, t]$, $0 \leq t_2 - t_1 \leq \epsilon$. Applying $(H3)$, we get

$$\|(v_1 x)(t_2) - (v_1 x)(t_1)\| \leq \int_0^{t_1} \|(U(t_2, s) - U(t_1, s))f(s, x(s))\|ds + \int_{t_1}^{t_2} \|U(t_2, s))f(s, x(s))\|ds$$

$$+ \int_0^t \|(U(t_2, s) - U(t_1, s))f(s, x(s))\|ds$$

$$+\epsilon N_1 \sup\{\|f(s, x(s))\| : x \in Y\}.$$

Hence, we derive the following inequality

$$\omega^t(v_1 Y, \epsilon)$$

$$\leq \sup\left\{\int_0^t \|(U(t_2, s) - U(t_1, s))f(s, x(s))\|ds : t_1, t_2 \in [0, t], \ 0 \leq t_2 - t_1 \leq \epsilon, \ x \in Y\right\}$$

$$+\epsilon N_1 \sup\{\|f(s, x(s))\| : x \in Y\}.$$

Letting $\epsilon \to 0+$ and hence the result.   □

Lemma 4. Assume that assumptions $(H1)$, $(H4)$ are satisfied and a set $Y \subset C(J, X)$ is bounded. Then $\omega_0^t(v_1 Y) \leq 2N_0(t)\beta(g(Y))$, for $t \in J$.

The simple proof is omitted.

Lemma 5. Assumptions $(H1) - (H4)$ are satisfied and a set $Y \subset C(J, X)$ is bounded. Then

$$\omega_0^t(v_3 Y) \leq 2bN_1 K_1\left(\|x_1\| + N_0\beta(g(Y)) + bN_1\beta(f(Q))\right), \ t \in J.$$

Proof: From the above Lemma 3 and Lemma 4, also fix $\epsilon > 0$ and t_1, $t_2 \in [0, t]$, $0 \leq t_2 - t_1 \leq \epsilon$. Applying $(H3)$ and $(H4)$, we get

$$\|(v_3 x)(t_2) - (v_3 x)(t_1)\|$$

$$\leq K_1 \int_0^{t_1} \|U(t_2, s) - U(t_1, s)\| \Big[\|x_1\| + \|U(b, 0)g(x)\| + \int_0^b \|U(b, \tau)f(\tau, x(\tau))d\tau\|\Big] ds$$

$$+ \epsilon K_1 N_1 \Big[\|x_1\| + N_0 \sup\{\|g(x)\| : x \in Y\} + N_1 \sup\{\|f(s, x(s))\| : x \in Y\}\Big].$$

Hence we derive the following inequality

$$\omega_0^t(v_3 Y) \leq \sup \Big\{ K_1 \int_0^{t_1} \|U(t_2, s) - U(t_1, s)\| \Big[\|x_1\| + \|U(b, 0)g(x)\|$$

$$+ \int_0^b \|U(b, \tau)f(\tau, x(\tau))d\tau\|\Big] ds : t_1, t_2 \in [0, b],\ 0 \leq t_2 - t_1 \leq \epsilon, x \in Y\Big\}$$

$$+ \epsilon K_1 N_1 \Big[\|x_1\| + N_0 \sup\{\|g(x)\| : x \in Y\} + N_1 \sup\{\|f(s, x(s))\| : x \in Y\}\Big].$$

Letting $\epsilon \to 0+$, we get $\omega_0^t(v_3 Y) \leq 2b N_1 K_1 \Big(\|x_1\| + N_0 \beta(g(Y)) + b N_1 \beta(f(Q))\Big)$. $\square$

Then we can calculate our main result as follows:

Theorem 1. If the Banach space X is separable under the assumptions $(H1) -$ $(H4)$, the system $(1) - (2)$, controllable on J.

Proof. Consider the operator $\mathcal{P}$ defined by

$$(\mathcal{P}x)(t) = U(t, 0)g(x) + \int_0^t U(t, s)f\big(s, x(s)\big)ds$$

$$+ \int_0^t U(t, s)BW^{-1}\Big[x_1 - U(b, 0)g(x) - \int_0^b U(b, s)f\big(s, x(\tau)\big)d\tau\Big](s)ds.$$

For an arbitrary $x \in \mathcal{C}(J, X)$ and $t \in J$, we get

$$\|(\mathcal{P}x)(t)\| \leq (1 + b N_1 K_1)\Big[N_0\|g(x)\| + N_1 \int_0^b \|f(\tau, x(\tau))\|d\tau\Big] + b N_1 K_1 \|x_1\| \leq r.$$

Now, we prove the operator $\mathcal{P}$ is continuous on $\mathbb{B}(\theta, r)$.

Let us fix $x \in \mathbb{B}(\theta, r)$ and take arbitrary sequence $\{x_n\} \in \mathbb{B}(\theta, r)$ such that $x_n \to x$ in $\mathcal{C}(J, X)$. Next we have

$$\|\mathcal{P}x_n - \mathcal{P}x\| \leq (1 + b N_1 K_1)\Big[N_0\|g(x_n) - g(x)\| + N_1 \int_0^b \|f(\tau, x_n(\tau)) - f(\tau, x(\tau))\|d\tau\Big].$$

Applying Lebesgue dominated convergence theorem and we derive that $\mathcal{P}$ is continuous on $\mathbb{B}(\theta, r)$.

Now, we consider the sequence of sets $\{\Omega_n\}$ defined by induction as follows:

$$\Omega_0 = \mathbb{B}(\theta, r),\quad \Omega_{n+1} = Conv(\mathcal{P}\Omega_n),\ \text{for } n = 1, 2, \dots.$$

This sequence is decreasing, that is, $\Omega_n \supset \Omega_{n+1}$, for $n = 1, 2, ...$

Further let us put $v_n(t) = \beta(\Omega_n([0, t]))$ and $w_n(t) = \omega_0^t(\Omega_n)$.

Observe that each of functions $v_n(t)$ and $w_n(t)$ is nondecreasing, while sequences $\{v_n(t)\}$ and $\{w_n(t)\}$ are nonincreasing at any fixed $t \in J$. Put

$$v_\infty(t) = \lim_{n \to \infty} v_n(t), \quad w_\infty(t) = \lim_{n \to \infty} w_n(t), \text{ for } t \in J.$$

Using Lemma 2, Lemma 4 and $(H4)$ we obtain

$$\beta(v_1\Omega_n([0, t])) \le \omega_0^t(v_1\Omega_n) + \sup_{s \le t} \beta(v_1\Omega_n(s)) = 3m_g N_0(t)v_n(b),$$

$$\text{that is, } \beta(v_1\Omega_n([0, t])) \le 3m_g N_0(t)v_n(b). \tag{13}$$

Moreover, $\beta(v_2\Omega_n([0, t])) \le \omega_0^t(v_2\Omega_n) + \sup_{s \le t} \beta(v_2\Omega_n(s))$

$$\le 2m_f b N_1(t)v_n(t) + m_f N_1(t)\int_0^t v_n(\tau)d\tau$$

and $\beta(v_3\Omega_n([0, t]))$

$$\le \omega_0^t(v_3\Omega_n) + \sup_{s \le t} \beta(v_3\Omega_n(s))$$

$$\le 3b N_1(t)K_1\Big(\|x_1\| + m_g N_0(t)v_n(b)\Big) + b m_f N_1(t)K_1\Big(2b N_1(t)v_n(t) + N_1\int_0^t v_n(\tau)d\tau\Big).$$

Linking this estimate with (13) we obtain

$$v_{n+1}(t) = \beta(\Omega_{n+1}([0, t])) = \beta(\mathcal{P}\Omega_n([0, t]))$$

$$\le 3m_g N_0(t)v_n(b) + 2m_f b N_1(t)v_n(t) + m_f N_1(t)\int_0^t v_n(\tau)d\tau$$

$$+ 3b N_1(t)K_1\Big(\|x_1\| + m_g N_0(t)v_n(b)\Big)$$

$$+ b m_f N_1(t)K_1\Big(2b N_1(t)v_n(t) + N_1\int_0^t v_n(\tau)d\tau\Big).$$

Letting $n \to \infty$ we get

$$v_\infty(t) \le 3m_g N_0(t)v_\infty(b) + 2m_f b N_1(t)v_\infty(t) + m_f N_1(t)\int_0^t v_\infty(\tau)d\tau$$

$$+ 3b N_1(t)K_1\Big(\|x_1\| + m_g N_0(t)v_\infty(b)\Big)$$

$$+ b m_f N_1(t)K_1\Big(2b N_1(t)v_\infty(t) + N_1\int_0^t v_\infty(\tau)d\tau\Big).$$

Puttting $t = b$, we get $v_\infty(b) = 0$. Moreover, applying Lemmas 3, 4, 5, we have

$$w_{n+1}(t) = \omega_0^t(\Omega_{n+1}) = \omega_0^t(\mathcal{P}\Omega_n)$$

$$\le (2 + b N_1 K_1)[m_g N_0 v_n(b) + m_f b N_1 v_n(t)] + 2b N_1 K_1\|x_1\|.$$

Letting $n \to \infty$ we get

$$w_\infty(t) \le (2 + bN_1K_1)[m_g N_0 v_\infty(b) + m_f b N_1 v_\infty(t)] + 2b N_1 K_1 \|x_1\|.$$

Putting $t = b$ and applying $v_\infty(b) = 0$, we conclude that $w_\infty(b) = 0$.

This fact together with $v_\infty(b) = 0$ implies that $\lim_{n \to \infty} \chi(\Omega_n) = 0$. Hence, in view of Remark, we deduce that the set $\Omega_\infty = \bigcap_{n=0}^{\infty} \Omega_n$ is nonempty, compact and convex. Finally linking above obtained facts concerning the set Ω_∞ and the operator $\mathcal{P} : \Omega_\infty \to \Omega_\infty$ and using the classical Schauder fixed point theorem we infer that the operator $\mathcal{P}$ has at least one fixed point x in the set Ω_∞. Obviously the function $x = x(t)$ is a mild solution of (1) and (2) satisfying $x(b) = x_1$. Hence the given system is controllable on J. $\qquad\qquad\square$

Acknowledgments. The author thankful to the referee for the improvement of the paper. The author was supported by University Grant Commission (UGC), India (No. G2/1287/UGC SAP DRS/2009).

References

1. Balachandran, K., Dauer, J.P.: Controllability of nonlinear systems via Fixed Point Theorems. J. Optim. Theory Appl. 53, 345–352 (1987)
2. Balachandran, K., Dauer, J.P.: Controllability of nonlinear systems in Banach spaces: A survey. J. Optim. Theory Appl. 115, 7–28 (2002)
3. Banas, J., Goebel, K.: Measure of Noncompactness in Banach Space. Lecture Notes in Pure and Applied Matyenath. Dekker, New York (1980)
4. Byszewski, L.: Theorems about the existence and uniqueness of solutions of a semilinear evolution nonlocal Cauchy problem. J. Math. Anal. Appl. 162, 494–505 (1991)
5. Chang, Y.K., Nieto, J.J., Li, W.S.: Controllability of semilinear differential systems with nonlocal initial conditions in Banach spaces. J. Optim. Theory Appl. 142, 267–273 (2009)
6. Chukwu, E.N., Lenhart, S.M.: Controllability questions for nonlinear systems in abstract spaces. J. Optim. Theory Appl. 68, 437–462 (1991)
7. Curtain, R.F., Zwart, H.: An Introduction to Infinite Dimensional Linear Systems Theory. Springer, Berlin (1995)
8. Klamka, J.: Controllability of Dynamical Systems. Kluwer Academic, Dordrecht (1993)
9. Lin, Y., Liu, J.H.: Semilinear integrodifferential equations with nonlocal Cauchy problem. J. Integral Eqn. Appl. 15, 79–93 (2003)
10. Pazy, A.: Semigroups of Linear Operators and Applications to Partial Differential Equations. Springer, New York (1983)
11. Radhakrishnan, B., Balachandran, K.: Controllability results for semilinear impulsive integrodifferential evolution systems with nonlocal conditions. J. Cont. Theory Appl. 10, 28–34 (2012)
12. Xue, X.: Nonlocal nonlinear differential equations with measure of noncompactness in Banach spaces. Nonlinear Anal.: Theory Methods & Appl. 70, 2593–2601 (2009)
13. Yan, Z.: Controllability of semilinear integrodifferential systems with nonlocal conditions. International J. Comp. Appl. Math. 3, 221–236 (2007)

Approximate Controllability of Fractional Stochastic Integral Equation with Finite Delays in Hilbert Spaces

C. Rajiv Ganthi and P. Muthukumar*

Department of Mathematics
Gandhigram Rural Institute - Deemed University,
Gandhigram - 624 302, Tamilnadu, India
{mathsrajivgandhi,pmuthukumargri}@gmail.com

Abstract. In this work the approximate controllability of fractional stochastic integral equation with finite delays in Hilbert spaces has been addressed. The results are obtained by using the assumption that the corresponding linear integral equation is approximately controllable and a stochastic version of the well known Banach fixed point theorem.

Keywords: Analytic semigroup, Approximate controllability, Banach fixed point theorem, Fractional stochastic integral equation, Hilbert space.

1 Introduction

Fractional order nonlinear equations, with or without delay, are abstract formulations for many control problems arising in engineering, physics and so on (see [1]); the study of controllability for such systems is important for many applications. The approximate controllability of nonlinear fractional order deterministic and stochastic differential systems in infinite dimensional spaces has been studied by several authors, (see [2, 3] and the references therein). Recently El-Borai, et al. [4] studied the existence property for semigroup and some fractional stochastic integral equations. Ahmed [5] studied controllability of fractional stochastic delay equations. If the semigroup is compact, then assumption (iv) in section 2 of [5] is valid only if the state space is finite dimensional. As a result, the applications are restricted to ordinary control systems. Motivated by these papers we extend the results to obtain the approximate controllability of fractional stochastic integral equation without assuming this condition.

The purpose of this paper is to obtain the sufficient conditions of approximate controllability results of fractional stochastic integral equation with finite delays in Hilbert space of the general form

$$x(t) = \psi(0) + \frac{1}{\Gamma(\alpha)} \int_0^t \frac{Ax(s)}{(t-s)^{1-\alpha}} ds + \frac{1}{\Gamma(\alpha)} \int_0^t \frac{Bu(s)}{(t-s)^{1-\alpha}} ds + \frac{1}{\Gamma(\alpha)}$$

---

* Corresponding author.

P. Balasubramaniam and R. Uthayakumar (Eds.): ICMMSC 2012, CCIS 283, pp. 302–309, 2012.

$$\times \int_0^t \frac{F(s,x_s)}{(t-s)^{1-\alpha}}ds + \frac{1}{\Gamma(\alpha)} \int_0^t \frac{G(s,x_s)}{(t-s)^{1-\alpha}}dW(s), \quad t \in J = [0,b],$$

$$x(t) = \psi(t), \quad t \in [-r,0], \tag{1}$$

where $0 < \alpha \leq 1$, $b > 0$. The state variable $x(\cdot)$ takes values in Hilbert space H with the inner product $\langle \cdot, \cdot \rangle$ and $\| \cdot \|$ and the control function $u(\cdot)$ is given in $L_2(J, U)$, a Banach space of admissible control functions with U as a Banach space. B is a bounded linear operator from U into H. A is a closed linear operator in H, it is assume that A generates an analytic semigroup $T(t), t \geq 0$. Let K be a another separable Hilbert space. Suppose $\{W(t)\}_{t\geq 0}$ is a given K-valued Brownian motion or Wiener process with a finite trace nuclear covariance operator $Q \geq 0$. We are also employing the same notation $\| \cdot \|$ for the norm of $L(K, H)$, where $L(K, H)$ denotes the space of all bounded operators from K into H, simply $L(H)$ if $K = H$. $F : J \times \mathcal{C} \to H$ and $G : J \times \mathcal{C} \to L_Q(K, H)$ are given functions. Here $L_Q(K, H)$ denotes the space of all Q- Hilbert schmidt operators from K into H. $\psi \in \mathcal{C} := C([-r, 0], H)$ and $C([-r, 0], H)$ denotes the space of continuous functions from $[-r, 0]$ to H and the histories x_t are defined by $x_t(s) = x(t + s), s \in [-r, 0]$. In this article, we study the approximate controllability of equation (1) by using the results [2, 3, 6, 7].

2   Preliminaries

For more details of this section, the reader may refer [3 - 7], and the references therein). Let $(\Omega, \mathfrak{F}, P)$ be a complete probability space furnished with complete family of right continuous increasing sub σ - algebras $\{\mathfrak{F}_t, t \in J\}$ satisfying $\mathfrak{F}_t \subset \mathfrak{F}$. We assume that $\mathfrak{F}_t = \sigma(W(s) : 0 \leq s \leq t)$ is the σ-algebra generated by W and $\mathfrak{F}_t = \mathfrak{F}$. Let $\varphi \in L(K, H)$ and define $\|\varphi\|_Q^2 = Tr(\varphi Q \varphi^*) = \sum_{n=1}^\infty \|\sqrt{\lambda_n}\varphi \zeta_n\|^2$. If $\|\varphi\|_Q < \infty$, then φ is called a Q-Hilbert Schmidt operator. Let $L_Q(K, H)$ denote the space of all Q-Hilbert Schmidt operators $\varphi : K \to H$. The collection of all strongly measurable, square integrable H valued random variables, denoted by $L_2(\Omega, \mathfrak{F}, P; H) \equiv L_2(\Omega; H)$, is a Banach space equipped with norm $\|x(\cdot)\|_{L_2} = (E\|x(\cdot; w)\|_H^2)^{\frac{1}{2}}$, where the expectation, E is defined by $E(\rho) = \int_\Omega \rho(w)dP$.

Throughout this paper we impose the following hypotheses:

(H_1) B is a bounded linear operator from U into H, such that $\|B\| = N$, for a constant $N > 0$,

(H_2) For $0 \leq \gamma < 1$, $H_\gamma = [D(A^\gamma)]$ is a Banach space with respect to the graph topology induced by the graph norm (see [8])

$$\|x\|_\gamma = \|A^\gamma x\| + \|x\|, \quad \text{for} \quad x \in D(A^\gamma),$$

(H_3) The function F maps H_γ to H and satisfies the Lipschitz condition and linear growth condition, there exists a constant $C > 0$ such that

$$\|F(t, x_t) - F(t, y_t)\|_H^2 \leq C\|x_t - y_t\|_\gamma^2,$$

$$\|F(t, x_t)\|_H^2 \leq C\Big(1 + \|x_t\|_\gamma^2\Big), \quad \text{for every} \quad x_t, y_t \in H_\gamma,$$

(H_4) The function G maps H_γ to $L_Q(K, H)$ and satisfies the Lipschitz condition and linear growth condition, there exists a constant $C > 0$ such that

$$\|G(t, x_t) - G(t, y_t)\|_Q^2 \le C\|x_t - y_t\|_\gamma^2,$$

$$\|G(t, x_t)\|_Q^2 \le C\left(1 + \|x_t\|_\gamma^2\right), \quad \text{for every} \quad x_t, y_t \in H_\gamma.$$

Definition 1. *A continuous stochastic process* $x(\cdot) \in C(J, H_\gamma)$ *is a mild solution of equation (1) if*

(i) $x(t)$ *is* $\mathfrak{F}$ *- adapted for each* $0 \le t \le b$,
(ii) $\int_{-r}^b E\|x(s)\|_\gamma^2 ds < \infty$, *a.s.,*
(iii) for each $u \in L_2^{\mathfrak{F}}(J; U)$ *the process* x *satisfies the following equation:*

$$x(t) = \widehat{T}_\alpha(t)\psi(0) + \int_0^t (t - s)^{\alpha-1} T_\alpha(t - s)Bu(s)ds + \int_0^t (t - s)^{\alpha-1} T_\alpha(t - s)$$

$$\times F(s, x_s)ds + \int_0^t (t - s)^{\alpha-1} T_\alpha(t - s)G(s, x_s)dW(s), \quad t \in J,$$

$$x(t) = \psi(t), \quad -r \le t \le 0, \tag{2}$$

where $\widehat{T}_\alpha(t) = \int_0^\infty \xi_\alpha(\theta)T(t^\alpha \theta)d\theta$, $T_\alpha(t) = \alpha \int_0^\infty \theta \xi_\alpha(\theta)T(t^\alpha \theta)d\theta$,
$\xi_\alpha(\theta) = \frac{1}{\alpha}\theta^{-1-\frac{1}{\alpha}}\overline{w}_\alpha(\theta^{-\frac{1}{\alpha}}) \ge 0$, $\overline{w}_\alpha(\theta) = \frac{1}{\pi}\sum_{n=1}^\infty (-1)^{n-1}\theta^{-\alpha n-1}\frac{\Gamma(n\alpha+1)}{n!}$
$\sin(n\pi\alpha)$, $\theta \in (0, \infty)$, ξ_α is a probability density function defined on $(0, \infty)$, that is $\xi_\alpha(\theta) \ge 0$, $\theta \in (0, \infty)$ and $\int_0^\infty \xi_\alpha(\theta)d\theta = 1$.

Let $x_b(\psi; u)$ be the state value of (1) at terminal time b corresponding to the control u and the initial value ψ. Introduce the set $\mathfrak{R}_b(N) = \{x_b(\psi; u)(\cdot) : u(\cdot) \in L_2(J, U)\}$ is called the reachable set of the equation (1) at terminal time b.

Definition 2. *A control system (1) is said to be approximately controllable on* J *if* $\mathfrak{R}_b(N)$ *is dense in* H, *i.e.,* $\overline{\mathfrak{R}_b(N)} = H$.

Consider the following linear fractional differential system

$$\frac{d^\alpha x(t)}{dt^\alpha} = Ax(t) + Bu(t), \quad t \in J,$$

$$x(0) = \psi(0). \tag{3}$$

The approximate controllability for linear fractional control system (3) is a natural generalization of approximate controllability of linear first order control system [6, 7]. It is convenient at this point to introduce the controllability operator associated with (3) as

$$\Pi_0^b = \int_0^b (b - s)^{\alpha-1} T_\alpha(b - s)BB^* T_\alpha^*(b - s)ds,$$

where B^* denotes the adjoint of B and $T_\alpha^*(t)$ is the adjoint of $T_\alpha(t)$. The operator Π_0^b is a linear bounded operator, $R(\lambda, \Pi_0^b) = (\lambda I + \Pi_0^b)^{-1}$ for $\lambda > 0$.

Lemma 1. *The linear fractional control system (3) is approximately controllable on J if and only if $\lambda R(\lambda, \Pi_0^b) \to 0$ as $\lambda \to 0^+$ in the strong operator topology.*

The proof of this lemma is a straightforward adaptation of the proof of ([7], Theorem 2).

The integral equation of (3) is

$$x(t) = \psi(0) + \frac{1}{\Gamma(\alpha)} \int_0^t \frac{Ax(s)}{(t-s)^{1-\alpha}} ds + \frac{1}{\Gamma(\alpha)} \int_0^t \frac{Bu(s)}{(t-s)^{1-\alpha}} ds, \quad t \in J. \quad (4)$$

Therefore, the equation (4) is approximately controllable on J if and only if $\lambda R(\lambda, \Pi_0^b) \to 0$ as $\lambda \to 0^+$ in the strong operator topology.

3   Approximate Controllability

Theorem 1. *If hypotheses $(H_1) - (H_4)$ are hold and then the operator Φ has a fixed point in $C(J, H_\gamma)$ provided that $\overline{K} < 1$, where $\overline{K} = 3\Big(\alpha^2 C^2 C_\gamma^2 (b +$*
$Tr(Q)) \frac{b^{2\beta-1}}{2\beta-1} + \frac{2\alpha^6 bN^4 C_\gamma^6 C^2}{\lambda^2} (b + Tr(Q)) \frac{b^{2\beta-1}}{2\beta-1} \frac{b^{2\alpha-4\gamma\alpha-1}}{2\alpha-4\gamma\alpha-1}\Big).$

Proof. Let us define the operator $\Phi : C(J, H_\gamma) \to C(J, H_\gamma)$ defined by

$$(\Phi x)(t) = \int_0^\infty \xi_\alpha(\theta) T(t^\alpha \theta) \psi(0) d\theta + \alpha \int_0^t \int_0^\infty \theta(t-s)^{\alpha-1} \xi_\alpha(\theta) T((t-s)^\alpha \theta)$$

$$\times Bu(s) d\theta ds + \alpha \int_0^t \int_0^\infty \theta(t-s)^{\alpha-1} \xi_\alpha(\theta) T((t-s)^\alpha \theta) F(s, x_s) d\theta ds$$

$$+ \alpha \int_0^t \int_0^\infty \theta(t-s)^{\alpha-1} \xi_\alpha(\theta) T((t-s)^\alpha \theta) G(s, x_s) d\theta dW(s), \quad t \in J,$$

$$= \psi(t), \quad -r \leq t \leq 0,$$

for any $x_b \in H$ and $\lambda \in (0, 1]$, we define the control function $u(t)$ as follows

$$u(t) = B^* T_\alpha^*(b-t) R(\lambda, \Pi_0^b) \Big\{ x_b - \int_0^\infty \xi_\alpha(\theta) T(t^\alpha \theta) \psi(0) d\theta$$

$$- \alpha \int_0^b \int_0^\infty \theta(b-s)^{\alpha-1} \xi_\alpha(\theta) T((b-s)^\alpha \theta) F(s, x_s) d\theta ds$$

$$- \alpha \int_0^b \int_0^\infty \theta(b-s)^{\alpha-1} \xi_\alpha(\theta) T((b-s)^\alpha \theta) G(s, x_s) d\theta dW(s) \Big\}.$$

Now, it is shown that, the operator Φ has fixed point using this control function, this fixed point is a solution of (1). First, it must be shown that Φ maps $C(J, H_\gamma)$ into itself. Without loss of generality, assume that $0 \in \rho(A)$. Otherwise, if $0 \notin \rho(A)$, for the identity operator I add the term λI to A giving $A_\lambda = A + \lambda I$, then $0 \in \rho(A_\lambda)$. This simplifies the graph norm to $\|x\|_\gamma = \|A^\gamma x\|$ for $x \in D(A^\gamma)$. Since $T(t), t \geq 0$ is an analytic semigroup and A^γ is a closed operator, there exist numbers $C_1 \geq 1$ and C_γ such that $\sup_{t \in J} \|T(t)\|^2 \leq C_1$, $\|A^\gamma T(t)\| \leq C_\gamma t^{-\gamma}$, for $t \geq 0$. Hence, for $x \in C(J, H_\gamma)$, $E\left(\sup_{t \in [-r,0]} \|\psi(t)\|_H^2\right) < \infty$, for $-r \leq t \leq 0$ and for $t \in J$,

$$
\begin{aligned}
&E\|u(t)\|_H^2 \\
&\leq 4 \frac{N^2 \alpha^2}{\lambda^2} \left\| \int_0^\infty \theta \xi_\alpha(\theta) T((b-t)^\alpha \theta) d\theta \right\|_\gamma^2 \left\{ \|x_b\|^2 + E \left\| \int_0^\infty \xi_\alpha(\theta) T(t^\alpha \theta) \psi(0) d\theta \right\|_\gamma^2 \right. \\
&\quad + \alpha^2 E \left\| \int_0^b \int_0^\infty \theta(b-s)^{\alpha-1} \xi_\alpha(\theta) T((b-s)^\alpha \theta) F(s, x_s) d\theta ds \right\|_\gamma^2 \\
&\quad \left. + \alpha^2 E \left\| \int_0^b \int_0^\infty \theta(b-s)^{\alpha-1} \xi_\alpha(\theta) T((b-s)^\alpha \theta) G(s, x_s) d\theta dW(s) \right\|_\gamma^2 \right\}, \\
&\leq 4 \frac{N^2 \alpha^2}{\lambda^2} C_\gamma^2 (b-t)^{-2\alpha\gamma} \left\{ \|x_b\|^2 + C_1 \|\psi(0)\|^2 + \alpha^2 b C^2 C_\gamma^2 \int_0^b (b-s)^{2\alpha(1-\gamma)-2} \right. \\
&\quad \left. \times (1 + \|x_s\|^2) ds + Tr(Q) \alpha^2 C^2 C_\gamma^2 \int_0^b (b-s)^{2\alpha(1-\gamma)-2} (1 + \|x_s\|^2) ds \right\}.
\end{aligned}
$$

Let $\alpha(1 - \gamma) = \beta$ where $\frac{1}{2} < \beta \leq 1$, then,

$$
\begin{aligned}
E\|u(t)\|_H^2 \leq 4 \frac{N^2 \alpha^2}{\lambda^2} C_\gamma^2 (b-t)^{-2\alpha\gamma} &\left\{ \|x_b\|^2 + C_1 \|\psi(0)\|^2 + \alpha^2 C^2 C_\gamma^2 (1 + \|x_s\|^2) \right. \\
&\quad \left. \times \frac{b^{2\beta-1}}{2\beta - 1} (b + Tr(Q)) \right\},
\end{aligned}
$$

$$
\begin{aligned}
&E(\sup_{t \in J} \|(\Phi x)(t)\|_H^2) \\
&\leq 4E \left\| \int_0^\infty \xi_\alpha(\theta) T(t^\alpha \theta) \psi(0) d\theta \right\|_\gamma^2 + 4E \left\| \alpha \int_0^t \int_0^\infty \theta(t-s)^{\alpha-1} \xi_\alpha(\theta) T((t-s)^\alpha \theta) \right. \\
&\quad \times Bu(s) d\theta ds \Big\|_\gamma^2 + 4E \left\| \alpha \int_0^t \int_0^\infty \theta(t-s)^{\alpha-1} \xi_\alpha(\theta) T((t-s)^\alpha \theta) F(s, x_s) d\theta ds \right\|_\gamma^2 \\
&\quad + 4E \left\| \alpha \int_0^t \int_0^\infty \theta(t-s)^{\alpha-1} \xi_\alpha(\theta) T((t-s)^\alpha \theta) G(s, x_s) d\theta dW(s) \right\|_\gamma^2, \\
&\leq 4 C_1 \|\psi(0)\|^2 + 4\alpha^2 C^2 C_\gamma^2 (1 + \|x_s\|^2) \frac{b^{2\beta-1}}{2\beta - 1} (b + Tr(Q)) + 16 \frac{\alpha^4 b N^4 C_\gamma^4}{\lambda^2} \\
&\quad \times \frac{b^{2\alpha-4\alpha\gamma-1}}{2\alpha - 4\alpha\gamma - 1} \left\{ \|x_b\|^2 + C_1 \|\psi(0)\|^2 + \alpha^2 C^2 C_\gamma^2 (1 + \|x_s\|^2) \right. \\
&\quad \left. \times \frac{b^{2\beta-1}}{2\beta - 1} (b + Tr(Q)) \right\} < \infty.
\end{aligned}
$$

Therefore $E(\sup_{t \in J} \|(\Phi x)(t)\|_H^2) < \infty$ for every $x \in C(J, H_\gamma)$. Since $\psi(\cdot)$ is continuous in $[-r, 0]$, to complete the proof it remains to show that $\Phi \in C((0, b), L_2(\Omega, H_\gamma))$ is continuous. To accomplish that, let $t \in (0, b)$, for $h > 0$ and $t + h \in J$, we have

$$E\|(\Phi x)(t+h) - (\Phi x)(t)\|_\gamma^2$$

$$\leq 7\|(A^\gamma T((t+h)^\alpha\theta) - A^\gamma T(t^\alpha\theta))\psi(0)\|_\gamma^2 + 7\alpha^2 bN^2 \int_0^t \|(t+h-s)^{\alpha-1}A^\gamma$$

$$\times T((t+h-s)^\alpha\theta) - (t-s)^{\alpha-1}A^\gamma T((t-s)^\alpha\theta)\|_\gamma^2\|u\|^2 ds + 7\alpha^2 bN^2 C_\gamma^2\|u\|^2$$

$$\times\frac{h^{2\beta-1}}{2\beta-1} + 7\alpha^2 C^2(b+Tr(Q))\int_0^t \|(t+h-s)^{\alpha-1}A^\gamma T((t+h-s)^\alpha\theta)$$

$$-(t-s)^{\alpha-1}A^\gamma T((t-s)^\alpha\theta)\|_\gamma^2(1+\|x_s\|^2)ds + 7\alpha^2 C^2 C_\gamma^2(1+\|x_s\|^2)$$

$$\times\frac{h^{2\beta-1}}{2\beta-1}(b+Tr(Q)),$$

for $t \in (0,b)$. Thus, letting $h \to 0$, the desired continuity follows. Hence Φ maps $C(J, H_\gamma)$ into itself. Now it is shown that for sufficiently small b, defining the interval J leads to a contraction in $C(J, H_\gamma)$. Indeed, for $x, y \in C(J, H_\gamma)$ satisfying $x(t) = y(t) = \psi(t)$ for $-r \leq t \leq 0$ it can easily seen that

$$E\|(\Phi x)(t) - (\Phi y)(t)\|_H^2$$

$$\leq 3\alpha^2 E\|\int_0^t \int_0^\infty \theta(t-s)^{\alpha-1}\xi_\alpha(\theta)T((t-s)^\alpha\theta)(F(s,x_s) - F(s,y_s))d\theta ds\|_\gamma^2$$

$$+3\alpha^2 E\|\int_0^t \int_0^\infty \theta(t-s)^{\alpha-1}\xi_\alpha(\theta)T((t-s)^\alpha\theta)(G(s,x_s)$$

$$-G(s,y_s))d\theta dW(s)\|_\gamma^2 + 3\alpha^2 E\|\int_0^t \int_0^\infty \theta(t-s)^{\alpha-1}\xi_\alpha(\theta)T((t-s)^\alpha\theta)B$$

$$\times\Big[B^* T_\alpha^*(b-s)R(\lambda, \Pi_0^b)\Big\{x_b - \int_0^\infty \xi_\alpha(\theta)T(t^\alpha\theta)\psi(0)d\theta$$

$$-\alpha\int_0^b \int_0^\infty \theta(b-s)^{\alpha-1}\xi_\alpha(\theta)T((b-s)^\alpha\theta)F(s,x_s)d\theta ds$$

$$-\alpha\int_0^b \int_0^\infty \theta(b-s)^{\alpha-1}\xi_\alpha(\theta)T((b-s)^\alpha\theta)G(s,x_s)d\theta dW(s) + y_b + \int_0^\infty \xi_\alpha(\theta)$$

$$\times T(t^\alpha\theta)\psi(0)d\theta + \alpha\int_0^b \int_0^\infty \theta(b-s)^{\alpha-1}\xi_\alpha(\theta)T((b-s)^\alpha\theta)F(s,y_s)d\theta ds$$

$$+\alpha\int_0^b \int_0^\infty \theta(b-s)^{\alpha-1}\xi_\alpha(\theta)T((b-s)^\alpha\theta)G(s,y_s)d\theta dW(s)\Big\}\Big]d\theta ds\|_\gamma^2,$$

$$\leq 3\Big(\alpha^2 C^2 C_\gamma^2(b+Tr(Q))\frac{b^{2\beta-1}}{2\beta-1} + \frac{2\alpha^6 bN^4 C_\gamma^6 C^2}{\lambda^2}(b+Tr(Q))$$

$$\times\frac{b^{2\beta-1}}{2\beta-1}\frac{b^{2\alpha-4\gamma\alpha-1}}{2\alpha-4\gamma\alpha-1}\Big)sup_{-r\leq t\leq b}\|x(s) - y(s)\|^2,$$

$$\leq \overline{K}sup_{-r\leq t\leq b}\|x(s) - y(s)\|^2.$$

Thus, for sufficiently small b, $\overline{K} < 1$ and Φ is contraction in $C(J, H_\gamma)$ and so, by the Banach fixed point theorem Φ has a unique fixed point $x \in C(J, H_\gamma)$. This completes the proof.

Theorem 2. *Assume that the Theorem 1 and hypotheses* $(H_1) - (H_4)$ *are holds and the linear system (4) is approximately controllable on* J. *Then the fractional stochastic integral equation (1) is approximately controllable on* J.

Proof. Let $\overline{x}^\lambda(\cdot)$ be a fixed point of Φ in $C(J, H_\gamma)$ by Theorem 1, any fixed point of Φ is a mild solution of (1) under the control

$$
\overline{u}^\lambda(t) = B^* T_\alpha^*(b-t) R(\lambda, \Pi_0^b) \Big\{ x_b - \int_0^\infty \xi_\alpha(\theta) T(t^\alpha \theta) \psi(0) d\theta
$$
$$
-\alpha \int_0^b \int_0^\infty \theta(b-s)^{\alpha-1} \xi_\alpha(\theta) T((b-s)^\alpha \theta) F(s, \overline{x}_s^\lambda) d\theta ds
$$
$$
-\alpha \int_0^b \int_0^\infty \theta(b-s)^{\alpha-1} \xi_\alpha(\theta) T((b-s)^\alpha \theta) G(s, \overline{x}_s^\lambda) d\theta dW(s) \Big\},
$$

and satisfies

$$
\overline{x}^\lambda(b) = \int_0^\infty \xi_\alpha(\theta) T(t^\alpha \theta) \psi(0) d\theta + \alpha \int_0^b \int_0^\infty \theta(b-s)^{\alpha-1} \xi_\alpha(\theta) T((b-s)^\alpha \theta)
$$
$$
\times B\overline{u}^\lambda(s) d\theta ds + \alpha \int_0^b \int_0^\infty \theta(b-s)^{\alpha-1} \xi_\alpha(\theta) T((b-s)^\alpha \theta)
$$
$$
\times F(s, \overline{x}_s^\lambda) d\theta ds + \alpha \int_0^b \int_0^\infty \theta(b-s)^{\alpha-1} \xi_\alpha(\theta)
$$
$$
\times T((b-s)^\alpha \theta) G(s, \overline{x}_s^\lambda) d\theta dW(s),
$$

where $\widehat{T_\alpha}(t) = \int_0^\infty \xi_\alpha(\theta) T(t^\alpha \theta) d\theta$ and $T_\alpha(t) = \alpha \int_0^\infty \theta \xi_\alpha(\theta) T(t^\alpha \theta) d\theta$.

$$
\overline{x}^\lambda(b) = \widehat{T_\alpha}(t) \psi(0) + \int_0^b (b-s)^{\alpha-1} T_\alpha(b-s) BB^* T_\alpha^*(b-s) R(\lambda, \Pi_0^b) \Big\{ x_b
$$
$$
-\widehat{T_\alpha}(t) \psi(0) - \int_0^b (b-s)^{\alpha-1} T_\alpha(b-s) F(s, \overline{x}_s^\lambda) ds - \int_0^b (b-s)^{\alpha-1}
$$
$$
\times T_\alpha(b-s) G(s, \overline{x}_s^\lambda) dW(s) \Big\} ds + \int_0^b (b-s)^{\alpha-1} T_\alpha(b-s) F(s, \overline{x}_s^\lambda) ds
$$
$$
+ \int_0^b (b-s)^{\alpha-1} T_\alpha(b-s) G(s, \overline{x}_s^\lambda) dW(s),
$$
$$
= x_b - \lambda R(\lambda, \Pi_0^b) \Big\{ x_b - \widehat{T_\alpha}(t) \psi(0) - \int_0^b (b-s)^{\alpha-1} T_\alpha(b-s) F(s, \overline{x}_s^\lambda) ds
$$
$$
- \int_0^b (b-s)^{\alpha-1} T_\alpha(b-s) G(s, \overline{x}_s^\lambda) dW(s) \Big\}.
$$

Obviously, the sequence $F(s, \overline{x}_s^\lambda)$ and $G(s, \overline{x}_s^\lambda)$ are bounded in H and $L_Q(K, H)$. There are subsequence denoted by $F(s, \overline{x}_s^\lambda)$ and $G(s, \overline{x}_s^\lambda)$, that weakly converges to say $f(s)$ and $g(s)$ in H and $L_Q(K, H)$. Then it is not difficult to see that

$$
E\| \int_0^b (b-s)^{\alpha-1} T_\alpha(b-s) [F(s, \overline{x}_s^\lambda) - f(s)] ds \|^2
$$

$$\leq b \int_0^b (b-s)^{2(\alpha-1)} \|T_\alpha(b-s)\|^2 E\|F(s,\overline{x}_s^\lambda) - f(s)\|^2 ds \to 0,$$

$$E\| \int_0^b (b-s)^{\alpha-1} T_\alpha(b-s)[G(s,\overline{x}_s^\lambda) - g(s)]dW(s)\|_Q^2$$

$$\leq Tr(Q) \int_0^b (b-s)^{2(\alpha-1)} \|T_\alpha(b-s)\|^2 E\|G(s,\overline{x}_s^\lambda) - g(s)\|_Q^2 ds \to 0.$$

Therefore, the equation becomes

$$E\|\overline{x}^\lambda(b) - x_b\|^2$$

$$= E\|\lambda R(\lambda, \Pi_0^b)[x_b - \widehat{T_\alpha}(t)\psi(0) - \int_0^b (b-s)^{\alpha-1} T_\alpha(b-s)F(s,\overline{x}_s^\lambda)ds$$

$$- \int_0^b (b-s)^{\alpha-1} T_\alpha(b-s)G(s,\overline{x}_s^\lambda)dW(s)]\|^2 \to 0 \quad \text{as} \quad \lambda \to 0^+,$$

so $\overline{x}^\lambda(b) \to x_b$ holds in $C(J, H_\gamma)$ and hence we obtain the approximately controllability of the equation (1). The proof is completed.

4   Conclusion

Approximate controllability gives the possibility of steering the system to states which form the dense subspace in the states pace. This paper contains approximate controllability results for fractional stochastic integral equation with finite delay in Hilbert spaces.

References

1. Hilfer, R.: Applications of fractional calculus in physics. World Scientific, Singapore (2000)
2. Sakthivel, R., Ren, Y., Mahmudov, N.I.: On the approximate controllability of semilinear fractional differential systems. Comput. Math. Appl. 62, 1451–1459 (2011)
3. Sakthivel, R., Suganya, S., Anthoni, S.M.: Approximate controllability of fractional stochastic evolution equations. Comput. Math. Appl. 63, 660–668 (2012)
4. El-Borai, M.M., El-Said El-Nadi, K., Mostafa, O.L., Ahmed, H.M.: Semigroup and some fractional stochastic integral equations. Internat. J. Pure Appl. Math. Sci. 3, 47–52 (2006)
5. Ahmed, H.M.: Controllability of fractional stochastic delay equations. Internat. J. Nonlinear Sci. 8, 498–503 (2009)
6. Mahmudov, N.I.: Approximate controllability of semilinear deterministic and stochastic evolution equations in abstract Spaces. SIAM J. Cont. Optim. 42, 1604–1622 (2003)
7. Mahmudov, N.I., Denker, A.: On controllability of linear Stochastic systems. Internat. J. Cont. 73, 144–151 (2000)
8. Pazy, A.: Semigroups of linear operators and applications to partial differential equations. Springer, New York (1983)

Controllability of Semilinear Dispersion Equation

Simegne Tafesse and N. Sukavanam

Department of Mathematics, IIT Roorkee, Roorkee, India
{wtsimegne,nsukavanam}@gmail.com

Abstract. In this paper, the exact controllability of the third order semilinear dispersion equation is established under simple sufficient conditions on the nonlinear term which is assumed to be the sum of three nonlinear functions. We use the integral contractor condition which is weaker than Lipschitz condition on one of the functions and imposing certain range conditions on the Nemskii operators of the other two nonlinear parts. Moreover, the Co-semigroup operator is assumed to be bounded.

Keywords: Dispersion equation, Semilinear, Integral contractor, Nemtskii operator, Exact controllable, Third order.

1 Introduction

Let $X = L_2(0, 2\pi)$ be Hilbert spaces and $Z = L_2((0, T) \times (0, 2\pi)) = L_2((0, T : X)$ be function space.

Consider the third order dispersion equation given by

$$\frac{\partial w}{\partial t}(x, t) + \frac{\partial^3 w}{\partial x^3}(x, t) = (Gu)(x, t) + f(t, w(x, t)), t > 0 \qquad (1)$$
$$w(x, 0) = 0,$$

in the domain $x \in [0, 2\pi]$ with periodic boundary condition

$$\frac{\partial^k w}{\partial x^k}(0, t) = \frac{\partial^k w}{\partial x^k}(2\pi, t), k = 0, 1, 2. \qquad (2)$$

The state function w and the control function u are in Z. The bounded linear operator G is defined on Z by

$$(Gu)(x, t) = g(x)(u(x, t) - \int_0^{2\pi} g(s)u(s, t))ds, \qquad (3)$$

where $g(x)$ is a piecewise continuous nonnegative function on $[0, 2\pi]$ such that $[g] = 1$ where $[g]$ is defined as

$$[g] = \int_0^{2\pi} g(s)ds \qquad (4)$$

P. Balasubramaniam and R. Uthayakumar (Eds.): ICMMSC 2012, CCIS 283, pp. 310–315, 2012.

$f : [0, T] \times X \to X$ is defined by $f = f_1 + f_2 + f_3$ where $f_i : [0, T] \times X \to X$, is continuous nonlinear function for each i = 1, 2, 3.

Many researchers have worked on the controllability problems of third order dispersion equation. R.K. George et al. [3] studied and obtained sufficient conditions for the exact controllability of the nonlinear third order dispersion equation by assuming the two standard types of nonlinearity, namely, monotone and Lipschitz continuous. In that paper, under each condition, the existence and uniqueness of the solution has been shown first and then the controllability was proved. In [1] Chalishajar studied the controllability of nonlinear integro-differential third order dispersion system. David L. and Bing Yu Zhang [2] discussed the exact controllability and stabilizability of a system described by the Korteweg-de Vries (KdV) equation given by

$$\frac{\partial w}{\partial t}(x, t) + \alpha w(x, t)\frac{\partial w}{\partial x}(x, t) + \frac{\partial^3 w}{\partial x^3}(x, t) = u(x, t), \tag{5}$$

where u is a control function. D.L Russel and B.Y Zhang [6] studied the controllability and stabilizability of the third order linear dispersion equation on a periodic domain. The equation (1) is the same as KdV equation when $\alpha = 0$. In each of the above papers, the control problems are focussing on the conservation of the so called fluid-volume which is given by $\int_0^{2\pi} w(x, t)dx$. N.K. Tomar, [4] studied on the controllability of third order dispersion equation where some parts of the nonlinear term are Lipschitz continuous, and monotone. Almost all of the above mentioned researchers dealt with the exact controllability of third order dispersion system. However, recently R. Sakthivel et.al [7] studied the approximate controllability of nonlinear third order dispersion equation.

In our paper, the exact controllability of the third order dispersion equation (1) is studied by splitting the nonlinear part into three nonlinear functions and each function is assumed to have different conditions.

2   Preliminaries and Basic Assumption

This paper is aimed to discuss the exact controllability of semilinear dispersion equation by assuming some simple conditions on each terms of the nonlinear function f. Consider the semilinear dispersion equation (1)-(2). Let A be an operator on X defined by

$$Aw = -\frac{\partial^3 w}{\partial x^3} \tag{6}$$

with domain $D(A) \subseteq H^3(0, 2\pi)$ which is consisting of functions satisfying boundary condition (2). Then the abstract form of the dispersion equation (1)-(2) can be written as

$$\frac{\partial w}{\partial t}(x, t) = Aw(x, t) + (Gu)(x, t) + f(t, w(x, t)), t > 0 \tag{7}$$

$$w(x, 0) = 0.$$

By Lemma 8.5.2 of Pazy [5], A is infinitesimal generator of a Co-semigroup $\{T(t), t \geq 0\}$ of isometries on $L_2(0, 2\pi)$. Then for all $w \in D(A)$

$$\langle Aw, w \rangle_X = \langle -w''', w \rangle = \langle w, w''' \rangle = -\langle Aw, w \rangle$$

where the middle equality is achieved by applying integration by parts three times. Also there exists a constant $M > 0$ such that $\sup\{\|T(t)\| : t \in [0, \tau]\} \leq M$. We define the mild solution of (7) by

$$w(x, t) = \int_0^t T(t - s)(Gu)(s)ds + \int_0^t T(t - s)f(s, w(x, s))ds$$

Suppose the control operator G is defined as in (3). Then we have

$$\int_0^{2\pi} (Gu)(x, s)dx = \int_0^{2\pi} g(x)(u(x, t) - \int_0^{2\pi} g(s)u(s, t)ds)dx$$

$$= \int_0^{2\pi} g(x)u(x, t)dx - \int_0^{2\pi} g(s)(\int_0^{2\pi} g(x)u(s, t)dx)ds$$

$$= \int_0^{2\pi} g(x)u(x, t)dx(1 - \int_0^{2\pi} g(s)ds) = 0.$$

Now let $w(x, t)$ be the solution of (1)-(2) then

$$\frac{d}{dt} \int_0^{2\pi} w(x, s)dx = \int_0^{2\pi} \frac{\partial w}{\partial t}(x, t)dx$$

$$= \int_0^{2\pi} [-\frac{\partial^3 w}{\partial x^3}(x, t) + (Gu)(x, t) + f(t, w(x, t))]dx$$

$$= \int_0^{2\pi} -\frac{\partial^3 w}{\partial x^3}(x, t)dx + \int_0^{2\pi} (Gu)(x, t)dx + \int_0^{2\pi} f(t, w(x, t))dx$$

$$= -\int_0^{2\pi} d[\frac{\partial^2 w}{\partial x^2}(x, t)] + \int_0^{2\pi} (Gu)(x, t)dx + \int_0^{2\pi} f(t, w(x, t))dx$$

$$= -\frac{\partial^2 w}{\partial x^2}(2\pi, t) + \frac{\partial^2 w}{\partial x^2}(0, t) + \int_0^{2\pi} f(t, w(x, t))dx$$

$$= \int_0^{2\pi} f(t, w(x, t))dx.$$

From this, we can see that the volume is conserved with periodic boundary condition when $\int_0^{2\pi} f(t, w(x, t))dx = 0$. Let

$$V = \{x \in X : [x] = \int_0^{2\pi} x(s)ds = 0\}$$

Then V is a Hilbert space with respect to L_2-norm (see [1]). Clearly $Gu \in V$.

Definition 1. *The system (1)-(2) is said to be exactly controllable over a time interval $[0, \tau]$, if for any given $w_\tau \in X$ with $[w_\tau] = 0$, there exists a control $u \in Z$ such that the mild solution $w(x, t)$ of (1)-(2) with control u satisfies $w(x, \tau) = w_\tau$.*

Define the solution mapping $W : Z \to Z$ by $(Wu)(t) = w(x,t)$ where $w(x,t)$ is the unique mild solution of the system (1)-(2) corresponding to the control u. Let $C = C([0,\tau] : X)$ denote the Banach space of continuous functions on $J = [0,\tau]$ with the standard norm $\|w\|_C = \sup\{\|w(t)\|_X : 0 \le t \le \tau\}$ for $w \in C$.

Definition 2. *[3] Suppose $\Gamma : J \times X \to BL(C)$ is a bounded continuous operator and there exists a positive γ such that for any $w, y \in C$ we have*

$$\sup_{0 \le t \le \tau} \{\|f(t, w(t)) + y(t) + \int_0^t T(t-s)(\Gamma(s, w(s))y)(s)ds) \tag{8}$$
$$-f(t, w(t)) - (\Gamma(t, w(t))y)(t)\|_X\} \le \gamma\|y\|_C$$

Then we say that f has a bounded integral contractor $\{I + \int T\Gamma\}$ (we may say simply Γ) with respect to the Co-semigroup T.

Definition 3. *[3] A bounded integral contractor Γ is said to be regular if the integral equation*

$$x(t) = y(t) + \int_0^t T(t-s)(\Gamma(s, w(s))y)(s)ds \tag{9}$$

has a solution y in C for every $w, x \in C$.

We assume $\|\Gamma(t, w(t))\| \le \beta_1$ for all $t \in J$ and $w \in C$. Let the nonlinear function f is decomposed into three nonlinear functions written as

$$f(t, w(x,t)) = f_1(t, w(x,t)) + f_2(t, w(x,t)) + f_3(t, w(x,t))$$

First, let us consider the dispersion equation with the nonlinear function f_3 term which is given by

$$\frac{\partial w}{\partial t}(x,t) + \frac{\partial^3 w}{\partial x^3}(x,t) = (Gu)(x,t) + f_3(t, w(x,t)), t > 0 \tag{10}$$
$$w(x,0) = 0,$$

where $x \in [0, 2\pi]$ and periodic boundary condition

$$\frac{\partial^k w}{\partial x^k}(0,t) = \frac{\partial^k w}{\partial x^k}(2\pi, t), k = 0, 1, 2. \tag{11}$$

From [3] one can obtain the exact controllability of (10)($f = f_3$) by assuming that f has an integral contractor. We define the operator $S : Z \to V$ as

$$Sw = \int_0^t T(t-s)w(x,s)ds \tag{12}$$

Define the Nemytskii operator $F_i : Z \to Z, i = 1, 2$ by

$$(F_iw)(x,t) = f_i(t, w(x,t)), i = 1, 2$$

Notation: The range of F is denoted by $R(F)$. Now we consider the following conditions

1. There exists a constant M such that $\|T(t)\| \leq M, t \in [0, \tau]$
2. $R(F_1) \subseteq N(S)$
3. $R(F_2) \subseteq R(G)$
4. f_3 has a regular bounded integral contractor Γ with sufficiently small constant γ
5. f_3 satisfies the monotone condition, that is there exists a constant $\beta > 0$ such that for all $x, y \in \Re$

$$\langle f_3(t, x) - f_3(t, y), x - y \rangle \leq -\beta |x - y|^2$$

6. f_3 satisfies the growth condition, that is there exists a constant $a \geq 0$ and $b > 0$ such that for all $x \in \Re$

$$|f_3(t, x)| \leq a|x| + b.$$

Remark 1. The existence and uniqueness of the solution of the system $(10)-(11)$ have been shown by Theorem 5.6 of [3] if the conditions (1) and (4) are satisfied. Moreover, if conditions (5) and (6) are fulfilled, then Lemma 2.2 of [3] shows that the solution mapping W is well defined and the uniqueness of the solution follows.

Remark 2. Clearly $\int_0^{2\pi} f_2(t, w(x, t)) dx = 0$.

3   Main Results

In this section we assume the existence of the mild solution since it has been done by [3] and other researchers in detail.

Theorem 1. *Suppose the nonlinear terms f_1, f_2 and f_3 satisfy the conditions (2), (3) and (4) respectively, then the semilinear dispersion equation $(1) - (2)$ is exactly controllable if the corresponding linear dispersion equation is exactly controllable.*

Proof. Let w_τ be any given final state. From Theorem 6.1 of [3] the dispersion equation (10)-(11) is exactly controllable in V. That means there is a control $u \in Z$ such that the mild solution $w(x, t)$ given by

$$w(x, t) = \int_0^t T(t - s)(Gu)(s) ds + \int_0^t T(t - s) f_3(s, w(x, s)) ds \qquad (13)$$

satisfies $w(x, \tau) = w_\tau$. Using condition 2, (13) can be written as

$$w(x, t) = \int_0^t T(t - s)(Gu)(s) ds + \int_0^t T(t - s) f_1(s, w(x, s)) ds$$
$$+ \int_0^t T(t - s) f_3(s, w(x, s)) ds. \qquad (14)$$

By condition (3) there exists a control v in Z such that $f_2(t, w(x,t)) = (F_2 w)(x,t)$ $= (Gv)(x,t)$. Then using a simple technique the mild solution $w(x,t)$ in (14) can be written as

$$w(x,t) = \int_0^t T(t-s)(Gu - Gv)(s)ds + \int_0^t T(t-s)f_1(s, w(x,s))ds$$
$$+ \int_0^t T(t-s)f_2(s, w(x,s))ds + \int_0^t T(t-s)f_3(s, w(x,s))ds$$
$$= \int_0^t T(t-s)G(u-v)(s)ds + \int_0^t T(t-s)f(s, w(x,s))ds,$$

where $f(t, w(x,t)) = f_1(t, w(x,t)) + f_2(t, w(x,t)) + f_3(t, w(x,t))$ Hence the system (1)-(2) is exactly controllable with control $v' = u - v$.

Theorem 2. *Suppose the nonlinear terms f_1, f_2 and f_3 satisfy the conditions (2), (3) and (5) respectively. Moreover, f_3 satisfies condition (6) with sufficiently small growth constant a. Then the nonlinear system (1) − (2) is exactly controllable.*

Proof. From theorem 4.3 of [3] the system (10) − (11) is exactly controllable in V. By similar approach to the proof of theorem 3.1 we can conclude that the nonlinear dispersion system is exactly controllable.

Remark 3. If $f_1 = 0 = f_2$, then theorem 6.1 of [3] is a particular case of theorem 3.1. Similarly theorem 4.3 of [3] is exactly the same as theorem 3.2 when $f_1 = 0 = f_2$.

References

1. Chalishajar, D.N.: Controllability of nonlinear integro-differential third order dispersion system. J. Math. Anal. Appl. 348, 480–486 (2008)
2. David, L., Zhang, B.Y.: Exact controllability and stability of the kortwegede vries equation. Amer. Math. Soc. 348, 3643–3672 (1996)
3. George, R.K., Chalishajar, D.N., Nandakumaran, A.K.: Exact controllablity of the nonlinear third-order dispersion equation. J. Math. Anal. Appl. 332, 1028–1044 (2007)
4. Tomar, N.K.: A Note on controllability of semilinear system, Ph D Thesis (2008)
5. Pazy, A.: Semigroups of linear operators and applications to partial differential equations. Springer, Heidelberg (1983)
6. Russel, D.L., Zhang, B.Y.: Controllability and stabilizability of the third order linear dispersion equation on a periodic domain. SIAM J. Control Optim. 31(3), 659–672 (1993)
7. Sakthivel, R., Mohmudov, N.I., Ren, Y.: Approximate controllability of nonlinear third order dispersion equation. Appl. Math. Comput. 217, 8507–8511 (2011)

Estimation of Controllable Initial Fuzzy States of Linear Time-Invariant Dynamical Systems

Bhaskar Dubey* and Raju K. George

Department of Mathematics, Indian Institute of Space Science and Technology,
Thiruvananthapuram - 695547, India
{bhaskard,george}@iist.ac.in

Abstract. In this paper, we consider linear time-invariant dynamical systems with fuzzy initial condition. Dynamics of such systems is given by linear time-invariant fuzzy differential equations. First, we discuss the evolution of solution to such fuzzy differential dynamical systems and also establish the existence and uniqueness of the solution. We then provide an estimate of initial fuzzy state that can be controlled to a predefined target fuzzy state, that is, given any two crisp states x_0, x_1 in $\mathbb{R}^n$, and a fuzzy state X_1 around x_1, we compute a fuzzy state X_0 around x_0 so that X_0 is fuzzy-controllable to X_1. In a special case, when the plant matrix has non-negative entries, the fuzzy state X_0 is fuzzy-controllable to X_1 with the crisp control that steers x_0 to x_1. Examples are given to substantiate the results obtained.

Keywords: Fuzzy-controllability, Fuzzy Number, Fuzzy States, Fuzzy Differential Equations, Fuzzy Dynamical Systems.

1 Introduction

In many practical applications, precise measurement of parameters is always a challenge. This is mostly due to the fact that precise measurement is costly and sometimes practically impossible. For instance, in case of problems related to controllability of dynamical systems, the initial condition from which the dynamical system starts evolving is fuzzy in nature. Despite the fuzziness in the initial condition, we are interested in the controllers that can control an initial fuzzy state to a desired target state which of course has to be fuzzy in nature. The dynamics of such systems will be given by a fuzzy differential equation.

The theory of fuzzy differential equations has been developed in late seventies and early eighties. The term "fuzzy differential equation" was first used by Kandel and Byatt [1] in 1980. Since then so many authors have substantially contributed towards the theory of fuzzy differential equations, for example, Dubois and Prade [2–4], Puri and Ralescu [5], Kaleva [6], etc. Kaleva has established the existence and uniqueness of a solution to fuzzy initial value problem $\dot{x}(t) = f(x,t), x(0) = X_0$ under some smoothness conditions on f [6].

---

* Corresponding author.

P. Balasubramaniam and R. Uthayakumar (Eds.): ICMMSC 2012, CCIS 283, pp. 316–324, 2012.

In this paper, we discuss the evolution of a solution to linear time-invariant fuzzy dynamical system. Furthermore, we will provide an estimate of fuzzy initial condition starting with which the state(fuzzy) of system at a given time lies within the prescribed state(fuzzy). The organization of paper is as follows. In Section 2, we state some preliminary results on the fuzzy set theory. In Section 3, we establish the main results, that is, we explain the evolution of a solution to linear time-invariant dynamical system with fuzzy initial condition as well as fuzzy-input. We shall also prescribe a procedure for computing an initial fuzzy state of linear time-invariant dynamical system that can be controlled to a desired fuzzy state. In Section 4, we give examples to support the results obtained.

2   Preliminaries

By $\mathbb{R}$, we mean the set of all real numbers. $\mathbb{R}^+$ denotes the set of all non-negative real numbers. By $\mathbb{E}^n$, we mean the set of all n-dimensional vectors of fuzzy numbers on $\mathbb{R}$.

Definition 1. *By a linear dynamical system, we mean a linear differential equation of the form*

$$\dot{x}(t) = A(t)x(t) + B(t)u(t), \tag{1}$$

where $A(\cdot) \in C([t_0, t_1]; \mathbb{R}^{n \times n})$ and $B(\cdot) \in C([t_0, t_1]; \mathbb{R}^{n \times m})$, $t_0, t_1 \in \mathbb{R}^+$. The matrix $A(\cdot)$ is sometimes called the plant matrix and the matrix $B(\cdot)$ is called the control matrix. For each $t \in [t_0, t_1]$, $x(t) \in \mathbb{R}^n$ is called the state of system and $u(t) \in \mathbb{R}^m$ is called the input or the control for system. If the matrices $A(t)$ and $B(t)$ are matrices with constant entries then we call (1) as linear time-invariant dynamical system.

The following definition of fuzzy sets is due to Zimmermann [7].

Definition 2. *If X is a collection of objects denoted generically by x, then a fuzzy set A in X is a set of ordered pairs $A = \{(x, \mu_A(x)) \mid x \in X\}$, where $\mu_A(x)$ is called the membership function or grade of membership of x in A. The range of membership function is a subset of nonnegative real numbers whose supremum is finite.*

It can be easily shown that for every $a \in \mathbb{E}$, α-level set of a is closed and bounded interval which is denoted by $[a]_\alpha = [\underline{a}^\alpha, \overline{a}^\alpha]$, where $\underline{a}^\alpha$, $\overline{a}^\alpha$ are called lower and upper $\alpha-$cut of a, respectively. For two fuzzy numbers a, b in $\mathbb{E}$, we say $a \leq b$ if $\mu_a(s) \leq \mu_b(s)$ for all s in $\mathbb{R}$, where $\mu_a(\cdot)$, $\mu_b(\cdot)$ are the membership functions of a, b, respectively. Given two fuzzy numbers $X_0 = [X_{01}, X_{02}, \ldots, X_{0n}]^T$, $X_1 = [X_{11}, X_{12}, \ldots, X_{1n}]^T$ in $\mathbb{E}^n$, we say $X_0 \leq X_1$ if $\mu_{X_{0i}}(\cdot) \leq \mu_{X_{1i}}(\cdot)$, $1 \leq i \leq n$.

Following are the well known fundamental arithmetic operations on fuzzy numbers. Let $a, b \in \mathbb{E}$ and $\alpha \in (0, 1]$ then we have

(i) $[a + b]_\alpha = [\underline{a}^\alpha + \underline{b}^\alpha, \overline{a}^\alpha + \overline{b}^\alpha]$
(ii) $[a.b]_\alpha = [min\{\underline{a}^\alpha \underline{b}^\alpha, \underline{a}^\alpha \overline{b}^\alpha, \overline{a}^\alpha \underline{b}^\alpha, \overline{a}^\alpha \overline{b}^\alpha\}, max\{\underline{a}^\alpha \underline{b}^\alpha, \underline{a}^\alpha \overline{b}^\alpha, \overline{a}^\alpha \underline{b}^\alpha, \overline{a}^\alpha \overline{b}^\alpha\}]$
(iii) $[a - b]_\alpha = [\underline{a}^\alpha - \overline{b}^\alpha, \overline{a}^\alpha - \underline{b}^\alpha]$

3   Main Results

Consider the following linear time-invariant system with fuzzy initial condition

$$\begin{cases} \dot{x}(t) = Ax(t) + Bu(t) \\ x(t_0) = x_0, \end{cases} \tag{2}$$

where A, B are real matrices of size $n \times n$, $n \times m$, respectively. The initial state $x_0 \in \mathbb{E}^n$ and the control $u(\cdot) \in \mathbb{E}^m$. Clearly at any time t we have the state $x(t) \in \mathbb{E}^n$. In the results to follow, we will characterize the solution of above fuzzy differential equation (2). The idea to solve Eq. (2) is basically proposed by Seikkala [8]. He considered the fuzzy initial value problem

$$\dot{x}(t) = f(t, x(t)), x(0) = x_0, \tag{3}$$

where f is continuous mapping from $\mathbb{R}^+ \times \mathbb{R}$ into $\mathbb{R}$ and x_0 is a fuzzy number in $\mathbb{E}$ with $\alpha-$level intervals given by

$$[x_0]_\alpha = [x_0^\alpha, \overline{x_0^\alpha}]. \tag{4}$$

The Zadeh's extension principle leads to the following definition of $f(t, x)$ when x is a fuzzy number:

$$f(t, x)(s) = \sup_{\tau | f(t,\tau)=s} \{x(\tau)\}, s \in \mathbb{R}. \tag{5}$$

From above equation it follows that

$$[f(t, x)]_\alpha = [min\{f(t, u) : u \in [\underline{x^\alpha}, \overline{x^\alpha}]\}, max\{f(t, u) : u \in [\underline{x^\alpha}, \overline{x^\alpha}]\}]. \tag{6}$$

Since the derivative $\dot{x}(t)$ of a fuzzy process $x : \mathbb{R}^+ \to \mathbb{E}$ is defined by:

$$[\dot{x}(t)]_\alpha = [\dot{\underline{x}^\alpha}(t), \dot{\overline{x}^\alpha}(t)] \tag{7}$$

By taking $\alpha-$cuts of Eq.(3) both sides, it follows that a fuzzy process $x : \mathbb{R}^+ \to \mathbb{E}$ is a solution of Eq.(3) on an interval $I = [0, T)$ if it satisfies the following level-wise decomposed crisp system of differential equations.

$$\begin{cases} \dot{\underline{x}^\alpha}(t) = min\{f(t, u) : u \in [\underline{x^\alpha}(t), \overline{x^\alpha}(t)]\}, \underline{x}^\alpha(0) = x_0^\alpha, \\ \dot{\overline{x}^\alpha}(t) = max\{f(t, u) : u \in [\underline{x^\alpha}(t), \overline{x^\alpha}(t)]\}, \overline{x}^\alpha(0) = \overline{x_0^\alpha}, \end{cases} \tag{8}$$

where $\alpha \in (0, 1]$ and $t \in I$.

Using the idea given above we state without proof the following lemma which describes the evolution of system (2), though the proof will follow along the same lines as for one dimensional case which was explained above. The evolution of the system (2) is described by the following 2n-differential equations corresponding to the end points of $\alpha-$cuts.

Lemma 1. For $\alpha \in (0, 1]$, let $x_k^\alpha(t) = [\underline{x_k^\alpha}(t), \overline{x_k^\alpha}(t)]$ be the α-cut of $x_k(t)$ for $1 \le k \le n$ and $u_j^\alpha(t) = [\underline{u_j^\alpha}(t), \overline{u_j^\alpha}(t)]$ be the α-cut of $u_j(t)$ for $1 \le j \le m$ then the evolution of system (2) is given by following $2n$-differential equations:

$$\begin{cases} \underline{\dot{x}_k^\alpha}(t) = min((Az + Bw)_k : z_i \in [\underline{x_i^\alpha}(t), \overline{x_i^\alpha}(t)], w_j \in [\underline{u_j^\alpha}(t), \overline{u_j^\alpha}(t)]) \\ \overline{\dot{x}_k^\alpha}(t) = max((Az + Bw)_k : z_i \in [\underline{x_i^\alpha}(t), \overline{x_i^\alpha}(t)], w_j \in [\underline{u_j^\alpha}(t), \overline{u_j^\alpha}(t)]) \\ \underline{x_k^\alpha}(t_0) = \underline{x_{0k}^\alpha} \\ \overline{x_k^\alpha}(t_0) = \overline{x_{0k}^\alpha}, \end{cases} \tag{9}$$

where $(Az + Bw)_k = \Sigma_{i=1}^n a_{ki} z_i + \Sigma_{j=1}^m b_{kj} w_j$ is the kth row of $Az + Bw$.

We will introduce new variables for the sake of convenience which we will use in rest of the paper. Given any $x(t) \in \mathbb{E}^n$, we define

$$\underline{x^\alpha}(t) := [\underline{x_1^\alpha}(t), \underline{x_2^\alpha}(t), \dots, \underline{x_n^\alpha}(t)]^T,$$

$$\overline{x^\alpha}(t) := [\overline{x_1^\alpha}(t), \overline{x_2^\alpha}(t), \dots, \overline{x_n^\alpha}(t)]^T,$$

where $[\underline{x_k^\alpha}(t), \overline{x_k^\alpha}(t)]$ is the α-cut of $x_k(t)$ for $1 \le k \le n$. We denote $x_*^\alpha(t) := [\underline{x^\alpha}(t), \overline{x^\alpha}(t)]^T := [\underline{x_1^\alpha}(t), \underline{x_2^\alpha}(t), \dots, \underline{x_n^\alpha}(t), \overline{x_1^\alpha}(t), \overline{x_2^\alpha}(t), \dots, \overline{x_n^\alpha}(t)]^T$ a column vector of size $2n$. $\underline{u^\alpha}(t)$, $\overline{u^\alpha}(t)$, and $u_*^\alpha(t)$ are similarly defined.

Remark 1. The system (9) can be described by the following system: For $\alpha \in (0, 1]$, $\dot{x}_*^\alpha(t) = A^* x_*^\alpha(t) + B^* u_*^\alpha(t)$, $x_*^\alpha(t_0) = x_{0*}^\alpha$ in which A^* and B^* are defined as follows:

(i) If A has all its entries non-negative then $A^* = M$ and $B^* = N$, where

$$M = \begin{bmatrix} A & 0 \\ 0 & A \end{bmatrix} \qquad N = \begin{bmatrix} B & 0 \\ 0 & B \end{bmatrix}.$$

That is, M is a block diagonal matrix of size $2n \times 2n$ and N is a block diagonal matrix of size $2n \times 2m$.

(ii) If A has some of its entries negative then A^* is obtained by the following flip operations on the entries of M.
$m_{ij} \longleftrightarrow m_{i(j+n)}$ if $1 \le j \le n$ and $m_{ij} < 0$.
$m_{ij} \longleftrightarrow m_{i(j-n)}$ if $n < j \le 2n$ and $m_{ij} < 0$

(iii) If B has some of its entries negative then B^* is obtained by the following flip operations on the entries of N.
$n_{ij} \longleftrightarrow n_{i(j+m)}$ if $1 \le j \le m$ and $n_{ij} < 0$.
$n_{ij} \longleftrightarrow n_{i(j-m)}$ if $m < j \le 2m$ and $n_{ij} < 0$

It is clear from the remark above that for each $\alpha \in (0, 1]$, system (9) represents an initial value problem in $\mathbb{R}^{2n}$ which has unique solution because A^* and B^* are constant matrices. Furthermore, it can be shown that for each $1 \le k \le n$, the intervals $[\underline{x_k^\alpha}(t), \overline{x_k^\alpha}(t)]$ defines a fuzzy number $x_k(t)$ in $\mathbb{E}$. Hence $x(t) = [x_1(t), x_2(t), \ldots, x_n(t)]^T$ is the unique fuzzy solution of (2).

Xu et.al. [9] have also studied a non-homogeneous linear time-invariant dynamical system with fuzzy initial condition and established the existence and uniqueness of fuzzy solution in a slightly different manner. In their approach, they reduced the $2n$−dimensional level-wise decomposed system in to n-dimensional system by moving to the field of complex numbers.

3.1 Fuzzy-Controllability

Before defining fuzzy-controllability of system (2), we shall remind the definition of controllability of the linear time-invariant system:

$$\dot{x}(t) = Ax(t) + Bu(t), \tag{10}$$

where $x(t) \in \mathbb{R}^n, u(t) \in \mathbb{R}^m$ and A, B are matrices of size $n \times n$ and $n \times m$ over $\mathbb{R}$.

Definition 3. *System* (10) *with initial condition* $x(t_0) = x_0 \in \mathbb{R}^n$ *is said to be controllable to* $x_1 \in \mathbb{R}^n$ *at* $t_1(> t_0)$, *if there exists a control* $u(\cdot) \in L^2([t_0, t_1]; \mathbb{R}^m)$ *such that the solution of system* (10) *with this control* $u(\cdot)$ *satisfies* $x(t_1) = x_1$. *If the choices of* x_0 *and* x_1 *are arbitrary in* $\mathbb{R}^n$ *then system* (10) *is called completely controllable.*

There are many equivalent conditions for linear time-invariant system (10) to be completely controllable; one can refer Szidarovszky [10], Hespanha [11] for details. We say that the pair (A, B) is controllable if the system (10) is completely controllable. There may be more than one control $u(\cdot)$ steering x_0 to x_1 during time interval $[t_0, t_1]$. However, the control which minimizes the energy is given by (see [10])

$$u(t) = B^T \Phi^T(t_0, t) W^{-1}(t_0, t_1)(\Phi(t_0, t_1)x_1 - x_0), \tag{11}$$

where $\Phi(t, \tau), W(t_0, t_1)$ are transition matrix and controllability-grammian matrix for system (10), respectively. We will now define the fuzzy-controllability of system (2).

Definition 4. *System* (2) *with fuzzy initial condition* $x(t_0) = X_0 \in \mathbb{E}^n$ *is said to be fuzzy-controllable to a fuzzy state* $X_1 \in \mathbb{E}^n$ *at* $t_1(> t_0)$ *if there exists a fuzzy-integrable control* $u(\cdot) \in \mathbb{E}^m$ *such that the solution of system* (2) *with this control satisfies* $x(t_1) \leqslant X_1$. *(For fuzzy-integrability, see [8])*

By a fuzzy state $X \in \mathbb{E}^n$ around $x \in \mathbb{R}^n$, we mean that $X_*^1 = [x, x]^T$, that is, $\underline{X^1} = \overline{X^1} = x$. We will now show that given a target fuzzy state X_1 in $\mathbb{E}^n$ around x_1, we can always find a fuzzy state X_0 in $\mathbb{E}^n$ around x_0 so that X_0 is fuzzy-controllable to X_1. Following theorem gives an estimate of initial fuzzy state X_0 with which the system (2) is fuzzy-controllable to the target fuzzy state X_1.

Theorem 1. *Let the system pair (A^*, B^*) be controllable and x_0, $x_1 \in \mathbb{R}^n$. Assume further that there exists a control $u^*(\cdot) \in L^2([t_0, t_1]; \mathbb{R}^{2m})$ with the following properties:*

(H1) The crisp system $\dot{x}(t) = A^* x(t) + B^* u(t)$ with the control $u^*(\cdot)$ steers $[x_0, x_0]^T$ to $[x_1, x_1]^T$ during time interval $[t_0, t_1]$.

(H2) For $1 \leq k \leq m$, $u_k^*(\cdot) \leq u_{k+m}^*(\cdot)$ in which $u_k^*(\cdot)$ is the kth component of $u^*(\cdot)$.

Then given a fuzzy state $X_1 \in \mathbb{E}^n$ around x_1, there exist a fuzzy state $X_0 \in \mathbb{E}^n$ around x_0 and a control $u(\cdot) \in \mathbb{E}^m$ so that system (2) with fuzzy initial condition X_0 and the control $u(\cdot)$ is fuzzy-controllable to X_1 during time interval $[t_0, t_1]$.

Proof. Using hypothesis (H2), define $u(t) = [u_1(t), u_2(t), \ldots, u_m(t)]^T \in \mathbb{E}^m$ for $t \in [t_0, t_1]$ in which each $u_k(t)$ for $1 \leq k \leq m$ is a rectangular fuzzy number on $\mathbb{R}$ defined by the following α−cut representation. For $\alpha \in (0, 1]$,

$$[u_k(t)]_\alpha = [u_k^*(t), u_{k+m}^*(t)] \tag{12}$$

Clearly $u(\cdot)$ is fuzzy-integrable. Let $X_0 \in \mathbb{E}^n$ be an unknown fuzzy state around x_0, which we will characterize at the end of proof. Assume without loss of generality that $[X_1]_\alpha = [x_1 - \gamma^\alpha, x_1 + \gamma^\alpha]$, where $\gamma^\alpha = [\gamma_1^\alpha, \gamma_2^\alpha, \ldots, \gamma_n^\alpha]^T$ with $\gamma_k^\alpha \in \mathbb{R}^+$ for $1 \leq k \leq n$, $\alpha \in (0, 1]$, that is, α−cut of the kth component of X_1 is given by $[x_{1k} - \gamma_k^\alpha, x_{1k} + \gamma_k^\alpha]$. The evolution of system (2), with fuzzy initial condition X_0 and the control $u(\cdot)$ as defined above, is given by the following equations (see Remark 1). For $\alpha \in (0, 1]$,

$$\begin{cases} \dot{x}_*^\alpha(t) = A^* x_*^\alpha(t) + B^* u_*^\alpha(t) \\ x_*^\alpha(t_0) = X_{0*}^\alpha \end{cases} \tag{13}$$

The solution of (13) at time t_1 is given by

$$x_*^\alpha(t_1) = \Phi^*(t_1, t_0) x_*^\alpha(t_0) + \int_{t_0}^{t_1} \Phi^*(t_1, \tau) B^* u^*(\tau) d\tau, \tag{14}$$

where $\Phi^*(t, \tau)$ denotes the transition matrix for system $\dot{x}(t) = A^* x(t)$. Denote $[x_1, x_1]^T$, a vector in $\mathbb{R}^{2n}$, by x_*^1 and similarly $[x_0, x_0]^T$ by x_*^0. Setting $\alpha = 1$ in Eq. (14) and using hypothesis (H1), we have

$$x_*^1 = \Phi^*(t_1, t_0) x_*^0 + \int_{t_0}^{t_1} \Phi^*(t_1, \tau) B^* u^*(\tau) d\tau \tag{15}$$

By subtracting (15) from (14) we have

$$\|x_*^\alpha(t_1) - x_*^1\| \leq \|\Phi^*(t_1, t_0)\| \ \|x_*^\alpha(t_0) - x_*^0\|, \tag{16}$$

where $\|\Phi^*(t_1, t_0)\| = \sup\limits_{x \neq 0} \left(\frac{\|\Phi^*(t_1, t_0) x\|}{\|x\|} \right)$ and $\|.\|$ denotes the 2-norm. From equation (16) it follows that

$$\|x_*^\alpha(t_1) - x_*^1\|_\infty \leq \sqrt{2n} \|\Phi^*(t_1, t_0)\| \|x_*^\alpha(t_0) - x_*^0\|_\infty, \tag{17}$$

where $\|.\|_\infty$ denotes the infinity-norm. Let $\beta^\alpha = \dfrac{\gamma_{min}^\alpha}{\sqrt{(2n)}\,\|\Phi^*(t_1,t_0)\|}$, where $\gamma_{min}^\alpha = min[\gamma_1^\alpha, \gamma_2^\alpha, \ldots, \gamma_n^\alpha]$. Now define the fuzzy state $X_0 \in E^n$ by $[X_0]_\alpha = [x_0 - \beta^\alpha, x_0 + \beta^\alpha]$, where $\boldsymbol{\beta}^\alpha = [\beta^\alpha, \beta^\alpha, \ldots, \beta^\alpha]^T \in \mathbb{R}^n$, $\alpha \in (0,1]$. It can be easily shown that with the initial condition $x(t_0) = X_0$, the propagated fuzzy state at time t_1 satisfies $x(t_1) \leq X_1$. Hence X_0 is the required initial fuzzy state with which system (2) is fuzzy-controllable to X_1 with the control $u(\cdot)$ during time interval $[t_0, t_1]$.   □

Remark 2. If the entries of the matrix A are non-negative, then the control $u(\cdot) \in \mathbb{E}^m$ can be taken as a crisp control $\tilde{u}(\cdot)$ with values in $\mathbb{R}^m$ that steers x_0 to x_1 during time interval $[t_0, t_1]$. That is, $u(\cdot) \in \mathbb{E}^m$ is obtained from $\tilde{u}(\cdot) \in \mathbb{R}^m$ via the natural embedding of $\mathbb{R}^m$ into $\mathbb{E}^m$. Thus, for $1 \leq k \leq n$, $u_k(t) \in \mathbb{E}$ is defined as follows:

$$u_k(t)(s) = \begin{cases} 1 \text{ if } s = \tilde{u}_k(t) \\ 0 \text{ if } s \neq \tilde{u}_k(t), \end{cases} \tag{18}$$

where $u_k(\cdot)$ and $\tilde{u}_k(\cdot)$ are the kth component of $u(\cdot)$ and $\tilde{u}(\cdot)$, respectively.

Remark 3. In the case, when A has all its entries non-negative then the constant β^α can be taken as $\dfrac{\gamma_{min}^\alpha}{\sqrt{(n)}\,\|\Phi(t_1,t_0)\|}$ because of the block diagonal structure of A^*. Here $\Phi(t,\tau)$ denotes the transition matrix for the system $\dot{x}(t) = Ax(t)$.

4   Examples

The following example demonstrates the evolution of solution to fuzzy differential equation as prescribed in Lemma 1.

Example 1. Consider the following differential equation.

$$\begin{pmatrix} \dot{x}_1(t) \\ \dot{x}_2(t) \end{pmatrix} = \begin{pmatrix} 1 & 2 \\ -2 & 1 \end{pmatrix} \begin{pmatrix} x_1(t) \\ x_2(t) \end{pmatrix} + \begin{pmatrix} 1 \\ -1 \end{pmatrix} t \tag{19}$$

Initial membership functions, $x_1(0) = \mu_1(s)$ and $x_2(0) = \mu_2(s)$ at time t=0 are defined as below:

$$\mu_1(s) = \begin{cases} s & 0 \leq s \leq 1 \\ 2 - s & 1 \leq s \leq 2 \end{cases} \qquad \mu_2(s) = \begin{cases} 2s & 0 \leq s \leq 1/2 \\ 2 - 2s & 1/2 \leq s \leq 1 \end{cases}$$

The evolution of above system is given by the following level-wise decomposed differential equations as mentioned in Lemma 1. For α in $(0,1]$,

$$\begin{pmatrix} \underline{\dot{x}_1^\alpha}(t) \\ \overline{\dot{x}_1^\alpha}(t) \\ \underline{\dot{x}_2^\alpha}(t) \\ \overline{\dot{x}_2^\alpha}(t) \end{pmatrix} = \begin{pmatrix} 1 & 2 & 0 & 0 \\ 0 & 1 & -2 & 0 \\ 0 & 0 & 1 & 2 \\ -2 & 0 & 0 & 1 \end{pmatrix} \begin{pmatrix} \underline{x_1^\alpha}(t) \\ \overline{x_1^\alpha}(t) \\ \underline{x_2^\alpha}(t) \\ \overline{x_2^\alpha}(t) \end{pmatrix} + \begin{pmatrix} 1 & 0 \\ -1 & 0 \\ 0 & 1 \\ 0 & -1 \end{pmatrix} \begin{pmatrix} t \\ t \end{pmatrix}, \tag{20}$$

with the initial condition

$$[\underline{x_1^\alpha}(0), \overline{x_1^\alpha}(0), \underline{x_2^\alpha}(0), \overline{x_2^\alpha}(0)]^T = [\alpha, \alpha/2, 2 - \alpha, 1 - (\alpha/2)]^T.$$

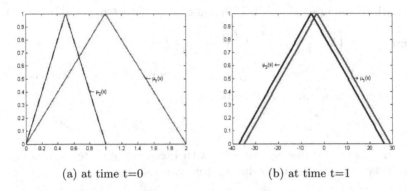

(a) at time t=0   (b) at time t=1

Fig. 1. Propagation of triangular-shaped membership functions for $x_1(\cdot)$ and $x_2(\cdot)$

The propagated fuzzy state at time t=1, starting from the initial fuzzy state at time $t = 0$, is shown in Fig.1.

In the next example, we compute a controllable initial fuzzy state that can be controlled to a desired target fuzzy state.

Example 2. Consider the following differential equation

$$\begin{pmatrix} \dot{x}_1(t) \\ \dot{x}_2(t) \end{pmatrix} = \begin{pmatrix} .5 & .1 \\ .1 & .3 \end{pmatrix} \begin{pmatrix} x_1(t) \\ x_2(t) \end{pmatrix} + \begin{pmatrix} 1 \\ -1 \end{pmatrix} u(t) \tag{21}$$

Let $x_0 = [2,3]^T$, $x_1 = [0,1]^T \in \mathbb{R}^2$ and $X_1 = [X_{11}, X_{12}]^T \in \mathbb{E}^2$, a fuzzy state around x_1, where X_{11}, X_{12} are given by the following membership functions:

$$X_{11}(s) = ee^{-\frac{1}{1-s^2}} \qquad X_{12}(s) = ee^{-\frac{1}{1-4(s-1)^2}}.$$

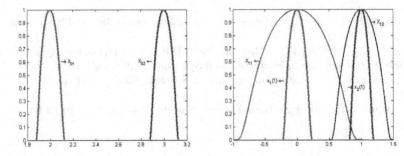

(a) Computed initial fuzzy state X_0 (b) Target fuzzy state X_1 and the propagated fuzzy state $x(t)$ at $t = 1$

Fig. 2. Initial fuzzy state X_0 (a) is fuzzy-controllable to a desired target fuzzy state X_1 (b)

Here the steering control $u(\cdot)$ is taken as minimum energy control that steers x_0 to x_1 during time interval $[0,1]$ (see Remark 2). In the setting of above example, we have $\gamma^\alpha = \left[\left(1 - \frac{1}{log\left(\frac{e}{\alpha}\right)} \right)^{\frac{1}{2}}, \frac{1}{2} \left(1 - \frac{1}{log\left(\frac{e}{\alpha}\right)} \right)^{\frac{1}{2}} \right]^T$.

Therefore, $\beta^\alpha = \frac{\gamma^\alpha_{min}}{\sqrt{(2)}\|e^A\|} = \frac{\gamma^\alpha_2}{3.6975}$. The fuzzy number $X_0 = [X_{01}, X_{02}]^T$ will be given by the following α-cut representation:

$$[X_{01}]_\alpha = [2 - \beta^\alpha, 2 + \beta^\alpha] \qquad [X_{02}]_\alpha = [3 - \beta^\alpha, 3 + \beta^\alpha]$$

It is clear from Fig.2 that system (21) with fuzzy initial state X_0 and control $u(\cdot)$ is fuzzy-controllable to X_1 during time interval $[0,1]$.

Acknowledgments. The authors wish to thank reviewers for their valuable comments and suggestions which resulted in improvement of this paper.

References

1. Kandel, A., Byatt, W.J.: Fuzzy Processes. Fuzzy Sets and Systems 4(2), 117–152 (1980)
2. Dubois, D., Prade, H.: Towards Fuzzy Differential Calculus Part 1: Integration of Fuzzy Mappings. Fuzzy Sets and Systems 8(1), 1–17 (1982)
3. Dubois, D., Prade, H.: Towards Fuzzy Differential Calculus Part 2: Integration on Fuzzy Intervals. Fuzzy Sets and Systems 8(2), 105–116 (1982)
4. Dubois, D., Prade, H.: Towards Fuzzy Differential Calculus Part 3: Differentiation. Fuzzy Sets and Systems 8(3), 225–233 (1982)
5. Puri, M.L., Ralescu, D.A.: Differentials of Fuzzy Functions. Journal of Mathematical Analysis and Applications 91(2), 552–558 (1983)
6. Kaleva, O.: Fuzzy Differential Equations. Fuzzy Sets and Systems 24(3), 301–317 (1987)
7. Zimmermann, H.-J.: Fuzzy Set Theory - and Its Applications, 4th edn. Springer, Heidelberg (2006)
8. Seikkala, S.: On the Fuzzy Initial Value Problem. Fuzzy Sets and Systems 24(3), 319–330 (1987)
9. Xu, J., Liao, Z., Hu, Z.: A class of Linear Differential Dynamical Systems with Fuzzy Initial Condition. Fuzzy Sets and Systems 158(21), 2339–2358 (2007)
10. Szidarovszky, F., Terry Bahill, A.: Linear Systems Theory, 2nd edn. CRC Press (1998)
11. Hespanha, J.P.: Linear Systems Theory. Princeton University Press (2009)

Fuzzy Modelling of S-Type Microbial Growth Model for Ethanol Fermentation Process and the Optimal Control Using Simulink

M.Z.M. Kamali[1], N. Kumaresan[2], Koshy Philip[3], and Kuru Ratnavelu[2]

[1] Centre for Foundation Studies in Science
University of Malaya
[2] Institute of Mathematical Sciences
University of Malaya
[3] Institute of Biological Sciences
University of Malaya
50603 Kuala Lumpur, Malaysia

Abstract. In this work, the fuzzy modelling of S(ubstrate)-type microbial growth model for ethanol fermentation process is built using the sector nonlinearity of Takagi-Sugeno (T-S) fuzzy system. The optimal control for the T-S fuzzy system is obtained using simulink. The motivation is to provide the optimal control by the solutions of the matrix Riccati differential equation (MRDE) obtained from an alternative approach. Accuracy of the solution of the simulink approach to the problem is qualitatively better. An illustrative numerical example is presented for the proposed method.

Keywords: Fuzzy modelling, Ethanol Fermentation process, Optimal control, Simulink.

1 Introduction

Fuzzy logic has shown to be the most suitable tools to represent biological system, analyzing biological data and capturing different uncertainties in biomedical and computational biology [1-3]. The potential of fuzzy control for biological processes was first recognized in the fermentation industry [1, 3] thus, in this work, the nonlinearity sector of T-S fuzzy model is used for representing the S-type microbial growth model for ethanol fermentation process. The S-type model assumed that (i) a linear relation is appropriate to describe the influence of sustrate depletion on the microbial growth, (ii) there is no substrate consumption for maintenance, (iii) there is no substrate breakdown in the medium, (iv) no additional susbtrate is added during the growth process and (v) there is only one limiting substrate. The details of the S-type model can be found in Ref [4]. In the T-S fuzzy model, there are two types of T-S fuzzy structures which are denoted as affine and linear T-S fuzzy system. The affine T-S fuzzy system is also called as the nonlinear part. Both demonstrate to be universal approximations to any nonlinear systems [5, 6]. The only difference between these two structures lies within the existence of a constant singleton in the fuzzy rule consequence for the affine T-S fuzzy model.

P. Balasubramaniam and R. Uthayakumar (Eds.): ICMMSC 2012, CCIS 283, pp. 325–332, 2012.

Under the topics of optimal control theory, the theory of the quadratic cost control problem has been treated as a more interesting problem and the optimal feedback with minimum cost control has been characterized by the solution of a Riccati equation [7]. The needs for solving such equations often arise in analysis and synthesis such as linear quadratic optimal control, stochastic filtering and control systems, model reduction, differential games etc. This equation, contributes significantly in optimal control problems, multivariable and large scale systems, scattering theory, estimation, transportation and radiative transfer [8]. The solution of this equation is difficult to obtain from two points of view. One is nonlinear and the other is in matrix form. Most general methods to solve MRDE with a terminal boundary condition are obtained by transforming MRDE into an equivalent linear differential Hamiltonian system [9]. Another approach is based on transforming MRDE into a linear matrix differential equation and then solving MRDE analytically or computationally [10-11].

The present paper implements the Simulink tool that runs as a companion to MATLAB software, for solving the matrix Riccati fuzzy differential equations in order to get the optimal solutions. This add-on package can be used to create a block of diagrams which can be translated into a system of ordinary differential equations. This paper is organized as follows: in section 2, the materials and methods are given. In section 3, the numerical example is discussed together with solution of the MRDE. Finally, the conclusion is given in section 4.

2   Materials and Methods

In the field of predictive microbiology, the mathematical models are developed to describe and predict the microbial evolution in foods. Single species microbial growth, whether in a (liquid) food product, normally undergoes three different phases. The first phase or the lag phase, the microbial cells adapt to their new environment and their numbers remain constant during this phase. In the exponential growth phase, the microbial cells multiply exponentially. Finally, in the stationary phase, the microbial cells cease multiplying and their total number remains constant at the maximum population density.

The S-type microbial growth model is given below:

$$\dot{N}(t) = \left(\frac{Q(t)}{1+Q(t)}\right).\mu_{\max}.S(t).N(t), \quad N(t=0) = N_0, \tag{1}$$

$$\dot{Q}(t) = .\mu_{\max}.Q(t), \quad Q(t=0) = Q_0, \tag{2}$$

$$\dot{S}(t) = -\left(\frac{Q(t)}{1+Q(t)}\right).\mu_{\max}.\frac{S(t)}{Y_{N/S}}.N(t), \quad S(t=0) = N_0. \tag{3}$$

The first differential equation describes the evolution of the microbial load $N(t)$ in time. It consists of the adjustment function which describes the lag phase by means of a variable representing the physiological state of the cells $Q(t)$, as well as the inhibition function which is a linear function of the substrate concentration $S(t)$. The second differential equation, describes the evolution of $Q(t)$, which increase exponentially, whereas the third differential equation represents the evolution of the substrate concentration $S(t)$. $Y_{N/S}$ refers as the yield coefficient.

2.1   Fuzzy Optimal Control Problem

The T-S fuzzy model can be derived from the above nonlinear system using sector nonlinearity approach [12]. Consider the linear dynamical fuzzy system [13] that can be expressed in the form:

R^i : If x_j is M_j, $i = 1,...,4$ and $j = 1, 2, 3$. then

$$\dot{x}(t) = A_i(t)x(t) + B_i u(t), \quad x(0) = 0, \quad t \in [0, t_f], \tag{4}$$

where R^i denotes the ith rule of the fuzzy model, M_j is membership function, $x(t) \in R^n$ is a generalized state space vector, $u(t) \in R^m$ is a control variable and it takes value in some Euclidean space,

$$\dot{x}(t) = \begin{bmatrix} \dot{x}_1(t) \\ \dot{x}_2(t) \\ \dot{x}_3(t) \end{bmatrix} = \begin{bmatrix} N(t) \\ Q(t) \\ S(t) \end{bmatrix}, \; A_i = \begin{bmatrix} z_1 & 0 & 0 \\ 0 & \mu_{max} & 0 \\ 0 & 0 & z_2 \end{bmatrix}, \; B_i = \begin{bmatrix} 0 \\ 0 \\ 1 \end{bmatrix},$$

$$z_1 = \mu_{max} \cdot \frac{Q(t)}{1+Q(t)} \cdot S(t), \; z_2 = -\mu_{max} \cdot \frac{Q(t)}{1+Q(t)} \cdot \frac{N(t)}{Y_{N/S}}. \tag{5}$$

Given a pair of $(x(t), u(t))$ and a product inference engine, the aggregate T-S fuzzy model can be inferred as

$$\dot{x} = \sum_{i=1}^{4} h_i(z(t))A_i x(t) + B_i u(t), \tag{6}$$

where

$$h_i(z(t)) = \frac{\Pi_{j=1}^{2} M_j^i(z_j(t))}{\sum_{i=1}^{4} \left(\Pi_{j=1}^{2} M_j^i(z_j(t)) \right)}.$$

for all t. The term $M_j^i(z_j(t))$ is the membership value of $z_j(t)$ in M_j^i. Here $h_i(z(t)) \geq 0$ and $\sum_{i=1}^{4} h_i(z(t)) = 1$ for all t and $i=1,2,3,4$. In order to minimize both state and control signals of the feedback control system, a quadratic performance index is usually minimized:

$$J = \frac{1}{2}x^T(t_f)Sx(t_f) + \frac{1}{2}\int_0^{t_f} [x^T(t)Qx(t) + u^T(t)Ru(t)]dt, \tag{7}$$

where the superscript T denotes the transpose operator, $S \in R^{n \times n}$ and $Q \in R^{n \times n}$ are symmetric and positive definite (or semi definite) weighting matrices for $x(t)$, $R \in R^{m \times m}$ is a symmetric and positive definite weighting matrix for $u(t)$. If all state variables are measurable, then linear state feedbacks control law [14, 15].

$$u(t) = -R^{-1}B_i^T \lambda_1(t), \tag{8}$$

can be obtained to the system described by eq.(4), where

$$\lambda_i(t) = K_i(t)x(t),$$

(9)

$K_i(t) \in R^{n \times n}$ is a symmetric matrix and the solution of MRDE. The relative MRDE for the linear T-S fuzzy system eq.(4) is

$$\dot{K}_i(t) + K_i(t)A_i + A_i^T K_i(t) + Q - K_i(t)B_i R^{-1} B_i^T K_i(t) = 0$$

(10)

with terminal condition(TC) $K_i(t_f)=S$. This equation is going to be solved for $K_i(t)$ in the next section for the optimal solution. After substituting the appropriate matrices in the above equation, they are transformed in to system of nonlinear differential equations.

2.2   Solution of MRDE

Consider the system of differential equation for eq.(10) in each rule of the fuzzy model.

$$\dot{k}_{ij}(t) = \phi_{ij}(k_{ij}(t)), \quad k_{ij}(t_f) = A_{ij} \quad (i, j = 1, 2,..., n).$$

(11)

2.3   Simulink Method

Procedure for simulink solution

   Step 1. Choose or select the required graphical block diagrams from the simulink Library.
   Step 2. Connect the appropriate blocks.
   Step 3. Set up the simulation parameters and run the simulink model to obtain the solution.

3   Numerical Examples

Consider the optimal control problem , where eq.(7) is minimized subject to the linear T-S fuzzy system R_i : If x_j is M_j, $i = 1,...,4$ and $j = 1,2,3$. Then following eq.(4) and for simplicity, the values of x_1, x_2 and x_3 are taken as $x_1 \in [1.68E-6, 0.478]$, x_2 is fixed and $x_3 \in [0.306, 0.542]$. The value of μ_{max} is given in [6] (i.e. μ_{max} = 9.006) whereas $Y_{N/S}=19.5147$. The   minimum and maximum values of z_1 and z_2 can be calculated and the membership functions can be obtained and shown in Fig. 1. Then, the nonlinear system is represented by the following fuzzy model.

Model Rule 1: IF $z_1(t)$ is Positive and $z_2(t)$ is Big, THEN $\dot{x}(t) = A_1 x(t) + Bu$,
Model Rule 2: IF $z_1(t)$ is Positive and $z_2(t)$ is Small, THEN $\dot{x}(t) = A_2 x(t) + Bu$,
Model Rule 3: IF $z_1(t)$ is Negative and $z_2(t)$ is Big, THEN $\dot{x}(t) = A_3 x(t) + Bu$,
Model Rule 4: IF $z_1(t)$ is Negative and $z_2(t)$ is Small, THEN $\dot{x}(t) = A_4 x(t) + Bu$,

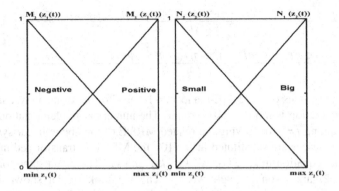

Fig. 1. Membership function of $z_1(t)$ and $z_2(t)$

where

$$S = \begin{bmatrix} 1 & 0 & 0 \\ 0 & 0 & 0 \\ 0 & 0 & 0 \end{bmatrix}, A_1 = \begin{bmatrix} 2.65 & 0 & 0 \\ 0 & 9.01 & 0 \\ 0 & 0 & 6.64E-4 \end{bmatrix}, A_2 = \begin{bmatrix} 2.65 & 0 & 0 \\ 0 & 9.01 & 0 \\ 0 & 0 & -4.2E-7 \end{bmatrix}$$

$$A_3 = \begin{bmatrix} -1.47E-2 & 0 & 0 \\ 0 & 9.01 & 0 \\ 0 & 0 & 6.64E-4 \end{bmatrix}, A_4 = \begin{bmatrix} -1.47E-2 & 0 & 0 \\ 0 & 9.01 & 0 \\ 0 & 0 & -4.2E-7 \end{bmatrix},$$

$$B = \begin{bmatrix} 0 \\ 0 \\ 1 \end{bmatrix}, R = 0, Q = \begin{bmatrix} 1 & 0 & 0 \\ 0 & 0 & 0 \\ 0 & 0 & 0 \end{bmatrix}.$$

If $x_1 = 1.0E-5$, $x_2 = 0.1$ and $x_3 = 0.31$, the T-S fuzzy modelling implication can be derived as in Table 1. Now the final values for $\dot{x}_1$ and $\dot{x}_3$, in T-S fuzzy defuzzification process, can be calculated as:

Table 1. T-S fuzzy model implication

Implication	Premise	Consequence	Truth Value
Rule 1	$M_1(z_1)=0.10,$	$\dot{x}_1 = 2.65E-5,$	$0.10 \wedge 2.19E-6 = 2.19E-6$
	$N_1(z_2)=2.19E-6$	$\dot{x}_2 = 0.90$	
		$\dot{x}_3 = 2E-4$	
Rule 2	$M_1(z_1)=0.10,$	$\dot{x}_1 = 2.65E-5,$	$0.10 \wedge 1.0 = 0.10$
	$N_2(z_2)=1.0$	$\dot{x}_2 = 0.90$	
		$\dot{x}_3 = -1.30E-7$	
Rule 3	$M_2(z_1)=0.90,$	$\dot{x}_1 = -1.47E-7,$	$0.90 \wedge 2.19E-6 = 2.19E-6$
	$N_1(z_2)=2.19E-6$	$\dot{x}_2 = 0.90$	
		$\dot{x}_3 = 2.06E-4$	
Rule 4	$M_2(z_1)=0.90,$	$\dot{x}_1 = -1.47E-7,$	$0.90 \wedge 1.00 = 0.90$
	$N_2(z_2)=1.00$	$\dot{x}_2 = 0.90$	
		$\dot{x}_3 = -1.30E-7$	

$$\dot{x}_I = \frac{(2.19E\text{-}6 \cdot 2.65E\text{-}5) + (1.01E\text{-}1 \cdot 2.65E\text{-}5) + (2.19E\text{-}6 \cdot -1.47E-7) + (0.90 \cdot -1.5E-7)}{(1.0)} = 2.54E-06$$

$$\dot{x}_3 = \frac{(2.19E\text{-}6 \cdot 2.06E\text{-}4) + (1.01E\text{-}1 \cdot -1.3E-7) + (2.19E\text{-}6 \cdot 2.06E\text{-}4) + (0.90 \cdot -1.3E-7)}{(1.0)} = -1.29E-7$$

Comparing the values of $\dot{x}_1 = 2.54E\text{-}6$ and $\dot{x}_3 = -1.30E\text{-}7$ of the original system, the T-S fuzzy approximation is more or less similar. The numerical implementation could be adapted by taking $t_f = 2$ for solving the related MRDE of the above linear system. The appropriate matrices are substituted in eq.(10), the MRDE is transformed into system of differential equation in k_{11}, k_{12}, k_{13}, k_{22}, k_{23} and k_{33}. The equidistant points in the interval [0, 2] are taken as input vector. The simulink model shown in Fig. 2 represents the systems of differential equations with the terminal conditions $k_{11} = 1.0$ and $k_{12} = k_{13} = k_{22} = k_{23} = k_{33} = 0.0$.

The numerical solutions of MRDE are calculated and displayed in Table 2. In the similar way, the MRDE can be solved for the matrices A_2, A_3 and A_4.

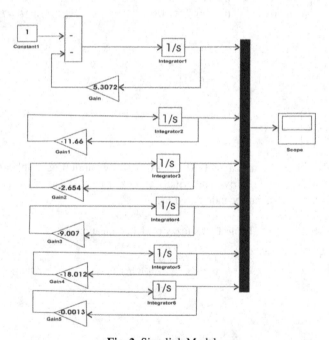

Fig. 2. Simulink Model

Table 2. Solutions for MRDE

t	Exact k_{11}	Simulink k_{11}
0	48389	48389
0.24	13690	13693
0.44	4736	4737
0.64	1639	1639
0.72	1072	1072
0.84	568	567
1.04	197	196
1.24	68	68
1.44	23	23
1.84	3	3
2	1	1

4   Conclusions

In this paper, we have illustrate the nonlinearity sector of T-S fuzzy modeling for ethanol fermentation process. The optimal control for the T-S fuzzy system was obtained by simulink approach. Thus, the simulink approach is expected to yield accurate solution of MRDE. A numerical example was given to demonstrate the derived results. The efficient approximations of the optimal solution are done in MATLAB on PC, CPU 2.0 GHz.

Acknowledgement. The funding for this work is the FRGS grant (Account No: FP048/2010B) is gratefully acknowledged.

References

1. Konstantinov, K., Yoshida, T.: Knowledge-based control of fermentation processes. Biotech. Bioengg. 39, 479–486 (1992)
2. Center, B., Verma, B.: A fuzzy photosynthesis model for tomato. Transactions of the ASABE 40(3), 815–821 (1997)
3. Dohnal, M.: Fuzzy bioengineering models. Biotech. Bioengg. 27, 1146–1151 (1985)
4. Van Impe, J.F., Poschet, F.P., Nicolai, B.M., Geeraerd, A.H.: S & P-type models a novel class of predictive microbial growth models. In: 13th World Congress of Food Sci. Tech., pp. 21–35 (2006)
5. Balasubramaniam, P., Abdul Samath, J., Kumaresan, N., Vincent Antony Kumar, A.: Neuro approach for solving matrix Riccati differential equation. Neural Parallel Scientific Comput. Dynam. 15, 125–135 (2007)
6. Baranyi, J., Roberts, T.A.: A dynamic approach to predicting bacterial growth in food. Internat. J. Food Microbiology 23, 277–294 (1994)
7. Da Prato, G., Ichikawa, A.: Quadratic control for linear periodic systems. Appl. Math. Optim. 18, 39–66 (1988)

8. Jamshidi, M.: An overview on the solutions of the algebraic matrix Riccati equation and related problems. Large Scale Syst. 1, 167–192 (1980)
9. Jodar, L., Navarro, E.: Closed analytical solution of Riccati type matrix differential equations. Indian J. Pure Appl. Math. 23, 185–187 (1992)
10. Lovren, N., Tomic, M.: Analytic solution of the Riccati equation for the homing missile linear quadratic control problem. J. Guidance Cont. Dynam. 17, 619–621 (1994)
11. Razzaghi, M.: A computational solution for a matrix Riccati differential equation. Numerische Mathematik 32, 271–279 (1979)
12. Kawamoto, S., Tada, K., Ishigame, A., Taniguchi, T.: An approach to stability analysis of second order fuzzy systems. IEEE Trans. Neural Netw. 4, 919–930 (1993)
13. Wu, S.J., Chiang, H.H., Lin, H.T., Lee, T.T.: Neural network based Optimal fuzzy controller design for nonlinear systems. Fuzzy Set Syst. 154, 182–207 (2005)
14. Ait Ram, M., Moore, J.B., Zhou, X.Y.: Indefinite stochastic linear quadratic control and generalized differential Riccati equation. SIAM J. Cont. Optim. 40, 1296–1311 (2001)
15. Zhu, J., Li, K.: An iterative method for solving stochastic Riccati differential equations for the stochastic LQR problem. Optim. Meth. Software 18, 721–732 (2003)

A Search for the Correlation Coefficient of Triangular and Trapezoidal Intuitionistic Fuzzy Sets for Multiple Attribute Group Decision Making

John Robinson P. and Henry Amirtharaj E.C.

PG and Research Department of Mathematics, Bishop Heber College
Tiruchirappalli – 620 017, Tamil Nadu, India
robijohnsharon@gmail.com, henry_23@rediff.com

Abstract. This paper introduces a new technique for defining the correlation coefficient of triangular and trapezoidal intuitionistic fuzzy sets for solving Multi-attribute Group Decision Making (MAGDM) problems. In situations where the information or the data is of the form of triangular or trapezoidal intuitionistic fuzzy numbers, some arithmetic aggregation operators, namely the trapezoidal intuitionistic fuzzy ordered weighted averaging (TzIFOWA) operator and the trapezoidal intuitionistic fuzzy hybrid aggregation (TzIFHA) operator are utilized. A new model is developed to solve the MAGDM problems using a new type of correlation coefficient defined for trapezoidal intuitionistic fuzzy sets based on the trapezoidal intuitionistic fuzzy weighted arithmetic averaging (TzIFWAA) operator and the TzIFHA operator.

Keywords: Multiple Attribute Group Decision Making, Correlation of Triangular Intuitionistic Fuzzy Number, Trapezoidal Intuitionistic Fuzzy Ordered Weighted Averaging Operator, Trapezoidal Intuitionistic Fuzzy Hybrid Aggregation Operator.

1 Introduction

Out of several higher-order fuzzy sets, Intuitionistic Fuzzy Set (IFS) was first introduced by Atanassov [1], [2] had been found to be compatible to deal with vagueness. With best of our knowledge, Burillo [3], [4] proposed the definition of intuitionistic fuzzy number (IFN) and studied the perturbations of IFN and the first properties of the correlation [4] between these numbers. Triangular IFS is a special case of IFSs, with two characterizations, namely the triangular fuzzy characterization and the intuitionistic fuzzy characterization [11], [13]. A MAGDM problem is to find a desirable solution from a finite number of feasible alternatives assessed on multiple attributes, both quantitative and qualitative ([9], [14] and [15]). In order to choose a desirable solution, the decision maker often provides his/her preference information which takes the form of numerical values, such as exact values, interval number values and fuzzy numbers [5]. Hence, MAGDM problems under a intuitionistic fuzzy or a triangular intuitionistic fuzzy environment is an interesting area of study for

P. Balasubramaniam and R. Uthayakumar (Eds.): ICMMSC 2012, CCIS 283, pp. 333–342, 2012.
© Springer-Verlag Berlin Heidelberg 2012

researchers in the recent days. Herrera et al. [8] developed an aggregation process for combining numerical, interval valued and linguistic information, and then proposed different extensions of this process to deal with contexts in which can appear information such as IFSs or multi-granular linguistic information. Xu and Yager [17] developed some geometric aggregation operators for MADM problems. D.F Li [12] investigated MADM problems with intuitionistic fuzzy information and constructed several linear programming models to generate optimal weights for attributes.

The method to measure the correlation between two variables involving fuzziness is a challenge to classical statistical theory. In this paper a new method to find the correlation coefficient of triangular and trapezoidal IFS is developed based on the method of zeng and Li [16] for the intuitionistic fuzzy information. At present, in the literature there is no formula for calculating the correlation coefficient of triangular and trapezoidal IFS. We utilize some aggregation operations and the proposed correlation coefficient of triangular and trapezoidal IFS to make a better decision on the given decision information. Since the decision making procedure followed in this paper is similar for both triangular IFS and trapezoidal IFS, aggregation operators namely *TzIFOWA* and *TzIFHA* operators are discussed and the procedure for decision making using the correlation coefficient of trapezoidal IFS is presented. A numerical example is presented to explain the developed decision making model.

2   Basic Concepts of Intuitionistic Fuzzy Sets (IFSs)

Definition. Intuitionistic Fuzzy Set.

Let a set X be the universe of discourse. An IFS A in X is an object having the form $A = \left\{ \left\langle x, \mu_A(x), v_A(x) \right\rangle / x \in X \right\}$, where μ_A: $X \rightarrow [0,1]$, v_A: $X \rightarrow [0,1]$ define the degree of membership and the degree of non-membership respectively ([1], [2]), of the element $x \in X$ to the set A, which is a subset of X, and for every element of $x \in X$, $0 \le \mu_A(x) + v_A(x) \le 1$.

Definition. Triangular intuitionistic fuzzy number TIFN [13].

A TIFN is an IFS in R with the following membership function $\mu_A(x)$ and non-membership function $v_A(x)$:

$$\mu_A(x) = \begin{cases} \dfrac{x - a_1}{a_2 - a_1} & \text{for } a_1 \le x \le a_2 \\ \dfrac{a_3 - x}{a_3 - a_2} & \text{for } a_2 \le x \le a_3 \\ 0 & \text{otherwise} \end{cases} \qquad v_A(x) = \begin{cases} \dfrac{x - a_1}{a_2 - a_1'} & \text{for } a_1' \le x \le a_2 \\ \dfrac{a_3 - x}{a_3' - a_2} & \text{for } a_2 \le x \le a_3' \\ 1 & \text{otherwise} \end{cases} \qquad (1)$$

where $a_1' \le a_1 \le a_2 \le a_3 \le a_3'$ and $\mu_A(x)$, $v_A(x) \le 0.5$ for $\mu_A(x) = v_A(x)$. This TIFN is denoted by $\left(a_1, a_2, a_3; a_1', a_2, a_3' \right)$. This TIFN is also denoted as:

$$A = \left\langle ([a_1, a_2, a_3]; \mu_A), ([a_1', a_2, a_3']; v_A) \right\rangle.$$

Definition. Trapezoidal intuitionistic fuzzy number TzIFN [7].

A TzIFN is an IFS in R with the following membership function $\mu_A(x)$ and non-membership function $v_A(x)$:

$$\mu_A(x) = \begin{cases} \dfrac{x - a_1}{a_2 - a_1} & \text{for } a_1 \le x < a_2 \\ 1 & \text{for } a_2 \le x < a_3 \\ \dfrac{a_3 - x}{a_4 - a_3} & \text{for } a_3 \le x < a_4 \\ 0 & \text{otherwise} \end{cases} \qquad v_A(x) = \begin{cases} \dfrac{a_2 - x}{a_2 - a_1'} & \text{for } a_1' \le x < a_2 \\ 0 & \text{for } a_2 \le x < a_3 \\ \dfrac{x - a_3}{a_4' - a_3} & \text{for } a_3 \le x < a_4' \\ 1 & \text{otherwise} \end{cases} \qquad (2)$$

where $a_1' \le a_1 \le a_2 \le a_3 \le a_4 \le a_4'$ and $\mu_A(x)$, $v_A(x) \le 0.5$ for $\mu_A(x) = v_A(x)$. This TzIFN is denoted by $(a_1, a_2, a_3, a_4; a_1', a_2, a_3, a_4')$. This TzIFN is also denoted as:

$$A = \left\langle ([a_1, a_2, a_3, a_4]; \mu_A), ([a_1', a_2, a_3, a_4']; v_A) \right\rangle.$$

3   Correlation Coefficient of Triangular and Trapezoidal Intuitionistic Fuzzy Sets

Since a triangular intuitionistic fuzzy set is characterized by intuitionistic fuzzy information and the triangular fuzzy information, any previous method proposed in the literature for finding correlation coefficient for intuitionistic fuzzy set will not be sufficient for working with the correlation coefficient of triangular intuitionistic fuzzy sets. Hence we need to define a new type of correlation coefficient for triangular intuitionistic fuzzy sets which will uniquely deal with both the intuitionistic fuzzy information and the triangular fuzzy information, combining these two characterizations to derive a single correlation coefficient. We have worked on correlation coefficient of vague sets and interval vague sets in the recent years [10], [11]. In this paper we propose a new and unique method for calculating the correlation coefficient of triangular and trapezoidal intuitionistic fuzzy sets.

Definition. Given a triangular fuzzy number $\widetilde{A} = (a, b, c)$, the graded mean integration representation (GMIR) of $\widetilde{A}$ is defined as [6]:

$$P(\widetilde{A}) = {(a + 4b + c)} \big/ {6} \qquad (3)$$

and given a trapezoidal fuzzy number $\widetilde{A} = (a, b, c, d)$, the graded mean integration representation (GMIR) of $\widetilde{A}$ is defined as [6]:

$$P(\widetilde{A}) = {(a + 2b + 2c + d)} \big/ {6}. \qquad (4)$$

A triangular IFS is represented as $A = ([a,b,c]; \mu, \gamma)$, where $[a,b,c]$ represent the triangular fuzzy entries and μ, γ represent the membership, non-membership grades respectively. Utilizing the GMIR for a triangular fuzzy entry together with the membership, non-membership grades of the triangular IFS, we define the correlation of triangular IFS as follows:

Let $A = ([a_1, b_1, c_1]; \mu_A, \gamma_A)$, $B = ([a_2, b_2, c_2]; \mu_B, \gamma_B)$ be two triangular IFSs.

Then for each A, B $\in$ TIFS(X), we define the informational triangular intuitionistic energy of A as follows:

$$E_{TIFS}(A) = \frac{1}{n} \sum_{i=1}^{n} \left[\frac{a_1 + 4b_1 + c_1}{6} \right]^2 \left(\mu_A^2(x_i) + \gamma_A^2(x_i) + \pi_A^2(x_i) \right) \tag{5}$$

and $$E_{TIFS}(B) = \frac{1}{n} \sum_{i=1}^{n} \left[\frac{a_2 + 4b_2 + c_2}{6} \right]^2 \left(\mu_B^2(x_i) + \gamma_B^2(x_i) + \pi_B^2(x_i) \right). \tag{6}$$

Now we define the correlation of A and B as:

$$C_{TIFS}(A,B) = \frac{1}{n} \sum_{i=1}^{n} \left[\frac{a_1 + 4b_1 + c_1}{6} \right]\left[\frac{a_2 + 4b_2 + c_2}{6} \right](\mu_A(x_i)\mu_B(x_i) + \gamma_A(x_i)\gamma_B(x_i) + \pi_A(x_i)\pi_B(x_i)). \tag{7}$$

Then we define the correlation coefficient between A and B as:

$$K_{TIFS}(A,B) = \frac{C_{TIFS}(A,B)}{\sqrt{E_{TIFS}(A).E_{TIFS}(B)}}. \tag{8}$$

Theorem 1. For A, B $\in$ TIFS(X), then $0 \le K_{TIFS}(A,B) \le 1$.

Theorem 2. $K_{TIFS}(A,B) = 1 \Leftrightarrow A = B$.

Theorem 3. $C_{TIFS}(A,B) = 0 \Leftrightarrow$ A and B are non-fuzzy sets and satisfy the condition $\mu_A(x_i) + \mu_B(x_i) = 1$ or $\gamma_A(x_i) + \gamma_B(x_i) = 1$ or $\pi_A(x_i) + \pi_B(x_i) = 1$, $\forall x_i \in$ X.

Theorem 4. $E_{TIFS}(A) = 1 \Leftrightarrow A$ is a non-fuzzy set.

Similarly the correlation coefficient of trapezoidal intuitionistic fuzzy sets can be defined. Let $A = ([a_1, b_1, c_1, d_1]; \mu_A, \gamma_A)$, $B = ([a_2, b_2, c_2, d_2]; \mu_B, \gamma_B)$ be two trapezoidal IFSs (TzIFS). Then for each A, B $\in$ TzIFS(X), we define the informational trapezoidal intuitionistic energy of A as follows:

$$E_{TzIFS}(A) = \frac{1}{n} \sum_{i=1}^{n} \left[\frac{a_1 + 2b_1 + 2c_1 + d_1}{6} \right]^2 \left(\mu_A^2(x_i) + \gamma_A^2(x_i) + \pi_A^2(x_i) \right) \tag{9}$$

and $$E_{TzIFS}(B) = \frac{1}{n} \sum_{i=1}^{n} \left[\frac{a_2 + 2b_2 + 2c_2 + d_2}{6} \right]^2 \left(\mu_B^2(x_i) + \gamma_B^2(x_i) + \pi_B^2(x_i) \right). \tag{10}$$

Now we define the correlation of A and B as:

$$C_{TzIFS}(A,B) = \frac{1}{n} \sum_{i=1}^{n} \left[\frac{a_1 + 2b_1 + 2c_1 + d_1}{6} \right]\left[\frac{a_2 + 2b_2 + 2c_2 + d_2}{6} \right](\mu_A(x_i)\mu_B(x_i) + \gamma_A(x_i)\gamma_B(x_i) + \pi_A(x_i)\pi_B(x_i)) \tag{11}$$

Then we define the correlation coefficient between A and B as:

$$K_{TzIFS}(A,B) = \frac{C_{TzIFS}(A,B)}{\sqrt{E_{TzIFS}(A).E_{TzIFS}(B)}}. \tag{12}$$

Proposition. For A, B $\in$ TzIFS(X), we have:

i) $0 \le K_{TzIFS}(A,B) \le 1$,

ii) $C_{TzIFS}(A,B) = C_{TzIFS}(B,A)$,

iii) $K_{TzIFS}(A,B) = K_{TzIFS}(B,A)$,

iv) $K_{TzIFS}(A,B) = 1$, if A = B.

4   Aggregation Operators for Decision Making

For a normalized trapezoidal intuitionistic fuzzy decision making matrix, $\tilde{R} = (\tilde{r}_{ij})_{m \times n} = \left(\left[a_{ij}, b_{ij}, c_{ij}, d_{ij} \right]; u_{ij}, v_{ij} \right)_{m \times n}$ where $0 \le a_{ij} \le b_{ij} \le c_{ij} \le d_{ij} \le 1$, $0 \le u_{ij} + v_{ij} \le 1$. The trapezoidal intuitionistic fuzzy positive ideal solution and trapezoidal intuitionistic fuzzy negative ideal solution are defined as follows: $\tilde{r}^+ = \left(\left[a^+, b^+, c^+, d^+ \right]; u^+, v^+, \right) = \left([\,1,1,1,1]; 1,0 \right)$, $\tilde{r}^- = \left(\left[a^-, b^-, c^-, d^- \right]; u^-, v^- \right) = \left([\,0,\,0,\,0,0]; 0,1 \right)$.

Definition. Let $\tilde{a}_j$ $(j = 1,2,\ldots,n)$ be a collection of trapezoidal intuitionistic fuzzy numbers and let $TzIFWAA:Q^n \to Q$, if $TzIFWAA\left(\tilde{a}_1, \tilde{a}_2, \ldots, \tilde{a}_n \right) = \sum\limits_{j=1}^{n} \tilde{a}_j \omega_j$

$$= \left(\left[\sum_{j=1}^{n} a_j \omega_j, \sum_{j=1}^{n} b_j \omega_j, \sum_{j=1}^{n} c_j \omega_j, \sum_{j=1}^{n} d_j \omega_j \right]; 1 - \prod_{j=1}^{n}(1 - \mu_{a_j})^{\omega_j}, \prod_{j=1}^{n}(\gamma_{a_j})^{\omega_j} \right) \tag{13}$$

where, $\omega = (\omega_1, \omega_2, \ldots, \omega_n)^T$ be the weight vector of $\tilde{a}_j$ $(j = 1,2,\ldots,n)$ and for $\omega_j > 0$, $\sum\limits_{j=1}^{n} \omega_j = 1$.

Definition. Let $\tilde{a}_j$ $(j = 1,2,\ldots,n)$ be a collection of TzIFNs. A trapezoidal intuitionistic fuzzy ordered weighted averaging *(TzIFOWA)* operator of decision n is a mapping $TzIFOWA: Q^n \to Q$, that has an associated vector $w = (w_1, w_2, \ldots, w_n)^T$ such that $w_j > 0$, and $\sum\limits_{j=1}^{n} w_j = 1$. Furthermore, $TzIFOWA_w\left(\tilde{a}_1, \tilde{a}_2, \ldots, \tilde{a}_n \right) = \sum\limits_{j=1}^{n} \tilde{a}_{\sigma(j)} w_j$

$$= \left(\left[\sum_{j=1}^{n} a_{\sigma(j)} w_j, \sum_{j=1}^{n} b_{\sigma(j)} w_j, \sum_{j=1}^{n} c_{\sigma(j)} w_j, \sum_{j=1}^{n} d_{\sigma(j)} w_j \right]; 1 - \prod_{j=1}^{n}(1 - \mu_{a_{\sigma(j)}})^{w_j}, \prod_{j=1}^{n}(\gamma_{a_{\sigma(j)}})^{w_j} \right) \tag{14}$$

where $(\sigma(1), \sigma(2), \ldots, \sigma(n))$ is a permutation of $(1,2,\ldots,n)$ such that $\tilde{\alpha}_{\sigma(j-1)} \ge \tilde{\alpha}_{\sigma(j)}$ for all j=2,\ldots,n. The *TzIFOWA* operator has the following properties:

Definition. A trapezoidal intuitionistic fuzzy hybrid aggregation *(TzIFHA)* operator of dimension n is the mapping *TzIFHA:* $Q^n \rightarrow Q$ that has an associated vector $w = (w_1, w_2,...., w_n)^T$ such that $w_j > 0$, and $\sum_{j=1}^{n} w_j = 1$. Furthermore,

$$TzIFHA_{\omega,w} \left(\widetilde{a}_1, \widetilde{a}_2,...., \widetilde{a}_n \right) = \sum_{j=1}^{n} \dot{\widetilde{a}}_{\sigma(j)} \, w_j$$

$$= \left(\left[\sum_{j=1}^{n} \dot{a}_{\sigma(j)} w_j , \sum_{j=1}^{n} \dot{b}_{\sigma(j)} w_j , \sum_{j=1}^{n} \dot{c}_{\sigma(j)} w_j , \sum_{j=1}^{n} \dot{d}_{\sigma(j)} w_j \right] ; 1 - \prod_{j=1}^{n} (1 - \dot{\mu}_{\ddot{a}_{\sigma(j)}})^{w_j} , \prod_{j=1}^{n} (\dot{\gamma}_{\ddot{a}_{\sigma(j)}})^{w_j} \right) \qquad (15)$$

where $\dot{\widetilde{a}}_{\sigma(j)}$ is the jth largest of the weighted *TzIFNs* $\dot{\widetilde{a}}_j (\dot{\widetilde{a}}_j = \widetilde{a}_j^{n\omega_j}, \; j = 1, 2......n)$, $\omega = (\omega_1, \omega_2,...., \omega_n)^T$ is the weight vector of $\widetilde{a}_j$ $(j = 1, 2,....,n)$ and for $\omega_j > 0$, $\sum_{j=1}^{n} \omega_j = 1$. n is the balancing co-efficient.

5   Proposed Method for the Decision Making Problem

Let $A = \{A_1, A_2....A_m\}$ be a discrete set of alternatives, and $G = \{G_1, G_2,...,G_n\}$ be the set of attributes, $w = (w_1, w_2,...., w_n)$ is the weighting vector of the attribute G_j $(j=1, 2,....,n)$, where $w_j \in [0,1], \sum_{j=1}^{n} w_j = 1$. Let $D = \{ D_1, D_2,....,D_t \}$ be the set of decision makers, $v = (v_1, v_2,...., v_n)$ be the weighting vector of decision makers, with $v_k \in [0,1], \sum_{k=1}^{t} v_k = 1$; Suppose that, $\tilde{R}_k = (\tilde{r}_{ij}^{(k)})_{m \times n}$ is the trapezoidal intuitionistic fuzzy decision matrix, with $u_{ij}^{(k)} \in [0,1], v_{ij}^{(k)} \in [0,1], u_{ij}^{(k)} + v_{ij}^{(k)} \leq 1$, where $i=1, 2,...,m; \; j=1,2,...,n; \; k=1,2,...,t$. In the following, we apply the *TzIFWAA* and *TzIFHA* operator to multiple attribute group decision making based on trapezoidal intuitionistic fuzzy nformation. The method involves the following steps:

Step 1. Utilize the decision information given in the trapezoidal intuitionistic fuzzy decision matrix $\tilde{R}_k$ and the *TzIFWAA* operator,

$$\tilde{r}_i^{(k)} = \left(\left[a_i^{(k)}, b_i^{(k)}, c_i^{(k)}, d_i^{(k)} \right] ; u_i^{(k)}, v_i^{(k)} \right)$$

$$= TrIFWAA_\omega \left(\tilde{r}_{i1}^{(k)}, \tilde{r}_{i2}^{(k)},..., \tilde{r}_{in}^{(k)} \right)$$

where $i = 1, 2,...,m, \; k = 1, 2,...,t$, to derive the individual overall preference trapezoidal intuitionistic fuzzy values $\tilde{r}_i(k)$.

Step 2. Utilize the *TzIFHA* operator, $\tilde{r}_i = \left(\left[a_i, b_i, c_i, d_i \right]; u_i, v_i \right)$

$$= TzIFHA_{v,w} \left(\tilde{r}_i^{(1)}, \tilde{r}_i^{(2)}, \ldots \ldots \tilde{r}_i^{(t)} \right), \quad i = 1, 2, \ldots, m$$

to derive the collective overall preference trapezoidal intuitionistic fuzzy values $\tilde{r}_i$ (i=1,2,....,m) of the alternative A_i, where $v = (v_1, v_2, ..., v_n)$ be the weighting vector of decision makers, with $v_k \in [0,1]$, $\sum_{k=1}^{t} v_k = 1$; and $w = (w_1, w_2, ..., w_n)$ is the associated weighting vector of the *TzIFHA* operator with $w_j > 0$, and $\sum_{j=1}^{n} w_j = 1$.

Step 3. To calculate the correlation coefficient using (12) between collective overall values $\tilde{r}_i = \left([a_i, b_i, c_i, d_i]; u_i, v_i \right)$ and the trapezoidal intuitionistic fuzzy positive ideal solution $\tilde{r}^+$.

$$K_{TzIFS}(A, B) = \frac{C_{TzIFS}(A, B)}{\sqrt{E_{TzIFS}(A).E_{TzIFS}(B)}}$$

Step 4. Rank all the alternatives A_i (i=1,2,...,m) and select one in accordance with $K_{TzIFS}(A, B)$, i=1,2,...,m. The greater values of $K_{TzIFS}(A, B)$ will be the better alternatives A_i, when positive ideal solution is taken.

6   Numerical Illustration

An investment company wants to invest the sum of money in the best option where there is a panel with five possible alternatives is considered. To invest the money, the investment company must take a decision according to the following four attributes: G_1 is the Past Market Trend, G_2 is the Risk Appetite, G_3 is the Investment Horizon, G_4 is the Expected Returns. The five possible alternatives A_i are to be evaluated using the trapezoidal intuitionistic fuzzy numbers by the three decision makers whose weighting vector is $\gamma=(0.45, 0.20, 0.35)$, and under the four attributes whose weighting vector is w=(0.4, 0.1, 0.2, 0.3). The decision matrices (5x4) are respectively:

$$\tilde{R} = \begin{bmatrix} ([0.3,0.4,0.5,0.6];0.3,0.5) & ([0.4,0.5,0.6,0.7];0.3,0.4) & ([0.3,0.4,0.5,0.6];0.2,0.6) & ([0.6,0.7,0.8,0.9];0.4,0.4) \\ ([0.4,0.5,0.6,0.7];0.2,0.7) & ([0.2,0.4,0.5,0.6];0.4,0.5) & ([0.2,0.4,0.6,0.8];0.2,0.5) & ([0.7,0.8,0.9,1.0];0.1,0.7) \\ ([0.6,0.7,0.8,0.9];0.4,0.5) & ([0.2,0.3,0.4,0.5];0.4,0.5) & ([0.6,0.7,0.9,1.0];0.2,0.7) & ([0.5,0.6,0.7,0.8];0.1,0.8) \\ ([0.7,0.8,0.9,1.0];0.4,0.6) & ([0.6,0.7,0.8,0.9];0.3,0.6) & ([0.5,0.6,0.8,0.9];0.3,0.6) & ([0.6,0.7,0.8,1.0];0.3,0.6) \\ ([0.2,0.3,0.5,0.6];0.5,0.5) & ([0.3,0.4,0.6,0.7];0.4,0.5) & ([0.6,0.7,0.8,0.9];0.4,0.4) & ([0.4,0.6,0.8,1.0];0.4,0.5) \end{bmatrix}$$

$$\tilde{R_2} = \begin{bmatrix} ([0.2,0.3,0.4,0.5];0.2,0.6) & ([0.3,0.4,0.5,0.6];0.3,0.4) & ([0.2,0.3,0.4,0.5];0.1,0.7) & ([0.5,0.6,0.7,0.8];0.1,0.3) \\ ([0.3,0.4,0.5,0.6];0.1,0.8) & ([0.1,0.3,0.4,0.5];0.3,0.6) & ([0.1,0.3,0.5,0.7];0.3,0.4) & ([0.6,0.7,0.8,0.9];0.2,0.6) \\ ([0.5,0.6,0.7,0.8];0.4,0.5) & ([0.1,0.2,0.3,0.4];0.3,0.6) & ([0.5,0.6,0.8,0.9];0.3,0.6) & ([0.4,0.5,0.6,0.7];0.2,0.7) \\ ([0.6,0.7,0.8,0.9];0.3,0.7) & ([0.5,0.6,0.7,0.8];0.2,0.7) & ([0.4,0.5,0.7,0.8];0.2,0.7) & ([0.5,0.6,0.7,0.9];0.4,0.5) \\ ([0.1,0.2,0.4,0.5];0.4,0.6) & ([0.2,0.3,0.5,0.6];0.4,0.5) & ([0.5,0.6,0.7,0.8];0.3,0.5) & ([0.3,0.5,0.7,0.9];0.2,0.3) \end{bmatrix}$$

$$\tilde{R_3} = \begin{bmatrix} ([0.1,0.2,0.3,0.4];0.1,0.5) & ([0.2,0.3,0.4,0.5];0.2,0.3) & ([0.1,0.2,0.3,0.4];0.2,0.6) & ([0.4,0.5,0.6,0.7];0.2,0.4) \\ ([0.2,0.3,0.4,0.5];0.1,0.7) & ([0.1,0.2,0.3,0.5];0.2,0.5) & ([0.1,0.2,0.4,0.6];0.2,0.3) & ([0.5,0.6,0.7,0.8];0.1,0.5) \\ ([0.4,0.5,0.6,0.7];0.3,0.4) & ([0.1,0.2,0.3,0.4];0.2,0.5) & ([0.4,0.5,0.7,0.8];0.2,0.5) & ([0.3,0.4,0.5,0.6];0.1,0.6) \\ ([0.5,0.6,0.7,0.8];0.2,0.6) & ([0.4,0.5,0.6,0.7];0.1,0.6) & ([0.3,0.4,0.6,0.7];0.1,0.6) & ([0.4,0.5,0.6,0.8];0.3,0.4) \\ ([0.1,0.2,0.3,0.4];0.3,0.5) & ([0.1,0.2,0.4,0.5];0.3,0.4) & ([0.4,0.5,0.6,0.7];0.2,0.4) & ([0.2,0.4,0.6,0.8];0.2,0.5) \end{bmatrix}$$

Step 1. Utilize the decision information given in the trapezoidal intuitionistic fuzzy decision matrix $\tilde{R}_k$ and the *TzIFWAA* operator to derive the individual overall preference intuitionistic triangular fuzzy values $\tilde{r}_i(k)$ of the alternative A_i .

$\tilde{r_1}^{(1)} = ([0.43,0.57,0.67,0.73];0.3150,0.4889), \tilde{r_2}^{(1)} = ([0.44,0.58,0.71,0.84];0.1852,0.6118)$

$\tilde{r_3}^{(1)} = ([0.52,0.62,0.75,0.85];0.2308,0.6675), \tilde{r_4}^{(1)} = ([0.59,0.69,0.82,0.96];0.3213,0.6000)$

$\tilde{r_5}^{(1)} = ([0.41,0.55,0.72,0.86];0.4215,0.4676),$

$\tilde{r_1}^{(2)} = ([0.33,0.43,0.53,0.63];0.1428,0.4573), \tilde{r_2}^{(2)} = ([0.34,0.48,0.61,0.74];0.2235,0.5627)$

$\tilde{r_3}^{(2)} = ([0.42,0.52,0.65,0.75];0.2841,0.6153), \tilde{r_4}^{(2)} = ([0.49,0.59,0.72,0.86];0.3057,0.6118)$

$\tilde{r_5}^{(2)} = ([0.31,0.45,0.62,0.76];0.2950,0.4227)$

and

$\tilde{r_1}^{(3)} = ([0.23,0.33,0.43,0.53];0.1809,0.4589), \tilde{r_2}^{(3)} = ([0.28,0.38,0.51,0.65];0.1414,0.4588)$

$\tilde{r_3}^{(3)} = ([0.33,0.43,0.56,0.66];0.1835,0.5143), \tilde{r_4}^{(3)} = ([0.35,0.45,0.58,0.68];0.2051,0.5102)$

$\tilde{r_5}^{(3)} = ([0.23,0.37,0.52,0.66];0.2314,0.4573).$

Step 2. Utilize the *TzIFHA* operator to derive the collective overall preference trapezoidal intuitionistic fuzzy values $\tilde{r}_i$ of the alternative A_i . We have,

$\gamma = (0.45,0.20,0.35)$ and $w=(0.4,0.4,0.2)$. Then we get,

$\tilde{r_1} = ([0.3518,0.4731,0.5746,0.6558]; 0.2381,0.4610), \tilde{r_2} = ([0.3703,0.5033,0.6354,0.7701]; 0.1863,0.5588),$

$\tilde{r_3} = ([0.4447,0.5458,0.6780,0.7804]; 0.2301,0.6145), \tilde{r_4} = ([0.5057,0.6078,0.7411,0.8790]; 0.2857,0.5793),$

$\tilde{r_5} = ([0.3361,0.4768,0.6456,0.7890]; 0.3419,0.4375).$

Step 3. To calculate the correlation coefficient between collective overall values $\tilde{r}_i = ([a_i, b_i, c_i, d_i]; u_i, v_i)$ and the trapezoidal intuitionistic fuzzy positive ideal solution $\tilde{r}^+ = ([1,1,1,1]; 1,0)$, is

$$K_{TzIFS}(\tilde{r}_i, \tilde{r}^+) = \frac{C_{TzIFS}(\tilde{r}_i, \tilde{r}^+)}{\sqrt{E_{TzIFS}(\tilde{r}_i).E_{TzIFS}(\tilde{r}^+)}}.$$

Hence the calculated values are given as follows:

$C_{TzIFS}(\tilde{r}_1, \tilde{r}^+) = 0.1231, C_{TzIFS}(\tilde{r}_2, \tilde{r}^+) = 0.1061, C_{TzIFS}(\tilde{r}_3, \tilde{r}^+) = 0.1408, C_{TzIFS}(\tilde{r}_4, \tilde{r}^+) = 0.1944, C_{TzIFS}(\tilde{r}_5, \tilde{r}^+) = 0.1920.$

$K_{TzIFS}(\tilde{r}_1, \tilde{r}^+) = 0.3969, K_{TzIFS}(\tilde{r}_2, \tilde{r}^+) = 0.2903, K_{TzIFS}(\tilde{r}_3, \tilde{r}^+) = 0.3411, K_{TzIFS}(\tilde{r}_4, \tilde{r}^+) = 0.4286, K_{TzIFS}(\tilde{r}_5, \tilde{r}^+) = 0.5724.$

Step 4. Rank all the alternatives A_i (i=1,2,3,4,5) based on the values of

$$K_{TzIFS}(A, B).$$ Then we have:
$$A_5 > A_4 > A_1 > A_3 > A_2.$$

Hence, the best alternative is A_5.

This same procedure of decision making is adopted with triangular intuitionistic fuzzy numbers which has the same effect of deciding the best alternative when trapezoidal intuitionistic fuzzy number is replaced by triangular intuitionistic fuzzy number.

7   Conclusion

In this paper, a novel approach of decision making is proposed with respect to MAGDM problems, where both the attribute weights and the expert weights take the form of real numbers, and the attribute values take the form of a trapezoidal intuitionistic fuzzy numbers (TzIFNs). The new method for finding the correlation coefficient of triangular and trapezoidal intuitionistic fuzzy numbers was introduced in this paper because there is no separate formula defined in the literature for finding correlation coefficient of these numbers. None of the earlier methods of correlation of IFSs can be utilized for triangular and trapezoidal intuitionistic fuzzy numbers. In this paper we have developed the formula for the correlation coefficient of triangular and trapezoidal intuitionistic fuzzy numbers. The newly developed correlation coefficient is used in choosing the best alternative out of many, in the decision making process. Finally, an illustration was presented to demonstrate and validate the effectiveness of our proposed method. We shall continue to work in the extension and application of the developed method in some of the complicated domains in future.

References

1. Atanassov, K.T.: Intuitionistic Fuzzy Sets. Fuzzy Sets and Systems 20(1), 87–96 (1986)
2. Atanassov, K.T.: More on Intuitionistic Fuzzy Sets. Fuzzy Sets and Systems 33, 37–45 (1989)

3. Burillo, P., Bustince, H., Mohedano, V.: Some Definitions of Intuitionistic Fuzzy Number. In: Lakov, D. (ed.) Proceedings of the 1st Workshop on Fuzzy Based Expert Systems, Sofia, Bulgaria, pp. 28–30 (1994)
4. Bustince, H., Burillo, P.: Correlation of Interval-valued Intuitionistic Fuzzy Sets. Fuzzy Sets and Systems 74(2), 237–244 (1995)
5. Chen, S.M., Tan, J.M.: Handling Multi-criteria Fuzzy Decision Making Problems Based on Vague Sets. Fuzzy Sets and Systems 67(2), 163–172 (1994)
6. Chen, S.H., Hsieh, C.H.: Graded Mean Integration Representation of Generalized Fuzzy Number. Journal of the Chinese Fuzzy System Association 5(2), 1–7 (1999)
7. Wei, G.: Some Arithmetic Aggregation Operators with Intuitionistic Trapezoidal Fuzzy Numbers and Their Application to Group Decision Making. Journal of Computers 5(3), 345–351 (2010)
8. Herrera, F., Martinez, L., Sanchez, P.J.: Managing Non-homogenous Information in Group Decision Making. European Journal of Operational Research 166, 115–132 (2005)
9. Hwang, C.L., Yoon, K.: Multiple Attributes Decision Making Methods and Applications. Springer, Heidelberg (1981)
10. John Robinson, P., Henry Amirtharaj, E.C.: A Short Primer on the Correlation Coefficient of Vague Sets. International Journal of Fuzzy System Applications 1(2), 55–69 (2011)
11. John Robinson, P., Henry Amirtharaj, E.C.: Extended TOPSIS with Correlation Coefficient of Triangular Intuitionistic Fuzzy Sets for Multiple Attribute Group Decision Making. International Journal of Decision Support System Technology 3(3), 15–40 (2011)
12. Li, D.F.: Multi Attribute Decision Making Models and Methods Using Intuitionistic Fuzzy Sets. Journal of Computer and System Sciences 70, 73–85 (2005)
13. Mahapatra, G.S., Roy, T.K.: Reliability Evaluation Using Triangular Intuitionistic Fuzzy Numbers Arithmetic Operations. World Academy of Science, Engineering and Technology 50 (2009)
14. Szmidt, E., Kacprzyk, J.: Using Intuitionistic Fuzzy Sets in Group Decision Making. Control and Cybernetics 31, 1037–1053 (2002)
15. Szmidt, E., Kacprzyk, J.: A Consensus-reaching Process under Intuitionistic Fuzzy Preference Relations. International Journal of Intelligent Systems 18, 837–852 (2003)
16. Zeng, W., Li, H.: Correlation Coefficient of Intuitionistic Fuzzy sets. Journal of Industrial Engineering International 3, 33–40 (2007)
17. Xu, Z.S., Yager, R.R.: Some Geometric Aggregation Operators Based on Intuitionistic Fuzzy Sets. International Journal of General Systems 35, 417–433 (2006)

A Method for Reduction of Fuzzy Relation in Fuzzy Formal Context

Prem Kumar Singh and Ch. Aswani Kumar*

School of Information Technology and Engineering
VIT University, Vellore-632014, Tamil Nadu, India
cherukuri@acm.org

Abstract. Recently, fuzzy set theory in Formal Concept Analysis (FCA) has become an effective tool for knowledge representation and discovery. In this paper we present a method, for reducing the fuzzy relation for a given fuzzy context. The proposed method provides an alternative approach towards reducing the knowledge in the formal context in contrast to existing object, attribute and formal context reduction. Through this method, fuzzy relation can be projected with regards to both objects and attributes and provides two projected fuzzy formal contexts. From these two contexts, we can generate the fuzzy concepts and visualize them in a lattice structure.

Keywords: Formal Concept Analysis (FCA), Formal Context, Fuzzy Concept, Fuzzy Formal Context, Fuzzy Relation.

1 Introduction

FCA proposed by Wille in 80's provided us a mathematical as well as theoretical framework for the knowledge discovery and concept lattice structure in information system [1]. The important notions of this theory are formal context, concepts and complete lattice. The FCA generates set of formal concepts from a given formal context and shows them in a lattice structure called as complete lattice, which reflects generalization and specialization property between the concepts [2]. Knowledge reduction is major concern for researchers in FCA. Some methods are available in literature for knowledge reduction [3-6]. Ganter and Wille discussed about removing the reducible objects and attributes [1], Elloumi presented multilevel reduction method for fuzzy context [7], Li and Zhang considered the T- implication replacing the Lukasiewicz operator[8], Aswani Kumar and Srinivas considered fuzzy K-means clustering, SVD decomposition[9-10]. Aswani Kumar discussed random projections, matrix factorization and fuzzy clustering in FCA [11-13]. Belohalavek defined fuzzy concept similarity for large context [14] and Formica extended the concept similarity method in fuzzy FCA [15]. Dubois and Parade discussed possibility theory in formal concept analysis for reduction of knowledge [16] and recently, Ghosh, Kundu and Sarkar related the fuzzy graph in FCA for reducing the time complexity [17].

---

* Corresponding author.

P. Balasubramaniam and R. Uthayakumar (Eds.): ICMMSC 2012, CCIS 283, pp. 343–350, 2012.
© Springer-Verlag Berlin Heidelberg 2012

In this paper, we present a method for reducing the fuzzy relation of a given fuzzy formal context. The proposed method compute the projection of fuzzy relation with regards to both objects and attributes of given fuzzy formal context [18]. Through these two projected fuzzy contexts we can generate the fuzzy concepts and visualize them in the concept lattice. This method generate two lattices, one for fuzzy relation projected with respect to objects, and other for, with respect to attributes.

Rest of the paper is organized as follows. Section 2 provides a brief background of fuzzy FCA. We discuss the proposed method and its illustration in section 3 and 4, respectively. Section 5 provides the conclusions, followed by acknowledgements and references.

2   Fuzzy Formal Concept Analysis

A fuzzy formal context is a triplet $\mathbf{K}{=}(\mathbf{G}, \mathbf{M}, R)$, where $\mathbf{G}$ is finite set of fuzzy objects, $\mathbf{M}$ is finite set of fuzzy attributes, and R is a fuzzy relation on $\mathbf{G}{\times}\mathbf{M}(R \subseteq \mathbf{G} \times \mathbf{M})$. Each relation $(o, a) \in R$ represents that fuzzy object $o \in \mathbf{G}$ has membership value $\mu(o, a)$ with fuzzy attribute $a \in \mathbf{M}$ in $[0, 1]$ [19] .

A fuzzy formal concept is a maximal rectangle of a given fuzzy context $\mathbf{K}$ filled with membership value between $[0, 1]$, which is an ordered pair of two sets (O, A), where $O{\subseteq} \mathbf{G}$ called as extent, and $A{\subseteq} \mathbf{M}$ is called as intent and they satisfy the fuzzy Galois closure property in residuated lattice for truth degrees as defined below.

A fuzzy Galois closure operators $\phi(\uparrow\downarrow) : \mathbf{L}^{\mathbf{G}} \to \mathbf{L}^{\mathbf{G}}$ and $\Psi(\downarrow\uparrow) : \mathbf{L}^{\mathbf{M}} \to \mathbf{L}^{\mathbf{M}}, \forall O_1, O_2 \in \mathbf{L}^{\mathbf{G}}$ and $A_1, A_2 \in \mathbf{L}^{\mathbf{M}}$ satisfy following property[20]:

(1). $A_1 \subseteq A_2 \to \phi(A_1) \subseteq \phi(A_2)$ and $O_1 \subseteq O_2 \to \phi(O_1) \subseteq \phi(O_2)$
(2). $A \subseteq \phi(A)$ and $O \subseteq \psi(O)$
(3). $\phi(\phi(A)) = \phi(A)$ and $\psi(\psi(O)) = \psi(O)$

The pair (O, A), where $O{\subseteq} \mathbf{G}$ and $A{\subseteq} \mathbf{M}$ is a fuzzy concept iff $\psi(O) = A$ and $\phi(A) = O$. Then O is called as extension and A is called as intension. Through this Galois closure property one can neither enlarge the attributes nor the objects of a concept.

A fuzzy set with binary relation $\leq$ on a set R is a partial order relation. A fuzzy lattice is a partial ordered set of $(R; \leq)$ in which for every pairs of $(x, y), \exists$ sup$=x \vee y$ and inf$=x \wedge y$. A residuated lattice $\mathbf{L}{=}(L, \wedge, \vee, \otimes, \to, 0, 1)$ is the finite structure of truth values of object and its properties. In [21], $\mathbf{L}$ is complete residuated lattice iff,

(1). $(L, \wedge, \vee, 0, 1)$ is a complete lattice.
(2). $(L, \otimes, 1)$ is commutative monoid. (i.e., $\otimes$ is commutative and associative means a $\otimes 1 = a = 1 \otimes a, \forall a \in \mathbf{L}$)
(3). $\otimes$ and $\to$ are adjoint operators and $a \otimes b \leq c$ iff $a \leq b \to c, \forall a, b, c \in \mathbf{L}$.

The $\otimes$ and $\to$ operators are defined distinctly by Lukaswiech, Godel, and Goguen t-norms and their residua.

For a fuzzy formal context if $O \in \mathbf{L}^\mathbf{G}$, and $A \in \mathbf{L}^\mathbf{M}$, then the fuzzy set $O^\uparrow \in \mathbf{L}^\mathbf{M}$ and $A^\downarrow \in \mathbf{L}^\mathbf{G}$ defined as below [21]:

(A) $O^\uparrow(a) = \wedge_{o \in \mathbf{G}}(O(o) \rightarrow R(o, a))$.
(B) $A^\downarrow(o) = \wedge_{a \in \mathbf{M}}(A(a) \rightarrow R(o, a))$.

The $O^\uparrow(a)$ is truth degree of attribute $a \in \mathbf{M}$ is shared by all objects from $o \in \mathbf{G}$ and $A^\downarrow(o)$ is the truth degree of object $o \in \mathbf{G}$ has all attributes from $a \in \mathbf{M}$. The fuzzy couple of $(O, A) \in \mathbf{L}^\mathbf{G} \times \mathbf{L}^\mathbf{M}$ is called as fuzzy formal concept iff $O^\uparrow(a) = A$ and $A^\downarrow(o) = O$. Also, they satisfy the fuzzy Galois closure property defined as above. If set of fuzzy formal concepts $FC_\mathbf{k}$, generated from a given fuzzy formal context $\mathbf{K}$, follows the ordering principle of set using the super and sub hierarchy properties of lattice i.e. $(O_2, A_2) \leq (O_1, A_1) \Longleftrightarrow O_2 \subseteq O_1 (\Longleftrightarrow A_1 \subseteq A_2)$ for every fuzzy concepts, then it forms a complete fuzzy Galois lattice $L_{FC_\mathbf{k}} = (FC_\mathbf{k}, \leq)$. In fuzzy concept lattice $\exists$, a inf $(0, 0, \cdots, 0)$ and sup $(1, 1, \cdots, 1)$ exists for every fuzzy concepts [20].

3   Proposed Method

We propose a method that, project the given fuzzy relation $(R \subseteq \mathbf{G} \times \mathbf{M})$ with respect to objects and attributes of a given fuzzy formal context. Through this method we can reduce the fuzzy relation of a given fuzzy context into two fuzzy contexts. This method can projects the fuzzy relation R with respect to $\mathbf{G}$ and $\mathbf{M}$ as follows:-

Step(1). For a given fuzzy relation R, let $(R \downarrow \mathbf{G})$ denote the projection of R on $\mathbf{G}$ defined as below:

$[R \downarrow \mathbf{G}](o) = \{(o), \max_A \mu_R(o, a) \in \mathbf{G} \times \mathbf{M}\}$
assigned the highest degree of membership value from the tuples$(O, A_i), \forall i \in |\mathbf{M}|$.

It shows the projection of the fuzzy relation on highest degree of membership value of all the tuples$(O, A_i), \forall i \in |\mathbf{M}|$.

Step (2). Let $(R \downarrow \mathbf{M})$ denote the projection of R on $\mathbf{M}$ defined as below:

$[R \downarrow \mathbf{M}](a) = \{(a), \max_O \mu_R(o, a) \in \mathbf{G} \times \mathbf{M}\}$
assigned the highest degree of membership value from the tuples $(O_i, A), \forall i \in |\mathbf{G}|$.

It shows the projection of the fuzzy relation on highest degree of membership value of all the tuples$(O_i, A), \forall i \in |\mathbf{G}|$.

Step (3). For both step 1 and step 2, we represent the fuzzy context with respect to the object and attribute in tabular form and generate the formal concepts from existing algorithms[17,19–22], and visualize them in a lattice structure. For study of algorithms for generating the fuzzy concepts from a given fuzzy formal context readers can refer some references [17,19–22].

4   Illustration

For illustration of proposed method, we considered a fuzzy formal context shown in Table 1 [17]. The projection of fuzzy relation with respect to the object (**G**) is calculated as below :- $(O_1, p_2)=1.0$, $(O_1, p_5)=1.0$, $(O_2, p_1)=1.0$, $(O_2, p_2)=1.0$, $(O_2, p_3)=1.0$, $(O_3, p_6)=1.0$, $(O_4, p_4)=1.0$, $(O_5, p_3)=1.0$, $(O_6, p_1)=0.5$. These relations are shown in Table 2. The projection of fuzzy relation with respect to attributes (**M**) is calculated as below :- $(O_2, p_1)=1.0$, $(O_1, p_2)=1.0$, $(O_2, p_2)=1.0$, $(O_2, p_3)=1.0$, $(O_5, p_3)=1.0$, $(O_4, p_4)=1.0$, $(O_1, p_5)=1.0$, $(O_3, p_6)=1.0$. These relations are shown in Table 3. We can observe that fuzzy relation shown in Tables 2 and 3 are reduced form of fuzzy relation shown in Table 1. Since, Tables 2 and 3 are representing a fuzzy formal context, so we can generate the fuzzy concepts and visualize them in the lattice structure.

All generated concepts from a fuzzy formal context shown in Table 1 are:

Table 1. Fuzzy Formal Context

	p_1	p_2	p_3	p_4	p_5	p_6
O_1	0.0	1.0	0.5	0.5	1.0	0.0
O_2	1.0	1.0	1.0	0.0	0.0	0.0
O_3	0.5	0.5	0.0	0.0	0.0	1.0
O_4	0.0	0.0	0.0	1.0	0.5	0.0
O_5	0.0	0.0	1.0	0.5	0.0	0.0
O_6	0.5	0.0	0.0	0.0	0.0	0.0

Table 2. Projected Context on **G**

	p_1	p_2	p_3	p_4	p_5	p_6
O_1	0.0	1.0	0.0	0.0	1.0	0.0
O_2	1.0	1.0	1.0	0.0	0.0	0.0
O_3	0.0	0.0	0.0	0.0	0.0	1.0
O_4	0.0	0.0	0.0	1.0	0.0	0.0
O_5	0.0	0.0	1.0	0.0	0.0	0.0
O_6	0.5	0.0	0.0	0.0	0.0	0.0

1.$\{\phi, 1.0/p_1 + 1.0/p_2 + 1.0/p_3 + 1.0/p_4 + 1.0/p_5 + 1.0/p_6\}$
2.$\{0.5/O_1, 1.0/p_2 + 1.0/p_3 + 1.0/p_4 + 1.0/p_5\}$
3.$\{1.0/O_2, 1.0/p_1 + 1.0/p_2 + 1.0/p_3\}$
4.$\{0.5/O_3, 1.0/p_1 + 1.0/p_2 + 1.0/p_6\}$
5.$\{0.5/O_1 + 0.5/O_5, 1.0/p_3 + 1.0/p_4\}$
6.$\{0.5/O_1 + 0.5/O_4, 1.0/p_4 + 1.0/p_5\}$
7.$\{1.0/O_1, 1.0/p_2 + 0.5/p_3 + 0.5/p_4 + 1.0/p_5\}$
8.$\{0.5/O_1 + 1.0/O_2, 1.0/p_2 + 1.0/p_3\}$

Table 3. Projected Context on **M**

	p_1	p_2	p_3	p_4	p_5	p_6
O_1	0.0	1.0	0.0	0.0	1.0	0.0
O_2	1.0	1.0	1.0	0.0	0.0	0.0
O_3	0.0	0.0	0.0	0.0	0.0	1.0
O_4	0.0	0.0	0.0	1.0	0.0	0.0
O_5	0.0	0.0	1.0	0.0	0.0	0.0
O_6	0.0	0.0	0.0	0.0	0.0	0.0

9. $\{1.0/O_2 + 0.5/O_3, 1.0/p_1 + 1.0/p_2\}$
10. $\{1.0/O_3, 0.5/p_1 + 0.5/p_2 + 1.0/p_6\}$
11. $\{0.5/O_1 + 1.0/O_5, 1.0/p_3 + 0.5/p_4\}$
12. $\{0.5/O_1 + 1.0/O_4, 1.0/p_4 + 0.5/p_5\}$
13. $\{1.0/O_1 + 0.5/O_4, 0.5/p_4 + 1.0/p_5\}$
14. $\{1.0/O_1 + 1.0/O_5, 0.5/p_3 + 0.5/p_4\}$
15. $\{0.5/O_1 + 1.0/O_2 + 1.0/O_5, 1.0/p_3\}$
16. $\{1.0/O_1 + 1.0/O_2, 1.0/p_2 + 0.5/p_3\}$
17. $\{1.0/O_2 + 0.5/O_3 + 0.5/O_6, 1.0/p_1\}$
18. $\{1.0/O_2 + 1.0/O_3, 0.5/p_1 + 0.5/p_2\}$
19. $\{0.5/O_1 + 1.0/O_4 + 0.5/O_5, 1.0/p_4\}$
20. $\{1.0/O_1 + 1.0/O_4, 0.5/p_4 + 0.5/p_5\}$
21. $\{1.0/O_1 + 1.0/O_2 + 1.0/O_5, 0.5/p_3\}$
22. $\{1.0/O_1 + 1.0/O_2 + 0.5/O_3, 1.0/p_2\}$
23. $\{1.0/O_1 + 1.0/O_4 + 1.0/O_5, 0.5/p_4\}$
24. $\{1.0/O_1 + 1.0/O_2 + 1.0/O_3, 0.5/p_2\}$
25. $\{1.0/O_2 + 1.0/O_3 + 1.0/O_6, 0.5/p_1\}$
26. $\{1.0/O_1 + 1.0/O_2 + 1.0/O_3 + 1.0/O_4 + 1.0/O_5 + 1.0/O_6, \phi\}$

The lattice for the fuzzy concepts generated from the Table 1 is shown in Fig. 1 [17].

All generated fuzzy concepts from a projected fuzzy relation with respect to **G** shown in Table 2 are:-

1. $\{\phi, 1.0/p_1 + 1.0/p_2 + 1.0/p_3 + 1.0/p_4 + 1.0/p_5 + 1.0/p_6\}$
2. $\{1.0/O_1, 1.0/p_2 + 1.0/p_5\}$
3. $\{1.0/O_2, 1.0/p_1 + 1.0/p_2 + 1.0/p_3\}$
4. $\{1.0/O_3, 1.0/p_6\}$
5. $\{1.0/O_4, 1.0/p_4\}$
6. $\{1.0/O_2 + 1.0/O_5, 1.0/p_3\}$
7. $\{1.0/O_1 + 1.0/O_2, 1.0/p_2\}$
8. $\{1.0/O_2 + 0.5/O_6, 1.0/p_1\}$

9.$\{1.0/O_2 + 1.0/O_6, 0.5/p_1\}$

10.$\{1.0/O_1 + 1.0/O_2 + 1.0/O_3 + 1.0/O_4 + 1.0/O_5 + 1.0/O_6, \phi\}$

The lattice for the fuzzy concepts generated from the Table 2 is shown in Fig. 2.

All generated fuzzy concepts from a projected fuzzy relation with respect to **M** shown in Table 3 are:-

1.$\{\phi, 1.0/p_1 + 1.0/p_2 + 1.0/p_3 + 1.0/p_4 + 1.0/p_5 + 1.0/p_6\}$

2.$\{1.0/O_1, 1.0/p_2 + 1.0/p_5\}$

3.$\{1.0/O_2, 1.0/p_1 + 1.0/p_2 + 1.0/p_3\}$

4.$\{1.0/O_3, 1.0/p_6\}$

5.$\{1.0/O_4, 1.0/p_4\}$

6.$\{1.0/O_2 + 1.0/O_5, 1.0/p_3\}$

7.$\{1.0/O_1 + 1.0/O_2, 1.0/p_2\}$

8.$\{1.0/O_1 + 1.0/O_2 + 1.0/O_3 + 1.0/O_4 + 1.0/O_5 + 1.0/O_6, \phi\}$

The lattice for the fuzzy concepts generated from the Table 3 is shown in Fig. 3. Here we can observe that the number of concepts generated from Tables 2 and 3 are reduced, with respect concepts generated from Table 1.

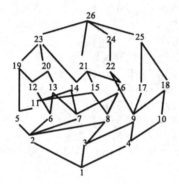

Fig. 1. Fuzzy Lattice for context of Table 1

We can observe that the number of generated concepts from Table 2 and Table 3 are lesser than the generated concepts from Table 1. Also, lattice structure shown in Fig. 1 is reduced into the Figs. 2 and 3 through presented method. We believe that, presented method may helpful for researchers in the field of (a) Language modeling, (b) Information retrieval, (c) Finding frequent item set (d) Reducing the concepts and other context related fields, e.g. stock exchange, economics, etc.

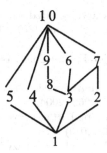

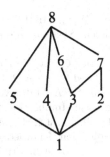

Fig. 2. Fuzzy Lattice for context of Table 2

Fig. 3. Fuzzy Lattice for context of Table 3

5   Conclusion

Reduction of fuzzy relation in fuzzy formal context through projection method is proposed in this study. In this method we have considered, fuzzy relation for reduction of fuzzy formal context in contrast to available methods in FCA about reducing the objects, attributes, finding similarity and matrix decomposition. From this method we can generate two lattices, one is from projection of fuzzy relation with respect to objects and other is from projection of fuzzy relation with respect to attributes. Fuzzy formal context shown in Table 1 is reduced into Table 2 and Table 3 through presented method.

In future this method may extend for other t-conorm operators at the place of max operator. Also, this method may helpful for the researchers to reduce the computational time for generating the concepts in fuzzy FCA and knowledge reduction.

Acknowledgement. Authors sincerely acknowledge the financial support from National Board of Higher Mathematics, Dept. of Atomic Energy, Govt. of India under the grant number 2/48(11)/2010-R&D II/10806.

References

1. Ganter, B., Wille, R.: Formal Concept Analysis: Mathematical Foundation. Springer, Heidelberg (1999)
2. Carpineto, C., Romano, G.: Concept Data Analysis: Theory and Application. John Willey and Sons Ltd., England (2004)
3. Wolff, K.E.: Concepts in Fuzzy Scaling Theory: Order and Granularity. Fuzzy Sets and Systems 132(1), 63–75 (2002)
4. Wu, W.Z., Leung, Y., Mi, J.S.: Granular Computing and Knowledge Reduction in Formal Contexts. IEEE Transactions on Knowledge and Data Engineering 21(10), 1461–1474 (2009)

5. Zhang, W.X., Mi, J.S., Wu, W.Z.: Approaches to Knowledge Reductions in Inconsistent Systems. International Journal of Intelligent Systems 18(9), 989–1000 (2003)
6. Li, J., Mei, C., Lv, Y.: Knowledge Reduction in Decision Formal Contexts. Knowledge Based Systems (2011)
7. Elloumi, S.: A Multilevel Conceptual Data Reduction Approach Based on the Lukasiewicz Implication. Information Sciences 163, 253–262 (2004)
8. Li, L., Jhang, J.: Attribute Reduction in Fuzzy Concept Lattices Based on the T-Implication. Knowledge Based Systems 23, 497–503 (2010)
9. Aswani Kumar, Ch., Srinivas, S.: Concept Lattice Reduction Using Fuzzy K-means Clustering. Expert Systems with Applications 37(3), 2696–2704 (2010)
10. Aswani Kumar, Ch., Srinivas, S.: Mining Associations in Health Care Data Using Formal Concept Analysis and Singular Value Decomposition. Journal of Biological Systems 18(4), 787–807 (2010)
11. Aswani Kumar, Ch.: Knowledge Discovery in Data Using Formal Concept Analysis and Random Projections. International Journal of Applied Mathematics and Computer Science 21(4), 745–756 (2011)
12. Aswani Kumar, Ch.: Mining Association Rules Using Non-Negative Matrix Factorization and Formal Concept Analysis. In: Venugopal, K.R., Patnaik, L.M. (eds.) ICIP 2011. CCIS, vol. 157, pp. 31–39. Springer, Heidelberg (2011)
13. Aswani Kumar, Ch.: Fuzzy Clustering Based Formal Concept Analysis for Association Rules Mining. Applied Artificial Intelligence (2012) (accepted for publication)
14. Belohlavek, R., Krupka, M.: Grouping Fuzzy Sets by Similarity. Information Sciences 179, 2656–2661 (2009)
15. Formica, A.: Concept Similarity in Fuzzy Formal Concept Analysis for Semantic Web. International Journal of Uncertainty, Fuzziness, and Knowledge Based System 18(2), 153–167 (2010)
16. Dubois, D., Parade, H.: Possibility Theory and Formal Concept Analysis: Characterizing Independent Sub-Contexts. Fuzzy Sets and Systems (2011)
17. Ghosh, P., Kundu, K., Sarkar, D.: Fuzzy Graph Representation of a Fuzzy Concept Lattice. Fuzzy Sets and Systems 161, 1669–1675 (2010)
18. Klir, G.J., Yuan, B.: Fuzzy Sets and Fuzzy Logic: Theory and Applications. Prentice Hall PTR, New Jersey (2008)
19. Burusco, A., Fuentes- Gonzales, R.: The Study of the L-Fuzzy Concept Lattice. Matheware and Soft Computing 3, 209–218 (1994)
20. Pocs, J.: Note on Generating Fuzzy Concept Lattices via Galois Connections. Information Sciences 185, 128–136 (2012)
21. Belohlavek, R., Vychodil, V.: What is Fuzzy Concept Lattice. In: Proc. CLAV Olomuc, Czech Republic, pp. 34–45 (2005)
22. Ayouni, S., Yahia, S.B., Laurent, A.: Extracting Compact and Information Lossless Sets of Fuzzy Association Rules. Fuzzy Sets and Systems 183, 1–25 (2011)

High-Order Method for a Singularly Perturbed Second-Order Ordinary Differential Equation with Discontinuous Source Term Subject to Mixed Type Boundary Conditions

R. Mythili Priyadharshini[1] and N. Ramanujam[2]

[1] Department of Mathematical and Computational Sciences,
National Institute of Technology, Karnataka, India
[2] Department of Mathematics,
Bharathidasan University, Tamilnadu, India
http://www.bdu.ac.in/depa/science/ramanujam.htm

Abstract. In this paper, a singularly perturbed second-order ordinary differential equation with discontinuous source term subject to mixed type boundary conditions is considered. A robust-layer-resolving numerical method of high-order is suggested. An ε-uniform error estimates for the numerical solution and also to the numerical derivative are derived. Numerical results are presented, which are in agreement with the theoretical results.

Keywords: Singular perturbation problem, mixed type boundary conditions, discontinuous source term, Shishkin mesh, discrete derivative.

1   Introduction

Motivated by the works given in [2] - [5], the present paper considers the following singularly perturbed mixed type boundary value problem for second-order ordinary differential equation with discontinuous source term:

$$Lu \equiv \varepsilon u'' + a(x)u' = f(x), \quad \text{for all} \quad x \in \Omega^- \cup \Omega^+ \quad (1)$$

$$B_0 u(0) \equiv \beta_1 u(0) - \varepsilon \beta_2 u'(0) = A, \quad B_1 u(1) \equiv \gamma_1 u(1) + \gamma_2 u'(1) = B, \quad (2)$$

where $a(x) \geq \alpha > 0$, for $x \in \bar{\Omega}$, $|[f](d)| \leq C$, $\beta_1, \beta_2 \geq 0$, $\beta_1 + \varepsilon\beta_2 \geq 1$, $\gamma_1 > 0$, $\gamma_1 - \gamma_2 \geq 1$, the constants A, B, β_1, β_2, γ_1 and γ_2 are given and $0 < \varepsilon \ll 1$. We assume that $a(x)$ is sufficiently smooth function on $\bar{\Omega}$ and $f(x)$ is sufficiently smooth on $\Omega^- \cup \Omega^+$; f and its derivatives have jump discontinuity at $x = d$. Since f is discontinuous at $x = d$ the solution u of (1), (2) does not necessarily have a continuous second derivative at the point d. Thus, $u(x)$ need not belong to the class of functions $C^2(\Omega)$ and $u \in Y \equiv C^1(\bar{\Omega}) \cap C^2(\Omega^- \cup \Omega^+)$. The novel aspect of the problem under consideration is that we take a source term in the differential equation which has a jump discontinuity at one or more points in the interior of the domain. This gives raise to a weak interior layer in the

P. Balasubramaniam and R. Uthayakumar (Eds.): ICMMSC 2012, CCIS 283, pp. 351–359, 2012.
© Springer-Verlag Berlin Heidelberg 2012

exact solution of the problem, in addition to the boundary layer at the outflow boundary point. In this paper, we constructed hybrid difference scheme (central finite difference scheme in the fine mesh region and mid-point difference scheme) for the BVP (1), (2) on Shishkin type meshes.

Note: Through out this paper, C denotes a generic constant that is independent of the parameter ε and the dimension of the discrete problem N. Let $u : D \to \mathbb{R}$, $(D \subset \mathbb{R})$. An appropriate norm for studying the convergence of numerical solution to the exact solution of a singular perturbation problem is the maximum norm $\| u \|_D = \max_{x \in D} |u(x)|$, [1]. We assume that $\varepsilon \leq CN^{-1}$ is generally the case for discretization of convection-diffusion equations.

Theorem 1. [5] *The problem (1), (2) has a solution* $u \in C^1(\bar{\Omega}) \cap C^2(\Omega^- \cup \Omega^+)$.

Theorem 2. *(Minimum Principle) [7] Let L be the differential operator in (1) and* $u \in Y$. *If* $B_0u(0) \geq 0$, $B_1u(1) \geq 0$, $Lu(x) \leq 0$, *for all* $x \in \Omega^- \cup \Omega^+$ *and* $[u'](d) \leq 0$, *then* $u(x) \geq 0$, *for all* $x \in \bar{\Omega}$.

Lemma 1. *If* $u \in Y$ *then* $\| u \|_{\bar{\Omega}} \leq C \max\{|B_0u(0)|, |B_1u(1)|, \| Lu \|_{\Omega^- \cup \Omega^+}\}$.

2 Solution Decomposition and Mesh Discretization

The solution u of (1), (2) can be decomposed into regular and singular components $u(x) = v(x) + w(x)$, where $v(x) = v_0(x) + \varepsilon v_1(x) + \varepsilon^2 v_2(x) + \varepsilon^3 v_3(x)$ and $v \in C^0(\Omega)$ is the solution of

$$Lv(x) = f(x), x \in \Omega^- \cup \Omega^+, \ B_0v(0) = B_0v_0(0) + \varepsilon B_0v_1(0) + \varepsilon^2 B_0v_2(0), \text{ (3)}$$
$$v(d) = v_0(d) + \varepsilon v_1(d) + \varepsilon^2 v_2(d), \quad B_1v(1) = B_1u(1). \tag{4}$$

Further, decompose $w(x) = w_0(x) + w_d(x)$, where $w_0 \in C^2(\Omega)$, is the solution of

$$Lw_0(x) = 0, \ x \in \Omega, \ B_0w_0(0) = B_0u(0) - B_0v(0), \quad B_1w_0(1) = 0, \tag{5}$$

and $w_d \in C^0(\Omega)$, is the interior layer function satisfying

$$Lw_d(x) = 0, \ x \in \Omega^- \cup \Omega^+, \ B_0w_d(0) = 0, \ [w_d'](d) = -[v'](d), \ B_1w_d(1) = 0.(6)$$

Lemma 2. [2,5] *(Derivative Estimate) For each integer k, satisfying* $0 \leq k \leq 4$, *the derivatives of the solutions* $v(x)$, $w_0(x)$ *and* $w_d(x)$ *of (3),(4), (5) and (6) respectively, satisfy the following bounds*

$$\| v^{(k)} \|_{\Omega^- \cup \Omega^+} \leq C(1 + \varepsilon^{3-k}), \ |[v](d)|, |[v'](d)|, |[v''](d)| \leq C$$
$$|w_0^{(k)}(x)| \leq C\varepsilon^{-k}e^{-\alpha x/\varepsilon}, \ x \in \bar{\Omega}, \ |w_d(x)| \leq C\varepsilon,$$
$$|w_d^{(k)}(x)| \leq \begin{cases} C(\varepsilon^{1-k}e^{-\alpha x/\varepsilon}), \ x \in \Omega^- \\ C(\varepsilon^{1-k}e^{-\alpha(x-d)/\varepsilon}), \ x \in \Omega^+, \end{cases}$$

where C is a constant independent of ε.

On Ω a piecewise uniform mesh of N mesh interval is constructed as follows. The domain $\bar{\Omega}$ is subdivided into the four subintervals $[0, \sigma_1] \cup [\sigma_1, d] \cup [d, d + \sigma_2] \cup [d + \sigma_2, 1]$ for some σ_1, σ_2 that satisfy $0 < \sigma_1 \leq \frac{d}{2}, \quad 0 < \sigma_2 \leq \frac{1-d}{2}$. On each subinterval a uniform mesh with $N/4$ mesh-intervals is placed. The interior mesh points are denoted by $\Omega^N = \{x_i : 1 \leq i \leq \frac{N}{2} - 1\} \cup \{x_i : \frac{N}{2} + 1 \leq i \leq N - 1\}$. Clearly, $x_{N/2} = d$ and $\bar{\Omega}^N = \{x_i\}_0^N$. It is fitted to the singular perturbation problem (1) - (2) by choosing σ_1 and σ_2 to be the following functions of N and ε $\sigma_1 = \min\{\frac{d}{2}, \frac{2\varepsilon}{\alpha} \ln N\}$ and $\sigma_2 = \min\{\frac{1-d}{2}, \frac{2\varepsilon}{\alpha} \ln N\}$. We now introduce the following notation for the four mesh widths

$$
h_i = x_i - x_{i-1} = \begin{cases} H_1 = \frac{4\sigma_1}{N}, & i = 1, ..., N/4 - 1, \\ H_2 = \frac{4(d-\sigma_1)}{N}, & i = N/4, ..., N/2, \\ H_3 = \frac{4\sigma_2}{N}, & i = N/2 + 1, ..., 3N/4, \\ H_4 = \frac{4(1-d-\sigma_2)}{N}, & i = 3N/4 + 1, ..., N. \end{cases}
$$

3   Hybrid Difference Scheme (HDS)

For the analysis of HDS, we assume that $\sigma = \sigma_1 = \sigma_2 = \dfrac{2\varepsilon}{\alpha} \ln N$, since otherwise N^{-1} is exponentially small compared with ε. In this scheme, we discretize (1) using the central difference scheme

$$
L_c^N U_i \equiv \frac{2\varepsilon}{h_i + h_{i+1}} \left(\frac{U_{i+1} - U_i}{h_{i+1}} - \frac{U_i - U_{i-1}}{h_i} \right) + a(x_i) \frac{U_{i+1} - U_{i-1}}{h_i + h_{i+1}} = f(x_i), \quad (7)
$$

in the fine mesh region that is, whenever the local mesh size allows us to do this without losing stability, but we employ a second order upwind scheme

$$
L_u^N U_i \equiv \frac{2\varepsilon}{h_i + h_{i+1}} \left(\frac{U_{i+1} - U_i}{h_{i+1}} - \frac{U_i - U_{i-1}}{h_i} \right) + a_{i+1/2} \frac{U_{i+1} - U_i}{h_{i+1}} = f_{i+1/2} \quad (8)
$$

in the coarse mesh region where $U_i = U(x_i)$, $a_{i-1/2} \equiv a((x_{i-1} + x_i)/2)$ and $a_{i+1/2} \equiv a((x_i + x_{i+1})/2)$; similarly for $f_{i-1/2}, f_{i+1/2}$. At the point $x_{N/2} = d$, we shall use the difference operator

$$
L_t^N U_{N/2} \equiv \frac{-U_{N/2+2} + 4U_{N/2+1} - 3U_{N/2}}{h_{N/2+1} + h_{N/2}} - \frac{U_{N/2-2} - 4U_{N/2-1} + 3U_{N/2}}{h_{N/2} + h_{N/2-1}} = 0. \quad (9)
$$

The first order derivative in the boundary conditions (2) are approximated by the central difference operator and we have

$$
U_{-1} = -\frac{2H_1\beta_1}{\varepsilon\beta_2}U_0 + U_1 + \frac{2H_1}{\varepsilon\beta_2}A \quad \text{and} \quad U_{N+1} = -\frac{2H_4\gamma_1}{\gamma_2}U_N + U_{N-1} + \frac{2H_4}{\gamma_2}B, \quad (10)
$$

where U_{-1} and U_{N+1} are the functional values at x_{-1} and x_{N+1}. The values U_{-1} and U_{N+1} may be eliminated by assuming that the difference equations (7) and (8) holds also for $i = 0$ and $i = N$. Substituting the values U_{-1} and U_{N+1} from (10) into the equations (7) and (8) for $i = 0$ and $i = N$, we get respectively

$$
\left(\frac{2\varepsilon}{H_1^2} + \frac{2H_1\beta_1}{\varepsilon\beta_2} \left(\frac{\varepsilon}{H_1^2} - \frac{a(x_0)}{2H_1} \right) \right) U_0 - \left(\frac{2\varepsilon}{H_1^2} \right) U_1 = \frac{2H_1 A}{\varepsilon\beta_2} \left(\frac{\varepsilon}{H_1^2} - \frac{a(x_0)}{2H_1} \right) - f_0, \quad (11)
$$

$$(\frac{2\varepsilon}{H_4^2} + \frac{a(x_N)}{H_4} + \frac{2H_4\gamma_1}{\gamma_2}(\frac{\varepsilon}{H_4^2} + \frac{a(x_N)}{H_4}))U_N - (\frac{2\varepsilon}{H_4^2} + \frac{a(x_N)}{H_4})U_{N-1}$$

$$= \frac{2H_4B}{\gamma_2}(\frac{\varepsilon}{H_4^2} + \frac{a(x_N)}{H_4}) - f_{N+1/2}. \tag{12}$$

The matrix associated with (7) - (9),(11) and (12) is not a M-matrix. We transform the equation so that the new equations do have the monotonicity property. From equations (7) and (8), we can get the expressions for $U_{N/2-2}$ and $U_{N/2+2}$ and substituting in (9) gives

$$L_T^N U_{N/2} = (\frac{4}{h_{N/2-1} + h_{N/2}} - \frac{2\varepsilon + a_{N/2-1/2}h_{N/2-1}}{2\varepsilon h_{N/2}})U_{N/2-1} - (\frac{3}{h_{N/2+1} + h_{N/2+2}}$$

$$+ \frac{3}{h_{N/2-1} + h_{N/2}} + \frac{1}{2\varepsilon + a_{N/2+1}h_{N/2+2}}(\frac{2\varepsilon - a_{N/2+1}h_{N/2+1}}{(h_{N/2+1} + h_{N/2+2})}) + \frac{2\varepsilon + a_{N/2-1/2}h_{N/2-1}}{2\varepsilon})U_{N/2}$$

$$+ (\frac{4}{h_{N/2+1} + h_{N/2+2}} - \frac{\varepsilon}{2\varepsilon + a_{N/2+1}h_{N/2+2}})U_{N/2+1} = \frac{f_{N/2-1}h_{N/2-1}}{2\varepsilon} + \frac{f_{N/2+1}h_{N/2+2}}{2\varepsilon + a_{N/2+1}h_{N/2+2}}.$$

Thus the HDS for the boundary value problem (1) - (2) is

$$L_H^N U_i = f_{H,i}, \quad \text{for} \quad i = 1, 2, ..., N-1, \tag{13}$$

$$B_{H0}^N U_0 = A_0, \quad B_{H1}^N U_N = B_N, \tag{14}$$

where $A_0 = \dfrac{2H_1A}{\varepsilon\beta_2}(\dfrac{\varepsilon}{H_1^2} - \dfrac{a(x_0)}{H_1}) - f_0$, $\quad B_N = \dfrac{2H_4B}{\gamma_2}(\dfrac{\varepsilon}{H_4^2} + \dfrac{a(x_N)}{H_4}) - f_{N+1/2}$,

$$L_H^N U_i = \begin{cases} L_c^N U_i, & \text{for} \quad i = 1, ..., N/4-1, N/2+1, ..., 3N/4-1, \\ L_u^N U_i, & \text{for} \quad i = N/4, ..., N/2-1, 3N/4, ..., N-1, \\ L_T^N U_i, & \text{for} \quad i = N/2, \end{cases}$$

$$f_{H,i} = \begin{cases} f_i, & \text{for} \quad i = 1, ..., N/4-1, N/2+1, ..., 3N/4-1, \\ f_{i+1/2}, & \text{for} \quad i = N/4, ..., N/2-1, 3N/4, ..., N-1, \\ \frac{hf_{i-1}}{2\varepsilon - ha_{i-1}} + \frac{hf_{i+1}}{2\varepsilon + ha_{i+1}}, & \text{for} \quad i = N/2. \end{cases}$$

Remark 1. The truncation error for (11) is given by
$|B_{H0}^N(U-u)(x_0)| \leq C\varepsilon H_1|u^{(3)}(x_0)| \leq CH_1^2|u^{(3)}(x_0)|$. Similarly, the truncation error for (12) is given by $|B_{H1}^N(U-u)(x_N)| \leq C_{\|a\|}H_4^2|u^{(3)}(x_N)|$. Further, we have from [8],

$$|L_H^N(U-u)(x_i)| \leq \begin{cases} \varepsilon h_i^2 \parallel u^{(4)} \parallel + h_i^2 \parallel a \parallel \parallel u^{(3)} \parallel, i = 1, ..., N/4-1, N/2+1, ..., 3N/4-1, \\ \varepsilon h_i \parallel u^{(3)} \parallel + C_{(\|a\|, \|a'\|)}h_i^2(\parallel u^{(3)} \parallel + \parallel u^{(2)} \parallel), i = N/4, ..., N/2-1, \\ 3N/4, ..., N-1. \end{cases}$$

To guarantee the monotonicity property of the difference operator L_H^N in HDS, we impose the following mild assumption on the minimum number of mesh points

$$\frac{N}{\ln N} \geq 4\frac{\max a(x)}{\min a(x)}. \tag{15}$$

Theorem 3. *[4, Lemma 2] Assume* (15). *If* $Z(x_i)$, $i = 0, 1, ..., N$ *is a mesh function that satisfies* $B_{H0}^N Z(x_0) \geq 0$, $B_{H1}^N Z(x_N) \geq 0$ *and* $L_H^N Z(x_i) \leq 0$, *for* $1, ..., N - 1$, *then* $Z(x_i) \geq 0$, *for all* $i = 0, ..., N$.

Lemma 3. *The solutions* $U(x_i)$ *satisfy* $\| U \|_{\bar{\Omega}^N} \leq \max\{|B_{H0}^N U(x_0)|, |B_{H1}^N U(x_N)| + \frac{1}{\alpha} \| f_H \|_{\Omega^N}\}$.

For the HDS, we define the mesh function $V(x_i), x_i \in \Omega^N$, to be the solution of

$$L_H^N V(x_i) = f_H(x_i), \ B_{H0}^N V(x_0) = B_0 v(0), \ V(d) = v(d), \ B_{H1}^N V(x_N) = B_1 v(1). \ (16)$$

Analogous to the continuous case we can further decompose $W(x_i)$ as $W(x_i) = W_0(x_i) + W_d(x_i)$, where $W_0(x_i)$ is defined as the solution of

$$L_H^N W_0(x_i) = 0, \text{ for } x_i \in \Omega^N \cup \{d\}, \ B_{H0}^N W_0(x_0) = B_0 w_0(x_0), \ B_{H1}^N W_0(x_N) = 0 \ (17)$$

and $W_d(x_i)$, is the solution of

$$L_H^N W_d(x_i) = 0, \quad \text{for} \quad x_i \in \Omega^N \tag{18}$$
$$B_{H0}^N W_d(x_0) = 0, B_{H1}^N W_d(x_N) = 0, L_T^N W_d(x_{N/2}) = -L_T^N W_0(x_{N/2}) - L_T^N V(x_{N/2})(19)$$

4   Error Estimates of Solution and Its Derivative

Lemma 4. *Assume* (15). *Let* $v(x)$ *and* $V(x_i)$ *be the solutions of* (3),(4) *and* (16) *respectively. Then at each mesh point* $x_i \in \bar{\Omega}^N$,

$$|(V - v)(x_i)| \leq \begin{cases} C(1 - x_i)N^{-2}, \text{ for } i = 0, ..., N/2 \\ C(2 - x_i)N^{-2}, \text{ for } i = N/2 + 1, ..., N, \end{cases}$$

$$|\varepsilon D^0 (V - v)(x_i)| \leq CN^{-2}.$$

Proof. Let us now consider the truncation error at the mesh points. As given in Remark 1, we have $|B_{H0}^N (V - v)(x_0)| \leq CN^{-2}$, $|B_{H1}^N (V - v)(x_N)| \leq CN^{-2}$, and $|L_H^N (V - v)(x_i)| \leq CN^{-2}$, for $i = 1, ..., N - 1$. Consider the two mesh functions

$$\Psi^{\pm}(x_i) = \Phi(x_i) \pm (V - v)(x_i), \text{ where } \Phi(x_i) = \begin{cases} C \dfrac{(1 - x_i)}{\alpha(1 - d)} N^{-2}, \text{ for } x_i \in \bar{\Omega}^N \cap [0, d], \\ C \dfrac{(2 - x_i)}{\alpha(2 - d)} N^{-2}, \text{ for } x_i \in \bar{\Omega}^N \cap [d, 1]. \end{cases}$$

We observe that $B_{H0}^N \Psi^{\pm}(x_0) > 0$, $L_H^N \Psi^{\pm}(x_i) \leq 0$ and $B_{H1}^N \Psi^{\pm}(x_N) > 0$. Thus, applying Theorem 3 to $\Psi^{\pm}(x_i)$, for $x_i \in \bar{\Omega}^N$, we get the required result.

Using the technique adapted in [6] and applying the argument separately on each of the subinterval $\Omega^N \cap [\sigma, d]$ and $\Omega^N \cap [d + \sigma, 1)$, we can prove that $|\varepsilon D^+ e(x_i)| \leq CN^{-2}$. Similarly we can prove that $|\varepsilon D^- e(x_i)| \leq CN^{-2}$. Now, we prove that $|\varepsilon D^+ e(x_i)| \leq CN^{-2}$, for all $x_i \in \Omega^N \cap ((0, \sigma) \cup (d, d + \sigma))$. Using the above result, we have $|\varepsilon D^+ e(x_{N/4})| \leq CN^{-2}$. To prove the result for $0 \leq i \leq N/4 - 1$, we rewrite the relation $\tau(x_i) = L^N e(x_i)$, in the form, $\varepsilon D^+ e(x_j) - \varepsilon D^+ e(x_{j-1}) + \frac{1}{4}(x_{j+1} - x_{j-1})a(x_j)(D^+ e(x_j) + D^+ e(x_{j-1})) = \frac{1}{2}(x_{j+1} - x_{j-1})\tau(x_j)$.

Summing and rearranging, we obtain $|\varepsilon D^+ e(x_i)| \leq |\varepsilon D^+ e(x_{N/4})| + \frac{1}{2} \sum\limits_{j=i+1}^{N/4} (x_{j+1} -$

$x_{j-1})|\tau(x_j)| + \frac{1}{4}| \sum\limits_{j=i+1}^{N/4} (x_{j+1} - x_{j-1})a(x_j)(D^+ e(x_j) + D^+ e(x_{j-1}))|.$

We now bound each term separately. We have already bounded the first term. We know that $|\tau(x_j)| \leq CN^{-2}$ and so the second term is also bounded by CN^{-2}.

To bound the last term we observe that $\sum\limits_{j=i+1}^{N/4} (x_{j+1} - x_{j-1})a(x_j)D^+ e(x_j) =$

$(\frac{x_{N/4+1}-x_{N/4-1}}{x_{N/4+1}-x_{N/4}}a(x_{N/4})e(x_{N/4+1}) - \frac{x_{i+1}-x_{i-1}}{x_{i+1}-x_i}a(x_i)e(x_{i+1})) - \sum\limits_{j=i+1}^{N/4} \frac{x_{j+1}-x_{j-1}}{x_{j+1}-x_j}(a(x_j) -$

$a(x_{j-1}))e(x_j) - (1 - \frac{H_1}{h})a(x_{N/4})e(x_{N/4-1})$ and similarly for

$\sum\limits_{j=i+1}^{N/4} (x_{j+1} - x_{j-1})a(x_j)D^+ e(x_{j-1})$. Using the fact that $|e(x_j)| \leq CN^{-2}$ and

$|a(x_j) - a(x_{j-1})| \leq \| a' \| (x_j - x_{j-1})$, we get $|\varepsilon D^+ e(x_i)| \leq CN^{-2}$, for $0 \leq i \leq N/4 - 1$. Similarly, we can prove that $|\varepsilon D^+ e(x_i)| \leq CN^{-2}$, for $3N/4 \leq i \leq N - 1$. Thus, we have $|\varepsilon D^+ e(x_i)| \leq CN^{-2}$, $x_i \in \Omega^N \cup \{d\}$. Similarly, we can prove that $|\varepsilon D^- e(x_i)| \leq CN^{-2}$, $x_i \in \Omega^N \cup \{d\}$. This implies that $|\varepsilon D^0 e(x_i)| \equiv |\frac{\varepsilon(D^+ + D^-)e(x_i)}{2}| \leq CN^{-2}$, $x_i \in \Omega^N \cup \{d\}$.

Lemma 5. *Assume (15). Let $w_0(x)$ and $W_0(x_i)$ be the solutions of (5) and (17) respectively. Then at each mesh point $x_i \in \bar{\Omega}^N$,*

$$|(W_0 - w_0)(x_i)| \leq CN^{-2}(\ln N)^2,$$
$$|\varepsilon D^0(W_0 - w_0)(x_i)| \leq CN^{-2}(\ln N)^2.$$

Proof. As given in Remark 1, we have the inequalities $|B^N_{H0}(W_0 - w_0)(x_0)| \leq CN^{-2}(\ln N)^2$, $|B^N_{H1}(W_0 - w_0)(x_N)| \leq CN^{-2}$. For $x_i \in \Omega^N \cap [\sigma, 1]$, using the triangle inequality we have $|(W_0 - w_0)(x_i)| \leq |W_0(x_i)| + |w_0(x_i)|$. Using Lemma 2, we have $|(W_0 - w_0)(x_i)| \leq CN^{-2}$, for $x_i \in \Omega^N \cap [\sigma, 1]$.

Now for $x_i \in [0, \sigma)$, we have $|L^N_H(W_0 - w_0)(x_i)| \leq C\varepsilon^{-1}N^{-2}(\ln N)^2 e^{-\alpha x_{i-1}/\varepsilon}$. For all i, $0 \leq i \leq N/4$, we introduce the mesh functions

$$\Psi^\pm(x_i) = CN^{-2} + CN^{-2}(\ln N)^2 e^{-\alpha x_{i-1}/\varepsilon} \prod_{j=1}^{i} (1 + \frac{\alpha h_j}{\varepsilon}) \pm (W_0 - w_0)(x_i).$$

We observe that $B^N_0 \Psi^\pm(x_0) \geq 0$, $B^N_1 \Psi^\pm(x_{N/4}) \geq 0$ and $L^N \Psi^\pm(x_i) \leq 0$. Thus, applying Theorem 3 to $\Psi^\pm(x_i)$, we get $|(W_0 - w_0)(x_i)| \leq CN^{-2}(\ln N)^2$, for $x_i \in \bar{\Omega}^N$.

For all $x_i \in \Omega^N \cap [\sigma, d) \cap [d+\sigma, 1)$, using triangle inequality we have $|\varepsilon D^0(W_0 - w_0)(x_i)| \leq |\varepsilon(D^0 W_0 - w'_0)(x_i)| + |\varepsilon(D^0 w_0 - w'_0)(x_i)|$. To bound the second term, we observe that $|\varepsilon(D^0 w_0(x_i) - w'_0(x))| \leq CN^{-2}\varepsilon \|w^{(3)}\|_{[\sigma, d]} + CN^{-2}\varepsilon \|w^{(3)}\|_{[d+\sigma, 1)}$ $\leq CN^{-2}(\ln N)^2$, since $|w^{(3)}_0(x)| \leq C\varepsilon^{-3}e^{-\alpha \sigma_1/\varepsilon} \leq C\varepsilon$, when $x \geq \sigma_1$. To bound the first term, first we consider $|\varepsilon(D^0 W_0 - w'_0)(x_i)|$ and using triangle inequality, we write it as $|\varepsilon(D^0 W_0 - w'_0)(x_i)| \leq |\varepsilon D^0 W_0(x_i)| + |\varepsilon w'_0(x_i)| \leq CN^{-2}$.

For $x_i = \sigma$, we write $L_H^N W_0(\sigma) = 0$ in the form $\varepsilon D^+ W_0(x_{N/4-1}) = (1 - \frac{H_2}{2\varepsilon} a(\sigma)) \varepsilon D^+ W_0(x_\sigma) + \frac{1}{2} a(\sigma)(W_0(x_{N/4+1}) - W_0(\sigma))$. Using the estimates obtained at the points σ, we obtain $\varepsilon D^+ W_0(x_{N/4-1}) \leq (1 + CN^{-1} \ln N)CN^{-2} + CN^{-2} \leq CN^{-2}$. Similarly for $x_i = d + \sigma$, we have $\varepsilon D^+ W_0(x_{3N/4-1}) \leq CN^{-2}$. Now consider $x_i \in \Omega^N \cap (0, \sigma)$. For convenience we introduce the notation $\hat{e}(x_i) = (W_0 - w_0)(x_i)$ and $\hat{\tau}(x_i) = L_H^N \hat{e}(x_i)$. We have already established that $|\hat{e}(x_i)| \leq CN^{-2}(\ln N)^2$ and $|\hat{\tau}(x_i)| \leq C\sigma\varepsilon^{-3}N^{-2}e^{-\alpha x_{i-1}/\varepsilon}$. Using the procedure adopted in Lemma 4 and using the above the results we have $\varepsilon D^+ \hat{e}(x_j) \leq CN^{-2}((\ln N)^2 + \frac{\sigma}{\varepsilon} \frac{\alpha h^2/\varepsilon^2}{1-e^{-\alpha H_1/\varepsilon}})$. But $y = \alpha H_1/\varepsilon = 4N^{-1} \ln N$ and $B(y) = \frac{y}{1-e^{-y}}$ are bounded and it follows that $|\varepsilon D^+ \hat{e}(x_i)| \leq CN^{-2}(\ln N)^2$, for $x_i \in \Omega^N \cap (0, \sigma)$, as required. Similarly applying the above argument on the interval $\Omega^N \cap (d, d+\sigma)$, we get $|\varepsilon D^+ e(x_i)| \leq CN^{-2}(\ln N)^2$. Also, we can prove that $|\varepsilon D^- e(x_i)| \leq CN^{-2}(\ln N)^2$, for $x_i \in \bar{\Omega}^N \cap (0, 1)$. This implies $|\varepsilon D^0 e(x_i)| \equiv |\frac{\varepsilon(D^+ + D^-)e(x_i)}{2}| \leq CN^{-2}(\ln N)^2$, for $x_i \in \bar{\Omega}^N \cap (0, 1)$.

Lemma 6. *Assume (15). Let $w_d(x)$ and $W_d(x_i)$ be the solutions of (6) and (18), (19) respectively. Then at each mesh point $x_i \in \bar{\Omega}^N$,*

$$|(W_d - w_d)(x_i)| \leq CN^{-2}(\ln N)^2,$$
$$|\varepsilon D^0(W_d - w_d)(x_i)| \leq CN^{-2}(\ln N)^2.$$

Proof. Using standard truncation error bounds and the bounds on the derivatives of w_d, we have the inequalities $| B_0^{HN}(W_d - w_d)(x_0) | \leq CN^{-2}(\ln N)^2$, $| B_1^{HN}(W_d - w_d)(x_N) | \leq CN^{-2}$, $|L_1^N(W_d - w_d)(x_i)| \leq CN^{-2}(\ln N)^2$, for $x_i \in \Omega^N \cap (0, d) \cap (d, 1)$. Now, we consider the magnitude of the truncation error at the point of discontinuity $x_{N/2} = d$.

$$|L_T^N W_d(x_{N/2}) - \frac{f(x_{N/2-1})h_{N/2-1}}{2\varepsilon} + \frac{f(x_{N/2+1})h_{N/2+2}}{2\varepsilon + a(x_{N/2+1})h_{N/2+2}}| \leq |L_t^N W_d(x_{N/2}) +$$

$$[w_d'](d)| + \frac{h_{N/2-1}}{2\varepsilon}|L_H^N W_d(x_{N/2-1}) - f(x_{N/2-1})| + \frac{h_{N/2+2}}{2\varepsilon + a(x_{N/2+1})h_{N/2+2}}$$

$$|L_H^N W_d(x_{N/2+1}) - f(x_{N/2+1})| \leq CN^{-2}(\ln N)^2.$$

Then applying Lemma 3 to $(W_d - w_d)(x_i)$, for all $x_i \in \bar{\Omega}^N$, we get $|(W_d - w_d)(x_i)| \leq CN^{-2}(\ln N)^2$. Using the technique adapted in the proof of Lemmas 4, 5, we get $|\varepsilon D^0(W_d - w_d)(x_i)| \leq CN^{-2}(\ln N)^2$.

Lemma 7. *[1, Lemma 3.13] Assume (15). For each $x_i \in \Omega^N \cup \{d\}$ and for all $x \in \bar{\Omega}_i = [x_i, x_{i+1}]$, we have $|\varepsilon(D^0 u(x_i) - u'(x))| \leq CN^{-2}(\ln N)^2$, where u is the solution of (1), (2).*

Theorem 4. *Assume that N is sufficiently large so that (15) is satisfied. Let $u(x)$ be the solution to the problem (1),(2) and let $U(x_i)$ be the corresponding numerical solution generated by HDS. Then for $x \in \bar{\Omega}_i = [x_i, x_{i+1}]$, we have*

$$\sup_{0<\varepsilon\leq 1} \| U - u \|_{\bar{\Omega}} \leq CN^{-2}(\ln N)^2,$$

$$\sup_{0<\varepsilon\leq 1} \| \varepsilon(D^0 U(x_i) - u') \|_{\bar{\Omega}_i} \leq CN^{-2}(\ln N)^2.$$

Proof. Proof follows immediately, if one applies the above Lemmas 4, 5, 6 and 7 to $|U - u| \leq |V - v| + |W - w|$ and $|\varepsilon(D^0 U(x_i) - u'(x))| \leq |\varepsilon D^0(V - v)(x_i)| + |\varepsilon D^0(W_0 - w_0)(x_i)| + |\varepsilon D^0(W_d - w_d)(x_i)| + |\varepsilon(D^0 u(x_i) - u'(x))|$.

5   Numerical Experiments

In this section, theoretical results obtained in the earlier sections are verified experimentally. For all integers N, satisfying N, $2N \in R_N = [64, 128, 256, 512, 1024]$. Let U_H^N and $\tilde{U}_H^{4096}$ denote respectively, the numerical solutions generated by HDS using N mesh points and the piecewise linear interpolant of the numerical solution obtained using 4096 mesh points. Similarly, for a finite set of values $\varepsilon \in R_\varepsilon = [2^{-26}, 2^{-12}]$, we compute the maximum pointwise errors for the solution and its first derivative respectively as $E_{H\varepsilon}^N = \| U_H^N - \tilde{U}_H^{4096} \|_{\Omega^N}$ $D_{H\varepsilon}^N = \max |\varepsilon(D^0 U_H^N - \tilde{D}^0 U_H^{4096})(x_i)|$, for $1 \leq i \leq N - 1$ and the ε-uniform maximum errors $E_H^N = \max_{\varepsilon \in R_\varepsilon} E_{H\varepsilon}^N$ and $D_H^N = \max_{\varepsilon \in R_\varepsilon} D_{H\varepsilon}^N$. Approximations of ε- uniform order of local convergence are defined, for all N, $4N \in R_N$, by $p_H^N = \log_2(\frac{E_H^N}{E_H^{2N}})$ and $s_H^N = \log_2(\frac{D_H^N}{D_H^{2N}})$. Consider the test problem :

$$\varepsilon u''(x) + (1 + x)u'(x) = \begin{cases} 2x, & x \leq 0.5, \\ 2(1 - x), & x \geq 0.5, \end{cases}$$
$$2u(0) - \varepsilon u'(0) = 0, \quad 3u(1) + u'(1) = 0.$$

Table 1 present values of the computed maximum pointwise error for the numerical solution E_H^N, p_H^N and the scaled discrete derivative throughout the domain D_H^N, s_H^N for the Test problem 1. The numerical results in Table 1 are clear

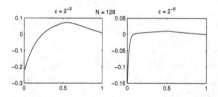
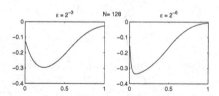

Fig. 1. Numerical solution and its scaled discrete derivative for $\varepsilon = 2^{-3}$ and $\varepsilon = 2^{-6}$, with $N = 128$ for the Test problem 1

Table 1. Values of E_H^N, p_H^N and D_H^N, s_H^N generated by HDS for the Test problem 1

N	64	128	256	512	1024
E^N	6.4376e-3	2.3491e-3	8.6916e-4	3.1509e-4	1.0150e-4
p^N	1.4544	1.4344	1.4639	1.6343	-
D_H^N	5.7466e-3	2.8257e-3	1.1911e-3	4.5423e-4	1.4246e-4
s_H^N	1.0241	1.2463	1.3908	1.6729	-

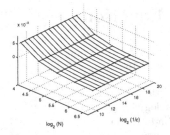 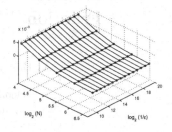

Fig. 2. Maximum pointwise errors $E_{H\varepsilon}^N$ and $D_{H\varepsilon}^N$ as functions of N and ε for the Test problem 1 generated by HDS

illustrations of the convergence estimate of the Theorem 4. In Figure 2 the maximum pointwise errors are plotted as functions of N and ε. Note that for all values of ε the error decreases steadily with increasing N.

Acknowledgement. The first author wishes to acknowledge the National Board for Higher Mathematics(NBHM), Mumbai(INDIA) for its financial support.

References

1. Farrell, P.A., Hegarty, A.F., Miller, J.J.H., O'Riordan, E., Shishkin, G.I.: Robust Computational Techniques for Boundary Layers. Chapman Hall/CRC, Boca Raton (2000)
2. Ansari, A.R., Hegarty, A.F.: Numerical Solution of a Convection Diffusion Problem with Robin Boundary Conditions. Appl. Math. Comput. 100, 27–48 (1999)
3. Kopteva, N., Stynes, M.: Approximation of Derivatives in a Convection-diffusion Two-point Bondary Value Problem. Appl. Math. Comput. 39, 47–60 (2001)
4. Cen, Z.: A Hybrid Difference Scheme for a Singularly Perturbed Convection-diffusion Problem with Discontinuous Convection Coefficient. Appl. Math. Comput. 169, 689–699 (2005)
5. Farrell, P.A., Hegarty, A.F., Miller, J.J.H., O'Riordan, E., Shishkan, G.I.: Singularly Perturbed Convection-diffusion Problem with Boundary and weak Interior Layers. J. Comput. Appl. Math. 166, 133–151 (2004)
6. Mythili Priyadharshini, R., Ramanujam, N.: Approximation of Derivative to a Singularly Perturbed Second Order Ordinary Differential Equation with Discontinuous Convection Coefficient using Hybrid Difference Scheme. Int. J. Comput. Math. 86(8), 1355–1369 (2009)
7. Mythili Priyadharshini, R., Ramanujam, N.: Approximation of Derivative for a Singularly Perturbed Second-order Ordinary Differential Equation of Robin Type with Discontinuous Convection Coefficient and Source Term. Numer. Math. Theor. Meth. Appl. 2(1), 100–118 (2009)
8. Mythili Priyadharshini, R., Ramanujam, N.: Approximation of Derivative to a Singularly Perturbed Reaction-convection-diffusion Problem with Two Parameters. J. Appl. Math. & Informatics 27(3-4), 517–529 (2009)

A Numerical Method for a Class of Linear Fractional Differential Equations

V. Balakumar* and K. Murugesan

Department of Mathematics
National Institute of Technology
Tiruchirappalli-620 015, Tamil Nadu, India
balakumar.math@gmail.com, murugu@nitt.edu

Abstract. A numerical method is proposed for the numerical solution of initial value problems of a certain class of linear Fractional Differential Equations (FDEs) with the Jumarie's modified Riemann-Liouville fractional derivative. Numerical examples are presented to justify the effectiveness of the method.

Keywords: Numerical Method, Fractional Differential Equations, Initial Value Problem, Jumarie Fractional Derivative.

1 Introduction

Fractional Differential Equations (FDEs) have been the subject of intense research in recent decades. There are various applications of fractional differential equations in, practically, all areas of science and engineering [1]. Only a few analytical methods are in the literature for the computation of solutions to FDEs. For most FDEs, we cannot provide methods to compute the exact solutions analytically. There is no generally applicable method to find an analytic solution to an arbitrarily given FDE [2]. Therefore, numerical methods plays a crucial role in solving FDEs. For Riemann-Liouville and Caputo fractional derivatives, some numerical methods are discussed in [2]-[5]. Recently, modified Riemann-Liouville fractional derivative has been proposed by Jumarie [6].

In this paper, we derive a numerical method for the solution of fractional ordinary differential equation, in the sense of Jumarie, of the form

$$D^\alpha y = \beta y + f(x) \text{ with } y(0) = y_0 \tag{1}$$

where α denotes the order of the derivative with $0 < \alpha \le 1$ and β is a constant.

This class of initial value problems occurs often in mechanics, diffusion processes, electromagnetics, electrochemistry, material science, the theory of ultra-slow processes and special functions [7].

This paper is organized as follows. Section 2 describes fractional integral and derivative operators. Section 3 is devoted for the mathematical preliminaries,

---

* Corresponding author.

P. Balasubramaniam and R. Uthayakumar (Eds.): ICMMSC 2012, CCIS 283, pp. 360–366, 2012.
© Springer-Verlag Berlin Heidelberg 2012

which are relevant for the derivation that follows. Section 4 is concerned with the derivation of the proposed numerical method. Section 5 demonstrates the efficiency of the method by numerical examples. Section 6 concludes the paper.

2 Fractional Derivatives

Fractional calculus deals with the investigation of integrals and derivatives of arbitrary noninteger order. A detailed survey is presented in [8]. There are various approaches for defining fractional order integral and derivative operators [1]. We give some useful definitions from the literature.

For a continuous real-valued function f, the Riemann-Liouville fractional integral operator of order α is defined as

$$J^\alpha f(x) = \frac{1}{\Gamma(\alpha)} \int_0^x (x - t)^{\alpha-1} f(t) dt, \text{ where } \alpha > 0, x > 0.$$

The Riemann-Liouville fractional derivative is defined as

$$D^\alpha f(x) = \frac{d}{dx} \left(J^{1-\alpha} f(x) \right), \text{ where } 0 < \alpha \leq 1.$$

The Caputo fractional derivative is defined as

$$D_*^\alpha f(x) = J^{1-\alpha} \left(\frac{d}{dx} f(x) \right), \text{ where } 0 < \alpha \leq 1.$$

Approximations of fractional integrals and Caputo fractional derivatives are presented in [9]. The Riemann-Liouville derivative of a constant is not zero. The Caputo derivative is defined for differentiable functions only.

To circumvent these difficulties, Jumarie [6] modified Riemann-Liouville definition via finite difference. Further results can be seen in [10]-[13]. For convenience, we denote Jumarie's fractional integral and fractional derivative operators by J_J^α and D_J^α respectively.

The Modified Riemann-Liouville fractional integral is defined as

$$J_J^\alpha f(x) = \frac{1}{\Gamma(\alpha)} \int_0^x (x - t)^{\alpha-1} [f(t) - f(0)] dt, \text{ where } \alpha > 0, x > 0. \tag{2}$$

The Modified Riemann-Liouville fractional derivative is defined as

$$D_J^\alpha f(x) = \frac{d}{dx} \left(J_J^{1-\alpha} f(x) \right), \text{ where } 0 < \alpha \leq 1. \tag{3}$$

3   Preliminaries

3.1   Numerical Approximation

Using product trapezoidal quadrature rule, a numerical approximation method for Jumarie's fractional integral has been deduced from [14]. Let the interval $[0, x_k]$ be partitioned into k subintervals $[x_j, x_{j+1}]$ of equal width $h = x_k/k$ by using the nodes $x_j = jh$, for $j = 0, 1, 2, \ldots, k$. Let $T(f, h, \alpha)$ denotes the numerical approximation for $J_J^\alpha f(x)$. Now,

$$T(f, h, \alpha) = \frac{h^\alpha}{\Gamma(\alpha + 2)} \left(F(x_k) + \sum_{j=1}^{k-1} [(k - j + 1)^{\alpha+1} + (k - j - 1)^{\alpha+1} - 2(k - j)^{\alpha+1}][F(x_j)] \right),$$

(4)

where $F(x_j) = f(x_j) - f(x_0)$. Further,

$$|J_J^\alpha f(x) - T(f, h, \alpha)| \le \frac{C_\alpha}{\Gamma(\alpha)} \|f''\| x_k^\alpha h^2 = \mathbf{O}(h^2),$$

(5)

for some constant C_α depending only on α. This represents that the error in this approximation is bounded.

3.2   Trapezoidal Method

Let $y = f(x)$ over $[x_k, x_{k+1}]$, where $x_{k+1} = x_k + h$. The classical Trapezoidal method is

$$\int_{x_k}^{x_{k+1}} f(x)dx = \frac{h}{2}\{f(x_k) + f(x_{k+1})\}.$$

This is the simplest method for numerical integration of definite integrals.

4   Derivation of the Method

In this section, we derive the proposed numerical method for solving the linear FDEs of the form (1).
    For,

$$D_J^\alpha y(x) = \beta y(x) + f(x)$$

$$\frac{d}{dx} J_J^{1-\alpha} y(x) = \beta y(x) + f(x)$$

$$J_J^{1-\alpha} y(x) \Big]_{x_k}^{x_{k+1}} = \int_{x_k}^{x_{k+1}} [\beta y(x) + f(x)] dx$$

$$J_J^{1-\alpha} y(x_{k+1}) - J_J^{1-\alpha} y(x_k) = \int_{x_k}^{x_{k+1}} [\beta y(x) + f(x)] dx.$$

(6)

Using the numerical approximation (4), we deduce

$$J_J^{1-\alpha} y(x_k) = \frac{h^{1-\alpha}}{\Gamma(3-\alpha)} \left(F(x_k) + \sum_{j=1}^{k-1} [(k-j+1)^{2-\alpha} + (k-j-1)^{2-\alpha} \right.$$
$$\left. -2(k-j)^{2-\alpha}][F(x_j)] \right)$$

$$J_J^{1-\alpha} y(x_{k+1}) = \frac{h^{1-\alpha}}{\Gamma(3-\alpha)} \left(F(x_{k+1}) + \sum_{j=1}^{k} [(k-j+2)^{2-\alpha} + (k-j)^{2-\alpha} \right.$$
$$\left. -2(k-j+1)^{2-\alpha}][F(x_j)] \right).$$

Using the Trapezoidal method, we obtain

$$\int_{x_k}^{x_{k+1}} [\beta y(x) + f(x)] dx = \frac{h}{2} [\beta y(x_k) + f(x_k) + \beta y(x_{k+1}) + f(x_{k+1})].$$

Substituting these approximations in (6), we get

$$\frac{2h^{1-\alpha} - h\beta\Gamma(3-\alpha)}{2\Gamma(3-\alpha)} y(x_{k+1}) = \frac{h}{2} \{\beta y(x_k) + f(x_k) + f(x_{k+1})\}$$
$$-\frac{h^{1-\alpha}}{\Gamma(3-\alpha)} \Big\{ -y(x_k)$$
$$+ \sum_{j=1}^{k} [(k-j+2)^{2-\alpha} + (k-j)^{2-\alpha} - 2(k-j+1)^{2-\alpha}][F(x_j)]$$
$$- \sum_{j=1}^{k-1} [(k-j+1)^{2-\alpha} + (k-j-1)^{2-\alpha} - 2(k-j)^{2-\alpha}][F(x_j)]\Big\}.$$

Further simplification yields

$$y(x_{k+1}) = \frac{h\Gamma(3-\alpha)}{2h^{1-\alpha} - h\beta\Gamma(3-\alpha)} \{\beta y(x_k) + f(x_k) + f(x_{k+1})\}$$
$$- \frac{2h^{1-\alpha}}{2h^{1-\alpha} - h\beta\Gamma(3-\alpha)} \Big\{ -y(x_k)$$
$$+ \sum_{j=1}^{k} [(k-j+2)^{2-\alpha} + (k-j)^{2-\alpha} - 2(k-j+1)^{2-\alpha}][F(x_j)]$$
$$- \sum_{j=1}^{k-1} [(k-j+1)^{2-\alpha} + (k-j-1)^{2-\alpha} - 2(k-j)^{2-\alpha}][F(x_j)]\Big\}. \quad (7)$$

This represents the proposed method. The error in this method is bounded as we use stable numerical approximation method and trapezoidal method. And so, the method is stable.

5 Numerical Examples

5.1 Example

Consider $D_J^\alpha y = \beta y + f(x)$, with $\beta = -1$, $f(x) = x^2 + \frac{2x^{2-\alpha}}{\Gamma(3-\alpha)}$ and $y(0) = 0$. This problem is discussed in [7], but with Riemann-Liouville sense. The exact solution is x^2. For,

$$D_J^\alpha(x^2) = \frac{1}{\Gamma(1-\alpha)} \frac{d}{dx} \int_0^x (x-t)^{-\alpha}[t^2 - 0]dt.$$

$$= \frac{2}{\Gamma(3-\alpha)} x^{2-\alpha}$$

The results are compared in Table 1 and Table 2.

Table 1. $\alpha = 0.5$, h=0.1

x	Exact solution	Numerical solution
0.1	0.0100	0.0100
0.2	0.0400	0.0400
0.3	0.0900	0.0900
0.4	0.1600	0.1600
0.5	0.2500	0.2500
0.6	0.3600	0.3600
0.7	0.4900	0.4900
0.8	0.6400	0.6400
0.9	0.8100	0.8100
1.0	1.0000	1.0000

Table 2. $\alpha = 0.75$, h=0.1

x	Exact solution	Numerical solution
0.1	0.0100	0.0100
0.2	0.0400	0.0400
0.3	0.0900	0.0900
0.4	0.1600	0.1600
0.5	0.2500	0.2500
0.6	0.3600	0.3600
0.7	0.4900	0.4900
0.8	0.6400	0.6400
0.9	0.8100	0.8100
1.0	1.0000	1.0000

5.2   Example

Consider $D_j^{\alpha} y = \beta y + f(x)$, with $\beta = 1$, $f(x) = 0$ and $y(0) = 1$.

The exact solution is $y(x) = \sum\limits_{k=0}^{\infty} \frac{x^{k\alpha}}{(k\alpha)!}$.

The results are compared in Table 3 and Table 4.

Table 3. $\alpha = 0.5$, h=0.1

x	Exact solution	Numerical solution
0.1	1.4868	1.5322
0.2	1.7990	1.7895
0.3	2.1077	2.1087
0.4	2.4300	2.4267
0.5	2.7743	2.7710
0.6	3.1462	3.1421
0.7	3.5508	3.5462
0.8	3.9928	3.9876
0.9	4.4772	4.4714
1.0	5.0090	5.0025

Table 4. $\alpha = 0.5$, h=0.01

x	Exact solution	Numerical solution
0.1	1.4868	1.4869
0.2	1.7990	1.7990
0.3	2.1077	2.1077
0.4	2.4300	2.4300
0.5	2.7743	2.7742
0.6	3.1462	3.1462
0.7	3.5508	3.5507
0.8	3.9928	3.9928
0.9	4.4772	4.4771
1.0	5.0090	5.0089

6   Conclusion

A new method is proposed to obtain the solution of initial value problems of a certain class of linear fractional differential equations. For smaller h values, the method gives better accuracy. The results show that the proposed method is quite reasonable when compare to exact solution.

References

1. Kilbas, A.A., Srivastava, H.M., Trujillo, J.J.: Theory and Applications of Fractional Differential Equations. North-Holland Mathematics Studies, vol. 204. Elsevier Science B.V., Amsterdam (2006)
2. Diethelm, K.: The Analysis of Fractional Differential Equations. Springer, Heidelberg (2010)
3. Diethelm, K., Ford, N.J., Freed, A.D., Luchko, Y.: Algorithms for the Fractional Calculus: A Selection of Numerical Methods. Comput. Methods Appl. Mech. Engg. 194, 743–773 (2005)
4. Li, C., Chen, A., Ye, J.: Numerical Approaches to Fractional Calculus and Fractional Ordinary differential Equation. J. Comput. Phys. 230, 3352–3368 (2011)
5. Podlubny, I.: Fractional Differential Equations. Academic Press, New York (1999)
6. Jumarie, G.: Modified Riemann-Liouville Derivative and Fractional Taylor Series of Non-differentiable Functions further Results. Comput. Math. Appl. 51, 1367–1376 (2006)
7. Diethelm, K.: An Algorithm for the Numerical Solution of Differential Equations of Fractional Order. Electron. Trans. Numer. Anal. 5, 1–6 (1997)
8. Tenreiro Machado, J., Kiryakova, V., Mainardi, F.: Recent History of Fractional Calculus. Commun. Nonlinear. Sci. Numer. Simulat. 16, 1140–1153 (2011)
9. Odibat, Z.: Approximations of Fractional Integrals and Caputo Fractional Derivatives. Appl. Math. Comput. 178, 527–533 (2006)
10. Jumarie, G.: Lagrangian Mechanics of Fractional Order, Hamilton–Jacobi fractional PDE and Taylor's Series of Non-differentiable Functions. Chaos Solitons Fractals 32, 969–987 (2007)
11. Jumarie, G.: Table of Some Basic Fractional Calculus Formulae Derived from a Modified Riemann-Liouville Derivative for Non-differentiable functions. Appl. Math. Lett. 22, 378–385 (2009)
12. Jumarie, G.: Laplace's Transform of Fractional Order via the Mittag-Leffler Function and Modified Riemann-Liouville Derivative. Appl. Math. Lett. 22, 1659–1664 (2009)
13. Jumarie, G.: Cauchy's Integral Formula via the Modified Riemann-Liouville Derivative for Analytic Functions of Fractional Order. Appl. Math. Lett. 23, 1444–1450 (2010)
14. Diethelm, K., Ford, N., Freed, A.: Detailed Error Analysis for a Fractional Adams Method. Numer. Algorithms 36, 31–52 (2004)

Oscillation of Second Order Nonlinear Ordinary Differential Equation with Alternating Coefficients

M.J. Saad, N. Kumaresan, and Kuru Ratnavelu

Institute of Mathematical Sciences
University of Malaya
50603 Kuala Lumpur, Malaysia

Abstract. In this paper, some the sufficient conditions for the oscillation of the solutions of the second order non-linear ordinary differential equation are obtained using Riccati Technique. The given results are the extension and improvement of the known oscillation results which was obtained before by many authors as Bihari [2] and Kartsatos [9]. These results are illustrated with examples that are solved using Runge Kutta method.

Keywords: Alternative coefficients, Nonlinear differential equations, Oscillatory solutions, Runge Kutta method.

1   Introduction

Consider the second order non-linear ordinary differential equation of the form

$$\left(r(t)\dot{x}(t) \right)^{\cdot} + q(t)\Phi\left(g(x(t)), r(t)\dot{x}(t) \right) = H(t, x(t)) \qquad (E)$$

where r and q are continuous functions on the interval $[t_0, \infty)$, $t_0 \geq 0$, $r(t)$ is a positive function, g is continuously differentiable function on the real line R except possibly at 0 with $xg(x) > 0$ and $g'(x) \geq k > 0$ for all $x \neq 0$, Φ is a continuous function on RxR with $u\Phi(u, v) > 0$ for all $u \neq 0$ and $\Phi(\lambda u, \lambda v) = \lambda \Phi(u, v)$ for any $(\lambda, u, v) \in R^3$ and H is a continuous function on $[t_0, \infty) \times$ R with $\dfrac{H(t, x(t))}{g(x(t))} \leq p(t)$ for all $x \neq 0$ and $t \geq t_0$. Throughout this study, our attention is only to the solutions of the differential equation (E) that exist on some ray $[t_x, \infty)$, where t_x may depend on the particular solution.

A solution $x(t)$ of the differential equation is said to be oscillatory if it has arbitrary large zeros. Otherwise it is said to be non-oscillatory. (E) is called oscillatory if all its solutions are oscillatory. Otherwise it is called non oscillatory. Particular cases of (E) have been considered by many authors for example [1-12]. Some of these particular cases can be classified as follows

P. Balasubramaniam and R. Uthayakumar (Eds.): ICMMSC 2012, CCIS 283, pp. 367–373, 2012.
© Springer-Verlag Berlin Heidelberg 2012

$$\ddot{x}(t) + q(t)x(t) = 0 \tag{1}$$

$$\ddot{x}(t) + q(t)\,\Phi(x(t), \dot{x}(t)) = 0 \tag{2}$$

$$\left(r(t)\dot{x}(t)\right)^{\cdot} + q(t)\,g(x(t)) = H(t, x(t)) \tag{3}$$

The oscillation of linear equation (1) has brought the attention of many authors since because of Fite [5]. He proved that if $q(t) > 0$ for all $t \geq t_0$ and $\int\limits_{t_0}^{\infty} q(s)\,ds = \infty$, then every solution of the (1) is oscillatory. Wintner [11] extended the result of Fite [5] to an equation in which q is of arbitrary sign and suppose that

$$\lim_{t \to \infty} \frac{1}{t} \int\limits_{t_0}^{t} (t - s)\,q(s)\,ds = \infty,$$

then, every solution of (1) is oscillatory. In the following, Kamenev [8] has proved a new integral criterion for the oscillation of the differential equation (1) based on the use of the n the primitive of the coefficient $q(t)$, which has Wintner's result [11] as a particular case. He has showed that (1) is oscillatory if

$$\lim_{t \to \infty} \sup \frac{1}{t^{n-1}} \int\limits_{t_0}^{t} (t-s)^{n-1} q(s)\,ds = \infty, \quad \text{for some integer } n \geq 3.$$

The oscillation of (2) has brought the attention of some authors because of Bihari [2], who has proved that If $q(t) > 0$ for all $t \geq t_0$ and $\int\limits_{t_0}^{\infty} q(s)ds = \infty$, then, every solution of (2) is oscillatory. The following result is extended the result of Bihari [2] to an equation in which q is of arbitrary sign. In this theorem, Kartsatos [9] has supposed that

(i) There exists a constant $C \in R = (-\infty, 0)$ such that

$$G(m) = \int\limits_{0}^{m} \frac{du}{\Phi(1, u)} \geq -C \text{ for all } m \in R$$

(ii) $\int\limits_{t_0}^{\infty} q(s)ds = \infty$,

then, every solution of (2) is oscillatory. Many authors are concerned with the oscillation criteria of solutions of the homogeneous second order nonlinear differential equations. Greaf, Rankin and Spikes [7] gave some theorems for the non-homogeneous (3). For example, they proved that if

(i) $r(t) \leq a_1,\ a_1 > 0,$

(ii) $\lim\limits_{t \to \infty} \frac{1}{t} \int\limits_{t_0}^{t} \int\limits_{t_0}^{s} (q(u) - p(u))duds = \infty,$

then, all solutions of equation (3) are oscillatory.

2   Main Results

In this section, Riccati technique is used to reduce the higher-order equations to the first-order Riccati equation or inequality to establish sufficient conditions for oscillation of (E). Comparisons between our results and the previously known are presented and some examples illustrate the main results.

Theorem 1: Suppose that

"1" $\dfrac{1}{\Phi(1,u)} < \dfrac{1}{C_0}, C_0 > 0,$

"2" $G(m) = \displaystyle\int_0^m \dfrac{ds}{\Phi(1,s)} > -B^*, B^* > 0$ for every $m \in R,$

"3" $\displaystyle\int_T^\infty [C_0 q(s) - p(s)]ds = \infty,$

where, $p : [t_0, \infty) \to (0, \infty)$, then, every solution of (E) is oscillatory.

Proof: Without loss of generality, we may assume that there exists a solution $x(t)$ of (E) such that $x(t) > 0$ on $[T, \infty)$ for some $T \geq t_0 \geq 0$.
Define

$$\omega(t) = \dfrac{r(t)\,\dot{x}(t)}{g(x(t))} \, , \, t \geq T$$

This and (E) imply

$$\dot{\omega}(t) \leq p(t) - q(t)\Phi(1, \omega(t)) \, , \, t \geq T$$

$$\dfrac{\dot{\omega}(t)}{\Phi(1, \omega(t))} \leq \dfrac{p(t)}{\Phi(1, \omega(t))} - q(t) \, , \, t \geq T$$

By the condition "1", we have

$$\dfrac{\dot{\omega}(t)}{\Phi(1, \omega(t))} \leq \dfrac{1}{C_0} p(t) - q(t) \, , \, t \geq T$$

Integrate from T to t, we have

$$\int_T^t [C_0 q(s) - p(s)]ds \leq -C_0 \int_{\omega(T)}^{\omega(t)} \dfrac{du}{\Phi(1,u)} = -C_0 G(\omega(t)) + C_0 G(\omega(T)) \leq C_0 B^* + C_0 G(\omega(T))$$

Thus, by condition "2", we have

$$\int_T^\infty (C_0 q(s) - p(s))ds < \infty,$$

which contradicts to the condition "3". Hence, the proof is completed.

2.1   Example 1

Consider the differential equation

$$\left(t\,\dot{x}(t) \right)^{\bullet} + \left(1 + 3\cos t \right) x(t) = \frac{x(t)\cos x(t)}{t^3}, \quad t > 0$$

Here $r(t)=t$, $q(t)=(1+3cost)$, $g(x)=x$, $\Phi(u,v)=u$ and

$$H(t,x(t)) = \frac{x(t)\cos(x(t))}{t^3}, \quad \frac{H(t,x(t))}{g(x(t))} = \frac{\cos(x(t))}{t^3} \le \frac{1}{t^3} = p(t).$$

All conditions of the Theorem 1 are satisfied and hence every solution of the given equation is oscillatory.

Table 1. Numerical solution of ODE 1

K	t_k	$x(t_k)$
1	1	1
2	1.198	0.837
3	1.396	0.5662
4	1.594	0.2258
5	1.792	-0.1412
6	1.99	-0.4917
7	2.188	-0.7824
8	2.386	-0.9733
9	2.584	-1.0352
10	2.782	-0.9578

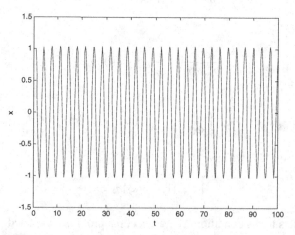

Fig. 1. Solution curve of ODE 1

To ensure that our result in Theorem 1 is true, the numerical solution of the given differential equation is also given in Example 1 using the Runge Kutta method of fourth order (RK4). We have

$\ddot{x}(t) = f(t, x(t), \dot{x}(t)) = x \cos x - 3.99x$ with initial conditions x(1)=1, $\dot{x}(1) =$ −0.5 on the chosen interval [1,100], finding values of the functions r, q and f where, we consider $H(t, x(t)) = f(t)l(x)$ at $t=1$, $n = 500$ and $h = 0.198$.

Remark 1. Theorem 1 is extension of the results of Bihari [2] and the results of Kartsatos [9].

Theorem 2: If the conditions "1" and "2" hold, assume that ρ be a positive continuous differentiable function on the interval $[t_0, \infty)$ with $\rho(t)$ is a decreasing function on the interval $[t_0, \infty)$ and such that

$$\text{"4"} \quad \int_T^\infty \rho(s)[C_0 q(s) - p(s)]ds = \infty,$$

where, p: $[t_0,\infty) \to (0,\infty)$, then every solution of (E) is oscillatory.

Proof: Without loss of generality, it is assumed that there exists a solution $x(t)$ of (E) such that $x(t) > 0$ on $[T,\infty)$ for some $T \geq t_0 \geq 0$ and
   Define

$$\omega(t) = \frac{\rho(t)r(t)\dot{x}(t)}{g(x(t))} , t \geq T$$

Thus and (E) imply

$$\left(\rho(t)\frac{\omega(t)}{\rho(t)}\right)^{\bullet} \leq \rho(t)p(t) - \rho(t)q(t)\Phi\left(1, \frac{\omega(t)}{\rho(t)}\right) + \frac{\dot{\rho}(t)}{\rho(t)}\omega(t) , t \geq T$$

$$\rho(t)\left(\frac{\omega(t)}{\rho(t)}\right)^{\bullet} \leq \rho(t)p(t) - \rho(t)q(t)\Phi\left(1, \frac{\omega(t)}{\rho(t)}\right) , t \geq T$$

Dividing the last inequality by $\Phi(1,\omega(t)/\rho(t)) > 0$ and by condition "1", we integrate from T to t, we obtain

$$\int_T^t \rho(s)[C_0 q(s) - p(s)]ds \leq -C_0 \int_T^t \frac{\rho(s)(\omega(s)/\rho(s))^{\bullet}}{\Phi(1, \omega(s)/\rho(s))}ds , t \geq T \tag{3}$$

By the Bonnet's theorem, we see that for each $t \geq T$, there exists $a_t \in [T, t]$ such that

$$-\int_T^t \frac{\rho(s)(\omega(s)/\rho(s))^{\bullet}}{\Phi(1, \omega(s)/\rho(s))}ds = -\rho(T)\int_T^{a_t} \frac{(\omega(s)/\rho(s))^{\bullet}}{\Phi(1, \omega(s)/\rho(s))}ds \tag{4}$$

From inequality (4) in inequality (3), we have

$$\int_T^t \rho(s)[C_0 q(s) - p(s)]ds \leq -\rho(T)\int_{\omega(T)/\rho(T)}^{\omega(a_t)/\rho(a_t)} \frac{du}{\Phi(1,u)} = C_0\rho(T)\left[-G\left(\frac{\omega(a_t)}{\rho(a_t)}\right) + G\left(\frac{\omega(T)}{\rho(T)}\right)\right]$$

$$\leq C_0\rho(T)B^* + C_0\rho(T)G\left(\frac{\omega(T)}{\rho(T)}\right) < \infty,$$

as $t \to \infty$, which contradicts to the condition "4". Hence the proof is completed.

Example 2. Consider the following differential equation

$$\left(t^3 \dot{x}(t)\right)^{\cdot} + \left(\frac{t^5 + 4t^5 \cos t}{t^5 + 1}\right)\left(x^9(t) + \frac{x^{27}(t)}{x^{27}(t) + (t^3 \dot{x}(t))^2}\right) = \frac{2t^5 x^9(t)\sin((x(t))}{\left(t^{10} + t^5\right)} \quad ,t > 0$$

Here $r(t)=t^3>0$, $q(t) = \frac{t^5+4t^5\cos t}{t^5+1}$, $g(x)=x^9$, $\Phi(u,v)=u$, $H(t,x(t)) = \frac{2t^5 x^9(t)\sin(x(t))}{\left(t^{10}+t^5\right)\left(x^2(t)+1\right)}$

and $\dfrac{H(t,x(t))}{g(x(t))} = \dfrac{\sin(x(t))}{\left(t^{10}+t^5\right)} \leq \dfrac{1}{\left(t^{10}+t^5\right)} = p(t)$.

Taking $\rho(t) = \dfrac{t^5+1}{t^5}$ such that

$$\int_T^t \rho(s)[C_0 q(s) - p(s)]ds = \infty.$$

We get all conditions of Theorem 2 are satisfied and hence every solution of the given equation is oscillatory. The numerical solutions of the given differential equation are found out using the Runge Kutta method of fourth order (RK4).
We have

$$\ddot{x}(t) = f(t, x(t), \dot{x}(t)) = x^9(t)\sin(x(t)) + 2.49\left(x^9(t) + \frac{x^{27}(t)}{x^{27}(t) + (t^3 \dot{x}(t))^2}\right) \quad \text{with}$$

initial conditions $x(1) = -0.5$, $\dot{x}(1) = 0$ on the chosen interval [1,500], finding values of the functions r, q and f where we consider $H(t,x(t))=f(t)l(x)$ at $t=1$, $n=500$ and $h=0.198$.

Table 2. Numerical solution of ODE 2

K	t_k	$x(t_k)$
1	1	-0.5
2	1.198	-0.302
3	1.396	-0.1039
4	1.594	0.0942
5	1.792	0.2922
6	1.99	0.4903
.	.	.
16	3.97	-0.1495
17	4.168	-0.3465
18	4.366	-0.5434
.	.	.
27	6.148	0.042
28	6.346	0.2334
29	6.544	0.4248

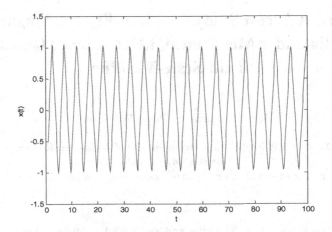

Fig. 2. Solution curve of ODE 2

Remark 2. Theorem 2 extends results of Bihari [2], Kartsatos [9], Grace and Lalli [6], E.M.El-Abbasy, T.S.Hassan and S.H.Saker [3] and Theorem 1 as well.

References

1. Atkinson, F.V.: On Second Order Non-linear Oscillations. Pacific. J. Math. 5, 643–647 (1955)
2. Bihari, I.: An Oscillation Theorem Concerning the Half Linear Differential Equation of the Second Order. Magyar Tud. Akad. Mat. Kutato Int. Kozl. 8, 275–280 (1963)
3. El-abbasy, E.M., Taher, S.H., Samir, H.S.: Oscillation of Second- order Non-linear Differential Equations with a Damping Term. Electron. J. Differential Equations 76, 1–13 (2005)
4. Elabbasy, E.M., Elhaddad, W.W.: Oscillation of Second Order Non-linear Differential Equations with a Damping Term. Electron. J. Qual. Theory Differ. Equ. 25, 1–19 (2007)
5. Fite, W.B.: Concerning the Zeros of the Solutions of Certain Differential Equations. Trans. Amer. Math. Soc. 19, 341–352 (1918)
6. Grace, S.R., Lalli, B.S.: Oscillation Theorems for Certain Second Perturbed Differential Equations. J. Math. Anal. Appl. 77, 205–214 (1980)
7. Greaf, J.R., Rankin, S.M., Spikes, P.W.: Oscillation Theorems for Perturbed Non-linear Differential Equations. J. Math. Anal. Appl. 65, 375–390 (1978)
8. Kamenev, V.: Integral Criterion for Oscillation of Linear Differential Equations of Second Order. Math. Zametki. 23, 249–251 (1978)
9. Kartsatos, A.G.: On Oscillations of Non-linear Equations of Second Order. J. Math. Anal. Appl. 24, 665–668 (1968)
10. Waltman, P.: An Oscillation Criterion for a Non-linear Second Order Equation. J. Math. Anal. Appl. 10, 439–441 (1965)
11. Wintner, A.: A Criterion of Oscillatory Stability. Quart. Appl. Math. 7, 115–117 (1949)
12. Yeh, C.C.: Oscillation Theorems for Non-linear Second Order Differential Equations with Damped Term. Proc. Amer. Math. Soc. 84, 397–402 (1982)

A Crack Arrest Study for Piezo-Electro-Magnetic Media under Mechanical, Electric and Magnetic Small-Scale-Yielding

R.R. Bhargava and Pooja Raj Verma

Department of Mathematics, Indian Institute of Technology Roorkee,
Roorkee 247667, India
rajrbfma@iitr.ernet.in, poojarajvs@gmail.com

Abstract. An anti-plane shear stress problem of crack arrest for a single crack weakening a piezo-electro-magnetic, unbounded, poled plate is investigated under remotely applied anti-plane mechanical loads and in-plane electromagnetic loads. Under applied loads the crack opens in self-similar fashion yielding mechanically, electrically and magnetically. The crack opening is arrested by prescribing yield-point mechanical, electrical and magnetic cohesive loads over the developed mechanical, saturation and induction zones rims, respectively. Two cases are considered when developed electrical zone is (i) bigger and (ii) equal to developed induction zone. The piezo-electro-magnetic ceramic being brittle the developed mechanical zones are assumed to be smallest. Solution is obtained using Fourier integral transforms. Closed-form analytic expressions are obtained for crack sliding displacement, crack opening potential drop, crack opening induction drop and energy release rate. Expressions are also obtained for determining length of each of developed mechanical, saturation and induction zones.

Keywords: Crack arrest, Induction zone, Piezo-electro-magnetic ceramic, Saturation zone, Slide yield zone.

1   Introduction

Piezo-electro-magnetic materials have many applications in engineering structure due to their electrical, magnetic and mechanical coupling effects. In recent years, the fracture of magnetoelectroelastic ceramics has been the focus area of researcher. Gao et al. [1], Zhao et al. [2] obtained the solutions for an elliptic cavity in a magneto-electro-elastic solid using Stroh formalism. The closed-form solution was obtained for an anti-plane crack in an infinite magneto-elastic composite using a conservative integral by Wang and Mai [3]. They [4] also discussed different electromagnetic boundary conditions on crack faces in magnetoelectroelastic ceramics which possessed coupled piezoelectric, piezomagnetic and electromagnetic effect. An interface crack problem for a layered magnetoelectroelastic strip of finite width was also addressed by them [5]. Mode II crack problem for weakening transversely isotropic magnetoelectroelastic media was investigated by Tupholme

P. Balasubramaniam and R. Uthayakumar (Eds.): ICMMSC 2012, CCIS 283, pp. 374–383, 2012.
© Springer-Verlag Berlin Heidelberg 2012

[6, 7]. Singh et al. [8] developed closed-form solutions for two collinear cracks in magneto-electro-elastic layer of finite thickness under anti-plane mechanical and in-plane electric and magnetic fields. Gao et al. [9] obtained a general solution for collinear cracks in a magneto-electro-elastic material under the permeable boundary conditions. Li et al. [10] developed mathematical model to find the exact solution for elliptical inclusion in magnetoelectroelastic ceramics. Zhong [11] obtained closed-form solutions for two collinear dielectric cracks in a magneto-electro-elastic solid.

The research done till date on piezo-electro-magnetic crack problems address the piezoelectric/piezomagnetic coupling effect, energy release rate when media is weakened by a single or double collinear straight cracks. Very little work is done on the crack arrest of such materials. The present work is an attempt to address this paucity.

A mathematical model is proposed to arrest crack opening in a piezo-electro-magnetic solid. The solution is obtained, using Fourier-transform technique. Expressions are derived to compute length of each of developed induction, saturation and yield zones. Closed form expression are derived for the crack opening displacement (COD), crack opening potential drop (COP) and crack opening induction drop (COI) at the crack tip for the first time. Energy release rate (ERR), for the first time, has also been computed. Cases examined: when developed saturation zone is bigger and other when it is equal to the developed induction zone. A case study is presented for both the case (i & ii) considered for $BaTO_3 - CoFe_2O_4$. The results confirm that the crack is arrested.

2   Fundamental Formulation

As are well-known the out-of-plane displacement, u_i, and in-plane electric field, E_i, and magnetic field , H_i, $(i=x, y, z)$ be defined as

$$u_x = 0, u_y = 0 \text{ and } u_z = w(x, y), \tag{1}$$

$$E_x = E_x(x, y), E_y = E_y(x, y) \text{ and } E_z = 0, \tag{2}$$

$$H_x = H_x(x, y), H_y = H_y(x, y) \text{ and } H_z = 0 . \tag{3}$$

The constitutive equations for transversely isotropic magneto-electro-elastic medium may be written as

$$\tau_{iz} = c_{44}w_{,i} + e_{15}\phi_{,i} + h_{15}\varphi_{,i} , \qquad D_i = e_{15}w_{,i} - \varepsilon_{11}\phi_{,i} - \beta_{11}\varphi_{,i},$$

$$B_i = h_{15}w_{,i} - \beta_{11}\phi_{,i} - \gamma_{11}\varphi_{,i}, \tag{4}$$

where τ_{iz}, D_i and B_i $(i = x, y)$ denote stress, electric-displacement and magnetic-induction, respectively. $c_{44}, e_{15}, h_{15}, \varepsilon_{11}, \beta_{11}$ and γ_{ij} denote elastic constant, piezoelectric coefficient, piezomagnetic coefficient, magnetic permeability, dielectric coefficient, respectively.

The stress equilibrium equations in the absence of body force, electric-displacement equilibrium equations in absence of charge and magnetic induction equilibrium equations in absence of magnetic force, respectively, are given as

$$\tau_{ij,i} = 0, \quad D_{i,i} = 0 \quad \text{and} \quad B_{i,i} = 0. \tag{5}$$

Divergence and gradient equations for strain, electric-displacement and magnetic induction are written as

$$\gamma_{ij} = \frac{1}{2}(u_{i,j} + u_{j,i}),\ E_i = -\phi,_i \text{ and } H_i = -\varphi,_i,\tag{6}$$

where $i, j = x, y$.

The governing equations may be written as

$$\nabla^2 u = 0.\tag{7}$$

where $u = [w, \phi, \varphi]^T$ and superscript T denotes the transpose of the matrix and ∇^2 is the Laplacian operator in two-dimension.

3   Methodology

Taking Fourier transform of the eq. (7), one obtains

$$\tilde{u}_{,yy} + \tilde{u} = 0,$$

where a subscripts comma denotes the partial differentiation with respect to variables following it and $\sim$ over the function denotes its Fourier transform and $q = -isy$, where s be the Fourier transform variable.

The solution of which may be written as

$$\tilde{u}^- = \begin{cases} Ce^{sy}, & s > 0 \\ Fe^{-sy}, & s < 0 \end{cases}, \tilde{u}^+ = \begin{cases} Ce^{sy}, & s < 0 \\ Fe^{-sy}, & s > 0 \end{cases},\tag{8}$$

where C and F are three component column, superscript $+$ and $-$ denote the value of the function on the rims of the crack as it is approached from $y > 0$ and $y < 0$ planes, respectively.

Denoting by $t = [\tau_{yz}, D_y, B_y]^T$ and taking Fourier transform of eq. (4) one arrives at

$$\tilde{t}(s, q) = -isB\tilde{u}(s, q),_q = -iB\tilde{u}(s, y),_y,\tag{9}$$

where

$$B = i\begin{bmatrix} c_{44} & e_{15} & h_{15} \\ e_{15} & -\varepsilon_{11} & -\beta_{11} \\ h_{15} & -\beta_{11} & -\gamma_{11} \end{bmatrix}.\tag{10}$$

The solution of eq. (9) may be written as

$$\tilde{t}^-(s, y) = \begin{cases} isBCe^{sy}, & s > 0 \\ -isBFe^{-sy}, & s < 0 \end{cases}, \tilde{t}^+(s, y) = \begin{cases} isBCe^{sy}, & s < 0 \\ -isBFe^{-sy}, & s > 0 \end{cases}\tag{11}$$

The traction continuity condition on $y = 0$ and $|x| < \infty$ using eq. (8) to eq. (11) yields

$$k\tilde{u}^+(s, 0) + k\tilde{u}^-(s, 0) = 0,\tag{12}$$

where $k = \begin{cases} isB, & s > 0 \\ -isB, & s < 0 \end{cases}$.

Introducing jump, $\Delta u(x)$, in displacement, electric displacement and magnetic induct-ion as

$$\Delta u(x) = u^+(x, 0) - u^-(x, 0),\tag{13}$$

and dislocation, dipole density and magnetic induction vector as

$$f(x) = \{f_1(x), f_2(x), f_3(x)\}^T = \frac{d}{dx}\Delta u(x),$$ (14)

and taking Fourier transform of eq. (13) and using eq. (14), one obtains

$$\Delta u(x) = \frac{i}{s}\int_{-a}^{a} f(x)e^{isx}dx.$$ (15)

eq.(14) and eq. (15) may also be written as

$$\tilde{t}(s,0) = -\frac{1}{2}k\Delta\tilde{u}(x),$$ (16)

Inverse Fourier transform of the above equation yields a system of integral equations to determine $f(x)$

$$-\frac{B}{\pi}\int_{-a}^{a}\frac{f(t)}{t-x}dt = t(x), \qquad |x| < \infty, y = 0.$$ (17)

4   The Problem

An unbounded transversely isotropic piezo-electro-magnetic plate occupies the xoy-plane and is thick enough along oz direction to allow anti-plane strain case. The plate is electrically and magnetically poled along oz-direction. A thorough internal hairline, quasi-static straight crack weakens the plate and occupies the interval [-a, a] on x-axis. The crack is assumed to be electrically and magnetically impermeable. The mechanical load $\tau_{yz} = \tau_\infty$,electrical-displacement load $D_y = D_\infty$ and magnetic induction $B_y = B_\infty$ prescribed at remote boundary open the crack in self-similar fashion. Consequently, under small-scale yielding a mechanical yield–slide zone, a electric-saturation zone and a magnetic induction zone protrude ahead of each tip of the crack. All the zones developed are assumed to lie in a thin strip along the length of the crack. The crack sliding zones occupy the plane $y=0$ and $a \le |x| \le b$, the saturation zones occupy the intervals $a \le |x| \le d$ and induction zones occupy the intervals $a \le |x| \le c$ all on x-axis. To arrest the crack from further opening the developed crack-slide yield zones are subjected to cohesive yield point anti-plane shear stress $\tau_{yz} = \tau_s$; the rims of saturation zones are subjected to normal closing saturation limit electrical displacement $D_y = D_s$ and induction zone rims are prescribed normal cohesive polarization limit magnetic induction $B_y = B_s$. For the present problem it is assumed that $b < c < d$ and c = d.

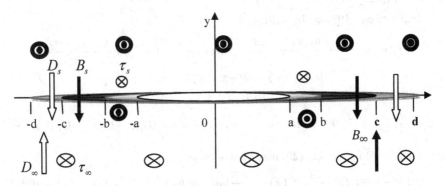

Fig. 1. Schematic representation of the configuration of the problem

5   The Solution of the Problem

The solutions obtained for two cases considered, are presented below.

5.1   Case (i) When Saturation Zone Exceeds the Developed Zones ($d > c > b$)

For this case $f_1(x) = 0$ for $|x| > b$; $f_2(x) = 0$ for $|x| > d$ and $f_3(x) = 0$ for $|x| > c$. The solution for the case may be written as follows using eq. (17) as

$$\frac{1}{\pi} \int_{-b}^{b} \frac{f_1(t)}{t-x} dt = G_{1j} t_j(x) \quad \text{for } |x| > b, \tag{18}$$

$$\frac{1}{\pi} \int_{-c}^{c} \frac{f_3(t)}{t-x} dt = G_{3j} t_j(x) \quad \text{for } |x| > c, \tag{19}$$

and

$$\frac{1}{\pi} \int_{-d}^{d} \frac{H_{2j} f_j(t)}{t-x} dt = t_2(x) \quad \text{for } |x| > d, \quad j = 1,2,3, \tag{20}$$

and
$$G = H^{-1} = \begin{bmatrix} C_{44} & e_{15} & h_{15} \\ e_{15} & -\varepsilon_{11} & -\beta_{11} \\ h_{15} & -\beta_{11} & -\gamma_{11} \end{bmatrix}^{-1}.$$

Since $f_1(x)$ is non-singular at $|x| = b$ consequently eq. (18) will have solution, (using Barnett and Asaro [12]), if

$$\int_{-b}^{b} \frac{G_{1j} t_j(t)}{\sqrt{b^2 - t^2}} dt = 0.$$

Solving this leads to a relation for determining yield-slide zone length as

$$\frac{b}{a} = \sec\left(\frac{\pi}{2} \frac{G_{11}\tau_\infty + G_{12}D_\infty + G_{13}B_\infty}{G_{11}\tau_s + G_{12}D_s + G_{13}B_s}\right). \tag{21}$$

Using this condition the solution of eq. (18) may be written as

$$f_1(x) = \frac{G_{11}\tau_s + G_{12}D_s + G_{13}B_s}{\pi} [\omega(x,a,b) - \omega(-x,a,b)], \tag{22}$$

where $\omega(x,a,b) = \cosh^{-1}\left(\frac{b^2 - ax}{b(a-x)}\right)$.

Solution of eq. (19) may be written as

$$f_3(x) = \frac{G_{31}\tau_s + G_{32}D_s + G_{33}B_s}{\pi} [\omega(x,a,c) - \omega(-x,a,c)] + \frac{G_{31}(\tau_\infty - \tau_s)}{\pi}$$
$$[\omega(x,b,c) - \omega(-x,b,c)]. \tag{23}$$

provided

$$\cos^{-1}\frac{a}{c}(G_{31}\tau_s + G_{32}D_s + G_{33}B_s) + G_{31}(\tau_\infty - \tau_s)\cos^{-1}\frac{b}{c} = \frac{\pi}{2}(G_{31}\tau_\infty + G_{32}D_\infty + G_{33}B_\infty) \cdot \tag{24}$$

Analogously solution of eq. (20) may be written as

$$f_2(x) = -\frac{H_{21}}{H_{22}} f_1(x) - \frac{H_{31}}{H_{22}} f_3(x) + \frac{D_s}{\pi H_{22}} [\omega(x,a,b) + \omega(-x,a,b)], \quad |x| < d. \tag{25}$$

Under the constrain

$$\frac{d}{a} = \sec\left(\frac{\pi}{2}\frac{D_\infty}{D_s}\right). \tag{26}$$

5.1.1   Applications for Case(i)

The crack sliding displacement,$\Delta w(x)$, across the crack is given by

$$\Delta w(x) = -\int_{-b}^{x} f_1(x)dx \tag{27}$$

$$= \frac{1}{\pi}(G_{11}\tau_s + G_{12}D_s + G_{13}B_s)[(a-x)\omega(x,a,b) + (a+x)\omega(-x,a,b)]. \tag{28}$$

Analogously the magnetic induction drop ,$\Delta\varphi(x)$, across the crack face is calculated form

$$\Delta\varphi(x) = -\int_{-d}^{x} f_3(x)dx \tag{29}$$

$$= \frac{1}{\pi}[(G_{31}\tau_s + G_{32}D_s + G_{33}B_s)\{(a-x)\omega(x,a,b) + (a+x)\omega(-x,a,b)\}] + (G_{31}(\tau_\infty - \tau_s))\{(b-x)\omega(x,b,c) + (b+x)\omega(-x,b,c)\}, \tag{30}$$

and electric potential drop, $\Delta\phi(x)$, is calculated as

$$\Delta\phi(x) - \int_{-c}^{x} f_2(x)dx \tag{31}$$

$$= -\frac{H_{21}}{H_{22}}\Delta w(x) - \frac{H_{31}}{H_{22}}\Delta\varphi(x) + \frac{D_s}{\pi H_{22}}[(a-x)\omega(x,a,b) + (a+x)\omega(-x,a,b)]. \tag{32}$$

The energy release rate (ERR),J_a, at the crack tip $x = 0$ is calculated using the definition

$$J_a = \tau_s\Delta w(a) + D_s\Delta\phi(x) + B_s\Delta\varphi(a). \tag{33}$$

5.2   Case (ii) When Developed Saturation and Induction Zones Are of Equal Lengths (c=d >b)

For this case $f_1(x) = 0$ for $|x| > b$ and each of $f_2(x)$and $f_3(x)$are zero for $|x| > c$.

The equation for determining $f_1(x)$ remains the same as in cases (i) i.e. eq. (22). Hence the solution, the yield – slide zone and the crack-sliding displacement remains same as derived in eq. (21) and eq. (28), respectively.

And the equation for determining $f_2(x)$ and $f_3(x)$ are given as

$$\frac{1}{\pi}\int_{-c}^{c}\frac{H_{2j}f_j(t)}{t-x}dt = t_2(x), \quad |x| < c, \tag{34}$$

$$\frac{1}{\pi}\int_{-c}^{c}\frac{H_{3j}f_j(t)}{t-x}dt = t_3(x), \quad |x| < c, j = 1,2,3. \tag{35}$$

The solution of eq. (34), eq. (35) may be obtained under the condition

$$\frac{c}{a} = \sec\left(\frac{\pi}{2}\frac{D_\infty}{D_s}\right) = \sec\left(\frac{\pi}{2}\frac{B_\infty}{B_s}\right), \tag{36}$$

and may be written as; for dipole density function, $f_2(x)$,

$$f_2(x) = \left(\frac{H_{31}H_{23} - H_{21}H_{33}}{H_{33}H_{23} - H_{32}H_{33}}\right) f_1(x) + \left(\frac{D_s H_{33} - B_s H_{23}}{\pi(H_{33}H_{23} - H_{32}H_{33})}\right) \times$$
$$\{\omega(x, a, c) - \omega(-x, a, c)\}. \tag{37}$$

And the magnetic induction function, $f_3(x)$, may be expressed as

$$f_3(x) = \left\{\frac{B_s}{\pi H_{33}} - \left(\frac{H_{32}}{H_{33}\pi}\left(\frac{D_s H_{33} - B_s H_{23}}{(H_{33}H_{23} - H_{32}H_{33})}\right)\right)\right\}[\omega(x, a, c) - \omega(-x, a, c)]$$
$$- \left\{\frac{H_{31}}{H_{33}} + \frac{H_{32}}{H_{33}}\left(\frac{H_{31}H_{23} - H_{21}H_{33}}{H_{33}H_{23} - H_{32}H_{33}}\right)\right\} f_1(x). \tag{38}$$

5.2.1   Applications for Case (ii)

The crack opening potential drop, $\Delta\phi(x)$, for this case across the crack face is given using eq. (29)

$$\Delta\phi(x) = \left(\frac{H_{31}H_{23} - H_{21}H_{33}}{H_{33}H_{23} - H_{32}H_{33}}\right) \Delta w(x) + \left(\frac{D_s H_{33} - B_s H_{23}}{\pi(H_{33}H_{23} - H_{32}H_{33})}\right) \times$$
$$[(a - x)\omega(x, a, c) + (a + x)\omega(-x, a, c)]. \tag{39}$$

and crack opening induction drop, $\Delta\varphi(x)$, across the crack face may be written using eq. (31) as

$$\Delta\varphi(x) = \left\{\frac{B_s}{\pi H_{33}} - \left(\frac{H_{32}}{H_{33}\pi}\left(\frac{D_s H_{33} - B_s H_{23}}{(H_{33}H_{23} - H_{32}H_{33})}\right)\right)\right\}[(a - x)\omega(x, a, c)$$
$$+ (a + x)\omega(-x, a, c)] - \left\{\frac{H_{31}}{H_{33}} + \frac{H_{32}}{H_{33}}\left(\frac{H_{31}H_{23} - H_{21}H_{33}}{H_{33}H_{23} - H_{32}H_{33}}\right)\right\}\Delta w(x). \tag{40}$$

The energy release rate at the crack tip is calculated using eq. (33) and substituting from eq. (28), eq. (39) and eq. (40) respectively, the values of $\Delta w, \Delta\phi$ and $\Delta\varphi$ at the crack tip x = a.

6   Case Study

The analytic results obtained above are used to demonstrate the crack opening arrest for $BaTiO_3 - CoFe_2O_4$. The material constants of which are listed below in Table 1. These are taken from [13].

The study has been carried out for the case when a crack of length $2a = 1$ weakens the plate. In every figure the graphs have been plotted for case (i) when saturation zone is bigger than induction zone (shown by a continuous line) and case (ii) when both saturation and induction zones are of equal length (shown by a dotted line).

Table 1. Material constant for $BaTiO_3 - CoFe_2O_4$

$c_{44}(10^9 N/m^2)$	$e_{15}(C/m^2)$	$h_{15}N/Am$	$\varepsilon_{11}(10^{-10}C/Nm^2)$	β_{11}	$\gamma_{11}(10^{-6}NS^2/C^2)$
44	5.8	275	56.4	0	297

Fig. 2 depicts the normalized crack opening potential drop (COP) versus saturation zone to half-crack length ratio. It may be observed from the figure as the saturation zone is increased then COP reduces. Also for case (ii) COP has higher values but the trend remains the same.

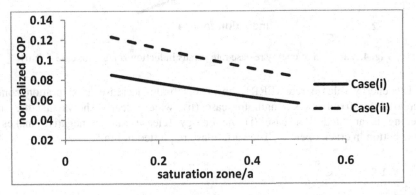

Fig. 2. Variation of crack opening potential drop versus saturation zone/a for case (i & ii)

Variation of crack opening induction drop (COI) with respect to induction zone to half-crack length ratio is plotted in Fig. 3. For case (i) the COI shows a negative decrease although for case (ii) it shows an increase, as expected.

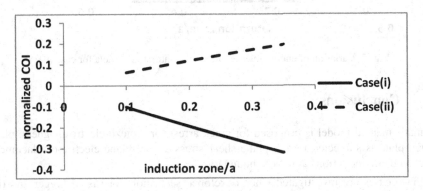

Fig. 3. Variation of crack opening induction drop versus induction zone/a for case (i & ii)

Fig. 4 show a sharp drop in ERR for case (i) and for case (ii) although ERR is positive but shows a decline as induction zone length is increased for fixed crack length. Thus it assists in crack opening arrest.

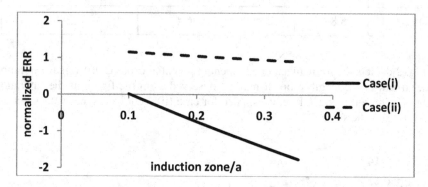

Fig. 4. Variation of energy release rate versus induction zone/a for case (i & ii)

The energy release rate (ERR) shows a definite decrease as saturation zone is increased for fixed crack length for case (ii), which clearly shows that the crack opening is arrested. For case (i) the energy release rate is negative makes no contribution in crack opening. The variations are plotted in Fig.5.

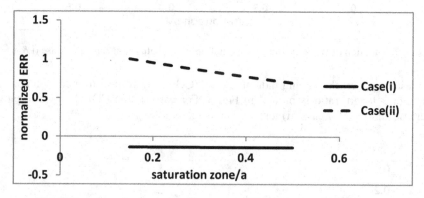

Fig. 5. Variation of energy release rate versus saturation zone/a for case (i & ii)

7   Conclusions

A mathematical model is proposed for crack arrest for a piezo-electro-magnetic plate when plate is subjected to anti-plane shear stress and in-plane electric-displacement and magnetic inductions at remote boundaries.

- Two cases are investigated when developed saturation zone is (i) bigger and (ii) equal to developed induction zone while developed slide zone is always considered being the smallest.

- The solutions are obtained analytically and closed form expression are also derived for saturation zone, induction zone, CSD, COP, COI, ERR.
- Quantities: COP, COI, ERR effecting the crack arrest are plotted against affecting parameters namely saturation zone, induction zone. The results confirm that present model is capable of crack arrest under small-scale yielding.

Acknowledgements. The authors are grateful to **Prof. R. D. Bhargava** (Sr. Prof. and Head (retd.), Indian Institute of Technology Bombay, Mumbai) for continuous encouragement during course of this work. The second author is thankful to MHRD for the financial support.

References

1. Gao, C.F., Tong, P., Zhang, T.Y.: Fracture mechanics for a mode III crack in a magnetoelectroelastic solid. Internat. J. Solids Struct. 41, 6613–6629 (2004)
2. Zhao, M.H., Wang, H., Yang, F., Liu, T.: A magnetoelectroelastic medium with an elliptical cavity under combined mechanical–electric–magnetic loading. Theoretical Appl. Fracture Mech. 45, 227–237 (2006)
3. Wang, B.L., Mai, Y.W.: Fracture of piezo-electro-magnetic materials. Mech. Res. Commun. 35, 65–73 (2004)
4. Wang, B.L., Mai, Y.W.: Applicability of the crack-face electromagnetic boundary conditions for fracture of magnetoelectroelastic materials. Internat. J. Solids Struct. 44, 387–398 (2007)
5. Wang, B.L., Mai, Y.W.: An Exact analysis for mode III cracks between two dissimilar magneto-electro-elastic layers. Mech. Composite Materials 44, 533–548 (2008)
6. Tupholme, G.E.: Crack in a transversely isotropic magnetoelectroelastic half-space. Mech. Res. Commun. 35, 466–474 (2008)
7. Tupholme, G.E.: Moving anti-plane shear crack in transversely isotropic magnetoelectroelastic media. Acta Mechanica 202, 153–162 (2009)
8. Singh, B.M., Rokne, J., Dhaliwal, R.S.: Closed-form solutions for two anti-plane collinear cracks in a magnetoelectroelastic layer. European J. Mech. A Solids 28, 599–609 (2009)
9. Gao, C.F., Kessler, H., Balke, H.: Crack problems in magnetoelectroelastic solids. Part II: general solution of collinear cracks. Internat. J. Engg. Sci. 41, 983–994 (2003)
10. Li, G., Wang, B.L., Han, J.C.: Exact solution for elliptical inclusion in magnetoelectroelastic materials. Internat. J. Solids Struct. 47, 419–426 (2010)
11. Zhong, X.C.: Closed-form solutions for two collinear dielectric cracks in a magnetoelectroelastic solid. Appl. Math. Modelling 35, 2930–2940 (2011)
12. Barnett, D.M., Asaro, R.J.: The fracture mechanics of slit-like cracks in anisotropic elastic media. J. Mech. Phys. Solids 20, 353–366 (1972)
13. Sih, G.C., Song, Z.F.: Magnetic and electric poling effects associated with crack growth in composite $BaTiO_3 - CoFe_2O_4$. Theoretical Appl. Fracture Mech. 39, 209–222 (2003)

Solution and Stability of n-Dimensional Quadratic Functional Equation

M. Arunkumar[1], S. Murthy[2], and G. Ganapathy[3]

[1] Department of Mathematics, Government Arts College,
Tiruvannamalai - 606 603, Tamil Nadu, India
[2] Department of Mathematics, Government Arts College for Men,
Krishnagiri-635 001, Tamil Nadu, India
[3] Department of Mathematics, R.M.D. Engineering College,
Kavaraipettai - 601 206, Tamil Nadu, India
{annarun2002,smurthy07,ganagandhi}@yahoo.co.in

Abstract. In this paper, we investigate the general solution and generalized Ulam-Hyers stability of a $n-$dimensional quadratic functional equation

$$\sum_{i=1}^{n} g\left(\sum_{j=1}^{i} x_j\right) - \sum_{i=1}^{n}(n-i+1)g\left(x_i\right)$$

$$= \frac{1}{2}\sum_{i=1}^{n-1}(n-i)\sum_{j=1}^{i}\left(g(x_j + x_{i+1}) - g(x_j - x_{i+1})\right)$$

with $n \geq 2$. Counterexamples for non-stability cases are also studied through out this paper.

Keywords: quadratic functional equation, Generalized Ulam-Hyers stability, JM Rassias stability.

1 Introduction

S.M. Ulam, in his famous lecture in 1940 to the Mathematics Club of the University of Wisconsin, presented a number of unsolved problems. This is the starting point of the theory of the stability of functional equations. One of the questions led to a new line of investigation, nowadays known as the stability problems. Ulam [16] discusses:
... the notion of stability of mathematical theorems considered from a rather general point of view: When is it true that by changing a *little* the hypothesis of a theorem one can still assert that the thesis of the theorem remains true or *approximately* true? ...

For very general functional equations one can ask the following question. When is it true that the solution of an equation differing slightly from a given one, must of necessity be close to the solution of the given equation? Similarly, if we replace a given functional equation by a functional inequality, when can one

P. Balasubramaniam and R. Uthayakumar (Eds.): ICMMSC 2012, CCIS 283, pp. 384–394, 2012.
© Springer-Verlag Berlin Heidelberg 2012

assert that the solutions of the inequality lie near to the solutions of the strict equation?

Suppose G is a group, $H(d)$ is a metric group, and $f : G \to H$. For any $\epsilon > 0$, does there exist a $\delta > 0$ such that

$$d(f(xy), f(x)f(y)) < \delta$$

holds for all $x, y \in G$ and implies there is a homomorphism $M : G \to H$ such that

$$d(f(x), M(x)) < \epsilon$$

for all $x \in G$?

If the answer is affirmative, then we say that the Cauchy equation M is stable. These kinds of questions form the basics of stability theory, and D.H. Hyers [8] obtained the first important result in this field. Many examples of this have been solved and many variations have been studied since.

In 1941, D. H. Hyers [8] gave an affirmative answer to the question of S.M. Ulam for Banach spaces. In 1950, T. Aoki [2] was the second author to treat this problem for additive mappings. In 1978, Th.M. Rassias [13] succeeded in extending Hyers' Theorem by weakening the condition for the Cauchy difference controlled by $(||x||^p + ||y||^p)$, $p \in [0, 1)$, to be unbounded.

In 1982, J.M. Rassias [11] replaced the factor $||x||^p + ||y||^p$ by $||x||^p ||y||^q$ for $p, q \in R$. A generalization of all the above stability results was obtained by P. Gavruta [7] in 1994 by replacing the unbounded Cauchy difference by a general control function $\varphi(x, y)$.

In 2008, a special case of Gavruta's theorem for the unbounded Cauchy difference was obtained by Ravi etal., [14] by considering the summation of both the sum and the product of two $p-$ norms. The stability problems of several functional equations have been extensively investigated by a number of authors and there are many interesting results concerning this problem (see [3], [10], [14]) and reference cited there in. Several investigations followed, and almost all functional equations are *stabilized*.

The basic algebraic (norm) condition that makes the normed linear space an inner product space is the parallelogram identity, also known as the Jordan.von Neumann identity (or the Appolonius law or norm equation),

$$||x + y||^2 + ||x - y||^2 = 2 ||x||^2 + 2 ||y||^2$$

for $x, y \in V$, where V is an normed linear space. This translates into a functional equation well known as the quadratic functional equation,

$$g(x + y) + g(x - y) = 2g(x) + 2g(y) \tag{1}$$

for $x, y \in V$, where V is a linear space.

A function $g : E_1 \to E_2$ between two vector spaces is quadratic if and only if there exists a unique symmetric biadditive function $g(x) = B(x, x)$ for any $x \in E_1$ (see [1],[10]). The biadditive function B is given by

$$B(x, y) = \frac{1}{4}[g(x + y) + g(x - y)].$$

The Hyers - Ulam stability of the quadratic functional equation (1) was proved by F. Skof [15] in 1983 for the functions $g : E_1 \to E_2$ where E_1 is a normed space and E_2 is a Banach space. In 1984, P.W. Cholewa [5] demonstrated that the result of Skofs theorem is still valid if the relevant domain E_1 is replaced by an Abelian group G. Skof's result was further extended by S. Czerwik [6] to the Hyers-Ulam-Rassias stability of the quadratic functional equation (1).

K.W. Jun and H.M. Kim [9] introduced the following generalized **quadratic and additive type functional equation**

$$g\left(\sum_{i=1}^{n} x_i\right) + (n-2)\sum_{i=1}^{n} g(x_i) = \sum_{1 \le i < j \le n} g(x_i + x_j) \qquad (2)$$

in the class of function between real vector spaces. For $n = 3$, Pl. Kannappan proved that a function g satisfies the functional equation (2) if and only if there exists a symmetric bi-additive function A and additive function B such that $g(x) = B(x,x) + A(x)$ for all x (see [10]). The Hyers-Ulam stability for the equation (2) when $n = 3$ was proved by S.M. Jung [12]. The Hyers-Ulam-Rassias stability for the equation (2) when $n = 4$ was also investigated by I.S. Chang et al., [4].

Consider the following elementary equality

$$\|x_1\|^2 + \|x_1 + x_2\|^2 + \cdots + \|x_1 + x_2 + x_3 + \cdots + x_n\|^2 - \left(n\|x_1\|^2 + (n-1)\|x_2\|^2\right.$$
$$\left. + \cdots + \|x_n\|^2\right) = \frac{1}{2}\sum_{i=1}^{n-1}(n-i)\sum_{j=1}^{i}\left(\|x_j + x_{i+1}\|^2 - \|x_j - x_{i+1}\|^2\right). \qquad (3)$$

The above equality (3) can be formulated by the following functional equation

$$\sum_{i=1}^{n} g\left(\sum_{j=1}^{i} x_j\right) - \sum_{i=1}^{n}(n-i+1)g(x_i)$$
$$= \frac{1}{2}\sum_{i=1}^{n-1}(n-i)\sum_{j=1}^{i}(g(x_j + x_{i+1}) - g(x_j - x_{i+1})) \qquad (4)$$

with $n \ge 2$, having the solution $g(x) = cx^2$.

In this paper, we study the solution and stability of the above functional equation (4). In Section 2 authors present the general solution of the quadratic functional equation (4). The generalized Ulam-Hyers stability of the quadratic functional equation (4) is given in Section 3. Counter-examples for non stability case is also discussed in this section.

2   General Solution

In this section, we discuss the general solution of the functional equation (4) by considering X and Y are real vector spaces.

Theorem 1. If $g : X \to Y$ satisfies the functional equation (1) for all $x, y \in X$ if and only if g satisfies the functional equation (4) for all $x_1, x_2, x_3, \ldots, x_n \in X$ and $n \geq 2$.

Proof. Assume $g : X \to Y$ satisfies the functional equation (1). Letting $x = y = 0$ in (1), we get $g(0) = 0$. Setting y by x and $-x$ in (1) respectively, we arrive $g(2x) = 4g(x)$ and $g(-x) = g(x)$ for all $x \in X$. Replacing y by $2x$, we get $g(3x) = 9g(x)$ for all $x \in X$. In general, for any positive integer n, $g(nx) = n^2 g(x)$ for all $x \in X$. Replacing $x = x_1, y = x_2$ in equation (1) we obtain

$$g(x_1 + x_2) + g(x_1 - x_2) = 2g(x_1) + 2g(x_2) \tag{5}$$

for all $x_1, x_2 \in X$. Adding $g(x_1)$ on both side and rearranging we get,

$$g(x_1) + g(x_1 + x_2) - 2g(x_1) - g(x_2) = g(x_1) + g(x_2) - g(x_1 - x_2)$$

for all $x_1, x_2 \in X$. Using (1) we obtain,

$$g(x_1) + g(x_1 + x_2) - 2g(x_1) - g(x_2) = \frac{1}{2}\left(g(x_1 + x_2) + g(x_1 - x_2)\right) - g(x_1 - x_2) \tag{6}$$

for all $x_1, x_2 \in X$. Then we get,

$$\sum_{i=1}^{2} g\left(\sum_{j=1}^{i} x_j\right) - \sum_{i=1}^{2}(2 - i + 1)g(x_i)$$

$$= \frac{1}{2}\sum_{i=1}^{2-1}(2 - i)\sum_{j=1}^{i}\left(g(x_j + x_{i+1}) - g(x_j - x_{i+1})\right) \tag{7}$$

for all $x_1, x_2 \in X$. Since g satisfies equation (2)(see Theorem 2.1[12]), keeping $n = 3$ in equation (2) and rearranging we have,

$$g(x_2 + x_2 + x_3) - g(x_1) - g(x_2) - g(x_3) = \frac{1}{2}\left[(g(x_1 + x_2) - g(x_1 - x_2))\right.$$
$$\left. + (g(x_1 + x_3) - g(x_1 - x_3)) + (g(x_2 + x_3) - g(x_2 - x_3))\right] \tag{8}$$

for all $x_1, x_2, x_3 \in X$. Adding (7) and (8) we arrive,

$$\sum_{i=1}^{3} g\left(\sum_{j=1}^{i} x_j\right) - \sum_{i=1}^{3}(3 - i + 1)g(x_i)$$

$$= \frac{1}{2}\sum_{i=1}^{3-1}(3 - i)\sum_{j=1}^{i}\left(g(x_j + x_{i+1}) - g(x_j - x_{i+1})\right) \tag{9}$$

for all $x_1, x_2, x_3 \in X$. Proceeding in the same manner, we arrive the equation (4) for all $x_1, x_2, x_3, \ldots, x_n \in X$ and $n \geq 2$. Hence g satisfies the functional equation (4) for all $x_1, x_2, x_3, \ldots, x_n \in X$.

Conversely, assume $g : X \to Y$ satisfies the functional equation (4). Letting $x_1 = x_2 = x_3 = \cdots = x_n = 0$ in (4), we get $g(0) = 0$. Setting $(x_1, x_2, x_3, \ldots, x_n,)$ by $(x, x, 0, \ldots, 0)$ and $(x, -x, 0, \ldots, 0)$ in (4), we obtain $g(2x) = 4g(x)$ and $g(-x) = g(x)$, respectively for all $x \in X$. Setting $(x_1, x_2, x_3, \ldots, x_n,)$ by $(x, 2x, 0, \ldots, 0)$ in (4), we obtain $g(3x) = 9g(x)$ for all $x \in X$. In general, for any positive integer n, $g(nx) = n^2 g(x)$ for all $x \in X$. Replacing $(x_1, x_2, x_3, \ldots, x_n,)$ by $(x, y, 0, \ldots, 0)$ in (4), we arrive (1) for all $x, y \in X$. Therefore g satisfies the functional equation (1) for all $x, y \in X$.

3   Stability Results

In this section, let X be a normed space and Y be a Banach space. We investigate the generalized Ulam -Hyers stability of the functional equation (4). Define a mapping $g : X \to Y$ by

$$D\, g(x_1, x_2, x_3, \ldots, x_n) = \sum_{i=1}^{n} g\left(\sum_{j=1}^{i} x_j\right) - \sum_{i=1}^{n}(n - i + 1)g\,(x_i)$$

$$-\frac{1}{2}\sum_{i=1}^{n-1}(n - i)\sum_{j=1}^{i}\left(g(x_j + x_{i+1}) - g(x_j - x_{i+1})\right)$$

for all $x_1, x_2, x_3, \ldots, x_n \in X$ and $n \geq 2$.

Theorem 2. Let $j = \pm 1$. Let $\phi : X^n \to [0, \infty)$ be a function such that

$$\sum_{\ell=0}^{\infty} \frac{\phi\left(2^{\ell j}x_1, 2^{\ell j}x_2, 2^{\ell j}x_3, \ldots, 2^{\ell j}x_n\right)}{4^{\ell j}} \quad converges$$

$$and \quad \lim_{\ell \to \infty} \frac{\phi\left(2^{\ell j}x_1, 2^{\ell j}x_2, 2^{\ell j}x_3, \ldots, 2^{\ell j}x_n\right)}{4^{\ell j}} = 0 \tag{10}$$

for all $x_1, x_2, x_3, \ldots, x_n \in X$, $n \geq 2$ and let $g : X \to Y$ be a function satisfying the inequality

$$\|D\, g(x_1, x_2, x_3, \ldots, x_n)\| \leq \phi\,(x_1, x_2, x_3, \ldots, x_n) \tag{11}$$

for all $x_1, x_2, x_3, \ldots, x_n \in X$. Then there exists a unique quadratic function $Q : X \to Y$ such that

$$\|g(x) - Q(x)\| \leq \frac{1}{2(n - 1)} \sum_{k=\frac{1-j}{2}}^{\infty} \frac{\beta(2^{kj}x)}{4^{kj}} \tag{12}$$

where $\beta(2^{kj}x) = \phi(2^{kj}x, 2^{kj}x, \underbrace{0, \ldots, 0}_{n-2\ times})$ for all $x \in X$. The function $Q(x)$ is defined by

$$Q(x) = \lim_{\ell \to \infty} \frac{g(2^{\ell j}x)}{4^{\ell j}} \tag{13}$$

for all $x \in X$.

Proof. Assume $j = 1$. Replacing $(x_1, x_2, x_3, \ldots, x_n)$ by $(x, x, \underbrace{0, \ldots, 0}_{n-2\ times})$ in (11)

and dividing by $2(n-1)$, we get

$$\left\| \frac{g(2x)}{4} - g(x) \right\| \leq \frac{\phi\left(x, x, \underbrace{0, \ldots, 0}_{n-2\ times}\right)}{2(n-1)} \tag{14}$$

for all $x \in X$. Letting $\beta(x) = \phi\left(x, x, \underbrace{0, \ldots, 0}_{n-2\ times}\right)$ in (14), we arrive

$$\left\| \frac{g(2x)}{4} - g(x) \right\| \leq \frac{\beta(x)}{2(n-1)} \tag{15}$$

for all $x \in X$. Now replacing x by $2x$ and dividing by 4 in (15) and adding the resultant inequality with (15), we obtain

$$\left\| \frac{g(2^2 x)}{4^2} - g(x) \right\| \leq \frac{1}{2(n-1)} \left[\beta(x) + \frac{\beta(2x)}{4} \right] \tag{16}$$

for all $x \in X$. In general for any positive integer ℓ, we have

$$\left\| \frac{g(2^\ell x)}{4^\ell} - g(x) \right\| \leq \frac{1}{2(n-1)} \sum_{k=0}^{\ell-1} \frac{\beta(2^k x)}{4^k} \tag{17}$$

$$\leq \frac{1}{2(n-1)} \sum_{k=0}^{\infty} \frac{\beta(2^k x)}{4^k}$$

for all $x \in X$. In order to prove the convergence of the sequence $\left\{ \frac{g(2^\ell x)}{4^\ell} \right\}$,

replace x by $2^m x$ and dividing by 4^m in (17), for any $m, \ell > 0$, we arrive

$$\left\| \frac{g(2^{\ell+m} x)}{4^{(\ell+m)}} - \frac{g(2^m x)}{4^m} \right\| = \frac{1}{4^m} \left\| \frac{g(2^\ell 2^m x)}{4^\ell} - g(2^m x) \right\|$$

$$\leq \frac{1}{2(n-1)} \sum_{k=0}^{n-1} \frac{\beta(2^{k+m} x)}{4^{k+m}}$$

$$\leq \frac{1}{2(n-1)} \sum_{k=0}^{\infty} \frac{\beta(2^{k+m} x)}{4^{k+m}}$$

$$\to 0 \quad as \quad m \to \infty \tag{18}$$

for all $x \in X$. Hence the sequence $\left\{ \frac{g(2^\ell x)}{4^\ell} \right\}$ is Cauchy sequence. Since Y is complete, there exists a mapping $Q : X \to Y$ such that

$$Q(x) = \lim_{\ell \to \infty} \frac{g(2^\ell x)}{4^\ell} \quad \forall \; x \in X.$$

Letting $\ell \to \infty$ in (17), we see that (12) holds for all $x \in X$. To prove Q satisfies (4), replacing $(x_1, x_2, x_3, \ldots, x_n)$ by $(2^\ell x_1, 2^\ell x_2, 2^\ell x_3, \ldots, 2^\ell x_n)$ and divided by 4^ℓ in (11), we arrive

$$\frac{1}{4^\ell} \left\| Dg \left(2^\ell x_1, 2^\ell x_2, 2^\ell x_3, \ldots, 2^\ell x_n \right) \right\| \le \frac{1}{4^\ell} \phi \left(2^\ell x_1, 2^\ell x_2, 2^\ell x_3, \ldots, 2^\ell x_n \right)$$

for all $x_1, x_2, x_3, \ldots, x_n \in X$. Letting $\ell \to \infty$ and in the above inequality, we see that

$$\| DQ \left(x_1, x_2, x_3, \ldots, x_n \right) \| = 0$$

Hence Q satisfies (4) for all $x_1, x_2, x_3, \ldots, x_n \in X$ and $n \ge 2$. To prove Q is unique, let $R(x)$ be another quadratic mapping satisfying (4) and (12). Then

$$\| Q(x) - R(x) \| = \frac{1}{4^n} \| Q(2^n x) - R(2^n x) \|$$

$$\le \frac{1}{4^\ell} \left\{ \| Q(2^\ell x) - g(2^\ell x) \| + \| g(2^\ell x) - R(2^\ell x) \| \right\}$$

$$\le \sum_{k=0}^{\infty} \frac{\beta(2^{k+\ell} x)}{4^{(k+\ell)}}$$

$$\to 0 \quad as \quad \ell \to \infty$$

for all $x \in X$. Hence Q is unique.

For $j = -1$, we can prove the similar type of stability result. This completes the proof of the theorem.

The following corollary is a immediate consequence of Theorems 2 concerning the stability of (4).

Corollary 3. Let λ and s be nonnegative real numbers. If a function $g : X \to Y$ satisfying the inequality

$$\| D\, g(x_1, x_2, x_3, \cdots, x_n) \| \le \begin{cases} \lambda, & \\ \lambda \sum\limits_{k=1}^{n} \|x_k\|^s, & s < 2 \quad or \quad s > 2; \\ \lambda \left\{ \sum\limits_{k=1}^{n} \|x_k\|^{ns} + \prod\limits_{k=1}^{n} \|x_k\|^s, \right\}, & s < \frac{2}{n} \quad or \quad s > \frac{2}{n}; \end{cases} \tag{19}$$

for all $x_1, x_2, x_3, \cdots, x_n \in X$, then there exists a unique quadratic function $Q : X \to Y$ such that

$$\| g(x) - Q(x) \| \le \begin{cases} \dfrac{2\lambda}{3(n-1)}, & \\ \dfrac{4\,\lambda \|x\|^s}{(n-1)|4 - 2^s|}, & \\ \dfrac{4\lambda \|x\|^{ns}}{(n-1)|4 - 2^{ns}|}, & \end{cases} \tag{20}$$

for all $x \in X$.

Now we will provide an example to illustrate that the functional equation (4) is not stable for $s = 2$ in condition (ii) of Corollary 3.

Example 4. Let $\phi : \mathbb{R} \to \mathbb{R}$ be a function defined by

$$\phi(x) = \begin{cases} \mu x^2, & \text{if } |x| < 1 \\ \mu, & \text{otherwise} \end{cases}$$

where $\mu > 0$ is a constant, and define a function $g : \mathbb{R} \to \mathbb{R}$ by

$$g(x) = \sum_{\ell=0}^{\infty} \frac{\phi(2^\ell x)}{4^\ell} \qquad \text{for all} \quad x \in \mathbb{R}.$$

Then g satisfies the functional inequality

$$|D\ g(x_1, x_2, x_3, \ldots, x_n)| \le \frac{8(n^3 + 3n^2 + 8n)}{18} \mu \left(\sum_{i=1}^{n} |x_i|^2 \right) \tag{21}$$

for all $x_1, x_2, x_3, \ldots, x_n \in \mathbb{R}$. Then there do not exist a quadratic function $Q : \mathbb{R} \to \mathbb{R}$ and a constant $\beta > 0$ such that

$$|g(x) - Q(x)| \le \beta |x|^2 \qquad \text{for all} \quad x \in \mathbb{R}. \tag{22}$$

Proof. Now

$$|g(x)| \le \sum_{\ell=0}^{\infty} \frac{|\phi(2^\ell x)|}{|4^\ell|} = \sum_{\ell=0}^{\infty} \frac{\mu}{4^\ell} = \frac{4}{3}\mu.$$

Therefore we see that g is bounded. We are going to prove that g satisfies (21). If $x_1 = x_2 = x_3 = \cdots = x_n = 0$ then (21) is trivial. If $\sum_{i=1}^{n} |x_i|^2 \ge \frac{1}{4}$ then the left hand side of (21) is less than $\dfrac{4(n^3 + 3n^2 + 8n)}{18} \mu$. Now suppose that $0 < \sum_{i=1}^{n} |x_i|^2 < \frac{1}{4}$. Then there exists a positive integer k such that

$$\frac{1}{4^{k+1}} \le \sum_{i=1}^{n} |x_i|^2 < \frac{1}{4^k}, \tag{23}$$

so that $2^{k-1} x_1 < \dfrac{1}{2}$, $2^{k-1} x_2 < \dfrac{1}{2}$, $2^{k-1} x_3 < \dfrac{1}{2}$, $\cdots$, $2^{k-1} x_n < \dfrac{1}{2}$ and consequently

$$2^{k-1}(x_1), 2^{k-1}(x_1 + x_2), \cdots, 2^{k-1}(x_1 + x_2 + \cdots + x_n),$$
$$2^{k-1}(x_j + x_{i+1}), 2^{k-1}(x_j - x_{i+1}) \in (-1, 1).$$

Therefore for each $\ell = 0, 1, \ldots, k-1$, we have

$$2^\ell(x_1), 2^\ell(x_1 + x_2), \cdots, 2^\ell(x_1 + x_2 + \cdots + x_n),$$
$$2^\ell(x_j + x_{i+1}), 2^\ell(x_j - x_{i+1}) \in (-1, 1).$$

and

$$\sum_{i=1}^{n} \phi\left(2^\ell \sum_{j=1}^{i} x_j\right) - \sum_{i=1}^{n}(n-i+1)\phi\left(2^\ell x_i\right)$$

$$-\frac{1}{2}\sum_{i=1}^{n-1}(n-i)\sum_{j=1}^{i}\left[\phi(2^\ell(x_j+x_{i+1})) - \phi(2^\ell(x_j-x_{i+1}))\right] = 0$$

for $\ell = 0, 1, \ldots, k-1$. From the definition of g and (23), we obtain that

$$|D\, g(x_1, x_2, x_3, \ldots, x_n)|$$

$$\leq \sum_{\ell=0}^{\infty}\frac{1}{4^\ell}\left|\sum_{i=1}^{n}\phi\left(2^\ell \sum_{j=1}^{i} x_j\right) - \sum_{i=1}^{n}(n-i+1)\phi\left(2^\ell x_i\right)\right.$$

$$\left.-\frac{1}{2}\sum_{i=1}^{n-1}(n-i)\sum_{j=1}^{i}\left[\phi(2^\ell(x_j+x_{i+1})) - \phi(2^\ell(x_j-x_{i+1}))\right]\right|$$

$$\leq \sum_{\ell=k}^{\infty}\frac{1}{4^\ell}\left|\sum_{i=1}^{n}\phi\left(2^\ell \sum_{j=1}^{i} x_j\right) - \sum_{i=1}^{n}(n-i+1)\phi\left(2^\ell x_i\right)\right.$$

$$\left.-\frac{1}{2}\sum_{i=1}^{n-1}(n-i)\sum_{j=1}^{i}\left[\phi(2^\ell(x_j+x_{i+1})) - \phi(2^\ell(x_j-x_{i+1}))\right]\right|$$

$$\leq \sum_{\ell=k}^{\infty}\frac{1}{4^\ell}\mu\left(n+\frac{n(n+1)}{2}+\sum_{i=1}^{n-1}(n-i)i\right)$$

$$= \sum_{\ell=k}^{\infty}\frac{1}{4^\ell}\mu\left(n+\frac{n(n+1)}{2}+n\sum_{i=1}^{n-1}i-\sum_{i=1}^{n-1}i^2\right)$$

$$= \sum_{\ell=k}^{\infty}\frac{1}{4^\ell}\mu\left\{n+\frac{n(n+1)}{2}+\frac{(n-1)n^2}{2}-\frac{(n-1)n(2n-1)}{6}\right\}$$

$$= \frac{8(n^3+3n^2+8n)}{18}\mu \times \frac{1}{4^{k+1}} \leq \frac{8(n^3+3n^2+8n)}{18}\mu\left(\sum_{i=1}^{n}|x_i|^2\right).$$

Thus g satisfies (21) for all $x_1, x_2, x_3, \ldots, x_n \in \mathbb{R}$ with $0 < \sum_{i=1}^{n}|x_i|^2 < \frac{1}{4}$.

We claim that the quadratic functional equation (4) is not stable for $s = 2$ in Corollary 3. Suppose on the contrary that there exist a quadratic mapping $Q : \mathbb{R} \to \mathbb{R}$ and a constant $\beta > 0$ satisfying (22). Since g is bounded and continuous for all $x \in \mathbb{R}$, Q is bounded on any open interval containing the origin and continuous at the origin. In view of Theorem 2, Q must have the form $Q(x) = cx^2$ for any x in $\mathbb{R}$. Thus we obtain that

$$|g(x)| \leq (\beta + |c|)\,|x|^2. \tag{24}$$

But we can choose a positive integer m with $m\mu > \beta + |c|$.

If $x \in \left(0, \frac{1}{2^{m-1}}\right)$, then $2^n x \in (0,1)$ for all $n = 0, 1, \ldots, m-1$. For this x, we get

$$g(x) = \sum_{n=0}^{\infty} \frac{\phi(2^n x)}{4^n} \geq \sum_{n=0}^{m-1} \frac{\mu(2^n x)^2}{4^n} = m\mu x^2 > (\beta + |c|)\, x^2$$

which contradicts (24). Therefore the quadratic functional equation (4) is not stable in sense of Ulam-Hyers if $s = 2$, assumed in the inequality (20).

Now we will provide an example to illustrate that the functional equation (4) is not stable for $s = \dfrac{2}{n}$ in condition (iii) of Corollary 3.

Example 5. Let $\phi : \mathbb{R} \to \mathbb{R}$ be a function defined by

$$\phi(x) = \begin{cases} \mu\, x^2, & |x| < \frac{2}{n} \\ \mu, & otherwise \end{cases}$$

where $\mu > 0$ is a constant, and define a function $g : \mathbb{R} \to \mathbb{R}$ by

$$g(x) = \sum_{n=0}^{\infty} \frac{\phi(2^n x)}{4^n} \quad for \quad all \quad x \in \mathbb{R}.$$

Then h satisfies the functional inequality

$$|D\, g(x_1, x_2, x_3, \ldots, x_n)| \leq \frac{8(n^3 + 3n^2 + 8n)}{18}\, \mu \left(\sum_{i=1}^{n} |x_i|^2 + \prod_{i=1}^{n} ||x_i||^{\frac{2}{n}} \right) \quad (25)$$

for all $x_1, x_2, x_3, \ldots, x_n \in \mathbb{R}$. Then there do not exist a quadratic function $Q : \mathbb{R} \to \mathbb{R}$ and a constant $\beta > 0$ such that

$$|g(x) - Q(x)| \leq \beta\, |x|^2 \quad for \quad all \quad x \in \mathbb{R}. \quad (26)$$

References

1. Aczel, J., Dhombres, J.: Functional Equations in Several Variables. Cambridge University Press (1989)
2. Aoki, T.: On the stability of the linear transformation in Banach spaces. J. Math. Soc. 2, 64–66 (1950)
3. Arunkumar, M.: Solution and stability of bi-additive functional equations. Internat. J. Math. Sci. Engg. Appl. 4, 33–46 (2010)
4. Chang, I.S., Lee, E.H., Kim, H.M.: On the Hyers-Ulam-Rassias stability of a quadratic functional equations. Math. Ineq. Appl. 6, 87–95 (2003)
5. Cholewa, P.W.: Remarks on the stability of functional equations. Aequationes Math. 27, 76–86 (1984)
6. Czerwik, S.: On the stability of the quadratic mappings in normed spaces. Abh. Math. Sem. Univ. Hamburg 62, 59–64 (1992)
7. Gavruta, P.: A generalization of the Hyers-Ulam-Rassias stability of approximately additive mappings. J. Math. Anal. Appl. 184, 431–436 (1994)

8. Hyers, D.H.: On the stability of the linear functional equation. Proc. Nat. Acad. Sci. USA 27, 222–224 (1941)

9. Jun, K.W., Kim, H.M.: On the stability of an n-dimensional quadratic and additive type functional equation. Math. Ineq. Appl. 9, 153–165 (2006)

10. Kannappan, P.: Quadratic functional equation inner product spaces. Results Math. 27, 368–372 (1995)

11. Rassias, J.M.: On approximately of approximately linear mappings by linear mappings. J. Funct. Anal. 46, 126–130 (1982)

12. Jung, S.M.: On the Hyers-Ulam stability of the functional equations that have the quadratic property. J. Math. Anal. Appl. 222, 126–137 (1998)

13. Rassias, T.M.: On the stability of the linear mapping in Banach spaces. Proc. Amer. Math. Soc. 72, 297–300 (1978)

14. Ravi, K., Arunkumar, M., Rassias, M.: On the Ulam stability for the orthogonally general Euler-Lagrange type functional equation. Internat. J. Math. Sci. Autumn 3, 36–47 (2008)

15. Skof, F.: Proprieta locali e approssimazione di operatori. Rend. Sem. Mat. Fis. Milano. 53, 113–129 (1983)

16. Ulam, S.M.: Problems in Modern Mathematics, Science edn. Wiley, New York (1964) (Chapter VI, Some Questions in Analysis: 1, Stability)

Monotone Iterative Scheme for System of Riemann-Liouville Fractional Differential Equations with Integral Boundary Conditions

J.A. Nanware and D.B. Dhaigude

Department of Mathematics,
Dr.Babasaheb Ambedkar Marathwada University,
Aurangabad - 431 004 (M.S.), India
jag_skmg91@rediffmail.com, dnyanraja@gmail.com

Abstract. Monotone iterative scheme is developed for weakly coupled system of Riemann-Liouville fractional differential equation with integral boundary condition using the method of lower and upper solutions. Two monotone convergent sequences converging to extremal solutions are obtained. Existence and uniqueness of solution of weakly coupled system of Riemann-Liouville fractional differential equations with integral boundary conditions is obtained by applying monotone iterative scheme.

Keywords: Fractional differential equation, Integral boundary condition, Monotone iterative method, Existence and uniqueness.

1 Introduction

The origin of fractional calculus [13] go back to seventeenth century when Liebnitz considered the derivatives of order one-half. Since then many mathematicians have contributed upto the middle of last century includes Laplace, Fourier, Abel, Liouville, Riemann, Grunwald etc. However the theory of fractional calculus was developed mainly as a purely theorotical field of mathematics. A wide class of applications of fractional calculus in science and technology are discussed in Debnath [1], Kilbas [4] and Podlubny [14]. The qualitative properties of solution of fractional differential equation [7] parallel to the well known theory of ordinary differential equation [5] has been growing recently. Amongst various techniques [14] to study theory of fractional differential equations one of the widely used technique is the monotone iterative method. The method is well developed and extensively employed in the study of differential equation [5], [9] which arise in biological and physical problems. The basic theory of fractional differential equation with Riemann-Liouville fractional derivative is well developed in [7], [10]. The local and global existence of solution of Riemann-Liouville fractional differential equation and uniqueness of solution are obtained by Lakshmikantham and Vatsala [6], [8]. Recently, McRae [11] developed monotone method for Riemann-Liouville fractional differential equation with initial conditions and studied the qualitative properties of solutions of initial value

P. Balasubramaniam and R. Uthayakumar (Eds.): ICMMSC 2012, CCIS 283, pp. 395–402, 2012.
© Springer-Verlag Berlin Heidelberg 2012

problem. The monotone method for system of Riemann-Liouville fractional differential equation with initial conditions when the function is quasimonotone nondecreasing is developed by Dhaigude et al. [2]. Comparision results and existence and uniqueness of solution of ordinary differential equation with integral boundary condition was firstly obtained by Jankowski [3]. Recently, Nanware and Dhaigude [12] developed the monotone method for Riemann-Liouville fractional differential equation with integral boundary condition when the right hand side function is splited as sum of two nondecreasing and nonincreasing function. Wang and Xie [15] developed the monotone method and obtained existence and uniqueness of solution of fractional differential equation with integral boundary condition.

We organize the paper in the following manner:

In section 2, we give some definitions and basic results. In section 3, we develop monotone iterative scheme and obtain existence and uniqueness results for the system of Riemann-Liouville fractional differential equation with integral boundary condition.

2   Preliminaries

The Riemann-Liouville fractional derivative [14] of order q $(0 < q < 1)$ of the function $u(t)$ is defined as

$$D^q u(t) = \frac{1}{\Gamma(1-q)} \frac{d}{dt} \int_0^t \frac{u(s)}{(t-s)^q} ds. \tag{1}$$

We consider the following Riemann-Liouville fractional differential equation with integral boundary condition

$$D^q u(t) = f(t,u), \quad t \in J = [0,T], \quad T \geq 0, \quad u(0) = \lambda \int_0^T u(s)ds + d, \quad d \in R. \tag{2}$$

where $0 < q < 1, \lambda$ is 1 or -1 and $f \in C[J \times R, R]$, for which monotone method is well developed by Wang and Xie.

Here we develop the monotone iterative scheme for following weakly coupled system of Riemann-Liouville fractional differential equation with integral boundary condition

$$D^q u_i(t) = f_i(t, u_1(t), u_2(t)), \quad u_i(0) = \int_0^T u_i(s)ds + d \tag{3}$$

where $i = 1, 2, d \in R, \quad t \in [0,T], \quad f_1, f_2$ in $C(J \times R^2, R), J = [0,T], 0 < q < 1$.

We employ this scheme to obtain the existence and uniqueness of solution of the system of Riemann-Liouville fractional differential equation with integral boundary condition (3).

Definition 1. *A pair of functions* $v = (v_1, v_2)$ *and* $w = (w_1, w_2)$ *in* $C_p(J, R)$ *are locally Hölder continuous with exponent* $\lambda > q$. *We say that* v *and* w *are ordered lower and upper solutions* $(v_1, v_2) \leq (w_1, w_2)$ *of* (3) *if*

$$D^q v_i(t) \leq f_i(t, v_1(t), v_2(t)), \qquad v_i(0) \leq \int_0^T v_i(s)ds + d$$

and

$$D^q w_i(t) \geq f_i(t, w_1(t), w_2(t)), \qquad w_i(0) \geq \int_0^T w_i(s)ds + d.$$

Definition 2. *A function* $f_i = f_i(t, u_1, u_2)$ *in* $C(J \times R^2, R)$ *is said to be quasi-monotone nondecreasing if*

$$f_i(t, u_1(t), u_2(t)) \leq f_i(t, v_1(t), v_2(t)) \quad if \quad u_i = v_i \quad and \quad u_j \leq v_j,$$

$i \neq j$, $i = j = 1, 2$.

Lemma 1. *[6] Let* $m \in C_p(J, R)$ *be locally Hölder continuous with exponent* $\lambda > q$ *and for any* $t_1 \in (0, T]$ *we have* $m(t_1) = 0$ *and* $m(t) \leq 0$ *for* $0 \leq t \leq t_1$. *Then it follows that* $D^q m(t_1) \geq 0$.

Lemma 2. *[15] If* $f \in C(J \times R, R)$ *and* $v_0(t), w_0(t) \in C^1(J, R)$ *are lower and upper solutions of* (2) *and there exists* $L \in \left(0, \frac{1}{T^q \Gamma(1-q)}\right)$ *such that* $f(t, x) - f(t, y) \geq -M(x - y)$ *for* $x \geq y$. *Then* $v(0) \leq w(0)$ *implies* $v(t) \leq w(t)$.

Theorem 1. *[15] Assume that*

(i) *functions* v *and* w *in* $C_p(J, R)$ *are ordered lower and upper solutions of* (2)
(ii) *function* $f(t, u(t))$ *satisfy one-sided Lipschitz condition:*

$$f(t, u(t)) - f(t, \overline{u}(t)) \leq \frac{L}{1 + t^q}(u - \overline{u})$$

Then $v(0) \leq w(0)$ *implies that* $v(t) \leq w(t)$, $0 \leq t \leq T$, *provided that* $LT^q \leq \frac{1}{\Gamma(1-q)}$.

3   Main Results

This section is devoted to develop monotone iterative scheme for weakly coupled system of fractional differential equation with integral boundary condition and to obtain existence and uniqueness of solution of weakly coupled system (3).

Theorem 2. *Assume that*

(i) *a function* $f_i = f_i(t, u_1, u_2)$ *i=1,2 in* $C[J \times R, R^2]$ *is quasimonotone nondecreasing,*

(ii) *functions* $v^0 = (v_1^0, v_2^0)$ *and* $w^0 = (w_1^0, w_2^0)$ *in* $C(J, R)$, *are ordered lower and upper solutions of (3) such that* $v_1^0(0) \leq w_1^0(0), v_2^0(0) \leq w_2^0(0)$ *on* $[0, T]$
(iii) *function* $f_i \equiv f_i(t, u_1, u_2)$ *satisfies one-sided Lipschitz condition,*

$$f_i(t, u_1, u_2) - f_i(t, \overline{u}_1, \overline{u}_2) \geq -M_i[u_i - \overline{u}_i] \qquad (4)$$

whenever $v_i^0(0) \leq u_i \leq w_i^0(0), M_i \geq 0, \quad v_i^0(0) \leq \overline{u}_i \leq u_i \leq w_i^0(0)$

Then there exists monotone sequences $\{v^n(t)\} = (v_1^n, v_2^n)$ *and* $\{w^n(t)\} = (w_1^n, w_2^n)$ *such that*

$$\{v^n(t)\} \to v(t) = (v_1, v_2) \quad and \quad \{w^n(t)\} \to w(t) = (w_1, w_2) \text{ as } n \to \infty$$

Proof. For any $\eta(t) = (\eta_1, \eta_2)$ in $C(J, R)$ such that for $v_1^0(0) \leq \eta_1, v_2^0(0) \leq \eta_2$ on J, we consider the following system of linear fractional differential equations

$$D^q u_i(t) = f_i(t, \eta_1(t), \eta_2(t)) - M_i[u_i(t) - \eta_i(t)], \quad u_i(0) = \int_0^T u_i(s)ds + d. \quad (5)$$

It is clear that for every $\eta(t)$ there exists a unique solution $u(t) = (u_1(t), u_2(t))$ of system (5) on J. For each $\eta(t)$ and $\mu(t)$ in $C(J, R)$ such that $v_i^0(0) \leq \eta_i(t), w_i^0(0) \leq \mu_i(t)$, we define a mapping A by $A[\eta, \mu] = u(t)$ where $u(t)$ is the unique solution of system (5). This mapping defines the sequences $\{v^n(t)\}$ and $\{w^n(t)\}$.

First, we prove that

(A_1) $v^0 \leq A[v_0, w_0], \quad w^0 \geq A[w_0, v_0]$
(A_2) A possesses the monotone property on the segment

$$[v_0, w_0] = \left\{ (u_1, u_2) \in C(J, R^2) : v_1^0 \leq u_1 \leq w_1^0, v_2^0 \leq u_2 \leq w_2^0 \right\}.$$

Set $A[v_0, w_0] = v^1(t)$, where $v^1(t) = (v_1^1, v_2^1)$ is the unique solution of system (5) with $\eta_i = v_i^0(0)$.
Setting $p_i(t) = v_i^0(t) - v_i^1(t)$ we see that

$$\begin{aligned}
D^q p_i(t) = D^q v_i^0(t) - D^q v_i^1(t) \\
\leq f_i(t, v_i^0(t), u_2(t)) - f_i(t, v_i^1(t), u_2(t)) \\
\leq -M_i[v_i^0(t) - v_i^1(t)] \\
\leq -M_i p_i(t)
\end{aligned}$$

$$\text{and} \quad p_i(0) \leq 0.$$

Thus by Theorem 1, it follows that $p_i(t) \leq 0$ on $0 \leq t \leq T$ and hence $v_i^0(t) - v_i^1(t) \leq 0$ which implies $v_i^0 \leq A[v_0, w_0]$. Set $A[v_0, w_0] = w^1(t)$, where $w^1(t) = (w_1^1, w_2^1)$ is the unique solution of system (5) with $\mu_i = w_i^0(t)$.
Setting $p_i(t) = w_i^0(t) - w_i^1(t)$ we see that

$$\begin{aligned}
D^q p_i(t) = D^q w_i^0(t) - D^q w_i^1(t) \\
\geq f_i(t, w_i^0(t), u_2(t)) - f_i(t, w_i^1(t), u_2(t)) \\
\geq -M_i[w_i^0(t) - w_i^1(t)] \\
\geq -M_i p_i(t)
\end{aligned}$$

$$\text{and} \quad p_i(0) \geq 0.$$

By Theorem 1, it follows that $w_i^0(t) \geq w_i^1(t)$. Hence $w^0 \geq A[v_0, w_0]$.

Let $\eta_1(t), \eta_2(t), \mu(t) \in [v_0, w_0]$ be such that $\eta_1(t) \leq \eta_2(t)$. Suppose that $A[\eta_1, \mu] = u_1(t) = (u_1^1, u_2^1), A[\eta_2, \mu] = u_2(t) = (u_1^2, u_2^2)$. Then setting $p_i(t) = u_1^i(t) - u_2^i(t)$ we find that

$$D^q p_i(t) = D^q u_1^i(t) - D^q u_2^i(t)$$
$$= f_i(t, \eta_1(t), u_2(t)) - f_i(t, \eta_2(t), u_2(t)) - M_i[u_1^i(t) - \eta_1(t)]$$
$$+ M_i[u_2^i(t) - \eta_2(t)]$$
$$\leq -M_i[\eta_1(t) - \eta_2] - M_i[u_1^i(t) - \eta_1(t)] + M_i[u_2^i(t) - \eta_2(t)]$$
$$\leq -M_i[u_1^i(t) - u_2^i(t)]$$
$$\leq -M_i p_i(t)$$

and $p_i(0) \leq 0$.

As before in (A_1), we have $A[\eta_1, \mu] \leq A[\eta_2, \mu]$.

Similarly, if we let $\eta(t), \mu_1(t), \mu_2(t) \in [v_0, w_0]$ be such that $\mu_1(t) \leq \mu_2(t)$. Suppose that $A[\eta, \mu_i] = u_i(t)$, we can prove that $A[\eta, \mu_1] \geq A[\eta, \mu_2]$. Thus it follows that the mapping A possesses monotone property on the segment $[v_0, w_0]$.

Now in view of (A_1) and (A_2), we define the sequences

$$v_i^n(t) = A[v_i^{n-1}, w_i^{n-1}], \quad w_i^n(t) = A[w_i^{n-1}, v_i^{n-1}]$$

on the segment $[v_0, w_0]$ by

$$D^q v_i^n(t) = f_i(t, v_1^{n-1}, v_2^{n-1}) - M_i[v_i^n - v_i^{n-1}], \quad v_i^n(0) = \int_0^T v_i^{n-1}(s)ds + d$$

$$D^q w_i^n(t) = f_i(t, w_1^{n-1}, w_2^{n-1}) - M_i[w_i^n - w_i^{n-1}], \quad w_i^n(0) = \int_0^T w_i^{n-1}(s)ds + d.$$

From (A_1), we have $v_i^0(t) \leq v_i^1(t), \quad w_i^0(t) \geq w_i^1(t)$.

Assume that $v_i^{k-1}(t) \leq v_i^k(t), \quad w_i^{k-1}(t) \geq w_i^k(t)$. In order to prove $v_i^k(t) \leq v_i^{k+1}(t), \quad w_i^k(t) \geq w_i^{k+1}(t)$ and $v_i^k(t) \geq w_i^k(t)$, we define $p_i(t) = v_i^k(t) - v_i^{k+1}(t)$. Thus

$$D^q p_i(t) = D^q v_i^k(t) - D^q v_i^{k+1}(t)$$
$$= f_i(t, v_1^{k-1}(t), v_2^{k-1}(t)) - M_i[v_i^k(t) - v_i^{k-1}(t)]$$
$$- \left\{ f_i(t, v_1^k(t), v_2^k(t)) - M_i[v_i^{k+1}(t) - v_i^k(t)] \right\}$$
$$\leq M_i[v_i^k(t) - v_i^{k-1}(t)] - M_i[v_i^k(t) - v_i^{k-1}(t)] + M_i[v_1^{k+1}(t) - v_1^k(t)]$$
$$\leq -M_i p_i(t)$$

and $p_i(0) \leq 0$.

It follows from Theorem 1 that $p_i(t) \leq 0$, which gives $v_i^k(t) \leq v_i^{k+1}(t)$. Similarly we can prove $w_i^k(t) \geq w_i^{k+1}(t)$ and $v_i^k(t) \geq w_i^k(t)$.

By induction, it follows that

$$v_i^0(t) \leq v_i^1(t) \leq v_i^2(t) \leq \ldots \leq v_i^n(t) \leq w_i^n(t) \leq w_i^{n-1}(t) \leq \ldots \leq w_i^1(t) \leq w_i^0(t). \tag{6}$$

Thus the sequences $v^n(t)$ and $w^n(t)$ are bounded from below and bounded from above respectively and monotonically nondecreasing and monotonically nonincreasing on $[0, T]$. Hence pointwise limit exists and are given by $\lim_{n \to \infty} v_i^n(t) = v_i(t)$, $\lim_{n \to \infty} w_i^n(t) = w_i(t)$ on $[0, T]$.

Using corresponding Volterra fractional integral equations

$$v_i^n(t) = v_i^0 + \frac{1}{\Gamma(q)} \int_0^T (t-s)^{q-1} \left\{ f_i(s, v_1^n(s), v_2^n(s)) - M_i[v_i^n - v_i^{n-1}] \right\} ds$$

$$w_i^n(t) = w_i^0 + \frac{1}{\Gamma(q)} \int_0^T (t-s)^{q-1} \left\{ f_i(s, w_1^n(s), w_2^n(s)) - M_i[w_1^n - w_1^{n-1}] \right\} ds, \tag{7}$$

it follows that $v(t)$ and $w(t)$ are solutions of system (5).

Finally, we prove that $v(t)$ and $w(t)$ are the minimal and maximal solution of the system (5). Let $u(t) = (u_1, u_2)$ be any solution of (3) different from $v(t)$ and $w(t)$, so that there exists k such that $v_i^k(t) \leq u_i(t) \leq w_i^k(t)$ on $[0, T]$ and set $p_i(t) = v_i^{k+1}(t) - u_i(t)$ so that

$$D^q p_i(t) = f_i(t, v_1^k(t), v_2^k(t)) - M_i[v_i^{k+1}(t) - v_i^k(t)] - f_i(t, u_1(t), u_2(t))$$
$$\geq -M_i[v_i^k(t) - u_i(t)] - M_i[v_i^{k+1}(t) - v_i^k(t)]$$
$$\geq -M_i[v_i^{k+1}(t) - u_i(t)]$$
$$\geq -M_i p_i(t)$$

and $p_i(0) \geq 0$.

Thus $v_i^{k+1}(t) \leq u_i(t)$ on $[0, T]$. Since $v_i^0(t) \leq u_i(t)$ on $[0, T]$, by induction it follows that $v_i^k(t) \leq u_i(t)$ for all k. Similarly, we can prove $u_i(t) \leq w_i^k(t)$ for all k on $[0, T]$. Thus $v_i^k(t) \leq u_i(t) \leq w_i^k(t)$ on $[0, T]$. Taking limit as $n \to \infty$, it follows that $v_i(t) \leq u_i(t) \leq w_i(t)$ on $[0, T]$. Thus the theorem.

We now obtain the uniqueness of solution of the system (3) in the following theorem.

Theorem 3. *Assume that*

(i) *a function $f_i = f_i(t, u_1, u_2)$ in $C[J \times R, R^2]$ is quasimonotone nondecreasing,*
(ii) *functions $v = (v_1, v_2)$ and $w = (w_1, w_2)$ in $C(J, R)$ are ordered lower and upper solutions of (3) on $[0, T]$*
(iii) *function $f_i = f_i(t, u_1, u_2)$ satisfies Lipschitz condition,*

$$|f_i(t, u_1, u_2) - f_i(t, \overline{u}_1, \overline{u}_2)| \leq M_i |u_i - \overline{u}_i| \tag{8}$$

(iv) *$\lim_{n \to \infty} ||w^n - v^n|| = 0$, where the norm is defined by $||f|| = \int_0^T |f(s)| ds$.*

then the solution of system (3) is unique.

Proof. Since $v(t) \leq w(t)$, we need to prove only $v(t) \geq w(t)$. To prove this, setting $p_i(t) = w_i(t) - v_i(t)$, we find that

$$D^q p_i(t) = D^q w_i(t) - D^q v_i(t)$$
$$= f_i(t, w_1(t), w_2(t)) - f_i(t, v_1(t), v_2(t))$$
$$\leq -M_i[w_i(t) - v_i(t)]$$
$$\leq -M_i p_i(t)$$

and $p_i(0) \leq 0$.

By Theorem 1, it follows that $p_i(t) \leq 0$. This gives $w_i(t) \leq v_i(t)$. Hence $v(t) = w(t)$ is the unique solution of system (3) on $[0, T]$. This proves the uniqueness of solution of the system (3) on $[0, T]$.

Acknowledgement. The first author is grateful to UGC India for awarding "Teacher Fellowship" under Faculty Development Programme, vide letter no 31-12/08(WRO). He also wish to put on record thanks to authorities of Shrikrishna Education Society, Gunjoti for sactioning study leave.

References

1. Debnath, L., Bhatta, D.: Integral Transforms and Their Applications, 2nd edn. Taylor and Francis Group, New York (2007)
2. Dhaigude, D.B., Nanware, J.A., Dhaigude, C.D.: Monotone Iterative Technique for Fractional Differential Equations. Bull. Marathwada Math. Soc. (Accepted)
3. Jankwoski, T.: Differential Equations with Integral Boundary Conditions. J. Comp. Appl. Math. 147, 1–8 (2002)
4. Kilbas, A.A., Srivastava, H.M., Trujillo, J.J.: Theory and Applications of Fractional Differential Equations. North Holland Mathematical Studies, vol. 204. Elsevier (North-Holland) Sciences Publishers, Amsterdam (2006)
5. Ladde, G.S., Lakshmikantham, V., Vatsala, A.S.: Monotone Iterative Techniques for Nonlinear Differential Equations. Pitman Advanced Publishing Program, London (1985)
6. Lakshmikantham, V., Vatsala, A.S.: Theory of Fractional Differential Equations and Applications. Commun. Appl. Anal. 11, 395–402 (2007)
7. Lakshmikantham, V., Vatsala, A.S.: Basic Theory of Fractional Differential Equations and Applications. Nonlinear Anal. 69(8), 2677–2682 (2008)
8. Lakshmikantham, V., Vatsala, A.S.: General Uniqueness and Monotone Iterative Technique for Fractional Differential Equations. Appl. Math. Lett. 21(8), 828–834 (2008)
9. Lakshmikantham, V., Leela, S.: Differential and Integral Inequalities, vol. I. Academic Press, New York (1969)
10. Lakshmikantham, V., Leela, S., Devi, J.V.: Theory and Applications of Fractional Dynamical Systems. Cambridge Scientific Publishers Ltd. (2009)

11. McRae, F.A.: Monotone Iterative Technique and Existence Results for Fractional Differential Equations. Nonlinear Anal. 71(12), 6093–6096 (2009)
12. Nanware, J.A., Dhaigude, D.B.: Existence and Uniqueness of Solution of Riemann-Liouville Fractional Differential Equation with Integral Boundary Condition (Communicated)
13. Oldham, K.B., Spanier, J.: The Fractional Calculus. Dover Publications, Inc., New York (2002)
14. Podlubny, I.: Fractional Differential Equations. Academic Press, San Diego (1999)
15. Wang, T., Xie, F.: Existence and Uniqueness of Fractional Differential Equations with Integral Boundary Conditions. J. Nonlinear Sci. Appl. 1(4), 206–212 (2008)

Some New Fractional Difference Inequalities

G.V.S.R. Deekshitulu[1] and J. Jagan Mohan[2,*]

[1] Fluid Dynamics Division, School of Advanced Sciences, VIT University,
Vellore - 632014, Tamil Nadu, India
[2] Department of Mathematics, Manipal Institute of Technology, Manipal University,
Manipal - 576104, Karnataka, India
j.jaganmohan@hotmail.com

Abstract. In this paper we present some new fractional order difference inequalities which can be used in the theory of fractional order difference equations.

Keywords: Difference equation, Fractional order, Inequalities.

1 Introduction

In the study of many problems concerning with behaviour of solutions of finite difference equations, we deal with certain inequalities and comparison principles involving sequence of real numbers. The inequalities and comparison principles which provide explicit bounds on unknown functions play a very important role in the study of qualitative and quantitative properties of solutions of nonlinear difference equations. In the theory of finite difference inequalities [8], the result on the discrete analogue of Gronwall-Bellman inequality is the fundamental inequality. In 1969, Lakshmikantham et al. [7] proved the following most precise and complete discrete analogue of the well known Gronwall - Bellman inequality which find numerous applications in the theory of finite difference equations.

Theorem 1. *(Discrete Gronwall - Bellman Inequality) Let $u(n)$, $p(n)$ and $q(n)$ be real valued nonnegative functions defined on $\mathbb{N}_0^+$. If, for $n \in \mathbb{N}_0^+$,*

$$u(n) \leq u(0) + \sum_{j=0}^{n-1}[p(j)u(j) + q(j)] \tag{1}$$

then, for $n \in \mathbb{N}_0^+$,

$$u(n) \leq u(0) \prod_{j=0}^{n-1}[1 + p(j)] + \sum_{j=0}^{n-1} q(j) \prod_{k=j+1}^{n-1}[1 + p(k)] \tag{2}$$

$$\leq u(0)\, exp\left[\sum_{j=0}^{n-1} p(j)\right] + \sum_{j=0}^{n-1} q(j)\, exp\left[\sum_{k=j+1}^{n-1} p(k)\right].$$

---

* Corresponding author.

P. Balasubramaniam and R. Uthayakumar (Eds.): ICMMSC 2012, CCIS 283, pp. 403–412, 2012.
© Springer-Verlag Berlin Heidelberg 2012

Several mathematicians established various important finite difference inequalities that have a wide range of applications in the theory of finite difference equations [1].

The notions of fractional differential equations [9] may be traced back to the works of Euler, but the idea of fractional difference equations is very recent. Diaz and Osler [5] defined the fractional difference by the rather natural approach of allowing the index of differencing, in the standard expression for the nth difference, to be any real or complex number. Later, Hirota [6], defined the fractional difference using Taylor's series. In 2002, Atsushi Nagai [2] introduced another definition of fractional difference which is a slight modification of Hirotas [6] definition. In 2010, Deekshitulu, G.V.S.R. and Jagan Mohan, J. [3] modified the definition of Atsushi Nagai [2] in such a way that the expression for fractional difference does not involve any difference operator and derived some basic inequalities and comparison theorems. Using the same definition, the authors have established the discrete fractional analogue of Bihari's inequality recently [4]. The main objective of the present paper is to present some new and interesting fractional order difference inequalities.

2   Notations and Terminology

Throughout the article, we shall use the following notations and definitions.

Let $\mathbb{Z}$ and $\mathbb{R}$ denote the set of all integers and the set of all real numbers respectively and $\mathbb{R}^+ = [0, \infty)$. $\mathbb{N} = \{0, 1, 2, ...\}$ be the set of all natural numbers including zero and $\mathbb{N}_a^+ = \{a, a+1, a+2, ...\}$ where $a \in \mathbb{N}$. Let $u(n) : \mathbb{N}_0^+ \to \mathbb{R}$ then for all a, $b \in \mathbb{N}_a^+$ and $a > b$, $\sum_{j=a}^{b} u(j) = 0$ and $\prod_{j=a}^{b} u(j) = 1$, i.e. empty sums and products are taken to be 0 and 1 respectively. If $n-1$ and n are in $\mathbb{N}_a^+$, then for this function $u(n)$, the backward difference operator ∇ is defined as $\nabla u(n) = u(n) - u(n-1)$. The extended binomial coefficient $\binom{a}{n}$, $(a \in \mathbb{R}, n \in \mathbb{Z})$ is defined by

$$\binom{a}{n} = \begin{cases} \frac{\Gamma(a+1)}{\Gamma(a-n+1)\Gamma(n+1)} & n > 0 \\ 1 & n = 0 \\ 0 & n < 0. \end{cases} \qquad (3)$$

In 2002, Atsushi Nagai [2] gave the following definition for fractional order difference operator.

Definition 1. *Let $\alpha \in \mathbb{R}$ and m be an integer such that $m - 1 < \alpha \le m$. The difference operator of order α, with step length ε, is defined as*

$$\nabla^\alpha u(n) = \varepsilon^{m-\alpha} \sum_{j=0}^{n-1} \binom{\alpha - m}{j} (-1)^j \nabla^m u(n - j). \qquad (4)$$

The above definition of $\nabla^\alpha u(n)$ given by Atsushi Nagai [2] contains ∇ operator and the term $(-1)^j$ inside the summation index and hence it becomes difficult to study the properties of solution. To avoid this, Deekshitulu and Jagan Mohan in [3] gave the following definition, for $\varepsilon = m = 1$.

Definition 2. *Let $\alpha \in \mathbb{R}$ such that $0 < \alpha < 1$. The difference operator of order α is defined as*

$$\nabla^\alpha u(n) = \sum_{j=0}^{n-1} \binom{j-\alpha}{j} \nabla u(n-j) \tag{5}$$

and the sum operator of order α is defined as

$$\nabla^{-\alpha} u(n) = \sum_{j=0}^{n-1} \binom{j+\alpha}{j} \nabla u(n-j). \tag{6}$$

Remark 1. *Let $v(n) : \mathbb{N}_0^+ \to \mathbb{R}$; α, β, $\gamma \in \mathbb{R}$ such that $0 < \alpha < 1$ and c, d are scalars. Then the fractional order difference operator satisfies the following properties.*

1. $\nabla^\beta \nabla^\gamma u(n) = \nabla^{\beta+\gamma} u(n)$.
2. $\nabla^\beta [cu(n) + dv(n)] = c\nabla^\beta u(n) + d\nabla^\alpha v(n)$.
3. $\nabla^\beta (u(n)v(n)) = \sum_{m=0}^{n-1} \binom{\beta}{m} [\nabla^{\beta-m} u(n-m)][\nabla^m v(m)]$.
4. $\nabla^\alpha \nabla^{-\alpha} u(n) = \nabla^{-\alpha} \nabla^\alpha u(n) = u(n) - u(0)$.
5. $\nabla^\alpha u(0) = 0$ and $\nabla^\alpha u(1) = u(1) - u(0) = \nabla u(1)$.

Definition 3. *Let $f(n,r)$ be any function defined for $n \in \mathbb{N}_0^+$, $0 \le r < \infty$. Then a nonlinear difference equation of order $\alpha \in \mathbb{R}$, $0 < \alpha < 1$, together with an initial condition is of the form*

$$\nabla^\alpha u(n+1) = f(n, u(n)), \quad u(0) = u_0. \tag{7}$$

Then the solution of (7) is expressed as a recurrence relation involving the values of the unknown function at the previous arguments as follows.

$$u(n) = u_0 + \sum_{j=0}^{n-1} \binom{n-j+\alpha-2}{n-j-1} f(j, u(j)) = u_0 + \sum_{j=0}^{n-1} B(n-1, \alpha; j) f(j, u(j)) \tag{8}$$

where $B(n, \alpha; j) = \binom{n-j+\alpha-1}{n-j}$ for $0 \le j \le n$.

3   Main Results

In this section we state and prove some useful finite difference inequalities of fractional order. Throughout the section we assume that $\alpha \in (0,1) \subseteq \mathbb{R}$. We begin with the following basic difference inequality.

Theorem 2. *Let $u(n)$, $a(n)$ and $b(n)$ be real valued nonnegative functions defined on $\mathbb{N}_0^+$. If*

$$\nabla^\alpha u(n+1) \le a(n)u(n) + b(n) \tag{9}$$

$$\text{then} \quad u(n) \le u(0) \prod_{j=0}^{n-1} \left[1 + B(n-1, \alpha; j)a(j) \right] + \sum_{j=0}^{n-1} B(n-1, \alpha; j)b(j)$$

$$\times \prod_{k=j+1}^{n-1} \left[1 + B(n-1, \alpha; j)a(k) \right] \tag{10}$$

for $n \in \mathbb{N}_0^+$.

Proof. Using (7) and (8), we have

$$u(n) = u(0) + \sum_{j=0}^{n-1} B(n-1, \alpha; j) \nabla^\alpha u(j+1). \tag{11}$$

Using (9) in (11), we get

$$u(n) \le u(0) + \sum_{j=0}^{n-1} B(n-1, \alpha; j)[a(j)u(j) + b(j)] \tag{12}$$

which is of the form

$$u(n) \le u(0) + \sum_{j=0}^{n-1} [p(j)u(j) + q(j)] \tag{13}$$

where $p(j) = B(n-1, \alpha; j)a(j)$ and $q(j) = B(n-1, \alpha; j)b(j)$ are real valued nonnegative functions defined on $\mathbb{N}_0^+$. Using discrete Gronwall - Bellman inequality (i.e. Theorem 1), we get the required inequalities in (10).

Now we state the discrete fractional Gronwall - Bellman inequality.

Corollary 1. *Let $u(n)$, $a(n)$ and $b(n)$ be real valued nonnegative functions defined on $\mathbb{N}_0^+$. If*

$$u(n) \le u(0) + \sum_{j=0}^{n-1} B(n-1, \alpha; j)a(j)[a(j)u(j) + b(j)] \tag{14}$$

$$\text{then} \quad u(n) \le u(0) \prod_{j=0}^{n-1} \left[1 + B(n-1, \alpha; j)a(j) \right] + \sum_{j=0}^{n-1} B(n-1, \alpha; j)b(j)$$

$$\times \prod_{k=j+1}^{n-1} \left[1 + B(n-1, \alpha; j)a(k) \right] \tag{15}$$

for $n \in \mathbb{N}_0^+$.

Theorem 3. *Let $u(n)$, $\nabla^{\alpha} u(n)$, $a(n)$, $b(n)$ and $c(n)$ be real valued nonnegative functions defined on $\mathbb{N}_0^+$ for which the inequality*

$$\nabla^{\alpha} u(n+1) \leq a(n) \left[c(n) + u(n) + \sum_{j=0}^{n-1} B(n-1, \alpha; j)[b(j)\nabla^{\alpha} u(j+1)] \right] \quad (16)$$

holds. Then, for $n \in \mathbb{N}_0^+$,

$$u(n) \leq u(0) + \sum_{j=0}^{n-1} B(n-1, \alpha; j)a(j)[c(j) + D(j)] \quad (17)$$

where $\quad D(n) = u(0) \prod_{j=0}^{n-1} A(n, j) + \sum_{j=0}^{n-1} \left[c(j)[A(n, j) - 1] \prod_{k=j+1}^{n-1} A(n, k) \right] \quad (18)$

and $\quad A(n, j) = \left[1 + B(n-1, \alpha; j)a(j)[1 + b(j)] \right]. \quad (19)$

Proof. Define a function $z(n)$ by

$$z(n) = u(n) + \sum_{j=0}^{n-1} B(n-1, \alpha; j)[b(j)\nabla^{\alpha} u(j+1)]. \quad (20)$$

Then $z(0) = u(0)$, $\nabla^{\alpha} u(n+1) \leq a(n)[c(n) + z(n)]$ and

$$\nabla^{\alpha} z(n+1) \leq a(n)c(n)[1 + b(n)] + [1 + b(n)]a(n)z(n). \quad (21)$$

Now by application of Theorem 2 to (21) yields

$$z(n) \leq u(0) \prod_{j=0}^{n-1} A(n, j) + \sum_{j=0}^{n-1} \left[c(j)[A(n, j) - 1] \prod_{k=j+1}^{n-1} A(n, k) \right] = D(n). \quad (22)$$

Using (22) in $\nabla^{\alpha} u(n+1) \leq a(n)[c(n) + z(n)]$, we get

$$\nabla^{\alpha} u(n+1) \leq a(n)[c(n) + D(n)]. \quad (23)$$

Now again by application of Theorem 2 to (23) gives the required inequality in (17).

Corollary 2. *Let $u(n)$, $\nabla^{\alpha} u(n)$, $a(n)$ and $b(n)$ be real valued nonnegative functions defined on $\mathbb{N}_0^+$ for which the inequality*

$$\nabla^{\alpha} u(n+1) \leq a(n) \left[u(n) + \sum_{j=0}^{n-1} B(n-1, \alpha; j)[b(j)\nabla^{\alpha} u(j+1)] \right] \quad (24)$$

holds. Then, for $n \in \mathbb{N}_0^+$,

$$u(n) \leq u(0) \left[1 + \sum_{j=0}^{n-1} B(n-1, \alpha; j) \prod_{k=0}^{j-1} \left[1 + B(j-1, \alpha; k)a(k)[1 + b(k)] \right] \right]. \quad (25)$$

Theorem 4. *Let $u(n)$, $\nabla^{\alpha} u(n)$, $a(n)$ and $b(n)$ be real valued nonnegative functions defined on $\mathbb{N}_0^+$ for which the inequality*

$$\nabla^{\alpha} u(n+1) \leq a(n) \left[\sum_{j=0}^{n-1} B(n-1, \alpha; j) b(j) [u(j) + \nabla^{\alpha} u(j+1)] \right] \quad (26)$$

holds. If $\left[1 + B(n-1, \alpha; j)[a(j) - 1]b(j) \right] \geq 0$ for all $0 \leq j \leq (n-1)$, then

$$u(n) \leq u(0) \left[1 + \sum_{j=0}^{n-1} B(n-1, \alpha; j) a(j) \left[\sum_{k=0}^{j-1} B(j-1, \alpha; k) b(k)[1 + E(k)] \right. \right.$$

$$\left. \left. \prod_{l=k+1}^{j-1} \left[1 + B(j-1, \alpha; l)[a(l) - 1]b(l) \right] \right] \right] \quad (27)$$

where

$$E(n) = \sum_{j=0}^{n-1} B(n-1, \alpha; j) b(j) \prod_{k=j+1}^{n-1} \left[1 + B(n-1, \alpha; k)[a(k) + b(k) + a(k)b(k)] \right]$$

$$(28)$$

Proof. Define a function $z(n)$ by

$$z(n) = \sum_{j=0}^{n-1} B(n-1, \alpha; j) b(j) [u(j) + \nabla^{\alpha} u(j+1)]. \quad (29)$$

Then $z(0) = 0$, $\nabla^{\alpha} u(n+1) \leq a(n) z(n)$ and

$$\nabla^{\alpha} z(n+1) = b(n)[u(n) + \nabla^{\alpha} u(n+1)] \leq b(n)[u(n) + a(n) z(n)]. \quad (30)$$

Since $\nabla^{\alpha} u(n+1) \leq a(n) z(n)$, an application of Theorem 2 to (30) yields

$$u(n) \leq u(0) + \sum_{j=0}^{n-1} B(n-1, \alpha; j) a(j) z(j). \quad (31)$$

Using (31) in (30), we get

$$\nabla^{\alpha} z(n+1) \leq b(n) \left[u(0) + \sum_{j=0}^{n-1} B(n-1, \alpha; j) a(j) z(j) + a(n) z(n) \right]. \quad (32)$$

Adding $b(n) z(n)$ on both sides of the above inequality, we have

$$\nabla^{\alpha} z(n+1) + b(n) z(n) \leq u(0) b(n) + a(n) b(n) z(n)$$

$$+ b(n) \left[z(n) + \sum_{j=0}^{n-1} B(n-1, \alpha; j) a(j) z(j) \right]. \quad (33)$$

Let,

$$v(n) = z(n) + \sum_{j=0}^{n-1} B(n-1, \alpha; j) a(j) z(j). \tag{34}$$

Then $v(0) = z(0)$, $z(n) \leq v(n)$ and

$$\nabla^\alpha v(n+1) = \nabla^\alpha z(n+1) + a(n) z(n). \tag{35}$$

Using (33) and (34) in (35), we get

$$\nabla^\alpha v(n+1) \leq u(0) b(n) + a(n) b(n) z(n) + b(n) v(n) + a(n) z(n)$$
$$\leq u(0) b(n) + [a(n) + b(n) + a(n) b(n)] v(n). \tag{36}$$

Now an application of Theorem 2 to (36) yields

$$v(n) \leq u(0) \sum_{j=0}^{n-1} B(n-1, \alpha; j) b(j) \prod_{k=j+1}^{n-1} \left[1 + B(n-1, \alpha; k)[a(k) + b(k) + a(k) b(k)] \right]$$
$$= u(0) E(n). \tag{37}$$

Using (37) in (32), we get

$$\nabla^\alpha z(n+1) \leq [a(n) - 1] b(n) z(n) + u(0) b(n)[1 + E(n)]. \tag{38}$$

Now again by application of Theorem 2 to (38) yields

$$z(n) \leq u(0) \sum_{j=0}^{n-1} B(n-1, \alpha; j) b(j)[1 + E(j)] \prod_{k=j+1}^{n-1} \left[1 + B(n-1, \alpha; k)[a(k) - 1] b(k) \right]. \tag{39}$$

Using (39) in $\nabla^\alpha u(n+1) \leq a(n) z(n)$, we get

$$\nabla^\alpha u(n+1) \leq u(0) a(n) \left[\sum_{j=0}^{n-1} B(n-1, \alpha; j) b(j)[1 + E(j)] \right.$$
$$\left. \times \prod_{k=j+1}^{n-1} \left[1 + B(n-1, \alpha; k)[a(k) - 1] b(k) \right] \right]. \tag{40}$$

Now again by application of Theorem 2 to (40), we get the required inequality in (27).

Theorem 5. *Let* $u(n)$, $\nabla^\alpha u(n)$, $a(n)$ *and* $b(n)$ *be real valued nonnegative functions defined on* $\mathbb{N}_0^+$ *for which the inequality*

$$\nabla^\alpha u(n+1) \leq a(n) \left[u(n) + \sum_{j=0}^{n-1} B(n-1, \alpha; j) b(j)[u(j) + \nabla^\alpha u(j+1)] \right] \tag{41}$$

holds. If $H(n, j) \geq 0$ for all $0 \leq j \leq (n-1)$, then for $n \in \mathbb{N}_0^+$,

$$
\begin{aligned}
u(n) \leq u(0)\Big[1 + \sum_{j=0}^{n-1} B(n-1, \alpha; j)a(j)\Big[\prod_{k=0}^{j-1} H(j, k) \\
+ \sum_{k=0}^{j-1} B(j-1, \alpha; k)b(k)[1 + F(k)] \prod_{l=k+1}^{j-1} H(j, l)\Big]\Big]
\end{aligned}
\tag{42}
$$

$$
\text{where} \quad F(n) = \Big[\prod_{j=0}^{n-1} G(n, j) + \sum_{j=0}^{n-1} B(n-1, \alpha; j)b(j) \prod_{k=j+1}^{n-1} G(n, k)\Big], \tag{43}
$$

$$
G(n, j) = \Big[1 + B(n-1, \alpha; j)[2a(j) + b(j) + a(j)b(j)]\Big], \tag{44}
$$

$$
and \quad H(n, j) = \Big[1 + B(n-1, \alpha; j)[a(j) - b(j) + a(j)b(j)]\Big]. \tag{45}
$$

Proof. Define a function $z(n)$ by

$$
z(n) = u(n) + \sum_{j=0}^{n-1} B(n-1, \alpha; j)b(j)[u(j) + \nabla^\alpha u(j+1)]. \tag{46}
$$

Then $z(0) = u(0)$, $\nabla^\alpha u(n+1) \leq a(n)z(n)$ and

$$
\nabla^\alpha z(n+1) \leq b(n)u(n) + [a(n) + a(n)b(n)]z(n). \tag{47}
$$

Since $\nabla^\alpha u(n+1) \leq a(n)z(n)$, an application of Theorem 2 to (47) yields

$$
u(n) \leq u(0) + \sum_{j=0}^{n-1} B(n-1, \alpha; j)a(j)z(j). \tag{48}
$$

Using (48) in (47), we get

$$
\nabla^\alpha z(n+1) \leq b(n)\Big[u(0) + \sum_{j=0}^{n-1} B(n-1, \alpha; j)a(j)z(j)\Big] + [a(n) + a(n)b(n)]z(n). \tag{49}
$$

Adding $b(n)z(n)$ on both sides of the above inequality we have

$$
\begin{aligned}
\nabla^\alpha z(n+1) + b(n)z(n) \leq u(0)b(n) + [a(n) + a(n)b(n)]z(n) \\
+ b(n)\Big[z(n) + \sum_{j=0}^{n-1} B(n-1, \alpha; j)a(j)z(j)\Big]. \tag{50}
\end{aligned}
$$

Let,

$$
v(n) = z(n) + \sum_{j=0}^{n-1} B(n-1, \alpha; j)a(j)z(j). \tag{51}
$$

Then $v(0) = z(0)$, $z(n) \leq v(n)$ and

$$\nabla^\alpha v(n+1) = \nabla^\alpha z(n+1) + a(n)z(n). \tag{52}$$

Using (51) and (52) in (50), we get

$$\nabla^\alpha v(n+1) \leq u(0)b(n) + [2a(n) + b(n) + a(n)b(n)]v(n). \tag{53}$$

Now an application of Theorem 2 to (53) yields

$$v(n) \leq u(0)\left[\prod_{j=0}^{n-1} G(n,j) + \sum_{j=0}^{n-1} B(n-1,\alpha;j)b(j) \prod_{k=j+1}^{n-1} G(n,k)\right] = u(0)F(n). \tag{54}$$

Using (54) in (50), we get

$$\nabla^\alpha z(n+1) \leq [a(n) - b(n) + a(n)b(n)]z(n) + u(0)b(n)[1 + D(n)]. \tag{55}$$

Now again by application of Theorem 2 to (55) yields

$$z(n) \leq u(0)\left[\prod_{j=0}^{n-1} H(n,j) + \sum_{j=0}^{n-1} B(n-1,\alpha;j)b(j)[1 + F(j)] \prod_{k=j+1}^{n-1} H(n,k)\right]. \tag{56}$$

Using (56) in $\nabla^\alpha u(n+1) \leq a(n)z(n)$, we get

$$\nabla^\alpha u(n+1) \leq u(0)a(n)\left[\prod_{j=0}^{n-1} H(n,j) + \sum_{j=0}^{n-1} B(n-1,\alpha;j)b(j)\right.$$

$$\left. \times [1 + F(j)] \prod_{k=j+1}^{n-1} H(n,k)\right]. \tag{57}$$

Now again by application of Theorem 2 to (57), we get the required inequality in (42).

4   Application

In this section we apply the fractional order difference inequality established in Corollary 2 to obtain a bound for the solution of a fractional order difference equation of the form

$$\nabla^\alpha u(n+1) = H\left(n, u(n), \sum_{j=0}^{n-1} B(n-1,\alpha;j)K(n,j,\nabla^\alpha u(j+1))\right) \tag{58}$$

where H and K be any real valued functions defined on $\mathbb{N}_0^+ \times \mathbb{R} \times \mathbb{R}$ and $\mathbb{N}_0^+ \times \mathbb{N}_0^+ \times \mathbb{R}$ respectively and $u(n) : \mathbb{N}_0^+ \longrightarrow \mathbb{R}$ such that

$$|H(n, u(n), v(n))| \leq a(n)[|u(n)| + |v(n)|] \tag{59}$$

$$and \quad |K(n, j, \nabla^\alpha u(j+1))| \leq b(n)|\nabla^\alpha u(j+1)| \tag{60}$$

for $n \in \mathbb{N}_0^+$, where $a(n)$, $b(n)$ and $v(n)$ be real valued nonnegative functions defined on $\mathbb{N}_0^+$. Let $u(n)$ be the solution of (58) for $n \in \mathbb{N}_0^+$. Using (59) and (60) in (58), we get

$$|\nabla^\alpha u(n+1)| \le a(n)\left[|u(n)| + \sum_{j=0}^{n-1} B(n-1,\alpha;j)b(j)|\nabla^\alpha u(j+1)|\right]. \qquad (61)$$

Now a suitable application of Corollary 2 to (61) yields

$$|u(n)| \le u(0)\left[1 + \sum_{j=0}^{n-1} B(n-1,\alpha;j) \prod_{k=0}^{j-1}\left[1 + B(j-1,\alpha;k)a(k)[1+b(k)]\right]\right] \qquad (62)$$

for $n \in \mathbb{N}_0^+$. The right hand side of (62) gives the bound on the solution of (58) in terms of the known functions.

References

1. Agarwal, R.P.: Difference equations and inequalities. Marcel Dekker, New York (1992)
2. Atsushi, N.: Fractional logistic map. arXiv: nlin/0206018 (2002)
3. Deekshitulu, G.V.S.R., Jagan Mohan, J.: Fractional difference inequalities. Commun. Appl. Anal. 14, 89–98 (2010)
4. Deekshitulu, G.V.S.R., Jagan Mohan, J.: Fractional difference inequalities of Bihari type. Commun. Appl. Anal. 14, 343–354 (2010)
5. Diaz, J.B., Osler, T.J.: Differences of fractional order. Math. Comp. 28, 185–201 (1974)
6. Hirota, R.: Lectures on difference equations. Science-sha (2000) (in Japanese)
7. Lakshmikantham, V., Leela, S.: Differential and integral inequalities, vol. I. Academic Press, New York (1969)
8. Pachpatte, B.G.: Inequalities of finite difference equations. Marcel Dekker, New York (2002)
9. Podlubny, I.: Fractional differential equations. Academic Press, San Diego (1999)

SaddleSURF:
A Saddle Based Interest Point Detector

Sajith Kecheril S., Arathi Issac, and C. Shunmuga Velayutham

Computer Vison and Image Processing, Department of Computer Science,
Amrita Vishwa Vidyapeetham University, Coimbatore, 641105, India
{isajith,arathiissac}@gmail.com, cs_velayutham@cb.amrita.edu

Abstract. This paper presents a modified Speeded Up Robust Features (SURF) with feature point detector based on scale space saddle points. Most of the feature detectors like Scale Invariant Feature Transform (SIFT), Principal Component Analysis (PCA)–SIFT and SURF are based on extrema points i.e. local maxima and minima. This work aims at utilizing the saddle points for panorama stitching which is a common and direct application for feature matching. Here Euclidean distance of descriptor is used to find the correct matches. Experiments to test the performance are done on Oxford affine covariant dataset and compared the performance with that of SURF.

Keywords: SURF, Saddle, SaddleSURF, Scale Space, Feature points, Panorama.

1   Introduction

Interest point detectors enjoy a distinct position in most of the computer vision applications like image matching, panorama stitching, image registration and others. Based on the scale space theory by Witkin [1] and Koenderink [2] several feature point detectors were found. With the idea of automated scale selection by Lindeberg [3] scale invariant blobs were used to identify objects of significance. Feature or interest points correspond to specific pixels/ region in image that reflects the local property on which the detector is based. Feature point along with its descriptor uniquely identifies a particular location in an image. This combination is used to find the position of a point in one image to the corresponding point in the next image. In this paper, a new class of feature point detector is introduced, which is based on saddle points in the image. We compare its performance with that of the open source implementation of SURF using oxford affine covariant image dataset. To further analyze, its real world application image panorama stitching is used.

The organization of the paper is as follows section 2 is about the related works, section 3 deals with the methodology to find the saddle based interest points, section 4 is about the experiments done to test the efficacy of the proposed method with a practical application on panorama stitching followed by section 5 on conclusion and discussion.

P. Balasubramaniam and R. Uthayakumar (Eds.): ICMMSC 2012, CCIS 283, pp. 413–420, 2012.

2   Related Works

One of the early interest point detectors is Harris corner detector [4] which is based on the Eigen values of the Hessian matrix. Lindeberg [3] used Laplacian maxima in 3 consecutive scale levels, Lowe [5] found difference of Gaussian in SIFT, Mikolajczyk et al [6] used combination of Harris and Laplacian to find the interest points. Harris-Affine [7] and Hessin-Affine[8] regions are invariant to affine transformations. Shape context by Belongie et al [9] uses edge based features to find correspondence points. Bay et al [12] used hessian maxima for interest points in SURF. Surya Prasanth [16] used total variation based partial differential equation to create non-linear scale space and color similarity is used to link the corresponding regions in the scale space stack.

3   Interest Point Detection

The image stitching application requires finding positions in images which are overlapping. These points are called correspondence points which act as control points for the stitching algorithm. For a point to be considered as a point of importance it has to satisfy a condition like the specified point should not change its differential structure with respect to rotation, scaling or affine transformations. Otherwise, the points need to be stable and be accurately detected in both images. The normalized derivatives attain local maxima at a specific scale and this scale is comparable to the size of objects present [3]. These stable local maxima points can be efficiently used as interest points [10].

3.1   Scale Space Images

The Gaussian scale space images are the images obtained from the convolution of an image with a Gaussian kernel of varying size. i.e. $L: R^2 \to R$

$$L(x, \sigma) = g(x, \sigma) \otimes L(x).$$ (1)

Where, $L(x, 0) = L(x)$, is the original image.

$$g(x, \sigma) = \frac{1}{\left(2\pi\sigma^2\right)^{D/2}} e^{-\frac{\left(x_1^2 + \cdots + x_D^2\right)}{2\sigma^2}}.$$ (2)

$D=2$, the dimension of image and σ is the standard deviation of Gaussian kernel.

3.2   Deep Structure

When the Gaussian of increasing scale is convolved with the image, it gets more and more blurred. Thus, as we move up the scale space image stack the features are decreased or specifically the spatial critical points are annihilated in pairs. The spatial critical points are points where the gradient vanishes, i.e. $\nabla_x L(x,t) = 0$. There are 3 types of critical points depending up on the hessian matrix. A point is maxima or minima if all Eigen values are negative or positive respectively. In case, if they are of both positive and negative sign then the point is a saddle. If one of the Eigen value is zero then the point is called a catastrophe point and is not often encountered in applications. Catastrophe points are defined for $n + 1$ dimension, if the image is of n dimension, so that they can be detected only by analyzing the scale space stack as a whole. In 2 dimensional images the fold catastrophe is the only generic event and occurs when annihilation or creation of extrema and saddle as pointed by Kuijper [11].

3.3   Saddle Interest Points

The extrema saddle annihilation occurs as the scale increases, and finally one extrema remains at the last level [11]. So theoretically the number of saddle is one less than the number of extremas. This makes saddles potential candidate as feature points. Most of the previous works concentrated on finding the extrema points and using them as feature points [3, 12]. In this work the potential of using saddles as features points are considered and its application on to panorama stitching.

This work concentrates on annihilation, the generic event, only. In the scale space stack, saddle points in each scale is tracked and is called saddle branch in the scale space stack, similarly tracking extrema gives extrema branch. In previous works, the extrema were taken as interest points and the scale was chosen in such a way that it was a local extrema in the scale space stack [3, 12], which gives an approximate size of the blob. In saddle branch the behavior is different from that of extrema branch. In this case at annihilation, the intensity of saddle and the extrema becomes same.

3.4   Finding Saddle Points

The rectangular grid has some fundamental disadvantages as 4 neighborhoods to be used for object and 8 neighborhood to be used for background to satisfy the condition of Euler number [13]. In this work, hexagonal neighborhood is used to find the saddle points. The rectangular grid is converted to hexagonal grid using Blom's technique as used by Kuijper [13]. Here, even rows in rectangular grid are shifted half pixel to the right to form the hexagonal grid. In practical case this turns out to be taking the 8 neighborhood and discarding top right and bottom right neighbors for even rows and discarding top left and bottom left elements for odd rows as shown in Fig. 1.

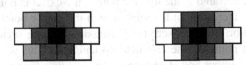

Fig. 1. Center pixel (*black*), hex neighborhood (*medium gray*), excluded pixels (*light gray*) from 8 neighborhoods for odd (*left*) and even (*right*)

For these 6 neighbors the sign of the difference with center pixel is found. This leads to 4 cases [13] as shown in Fig. 2 with

1. No sign change, the given point to be an extrema
2. Two sign changes, the point is a regular point
3. Four sign changes, the point is a saddle.
4. Six sign changes, the point is a monkey saddle

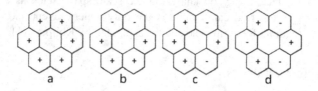

Fig. 2. Sign of the hex neighbor after subtracting center pixel a) extremum b) regular c) saddle d) monkey saddle

3.5   Finding Saddle Maxima Points

After the saddle points are found on various scales next the local maximum in saddle branch is found. The local maxima in 3 consecutive scales, σ_{n-1}, σ_n and σ_{n+1} is found using a $3\times3\times3$ neighborhood as shown in Fig. 3. The scale at which the middle layer gets the maximum value is taken and that particular scale is assigned for the respective saddle. This process is repeated for the entire scale space stack.

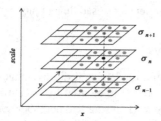

Fig. 3. Local maxima in three consecutive scale levels

3.6   SURF Descriptor

The SURF descriptor is found at the interest points, the extrema points, in the original algorithm. Here, this interest point is replaced with the saddle point. After finding the reproducible orientation SURF descriptor is found. A square region of size 20σ is placed at the interest points in the calculated orientation. This square region is then divided into 4×4 sub regions and Haar response is calculated for each of these regions at 5×5 equally separated positions. The x and y Haar responses found are weighted with a Gaussian of size 3.3σ. From this four responses are derived as the sums and absolute sums in x and y directions, i.e., $\Sigma d_x, \Sigma|d_x|, \Sigma d_y, \Sigma|d_y|$. Thus the response at all the 4×4 sub regions is taken to obtain a descriptor of dimension $4\times4\times4$. The extended version is obtained by further sub dividing the responses and calculating d_x and $|d_x|$ separately for $d_y < 0$ and $d_y \geq 0$. Similarly for d_y also, thus obtaining a descriptor of dimension 128.

4   Experiments and Results

The variant of the original SURF detector/descriptor developed is named as SaddleSURF. For implementation the open source implementation of SURF, OpenSurf, Matlab version by Dirk-Jan is used. The feature detector part in the original SURF is replaced with the proposed algorithm. The SaddleSURF is tested using oxford dataset for rotation, blur, illumination, affine changes and jpeg compression [17].

Repeatability score, used by Lou et.al [14] and Mikolajczyk [6], is used to analyze the performance and compared the same with SURF algorithm. Repeatability is a measure used to find the accuracy of the interest point detector. It is measured with respect to two images as the ratio of the number of corresponding matches found to the minimum no. of features detected in the two images.

$$r_{12} = \frac{c(I_1, I_2)}{\min(n_1, n_2)} \tag{3}$$

r_{12} is the repeatability score obtained for images I_1 and I_2, c is the corresponding matches obtained using SaddleSURF, n_1 and n_2 are the number of feature points detected in I_1 and I_2 respectively.

4.1   Image Blur

The performance under image degradation such as blur is crucial for real world applications like stabilizing image against motion blur. The blurred image set trees and bikes are used for the purpose. The repeatability score is used to find the performance. The Fig. 4 shows the performance. The average repeatability for SURF is 54.06% and 67.04% and for SaddleSURF it is 48.99% and 59.52% for tree and bike respectively.

4.2   Image Scale Change and Rotation

The second experiment tests for the performance under rotation. The real world objects under goes various kinds of variations with respect to the observer. So to identify objects under rotational and scaling transformation is important for object recognition. For this test, boat and bark are used, the results in Fig. 5. The average repeatability for boat is 25.05% for SURF and 26.66% for SaddleSURF and for bark it is 17.02% and 20.38% for the respective cases.

4.3   View Point Variation

The third experiment is on variation in view point on graffiti and wall with results in Fig. 6. For graffiti image the average repeatability is 25.20% for SURF, 26.30% for SaddleSURF, 46.55% and 42.65% for wall image to the respective cases.

4.4   Illumination Changes

The fourth experiment tests the performance variation under changes in lighting condition. Illumination variation is real challenge for computer vision applications especially in robotic vision. The leuven image set used and result is shown in Fig. 7. The average repeatability is 65.35% for SURF and 59.93% for SaddleSURF.

4.5   JPEG Compression

The fifth experiment is on changes in repeatability score due to lossy JPEG compression on UBC and result is shown in Fig. 7. The average repeatability is 66.78% for SURF and 55.27% for SaddleSURF.

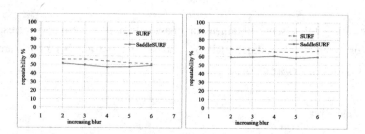

Fig. 4. Blur comparison using trees (*left*) and bikes (*right*)

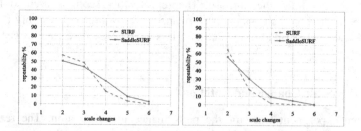

Fig. 5. Scale change and rotation comparison using boat (*left*) and bark (*right*)

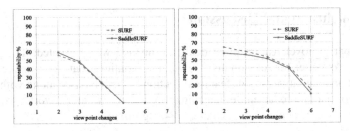

Fig. 6. View point variation comparison using graffiti (*left*) and wall (*right*)

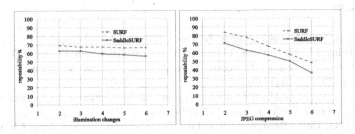

Fig. 7. Ilumination variation comparison using leuven (*left*) and JPEG compression comparison using UBC (*right*)

4.6 Panorama Stitching

The experimental results show that for rotation and view point variations SaddleSURF performs better than SURF algorithm. So image panorama stitching was done using the SaddleSURF. Once the overlapping images are obtained, the stable saddle points are found on both the images. At these points, the SURF features are obtained and nearest neighbor matching is done. The detected points on the right half of the left image and left half of the right image are taken so that the region in the overlapping parts alone will be matched. Then, to eliminate error matches median filtering is applied. Once the matching points are obtained the transformation matrix from the input control points is found, after this the second image is transformed and stitched with the first image. For panorama stitching, Matlab code by Michael Carrol and Andrew Davidson [15] is used.

Fig. 8. Stitched image using SaddleSURF (*left*) zoomed in image at the seam (*right*)

5   Conclusion and Discussion

In this paper, a novel feature point detection method was presented. The proposed method works better for rotation and view point variations than the existing method. This is not an alternative way of feature detection but can be used along with existing methods to find more accurate matching points. As a future work, we plan to implement a faster version of the SaddleSURF algorithm.

References

1. Witkin, A.P.: Scale-space Filtering. In: Proc. of IJCAI, Karlsruhe, pp. 1019–1021 (1983)
2. Koenderink, J.J.: The Structure of Images. Biological Cybernetics 50, 363–370 (1984)
3. Lindeberg, T.: Feature Detection with Automatic Scale Selection. J. Computer Vision 30(2), 79–116 (1998)
4. Harris, C., Stephens, M.: A Combined Corner and Edge Detector. In: Proceedings of the 4th Alvey Vision Conference, pp. 147–151 (1988)
5. Lowe, D.G.: Object Recognition from Local Scale-invariant Features. In: ICCV, pp. 1150–1157 (1999)
6. Mikolajczyk, K., Schmid, C.: Indexing Based on Scale Invariant Interest Points. In: Proc. Eighth International Conference on Computer Vision, pp. 525–531 (2001)
7. Mikolajczyk, K., Schmid, C.: Scale and Affine Invariant Interest Point Detectors. J. Computer Vision 1(60), 63–86 (2004)
8. Mikolajczyk, K., Tuytelaars, T., Schmid, C., Zisserman, A., Matas, J., Schaffalitzky, F., Kadir, T., Gool, L.V.: A Comparison of Affine Region Detectors. J. Computer Vision 65(2), 43–72 (2005)
9. Belongie, S., Malik, J., Puzicha, J.: Shape Matching and Object Recognition using Shape Contexts. IEEE Trans. Pattern Analysis. Machine Intelligence 2(4), 509–522 (2002)
10. Bay, H., Tuytelaars, T., Van Gool, L.J.: SURF: Speeded Up Robust Features. In: Leonardis, A., Bischof, H., Pinz, A. (eds.) ECCV 2006. LNCS, vol. 3951, pp. 404–417. Springer, Heidelberg (2006)
11. Kuijper, A.: Exploring and Exploiting the Structure of Saddle Points in Gaussian Scale Space. J. Computer Vision and Image Understanding 112, 337–349 (2008)
12. Bay, H., Ess, A., Tuytelaars, T., Van Gool, L.J.: Speeded-Up Robust Features (SURF). J. Computer Vision and Image Understanding 110(3), 346–359 (2008)
13. Kuijper, A.: On Detecting All Saddle Points in 2D Images. Pattern Recognition Lett. 25, 1665–1672 (2004)
14. Juan, L., Gwun, O.: A Comparison of SIFT, PCA-SIFT and SURF. J. Image Processing 3(4), 143–152 (2009)
15. Michael Carroll projects, http://michaelroyce.org/projects2.html
16. Surya Prasanth, V.B.: Color Image Segmentation Based on Vectorial Multiscale Diffusion with Inter-scale Linking. In: Chaudhury, S., Mitra, S., Murthy, C.A., Sastry, P.S., Pal, S.K. (eds.) PReMI 2009. LNCS, vol. 5909, pp. 339–344. Springer, Heidelberg (2009)
17. Affine Covariant Features, http://www.robots.ox.ac.uk/~vgg/research/affine/

Modified Color Layout Descriptor
for Gradual Transition Detection

Lakshmi Priya G.G. and Domnic S.

Department of Computer Applications, National Institute of Technology,
Tiruchirappalli, India 620015
gg_lakshmipriya@yahoo.co.in, domnic@nitt.edu

Abstract. Shot transition detection is a basic step in video content analysis and retrieval. A number of automatic detection techniques exist but they leak their performance in the detection of gradual transitions. These techniques often suffer from high false detection rates due to the presence of illumination, camera or object movements in the video sequences. In order to reduce the influences of these effects in the detection process and to detect long dissolve and fade regions in the video sequences, we propose a modified color layout descriptor with a new similarity metric; suits well for the extracted features to identify and classify the gradual transitions. Experiments are carried out on various video sequences taken from TRECVid dataset and publically available dataset. The results show that the proposed method yields better result compared to that of the existing gradual transition detection methods.

Keywords: Gradual transition detection, Modified color layout descriptor, Similarity metric, Decision making, Performance evaluation.

1 Introduction

Recent advances in multimedia technology have led to the extensive availability and use of the digital video. Many applications generate and use large collections of video. This has created a need for tools that can efficiently index, search, browse and retrieve the relevant data. Video shot boundary detection is the initial step for automatic annotation of the digital video sequences. Its goal is to divide the video stream into a set of meaningful and manageable segments called shots, which are the basic elements for indexing. A shot is an uninterrupted and continuous segment in a video sequence that defines the building blocks of video content. It is comprised of a number of consecutive frames filmed with a single camera with variable durations.

Depending on the nature of transitions between two consecutive shots, shot boundaries are classified into two types, abrupt (cut) and gradual transitions. The common way for detecting the transitions is to evaluate the difference between consecutive frames represented by certain feature. For shot boundary detection, different algorithms have been proposed based on pixel-wise differences [1], histogram [2], edge [3], motion vectors [4], DC frame [5], DCT coefficients [5, 6], bit-rate information [7], Macro block information [8], Visual descriptor based [9,10] and so on.

P. Balasubramaniam and R. Uthayakumar (Eds.): ICMMSC 2012, CCIS 283, pp. 421–428, 2012.
© Springer-Verlag Berlin Heidelberg 2012

In paper [10], Color Layout Descriptor (CLD) [11] is considered as feature and the difference between the consecutive frames are computed. In the MPEG-7 standard [9], it is recommended to use a total of 12 coefficients, 6 for luminance and 3 for each chrominance. The detection of shot boundaries are performed by using adaptive threshold technique as followed in [12]. As stated by the author [10] this method works well for hard cuts but lacks its performance in the presence of long dissolve and fade regions in the video stream. Our objective is to improve the detection process in the presence of long dissolve and fades. In order to achieve this objective, we propose to modify the existing color layout descriptor [10] and a new similarity metric which suits well for the extracted features. Also, a different identification and classification strategy is followed in this paper. The structure of the paper is as follows: proposed feature extraction using Modified Color Layout Descriptor, similarity metric and gradual transition identification is described in section 2, experimental results are discussed in section 3 and conclusions are narrated in section 4.

2   The Proposed Method

In this section, we present a general framework for efficient feature extraction from the video sequences. In this paper, we have modified the color layout descriptor [10] and named as Modified Color Layout Descriptor (MCLD), which contains characteristics of color based features. The overall block diagram of the proposed method is shown in Fig. 1.

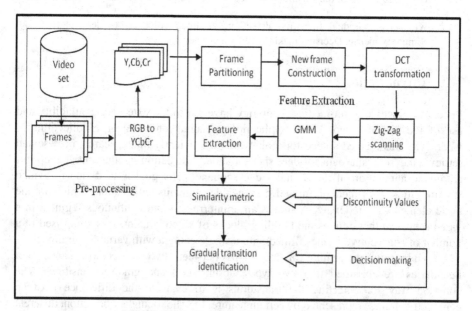

Fig. 1. Block diagram of the proposed method

2.1   Feature Extraction

As a pre-processing stage, frames are resized to 256 x 256 and RGB frames are converted to YCbCr frames. This is because RGB color space often not suited for frame processing purposes due to a number of reasons [14]. In our method, RGB frames are converted into Y, Cb, Cr component frames using the equation given below

$$Y = 0.299 * R + 0.587 * G + 0.114 * B$$
$$Cb = 0.169 * R - 0.331 * G + 0.500 * B \qquad (1)$$
$$Cr = 0.500 * R - 0.419 * G - 0.081 * B$$

The extraction process of the MCLD from a frame consists of six stages where the first three stages are similar to that of the CLD feature extraction: frame-region partitioning, representative color extraction, Child frame, DCT transformation [13], zigzag-scanned coefficients, and coefficients clustering.

For each color component perform the following: Partition the frames into equal sized non overlapping regions or blocks of size (8 x 8). On doing so, the number of blocks per frame is 1024. For each block, a single representative color is selected by applying average of the pixel intensities. As a result, a new child frame of size 32 x 32 is obtained as shown in Fig. 2, where each pixel contains the representative color of its parent blocks.

The child frame is further partitioned into 8 x 8 blocks and DCT is performed over each block. For a standard 8 x 8 block, the DCT $D(u,v)$ is defined as:

$$D(u,v) = \frac{1}{4} V(u) V(v) \sum_{x=0}^{7} \sum_{y=0}^{7} P(x,y) \cos\left[\frac{(2x+1)u\pi}{16}\right] \cos\left[\frac{(2y+1)v\pi}{16}\right] \quad (2)$$

$$\text{Where } V(l) = \begin{Bmatrix} \frac{1}{\sqrt{2}} & if \ l = 0 \\ 1 & if \ l > 0 \end{Bmatrix}$$

$P(x,y)$ is the x,y^{th} element of the block. Now from the child frame, number of blocks obtained is 16. From each transformed block, for Y - 6 coefficients, for Cb – 3 coefficients and for Cr - 3 coefficients where as overall 12 coefficients are extracted by Zig-zag scanning as shown in Fig. 3.

From Fig. 3, the coefficients that are specified with '**' are selected for Y component, '@' for Cb and Cr components. On the whole, 16 x 12 (192) coefficients will be generated per frame. When compared to that of the CLD these coefficients are too large in number and this can be reduced. In order to reduce the number of coefficients, new components are extracted using Gaussian Mixture model (GMM) [15] based clustering and the cluster mean values are considered as the feature representative of the whole child frame. For each color component, cluster C_i is defined as follows:

$$c_i^D = \{D_{ij}, \lambda_i^D\} \qquad (3)$$

where D is the color component, 'i' is the number of cluster; 'j' is the number of coefficients that varies depending on the color component i.e. 6 for Y, 3 for Cb and Cr.

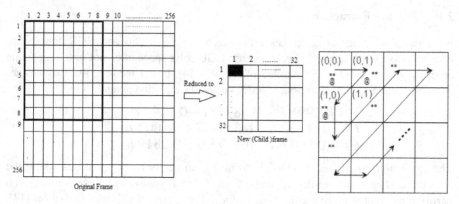

Fig. 2. New Child frame construction   **Fig. 3.** Zig-Zag scanning

λ_i^D is the fraction of the blocks in the cluster as compared to the total number of blocks in the frame of size m x n.

$$\lambda_i^D = \frac{\text{number of blocks} \in C_i^D}{N} \qquad (4)$$

N is the number of blocks in the child frame. If two clusters are considered then 12 coefficients for Y, 6 for *Cb* and *Cr* which is totally of 24 coefficients + 6 (λ_i) features are extracted from each frame. This can be further reduced by considering either of the *Cb* or *Cr* chrominance components for feature extraction. On doing so, the total number of features extracted from the child frame will be reduced to 18 coefficients + 4 (λ_i) features. Based on the features extracted from the video frames, the similarity between the consecutive frames is calculated using a new similarity metric.

2.2   Similarity Metric

As a summary, each frame has 22 (18+4) feature coefficients as a representative for the whole frame. λ_i holds the information about ratio of blocks in the cluster *i*. The similarity between the consecutive frames is calculated by finding the relation between the clusters of the frames *k* and *k+1*. For finding the similarity/dissimilarity we use the weighted distance measure:

$$S(k, k+1) = \sum_D \omega_D \sum_{i=1,2} \sqrt{(C_{ik}^D - C_{i,k+1}^D)^2} \qquad (5)$$

where ω_D is the weight value calculated as

$$\omega^D = \max(\lambda_{ik}^D, \lambda_{ik+1}^D) \qquad (6)$$

D is the Y and Cb components, i is the number of cluster for each component. The similarity metric $S(k,k+1)$ holds information about the relative frequency of the blocks of the corresponding clusters as well as the color information between the two clusters. On calculating the similarity values the next step is to classify the long dissolve and fade regions along with camera / object motions and zooming effects that occur in the video sequences.

2.3   Gradual Transition Identification

The gradual transition from one shot to another takes place progressively. In general these transitions lost for 1s -2s and in some situations the duration of these transitions are long i.e. exceeds the range. As mentioned the performance of CLD [10], its corresponding similarity metric and the adaptive threshold technique leak its performance in the presence of long dissolve and fade. To improve its performance, we mainly focused on (i) a new feature that yields better representation of the whole frame (feature extraction), (ii) new metric to find the similarity between the consecutive frames (similarity metric) and (iii) the process of detecting the transition occurrence (identification and classification). So far, in the previous section, we have discussed about the first two steps. Here, in this section our focus is mainly on identification of transitions and the occurrence of camera and object motions which may lead to false detection. The following steps are involved in the gradual transition identification which focuses on all the issues discussed. The steps are

Step 1: Calculate the mode, lower bound and upper bound using

$$upperbound = \bar{x} + \sqrt{\frac{1}{n-1}\sum_{i=1}^{n-1}(S_i - \bar{x})^2}$$

$$\text{where } \bar{x} = \frac{1}{n-1}\sum_{i=1}^{n-1}(S_i)$$

$$lowerbound = \bar{x} - \sqrt{\frac{1}{n-1}\sum_{i=1}^{n-1}(S_i - \bar{x})^2} \tag{7}$$

where S_i denotes the similarity value $S(k,k+1)$ between the frames k and $k+1$, n is the total number of frames in the video sequence.

Step 2: if $S_i > upperbound$ then discard those values because those regions have more similarity values which is due to the occurrence of camera / object motions or zooming effects in videos.

Step 3: On considering the values which lie between the '*lowerbound*' and '*upperbound*' values calculated using equation (7) the gradual transitions are identified by finding the two consecutive peaks which indicate the start and end of the transitions. Later, the frame count (α) between the start peak and end peak are calculated and if the α value is equivalent to number of gradual transition frames (30-60) then the region is declared as the gradual shot region.

3   Experimental Results and Discussions

To evaluate our proposed method, we need to consider test data set and some related existing works for comparison purpose. Test data are taken from TRECvid data set [16] and publically available data [17, 18]. Here, data is selected which contains long gradual transitions, fast object and camera movements. The description of the test data sets are summarized in Table 1.

Table 1. Test Sequences

S.No	Test Data	Duration (s)	Number of frames	Number of transitions
1	BG_35917	14	350	5
2	Anni001	30	1170	8
3	F.Mpg	7	236	3
4	Movie songs I	292	7592	43
5	Paycheck	420	11760	25
6	Movie songs II	323	9044	36
Total		1086	30152	117

The performance of the whole transition detection process is evaluated by using Precision, Recall and combined measure as given in equation (8). Higher these ratio, the result obtained is better. The proposed method is implemented in Matlab 7.6.

$$\text{Precision (P)} = \frac{A}{A + B}$$
$$\text{Recall (V)} = \frac{A}{A + C} \tag{8}$$
$$\text{Combined Measure (F1)} = \frac{2.P.V}{(P + V)}$$

where A is the number of correctly detected hits, B is the number of false hits and C is the number of missed hits. The precision, recall and combined measures of the proposed method and CLD are listed in Table 2. Since our objective is to improve the performance of the detection process during the occurrence of long gradual transitions, we have considered [10] for comparison purpose. As described in the section 2.1, the proposed method can consider Y, Cb, Cr or Y, Cb or Y, Cr components for feature extraction step where the features count also varies accordingly. The experimental comparison of the proposed method and [10] is listed in Table 2.

From the Table 2, it is observed that the performance of our algorithm on considering 18+4 (22) coefficients yields better results than CLD. So it is better to have 22 coefficients for our further consideration. The Fig. 4 shows the occurrence of the gradual transitions and the camera, object motions. In order to visualize possible editing effects in the Fig. 4, frame skipping technique is used and the frame count between the gradual transitions appears to be less.

Table 2. Performance Evaluation

Test Data	Proposed Method (22 coefficients)			Color layout Descriptor[10]		
	Precision (P)	Recall (V)	Combined Measure (F1)	P	V	F1
BG_35917	100	100	100	80	80	80
Anni001	88.88	100	94.11	87.5	87.5	87.5
F.Mpg	100	100	100	100	66.66	80
Movie songs I	95.34	95.34	95.34	92.68	88.37	90.47
Paycheck	95.65	88	91.66	83.33	80	81.63
Movie songs II	97.14	97.14	97.14	93.93	86.11	89.85
Average	**96.16**	**96.74**	**96.37**	**89.57**	**81.44**	**84.90**

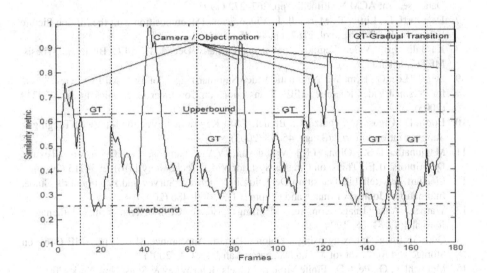

Fig. 4. Occurrence of gradual transitions, camera / object motion and zooming effects

4   Conclusions

In this paper, we have focused mainly on detection of gradual transitions, more specifically long transition effects in the video sequences. Our proposed method is capable of detecting the transitions in the presence of object/ camera motion and zooming effects. Evidence for this statement is depicted through the performance evaluation where the proposed method yields overall 96.37 % of results which are better than the existing CLD based video shot boundary detection methods.

References

1. Yeo, B., Liu, B.: Rapid Scene Analysis on Compressed Video. IEEE Trans. Circuits and Systems for Video Technology 5, 533–544 (1995)
2. Han, S.H., Yoon, K.J., Kweon, I.S.: A New Technique for Shot Detection and Key Frame Selection in Histogram Space. In: 12th Workshop on Image Processing and Image Understanding, pp. 475–479 (2000)
3. Zabih, R., Miller, J., Mai, K.: A Feature-based Algorithm for Detecting Cuts and Classifying Scene Breaks. In: ACM Multimedia, San Francisco, pp. 189–200 (1995)
4. Fernando, W.A.C., Canagarajah, C.N., Bull, D.R.: Video Segmentation and Classification for Content Based Storage and Retrieval Using Motion Vectors. In: SPIE Conference on Storage and Retrieval for Image and Video Databases, vol. 3656, pp. 687–698 (1999)
5. Truong, B.T.: Shot Transition Detection and Genre Identification for Video Indexing and Retrieval, Honours, School of Computing, Curtin University of Technology (1999)
6. Arman, F., Hsu, A., Chiu, M.Y.: Image Processing on Compressed Data for Large Video Databases. In: ACM Multimedia, pp. 267–272 (1993)
7. Deardorff, E., Little, T., Marshall, J.: Video Scene Decomposition with the Motion Picture Parser. In: IS&T/SPIE, vol. 2187, pp. 44–45 (1994)
8. Kayaalp, I.B.: Video Segmentation using Partially Decoded MPEG Bitstream. Thesis, METU, Ankara (2003)
9. Lee, J., Lee, G., Kim, W.: Automatic Video Summarizing Tool using MPEG-7 Descriptors for Personal Video Recorder. IEEE Transactions on Consumer Electronics 49(3), 742–749 (2003)
10. Borth, D., Ulges, A., Schulze, C., Breuel, T.M.: Keyframe Extraction for Video Tagging & Summarization. Informatiktage, 45–48 (2003)
11. Manjunath, B.S., Ohm, J.R., Vasudevan, V.V., Yamada, A.: Color and Texture Descriptors. IEEE Trans. on Circuits Syst. for Video Technology 11(6), 703–715 (2001)
12. Lienhart, R.: Reliable Transition Detection in Videos: A Survey and Practitioner's Guide. International Journal of Image and Graphics 1(3), 469–486 (2001)
13. Watson, A.B.: Image compression using Discrete Cosine Transform. Mathematica Journal 4(1), 81–88 (1994)
14. Qiu, G.: Image Indexing using a Colored Pattern Appearance Model. In: SPIE Conf. on Storage and Retrieval for Media Databases, San Jose, CA (2001)
15. McLachlan, G., Peel, D.: Finite Mixture Models. John Wiley & Sons, New York (2000)
16. TRECVID Dataset, http://trecvid.nist.gov/
17. Carleton University, http://iv.csit.carleton.ca/~awhitehe/vidproc/data.html
18. The open video project, http://www.openvideo.com

Personal ID Image Normalization Using ISO/IEC 19794-5 Standards for Facial Recognition Improvement

K. Somasundaram[1] and N. Palaniappan[2]

[1] Image Processing Lab, Department of Computer Science and Applications
[2] Computer Centre,
The Gandhigram Rural Institute – Deemed University, Gandhigram – 624 302, India
somasundaramk@yahoo.com, naapala@gmail.com

Abstract. Facial recognition methods are the basis for security systems. Identification and verification are the Facial recognition methods are used to identify and authorize persons. For automated face recognition, the facial images should be normalized to improve the efficiency of recognition. In this paper we normalize personal ID images using ISO/IEC 19794-5 standards. Experimental results show that the proposed algorithm significantly improved face recognition efficiency.

Keywords: Facial identification and verification, Frontal image, Image normalization.

1 Introduction

Identification of personnel in an organization and authenticate them to access the vital resources are part of the security system. Authentication systems make use of some unique biological features of the person like personal ID image, finger print, iris print, etc. One of the most commonly used and reliable features is the personal ID image [1]. Personal ID images or facial images are used for identification or verification of persons. Facial recognition methods require facial images in a form, in which the features are identified as accurately as possible. Among the facial features, eyes are very prominent and easily identifiable. Eye position cannot be easily affected by various facial expressions, head rotation and scaling. Therefore face alignment is necessary. Face alignment process, spatially scales and rotates a face image with eyes as a reference point. Facial alignment results in normalized face image for matching [2].

Facial recognition methods find their application not only within the organizations but also in vital processes like issue of visa and e-passport. So, there is a necessity for a digital facial image with features in a standard format. In 1980, the International Civil Aviation Organization (ICAO) started a project on machine assisted biometric identification of persons. Based on this project, in 2004 the International Standard Organization (ISO) defined digital face image standard for machine readable travel documents [3]. This standard specifies the data format, scene constraints, photographic properties and digital image attributes.

P. Balasubramaniam and R. Uthayakumar (Eds.): ICMMSC 2012, CCIS 283, pp. 429–438, 2012.

The aim of our paper is to develop an algorithm to normalize the personal ID image of a person to a token frontal image adhering to the digital image attributes of ISO standards. This token frontal image format is size normalized with specific width and height and geometrically aligned which can give better results while verification or identification. The results of the proposed method are verified using Principal Component Analysis (PCA) face recognition algorithm. The remaining part of the paper is organized as follows. In section 2 ISO image features are given. In section 3, the proposed normalization process is explained. In section 4, results and discussions are given. Conclusion is given in section 5.

2   ISO/IEC 19794-5 Standards for Digital Image Attributes

ISO standard specifies the data format, scene constraints (pose, expression etc.), photographic properties (lighting, positioning, camera focus etc) and digital image attributes (image resolution, image size etc). The fundamental face image types are given below.

2.1   Basic Image

Facial image with the record format, including header and image data without any specific requirements.

2.2   Frontal Image

Facial image, as per the ISO requirements is appropriate for face recognition and human examination. Full frontal and token frontal are the two types of frontal images.

2.2.1 Full Frontal Image
Frontal image with required resolution for human and computer face recognition. This image includes full head with all hair, neck and shoulders as shown in Fig. 1.

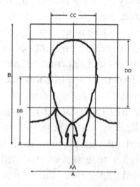

Fig. 1. Full frontal image

2.2.2 Token Frontal Image

Token frontal image is obtained by using the specification given in Table 1. Eye positioning is based on the dimensions of the image. Fig. 2 shows a token frontal image with parameters and sample values in pixels.

Table 1. Geometric characteristics of a token frontal image

Feature or Parameter	Parameter	Example (Pixels)
Image Width	W	240
Image Height	W/0.75	320
Y coordinate of Eyes	0.6 * W	144
X coordinate of right eye	(0.375 * W) -1	89
X coordinate of left eye	(0.625 * W) – 1	150
Width from eye to eye (inclusive)	0.25 W	60

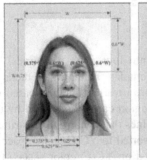

Fig. 2. A sample token frontal image

2.3 Pose Angles

The face view of an image may appear in different orientation. The head rotation is classified into three categories viz. yaw, pitch and roll as shown in Fig. 3. The yaw is the rotation about the vertical axis, pitch is the rotation about the horizontal side-to-side axis and the roll angle is the rotation about the horizontal back to front axis. Frontal images should have the yaw, pitch and roll angles as 0.

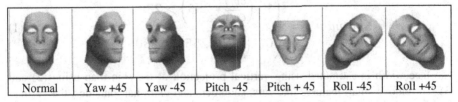

| Normal | Yaw +45 | Yaw -45 | Pitch -45 | Pitch + 45 | Roll -45 | Roll +45 |

Fig. 3. Pose angles – Yaw, Pitch and Roll

3   Normalization of Personal ID Image to Token Frontal

To use the given personal ID image, first it must be converted to a form that the recognition algorithm could produce the highest possible accuracy of recognition. This is done through the normalization process. The stages involved are pre-processing, face detection, eye location, eye localization, roll angle alignment, interpupillary distance adjustment and token frontal image segmentation.

3.1   Pre-processing

Personal ID images having single face in the image, with uniform white or light grey background are taken for this study. Pose angles should be 0 for yaw and pitch. An angle range of 30 to -30 is allowed for roll.   Red-eye removal is done using Photoshop.   Illumination is normalized using adaptive histogram equalization and median filter is applied to de-noise the image.

3.2   Face Detection

Face detection methods are classified into four main categories, knowledge-based methods, feature invariant approaches, template matching methods and appearance-based methods [4]. We propose a feature invariant method based on integration of skin color, size and shape. Skin-color region can be identified by the presence of a certain set of chrominance (C_b and C_r) values that is narrowly distributed in the YC_bC_r color space [5].   The test image is converted to YC_bC_r color space.   Hsu et. al. [6] proposed a face detection algorithm based on YC_bC_r color space in varying lighting conditions with complex backgrounds which involves more computation.   Chan and Abu-Bakar [7] used (1) for eyes' candidate location into four binary images of pixel values 255 – 231, 230 – 224, 223 – 216 and 215 - 0. Comparing the four binary images with each other manually, the best approximation of eye map is identified. Trial experiments on number of personal ID images with uniform background calculated using (1) gives the skinmap with eye locations in the pixel range 231 – 255. Non-skin features like eyes and eyebrows within the face area are left as holes in the skinmap. Using morphological operations other small features like light reflection on the face are removed.

$$Skinmap_{i,j} = \begin{cases} 1 & \left(\frac{Cb_{i,j}}{Cr_{i,j}} * 255\right) > 230 \\ 0 & \left(\frac{Cb_{i,j}}{Cr_{i,j}} * 255\right) \leq 230 \end{cases} \tag{1}$$

The face width is estimated using vertical projection of the skin map using (2) and (3). To remove small features like ears, shoulders and extra hair on sides, the vertical mean of the skin map is subtracted from the vertical sum.

$$VP_{j=1..C} = \left(\sum_{i=1}^{R} Skinmap_{i,j}\right) - \frac{1}{C}\left(\sum_{j=1,i=1}^{C,R} Skinmap_{i,j}\right), \tag{2}$$

$$VP_{j=1..C} = \begin{cases} VP_x & VP_x > 0 \\ 0 & VP_x \le 0 \end{cases} 1 \le x \le C, \tag{3}$$

where VP is the vertical sum, R,C are the number of rows and columns of the skin map. The left and right edges of the projections y_1 and y_2 estimate the left most and right most boundaries of the face area. The difference of $y_2 - y_1$ gives the face width. Fig. 4 shows the estimation of left and right boundaries.

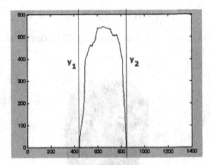

Fig. 4. Face width estimation

The face height is calculated by horizontal projection (HP) of the skin map given by (4) and (5). Horizontal mean of the skin map is subtracted from the horizontal sum to eliminate small features.

$$HP_{i=1..R} = \left(\sum_{j=1}^{C} Skinmap_{i,j}\right) - \frac{1}{R}\left(\sum_{i=1,j=1}^{R,C} Skinmap_{i,j}\right) \tag{4}$$

$$HP_{i=1..R} = \begin{cases} HP_x & HP_x > 0 \\ 0 & HP_x \le 0 \end{cases} 1 \le x \le R \tag{5}$$

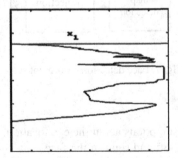

Fig. 5. Estimation of upper most face boundary

The upper edge of the projection gives the upper most boundary of the face x_1. Fig. 5 shows the upper most boundary of the facial area. Normally personal ID images will cover up to the chest or shoulders at the bottom. So, the lower edge of the projection cannot be taken as the lower most boundary of the face area. The golden ratio (Φ) for an ideal face based on anthropometric facial proportions [8] is given by

$$\Phi = \frac{height}{width} \equiv \frac{1+\sqrt{5}}{2} \cong 1.618 \qquad (6)$$

Since all the faces are not ideal, the lower most boundary is estimated to the chin or neck of the images. The lower most boundary x_2 of the face area is estimated using (7).

$$x_2 = ((y_2 - y_1) * 1.618) + x_1 \qquad (7)$$

The complement of the skin map area covered by the coordinates $(x_1, y1) - (x_2, y2)$ is the face area as show in Fig. 6.

Fig. 6. Detected face area

The different stages of face detection system is shown in Fig. 7.

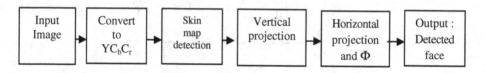

Fig. 7. Face detection – different stages

3.3   Eye Location

We make the following assumptions about the eye location. The eyes are in the upper half of the face, placed on left and right of the face centers; height of the eyes is less than their width and can be verified as a pair. If more than one pair is found, then the upper pair will be ignored as eye brows. Using connected component analysis the available components are selected and verified with the above assumptions and non-matching components are removed. The eyebrows are removed using the horizontal projection. Vertical projection is also used to remove other components like nose and mouth. Fig. 8 shows the detection of eye locations. Let (x_r, y_r) and (x_l, y_l) be the centers of the detected rough eye locations.

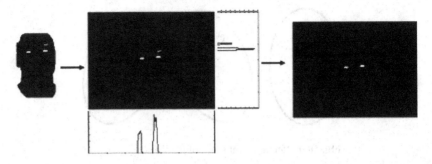

Fig. 8. Detection of eye locations

3.4 Eye Localization

The eye localization is done to find the exact centre of the iris to calculate the interpupillary distance. Many research works are available for eye localization based on shape and template matching [9], AdaBoost classifiers [10], SVM techniques [11] and Haar features [12]. We adopted a simple algorithm proposed by Bonney et.al. [13] which uses the least significant bit-plane and morphological operations.

Two windows of size $(x_r-75, y_r-50) - (x_r+75, y_r+50)$ and $(x_l-75, y_l-50) - (x_l+75, y_l+50)$ for right and left eyes of the Y component of the input image are used for eye localization. Image contrast is adjusted using gamma correction. The pixel values in the ranges 0 - 60 and 241 - 254 are replaced with 255. This enables the pupil to binary '1' in all bit-planes. Then, the least significant bit-plane is extracted. After applying morphological erosion, filling and connected component analysis techniques, the resultant blob represents the middle part of the iris. By finding the end points in the upper, lower, left and right edges the refined centre of the iris can be computed. Fig. 9 shows the various stages of eye localization.

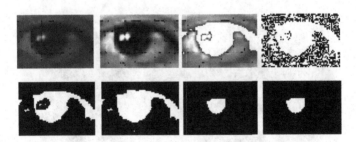

Fig. 9. Various stages of eye localization

3.5 Roll Angle Alignment

The roll angle is determined by a horizontal line passing through the centre of the eye. The orientation of the line $(x_r,y_r)-(x_l,y_l)$ defines the roll angle. Fig.10 shows the different orientations of the horizontal line. To bring the correct position, the image is rotated over the y-axis by an angle θ given by

$$\Theta = atan\left(\frac{y_r-y_l}{x_r-x_l}\right) \tag{8}$$

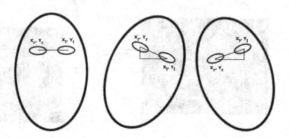

Fig. 10. Different orientations of the line (x_r,y_r)-(x_l,y_l)

3.6   Interpupillary Distance Adjustment

As per the ISO standard, the interpupillary distance should be one fourth of the resultant image's width. For an image of width 240 pixels, the interpupillary distance should be 60 pixels. The distance between the eye centers (x_l,y_l) and (x_r,y_r) are measured and adjusted to 60 pixels by resizing the image using bicubic interpolation technique with a scaling factor d_{eye} given by

$$d_{eye} = \frac{60}{|y_r - y_l|} \tag{9}$$

3.7   Token Frontal Image Segmentation

The required token frontal image is segmented from the resampled, scaled, rotated and interpupillary distance adjusted image, by cutting window with co-ordinates $(x_r$-143, y_r-89) and $(x_r$+173, y_r+150) with (x_r,y_r) as the reference co-ordinate, as explained in Table 1. Fig. 11 shows an input image and an aligned token frontal image.

Fig. 11. (a) Preprocessed image (b) Aligned token frontal image

4   Experimental Results

We carried out experiments by using Matlab Version 7.8.0 on images in CVL face database [14], having images of 114 persons and our own Gandhigram Rural Institute (GRI) face database of 60 images. The performance of the algorithm is evaluated by finding eye locations. This algorithm failed on images having light reflections on their spectacles or half-closed eyes. Fig.12 shows few of the images for which our method failed. Even in the failed cases, the algorithm can still find a single eye if there is a moderate light reflection.

Fig. 12. Few of the images where our method failed due to (a) reflection, (b) reflection and (c) eye balls not visible / very small

We employed Principal Component Analysis (PCA) face recognition with euclidean distance algorithm to verify the efficiency of the facial image normalization. The facial images are well aligned with a fixed eye position, fixed size and zero angle of head rotation. Facial images that are not fulfilling the above requirements are excluded from the gallery images[1]. When tested with the probe images[2], our normalization algorithm gives comparable results. Even it can recognize partially occluded facial images.

Table 2. Comparison of facial image recognition before alignment and after alignment

Database	Before normalization	After normalization
CVL	77%	84%
GRI	28%	54%

Table 2 gives the comparison of the facial image recognition before alignment and after alignment. The source images from CVL database are of uniform size 640 x 480 pixels, with a uniform camera distance[3]. In the case of GRI database the images are un uniform in size, with different resolution and the camera distance also varies from image to image. The results clearly show that the image normalization increases the rate of face recognition.

5   Conclusion

In this paper, we have proposed an algorithm for personal ID image normalization using ISO/IEC 19794-5 standards. The resultant normalized images are having fixed eye position, fixed size and zero angle of head rotation. These normalized images

---

[1] Images included in training phase.
[2] Images used to test the algorithm.
[3] The distance between the person and the camara while taking photograph.

improve the face recognition efficiency. This algorithm finds it applications in identification and verification of personnel in organizations and vital activities like issue of visa and e-passports.

References

1. Somasundaram, K., Palaniappan, N.: Adaptive Low Bit Rate Facial Feature Enhanced Residual Image Coding Method using SPIHT for Compressing Personal ID Images. Int. J. Electron. Commun. 65, 589–594 (2011)
2. Wang, P., Tran, L.C., Ji, Q.: Improving Face Recognition by Online Image Alignment. In: 18th IEEE International Conference on Pattern Recognition, pp. 311–314. IEEE Press, New York (2006)
3. Biometric Data Interchange Formats – Part 5: Face Image Data Draft Revision 19794–5 (2004)
4. Yang, M.-H., Kriegman, D.J., Ahuja, N.: Detecting Faces in Images: A Survey. IEEE Trans. Pattern Anal. Mach. Intell. 24, 34–58 (2002)
5. Chai, D., Ngan, K.N.: Locating Facial Region of a Head-and-shoulders Color Image. In: IEEE International Conference on Automatic Face and Gesture Recognition, pp. 124–129. IEEE Press, New York (1998)
6. Hsu, R.L., Mottaleb, M.A., Jain, A.K.: Face Detection in Color Images. IEEE Trans. Pattern Anal. Mach. Intell. 24, 696–706 (2002)
7. Chan, Y.H., Abu-Baker, S.A.R.: Face Detection System Based on Feature-Based Chrominance Color Information. In: IEEE International Conference on Computer Graphics, pp. 153–158. IEEE Press, New York (2004)
8. Frakas, L.G., Munro, I.R.: Anthropometric Facial Proportion in Medicine. Charles C. Thomas Publisher, Springfield IL (1987)
9. Wang, Q., Yang, J.: Eye Detection in Facial Images with Unconstrained Background. J. Pattern Recognition Research 1, 55–62 (2006)
10. Viola, P., Jones, M.: Rapid Object Detection using a Boosted Cascade of Simple Features. In: IEEE Conference on Computer Vision and Pattern Recognition, vol. 1, pp. 511–518. IEEE Press, New York (2001)
11. Tang, X., Ou, Z., Su, T., Sun, H., Zhao, P.: Robust Precise Eye Location by AdaBoost and SVM Techniques. In: International Symposium on Neural Networks, pp. 93–98 (2005)
12. Friedman, J., Hastie, T., Tibshirani, R.: Additive Logistic Regression: a Statistical View of Boosting, Technical Report, Stanford University (1998)
13. Bonney, B., Ives, R., Etter, D., Du, Y.: Iris Pattern Extraction using Bit Planes and Standard Deviations. In: Thirty-Eighth Asilomar Conference on Signals, Systems and Computers, vol. 1, pp. 582–586. IEEE Press, New York (2004)
14. Computer Vision Laboratory, Faculty of Computer and Information Science, University of Ljubljana, Slovenia, http://lrv.fri.uni-lj.si/facedb.html

Skull Stripping Based on Clustering and Curve Fitting with Quadratic Equations

K. Somasundaram and R. Siva Shankar

Image Processing Lab, Department of Computer Science and Applications,
Gandhigram Rural Institute-Deemed University, Tamil Nadu, India
{ka.somasundaram,arjhunshankar}@gmail.com

Abstract. The segmentation of brain tissue finds an important role in many clinical applications. In this paper, we propose a new skull stripping method for magnetic resonance Images (MRIs) of human head scan. First, we find the clusters based on the intensity. Then the centroids of these clusters are found. These centroids are then connected, which form the brain boundary. We use Quadratic equations to fit the Curve to find the edges by connecting all the maxima values in the boundary. Then, we remove the skull area and the clusters which are outside the boundary, resulting in the segmentation of the brain portion.

Keywords: MRI Processing, Segmentation, K-Mean Clustering, Centroids, Quadratic equations, Curve fitting, Skull stripping.

1 Introduction

The segmentation of brain tissue from non-brain tissues in MRI Images is referred as Skull Stripping or Brain Extraction. It is an important pre-processing technique in MRI analysis. MRIs of the brain give the anatomy of brain that is helpful to diagnose the brain related diseases. It is a non-invasive and non-destructive method.

Region based methods view brain regions as a group of connected pixel data sets. These regions will have muscles, tissues, cavities, skin, optic nerves, etc. The extraction of the brain region from the non-brain region is done by methods like region growing, watershed and mathematical morphological methods.

In [1]-[8] mathematical morphology had been used for skull stripping. Hohne et al. [2] proposed a semi-automated segmentation algorithm based on region growing and morphological operations. Justice et al. [4] proposed a semi automated segmentation method using 3D seeded region growing (SRG) which is an extension of 2D SRG method proposed by Adams et al. [7]. One disadvantage of these methods is that the user has to select the seed region and the threshold values. Combination of Intensity based methods with mathematical formulas. [10] gives better results.

We use a method based on intensity with clustering and curve fitting technique for skull stripping. The remainder of the paper is organized as follows. In section 2, we give an outline of the basic principles used in our scheme. In section 3, we present our method. In section 4, results and discussions are given. The conclusion is given in section 5.

P. Balasubramaniam and R. Uthayakumar (Eds.): ICMMSC 2012, CCIS 283, pp. 439–444, 2012.
© Springer-Verlag Berlin Heidelberg 2012

2 Basic Principles Used in Skull Stripping

2.1 Clustering

Clustering is the assignment of a set of observations into several subsets usually called as clusters. The observations in the same cluster are similar in some property. Clustering is an unsupervised learning method used in statistical data analysis. It is used in many fields like machine learning, data mining, pattern recognition, image analysis and bioinformatics etc. In our work three main regions form the clusters. They are white matter (WM), grey matter (GM) and cerebrospinal fluid (CSF) present in MRI of head scans. They are the important regions in the study of brain diagnostic system.

2.2 K-Mean (KM) Algorithm

KM Algorithm is a hard segmentation procedure that generates a sharp classification. It assigns each tissue to one cluster definitely. This algorithm is an unsupervised classification technique. KM algorithm [9] is used to classify the given pixels to a certain number of clusters.

The steps involved in KM algorithm are:
1. Set the cluster size as K.
2. Initialize the centroid of each cluster $C_i=0$, $i=1,2...K$.
3. Process the observations and assign to a cluster.
4. Find the centroids of each cluster.

The centroid of a cluster is found by minimizing the objective function

$$J = \sum_{j=1}^{k} \sum_{i=1}^{n} \| P_i^{(j)} - C_j \|^2 \tag{1}$$

where $\| P_i^{(j)} - C_j \|^2$ is the distance between the centroid, $P_i^{(j)}$ is the jth pixel of the ith cluster C_j is the jth pixel of the centroid.

3 The Proposed Method

3.1 Fixing the Edges

The MRI image is clustered into different clusters using KM algorithm. We then find centroids of the clusters and randomly select 5 to 10 co-ordinate points for each row in the Image. This will form curves forming rough edges. We then use the intensity values at those centroid co-ordinate points to fit the quadratic equation:

$$Y = ax^2 + bx + c \tag{2}$$

$$Y'' = 2ax + b \tag{3}$$

$$Y'' = 2a \tag{4}$$

where a,b,c are intensity values in (1)-(3) are first and second derivations of its previous equations.

By setting (3) to 0, we can decide the Maxima and Minima values of x. We can select the maximum and the minimum values occurring as next co-ordinate values. This combination of these occurrences makes us to pick the co-ordinate points of the edges alone. X is the intensity value at current centroid values and comparing them with near by value sets and adding them to the finalized array. The inner most maxima values are connected. The row and the column of the point m is mid point of the image is considered as connecting point of all the curves. By connecting all the maxima values with the curve satisfying conditions, we will get the boundary (see Fig. 1). This is repeated all around the contour without pixel gaps (see Fig.2). The pixel intensity set as primary property of the pixel, will be compared by the surrounding pixels to decide whether they are connected or not definitely. The curve can be visible at all quarters by satisfying our condition. Each pixel will be connected to either horizontally or vertically near by pixel, with the property related to intensity values. This boundary is used as a mask and the area enclosed inside the mask gives the brain portion.

Using the combination of K-Mean algorithm and curve fitting using quadratic equation results the brain extraction.

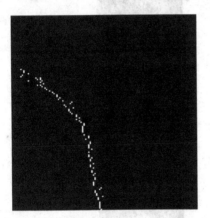

Fig. 1. Curve fixed by quadratic equation

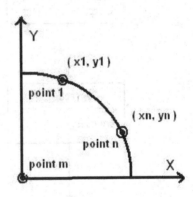

Fig. 2. Connected without pixel gaps

3.2   Extraction of the Brain Portion in the Middle Slice

We start this process from the middle slice of the given MRI volume and then proceed through the slices in the volume as it contains largest connected components. We apply the Curve fitting by quadratic equations method and get a brain portion in the middle slice. Brain portion in the mid slice is the largest area. We can store the co-ordinate points of the edge in the middle slice and we can process the inner portion further to the rest of the slices. After finding the edge of the next slices that will be marked co-ordinate points to be processed for the next slice until we reach the first and last slice of the volume. This makes our process robust and reliable. The time complexity is also reduced while processing the entire volume. The flow chart of out method is shown in Fig.3.

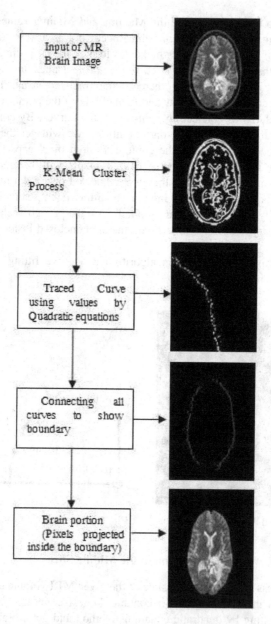

Fig. 3. Flow chart of the proposed method

4   Results and Discussions

We carried out experiments by applying our method on T2 weighted axial MRI of head scans taken from Whole Brain Atlas(WBA).[11] maintained by Department of

Radiology and Neurology at Brigham and women's hospital, Harward Medical school, Boston, USA. It contains 56 slices each of 256*256 pixel size. We computed Jaccard and Dice co-efficient using the hand segmented brain called ground truth images. We compare our results with existing Brain Extraction tool (BET) results.

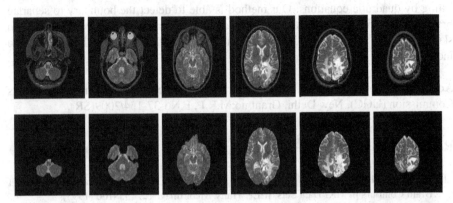

Fig. 4. Original slices are shown in first row and the extracted brain using our method is shown in second row

For Visual inspection the original images and the extracted images are shown in Fig.4. To quantitatively evaluate the performance of the proposed method, the Jaccard similarity and the Dice similarity were computed. The obtained result values vary between 0 to 1. The best results will be very close to 1. Our results were compared to existing Brain Extraction Tool (BET) method. The chosen values are from both agreement and both disagreement Measures. Consider that A, B are the set of the pixel values in our segmented Image and BET segmented Images. Then we compute the Jaccard and Dice values as follows.

The Jaccard Similarity index is given by:

$$J(A,B) = |A \cap B| \, / \, |A \cup B| \tag{5}$$

The Dice Similarity index is given by:

$$D(A,B) = 2 \, |A \cap B| \, / \, |A| + |B| \tag{6}$$

The recorded average values of Jaccard and Dice for the volume by the proposed method and BET method are shown in Table. 1. We note that the proposed method gives better result in terms of the Dice and Jaccard indices.

Table 1. Computed values of Jaccard, Dice Co-efficient for our method and BET

Date Set	Proposed Method		BET	
Type	Jaccard	Dice	Jaccard	Dice
Axial T2 Normal	.9688	.9841	.9531	.9760

5   Conclusions

In this paper, we have proposed a novel method to extract brain portion from MRI human head scan images. The proposed scheme is based on clustering and Curve fitting by quadratic equations. Our method is able to detect the boundary to separate brain portion and skull directly and thus avoids the processing of background and skull areas. The proposed method gives better result in terms of the Dice and Jaccard indices.

Acknowledgment. This work is supported by a Research grant by University Grants Commission (UGC), New Delhi, Grant no:M.R.P, F.No-37-154/2009(SR).

References

1. Brummer, M.E., Mersereau, R.M., Eisner, R.L., Lewine, R.R.J.: Automatic Detection of Brain Contours in MRI Data Sets. IEEE Trans. Med. Imag. 12, 153–166 (1993)
2. Hohne, K.H., Hanson, W.A.: Interactive 3D Segmentation of MRI and CT Volumes using Morphological Operations. J. of Comput. Assist. Tomogr. 16, 285–294 (1992)
3. John, C., Kevin, W., Emma, L., Chao, C., Barbara, P., Declan, J.: Statistical Morphological Skull Stripping of Adult and Infant MRI Data. Comput. Biol. Med. 37, 342–357 (2007)
4. Justice, R.K., Stokely, E.M., Strobel, J.S., Ideker, R.E., Smith, W.M.: Medical Image Segmentation using 3D Seeded Region Growing. In: Proc. SPIE Med. Imag., vol. 3034, pp. 900–910 (1997)
5. Lemieux, L., Hagmann, G., Krakow, K., Woermann, F.G.: Fast, Accurate, and Reproducible Automatic Segmentation of the Brain T1-Weighted Volume MRI Data. Mgn. Reson. Med. 12, 127–135 (1999)
6. Tsai, C., Manjunath, B.S., Jagadeesan, R.: Automated Segmentation of Brain MR Images. Pattern Recognition 28, 1825–1837 (1995)
7. Adams, R., Bischof, L.: Seeded Region Growing. IEEE Trans. Pattern Anal. Mach. Intell. 16, 641–646 (1994)
8. Somasundaram, K., Kalaiselvi, T.: Fully Automatic Brain Extraction Algorithm for Axial T2-weighted Magnetic Resonance Images. Computer and Biology in Medicine 40, 811–822 (2010)
9. Cheung, Y.M.: K*Means: A New Generalized K-Means Clustering Algorithm. Pattern Recognition Lett. 24, 2883–2893 (2003)
10. Somasundaram, K., Siva Shankar, R.: Skull Stripping of MRI Using Clustering and Resonance Method. Int. J. Knowledge Mana. & E-Learning 3, 19–23 (2011)
11. The Whole Brain Atlas (WBA), Department of Radiology and Neurology at Brigham and women's hospital, Harward Medical school, Boston, USA

Modified Otsu Thresholding Technique

K. Somasundaram and T. Genish

Image Processing Lab, Department of Computer Science and Applications,
Gandhigram Rural Institute – Deemed University, Gandhigram - 624302, Tamil Nadu, India
{ka.somasundaram,genishit}@gmail.com

Abstract. In this paper we propose a method to compute intensity threshold by modifying Otsu`s method. We replace the between-class mean by standard deviation. The proposed method produce binary images with sharp edges than Otsu`s method.

Keywords: Otsu's algorithm, Thresholding, Variance, Binarization of image.

1 Introduction

Image segmentation is one of the most important steps in the analysis of image. Its main goal is to divide an image into regions that have a strong correlation with objects or areas of the real world contained in the image. We usually segment regions by using some property found in the image. The simplest property that pixel in a region has the pixel intensity. So, a natural way to segment such regions is through intensity thresholding. Thresholding helps to create binary images from grey-level image by turning all pixels below a threshold to zero and the remaining pixels to one [1, 2]. There are several methods to binarize a given image [3]-[6]. In this paper we propose a modified Otsu`s method to binarize an image. We apply our method on few Magnetic Resonance Images (MRI) of human head scans. Experimental results show improved results over the original method. The remaining paper is organized as follows. In Section 2, we present the modified Otsu's method. In Section 3, we give the results and discussions. In Section 4, the conclusion is given.

2 Thresholding Technique

Otsu's thresholding method [1, 2] involves iterating through all the possible threshold values and calculating a measure of spread for the pixel levels each side of the threshold, i.e. the pixels that either fall in foreground or background. The aim is to find the threshold value where the sum of foreground and background is at its minimum. In image processing, Otsu's method is used to automatically compute the threshold and to convert a gray scale image in to a binary image. The algorithm assumes that the image to be thresholded contains two classes of pixels. We can define the within-class variance as the weighted sum of the variances of each cluster object (O) and background (B):

P. Balasubramaniam and R. Uthayakumar (Eds.): ICMMSC 2012, CCIS 283, pp. 445–448, 2012.

$$\sigma_{Within}^2(T) = n_B(T)\,\sigma_B^2(T) + n_O(T)\,\sigma_O^2(T) \tag{1}$$

where,

$n_B(T) = \sum_{i=0}^{T-1} p(i)$,

$n_O(T) = \sum_{i=T}^{N-1} p(i)$.

$\sigma_B^2(T)$ is the variance of the pixels in the background, and $\sigma_O^2(T)$ is the variance of the pixels in the foreground, p_i is the probability of occurring of pixel value x_i.

Computing this within-class variance for each of the two classes for each possible threshold involves a lot of computation, but this computation can be reduced by modifying eqn. (1). If we subtract the within-class variance in eqn. (1) from the total variance of the combined distribution, we obtain the between-class variance as:

$$\begin{aligned}\sigma_{Between}^2(T) &= \sigma^2 - \sigma_{Within}^2(T) \\ &= n_B(T)\,[\mu_B(T)-\mu]^2 + n_O(T)[\mu_O(T)-\mu]^2\end{aligned} \tag{2}$$

where, σ^2 is the combined variance and μ is the combined mean of the pixels x_i . Notice that the between-class variance is simply the weighted variance of the cluster means themselves around the overall mean. Substituting $\mu = n_B(T)\,\mu_B(T) + n_O(T)\,\mu_O(T)$ in eqn. (2) and simplifying, we get:

$$\sigma_{Between}^2(T) = n_B(T)\,n_O(T)\,[\mu_B(T) - \mu_O(T)]^2 \tag{3}$$

Instead of using mean value in eqn. (3) we use the standard deviation of background pixels subtracted from the standard deviation of object pixels. The standard deviation for any region with N pixels of intensity x_i , $i = 1, 2\ldots\ldots N$, is given by:

$$S = \sqrt{\sum_{i=1}^N (x_i - \bar{x})^2 / N} \tag{4}$$

where $\bar{x}$ is the mean of x_i. By substituting eqn. (4) in eqn. (3) we get:

$$\sigma_{Between}^2(T) = n_B(T)n_O(T)[S_B(T) - S_O(T)]^2 \tag{5}$$

where, S_B is the standard deviation for the background pixels and S_O is for the object pixels. To get an optimum threshold value, eqn. (5) is computed by varying T and finding the lowest value for $\sigma^2{}_{Between}$. The value of the pixel value x_i at which the minimum $\sigma^2{}_{Between}$ obtained is taken as the threshold value T. Using the threshold value T, the given input image f(x,y) is converted to a binary image g as:

$$\begin{aligned}g(x,y) &= 1 \text{ if } f(x,y) >= T \\ &= 0 \text{ otherwise.}\end{aligned} \tag{6}$$

Algorithm

1. Compute histogram and probabilities of each intensity level.
2. Step through all possible thresholds t=1,2,…,maximum intensity.
   a. Compute $n_B(T)$ and $n_O(T)$.
   b. Compute $S_B(T)$ and $S_O(T)$.
3. Desired threshold corresponds to the maximum $\sigma_{Between}^2$.

3   Results and Discussions

To evaluate the performance of our method, we applied it on a sample MRI of human head scans and binarized the image using our method. For comparison we used Otsu's method to compute the threshold T for the same image. The results obtained using Otsu's method and proposed method are shown in Fig. 1.

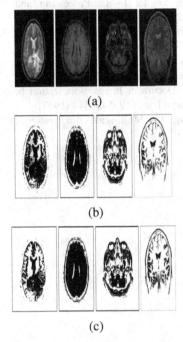

(a)

(b)

(c)

Fig. 1. Result obtained by using the proposed method. (a) Original MRI of human head scans. Binary image obtained using (b) Otsu's method (c) Proposed method.

Almost no-zero value pixels are present in an edge image, but small edge values correspond to non-significant gray-level changes resulting from quantization noise, small lighting irregularities. Hence, other thresholding methods need pre-processing the image. But for our method this pre-processing is not required.

By comparing the Fig.1 (b) and Fig.1 (c) we observe that the proposed method give sharp edges than that obtained for original Otsu's method. Such sharp edges help to segment various regions of an image accurately.

4   Conclusion

In this paper, we have proposed a novel method to find an intensity threshold by modifying the Otsu's method. Experimental results using our method on an MRI of human head scan, shows that it can produce binary images with sharp edges than with Otsu's method.

Acknowledgement. This work is supported by research grant by University Grants Commission, New Delhi, India. Grant No: M.R.P, F.No-37- 154/2009(SR).

References

1. Gonzalez, R.C., Woods, R.E.: Digital Image Processing, 2nd edn. Prentice Hall (2002)
2. Sonka, Hlavac, Boyle: Digital Image Processing and Computer Vision, India edn. CENGAGE Learning (2007)
3. Sauvola, J., Pietikainen, M.: Adaptive Document Image Binarization. Pattern Recognition 33, 225–236 (2000)
4. Gatos, B., Pratikakis, I., Perantonis, S.J.: Adaptive Degraded Document Image Binarization. Pattern Recognition 39, 317–327 (2006)
5. Liu, Y., Srihari, S.N.: Document Image Binarization based on Texture Features. IEEE Trans. Pattern Anal. Mach. Intell. 19, 540–544 (1997)
6. Tabbone, S., Wendling, L.: Multi-scale Binarization of Images. Pattern Recognition Lett. 24, 403–411 (2003)

A Brain Extraction Method Using Contour and Morphological Operations

K. Somasundaram and K. Ezhilarasan

Image Processing Lab, Department of Computer Science and Applications,
Gandhigram Rural Institute - Deemed University, Gandhigram – 624302, Tamil Nadu, India
{ka.somasundaram,ezhilarasankc}@gmail.com

Abstract. In this paper, we prepare segmenting the brain portion Magnetic Resonance Image (MRI). In our method, we first find the threshold using Riddler's method and convert gray image to binary image. The head boundaries are detected by using contour of the head. By subtracting, we detect the scalp and remove it. We perform morphological operation to disconnect the non-brain region. Finally, by applying the largest connected component method we devise an exact brain mask. Using the brain mask the brain portion can be obtained.

Keywords: Brain MRI, Largest connected component, Morphological operations.

1   Introduction

Medical imaging using various techniques is invented to diagnose disease/ deformities in human. Few of them are X-ray, Computed Tomography and PET to quote. Magnetic Resonance Imaging (MRI) technique is employed to visualize detailed internal structure of soft tissues in our body. We can get three different types of images T1-weighted, T2-weighted and Proton Density (PD). The brain MRI is taken in three different orientations axial, coronal and saggital which gives some clear views of brain tissues. This method is non-ionizing and produces no side effects. MRI gives the clear view of soft tissues and it helps to diagnose the problems easily that affect soft tissues. An MRI scanner can take images of most parts of our body.

There are many algorithms available for the segmentation of brain portion from MRI head scan [1]-[6]. Each method has its own advantages and disadvantages. Brain image segmentation is a pre-process for several medical image analysis. Each algorithm makes use of some property of the image to remove non-brain tissues. Basically, there are three types of method used for segmentation. They are intensity based method, region based method and shape based method [7]. Most of them are three stage algorithm to segment the brain image. First and final stages are the priori and post operations on image and second stage is generate rough brain mask. For second stage they use nonlinear anisotropic diffusion to find active contour. Third stage is the refinement of rough result. In [3], geographical model Watersheds

P. Balasubramaniam and R. Uthayakumar (Eds.): ICMMSC 2012, CCIS 283, pp. 449–455, 2012.
© Springer-Verlag Berlin Heidelberg 2012

are used to segment the brain image. The brain tissues are detected with the use of center of gravity of brain image and tessellated sphere that are generated from the center of gravity and move outwards to brain surface [4] (BET). In BEM [8], it creates a label with the help of anatomical facts and encoding with morphological operation to perform segmentation, 3D concepts are used to detect the critical brain regions. In this paper we proposed a brain extraction method using active contour method and morphological operation. In Section 2, the proposed method is presented. Result and Discussion, are given in Section 3 and conclusion is given in Section 4 respectively.

2   Methods Used

The proposed method is a modified form of BEM [2]. The image is processed using a noise filter. Then, by using a thresholding, the area of interest in the head region is formed; the active contour method is used to find the edge points in the area of interest, followed by run length filling.

2.1   Thresholding

There are various methods available to find the threshold of an image. We use Riddler's method [7] for finding the threshold value. It is an iterative method. The initial threshold (I) is given by

$$T_0 = \frac{\sum_{i=0}^{n} x_i}{n} \quad , \qquad n = 0, 1, 2, 3, \ldots, N \tag{1}$$

Where, N is the total number of pixel in the image, x_i is the pixel values.

We consider T_0 value as the critical value T, which help to separate the two sets G and L, as shown below

$$x_i \in G \quad \text{if } x_i \geq T_0$$
$$x_i \in L \quad \text{if } x_i < T_0$$

We then compute the following parameters to compute an improved threshold value T_1.

$$sum = \frac{\sum_{i=0}^{n} G_i}{G_n} + \frac{\sum_{i=0}^{n} L_i}{L_n} \tag{2}$$

G_n and L_n are the total number of elements in the set G (White) and set L (Black).

The next threshold T_1 is computed as

$$T_1 = sum / 2 \tag{3}$$

The above process is repeated until the successive T_i and T_{i+1} are very close to each other, the final threshold value T is applied to the input image I and a binary image I_b is obtained. This binary image is used to segment the brain tissues. The whole process is shown in the flow chart in Fig. 1. The binary image I_b is obtained from the input image I as:

$$I_b(x, y) = 1 \text{ if } I(x, y) \geq T$$
$$= 0 \text{ otherwise.}$$

2.2   Head Contour

From the binary image, we can trace out the boundary of the head. For this, we scan each row of the image I_b, from left to right and from right to left. For each run, the first bright value having value 1 is identified. The collection of all such points forms the contour enclosing the head. The remaining pixels are set to 0. Thus an image I_c with the contour of the head is obtained. (Fig. 1).

2.3   Head Mask

From the head contour binary image, we create a head mask. For this, we again move through each row, if two bright points exists connect them by filling the entire pixels with value 1. By repeating this for all rows, we get an image I_h with white region (with value 1), covering the head region. This image I_h is used as head mask.

2.4   Rough Brain Region

We then subtract I_T from I_h, we get an image I_s which gives a contour enclosing the brain region and other non-brain region.

$$I_s = I_h - I_T \tag{4}$$

This I_s image contains several contours. We fill the gap between a pair of bright points by 1. The filled value result with several connected regions is given as I_{pb}. We also perform morphological operations to removes isolated points and weakly connected small regions.

2.5   Connected Components

The connected component analysis is done in the image I_{pb}. This image contains very few non-brain tissues that are spread over the image I_{pb}. The largest connected component in the image is brain tissue. It is taken as the brain region I_{br}. The other regions are discarded. Using the brain mask I_{br}, the brain portion can be extracted.

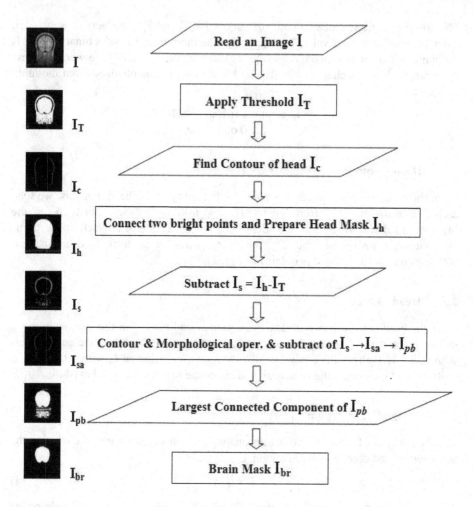

Fig. 1. Flow chart of our method

3   Results and Discussion

We applied our algorithm on one set of volume taken from the Internet Brain Segmentation Repository (IBSR). The segmented brain portion is evaluated with the help of Jaccard's coefficient or Jaccard's similarity. The Gold standard hand segmented images are taken from the IBSR 202 series. The Jaccard similarity is given by [9].

$$J (A, B) = |A \cap B| / |A \cup B| \tag{5}$$

Table 1. Jaccard index computed for 60 slice

Slice No	Jaccard similarity	Slice No	Jaccard similarity	Slice No	Jaccard similarity
202_3_1	-	202_3_21	0.9718	202_3_41	0.8687
202_3_2	-	202_3_22	0.9696	202_3_42	0.9606
202_3_3	-	202_3_23	0.9609	202_3_43	0.9685
202_3_4	0.9782	202_3_24	0.9638	202_3_44	0.9546
202_3_5	0.9810	202_3_25	0.8415	202_3_45	0.9593
202_3_6	0.9817	202_3_26	0.8410	202_3_46	0.9682
202_3_7	0.9891	202_3_27	0.9341	202_3_47	0.9700
202_3_8	0.9924	202_3_28	0.8676	202_3_48	0.9908
202_3_9	0.9874	202_3_29	0.9650	202_3_49	0.9916
202_3_10	0.9847	202_3_30	0.9671	202_3_50	0.9896
202_3_11	0.9827	202_3_31	0.9327	202_3_51	0.9911
202_3_12	0.9808	202_3_32	0.8667	202_3_52	0.9899
202_3_13	0.9823	202_3_33	0.8485	202_3_53	0.9913
202_3_14	0.9892	202_3_34	0.8154	202_3_54	0.9919
202_3_15	0.9825	202_3_35	0.8069	202_3_55	0.9891
202_3_16	0.9762	202_3_36	0.8031	202_3_56	0.9892
202_3_17	0.9809	202_3_37	0.8078	202_3_57	0.9890
202_3_18	0.9808	202_3_38	0.8209	202_3_58	0.9823
202_3_19	0.9820	202_3_39	0.8118	202_3_59	0.9700
202_3_20	0.9790	202_3_40	0.8520	202_3_60	0.9592
Avg.					0.9442

We get an average value of J= 0.9442 for the MRI volume.

The computed values of the Jaccard similarity index of the segmented images are given in Table. 1. For the complicated slices, we got very less values. Those slices have neck regions. The neck region tissues and brain region tissues have same intensity levels, so this algorithm is not able to differentiate them. This algorithm does not work well in these conditions. For visual inspection, the original and segmented brain marks are shown in Fig. 2.

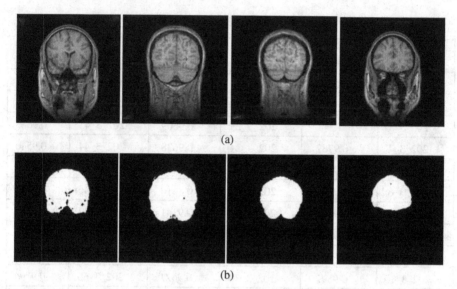

(a)

(b)

Fig. 2. a) Original Image MRI of head scans. b) Brain mask generated using our method.

4   Conclusion

In this paper, we have proposed a method for extracting the brain portion from MRI of head scan. It is based on simple thresholding and contouring. Experimental results on IBSR image data shows that the proposed method works satisfactory. With these techniques the time taken by the physician to analyze the brain MRI of the patient is reduced. The computed Jaccard index, which is the measure of accuracy, is 0.9442.

Acknowledgement. This work is funded by the University Grant Commission, New Delhi, and Grant No: FNo 37-154-2009(SR).

References

1. Atkins, M.S., Mackiewich, B.T.: Fully Automatic Segmentation of the Brain in MRI. IEEE Trans. Medical Imaging. 17, 98--107 (1998)
2. Soumasundaram, K., Kalaiselvi, T.: Automatic Brain Extraction Methods for T1 Magnetic Resonance Images using Region Labeling and Morphological Operations. Computers in Biology and Medicine. 41, 716--725 (2011)
3. Horst, K., Hahn, Heinz Otto Peitgen.: The Skull Stripping Problem in MRI Solved by a Single 3D Watershed Transform. Medical Image Computing and Computer-Assisted Intervention – MICCAI 2000 Lecture Notes in Computer Science. 12--145 (2000)
4. Smith, S.M.: Fast Robust Automated Brain Extraction. Human Brain Mapping. 17, 143--155 (2002)
5. David, E. Rex, David, W. Shattuck, Roger, P. Woods, Katheine, L. Narr, Eileen Luders, Kelly Rehm, Sarah, E. Stolzner, David, A. Rottenberg, Aruthur, W. Toga: A Meta – algorithm for Brain Extraction in MRI. NeuroImage. 23, 625--637 (2004)

6. Dzung, L. Pham, Jerry, L. Prince: Adaptive Fuzzy Segmentation of Magnetic Resonance Images". IEEE Trans. Medical Imaging. 18, 737--752 (1999)
7. Sonka, M., Hlavac, V., Boyal, R.: Image Processing Analysis and Machine vision. Brooks/cole Publishing Company (1999)
8. Collins, D. L., Zijdenbos, A. P., Kkollokian, V.Sled., Kabani, J.G., Holmes. N.J., Evans, A. C.: Design and Construction of Realistic Digital Brain Phanthom. IEEE Trans. Medical Imaging. 17, 463--468 (1998)
9. Somasundaram, K., Siva Shankar, R.: Skull Stripping of MRI using Clustering and Resonance Method. Int. J. Knowledge Management and e-Learning. 3, 19--23 (2011)

Image Data Hiding in Images Based on Interpolative Absolute Moment Block Truncation Coding

Kaspar Raj I.

Gandhigram Rural Institute – Deemed University
Gandhigram -624 302, Tamilnadu, India
kasparraj@gmail.com

Abstract. Absolute Moment Block Truncation Coding (AMBTC) is an efficient and fast lossy compression technique for still images. It is easy to implement compared to transform coding. Interpolative AMBTC is a modified method of AMBTC compression. In this paper image data hiding in images using Interpolative AMBTC is proposed. In this method the secret image is stored in the bit plane of the AMBTC compressed image and is recovered during decompression process. Experimental results show that the proposed method achieves good quality decompressed image with low computational complexity.

Keywords: Data hiding, Compression, Block truncation coding, AMBTC, Bit plane.

1 Introduction

Data a security have become inevitable during communication in this internet era. Data encryption and data hiding are two different methods for data protection during transmission of sensitive information. Data encryption refers to mathematical calculations and algorithmic schemes that transform plaintext into non-readable form to unauthorized parties. The recipient of an encrypted message uses a key which triggers the algorithm mechanism to decrypt the data, transforming it to the original plaintext version. Data hiding in images is process of embedding a secret data into a digital image which is called as cover image [1]. Data hiding can be used in different applications like e-commerce, confidential communication, authentication military data transmission and copyright protection. Hence, in the internet data hiding in images is becoming a necessary technique. A large number of techniques have been proposed to embed data into images [2]-[4].

Block Truncation Coding (BTC) is a simple image compression technique, introduced by Delp and Mitchell [5]. BTC is based on the conservation of statistical properties. Although it is a simple technique, BTC has played an important role in the history of image compression. Many image compression techniques have been developed based on BTC [6]. Block truncation coding is a lossy compression method. It achieves 2 bits per pixel (bpp) with low computational complexity. Lema and Mitchell [7] presented a variant of BTC called AMBTC. It preserves the higher mean and lower mean of the sub image blocks before and after compression. However, the bit rate achieved by both BTC and AMBTC is 2 bpp. Ramana and Eswaran [8]

P. Balasubramaniam and R. Uthayakumar (Eds.): ICMMSC 2012, CCIS 283, pp. 456–463, 2012.

proposed a simple predictive scheme for BTC bit plane coding. The idea of this scheme is based on the observation that a high correlation exists among neighbouring pixels in most digital images. Somasundaram and Kaspar Raj [9], proposed another modified version of the BTC. In this method eight bits in the bit plane are transmitted to the decoder while the other eight bits in the bit plane are dropped. The dropped bits are recovered by performing interpolation. The interpolation is done by taking the arithmetic mean of the adjacent gray values. So it is named as Interpolative AMBTC method.

In this paper, image data hiding in images using Interpolative AMBTC is proposed. In this method the secret image is stored in the bit plane of the compressed cover image. In this technique, interpolative AMBTC method is used for compression. The secret image is recovered during decompression process. The experimental results show that the secret image hiding is not affecting the quality of decompressed image. This method has a low computational complexity.

In Section 2, we briefly outline AMBTC and interpolative AMBTC methods. We describe the proposed data hiding method in Section 3. The experimental results are discussed in Section 4. Finally, we conclude this paper in Section 5.

2   Absolute Moment Block Truncation Coding

The basic concept of BTC is to perform moment preserving quantization for blocks of pixels of an image so that the quality of the image will remain acceptable and at the same time the image file size decreases. Even though the bit rate of the original BTC is relatively high when compared to other still image compression techniques such as JPEG or Vector Quantization, BTC has gained popularity due its practical usefulness.

In the BTC method, the image is divided into non-overlapping small blocks (normally 4 x 4 pixels). The moments are calculated for each block, i.e., the sample mean $\bar{x}$ and standard deviation σ . The mean $\bar{x}$ standard deviation σ are computed using

$$\bar{x} = \frac{1}{n} \sum_{i=1}^{n} x_i \tag{1}$$

$$\sigma = \sqrt{\frac{1}{n} \sum_{i=1}^{n} (x_i - \bar{x}_i)^2} \tag{2}$$

Where x_i represent the ith pixel value of the image block and n is the total number of pixels in the block. The two values $\bar{x}$ and σ are termed as quantizers of BTC.

Taking $\bar{x}$ as the threshold value a two-level bit plane is obtained by comparing each pixel value x_i with the threshold. If $x_i < \bar{x}$ then the pixel is represented by '0', otherwise by '1'. By this process each block is reduced as a bit plane. The bit plane along with $\bar{x}$ and σ forms the compressed data. For example a block of 4 x 4 pixels will give a 32 bit compressed data, amounting to 2 bpp.

In the decoder an image block is reconstructed by replacing by '1' s with H and the '0's by L, which are given by

$$H = \overline{x} + \sigma\sqrt{\frac{p}{q}} \tag{3}$$

$$L = \overline{x} - \sigma\sqrt{\frac{q}{p}} \tag{4}$$

where p and q are the number of 0's and 1's in the compressed bit plane respectively.

Lema and Mitchell [4] presented a simple and fast variant of BTC, named Absolute Moment BTC (AMBTC) that preserves the higher mean and lower mean of a block. However, the bit rate achieved with the AMBTC algorithm is also 2 bpp. The original AMBTC algorithm involves the following steps:

An image is divided into non-overlapping blocks. The size of a block could be (4 x 4) or (8 x 8), etc. Calculate the average gray level of the block (4x4) as :

$$\overline{x} = \frac{1}{16} \sum_{i=1}^{16} x_i \tag{5}$$

where x_i represents pixels in the block. Pixels in the image block are then classified into two ranges of values. The upper range is those gray levels which are greater than the block average gray level ($\overline{x}$) and the remaining brought into the lower range. The mean of higher range x_H and the lower range x_L are calculated as

$$x_H = \frac{1}{k} \sum_{x_i \geq \overline{x}}^{n} x_i \tag{6}$$

$$x_L = \frac{1}{16 - k} \sum_{x_i < \overline{x}}^{n} x_i \tag{7}$$

where k is the number of pixels whose gray level is greater than $\overline{x}$.

A binary block, denoted by b, is also used to represent the pixels. We can use "1" to represent a pixel whose gray level is greater than or equal to $\overline{x}$ and "0" to represent a pixel whose gray level is less than $\overline{x}$. The encoder writes x_H, x_L and b to a file. Assume that we use 8 bits to represent x_H, x_L respectively. Then, the total number of bits required for a block is 8+8+16 =32 bits. Thus, the bit rate for the AMBTC algorithm is 2 bpp . In the decoder, an image block is reconstructed by replacing the `1's with x_H and the '0''s by x_L In the AMBTC, we need 16 bits to code the bit plane which is same as in the BTC. But, AMBTC requires less computation than BTC.

2.1   Interpolative AMBTC Method (IAMBTC)

In this technique half (8 bits) of the bits in the bit plane of AMBTC is dropped at the time of encoding as in Fig. 1. In decoding the dropped bits are recovered by taking the arithmetic mean of the adjacent values as in (8).

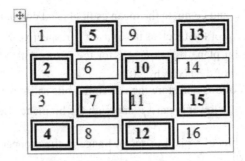

Fig. 1. The pattern of dropping bits. The bold faced bits are dropped.

3   Proposed Data Hiding Method

The proposed data hiding method makes use of AMBTC and interpolative technique. In this method, the cover image is compressed using the Interpolative AMBTC technique. In the Interpolative AMBTC , in the place of dropped bits the bits of the secret image will be stored. If the size of the cover image is MxM then the size of the secret image can be stored is NxN, where N is M/4. The secret image can be recovered during image decompression. The steps involved in the data hiding are as follows.

$$\hat{x}_i = \frac{1}{2}(x_{i-1} + x_{i+1} + x_{i+4}) \; for \; i = 2$$

$$\hat{x}_i = \frac{1}{2}(x_{i-1} + x_{i+4}) \; for \; i = 4$$

$$\hat{x}_i = \frac{1}{3}(x_{i-4} + x_{i+1} + x_{i+4}) \; for \; i = 5$$

$$\hat{x}_i = \frac{1}{4}(x_{i-4} + x_{i-1} + x_{i+1} + x_{i+4}) \qquad for$$
$$i = 7,10$$

$$\hat{x}_i = \frac{1}{3}(x_{i-4} + x_{i-1} + x_{i+4}) \; for \; i = 12$$

$$\hat{x}_i = \frac{1}{2}(x_{i-4} + x_{i+1}) \; for \; i = 13$$

$$\hat{x}_i = \frac{1}{3}(x_{i-1} + x_{i+1} + x_{i-4}) \; for \; i = 15$$

$$\left. \right\} \qquad (8)$$

Step 1: Divide the cover image X into blocks of size n = 4 x 4 pixel and convert the pixel values of secret image into binary values.

Step 2: Input a block X and compute the mean as in (1) and compute the lower mean $\overline{x}_L$ and higher mean $\overline{x}_H$ of the block as in (6) and (7) respectively.

Step 3 : Construct the bit plane by taking '1' for the pixels with values larger than the mean $\overline{X}$ and the rest of the pixels by '0'.

Step 4: Drop a pattern of bits and store the remaining bits as shown in Fig. 1.   Store the higher   mean $\overline{x}_H$, the lower mean $\overline{x}_L$ and with the bit plane after dropping eight bits.

Step 5: Store the eight bits of a pixel of secret image in the dropped places of the bit plane of the cover image.

Step 6: Repeat step 2 to 5 for all blocks in the image.

In the decoder, an image block is reconstructed using the bit plane transmitted by replacing the 1s with $\overline{X}_H$ and the 0s by $\overline{x}_L$. The pixels corresponding to the dropped bits of the cover image are estimated by computing the mean of the adjacent values by the (8) then by converting the binary values stored in the dropped places of the bit plane we can recover the secret image.

4   Results and Discussion

In order to evaluate the performance of our method, experiments were carried out on six standard monochrome images of size 512 X 512 as cover images "Couple", "Lena", "Jet", "Peppers", "Girl" and "Zelda" which are given in Fig. 2.

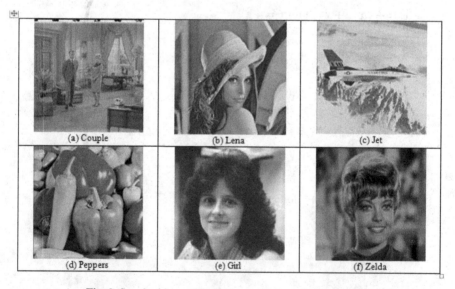

(a) Couple

(b) Lena

(c) Jet

(d) Peppers

(e) Girl

(f) Zelda

Fig. 2. Standard images used as cover images for experiment

Each digital image was partitioned into a set of non-overlapping image blocks of 4 X 4 pixels Since the size of the cover image is 512 X 512, the maximum size of the secret image can be 128 X 128 as per the proposed method. Hence, the standard monochrome image "barb" of size 128 X 128 is taken as secret image which is given in Fig. 3.

Fig. 3. Secret image used for experiment

For comparison, experiments using the AMBTC method and the method proposed by Ramana and Eswaran (RE Method) were also carried out. The PSNR values were computed for each of the reconstructed images and are given in Table 1.

Table 1. PSNR and BPP values for different methods on standard images

Image	AMBTC bpp=2.0	RE Method bpp=1.5	Interpolative AMBTC bpp=1.5
Couple	35.01	32.79	34.12
Lena	30.07	26.99	28.33
Jet	31.42	28.42	30.28
Peppers	33.44	30.33	31.27
Girl	33.95	31.45	32.35
Zelda	36.74	34.28	35.52

From Table 1, it can be seen that Interpolative AMBTC has better PSNR values than that of RE method. Hence, one can obtain better quality of reconstructed image and with the same the compression when compared to the RE method. It is also observed that interpolative AMBTC has closer PSNR value to the AMBTC method but with less bpp. Here, we can observe that data hiding in the compressed image is not affecting the quality of the decompressed image. So, we can take the secret image separately at the time of decompression.

Fig. 4. The reconstructed standard images using PIBTC

5   Conclusion

In this paper, image data hiding in images using Interpolative AMBTC is proposed. In this method, the secret image is stored in the bit plane of the AMBTC compressed image and is recovered during decompression process without much affecting the quality of the cover image. Experimental results show that the proposed method achieves quality of the image as per the decompression technique. This method has a low computational complexity.

References

1. Bender, W., Gruhl, D., Morimoto, A., Lu, A.: Techniques for Data Hiding. IBM Systems Journal 35, 313–336 (1996)
2. Chan, C., Cheng, L.: Hiding Data in Images by Simple LSB Substitution. Pattern Recognition 37(3), 469–474 (2004)
3. Du, W., Hsu, W.: Adaptive Data Hiding based on VQ Compressed Images. IEEE Proceedings Vision Image and Signal Processing 150(4), 233–238 (2003)
4. Chang, C., Lin, C., Tseng, C., Tai, W.: Reversible Hiding in DCT-based Compressed Images. Information Sciences 177, 2768–2786 (2007)
5. Delp, E.J., Mitchell, O.R.: Image Compression using Block Truncation Coding. IEEE Trans. Communications 27, 1335–1342 (1979)

6. Franti, P., Nevalainen, O., Kaukoranta, T.: Compression of Digital Images by Block Truncation Coding: a Survey. The Computer Journal 37(4), 308–332 (1994)
7. Lema, M.D., Mitchell, O.R.: Absolute Moment Block Truncation Coding and its Application to Color images. IEEE Trans. Communications 32, 1148–1157 (1984)
8. Ramana, Y.V., Eswaran, C.: A New Algorithm for BTC Image Bit Plane Coding. IEEE Trans. Communications 43, 2010–2011 (1995)
9. Somasundaram, K., Kaspar Raj, I.: Low Computational Image Compression Scheme based on Absolute Moment Block Truncation Coding. Enformatika Transactions on Engineering, Computing and Technology 13, 184–190 (2006) ISSN 1305-5313

A Novel Self Initiating Brain Tumor Boundary Detection for MRI

T. Kalaiselvi, K. Somasundaram, and S. Vijayalakshmi

Department of Computer Science and Applications
Gandhigram Rural Institute, Deemed University, Gandhigram–624 302, India
kalaivpd@gmail.com, somasundaramk@yahoo.com,
lakshmiviji.gru@gmail.com

Abstract. In this paper, we present a fusion based technique that produced robust and fully automatic tumor extraction for magnetic resonance images (MRI) of Head scans. This process constitutes the segmentation method which is based on a combination of spatial relations and deformable model. Three popular deformable methods: snake, level set and Distance regularized level set evolution were chosen for predicting their performance in generating the brain tumor boundaries. Generally, deformable methods require user-interaction for initialization. But the initial curve is automatically generated by our proposed method. We have combined boundary based techniques with our previous region based method and compared the results. Tumor slices were chosen from WBA and used for our experimental study. The results produced by our fully automatic technique are compared visually and parametrically. The evaluation is based on, how consistent is a specific deformable technique when compared with the rest by using Jaccard and William's index.

Keywords: Brain tumor, Deformable model, Boundary detection, MRI.

1   Introduction

Enormous and uncontrollable growth of cells leads to tumor in brain. In clinics, radiologists analyze a large numbers of MRI of head scans for detection of tumor. This manual process is very tedious and may easily slip-up when the radiologist doesn't have enough time. The automatic segmentation of tumors is essential to deal with large numbers of head scans. All medical treatments need information about edge of the tumor for successful surgery. Hence, finding a fully automated method to give perfect information to surgeons is necessary.

Tumor segmentation methods found in literature are classically divided into two categories, Region based and Boundary based methods. The first category "knowledge based fuzzy classification" [1], requires multi-channel images. "Statistical pattern recognition techniques" automatically detect small tumors [2]. The "outliers of normal voxel distribution method" [3] gives rough segmentation of brain. Four classes are segmented using "Segmented weighted aggregation" method [4]. The method based on

P. Balasubramaniam and R. Uthayakumar (Eds.): ICMMSC 2012, CCIS 283, pp. 464–470, 2012.

"evidential parametric model" [5] is sensitive to noise. A support vector machine (SVM) classification method has been reported recently [5]. In contour based methods, several works have been reported, such as "Neural Networks" based algorithms [1], "Histogram analysis" [2] based methods. The region-based method exploits only the local information and contour based methods suffer in determining the initial contour.

In our previous work [8], we proposed a region based technique with symmetric measure and maxima transform to detect and segment the brain tumor. In this paper, we extend and combine it with boundary based deformable methods. Usually in deformable model the user makes an initial guess of the contour which is then deformed by image driven forces to the boundary of desired objects. The accuracy of the final result depends on the initial curve. The resultant tumor curve of our proposed region based method [8] is given as initial curve to the deformable model which eliminates the user intervention. This fusion gives fine differentiation between pathological and normal tissues in slices. We have used three different boundary methods to compare and evaluate the performance. Validation and comparison metrics for segmentation of tumors are difficult due to lack of reliable ground truth. William's index was chosen as a performance measure for the selected boundary methods. It measures the common agreement among different classifiers [6, 7]. The notion of common agreement is useful and can be quantified directly through William's measure. In section 2, we present the basic algorithms used in each methods. Results and discussion are given in section 3. In section 4, the conclusion is given.

2   Methods

In MR images the brain portion is extracted first by eliminating the non brain tissues like fat, skin, muscles and background using BEA algorithm [12]. Then the extracted brain portion is segmented into four classes' cerebrospinal fluid (CSF), White matter (WM), grey matter (GM) and background. In MRI slices the tumor intensity characteristics are similar to CSF. So the CSF class is analyzed for symmetry property along the central vertical line. We make use of bilateral symmetry of the human brain to detect the abnormality in segmented CSF tissue. If no symmetry is found then the slice is tumor slice and further processed with ExM transform to segment the tumor. This resultant curve is nearby the actual border of the tumor. From the ExM transform the curve is drive to crisp border of tumor by incorporating boundary. Three boundary based methods based on deformable models were chosen for this. Parametric Deformable model (snakes), Geometric deformable model (Level set Function) and Distance Regularized Level set Method (DRLSM). Our current work consists of three stages.

Stage 1: Initial curve placement is made by ExM transform
Stage 2: Final curve segment is converged using boundary based deformable models.
Stage 3: Performance is compared using final curve region with each other.

2.1   Extended Maxima (ExM) Transform

ExM algorithm is a region based method that usually searches for connected regions of pixels with a similar intensity [8]. It performs the H-maxima transform followed by the regional maxima. H-maxima is achieved in two steps

$$HMAX_h(f) = R_f(f - h) \tag{1}$$

$$EMAX_h(f) = RMAX[HMAX_h(f)] \tag{2}$$

where $R_f(f-h)$ is the morphological reconstruction by dilation of image f with respect to $(f-h)$, h is threshold level. Fuzzy Symmetric Measure (FSM) is used as a threshold to discriminate between normal and abnormal CSF images. Symmetric property of human brain is used to detect the abnormality and the maxima transform, to locate the tumor region. This transform helps to separate the tumor region from the normal CSF region.

2.2   Parametric Deformable Model (Snakes)

An active contour or snake is an energy minimizing technique guided by external minimizing force and manipulated by gradient of image forces [9]. This method drives a contour to reach onto the boundary of object on an image. This method gets initial contour by manual or by other mechanism with close to object of interest. The evolution of our snake is described by the following usual dynamic force equation.

$$E_{snake} = \int_0^1 E_{int}(v(s)) + E_{image}(v(s)) + E_{con}(v(s))ds \tag{3}$$

$$E_{int} = \frac{(\alpha(s)|v_s(s)|^2 + \beta(s)|v_{ss}(s)|^2)}{2} \tag{4}$$

where E_{int} is internal spline energy, E_{con} is external energy constraints, E_{image} is image forces and $v(s)$ is the collection of contour points. In (4), $v_s(s)$ and $v_{ss}(s)$ are the first and second order derivatives. The internal spline energy is controlled by (s) and second order term controlled by $\beta(s)$. α, β are given as constant values.

2.3   Geometric Deformable Model (Level Set Function)

We use Level Set Function (LSF) devised by Osher and Fedkiw [10]. This is used for implementing the curve evolution under a combination of three forces, curvature-based forces, vector field-based forces and forces in normal direction. The mathematical form of level sets scheme is given as:

$$C = \{I(x, y) \mid f(x, y) = 0\} \tag{5}$$

where $I(x,y)$ is an image, C is curve. $f(x,y)$ is the initial curve, set to 0. The curve evolution given in (3), in terms of a parameterized contour, can be converted to a level set formulation by embedding the dynamic contour as zero level set of a time dependent LSF .Then, the curve evolution (3) is converted to the following partial differential equation (PDE):

$$\underbrace{\frac{\partial f}{\partial t}}+ \cdot \underbrace{\vec{s}.\nabla f}_{\substack{Vector\ Field \\ Based}} + \underbrace{v_n\ |\nabla f|}_{\substack{In\ normal \\ Direction}} = \underbrace{b_k\ |\nabla f|}_{\substack{Curvature \\ Based}} \tag{6}$$

where ∇ is the gradient operator. S is spline energy, v_n is curve evolution normal direction, b_k edge information. This Partial Differential Equation consists of edge function and vector fields used to stops the curve around the edge and pull the curve towards the edges.

2.4   Distance Regularized Level Set Evolution (DRLSM)

Distance regularized model is a new variation level set formulation in which the regularity of the level set function is intrinsically maintained during the level set evolution [11]. To demonstrate the effectiveness of the DRLSM formulation, we apply it to an edge-based active contour model for image segmentation, and provide a simple narrowband implementation to greatly reduce computational cost. The curve evolution f can be expressed as

$$\frac{\partial f}{\partial t} = -\frac{\partial \varepsilon}{\partial f} \tag{7}$$

$$\varepsilon\ (f) = \mu R_p\ (f) + \varepsilon_{ext}\ (f) \tag{8}$$

Where $R_p\ (f)$ is an internal energy penalizing deviation f . $\varepsilon_{ext}\ (f)$ is external energy drives motion of toward image boundary.

2.5   Williams Index

For quantitative results, we make use of William's index. Williams' index is a measure of common agreement, which estimates the performance of overall similarity between different techniques. If the index value is greater than one, the classifier under study agrees with other classifiers. The similarity of William's index requires Jaccard co-efficient similarity between the methods. WI and Jaccard similarity index (J) [13] are given by:

$$WI_n = \frac{(r-2)\ \sum_{n'\neq n}^{r} J(x_n,x_{n'})}{2\ \sum_{n'\neq n}^{r} \sum_{n''\neq n}^{n'} J(x_{n'},x_{n''})} \tag{9}$$

$$J(x_n, x_{n'}) = \frac{|(x_n \cap x_{n'})|}{|(x_n \cup x_{n'})|} \tag{10}$$

where r is the number of raters (methods), J $(x_n, x_{n'})$ the Jaccard similarity between rater n and n', x_n is resultant image of rater n and $x_{n'}$ is the resultant of n'.

Materials

We use some tumor affected slices T2 weighted axial head scans that are obtained from the whole brain atlas (WBA) maintained by Harward medical school [14] for our experiments.

3 Results and Discussions

We carried out our experiments by applying the above methods on the tumor slices and compared the performance of all the techniques with each other. First column of Fig. 1 shows the sample slices with irregular tumor boundaries. Second column shows the results of extracted tumor boundary by ExM transform. This is automatically used as an initial curve for the rest of the three boundary based techniques: Snakes, LSM and DRLSM. And their fixed parameters are shown in Table 1. The final boundary produced by these methods is shown in column III, IV, V of Fig. 1 respectively.

Table 1. Parameters used in existing methods

Method	Fixed Parameter	Value
Snakes	Sigma (σ)	1.00
	Alpha(α)	0.40
	Beta (β)	0.20
	Gamma(γ)	1.00
	Kappa	0.15
	ELine	0.30
	Eedge	0.40
	Eterm	0.70
	iterations	200
LSM	Alpha(α)	0.50
	Normal_evolve	1
	Vector_evolve	1
	Kappa_evolve	0
	Accuracy	3
	Iterations	60
DRLSM	Gamma(γ)	5.0
	Alpha(α)	0.3
	Mu(μ)	0.04
	Iterations	80

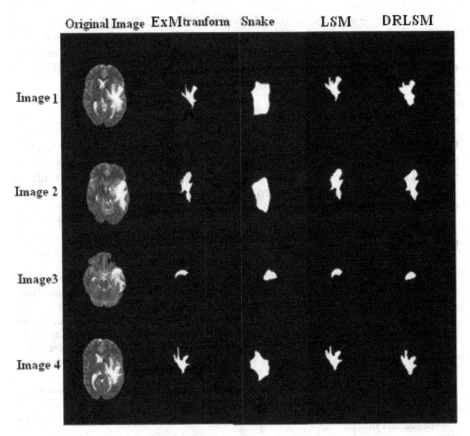

Fig. 1. Results of boundary detection methods. Column 1 – Original Image, Column 2 – initial curve by ExM transform method, Column 3 – final curve by snakes, Column 4 – final curve by LSM, Column 5 – final curve by DRLSM.

Table 2. Computed Jaccard coefficient and William's index for abnormal slices in a volume. A shows results by ExM transform technique, B by snake method, C by LSM, D by DRLSM.

Raters	Jaccard Coefficient in %						WI			
Image	AB	AC	AD	BC	BD	CD	A	B	C	D
Slice1	99.58	99.57	99.57	99.71	99.20	99.23	1.002	1	1.001	0.997
Slice2	99.71	99.63	99.13	99.47	99.39	99.30	1.001	1.002	1.001	0.997
Slice3	98.94	99.41	98.93	99.35	98.51	98.79	1.002	0.999	1.004	0.995
Slice4	95.52	98.28	98.69	97.12	94.56	96.70	1.014	0.978	1.012	0.997
Slice5	95.36	98.62	98.68	96.54	93.36	96.03	1.024	0.972	1.013	0.992
Slice6	95.41	98.64	97.59	96.52	93.36	95.44	1.022	0.978	1.015	0.986
Slice7	94.04	98.81	97.58	95.01	92.46	96.39	1.023	0.962	1.022	0.995
Slice8	96.22	98.95	97.90	97.04	94.63	96.99	1.015	0.98	1.015	0.991
Slice9	96.30	99.05	97.77	97.00	94.91	97.31	1.013	0.98	1.015	0.992
Slice10	96.60	99.43	98.59	97.10	96.22	98.70	1.009	0.977	1.013	1.001
						Avg	1.013	0.983	1.011	0.994

The quantitative analysis based on WI is given in Table 2. WI of ExM (A) is 1.013, snake (B) is 0.983, LSM (C) is 1.011 and DRLSM (D) is 0.994. Among the deformable methods WI is above one for LSM technique. This shows level set method agrees with the final results of all the selected techniques as well as the shape of initial curve as given in Fig. 1.

4   Conclusion

In this work, we proposed to detect final boundary of brain tumor. We have selected three popular deformable models by fixing the initial curves automatically by our proposed method. The performance of these three models is compared using Williams index for ten tumor slices. The experimental results show the LSM produces better results than other deformable models.

References

1. Clark, M.C., Lawrence, L.O., Golgof, D.B., Velthuizen, R., Murtagh, F.R., Silbiger, M.S.: Automatic Tumor Segmentation Using Knowledge-based Techniques. IEEE Trans. Medical Imaging 17, 187–201 (1998)
2. Kaus, M.R., Warfield, S.K., Nabavi, A., Chatzidakis, E., Black, P.M., Jolesz, F.A., Kikinis, R.: Segmentation of Meningiomas and Low Grade Gliomas in MRI. In: Taylor, C., Colchester, A. (eds.) MICCAI 1999. LNCS, vol. 1679, pp. 1–10. Springer, Heidelberg (1999)
3. Moon, N., Bullitt, E., Leemput, K.V., Gerig, K.V.: Model-based Brain and Tumor Segmentation. In: ICPR, Quebec, pp. 528–531 (2002)
4. Corso, J.J., Sharon, E., Yuille, A.: Multilevel Segmentation and Integrated Bayesian Model Classification with an Application to Brain Tumor Segmentation. In: Larsen, R., Nielsen, M., Sporring, J. (eds.) MICCAI 2006. LNCS, vol. 4191, pp. 790–798. Springer, Heidelberg (2006)
5. Capelle, A.S., Colot, O., Fernandez-Maloigne, C.: Evidential segmentation scheme of multi-echo MR images for the detection of brain tumors using neighborhood information. Information Fusion 5, 203–216 (2004)
6. Williams, G.W.: Comparing the joint agreement of several raters with another rater. Biometrics 32, 619–627 (2004)
7. Sylvain, B., Martin-Fernandez, M., Ungar, L., Nakamura, M., Carley, R., Shenton, E.: On evaluating brain tissue classifiers without a ground truth. Neuro Image 36, 1207–1224 (2007)
8. Somasundaram, K., Kalaiselvi, T.: Automatic Detection of Brain Tumor from MRI Scans using Maxima Transform. In: UGC Sponsored National Conference on Image Processing – NCIMP 2010, Gandhigram, India, pp. 136–141 (2010)
9. Kass, M., Witkin, A., Terzopoulos, D.: Snakes: Active contours models. International Journal of Computer Vision 1, 321–331 (1987)
10. Osher, S., Sethian, J.A.: Fronts propagating with curvature-dependent aped: Algorithms based on Hamilton- Jacobi formulations. J. Comput. Phys. 79, 12–49 (1988)
11. Chunming, L., Chenyang, X., Changfeng, G., Martin, D.: Fox, Distance Regularized Level Set Evolution and Its Application to Image Segmentation. IEEE Trans. on Image Processing 19, 3243–3254 (2010)
12. Somasundaram, K., Kalaiselvi, T.: Fully Automatic Brain Extraction Algorithm for Axial T2-Weighted Magnetic Resonance Images. Comp. in Bio. Med. 40, 811–822 (2010)
13. Jaccard, P.: The Distribution of Flora in Alpine Zone. New Phytol. 11, 37–50 (1912)
14. WBA (Whole Brain Atlas) MR brain image,
    http://www.med.harvad.edu/AANLIB/home.html

An Integrated Framework for Mixed Data Clustering Using Growing Hierarchical Self-Organizing Map (GHSOM)

D. Hari Prasad[1,*] and M. Punithavalli[2,**]

[1]Department of Computer Applications,
Sri Ramakrishna Institute of Technology, Coimbatore, India
[2] Department of Computer Applications, Sri Ramakrishna Engineering College,
Coimbatore, India

Abstract. Clustering plays an important role in data mining of large data and helps in analysis. This develops a vast importance in research field for providing better clustering technique. There are several techniques exists for clustering the similar kind of data. But only very few techniques exists for clustering mixed data items. The cluster must be such that the similarity of items within the clusters is increased and the similarity of items from different clusters must be reduced. The existing techniques possess several disadvantages. To overcome those drawbacks, Self-Organizing Map (SOM) and Extended Attribute-Oriented Induction (EAOI) for clustering mixed data type data can be used. This will take more time for clustering; the usage of SOM has the inability to capture the inherent hierarchical structure of data. To overcome this, a Growing Hierarchical Self-Organizing Map (GHSOM) is proposed in this paper. The experimentation is done by using UCI Adult Data Set.

Keywords: Attribute-Oriented Induction, Pattern Discovery, Self-Organizing Map (SOM), Growing Hierarchical Self-Organizing Map (GHSOM).

1 Introduction

One of the widely used techniques in data mining is clustering [10,11]. It focuses on grouping a whole data based on its similarity measures that depends on some distance measure. Most techniques of clustering comprise document grouping, and customer/market segmentation. In knowledge discovery, a basis data mining technique used is data clustering [14, 15, 17]. For investigative data, the clustering with the help of Gaussian mixture models is widely used. The six sequential, iterative steps of Data mining processes are problem definition, data acquisition, data preprocessing and survey, data modeling, evaluation and knowledge deployment.

Generally, clustering [16] includes the classification of the provided data that includes n points in m dimension into k clusters [7]. The clustering must be such that the data points in the respective cluster should be highly identical to one another. The

---

[*] Assistant Professor (Selection Grade).
[**] Director.

P. Balasubramaniam and R. Uthayakumar (Eds.): ICMMSC 2012, CCIS 283, pp. 471–479, 2012.

troubles involved clustering techniques are: identifying a likeness measure to guess the similarity among various data, it is hard to determine the appropriate techniques for identifying the identical data in unsupervised way and derive a description that can distinguish the data of a cluster in an efficient way. Euclidean distance measure is helpful in existing clustering techniques for identifying the similarity among various data. For the data gathered from banks, or health sector, which are categorical data require better clustering technique [13]. It is highly difficult to cluster the categorical data into meaningful category with good distance measure, capturing adequate data similarity and to utilize conjunction with an efficient clustering algorithm [8, 9].

For dealing with mixed numeric and categorical data [1, 4, 5, 12], only few techniques exist. One of the techniques is usage of SOM [2, 18] and EAOI for clustering mixed data type. This will take more time for clustering [3]. To overcome this, a GHSOM is proposed in this paper.

2   Methodology

2.1   Modified Self-Organizing Map

SOM is an unsupervised neural network that assigns high-dimensional data onto a low dimensional grid, generally two-dimensional and conserves the topological connection of the original data. Training an SOM usually involves two phase: the identifying and the adjusting steps. In the identifying phase, every training pattern contains the units of the map and finds the Better Matching Unit (BMU) that is highly identical to the training model. Next, in the adjusting phase, the BMU and its neighbors are updated to be similar to the training pattern.

Table 1. The conventional approach transforms the categorical attribute Favorite_Drink to three binary attributes with domain {0, 1} prior to training an SOM

Name	Favorite_Drink	Amt
Gary	Coke	60
John	Pepsi	70
Tom	Coffee	30

Name	Coke	Pepsi	Coffee	Amt
Gary	1	0	0	60
John	0	1	0	70
Tom	0	0	1	30

Formulas for identifying and the adjusting phases are as follows:

$$\|x - m_b\| = \min_c \{\|x - m_c\|\} \tag{1}$$

$$m_c(t + 1) = m_c(t) + \alpha(t)h_{bc}(t)[x(t) - m_c(t)] \tag{2}$$

where x represents the input vector. m_c and m_b represents the model vectors of unit c and BMU, correspondingly. $\alpha(t)$ and $h_{bc}(t)$ represents the learning rate and the neighborhood function, both decreasing steadily upon increasing the training step t.

Demerit of SOM is over fitting. For mixed data, binary transformation that changes each mixed data to a set of binary attributes (e.g., Table 1) is usually carried out before the training phase [6]. Over fitting is resulted because of the sparseness and high degree of non-linearity in the training material.

On the other hand, the binary transformation technique has at least four demerits:

1. Similarity details between categorical values are not conveyed.
2. When the domain of a categorical attribute is higher, the transformation maximizes the dimensionality of the transformed relation, resulting in wasting storage space and in maximizing the training time.
3. Maintenance is very complex; when the attribute domain is modified, the new relation scheme requires modifying also.
4. The names of binary attributes unsuccessful to conserve the semantics of the original categorical attribute.

In order to overcome these issues, GHSOM is used in this paper.

2.2 Growing Hierarchical Self-Organizing Map (GHSOM)

GHSOM consists of a hierarchical structure of multiple layers in which every layer contains numerous independent growing SOMs. Beginning from a top-level map, every map which is similar to the Growing Grid model, grows in size to symbolize a gathering of data at a particular level of detail [19]. The obtained GHSOM is completely adaptive to reflect, the hierarchical structure inherent in the data, assigning additional space for the illustrating of inhomogeneous areas in the input space. A graphical demonstration of a GHSOM is provided in Fig. 1.

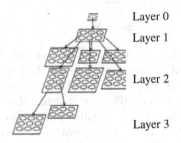

Layer 0
Layer 1
Layer 2
Layer 3

Fig. 1. Trained GHSOM

2.3 Training Algorithm

The advantage of the GHSOM architecture is its adjustment to the training data. The worth of this adjustment is deliberated by means of divergence among a unit's model vector and the input vectors indicates by this specific unit. The mean quantization error (mqe) of a unit i is computed based on (3) as the mean Euclidean distance among its model vector m_i and the n_C input vectors x_j that are elements of the set of input vectors C_i mapped onto this unit i:

$$mqe_i = \frac{1}{n_c} \sum_{x_j \in C_i} ||m_i - x_j||, n_c = |C_i|, C_i \neq \emptyset \qquad (3)$$

The initial point for the GHSOM training procedure is the computation of a mqe_0 of the unit forming the layer 0 map as represented in (4) in which n_I indicates the number of every input vectors x of the input data set I and m_0 represents the mean of the input data.

$$mqe_0 = \frac{1}{n_I} \sum_{x_i \in I} ||m_0 - x_i||, n_I = |I| \qquad (4)$$

The mqe computes the variation of each input data mapped onto a specific unit and will be utilized to manage the growth procedure of the neural network. Mainly, the minimum characteristic of data representation of each unit will be indicates as a fraction, specified by a parameter r_2, of mqe_0. Also, each unit must represent their individual subsets of data at a mqe lower than a fraction r_2 of mqe_0, i.e., convincing the global termination criterion indicated in (5):

$$mqe_i < r_2. mqe_0 \qquad (5)$$

For every unit not agreed with this condition, a more comprehensive data representation is needed, resulting in addition of extra units to afford more map space for data representation. But the quantization error of the unit as represented in (6), indicated by qe, may be utilized as an alternative of the mqe, resulting in a global termination condition as suggested in (7).

$$qe_i = \sum_{x_j \in C_i} ||m_i - x_j|| \qquad (6)$$

$$qe_i < r_2. qe_0 \qquad (7)$$

While utilizing the standard deviation of the data distribution as an overall quality measure for the GHSOM training procedure may be highly intuitive, utilizing the qe pursues more narrowly the basic features of SOM of providing more map space for more densely populated regions of the input space, also referred to as magnification factor. Therefore, with the help of (7) rather than (5) as a global stopping condition resulting in maps that more intuitively replicates the features of data distributions by means of capturing finer variations in highly compactly populated clusters. This result is particularly significant when the resulting maps shall be utilized as explorative interfaces to data sets, as it is regularly the case for SOM-like architectures. The starting size of this first-layer map is set to 2×2 units, with its model vectors being initialized to random values.

Generally, the growth procedure of a growing SOM can be explained as follows. Let C_i indicates the subset of vectors x_i of the input data that is mapped onto unit i, i.e., $C_i \subseteq I$; and m_i indicates the model vector of unit i. After that, the error unit e is computed as the unit with the maximum quantization error as represented in (8):

$$e = \arg\max_i \left(\sum_{x_j \in C_i} ||m_i - x_j|| \right), n_c = |C_i|, C_i \neq \emptyset \qquad (8)$$

After the error unit is chosen, its highly dissimilar neighbor d is identified as listed in (9), where N_e represents the set of neighboring units of the error unit e:

$$d = \arg\max_i (||m_i - x_j||), m_i \in N_e \qquad (9)$$

A row or column of units is included between d and e. To acquire a smooth positioning of the newly included units in the input space, their model vectors are initially set as the means of their individual neighbors. After including the learning rate and neighborhood range are reorganizing to their initial values, and training continues in a SOM-like fashion for the next λ iterations. This training procedure of single growing SOM is extremely similar to the Growing Grid model. The difference up to now is that a decreasing learning rate is utilized and a decreasing neighborhood range rather fixed values.

2.4   Extended Attribute-Oriented Induction

For the exploration of major values, a parameter majority threshold β is introduced. If some values (i.e., major values) take up a major portion (exceeding β) of an attribute, the EAOI preserves those major values and generalizes other non major values. If no major values exist in an attribute, the EAOI proceeds like the AOI, generating the same results as that of the conventional approach. If β is set to 1, the EAOI degenerates to the AOI.

The EAOI algorithm is outlined as follows:

Algorithm: An extended attribute-oriented induction algorithm for major values and alternative processing of numeric attributes
Input: A relation W with an attribute set A; a set of concept hierarchies; generalization threshold θ, and majority threshold β.
Output: A generalized relation P.
Method:
1.   Determine whether to generalize numeric attributes.
2.   For each attribute A_i to be generalized in W,
    2.1 Determine whether A_i should be removed, and if not, determine its minimum desired generalization level L_i in its concept hierarchy.
    2.2 Construct its major-value set Mi according to θ and β.
    2.3 For $v \in Dorn(A_i)$, if $v \in M_i$ construct the mapping pair as $(v, v_{L_i} - M_{L_i})$ otherwise, as (v, v).
3.   Derive the generalized relation P by replacing each value v by its mapping value and computing other aggregate values.

In Step 1, if numeric attributes are not to be generalized, their averages and deviations will be computed in Step 3. Step 2 aims at preparing the mapping pairs of attribute values for generalization. First, in Step 2.1, an attribute is removed either because there is no concept hierarchy defined for the attribute, or its higher-level concepts are expressed in terms of other attributes. In Step 2.2, the attribute's major-value set M_i is constructed, which consists of the first $\alpha(< \theta)$ count leading values if they take up a major portion $(\geq \beta)$ of the attribute, where θ is the generalization threshold that sets the maximum number of distinct values allowed in the generalized attribute.

In Step 2.3, if v is one of the major values, its mapping value remains the same, i.e., major values will not be generalized to higher-level concepts. Otherwise, v will be generalized by the concept at level L_i by excluding the values contained in both the major-value set and the leaf set of the v_{L_i} subtree (i.e., v_{L_i}- M_{L_i} where $M_{L_i} = \text{Leaf}(v_{L_i} \cap M_i)$. Note that, if there are no major values in A_i, M_i and M_{L_i} will be empty. Accordingly, the EAOI will behave like the AOI. In Step 3, aggregate values are computed, including the accumulated count of merged tuples, which have identical values after the generalization, and the averages and deviations of numeric attributes of merged tuples if numeric attributes are determined not to be generalized.

3   Experimental Results

The proposed clustering technique is experimented with UCI Adult Data Set. The data set contains 15 attributes that include eight categorical, six numerical, and one class attributes. 10,000 tuples from the 48,842 tuples are chosen randomly for the evaluation. For the attribute choosing, the method of relevance analysis based on information gain is utilized. The relevance threshold was set to 0.1, and seven qualified attributes are obtained: Marital-status, Relationship, Education, Capital_gain, Capital_loss, Age, and Hours_per_week. The first three are categorical, and the others are numeric.

Table 2. Number of Resultant Clusters for using SOM and GHSOM

Distance Criteria	GHSOM		SOM	
	Cluster	Outliers	Cluster	Outlier
d=0	75	-	88	-
$d \leq 1.414$	9	-	19	-
$d \leq 2.828$	4	-	9	-
$d \leq 3\&Adj$	5	5	14	1

The map size is 400 units. The training parameters are set to the same with that of the previous experiment. The number of resultant clusters by using SOM and modified SOM with different distance criteria is in table 2 and Fig. 2. It can be seen that the proposed technique results in better categorization. To evaluate how the clustering improves the likelihood of similar values falling in the identical cluster, the Average Categorical Utility (ACU) of clusters can be helpful. The higher the value of categorical utility, the better the clustering fares. The ACU is calculated as:

$$ACU = \frac{1}{K} \sum_k \left(\frac{|C_k|}{|D|} \sum_i \sum_j [P(A_i = V_{ij} \mid C_k)^2 - P(A_i = V_{ij})^2] \right)$$

where $P(A_i = V_{ij} \mid C_k)$ is the conditional probability that the attribute A_i has the values V_{ij} given the cluster C_k, and $P(A_i = V_{ij})$ is the overall probability of A_i having V_{ij} in the entire data set.

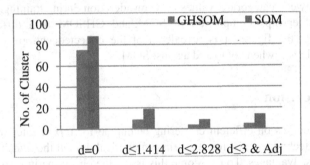

Fig. 2. Number of Resultant Clusters for using SOM and GHSOM

Table 3. Increased Rate of Average CU and the Expected Entropy of the Class Attribute Salary

	GHSOM			
	Leaf Level	Level1	Increased	Exp.Entropy
$d \leq 1.414$	0.085	0.099	17%	0.594
$d \leq 2.828$	0.149	0.085	19%	0.654
$d \leq 3\&Adj$	0.161	0.095	22%	0.582
	SOM			
$d \leq 1.414$	0.118	0.108	-.07%	0.613
$d \leq 2.828$	0.119	0.107	-0.08%	0.701
$d \leq 3\&Adj$	0.121	0.106	-0.09%	0.613

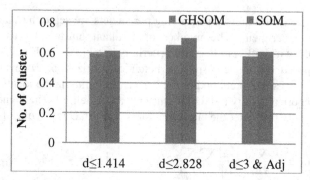

Fig. 3. Expected Entropy Comparison of SOM and GHSOM with Different Distance Criteria

The ACU of categorical values of clusters formed by the three clustering criteria are computed at the leaf level and Level 1 of the distance hierarchies, and the increased rate, as shown in Table 3. This also indicates the effect of taking the similarity between categorical values into consideration during training. Fig. 3. shows the expected entropy comparison of SOM and GHSOM with different distance criteria. From the figure, it is revealed that the expected entropy of GHSOM is comparatively low when compared against SOM.

4   Conclusion

This paper focuses on efficient clustering technique for mixed category data. There are different technique exist for clustering categorical, but all those technique resulted in several disadvantages. To overcome this issue, the clustering in mixed data can be performed based on SOM and EAOI. But this technique also takes more time for classification. To this issue, GHSOM is used in this paper. Experiment is performed using UCI Adult Data Set and it can be observed that the better classification result is obtained for the proposed technique when compared to the existing techniques.

References

[1] Roy, D.K., Sharma, L.K.: Genetic K-Means Clustering Algorithm for Mixed Numeric and Categorical Data Sets. Int. J. Artif. Intell. Appl. 1, 23–28 (2010)

[2] Vesanto, J., Alhoniemi, E.: Clustering of Self-Organizing Map. IEEE Trans. Neural Netw. 11, 586–600 (2000)

[3] Girolami, M.: Mercer Kernel-based Clustering in Feature space. IEEE Trans. Neural Netw. 13, 780–784 (2002)

[4] Li, S., Liu, J., Zhu, Y., Zhang, X.: A New Supervised Clustering Algorithm for Data Set with Mixed Attributes. In: Eighth ACIS Int. Conf. Softw. Engg. Artif. Intell. Netw. Parallel Distrib. Comput., vol. 2, pp. 844–849 (2007)

[5] Yin, J., Tan, Z.F., Ren, J.T., Chen, Y.Q.: An Efficient Clustering Algorithm for Mixed Type Attributes in Large Dataset. In: Proc. Int. Conf. Mach. Learn. Cybernet, vol. 3, pp. 1611–1614 (2005)

[6] Chiu, T., Fang, D.P., Chen, J., Wang, Y.: Christopher Jeris: A Robust and Scalable Clustering Algorithm for Mixed Type Attributes in Large Database Environment. In: Proc. Int. Conf. Data Min. Knowl. Discov., pp. 263–268 (2001)

[7] Liu, H., Yu, L.: Toward Integrating Feature Selection Algorithms for Classification and Clustering. IEEE Trans. Knowl. Data Engg. 17, 491–502 (2005)

[8] Aggarwal, C.C., Procopiuc, C., Wolf, J.L., Yu, P.S., Park, J.S.: Fast Algorithm for Projected Clustering. In: Proc. ACM SIGMOD, pp. 61–72 (1999)

[9] Agrawal, R., Gehrke, J., Gunopulos, D., Raghavan, P.: Automatic Subspace Clustering of High Dimensional Data. Data Min. Knowl. Discov. 11, 5–33 (2005)

[10] Yip, K.Y.L., Cheung, D.W., Ng, M.K., Cheung, K.: Identifying Projected Clusters from Gene Expression Profiles. J. Biomed. Inform. 37, 345–357 (2004)

[11] Yip, K.Y.L., Cheung, D.W., Ng, M.K.: On Discovery of Extremely Low-Dimensional Clusters using Semi-Supervised Projected Clustering. In: Proce. 21st Int. Conf. Data Engg., pp. 329–340 (2005)

[12] Bouguessa, M., Wang, S., Jiang, Q.: A K-Means-based Algorithm for Projective Clustering. In: Proce. 18th IEEE Int. Conf. Pattern Recognition, pp. 888–891 (2006)

[13] Cheng, C.H., Fu, A.W., Zhang, Y.: Entropy-Based Subspace Clustering for Mining Numerical Data. In: Proc. ACM SIGMOD, pp. 84–93 (1999)

[14] Kailing, K., Kriegel, H.-P., Kroger, P.: Density-Connected Subspace Clustering for High-Dimensional Data. In: Proce. Fourth SIAM Int'l Conf. Data Mining, pp. 246–257 (2004)

[15] Parsons, L., Haque, E., Liu, H.: Subspace Clustering for High Dimensional Data: A Review. ACM SIGKDD Explorations Newsletter 6, 90–105 (2004)

[16] Bouguessa, M., Wang, S., Sun, H.: An Objective Approach to Cluster Validation. Pattern Recognition Lett. 27, 1419–1430 (2006)

[17] Jain, A.K., Murty, M.N., Flynn, P.J.: Data Clustering: A Review. ACM Comput. Surv. 31, 264–323 (1999)

[18] Kaski, S., Kangas, J., Kohonen, T.: Bibliography of Self-Organizing Map (SOM). Neural Comput. Surv. 1, 1–176 (1998)

[19] Fritzke, B.: Growing Grid-A Self-Organizing Network with Constant Neighborhood Range and Adaption Strength. Neural Proce. Lett. 2, 1–5 (1995)

Design of Efficient Objective Function
for Stochastic Search Algorithm

Paulraj Vijayalakshmi[1] and Mahadevan Sumathi [2]

[1] Department of Information Technology,
Pandian Saraswathi Yadav Engineering College,
Arasanoor, Sivagangai-630561,Tamilnadu, India
[2] Department of Computer Science,
Sri Meenakshi Govt. College for Women,
Madurai-625002, Tamilnadu, India
vijikcet@gmail.com

Abstract. In this paper, an approach to design the objective function for stochastic search algorithm is presented. N-Queen problem is solved employing genetic algorithm. Fitness function is one of the most critical parts of a genetic algorithm and its purpose is for parent selection and a measure for convergence. Problems with fitness are premature convergence and slow finishing. In general, a chromosome's value is computed by the order of its genes; any change in the order results in different chromosome's value. Here, a weight is computed to each bit (gene) position to compute the fitness of the string. To improve system performance, a weakest bit in the string is selected for cross over. It reduces the   probability of dummy iterations and generation of recessive strings. Experimental results are compared with simple genetic algorithm, and enhanced improved genetic algorithm and adaptive genetic algorithm. Potential application includes search techniques and machine learning.

Keywords: N queen problem, Genetic algorithm, NP hard, Search techniques, Fitness evaluation.

1   Introduction

Genetic Algorithms have proved to be a powerful tool in various areas of computer science including machine learning, search, and optimization. It is a search technique used in computing to find exact or approximate solutions to optimization and search problems. In [1], a statistics-based adaptive non-uniform crossover (SANUX) was proposed. SANUX used the statistics information of the alleles to calculate the swapping probability for crossover operation. The crossover is an operation in GA that can not generate special off springs from their parents because it uses only acquired information. In [2], Parallel genetic algorithm (PGA) was used to solve N-Queen problem with custom genetic operators to increase GA speed. Speed gains were limited by master-slave communication overhead. In [3], a state of the art PGA was reviewed and a new taxonomy of PGA was proposed. The idea was to have one

P. Balasubramaniam and R. Uthayakumar (Eds.): ICMMSC 2012, CCIS 283, pp. 480–487, 2012.

individual for every processing element. It demanded more computational load, memory and high communication cost. In [4], a modification of GA, enhanced improvement of individuals was discussed. The approach was to maintain good - individuals by local search techniques, tabu search, simulated annealing, and iterated local search. After reproduction and improvement of offspring, again poor off springs were improved by additional iterations. In [5, 6], a selective mutation method for improving the performance of GA was proposed, individuals were ranked and then mutated one bit based on their ranks. This selective mutation helped GA to quickly escape local optima. The mutation is an aiding process that can not change many bits in the individuals and hence its performance. In [7], an approach to identify factors that affect the efficiency of GA was reviewed; optimal values were obtained using the approach. In [8], a detection system employing GAs and neural networks were presented. First, neural networks were trained to recognize the characters and then, a template matching was used. To control the size of the neural network inputs and template, a GA search was applied. The drawback was that a huge color database was to be created manually extracting colors from license plates. In [9], a modification of GA for shape detection using edge detection algorithm was presented. The use of GA had reduced the time needed for detection task. The time and storage complexity of the method were linear function of number of the detected features. In [10], a tree based Genetic Programming (GP) for classification methods were reviewed and analyzed. Strengths and weaknesses of various techniques were studied and a framework to optimize the task of GP based classification was provided. The major drawback of that approach was a conflict between more than one classifier needed to be handled. In [11], an approach using GP based object detection to locate small objects of multiple classes in large pictures was described. It used a feature set computed from a square input field that contained objects of interest. There were more errors in detection in complex images.

Simple genetic algorithm (SGA) takes more time and memory. The main idea of PGA is to distribute expensive tasks across slaves controlled by a master process that are executed in parallel. The master maintains a population and executes genetic operators and slaves perform fitness evaluation. Due to communication between the master and slaves, speed gains are limited. A Hybrid Genetic Algorithm has been designed by combining a number of heuristics. Two heuristics were applied to offspring to prevent getting struck up at local optimum. The hybrid genetic algorithm was designed to use heuristics for initialization of population and improvement of offspring produced by crossover. Adaptive genetic algorithms (AGA) use many parameters, such as the population size, crossing over probability and the mutation probability and they are varied while the algorithm is running. The draw backs of AGA includes crossing point is selected randomly and the population retains without any improvement for longer time [12]-[15].

In this paper, an objective function for N queen problem is designed assigning weights to each bit place forming the string. The remaining paper is organized as follows: In section 2, proposed design approach and algorithm are discussed. In section 3, experimental results are analyzed and compared. In section 4, conclusion and future work are presented.

2   System Overview

Genetic algorithm is adopted to solve the N Queen problem that has three variants: finding one solution, a family of solutions, and all solutions. Optimal solutions to small "N' values are found in reasonable time by classical search algorithms. However, to solve larger N values, it is very time consuming since the N-Queen problem is a constraint satisfaction problem (CSP).

2.1   Proposed Method

In N queen problem, N queens are to be placed without conflicting such that no two queens fall on the same row or column or diagonal. The fitness (objective) function is obtained by a simple combination of the different criteria that gives good solution the problem. Parameters that decide solution of the N queen problem are identified and the string is coded. Consider an initial population $\{P_1, P_2.....P_n\}$ strings, where 'n' is population size and P_i represents an individual string. In general strings over all fitness is computed based on objective function and is shown in Fig. 1(a). Here, numerals are used to define objective function. Also, a small weight is computed to each bit place to compute over all fitness of the string Fig. 1(b).

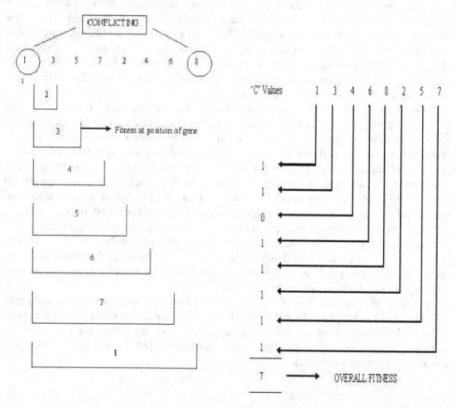

Fig. 1. a) weight for bit places in new approach b) Fitness calculation in classical algorithm

For the design of new objective function, weight for each bit place forming the string is computed using formula (1). Q_i, Q_j are queens and i and j are positions of the queens.

$$F_i = 1 + \sum_{j-1}^{i-1} c\left[Q_i, Q_j\right] \tag{1}$$

Fi-weight of ith bit place in the string, C $[Q_i, Q_j]$ is computed using formula (2) and (3).

$$c\left[Q_i, Q_j\right] = \begin{cases} 1 \; for \; Q_i \neq Q_j \; and \mid Q_i - Q_j \mid \neq \mid i - j \mid and \; i \neq j \\ \qquad\qquad 0 \qquad\qquad\qquad otherwise \end{cases} \tag{2}$$

i, j are the position of queen

Q_i, Q_j *are queens*

$$c\left[Q_i, Q_j\right] = \begin{cases} 1, strong \\ 0, weak \end{cases} \tag{3}$$

The objective function of the string OF_i is computed iteratively using eqns. (4) and (5). $POS(Q_i)$ is the position of queen Q_i and F_i is weight value assigned to that bit place.

$$P_i = \begin{cases} 1, F_i = POS(Q_i) \\ 0, \quad otherwise \end{cases} \tag{4}$$

$$OF_i = + \sum_{i=1}^{n} Q_i \tag{5}$$

OF_i is the overall fitness of ith string and n is the length of the string.

2.2   Cross over and Mutation Process

In cross over, chromosomes of two parents randomly recombine to form offspring. Some times chromosomes of the two parents are copied unmodified as offspring. The crossover generates normal off springs from their parents. To improve the convergence and avoid dummy iterations, cross over is applied at conflicting (weaker) positions in the string. Cross over operation is explained in eqn (6). Consider POSi ....POS_{sl} is the position of element in parent Q_i, Q_{i+1}, C_1............C_n are the conflicting position of parent Q_i then, $newQ_i$ is computed using equation (6).

$$newQ_i = \begin{cases} Q_i\left[POS_I\right] = Q_{i+1}\left[C_I\right]; POS_I = C_I \\ Q_i\left[POS_I\right] = Q_i\left[POS_I\right]; POS_I \neq C_I \end{cases} \tag{6}$$

Crossover explores combinations of the current gene pool. Mutation generates a new string from single string(parent). It can "generate" new genes. The mutation process is explained using eqn. (7). Consider $Cr_1, Cr_2...Cr_n$ are crossed positions; $e_i,......e_N$ are repeated elements in Q_i ; $ne_i,......ne_N$ are non redundant elements; $pos_1,........pos_{sl}$ are position of bits in Q_i, then new Q_i is computed using equation 7.

$$newQ_i = \begin{cases} Q_i[pos_i] = ne_i \; ; \; pos_i \neq cr_i \; \& \; Q_i[pos_i] = e_i \\ Q_i[pos_i] = Q_i[pos_i]; \; pos_i = cr_i \end{cases}$$

(7)

2.3   Algorithm for N Queen Problem

begin

       Collect solution parameters
       Design an objective function
       Code the string
       Compute objective function value
       Generate initial population

begin

       Initialize i as one;
       While (i not equal to population size)

begin

       Take P_i and P_{i+1} string from population;
       Apply crossover;
       Apply mutation;
       Evaluate objective function of new P_i;
       Add new string P_i; and construct a new population
       end{new population construction)

end{expected population}
end {algorithm}

3   Analysis of Experimental Results

The new approach is tested to solve N Queen Problem using genetic algorithm. Program in Turbo C developed on Intel Core to Duo Processor. Factors that affect performance of stochastic search algorithm are objective function, population size, cross over and mutation probability. Increasing population size, increases diversity and computation time. Rising crossover probability improves opportunity for recombination but disrupts good combinations and high mutation probability makes the search algorithm into a random search. Here, by designing objective function system performance is improved. Time and average fitness of the population is performance indicator of any genetic algorithm. Average fitness is computed as the ratio of total fitness to the population size. SGA, AGA and EIGA algorithms are tested in the same environment for analysis purpose. Parameters used for experimentation are tabulated in Table 1. Due to limitation of computational resources, experiments are conducted for values of 'N' 8 to 27 queens. Strings of population are formed using numerals instead of binary bits. For N=8, 1 2 ....

8 queens are considered and for N =27, 1 2 3. . . . . 27 queens are considered. Experiments are conducted for all solutions and single solution. For all solutions, the algorithm is made to converge only when all N queens are placed as per the constraints and iterations are repeated at least one objective solution has reached for a single solution. Experimental results of population fitness for single solution are tabulated in Table 2. For N=18, it is observed that population's fitness is 11 for SGA, 11 for AGA and 12 for EIGA and 14 for new approach. It is important to note in the calculations of string over all fitness, a weight of "1" is allotted uniformly for all non conflicting positions in the string irrespective of bit places and a "0"is assigned for conflicting positions in the classical approaches. Also, it is observed for certain iterations fitness value does not change, they are dummy iterations.

Time is another important performance indicator for any computational algorithm. It is clear that increasing crossover probability improves opportunity for recombination but disrupts good combinations. Increasing mutation probability makes search algorithm into a random search. Experimental results for computation efficiency are tabulated in Table 3. For "N= 22", time taken is 1200 seconds for SGA, 802 seconds for EIGA, 531 seconds for AGA and 306 seconds for new approach for all solutions. It is understood that fitness value of the objective function is retained over many iterations in classical algorithms that employs various cross over operators. This is due to cross over at improper site and demands more iteration to converge. It is observed that the new approach performs better when a weak point (conflicting point) is chosen for cross over along the string length than a random point selection .It also avoids dummy iteration and reduction in fitness value. It is also possible to determine number of iterations and time required to arrive at a particular solution which is hard to find in classical approaches. Though 'N queen problem' is NP hard and non

Table 1. Parameters used in Experimentation

Parameters	Method/ value
Parent Selection	Roulette wheel
Cross over probability	0.72
Mutation Probability	0.001
Population size	8-27
String length	8-27

deterministic in nature, the new iterative approach could make the algorithm to converge in deterministic time for large population size, N=100, N=500, N=1000. Moreover, for certain values of "N", it is difficult to decide a good algorithm and it could be done based on trade off between time factor and average fitness value.

Table 2. Comparison of population fitness

Number of Queen	Population's average fitness			
	SGA	EIGA Ref[4]	AGA Ref[1]	New Approach
8	5	6	5	7
9	5	6	5	7
10	6	6	6	8
11	6	7	6	11
12	6	8	6	11
13	7	8	7	13
14	7	9	7	12
15	8	9	8	13
16	9	10	9	13
17	10	11	10	14
18	11	12	11	14
19	11	13	12	16
20	12	12	12	17
21	12	15	13	17
22	13	16	15	18
23	14	15	12	19
24	14	18	16	20

Table 3. Comparison of Efficiency

No. of Queen	Time Taken in seconds(for all solutions)			
	SGA	EIGA Ref[4]	AGA Ref[1]	New approach
8	0.25	0.15	0.00	0.00
9	0.50	0.36	0.20	0.12
10	0.90	0.50	0.30	0.17
11	1.60	0.85	0.40	0.35
12	2.01	1.08	0.72	0.53
13	3.05	2.0	1.35	0.90
14	4.60	3.29	2.54	1.57
15	6.09	5.02	4.10	3.05
16	51.26	20.27	15.09	10.72
17	445	178	45. 09	27.18
18	703.98	230.43	131.05	78.25
19	603.35	370.12	203.36	108.27
20	890.34	432.92	306.72	162.00
21	1043.09	609.78	450.81	261.09
22	1200.99	802.46	531.72	306.54

4   Conclusion

An objective function design approach for stochastic search algorithm is formulated by computing weight for each (gene) bit in the string and selecting a weak point (conflicting position) along the string length for cross over. As a result, dummy iterations are avoided completely. Experimental results demonstrate that new design approach generates efficient solutions with better fitness value. The approach could be modified to solve optimization problems. Also, data base concepts could be employed to solve machine vision problems.

References

1. Yang, S., Abbass, Bedau: Adaptive Crossover in Genetic Algorithms Using Statistics Mechanism. In: Standish, Abbass, Bedau (eds.) Artificial Life VIII, pp. 182–185. MIT Press (2002)
2. Božikovic, M., Golub, M., Budin, L.: Solving N-Queen Problem using Global Parallel Genetic Algorithm. In: EUROCON, Ljubljana, Slovenia (2003)
3. Nowostawski, M., Poli, R.: Parallel Genetic Algorithm Taxonomy. In: Kes 1999, pp. 88–92 (1999)
4. Misevičius, A., Rubliauskas, D.: Enhanced Improvement of Individuals in Genetic Algorithms. Information Technology and Control (2008) ISSN 1392 – 124x
5. Jung, S.H.: Selective Mutation for Genetic Algorithms. World Academy of Science, Engineering and Technology 56 (2009)
6. Vrajitoru, D.: Crossover Improvement for the Genetic Algorithm in Information Retrieval. Information Processing & Management 34(4), 405–415 (1998)
7. Petrovski, A., Wilson, A., Mccall, J.: Statistical identification and optimisation of significant GA factors. In: Fifth Joint Conference on Information Sciences (JCIS 2000), Atlantic City, NJ (2000)
8. Karungaru, S., Fukumi, M., Akamatsu, N.: Detection and Recognition of Vehicle License Plates Using Template Matching, Genetic Algorithms And Neural Networks. Int. J. Inno. Comp. Inf. Control 5(7), 1975–1985 (2009)
9. Abdel-gaied, S.M.: Employing Genetic Algorithms for Qualitative Shapes Detection. ICGST-GVIP 8(4), 19–25 (2008) ISSN 1687-398X
10. Jabeen, H., Baig, A.R.: Review of Classification using Genetic Programming. Int. J. Eng. Sci. Tech. 2(2), 94–103 (2010)
11. Ludwig, O., Nunes, U.: Improving the Generalization Properties of Neural Networks: an Application to Vehicle Detection. In: 11th International IEEE Conference on Intelligent Transportation Systems, Beijing, pp. 310–315 (2008)
12. Man, K.-F., Tang, K.-S., Kwong, S.: Genetic Algorithms: Concepts and Designs (Advanced Textbooks in Control and Signal Processing). Springer, Heidelberg (1999) ISBN-10: 1852330724
13. Michalewicz, Z.: Genetic Algorithms + Data Structures = Evolution Programs. Springer, Heidelberg (1998) ISBN-13: 978-3540606765
14. Sivanandam, S.N., Deepa, S.N.: Introduction to Genetic Algorithms. Springer, Heidelberg (2008)
15. Goldberg, D.E.: Genetic Algorithms in Search, Optimization, and Machine Learning, 1st edn. Addison-Wesley Professional (1989) ISBN-10: 0201157675

Mining of Quantitative Association Rule on Ozone Database Using Fuzzy Logic

A.M. Rajeswari, M.S. Karthika Devi, and C. Deisy

Thiagarajar College of Engineering, Madurai,
Tamil Nadu, India
amrcse@tce.edu, karthikadevi88@gmail.com,
cdcse@tce.edu

Abstract. In this paper, we present a fuzzy data mining approach for extracting association rules from quantitative data using search tree technique. Fuzzy association rule is used to solve the high dimensional problem by allowing partial memberships to each different set. It suffers from exponential growth of search space, when the number of patterns and/or variables becomes high. This increased search space results in high space complexity. To overcome this problem, the proposed method uses search tree technique to list all possible frequent patterns from which the fuzzy association rules have been generated.

Keywords: Data Mining, Association Rule, Fuzzy Logic.

1 Introduction

Data mining is the process of extracting a hidden, predictive data from a very large database [1]. Data mining techniques support automatic exploration of data and attempts to source out the patterns and trends in data and also to infer rules from those patterns.

Association discovery is one of the most common data mining techniques used to extract interesting knowledge from a very large datasets. Once the frequent item sets from the transaction dataset has found, association rules can be generated. With the support threshold and confidence threshold, association rules will be generated [2], [3].

In reality, database not only consists of binary attributes but also quantitative attributes. Mining of such quantitative attributes are said to be quantitative association rule mining. With quantitative attributes, partitioning of data will have unnatural boundaries which lead to overestimate or underestimate the values [4], [5].

Fuzzy logic is most used in data mining technique because of its crisp and simplicity nature. It overcomes the problem of quantitative association rule by having natural boundary in partitioning of data. It facilitates the interpretation of rules in a most realistic way [6], [7], [8].

In this paper we present an approach to extract interesting rules from quantitative attributes using fuzzy association rule based on search tree technique.

P. Balasubramaniam and R. Uthayakumar (Eds.): ICMMSC 2012, CCIS 283, pp. 488–494, 2012.

2 Related Work

In quantitative association rule mining, partitioning of data will be with unnatural boundary. If there is unnatural boundary in partitioning of data, it leads to overestimate/underestimate of values [4]. Because of this drawback, it has been moved towards fuzzy association rule mining. Fuzzy data mining algorithm can be given by allocating dynamic membership functions for 2 tuple linguistic model [9]. In this algorithm, the rules generated are twice as of fuzzy association rules as remaining approaches. The drawback of this approach is more number of rules being generated [10]. For the purpose of predicting rainy/normal day, fuzzy association rules are generated using weighted counting algorithm. This approach proved to be efficient even when the linear statistical relation between the indices is weak [11]. Rules generated using fuzzy association rule mining will be more in number which in turn increases the search space. To reduce the search space, search tree technique [12] is used which uses static membership function.

3 Design Concept for Mining Fuzzy Association Rules

3.1 Architecture of the Proposed System

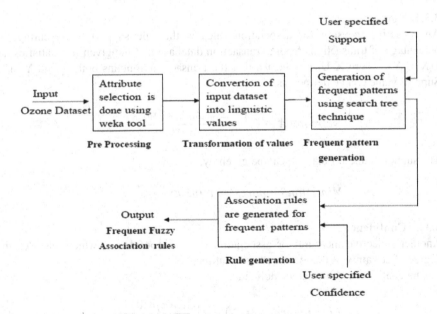

Fig. 1. Architecture of proposed system

3.2   Dataset Used

To evaluate the proposed approach, ozone level detection database which is taken from UCI repository [13], [14] is used. The description of dataset is as follows

  Number of instances: 2536
  Number of instances after data cleaning: 1838
  Number of attributes: 74

Dataset consists of temperature with 26 attributes (T0,......,T23, T_PK, T_AV), wind speed ratio with 26 attributes (WSR0,.....WSR23,WSR_PK, WSR_AV), relative humidity with 3 attributes (RH85,RH70,RH50), height with 3 attributes (HT85, HT70, HT50) and directions with 6 attributes (U85, U70, U50, V85, V70, V50) and sea level pressure. Even though the temperature and wind speed ratio are taken for different period of time, the values are with only slight variations. Hence attribute selection is done based on cfs subset level algorithm and best first algorithm using weka tool to avoid the redundancy and repetition. Based on this algorithm, the selected attributes are WSRAV, T_PK, RH50, HT50 and V70.

3.3   Measures Used

3.3.1 Support

An objective measure for association rules is the rule support, representing the percentage of transactions from a transaction database that the given rule satisfies i.e. P(X U Y), where X U Y indicates that the transaction contains both X and Y [12]. Support value can be defined as,

$$support(A \rightarrow C_j) = \frac{\Sigma_{x_p \in class C_j} \mu_A(x_p)}{|N|} \tag{1}$$

The support value for class C_j can be given by,

$$MinimumSupport_{C_j} = minSup * f_{C_j} \tag{2}$$

3.3.2 Confidence

Another objective measure for association rules is confidence, which assesses the degree of certainty of the identified association [12].

  The confidence value can be defined as

$$Confidence(A \rightarrow C_j) = \frac{\Sigma_{x_p \in class C_j} \mu_A(x_p)}{\Sigma_{x_p \in T} \mu_A(x_p)} \tag{3}$$

4   Algorithmic Design Concept

To reduce the search space, search tree technique is used. Algorithmic design concept for the proposed method is given in Fig. 2.

INPUT:
d - raw dataset
D - selected attributes
T - No of instances
Ø - Membership function
ψ - minimum support value
∂ - minimum confidence value

OUTPUT:
R - fuzzy association rule

METHOD:
    Step1: Data Pre processing
        Data cleaning : cleaning of missing values
        Attribute Selection: $d(a1, a2, a3, a4, a5, a6) \rightarrow D(a1, a2, a3, a4)$
        Discretization : $D(a1, a2, a3, a4) \rightarrow D(A1, A2, A3, A4)$
        For I=1 to T
    Step2: Convertion of input values to linguistic terms i.e fuzzy values Ø
    Step3: Frequent pattern generation
        Frequent patterns are generated using search tree technique and the patterns
        which fall below ψ are rejected.
    Step4: Fuzzy Association Rule Generation
        Rule generation R is performed with frequent patterns and the rules which
        fall below ∂ are   rejected.

Fig. 2. Algorithmic Design of the proposed system

4.1   Search Tree Technique

To reduce the search space, search tree technique is used[12]. In this search tree technique, root level or level 0 is an empty set. The attributes are assumed to have an order and the one item set corresponding to attributes are listed in first level of search tree. If an attribute has j possible outcomes, it will have j one itemsets   listed in the first level. The children of a one-item node for an attribute A are the two-item sets that include the one item set of attribute A and a one-item set for another attribute behind attribute A in the order, and so on. If an attribute has j > 2 possible outcomes, it can be replaced by j binary variables to ensure that no more than one of these j binary attributes can appear in the same node in a search tree. An itemset with a support higher than the minimum support is a frequent itemset. If the support of an n-item set in a node J is less than the minimum support, it does not need to be extended more.

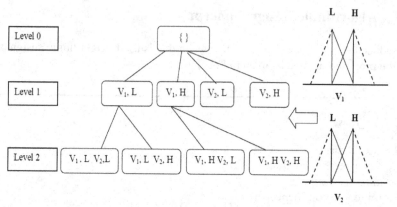

Fig. 3. Search tree for two quantitative attributes V_1 and V_2 with two linguistic terms L and H

5   Experimental Analysis

To evaluate the proposed approach we carried several experiments on a real world dataset named ozone dataset [14]. Those data were collected from 1998 to 2004 at the Houston, Galveston and Brazoria area. Ozone dataset contains 2536 transactions with 73 attributes.

5.1   Analysis of Search Space Complexity

To prove that the search tree technique reduces the search space, we compare this technique with conventional method. From Figs. 4 and 5, we can easily conclude that proposed work uses less search space when compared to conventional Apriori method.

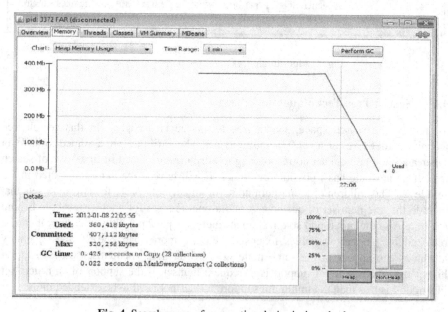

Fig. 4. Search space of conventional- Apriori method

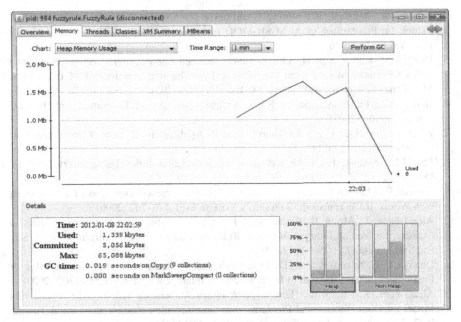

Fig. 5. Search space of proposed work

6   Conclusion and Future Work

In this proposed method, quantitative values are taken as an input. In quantitative values, unnatural boundary partitioning is overcome by using fuzzy logic. In fuzzy logic natural boundary partitioning is done and the values are changed into a most linguistic term. The search space is reduced using search tree technique by limiting the depth of the tree. With the search tree technique, number of frequent patterns generated has been reduced. Finally fuzzy association rules are generated to predict ozone/normal day from the ozone database. It has been found that, Search tree technique reduces the search space when compared to conventional method (Apriori).

In future, the proposed work can be extended to perform further reduction in search space. And also experiment the proposed work with different real time datasets.

References

1. Han, J., Kamber, M.: Data Mining: Concepts and Techniques, 2nd edn. Morgan Kaufmann Publishers, San Fransisco (2006)
2. Agrawal, R., Srikant, R.: Fast Algorithms for Mining Association Rules in Large Databases. In: Proceedings of 20th International Conference on Very Large Data Bases, pp. 487–499 (1994)
3. Agrawal, R., Imielinski, T., Swami, A.: Mining Association Rules between Sets of Items in Large Databases. In: Proceedings of ACM-SIGMOD International Conference on Management of Data, vol. 22(2), pp. 207–216 (1993)

4. Srikant, R., Agrawal, R.: Mining Quantitative Association Rules in Large Relational Tables. In: Proceedings of ACM-SIGMOD International Conference on Management of Data, vol. 25(2) (1996)
5. Hong, P.T., Kuo, C.S., Chi, S.C.: Trade-off between Computation Time and Number of Rules for Fuzzy Mining from Quantitative Data. International Journal of Information Management Fuzziness and Knowledge-Based Systems 9(5), 587–604 (2001)
6. Helm, B.L.: Fuzzy Association Rules. Vienna University of Economics and Business Administration (2007)
7. Zimmermann, H.-J.: Fuzzy Set Theory - and Its Applications, 2nd edn. Kluwer Academic Publishers, Boston (1991)
8. Lee, J.H., Kwang, H.L.: An Extension of Association Rules Using Fuzzy Sets. In: Proceedings of 7th International Fuzzy Systems (1997)
9. Herrera, F., Martnez, L.: A 2-Tuple Fuzzy Linguistic Representation Model for Computing with Words. IEEE Transactions on Fuzzy Systems 8(6), 746–752 (2000)
10. Alcalá-Fdez, J., Alcalá, R., Gacto, M.J., Herrera, F.: Learning the Membership Function Contexts for Mining Fuzzy Association Rules by Using Genetic Algorithms. Fuzzy Sets and Systems 160(7), 905–921 (2009)
11. Dhanya, C.T., Nagesh Kumar, D.: Data Mining for Evolving Fuzzy Association Rules for Predicting Monsoon Rainfall of India. Journal of Intelligent Systems, 1299–1309 (2009)
12. Alcalá-Fdez, J., Alcalá, R., Herrera, F.: A Fuzzy Association Rule-Based Classification Model for High-Dimensional Problems with Genetic Rule Selection and Lateral Tuning. IEEE Transactions on Fuzzy Systems 19(5), 857–872 (2011)
13. Zhang, K., Fan, W.: Forecasting Skewed Biased Stochastic Ozone Days: Analyses, Solutions and Beyond. Knowledge and Information Systems 14(3), 299–326 (2008)
14. UCI Repository of Machine Learning Databases, http://archieve.ics.uci.edu

JNLP Based Secure Software as a Service Implementation in Cloud Computing

M. Jaiganesh and A. Vincent Antony Kumar

Department of Information Technology,
PSNA College of Engineering and Technology,
Dindigul, Tamil Nadu, India
jailingam@gmail.com, vincypsna@rediffmail.com

Abstract. Cloud computing is starting a new era in getting information puddles through various internet connections by any connected device. It provides pay by use method for grasping the services on the claim required by the clients. Cloud service providers treat the client as a virtual client to accommodate a virtual environment of cloud computing. Virtual clients get the services and operate the software, hardware in the cloud service provider domain. The major issue in cloud is providing security against the unauthenticated accessibility of the cloud services. We propose a novel approach to provide Java network Protocol Launcher (JNLP) based secure model for cloud computing. This paper concentrates on designing simple cloud architecture by setting up PHP, Java for clients in cloud environment. The software is provided as a service on demand in this dynamic system. In addition to identification and evaluation of the design, we implement a real time text editor, calculator, Puzzle game. We fabricate an innovative Java Archive file (JAR) to prevent the attackers in different time intervals effectively.

Keywords: Cloud Computing, JNLP, JAVA, Security, PHP.

1 Introduction

Cloud computing is the deliverance of services over an internet. It offers utility based computing like resource sharing, software, platform and infrastructure [1]. Cloud service providers bring services through internet that is accessed by web browsers, desktop applications, and mobile apps. The pool of data is stored in data centre at cloud service provider location. The applications needed for business purposes can be developed by the use of web programming languages namely php and AJAX. The end user will not be able to know the details about the server location and the hardware software specifications of the cloud server. There are five important layers which are responsible for the cloud environment; those are Client, Application, Platform, Infrastructure and Server [13, 14]. The Client is the interface which is responsible for the delivery of applications. The application layer is needed for providing the support for running Software as a Service (SaaS) which removes the necessity to install the software in the client system. The platform layer helps in deploying the application

P. Balasubramaniam and R. Uthayakumar (Eds.): ICMMSC 2012, CCIS 283, pp. 495–504, 2012.

eliminating the need of the complex procedures involved in the use of hardware and software in the client machine [12]. The Platform as a Service (PaaS) provides the infrastructure required for sustaining the software which is deployed as cloud services. Infrastructure as a Service (IaaS) provides the infrastructure necessary for the environment to provide cloud services. The users need not buy the software required; instead they need to pay only for the service based on utility. The collection of services, hardware, data are stored in a centralized pool called data centre [4, 6, 10]. The data centre may include multi core processing capabilities and operating system specific to cloud. The interconnection of one cloud with another called the Intercloud is used in some applications. The engineering discipline of cloud called cloud engineering is used to provide consistency, supremacy and commercial approach to the systems providing cloud concepts.

Cloud computing security is more important in implementing the cloud based services [3, 8]. It comprises the inclusion of various policies and techniques which are used to provide the security to applications accessed via the cloud. Security issues concerned with cloud are broadly classified for the two classes such as service providers and customers. It is the responsibility of the data centre to provide the necessary security measures needed for both the data centre and clients. The clients must make sure that the data they receive and the applications and services are secure. The data protection mechanisms from the servers need to provide data isolation of various users [5, 12]. The process of running multiple virtual clients each with their own OS is called the isolation. It is one of the properties of virtualization. Virtual isolation is a problem in which the virtual clients are attacked by malware by injecting codes into other virtual clients [9]. Hypervisor also is vulnerable to these attacks. The virtual clients request their own resources and they run in their own application domain.

The cloud computing research is in the field of virtualization to represent virtual isolation problems. The contribution of this paper is as follows.

1. We propose a novel security model using JNLP to overcome virtual isolation problems in cloud computing, which can more effectively prevent the malware in virtual environment.

2. We construct an innovative Java Archive file structure to hide the archive file in virtual environment.

3. We design software as a service model to run applications in private cloud systems, which can be implemented and evaluated using PHP, J2EE environment.

2   Background

Simon Wardley, Etienne Goyer and Nick Barcet [7] proposed Ubuntu Enterprise Cloud gives an open source implementation standard. It is designed to simplify the process of establishing and organizing an internal cloud for business. It provides a UEC internal architecture having Cloud Controller (CLC), Walrus Storage Controller (WS3), Elastic Block Storage Controller (EBS), Cluster Controller (CC) and Node

Controller (NC). The users or administrators of the system which have specific rights to modify the system and start and stop instances, the components of the system (NC, CLC, CC) which need to trust each other when transmitting requests.

Jason Carolan and Steve Gaede [2] Sun Cloud computing architecture gives to increase the velocity with which applications are deployed, increase innovation, and lower costs, all while increasing business agility It takes an inclusive view of cloud computing that allows it to support every facet, including the server, storage, network, and virtualization technology that drives cloud computing environments to the software that runs in virtual appliances that can be used to assemble applications in minimal time.

Borja Sotomayor, Rub´en S. Montero, Ignacio M. Llorente, and Ian Foster presented Open Nebula [1] and Haizea, two open source projects that address these challenges. Open Nebula is designed from the outset to be easy to integrate with other components, such as the Haizea lease manager. When used together, Open Nebula and Haizea are the only virtual infrastructure management solution that provides leasing capabilities beyond immediate provisioning, including best-effort leases and advance reservation of capacity.

VM WARE [11] Solutions describes an overview of the term virtualization is a separation of a resource or request for a service from the underlying physical delivery of that service. With virtual memory, for example, computer software gains access to more memory than is physically installed, via the background swapping of data to disk storage. Similarly, virtualization techniques can be applied to other IT infrastructure layers - including networks, storage, laptop or server hardware, operating systems and applications.

3   Preliminaries and Design

3.1   Java Network Launching Protocol

Java Network Launching Protocol (JNLP), architecture developed by Sun Microsystems [15, 16]. It is possible to load and store the file as needed by the client, does not need any applet viewer. There are many minor extensions provided by the protocol. The applications need to be signed in order to have all permission levels. Using this protocol eliminates the need for the various Java Virtual Machine versions and different java plug ins required for the browsers. The java programs run in separate frames and do not require any web page. The java programs can be packaged as *jar* files and then can be converted as JNLP files which can be accessed across the web using browsers. The advantages of the java are also included in JNLP such as cross platform and open source implementation. The protocol is defined using Extensible Markup Language (XML) schema. It specifies how to initialize Web start. It consists of a set of rules that is used for the implementation of launching mechanism.

The files contain the information which is used to locate the *jar* package file and the main class in which the codes are included. The web browser then transfers the JNLP files to the Java Runtime Environment (JRE). The JRE then fetches the application from the server and stores it in the local machine. It is then executed. The main advantage of web start is to download and automatically install JRE if it is not found in the client machine. It is stored in the cache and can be accessed locally. When the user connects to the internet, the available downloads and updates needed for the application is fetched. The client's work is to simply install the JNLP client. The working is similar to that of the HTTP and HTML. The request for an HTML file is given by clicking the hyperlink. The URL is then passed to the web browser and the HTML file is returned to the client giving the request. The same principle is used in JNLP. By default, the web start applications are restricted in their access to the local resources, but this restriction can be removed by signing the apps with jar signer tool.

The JAR file which is the Java Archive file is used as a comprehension tool to accumulate all the necessary java class files and the resources and the data required for the application [17]. They have .jar extension. The JDK contains the jar command to create the jar file.

The security problems that occur in the virtual system environments mostly the same as that of the problems that occur in a physical system. The virtual world has exhausted the last decade hobbling on preventing networks from conventional security attack vectors. Some of many network firewalls and filters, ethical hacking software's, tools solved levels sophisticatedly sufficient to allow and pull and fall enforcement of security policies. In Table 1, we highlighted some of the attacks and their attempts to eavesdrop attributes in cloud environment.

Table 1. Tale of attacks and attempts

No. of attempts	Vulnerability type	Description of attacks
57 − 18	Novell e Directory Session Cookie	Virtual cookie attack
47 − 28	2Wire Cross-Site Password Reset	Reset admin password
11 − 64	Veritas Backup Exec	Backup arbitary files attack)
45 − 30	Iomega StorCenter Pro Bypass	Authentication bypass
29 − 46	FileZilla FTP Server Admin Interface	Denial of service attack
39 − 36	SAP Management Console OSExecute	Execution of OS attack
37 − 38	VxWorks WDB Agent Remote Memory Dump	Dump the system memory attack
55 − 20	TrendMicro Data Loss	Data loss attack
60 − 15	Samba Symlink Directory Traversal	Writeable share directory attack
67 − 08	Apache mod-isapi	Server attack

4   Proposed Method

The main aim of proposed method is to hide the *jar* file from attackers and hackers in cloud environment. It is followed by a six steps of procedure to keep JNLP with *jar* file handling and securing *jar* files.

Step1. Client Request Module

In this module, the end user (or) client requests the services from the cloud environment (or) cloud service provider. The cloud administrator initiates the services by affording registration form, with full details of client. After receiving the registration form, client fills up (or) includes entire information and submits to the cloud admin called publishing the services. The cloud computing services are like on-demand and open pool in nature. Receiving the registration form, the cloud admin updates the entire information of client in data centre, and subsequently it establishes a service connection and allocates resources according to the nature of the service request by the client.

The resource allocation includes memory, CPU or any hardware requirements. The client registration form is designed using PHP and HTML. The registration form is directly obtained from the browser. The client must fill their first name, last name, e-mail id, password, address, time and alternate e-mail id. This information stored in the data centre is given by the virtual client. The cloud admin is able to allocate the resources to the clients. The virtual client details are set aside in the database (MySql), by saving the user details we can surpass the virtual Isolation problems straightforwardly. The java network launching protocol is used for defining the xml schema that contains data on how to launch the application in the client machine. It is noted that these approaches help in triumphing the drawback of the virtual client problem and the security in the cloud organization.

Step 2. Initialization of JNLP Using Java Web Start

JNLP files are loaded in the java web start; it retrieves the JAR file from the server. In this Fig. 1, the JNLP architecture design is discussed.

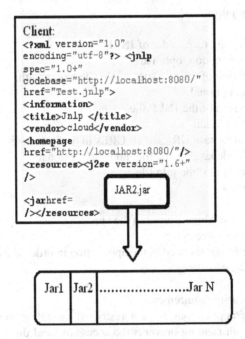

Fig. 1. JNLP Architecture – Schema

The above processes are implemented in the client. Server holds the *jar* files. By means of PHP we can copy the corresponding *jar* file from the root folder to fleeting folder. During this file transfer the *jar* file is loaded into the cloud admin and it is allocated to the virtual client. During this transformation the client details are updated in the database with new details of application which means the new timing will be added in the records. By using this conception we can leave behind the virtual isolation quandary. The number of clients and their details are continuously monitored and any changes will indicate the forge attackers who cause virtual isolation. The timing will be saved in client catalogue. This timing will be refreshed for each five minutes in the client side. Additionally the new timing will be updated in the database. For the issue of client side request the J2EE application is exploited and the package *openStream ( )* will open the URL for accomplishment of response from the server. This corresponding response will be saved in buffered folder of Java SDK. Subsequently the response will be displayed in the client using *bufferredwriter*. By means of *setTimeout ( )* the client is called for each five minutes. If the client time is equal to end time of database, the application will be blocked (or) killed by the admin.

JNLP tags

- <jnlp>...</jnlp>

The above is the only root folder of JNLP
Attributes: spec=version, optional
Indicates the version of the jnlp spec a jnlp file works with.
version=version, optional
Indicates the version of the JNLP file.
codebase=URL, optional
Codebase is used as base URL for all URLs in href attributes.
It describes the codebase for the file. href=url, optional
Contains the location of the jnlp file.

- <jar>...

Indicates the jar file which is used in the app.
<resources>...</resources>
Contains all the resources used in the application in ordered form.

- <security>...</security>

Indicates the security requirements.
The running of applications is in a restricted execution environment called the sandbox. This default running prevents the access to local disk. The jar files must be downloaded from the local host.

- <title>...</title>

Name of the application is enclosed.
...

Step 3. Creation of JAR Application

```
BufferredInputStream in = new
Java.net.URL("url").openStream();
```

The *jar* file is copied from the root directory to the temporary director. The Buffered Input Stream stores the input in the buffer. Internal buffer array is created using this command. The internal buffer is refilled as the bytes from the stream are read. The refilling is done from the contained input stream. Open stream command is used to open the URLl required to get response from the server.

Step 4. Accessing JAR Application

```
Process p=Runtime.get
Runtime().exec("javaws url");
```

The java application contains a single instance of Runtime class. This allows the interface of application with the environment. This Runtime class is used to load the *jnlp* file in Java web Start.

Step 5. JNLP Based Secure Connection Model

The client requests its *jar* file; the temporary file will be retrieved from the root directory. The COPY () function is used to copy the *jar* file from the root directory to the temporary directory. After copying the *jar* file from the root directory, the corresponding *jar* file will be used by the client. During this server side execution, the *Jnlp* file is loaded in the Java web start. It contains the *jar* package server address which will be retrieved and the web start application will be loaded in the client. Using the *getRuntime ()* we can load the *jnlp* in the web start. The *jar* file which is copied from the temporary directory is removed so that any attacks from the server side application can be prevented. Unlink () is the PHP function which is used to delete file from the temporary directory. If client needs the application (or) using of cache viewer if client wants to access the application the *Jar* file will not be loaded in the Java web start.

   PHP code for copy:
   COPY($efile,"/temp/".$nfile);
   This will copy the *JAR* file from the root directory to temp directory. When the server gets the request from the client, then the corresponding PHP file is executed.
   Unlink (file);
   This will destroy the file from the temporary directory.

Step 6. Execution of Server Files

The server files are accessed by the client to send requests to cloud administrator. In general the client requests the necessary *jar* to the cloud admin. The cloud admin is responsible for accessing the temporary folder and retrieving the jar files from the database as shown in Fig. 2. The temporary folder holds the *jnlp* files and the php codes to send a response to the client request. The cloud admin obtains the *jar* files from the *jar* folder. The requested jar is loaded in the temporary folder.

The core work of the temporary folder is to afford a corresponding *jnlp* file for the specific *jar* files. These *Jnlp* files are sent to the browser of the client with the assistance of *jnlp*. The figure shows the detailed information of the server execution. By using *jnlp* and PHP codes the corresponding *jar* file is loaded in the temporary folder. We have proposed this concept for providing security in both the client and server architecture. The hacker will not know where the *jar* file is located in the server, since the *jar* file is loaded in the temporary folder. These temporary *jar* files are deleted when the process is completed. This provides complete security from unauthorized users or hackers.

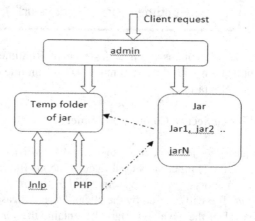

Fig. 2. Server Execution

5   Secure JNLP Real Time Implementation and Performance Evaluation

To evaluate proposed architecture we performed a simple text editor, calculator, Puzzle game. It is used software as a service in our cloud environment. We performed the experiment in real time by using a total number of 30 systems which were virtually connected to establish these applications. 29 Virtual clients and one server as a cloud administrator coupled with data centre and library.

Initially we included five virtual clients and continued incremented it in arithmetic progression as 10, 15, 20, 25, 30... and studied the performance results. We implemented this in real time with the following decision parameters plotted in Fig. 3 as graphs given below.

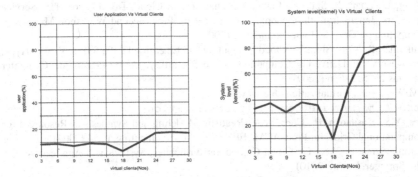

Fig. 3. User application, System Kernel grades in JNLP secure architecture

6   Conclusion

The Cloud computing world has exhausted the last decade hobbling on preventing networks from conventional security attack vectors. Some of many network firewalls and filters, ethical hacking software's, tools solved levels sophisticatedly sufficient to allow and pull and fall enforcement of security policies. We established a JNLP software based service implementation in cloud computing. In future work, we will construct private cloud architecture to provide software services as on demand in enterprises based environment.

References

1. Jinesh Varia, V.: Cloud Architectures. White Paper, Amazon Web Services, Amazon Inc. 1–14 (2008)
2. Carolan, J., Gaede, S.: Introduction to Cloud Computing Architecture. White Paper, 1st edn. Sun Micro Systems Inc. (2009)
3. Shacham, H., Waters, B.: Compact Proofs of Retrievability. In: Pieprzyk, J. (ed.) ASIACRYPT 2008. LNCS, vol. 5350, pp. 90–107. Springer, Heidelberg (2008)
4. Francesco, M.A., Gianni, F., Simone, S.: An Approach to a Cloud Computing Network. In: Proc. International Conf. on Application of Digital Information and Web Technologies, pp. 113–118. IEEE Press (2008)
5. Borja, S., Ruben, M.S., Ignacio, M.T., Ian, F.: An Open Source Solution for Virtual Infrastructure Management in Private and Hybrid Clouds. IEEE Internet Computing 1, 14–22 (2009)
6. Rajkumar, B.: Market-Oriented Cloud Computing: Vision, Hype and Reality of Delivering Computing as the 5th Utility. In: Proc. IEEE/ACM International Symposium on Cluster Computing and the Grid, pp. 1–13. IEEE Press/ACM (2009)

7. Simon. W., Etienne. G., Nick. B.: Ubuntu Enterprise Cloud Architecture. Linux Inc. (2009)
8. Wang, Q., Wang, C., Li, J., Ren, K., Lou, W.: Enabling Public Verifiability and Data Dynamics for Storage Security in Cloud Computing. In: Backes, M., Ning, P. (eds.) ESORICS 2009. LNCS, vol. 5789, pp. 355–370. Springer, Heidelberg (2009)
9. Wang, L., Zhan, J., Shi, W., Liang, Y., Yuan, L.: In Cloud, Do MTC or HTC Service Providers Benefit from the Economies of Scale? In: Proc. Second Workshop Many-Task Computing on Grids and Supercomputers (2009)
10. Rajkumar, B., Chee Shin, Y.: Cloud Computing and Emerging IT Platforms: Vision, Hype, and Reality for Delivering Computing as the 5th utility. Future Generation Computer Systems 25, 599–616 (2009)
11. VM Ware. Inc., Virtualization Overview,
    http://www.vmware.com/pdf/virtualization.pdf
12. Kee, Y.S., Casanova, H., Chen, A.A.: Realistic Modeling and Synthesis of Resources for Computational Grids. In: Proc. ACM/IEEE Conf. Supercomputing, p. 54 (2004)
13. Antonopoulos, N., Lee, G.: Cloud Computing: Principles, Systems and Applications, 1st edn. Springer, Berlin (2010)
14. Foster, I., Zhao, Y., Raicu, I., Lu, S.: Cloud Computing and Grid Computing 360-Degree Compared Grid Computing Environments Workshop. IEEE Press (2008)
15. Zukowski, J.: Deploying Software with JNLP and Java,
    http://java.sun.com/developer/TechnicalArticles/Programming/jnlp
16. JNLP File Syntax, java.sun.com,
    http://java.sun.com/javase/6/docs/technotes/guides/deelopers.html
17. Lacombe, C.: Dynamic JNLP,
    https://www.software.ibm.com/developerworks/education.pdf

Inter-journal Citations and Ranking Top Journals in Financial Research

Kuru Ratnavelu[1], Choong Kwai Fatt[2], and Ephrance Abu Ujum[1]

[1]Institute of Mathematical Sciences,
University of Malaya, Kuala Lumpur 50603, Malaysia
[2]Faculty of Business and Accountancy,
University of Malaya, Kuala Lumpur 50603, Malaysia

Abstract. This article uses the methodology developed by Borokhovich, Bricker and Simkins (1994) to determine the relative influence of seven prominent finance journals. The original analysis is expanded on a longitudinal basis for the years 1990 to 2006 inclusive. It is found that the relative influence rank produces some stable ordering over the study period with the Journal of Finance and Journal of Financial Economics occupying top spots. A change in the ordering of the relative influence rank indicates a shift in inter-journal communication trends.

Keywords: Bibliometrics, Finance literature, Interjournal communication, Influence ranking.

1   Introduction

There are a number of accepted measures to rank and measure the quality of journals. Nevertheless, the challenge of measuring journal influence is fraught with pitfalls. The common convention is to use the Journal Citation Reports (JCR) impact factor score which measures a journal's average number of citations per article within a specific time window (Garfield [1]; Adler, Ewing and Taylor [2]). It has become the convention to compute the impact factor over a period of two (or five) years. Shorter time windows give greater weight to rapidly changing fields. On the other hand, longer time windows take into account a larger number of citations and/or sources, but results in a less current measure of impact. This measure has received considerable attention, most notably because citation rates vary from one field to the next and therefore a standardized two-year time window across all fields may exaggerate the impact of some journals (especially multidisciplinary ones) while under estimating others. Clearly, a field specific treatment is required.

In 1994, Borokhovich, Bricker and Simkin [3] (hereon referred to as BBS) had presented a case study of inter-journal communication and influence between eight of the most prominent mainstream journals in finance during 1990-1991. The eight journals are: Financial Management (FM), Journal of Banking and Finance (JBF),

P. Balasubramaniam and R. Uthayakumar (Eds.): ICMMSC 2012, CCIS 283, pp. 505–513, 2012.
© Springer-Verlag Berlin Heidelberg 2012

Journal of Business (JBUS), Journal of Finance (JF), Journal of Financial Economics (JFE), Journal of Financial Research (JFR), Journal of Financial and Quantitative Analysis (JFQA), and the Review of Financial Studies (RFS). In their analysis, the concept of "self-citation index" was developed, defined as a measure of how frequently a journal cites itself compared to how frequently its articles are cited by other journals. Self-citations are instances where an article cites another article published within the same journal. According to BBS, there are a number of possible explanations for differences in self-citation rates across journals. On one hand, it might be argued that a journal, in order to promote itself, encourages self-citations. Then, differences in these self-citation rates reflect the extent of self-promotion. Alternatively, journals that more frequently publish important studies tend to be cited more frequently. In this case, the differences in self-citation rates reflect the relative importance of the journals' articles. Finally, it is reasonable to assert that journals publishing in narrower or more specialized research areas tend to cite themselves more frequently, simply because they are the principal source of knowledge in that area."

Following the BBS approach, this paper uses synchronous citation data to explore the inter-journal citation patterns between journals, using the field of financial research as a case study. We believe that this study will provide a further insight into the inter-journal communication and influence on the use of a larger set of data (1990-2006) from these core journals. We organize this paper in a similar manner: Section 1 describes the data used in this study, while Section 2 covers the analysis of inter-journal citations. We begin Section 2 by first reconstructing Table 1 in BBS to benchmark the results of our methods. We then analyze the time series for self-citation rates and self-citation index.

2   Data and Methodology

We select the following seven of the eight mainstream finance journals, as identified by BBS: FM, JBF, JBUS, JF, JFE, JFQA, and RFS. Publication and citation data for the eight journals were obtained from the Social Science Citation Index (SSCI), covering the period 1990-2006. The JFR, which was originally a part of the BBS dataset, was omitted from this study as it had been dropped from the SSCI from 1995 onwards. We also deliberately chose to end our study period at 2006 as the JBUS ceased publication after November 2006. Extra care was expended to handle typographical irregularities in the cited references of the journal articles sampled; e.g. "j finan" and "j fiance", as opposed to the correct abbreviated form for the JF, "j financ".

Table 1. Summary information for the publication data (1990-2006) used in this study

Journal	Source Publications	Number of Citations	Mean Citations per Publication
Financial Management (FM)	586	14,040	23.96
Journal of Banking and Finance (JBF)	1,695	43,485	25.65
Journal of Business (JBUS)	554	17,726	32.00
Journal of Finance (JF)	1,967	47,292	24.04
Journal of Financial Economics (JFE)	916	29,640	32.36
Journal of Financial and Quantitative Analysis (JFQA)	549	15,834	28.84
Review of Financial Studies (RFS)	627	21,131	33.70
Total	6,894	189,148	27.44

In order to put the present work into context, we have benchmarked the results of our methods with those obtained by Borokhovich, Bricker and Simkins in [2]. Discrepancies are used to identify possible errors. Errors resulting from the present computer codes are debugged accordingly. The numbers presented in the following Table 2 represent the number of times articles published in each of the eight finance journals cited articles in these journals during 1990 and 1991. The eight journals are FM, JBF, JBUS, JF, JFE, JFR, JFQA, and RFS. Additional entries are for the Journal of Political Economy (JPE), the American Economic Review (AER), Econometrica (ECMA), and an aggregate of other journals and nonjournals.

Table 2. A summary of the publication data (1990-1991) used in the study. Items in brackets indicate BBS values

Journal	Source Publications	Number of Citations	Mean Citations per Publication
Financial Management (FM)	99 (62)	1,681	16.98
Journal of Banking and Finance (JBF)	140 (130)	2,925	20.89
Journal of Business (JBUS)	58 (54)	1,250	21.55
Journal of Finance (JF)	210 (173)	4,590	21.86
Journal of Financial Economics (JFE)	74 (74)	1,998	27.00
Journal of Financial Research (JFR)	62 (62)	788	12.71
Journal of Financial and Quantitative Analysis (JFQA)	73 (73)	1,549	21.22
Review of Financial Studies (RFS)	63 (57)	1,601	25.41
Total	779 (685)	16,382	21.03

The self-citation rate is the percentage of a journal's citations attributable to its own articles. The self-citation index is a measure of how frequently a journal cites itself compared to how frequently the articles are cited by other journals. We had examined inter-journal communications by measuring the journal citation patterns within and outside the eight-journal set.

In summary, all journal datasets are within 10% of the BBS values except for JFQA. The reason for the extremely large discrepancy with JFQA is unknown at this point. We point out that the original table in BBS contained one typographical error, i.e. the value of citations from JBF to RFS is 15 and not 145 if the row sum is to equal 1,003 as indicated. It may be possible that there are further typographical errors yet to be identified. For this reason, we choose to include the present analyses for JFQA in case these values turn out to be more accurate estimates of journal citation patterns for these eight journals.

3   Inter-journal Citation Patterns: 1990-2006

A research article typically makes cited references to other research articles, thus creating a network of papers and journals that are connected through citation linkages. If one group cited the references of source articles by their respective source journals, the journal-to-journal citations can be split into two types: those that are directed internally and externally. The former corresponds to journal self-citation while the latter represents inter-journal communication. Since the total cited references made by a journal are proportional to its source article volume (total number of publications), and because the latter generally fluctuate from year to year, it is perhaps more appropriate to talk about inter-journal citation patterns through percentage of contributions.

3.1   Financial Management

From the present analysis, we find that the mean total citing frequency is 828, with a standard deviation of roughly 200 cited references. The total citing frequency ranges between 640 and 1,417 cited references. On average, FM cites JF the most (17.68 ± 3.59%), followed by JFE (16.55 ± 2.74%), FM (8.35 ± 3.75%), JFQA (2.78 ± 0.69%), JBUS (2.39 ± 0.98%), RFS (1.87 ± 1.01%) and JBF (1.05 ± 0.54%). This suggests that FM is primarily influenced by works in JF, JFE and FM itself. The self-citation rate for FM spikes considerably in 1997 at 0.1794 from 0.1076 in 1996 (See Table 3). This is largely due to a significant drop in citing frequency during that year for JF (by half) and JFE (by nearly a third), while FM experiences a considerable increase (22.3%).

3.2   Journal of Banking and Finance

For JBF, we find that the mean total citing frequency is 2572, with values each year ranging between 1,349 and 5,187 cited references, on the rise with annual publication volume. *JBF* contributes the most citations to *JF* (11.62 ± 2.11%), *JFE* (7.41 ± 1.71%), *JBF* (6.57 ± 1.35%), *JFQA* (2.48 ± 0.64%), *JBUS* (1.94 ± 0.56%), *RFS* (1.81 ± 0.85%) and *FM* (0.78 ± 0.28%). The self-citation rate for *JBF* swings between 0.0319 and 0.0831.

3.3   Journal of Business

For *JBUS*, we find that the mean total citing frequency during 1990-2003 is roughly 646, with a standard deviation of 108 cited references. The total citing frequency then doubled with publication volume in 2004 to 1,747 cited references. In 2005 and 2006, that value soared to 3,013 and 3,991 cited references, respectively. On average, *JBUS* cites *JF* the most (13.75 ± 3.57%), followed by *JFE* (10.66 ± 2.73%), *JBUS* (4.52 ± 1.79%), *RFS* (2.79 ± 1.44%), *JFQA* (1.66 ± 0.62%), *JBF* (0.72 ± 0.56%) and *FM* (0.56 ± 0.35%). The self-citation rate for *JBUS* swings between 0.0238 and 0.0833.

3.4   Journal of Finance

For *JF*, we find that the mean total citing frequency is 2,793. On average, *JF* cites itself the most (19.30 ± 1.95%), followed by *JFE* (14.59 ± 2.06%), *RFS* (3.97 ± 1.30%), *JBUS* (2.53 ± 0.45%), *JFQA* (2.16 ± 0.45%), *FM* (0.69 ± 0.20%) and *JBF* (0.63 ± 0.24%). The self-citation rate for *JF* swings between 0.1518 and 0.2228.

3.5   Journal of Financial Economics

For *JFE*, we find that the mean total citing frequency is 1,755. On average, *JFE* cites itself the most (20.22 ± 4.18%), followed by *JF* (16.61 ± 2.85%), *RFS* (3.31 ± 1.19%), *JBUS* (2.13 ± 0.56%), *JFQA* (1.98 ± 0.60%), *FM* (0.95 ± 0.46%) and *JBF* (0.66 ± 0.30%). The self-citation rate for *JFE* swings between 0.1389 and 0.2910.

3.6   Journal of Financial and Quantitative Analysis

For *JFQA*, we find that the mean total citing frequency is 935. On average, *JFQA* cites *JF* the most (19.97 ± 2.98%), followed by *JFE* (16.45 ± 2.47%), *JFQA* (4.93 ± 1.26%), *RFS* (4.21 ± 1.92%), *JBUS* (2.68 ± 0.89%), *FM* (1.04 ± 0.54%) and *JBF* (0.92 ± 0.45%). The self-citation rate for *JFQA* swings between 0.0315 and 0.0806.

3.7   Review of Financial Studies

For *RFS*, we find that the mean total citing frequency is 1,246. On average, *RFS* cites *JF* the most (16.60 ± 3.47%), followed by JFE (11.97 ± 2.22%), RFS (6.79 ± 1.16%), JFQA (2.33 ± 0.52%), JBUS (2.41 ± 0.73%), JBF (0.63 ± 0.35%) and FM (0.40 ± 0.20%). The self-citation rate for *RFS* swings between 0.0272 and 0.0811.

The self-citation index for each journal in year Y is computed as the (self-citation rate in year Y × 100) ÷ (normalized average citations from other journals in year Y). From Fig. 1, the two journals with the highest self-citation index are FM and JBF with values rising and dipping below 1.00 throughout 1990-2006. Fig. 2 excludes the FM and JBF plots to resolve annual variations for the other five journals. Evidently, the other five journals possess a self-citation index below 1.00 throughout the study period. Furthermore, RFS appears to be experiencing a decreasing growth trend, while the other four journals fluctuate more or less around their mean values.

Table 3. Self-citation rates (1990-2006) for the seven journals studied

				Self-citation rate			
Year	FM	JBF	JBUS	JF	JFE	JFQA	RFS
1990	0.0685	0.0319	0.0833	0.1518	0.2451	0.0679	0.0272
1991	0.0529	0.0831	0.0480	0.1814	0.2126	0.0806	0.0692
1992	0.0471	0.0576	0.0474	0.1751	0.2488	0.0488	0.0637
1993	0.0543	0.0691	0.0464	0.1576	0.2910	0.0540	0.0747
1994	0.0815	0.0562	0.0782	0.1946	0.1988	0.0556	0.0723
1995	0.0672	0.0738	0.0317	0.1796	0.2182	0.0633	0.0627
1996	0.1076	0.0539	0.0511	0.1954	0.2147	0.0379	0.0700
1997	0.1794	0.0759	0.0375	0.1826	0.2410	0.0510	0.0671
1998	0.1292	0.0746	0.0552	0.2141	0.2270	0.0427	0.0748
1999	0.1282	0.0830	0.0238	0.1971	0.2149	0.0405	0.0633
2000	0.1139	0.0606	0.0255	0.2117	0.1673	0.0353	0.0703
2001	0.0810	0.0814	0.0517	0.2042	0.1710	0.0492	0.0704
2002	0.0909	0.0557	0.0653	0.2109	0.1442	0.0478	0.0811
2003	0.0765	0.0742	0.0364	0.2029	0.1807	0.0478	0.0657
2004	0.0531	0.0707	0.0280	0.2056	0.1652	0.0315	0.0750
2005	0.0437	0.0572	0.0266	0.1941	0.1584	0.0361	0.0765
2006	0.0450	0.0575	0.0323	0.2228	0.1389	0.0483	0.0704

Panel B: Basic data descriptors

	FM	JBF	JBUS	JF	JFE	JFQA	RFS
Min.	0.0437	0.0319	0.0238	0.1518	0.1389	0.0315	0.0272
Median	0.0765	0.0691	0.0464	0.1954	0.2126	0.0483	0.0703
Mean	0.0835	0.0657	0.0452	0.1930	0.2022	0.0493	0.0679
Max.	0.1794	0.0831	0.0833	0.2228	0.2910	0.0806	0.0811
Range	0.1357	0.0512	0.0596	0.0710	0.1521	0.0491	0.0539
Std Dev	0.0375	0.0135	0.0179	0.0195	0.0418	0.0126	0.0116

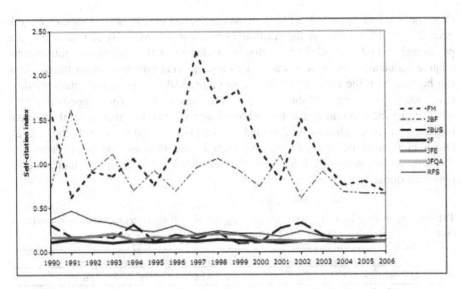

Fig. 1. Time evolution of self-citation index for the journals in the study dataset

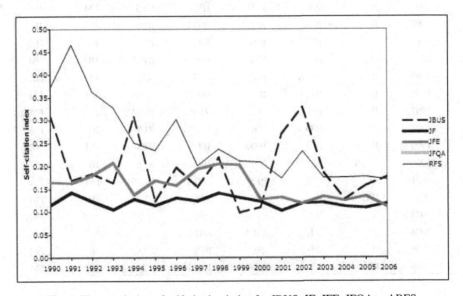

Fig. 2. Time evolution of self-citation index for JBUS, JF, JFE, JFQA and RFS

By sorting the self-citation index values in ascending order, we obtain the annual relative influence rank as shown in Table 4. This ranking reflects inter-journal citation patterns, with the highest rank representing the journal with either the smallest self-citation rate, and/or the largest annual citations contributed from other journals. This gives a practical and simple measure to gauge relative influence between journals, i.e. a journal is more influential to the development of other journals if it is cited more externally than internally. Although some permutations in the ordering occur throughout the study period, a few patterns are visible: (1) JF always appears at rank 1 or 2; (2) JFE maintains its position in the top 4 ranks; (3) JBF and FM are positioned at ranks 6 and 7. Accordingly, a change in the ordering of journals by relative influence rank indicates a shift in inter-journal communication trends. This can be seen with the RFS, which has shown a gradual decline in self-citation index, corresponding to an upward shift in relative influence ranking. This suggests that RFS is becoming more prominent in the finance literature. On the other hand, the Journal of Business can be seen shuffling around ranks 1 (1999-2000) to 5 (1994, 2001-2002). Drops in ranking occur when the self-citation rate increases or when there is a decrease in the normalized average citations from other journals (indicative of reduced external influence).

Table 4. The relative Influence Rank obtained by sorting self-citation index values in ascending order

	Relative Influence Rank						
Year	1	2	3	4	5	6	7
1990	JF	JFE	JBUS	JFQA	RFS	JBF	FM
1991	JF	JFE	JBUS	JFQA	RFS	FM	JBF
1992	JF	JFE	JBUS	JFQA	RFS	FM	JBF
1993	JF	JBUS	JFE	JFQA	RFS	FM	JBF
1994	JF	JFE	JFQA	RFS	JBUS	JBF	FM
1995	JF	JBUS	JFE	RFS	JFQA	FM	JBF
1996	JF	JFQA	JFE	JBUS	RFS	JBF	FM
1997	JF	JBUS	JFE	RFS	JFQA	JBF	FM
1998	JF	JFQA	JFE	JBUS	RFS	JBF	FM
1999	JBUS	JF	JFQA	JFE	RFS	JBF	FM
2000	JBUS	JF	JFE	JFQA	RFS	JBF	FM
2001	JF	JFE	RFS	JFQA	JBUS	FM	JBF
2002	JFE	JF	JFQA	RFS	JBUS	JBF	FM
2003	JF	JFE	RFS	JBUS	JFQA	JBF	FM
2004	JF	JFE	JBUS	JFQA	RFS	JBF	FM
2005	JF	JFE	JBUS	RFS	JFQA	JBF	FM
2006	JFE	JF	RFS	JBUS	JFQA	JBF	FM

4   Concluding Remarks

In this paper, we have quantified the inter-journal citation patterns for seven prominent finance journals. Our analysis suggests that these journals have a particular ordering in terms of relative influence rank, with JF and JFE occupying top positions throughout the period of study. This could be indicative of a significant number of influential works located within the two journals that are current (relevant) to the development of other works elsewhere. Incidentally, this creates a bias for older, more established journals which have a larger pool of works to cite from. This could explain the low rank for the younger, comparatively less self-citing, yet highly cited RFS. Despite beginning publication only in 1988, RFS averages 33.70 citations per publication during the period 1990-2006, the highest among the seven journals studied (see Table 1). To address this issue, one could tally citations to journals within a fixed time window, but this is exactly the approach utilized by the impact factor which we are trying to avoid. A more promising approach is to conduct a centrality analysis of the citation network for business/finance journals, from which a number of prominence scores can be constructed (reflecting different aspects of a journal's relative position within a structure of citation ties). This will be explored in future works.

Acknowledgements. The authors would like to acknowledge the UM-High Impact Research Grant No: F000013-21001 as well as the Fundamental Research Grant Scheme No: RG298-11HNE for support of this research work.

References

1. Garfield, E.: Agony and the ecstasy - the history and meaning of the journal impact factor. In: International Congress on Peer Review and Bibliomedical Publication, Chicago, September 16 (2005)
2. Adler, R., Ewing, J., Taylor, P.: Citation Statistics - A report from the International Mathematical Union (IMU) in cooperation with the International Council of Industrial and Applied Mathematics (ICIAM) and the Institute of Mathematical Statistics (IMS). Technical report, Joint Committee on Quantitative Assessment of Research (2008)
3. Borokhovich, K.A., Bricker, R.J., Simkins, B.J.: Journal Communication and Influence in Financial Research. J. Finance 49(2), 713–725 (1994)

A Key Sharing Scheme over $GF(2^5)$ to Use in ECC

A.R. Rishivarman[1,*], B. Parthasarathy[2,**], and M. Thiagarajan[3,**]

[1] Mathematics, Pauls Engineering College,
Villupuram, TN, India
`rishi_130777@yahoo.co.in`
[2] Mathematics, Mailam Engineering College,
Villupuram, TN, India
`mlampmlc@gmail.com`
[3] School of Computing, SASTRA University,
Tanjore, TN, India
`m_thiyagarajan@yahoo.com`

Abstract. Since the introduction of public-key cryptography by Diffe and Hellman in 1976, the potential for the use of the discrete logarithm problem in public-key cryptosystems has been recognized. Although the discrete logarithm problem as first employed by Diffe and Hellman was defined explicitly as the problem of finding logarithms with respect to a generator in the multiplicative group of the integers module a prime, this idea can be extended to arbitrary groups and in particular, to elliptic curve groups. The resulting public-key systems provide relatively small block size, high speed, and high security. In this paper, a vector space secrets sharing scheme is proposed in detail. Its security is based on the security of ECC. This scheme has the following characteristic: the precondition of (t, n)- threshold secret sharing scheme that all assignees purview must be same is generalized. A verifiable infrastructure is provided, which can be used to detect the cheaters from the dealers and assignees. The shared key distributed by dealer is encrypted based on ECC, which enhances the security. So this scheme is of less computation cost which is valuable in applications with limited memory, communications bandwidth or computing power.

Keywords: Secret sharing, Elliptic curve cryptography (ECC), $GF(2^5)$, Irreducible polynomial, ECDLP, Vector space.

1 Introduction

Modern cryptosystems are designed to ensure the security of any system by keys. The system will loose its security if the key is revealed. In other cases, losing of the key by accidents (for personal reason or not) may result in your being unable to rehabilitate encoded documents. All these are the problems that

---

* Assitant Professor.
** Professor.

P. Balasubramaniam and R. Uthayakumar (Eds.): ICMMSC 2012, CCIS 283, pp. 514–521, 2012.
© Springer-Verlag Berlin Heidelberg 2012

must be solved by cryptosystem designers. One method to ensure the security in the cryptosystem is to change the keys with a high frequency, however, it is not realistic to do so in today's age that needs large amounts of information. Therefore, problems occur, such as how to select, exchange keys and how to store and send keys safely-what are also called key management problems or shared control problems [1].

In 1979, Shamir [2] and Blakley [3] proposed a Threshold Secret Sharing Scheme respectively. Calculation of Shamir's scheme is based on polynomial interpolation while Blakely's system is based on finite geometry. The (t,n) Threshold Secret Sharing Scheme indicates that t participants will reconstruct keys, that is to say, purviews of all assignees are the same. But in reality, purviews of assignees are different due to different position and function of assignees. Hence a general secret sharing scheme should be taken into account. In 1990, Laih [4] and his followers suggested Dynamic threshold Secret Sharing Scheme in which keys can be changed at random while the value of sub key of participants are unchanged. However the security of Laih's scheme will diminish for the increasing times of renewing of keys. Compared to Laih's scheme, keys in Sun and Shieh's scheme can be renewed infinitely and can be used to detect dishonest participants. In 2004, Xu Chunxiang provided A Threshold Multiple Secret Sharing Scheme [5]. The security of Xu's scheme depends on the security of RSA digital signature. In this scheme, shared secrets of participants can be used repeatedly and infinite secrets can be shared by participants, meanwhile this can prevent dealer cheating and prevent mutual cheating between participants. In 2004, Ma Wenping put forward an Unconditionally security verifiable secret sharing system [6] in which anti-cheating function of Threshold Secret Sharing Scheme was discussed and a new unconditionally secure verifiable Secret Sharing Scheme based on unconditionally secure authentication code is constructed.

In 1989, Brickell [7] brought out Vector space Secret Sharing Scheme. The scheme stipulates the participant subset of shared keys through the concept of access structure. Threshold Secret Sharing Scheme is a special case in this scheme which means that the precondition of the threshold secret sharing scheme that all assignees, purview must be the same is generalized. In 2002, Xu and others [8] provided a secure vector space secret sharing scheme: a safe vector space secret sharing scheme whose key space is based on finite field $GF(q)$ is provided and quantitative study on security and information rate of the scheme is accomplished.

Elliptic curve calculation was not introduced to cryptography until 1985. Compared with RSA, the advantage of ECC lies in its ensuring the same security while the length of key of ECC is much less than RSA cryptography and its lessening operation load. Papers [9,10,11] help make multiple keys negotiation and exchange agreements based on ECC.

This article provides a safe and verifiable vector space dynamic secret sharing scheme based on elliptic curve key cryptography. The security of the scheme is based on elliptic curve discrete logarithm problem. Meanwhile this scheme provides a verifiable infrastructure to prevent participating and cheating of illegal members.

2 Mathematical Background

2.1 The Finite Field GF(2^m)

The finite field GF(2^m) is the characteristic 2 finite field containing 2^m elements. Although there is only one characteristic 2 finite field GF(2^m) for each power 2^m of 2 with $m \geq 1$, there are many different ways to represent the elements of GF(2^m).

Here the elements of GF(2^m) should be represented by the set of binary polynomials of degree $m - 1$ or less:

With addition and multiplication defined in terms of an irreducible binary polynomial $f(x)$ of degree m, known as the reduction polynomial, as follows:

Addition. If $a = a_{m-1}x^{m-1} + \cdots + a_0$, $b = b_{m-1}x^{m-1} + \cdots + b_0 \in$ GF(2^m), then $a+b = r \in$ GF(2^m), where $r = r_{m-1}x^{m-1} + \cdots + r_0$ with $r_i \equiv a_i + b_i$ (mod 2).

Multiplication. If $a = a_{m-1}x^{m-1} + \cdots + a_0$, $b = b_{m-1}x^{m-1} + \cdots + b_0 \in$ GF(2^m), then $a \cdot b = s$ in GF(2^m), where $s = s_{m-1}x^{m-1} + \cdots + s_0$ is the remainder when the polynomial ab is divided by $f(x)$ with all coefficient arithmetic performed modulo 2.

Addition and multiplication in GF(2^m) can be calculated efficiently using standard algorithms for ordinary integer and polynomial arithmetic. In this representation of GF(2^m), the additive identity is the polynomial 1.

Again it is convenient to define subtraction and division of field elements. To do so the additive inverse (or negative) and multiplicative inverse of a field element must be described.

Additive Inverse. If $a \in$ GF(2^m), then the additive inverse (-a) of a in GF(2^m) is the unique solution to the equation $a + x = 0$ in GF(2^m).

Multiplicative Inverse. If $a \in$ GF(2^m), $a \neq 0$, then the multiplicative inverse a^{-1} of a in GF(2^m) is the unique solution to the equation $a \cdot x = 1$ in GF(2^m).

Additive inverses and multiplicative inverses in GF(2^m) can be calculated effectively using the extended Euclidean algorithm. Division and subtraction are defined in terms of additive and multiplicative inverse:

$a - b$ in GF(2^m) is $a + (-b)$ in GF(2^m) and a/b in GF(2^m) is $a \cdot (b^{-1})$ in GF(2^m).

2.2 Elliptic Curves over GF(2^m)

Let GF(2^m) be a characteristic 2 finite field, and let $a, b \in$ GF(2^m) satisfy $b \neq 0$ GF(2^m). Then a (non-supersingular) elliptic curve E(GF(2^m)) over GF(2^m) defined by the parameters $a, b \in$ GF(2^m) consists of the set of solutions or points $P = (x, y)$ for $x, y \in$ GF(2^m) to the equation:

$$y^2 = x \cdot y = x^3 + a \cdot x^2 + b \text{ in GF}(2^m)$$

together with an extra point O called the point at infinity [11,12]. (Here the only elliptic curves over GF(2^m) of interest are non-supersingular elliptic curves.)

The number of points on $E(\mathrm{GF}(2^m))$ is denoted by $\#E(\mathrm{GF}(2^m))$. The Hasse theorem states that:

$$2^m + 1 - 2\sqrt{2^m} \leq \#E(\mathrm{GF}(2^m)) \leq 2^m + 1 + 2\sqrt{2^m}.$$

It is again possible to define an addition rule to add points on E as the addition rule is specified as follows:

1. Rule to add the point at infinty to itself:

$$O + O = O.$$

2. Rule to add the point at infinity to any other point:

$$(x, y) + O = O + (x, y) = (x, y) \text{ for all } (x, y) \in \mathrm{GF}(2^m).$$

3. Rule to add two points with the same coordinates when the points are either distinct or have x-coordinates 0:

$$(x, y) + (x, x + y) = O \text{ for all } (x, y) \in \mathrm{GF}(2^m).$$

4. Rule to add two points with different x-coordinates: Let $(x_1, y_1) \in \mathrm{GF}(2^m)$ and $(x_2, y_2) \in \mathrm{GF}(2^m)$ be two points such that $x_1 \neq x_2$. Then $(x_1, y_1) + (x_2, y_2) = (x_3, y_3)$, where:

$$x_3 = \lambda^2 + \lambda + x_1 + x_2 + a \text{ in } \mathrm{GF}(2^m),$$
$$y_3 = \lambda(x_1 + x_3) + x_3 + y_1 \text{ in } \mathrm{GF}(2^m), \text{ and}$$
$$\lambda \equiv \frac{y_1 + y_2}{x_1 + x_2} \text{ in } \mathrm{GF}(2^m).$$

5. Rule to add a point to itself (double a point): Let $(x_1 y_1) \in \mathrm{GF}(2^m)$ be a point with $x_1 \neq 0$. Then $(x_1, y_1) + (x_1, y_1) = (x_3, y_3)$, where:

$$x_3 = \lambda^2 + \lambda + a \text{ in } \mathrm{GF}(2^m),$$
$$y_3 = x_1^2 + (\lambda + 1) \cdot x_3 \text{ in } \mathrm{GF}(2^m), \text{ and}$$
$$\lambda = x_1 + \frac{y_1}{x_1} \text{ in } \mathrm{GF}(2^m).$$

The set of points on $E(\mathrm{GF}(2^m))$ forms an abelian group under this addition rule. Notice that the addition rule can always be computed efficiently using simple field arithmetic.

Cryptographic schemes based on ECC rely on scalar multiplication of elliptic curve points. As before given an integer k and a point $P \in \mathrm{GF}(2^m)$, scalar multiplication is the process of adding P to itself k times. The result of this scalar multiplication is denoted kP.

3 The Finite Field $GF(2^5)$ Arithmetic Operations

There is a representations for an element of a finite field $GF(2^5)$. The polynomial basis (PB), where $\{1, \omega, \omega^2, \omega^3, \omega^4\}$ where ω is a root of an irreducible polynomial $f(x) = x^5 + x^2 + 1$ over the field F_2. An element in $GF(2^5)$ can then be represented as a polynomial $\{c_0 + c_1\omega + c_2\omega^2 + c_3\omega^3 + c_4\omega^4 / c_i = 0 \text{ or } 1\}$ or in vector form $\{c_0, c_1, c_2, c_3, c_4\}$.

Hence it is easy to see the elements of $GF(2^5)$ and $[\mathrm{Res}_{F_{2^5}/F_2} GF(2^5)](F_2)$. Observe that there is a isomorphism of sets.

$$GF(2^5) \sim [\mathrm{Res}_{F_{2^5}/F_2} GF(2^5)](F_2).$$

The efficiency of elliptic curve algorithm heavily depends on the performance of the under lying field arithmetic operations. These operations include addition, Subtraction, Multiplication, and Invertion. Given two elements $[a_0, a_1, a_2, a_3, a_4]$ and $[b_0, b_1, b_2, b_3, b_4]$, these operations are defined in our earlier work [13], [14].

4 A Vector Space Secret Sharing Scheme

A vector space conformation is a method [12] aims at constructing an ideal scheme which is directed against access scheme which is directed against access structure. If $P = \{p_1, p_2, \ldots, p_n\}$ is a set of n participants, then, $\Gamma \in 2^P$, and if we consider $\forall A \in \Gamma$, participants in set A can calculate the key k, then we say Γ is named authorized subset $D(\notin P)$ is shared key distributor. If D selects $GF(2^5)$, is a finite field, K^5 represents 5-dimensional vector space based on $GF(2^5)$.

Definition. Γ is access structure, if there is a function: $\varphi : P \cup \{D\} \to K^5$ in the condition that

$$\varphi(D) = (1, 0, 0, 0, 0) \in < \varphi(p_i)$$
$$= (x_{1i}, x_{2i}, x_{3i}, x_{4i}, x_{5i}) : p_i \in A > \Leftrightarrow A \in \Gamma.$$

Then Γ is named A vector space access structure. In other words, vector $\varphi(b)$ can be represented as linear combination of vector of set $\{\varphi(p_i) : p_i \in A\}$ if and only if A is an authorized set. If Γ is a vector space access structure, an ideal secret sharing scheme is as follows:

provided given secret key
$K \in GF(2^5)$ selects any $v = (k, v_2, v_3, v_4, v_5) \in K^5$,
then obviously
$k = v \cdot \varphi(D)$, $k_i = v \cdot \varphi(p_i) = v_1 x_{1i} + v_2 x_{2i} + v_3 x_{3i} + v_4 x_{4i} + v_5 x_{5i}$
is distributed to number i participants, function $\varphi(p_i)$ is open, participants in authorized set can calculate k by utilizing distributed linear combination of shared keys. In fact, if we suppose $A = \{p_1, p_2, \ldots, p_l\}$ is an authorized subset, then we find:

$$k = v \cdot \varphi(D) = v \cdot \sum_{i=1}^{l} c_i \varphi(p_i) = \sum_{i=1}^{l} c_i v \cdot \varphi(p_i) = \sum_{i=1}^{l} c_i k_i$$

where $c_i \in \mathrm{GF}(2^5)$.

The scheme constructed by the above method is called vector space secret sharing scheme. The (t, n) threshold scheme is a special case of vector space secret sharing scheme. Because if we make sure that $\varphi(p_i) = (1, x_i, x_i^2, \ldots, x_i^{t-1})$ then here x_i is n different non–zero elements in GF(2^5).

5   A Publicly Verifiable Vector Space Secret Sharing Scheme

The dealer $D (\notin P)$ choose the finite field GF(2^5), K^5 denotes a five dimensional vector space on GF(2^5), E denotes a elliptic curve on GF(2^5) and the equation is $y^2 + xy = x^3 + ax + b \pmod{f(x)}$ where $f(x) = x^5 + x^2 + 1$: $a, b, \in \mathrm{GF}(2^5)$ and $4a^3 + 27b^2 \neq 0 \pmod{f(x)}$, α is the base point of E. Suppose $P = \{p_1, p_2, p_3, \ldots, p_n\}$ is the set of n participants, $\Gamma (\subset 2^P)$ is access structure of vector space K^5, D computes $\beta = a\alpha$, $(1 \leq a \leq \mathrm{ord}(\alpha))$.

a is private key α and β are published. $p_i \in P$ is the assignee of the secret key and each p_i calculate $\beta_i = a_i\alpha$ and publish it.

5.1   Dealing Sub Key

D finish following work:

(i) For a given secret s, D choose a vector $V = \{s, v_2, v_3, v_4, v_5\}$ randomly from K^5. Then D, calculates

$$s = v \cdot \varphi(D) = v_1\alpha_1 + v_2\alpha_2 + v_3\alpha_3 + v_4\alpha_4 + v_5\alpha_5 \text{ and}$$
$$s_i = v \cdot \varphi(p_i) = v_1 x_{i1} + v_2 x_{i2} + v_3 x_{i3} + v_4 x_{i4} + v_5 x_{i5} \text{ where}$$
$$\varphi(D) = \{1, 0, 0, 0, 0\} \text{ and } \varphi(p_i) = \{x_{i1}, x_{i2}, x_{i3}, x_{i4}, x_{i5}\};$$

   $i = 1, 2, \ldots, n$ then function φ is published.

(ii) Assume t_0 is immediate time, $m_i = (s_i, t_0 \bmod f(x))$.
    D calculates $u_i = m_i - \alpha_i\beta_i(\bmod f(x))$ then publish parameter u_i.

(iii) D calculates

$$r_j = v_j\alpha \ (j = 1, 2, 3, 4, 5)$$
$$\gamma_1 = s\alpha$$
$$\gamma_{i2} = s_i\alpha \ (i = 1, 2, 3, 4, 5).$$

And then $r_j \ (j = 1, 2, 3, 4, 5)$, $\gamma_1, \gamma_{i2} = s_i\alpha \ (i = 1, 2, 3, 4, 5)$ is published.

5.2   Key Recovery

Suppose authorization subset $A \in \Gamma$, and $A = \{p_1, p_2, \ldots, p_l\}$ ($l < n$). Firstly participants $p_i \in A$ computes $m_i = (s_i, t_0 \bmod f(x))$ from

$$m_i = u_i + \alpha \beta_i = u_i + a a_i \alpha = u_i + a_i \beta$$

and then issues it to each participants $p_i \in A$. After receiving all the m_i, p_i figures out $c_i \in \mathrm{GF}(2^5)$ from the equation $\varphi(D) = c_1 \varphi(p_1) + c_2 \varphi(p_2) + c_3 \varphi(p_3) + \cdots + c_l \varphi(p_1)$ and then figures out the secret $s = c_1 s_1 + c_2 s_2 + \cdots + c_l s_l$.

5.3   Key Verification

(i) Verification of s_i : $\gamma_{i2} = x_{i1} r_1 + x_{i2} r_2 + x_{i3} r_3 + x_{i4} r_4 + x_{i5} r_5$. In fact

$$\begin{aligned}
\gamma_{i2} = s_i \alpha &= v \cdot \varphi(p_i)\alpha = v_1 x_{i1} \alpha + v_2 x_{i2} \alpha + v_3 x_{i3} \alpha + v_4 x_{i4} \alpha + v_5 x_{i5} \alpha \\
&= x_{i1} v_1 \alpha + x_{i2} v_2 \alpha + x_{i3} v_3 \alpha + x_{i4} v_4 \alpha + x_{i5} v_5 \alpha \\
&= x_{i1} r_1 + x_{i2} r_2 + x_{i3} r_3 + x_{i4} r_4 + x_{i5} r_5.
\end{aligned}$$

(ii) Verification of s:

$$\gamma_1 = \alpha_i r_1 + \alpha_i r_2 + \alpha_i r_3 + \alpha_i r_4 + \alpha_i r_5.$$

If the face

$$\begin{aligned}
\gamma_1 = s\alpha &= v \cdot \varphi(D)\alpha \\
&= v_1 \alpha_1 \alpha + v_2 \alpha_2 \alpha + v_3 \alpha_3 \alpha + v_4 \alpha_4 \alpha + v_5 \alpha_5 \alpha \\
&= \alpha_1 v_1 \alpha + \alpha_2 v_2 \alpha + \alpha_3 v_3 \alpha + \alpha_4 v_4 \alpha + \alpha_5 v_5 \alpha \\
&= \alpha_1 r_1 + \alpha_2 r_2 + \alpha_3 r_3 + \alpha_4 r_4 + \alpha_5 r_5.
\end{aligned}$$

(iii) When receiving s_j, p_i can verify equations in (i) and to detect from dealer and participants.

5.4   Update of Secret Key

Suppose dealer D has issued $i - 1$ sub key and he want to issue the number i, what he need to do is just follow the steps described above. Because the time parameters is t_i and the time $t_0, t_1, t_2, \ldots, t_{i-1}$ corresponding to dealt secret $s_0, s_1, \ldots, s_{i-1}$ is different from each other, secret has nothing to do with each other.

6   Conclusion

The security of this scheme is based on elliptic curve discrete logarithm problem which is a really hard problem because there are no sub exponential algorithms for it known today. The issued data u_i is public, but s_i and t_0 is safe since it is an ECDLP parameter.

r_j ($j = 1, 2, \ldots, 5$), γ_i, γ_{i2} ($i = 1, 2, \ldots, 5$) is public, so everyone can find out whether the dealer is a cheater by verifying, γ_{i2} and γ_1. After receiving data from p_j, p_i can verify identity of p_j as follows; firstly, p_i check t_0. If t_0 is not right then the participant is an illegal invader. If t_0 is right, go to check s_i. If s_i is not right that means the participants is a cheater. In addition, data transmitted in communication is safe because the data is encrypted by ECC.

References

1. Mulan, L., Zhanfei, Z., Xiaoming, C.: Secret sharing scheme. Chin. Bull. 45(9), 897–906 (2000)
2. Shamir, A.: How to share a secret. Commun. ACM 22(11), 612–613 (1979)
3. Blakley, G.R.: Safeguarding cryptographic keys. In: Proce. AFIPS 1979 Nat. Comput. Conf., vol. 48, pp. 313–317 (1979)
4. Laih, C.-S., Harn, L., Lee, J.-Y., Hwang, T.: Dynamic Threshold Scheme Based on the Definition of Cross-Product in an N-dimensional Linear Space. In: Brassard, G. (ed.) CRYPTO 1989. LNCS, vol. 435, pp. 286–298. Springer, Heidelberg (1990)
5. Chunxiazng, X., Guozhen, X.: A threshold multiple secret sharing scheme. Acta Electronica Sinica 10(32), 1688–1689 (2004)
6. Wenping, M., Xinhai, W.: Unconditionally secure verifiable secret sharing system. J. China Inst. Commun. 4(25), 64–68 (2004)
7. Brickell, E.F.: Some Ideal Secret Sharing Schemes. In: Quisquater, J.-J., Vandewalle, J. (eds.) EUROCRYPT 1989. LNCS, vol. 434, pp. 468–475. Springer, Heidelberg (1990)
8. Chunxiang, X., Kai, C., Guozhen, X.: A secure Vector space secret sharing scheme. Acta Electronica Sinica 5(30), 715–718 (2002)
9. Aifen, S., Yixian, Y., Xinxin, N., Shoushan, L.: On the Authenticated key Agreement Protocol Based on Elliptic Curve Cryptography. J. Beijing Univ. Post. Telecomm. 3(27), 28–32 (2004)
10. Yajuan, Z., Yuefei, Z., Qiusheng, H.: Elliptic Curve Key-Exchange Protocol. J. Inf. Engg. University 4(5), 1–5 (2004)
11. Wenyu, Z., Qi, S.: The Elliptic Curves over Z_n and key Exchange Protocol. Acta Electronica Sinica 1(33), 83–87 (2005)
12. Brickell, E.F.: Some ideal secret sharing schemes. J. Combin. Math. Combin. Comput. 9, 105–113 (1989)
13. Rishivarman, A.R., Parthasarathy, B., Thiyagarajan, M.: An efficient performance of GF(2^5) arithmatic in an elliptic curve cryptosystem. Int. J. Comput. Appl. 4(2), 111–116 (2009)
14. Rishivarman, A.R., Parthasarathy, B., Thiyagarajan, M.: A Montgomery representation of elements in GF(2^5) for efficient arithmetic to use in ECC. Int. J. Adv. Netw. Appl. 1(5), 323–326 (2010)

Construction of Highly Nonlinear Plateaued Resilient Functions with Disjoint Spectra

Deep Singh

Department of Mathematics,
Indian Institute of Technology Roorkee, Roorkee 247667 India
deepsinghspn@gmail.com

Abstract. The nonlinearity of Boolean functions is one of the most important cryptographic criterion to provide protection against linear approximation attack. In this paper, we use technique suggested by Gao et al. to construct plateaued resilient functions. We provide some new constructions of highly nonlinear resilient Boolean functions on large number of variables with disjoint spectra by concatenating disjoint spectra functions on small number of variables. The nonlinearity of the constructed functions (for some functions) has improved upon the bounds obtained by Gao et al..

Keywords: Walsh-Hadamard transform, Parseval's theorem, Resiliency, Nonlinearity.

1 Introduction

Boolean functions are important in terms of their cryptographic and combinatorial properties for different kind of cryptosystems. Boolean functions used in different cryptosystems must satisfy some desired cryptographic properties such as balancedness, high nonlinearity, high algebraic degree, high order of resiliency and low autocorrelation to provide protection against different kinds of attacks. It is not possible to get best optimized values for each of them simultaneously. Sarkar and Maitra [4] provided weight divisibility results on correlation and resilient Boolean functions. The weight divisibility results have direct consequence for upper bound on nonlinearity for these Boolean functions and a bench-mark in design of such resilient functions. The nonlinearity is one of the most important cryptographic property to provide protection against linear approximation attack. For most recent papers on the construction of Boolean functions with high nonlinearity we refer [4,5,8,10,11]. For the first time, Maitra and Pasalic [6] have constructed 8-variable 1-resilient Boolean functions with nonlinearity 116. The construction of such type of function was posed as an open question by Sarkar and Maitra in [4].

Siegenthaler has proved in [13] that for an n-variable balanced Boolean function of degree d and order of correlation immunity $m(1 \leq m \leq n - 2)$, the inequality $m + d \leq n - 1$ holds and known as Siegenthaler's inequality. The construction of resilient Boolean functions has been discussed in [2,3,12]. Pasalic

P. Balasubramaniam and R. Uthayakumar (Eds.): ICMMSC 2012, CCIS 283, pp. 522–529, 2012.
© Springer-Verlag Berlin Heidelberg 2012

and Johannson [7] provide an important relation between resiliency and nonlinearity of Boolean functions. Recently, Zhang and Xiao [1] described a technique for construction of nonlinear resilient Boolean functions. Using this technique on several sets of disjoint spectra functions on a small number of variables, an almost optimal resilient function on a large even number of variables can be constructed. Also proved that for given m, one can construct infinitely many n-variable (n even), m-resilient functions with nonlinearity $> 2^{n-1} - 2^{n/2}$.

Recently, Gao et al. [9], provided a technique to construct plateaued resilient functions with disjoint spectra. In this paper, we use Gao et al.'s technique to construct plateaued resilient functions and provide some new constructions of highly nonlinear resilient Boolean functions on large number of variables with disjoint spectra by concatenating disjoint spectra functions on small number of variables. From Proposition 2 and Theorem 2, we observe that the nonlinearity of the concatenated functions (for some functions) as constructed in Theorem 1 has improved upon the bounds obtained in Proposition 2.

2   Preliminaries

A function from $\mathbb{F}_2^n$ to $\mathbb{F}_2$ is called a Boolean function where $\mathbb{F}_2 = \{0, 1\}$ is field of characteristic 2 and $\mathbb{F}_2^n = \{0, 1\}^n$ is n-dimensional vector space over field $\mathbb{F}_2$. The set of all n-variable Boolean functions is denoted by $\mathcal{B}_n$. The inner addition of two elements $x, y \in \mathbb{F}_2^n$ is defined as $x \oplus y = (x_1 \oplus y_1, x_2 \oplus y_2, \ldots, x_n \oplus y_n)$ and $\lambda \cdot x = \lambda_1 x_1 \oplus \lambda_2 x_2 \oplus \ldots \oplus \lambda_n x_n$ is the usual inner product on $\mathbb{F}_2^n$. The well known representation of Boolean function $f \in \mathcal{B}_n$ is a binary string of length 2^n given by $[f(0,0,\ldots,0), f(0,0,\ldots,1), \ldots, f(1,1,\ldots,1)]$. The representation of Boolean function in polar form is defined by $[(-1)^{f(0)}, \ldots, (-1)^{f(2^n-1)}]$. The Hamming weight $w_H(f)$ of a Boolean function f on $\mathbb{F}_2^n$ is also the size of its support $\{x \in \mathbb{F}_2^n | f(x) \neq 0\}$. The Hamming distance $d_H(f, g)$ between two functions f and g on $\mathbb{F}_2^n$ is the size of the set $\{x \in \mathbb{F}_2^n | f(x) \neq g(x)\}$, i.e. $d_H(f, g) = w_H(f \oplus g)$.

A function $f \in \mathcal{B}_n$ of the form $f(x) = \lambda \cdot x + \epsilon$ for all $x \in \mathbb{F}_2^n$, where $\lambda \in \mathbb{F}_2^n$ and $\epsilon \in \mathbb{F}_2$, is called an affine function. $\mathbb{A}_n$ denotes set of all affine functions on n-variables.

Definition 1. *The nonlinearity $nl(f)$ of $f \in \mathcal{B}_n$ is defined as the minimum Hamming distance of f from the set $\mathbb{A}_n$. Symbolically,*

$$nl(f) = min\{d_h(f, l) | l \in \mathbb{A}_n\}.$$

Definition 2. *Walsh-Hadamard transform $W_f(\lambda)$ of $f \in \mathcal{B}_n$ at $\lambda \in \mathbb{F}_2^n$ is defined as*

$$W_f(\lambda) = \sum_{x \in \mathbb{F}_2^n} (-1)^{f(x) + \lambda \cdot x}.$$

Parseval's identity,

$$\sum_{\lambda \in \mathbb{F}_2^n} W_f(\lambda)^2 = 2^{2n},$$

it can be shown that $max_{\lambda \in \mathbb{F}_2^n}\{|W_f(\lambda)|\} \geq 2^{n/2}$ which implies $nl(f) \leq 2^{n-1} - 2^{\frac{n}{2}-1}$. Nonlinearity in terms of Walsh-Hadamard transform is defined as

$$nl(f) = 2^{n-1} - \frac{1}{2}max_{\lambda \in \mathbb{F}_2^n}|W_f(\lambda)|.$$

In case n is even, the highest possible nonlinearity is $2^{n-1} - 2^{\frac{n}{2}-1}$. The Boolean functions achieving this bound are bent.

Definition 3. *Let n be a positive even integer. A Boolean function f on $\mathbb{F}_2^n$ is called bent if and only if $W_f(\lambda) = \pm 2^n$ for all $\lambda \in \mathbb{F}_2^n$.*

Definition 4. *A Boolean function f on $\mathbb{F}_2^n$ is called Plateaued if its Walsh-Hadamard values are only 0 and $\pm 2^\lambda$, where λ is some positive integer and known as amplitude of f .*

Definition 5. *The functions $f, g \in \mathcal{B}_n$ are called a pair of disjoint spectra functions if $W_f(\omega)W_g(\omega) = 0$ for all $\omega \in \mathbb{F}_2^n$.*

A function $f \in \mathcal{B}_n$ is said to be balanced if $w_H(f) = 2^{n-1}$. In this case $W_f(0) = 0$.

Definition 6. *A function $f \in \mathcal{B}_n$ is said to be correlation immune of order m if $W_f(\alpha) = 0 \; \forall \alpha \in \mathbb{F}_2^n$ such that $1 \leq w_H(\alpha) \leq m$. A function f is m resilient if and only if it is correlation immune of order m and balanced. Mathematically, f is m resilient if and only if $W_f(\alpha) = 0 \; \forall \alpha \in \mathbb{F}_2^n$ such that $0 \leq w_H(\alpha) \leq m$.*

Iwata and Kurosawa [14] provide the following result related to the nonlinearity of a Boolean function and its subfunctions.

Lemma 1. *[14] Suppose f be an n-variable Boolean function then if f_0 is the restriction of f to the linear hyperplane H of equation $x_n = 0$ and f_1 the restriction of f to the affine hyperplane H' of equation $x_n = 1$ then these two functions will be viewed as $(n-1)$-variable functions and r-th order nonlinearity of f is given as*

$$nl_r(f) \geq nl_r(f_0) + nl_r(f_1).$$

In this paper, we will use concatenation of Boolean functions [6]. The concatenation of two n-variable Boolean functions f_1 and f_2, is denoted by $f_1 \parallel f_2$, which means the output columns of the truth tables of these functions will be concatenated to provide the output column of the truth table of an $(n + 1)$-variable function f. The algebraic normal form of $f = f_1 \parallel f_2$ can be defined as

$$f(x_1, ..., x_{n+1}) = (1 + x_{n+1})f_1(x_1, ..., x_n) + x_{n+1}f_2(x_1, ..., x_n)$$

Define the profile (a, b, c, d), where the entries a, b, c, d denotes the number of variables, order of resiliency, algebraic degree and nonlinearity of Boolean function respectively.

The following result on resiliency is due to Maitra and Sarkar [4].

Lemma 2. *[4] Suppose f is a function with profile $(n, m, n - m - 1, nl)$, then the profile of function concatenated $f \parallel \overline{f}$ is $(n+1, m+1, n-m-1, 2nl)$, where $\overline{f}$ is the complement function of f, i.e., $\overline{f} = 1 + f$.*

Siegenthaler provide the result on resiliency of the function concatenation as

Lemma 3. *[13] Let f_1 and f_2, are any two m-resilient Boolean function then the concatenated function $f = f_1 \parallel f_2$, is also m-resilient.*

Suppose $v = (v_r, ..., v_1)$. We define

$$f_v(x_{n-r}, ..., x_1) = f(x_n = v_r, ..., x_{n-r+1} = v_1, x_{n-r}, ..., x_1).$$

Let $u = (u_r, ..., u_1) \in \mathbb{F}_2^r$ and $w = (w_{n-r}, ..., w_1) \in \mathbb{F}_2^{n-r}$. We define the vector concatenation of u and v as

$$uw = (u, w) = (u_r, ..., u_1, w_{n-r}, ..., w_1).$$

Definition 7. *Let $f \in \mathcal{B}_n$ and u be the string of certain length then $NZ(f)$ denotes spectrum characterization matrix of f, whose rows are the vector of $\mathbb{F}_2^n$ at which the Walsh-Hadamard value of f is non zero i.e.*

$$NZ(f) = \{\omega \in \mathbb{F}_2^n : W_f(\omega) \neq 0\} \text{ and } u \parallel NZ(f) = \{u\omega : \omega \in NZ(f)\}.$$

The following two results are due to Gao et al. [9].

Proposition 1. *[9, Theorem 1] Let $f \in \mathcal{B}_n$ and $w = (w_{n-r}, ..., w_1) \in \mathbb{F}_2^{n-r}$. Then a relation between Walsh-Hadamard spectrum of f and its subfunction can be defined as*

$$W_f(uw) = \sum_{y \in \mathbb{F}_2^r} W_{f_v}(w)(-1)^{u.y}.$$

Proposition 2. *[9, Theorem 3] Let $f_0, g_0 \in \mathcal{B}_n$, be pair of optimal plateaued disjoint spectra functions of profile $(n, m, n - m - 1, 2^{n-1} - 2^{m+1})$ and let $f = (x_{n+1} \oplus 1)f_0 \oplus x_{n+1}g_0$, $g = x_{n+1} \oplus f_0$ and $h = x_{n+1} \oplus g_0$ with profile of f as $(n+1, m, n-m, 2^n - 2^{m+1})$. Then $F = f \parallel \overline{f} \parallel \overline{f} \parallel f \parallel \overline{f} \parallel f \parallel f \parallel \overline{f}$ and $G = g \parallel h \parallel \overline{h} \parallel \overline{g} \parallel \overline{g} \parallel \overline{h} \parallel h \parallel g \in \mathcal{B}_{n+4}$ have profile $(n + 4, m + 3, n - m, 2^{n+3} - 2^{m+4})$.*

Some constructions of plateaued resilient functions with disjoint spectra and high nonlinearity are provided in the following.

3   Main Results

Theorem 1. *Suppose f_0 and g_0 are two disjoint spectra Boolean function on $\mathbb{F}_2^n$. Define f, g and h on $\mathbb{F}_2^{n+1}$ such that $f = (x_{n+1} \oplus 1)f_0 \oplus x_{n+1}g_0$, $g = x_{n+1} \oplus f_0$ and $h = x_{n+1} \oplus g_0$. Then $F = f \parallel \overline{f} \parallel f \parallel \overline{f} \parallel \overline{f} \parallel \overline{f} \parallel f \parallel f$ and $G = g \parallel h \parallel h \parallel g \parallel \overline{g} \parallel \overline{h} \parallel \overline{h} \parallel \overline{g}$ on $\mathbb{F}_2^{n+4}$ are also functions with disjoint spectra .*

Proof. Since $f_0, g_0 \in \mathcal{B}_n$ is pair of disjoint spectra functions so, for any $\omega \in \mathbb{F}_2^n$, we have $W_{f_0}(\omega)W_{g_0}(\omega) = 0$. If $\overline{f}$ denotes the complement of the Boolean function f, then $W_f(\omega) = -W_{\overline{f}}(\omega)$. Let $\overline{f_0} = f_0 \oplus 1$ and $\overline{g_0} = g_0 \oplus 1$. If $\omega \in NZ(f_0)$ or $\omega \in NZ(g_0)$, by Proposition 1, we have $W_f(0\omega) = W_{f_0}(\omega) + W_{g_0}(\omega) \neq 0$, $W_f(1\omega) = W_{f_0}(\omega) - W_{g_0}(\omega) \neq 0$. Hence

$$NZ(f) = \begin{pmatrix} 0 \parallel NZ(f_0) \\ 1 \parallel NZ(f_0) \\ 0 \parallel NZ(g_0) \\ 1 \parallel NZ(g_0) \end{pmatrix}$$

Since $F = f \parallel \overline{f} \parallel f \parallel \overline{f} \parallel \overline{f} \parallel \overline{f} \parallel f \parallel f$, by Proposition1, $W_F(000\omega) = 0$, $W_F(001\omega) = 4W_f(\omega)$, $W_F(010\omega) = -4W_f(\omega)$, $W_F(011\omega) = 0$, $W_F(100\omega) = 0$, $W_F(101\omega) = 4W_f(\omega)$, $W_F(110\omega) = 4W_f(\omega)$, $W_F(111\omega) = 0$. The spectrum characterization matrix of F can be written as

$$NZ(F) = \begin{pmatrix} 001 \parallel NZ(f) \\ 010 \parallel NZ(f) \\ 101 \parallel NZ(f) \\ 110 \parallel NZ(f) \end{pmatrix} = \begin{pmatrix} 0010 \parallel NZ(f_0) \\ 0011 \parallel NZ(f_0) \\ 0100 \parallel NZ(f_0) \\ 0101 \parallel NZ(f_0) \\ 1010 \parallel NZ(f_0) \\ 1011 \parallel NZ(f_0) \\ 1100 \parallel NZ(f_0) \\ 1101 \parallel NZ(f_0) \\ 0010 \parallel NZ(g_0) \\ 0011 \parallel NZ(g_0) \\ 0100 \parallel NZ(g_0) \\ 0101 \parallel NZ(g_0) \\ 1010 \parallel NZ(g_0) \\ 1011 \parallel NZ(g_0) \\ 1100 \parallel NZ(g_0) \\ 1101 \parallel NZ(g_0) \end{pmatrix}$$

Since $G = g \parallel h \parallel h \parallel g \parallel \overline{g} \parallel \overline{h} \parallel \overline{h} \parallel \overline{g}$, by Proposition 1, $W_G(000\omega) = 0$, $W_G(001\omega) = 0$, $W_G(010\omega) = 0$, $W_G(011\omega) = 0$, $W_G(100\omega) = 4W_g(\omega) + 4W_h(\omega)$, $W_G(101\omega) = 0$, $W_G(110\omega) = 0$, $W_G(111\omega) = 4W_g(\omega) - 4W_h(\omega)$. Hence, the spectrum characterization matrix of G can be written as

$$NZ(g) = [1 \parallel NZ(f_0)] \text{ and } NZ(h) = [1 \parallel NZ(g_0)]$$

$$NZ(G) = \begin{pmatrix} 100 \parallel NZ(g) \\ 111 \parallel NZ(g) \\ 100 \parallel NZ(h) \\ 111 \parallel NZ(h) \end{pmatrix} = \begin{pmatrix} 1001 \parallel NZ(f_0) \\ 1111 \parallel NZ(f_0) \\ 1001 \parallel NZ(g_0) \\ 1111 \parallel NZ(g_0) \end{pmatrix}$$

From the Walsh-Hadamard spectrum (spectrum characterization matrix) of concatenated functions F and G we observe that F and G are Boolean functions on $\mathbb{F}_2^{n+4}$ with disjoint spectra. $\qquad\square$

Now, in the following theorem, we show that F and G are resilient plateaued functions with high nonlinearity.

Theorem 2. *Suppose f_0 and g_0 are two disjoint spectra optimal plateaued resilient Boolean functions of profile $(n, m, n - m - 1, 2^{n-1} - 2^{m+1})$ and f, g, h are as defined in Theorem 1. Then the function F and G, constructed in Theorem 1 are disjoint spectra plateaued functions of profile $(n+4, m+1, n-m, 2^{n+3} - 2^{m+3})$ and $(n + 4, m + 2, n - m, 2^{n+3} - 2^{m+4})$ respectively.*

Proof. Since $f = f_0 \parallel g_0$, by Lemma 3, f is m-resilient function and by Lemma 2, $f \parallel \bar{f}$ is $(m + 1)$-resilient function with nonlinearity $nl(f \parallel \bar{f}) = 2nl(f)$. Let $P = f \parallel \bar{f} \parallel f \parallel \bar{f}$, then by Proposition 1, for any $\omega \in \mathbb{F}_2^n$, we have $W_P(00\omega) = 0$, $W_P(01\omega) = 4W_f(\omega)$, $W_F(10\omega) = 0$, $W_P(11\omega) = 0$. Clearly when $wt(01\omega) \le m+1$, then $wt(\omega) \le m$ and f is m-resilient, which implies $W_f(\omega) = 0$ and hence $W_P(01\omega) = 0$, i.e., P is an $(m + 1)$-resilient. Further by Lemma 1 the nonlinearity of P is $nl(P) = 4nl(f)$.

Now we have $F = f \parallel \bar{f} \parallel f \parallel \bar{f} \parallel \bar{f} \parallel \bar{f} \parallel f \parallel f$ and from Proposition 1, $W_F(000\omega) = 0$, $W_F(001\omega) = 4W_f(\omega)$, $W_F(010\omega) = -4W_f(\omega)$, $W_F(011\omega) = 0$, $W_F(100\omega) = 0$, $W_F(101\omega) = 4W_f(\omega)$, $W_F(110\omega) = 4W_f(\omega)$, $W_F(111\omega) = 0$. Now if $wt(001\omega), wt(010\omega) \le (m+1)$, then $wt(\omega) \le m$ which gives $W_F(001\omega) = 0$, $W_F(010\omega) = 0$, $W_F(101\omega) = 0$, $W_F(110\omega) = 0$. i.e, F is $(m + 1)$-resilient function.

From the profile of f it is clear that $max_{\omega \in \mathbb{F}_2^{n+1}} |W_f(\omega)| = 2^{m+2}$ and from spectrum of F, we have $max_{u \in \mathbb{F}_2^3, \omega \in \mathbb{F}_2^{n+1}} |W_F(u, \omega)| = 4max_{\omega \in \mathbb{F}_2^{n+1}} |W_f(\omega)| = 2^{m+4}$. Hence nonlinearity of F is $nl(F) = 2^{n+3} - 2^{m+3}$.

Since $g = f_0 \parallel \bar{f_0}$, $h = g_0 \parallel \bar{g_0}$. From construction of disjoint spectra functions g, h and Lemma 2, we can observe that g and h are functions of profile $(n + 1, m+1, n-m-1, 2^n - 2^{m+2})$. By interchanging variables x_{n+1} and x_{n+2} in the function $g \parallel h = f_0 \parallel \bar{f_0} \parallel g_0 \parallel \bar{g_0}$, we can obtain the function $f_0 \parallel g_0 \parallel \bar{f_0} \parallel \bar{g_0}$, which is equal to the concatenation $f \parallel \bar{f}$. By Lemma 2 we get nonlinearity $nl(g \parallel h) = 2nl_f$. Suppose $Q = g \parallel h \parallel h \parallel g$, then by Proposition 1, we get $W_Q(00\omega) = 2W_g(\omega) + 2W_h(\omega)$, $W_Q(01\omega) = 0$, $W_Q(10\omega) = 0$, $W_Q(11\omega) = 2W_g(\omega) - 2W_h(\omega)$. If $wt(00\omega) \le m + 1$, then $wt(\omega) \le m + 1$. Further g and h are $(m + 1)$-resilient functions, which implies $W_g(\omega) = 0$, $W_h(\omega) = 0$ and hence $W_Q(11\omega) = 0$. Hence Q is an $(m + 1)$-resilient function with nonlinearity $nl(Q) = 2nl(g \parallel h) = 4nl(f)$.

From construction we have $G = g \parallel h \parallel h \parallel g \parallel \bar{g} \parallel \bar{h} \parallel \bar{h} \parallel \bar{g}$ and from Proposition 1, $W_G(000\omega) = 0$, $W_G(001\omega) = 0$, $W_G(010\omega) = 0$, $W_G(011\omega) = 0$, $W_G(100\omega) = 4W_g(\omega) + 4W_h(\omega)$, $W_G(101\omega) = 0$, $W_G(110\omega) = 0$, $W_G(111\omega) = 4W_g(\omega) - 4W_h(\omega)$. If $wt(100\omega) \le (m + 2)$, then $wt(\omega) \le m + 1$. Since g and h are $(m + 1)$-resilient, which implies $W_g(\omega) = 0$ and $W_h(\omega) = 0$, which gives $W_G(100\omega) = 0$, $W_G(111\omega) = 0$, i.e, G is $(m + 2)$-resilient function.

From the profile of g and h it is clear that $max_{\omega \in \mathbb{F}_2^{n+1}} \{|W_g(\omega)|, |W_h(\omega)|\} = 2^{m+3}$. Since g and h are disjoint spectra Boolean functions, therefore, from spectrum of G we have

$$max_{u \in \mathbb{F}_2^3, \omega \in \mathbb{F}_2^{n+1}} |W_F(u, \omega)| = 4max_{\omega \in \mathbb{F}_2^{n+1}} \{|W_g(\omega)|, |W_h(\omega)|\} = 2^{m+5}.$$

Hence, the nonlinearity of G is $nl(G) = 2^{n+3} - 2^{m+4}$. Thus, the concatenated functions F and G are disjoint spectra plateaued resilient functions of profile $(n+4, m+1, n-m, 2^{n+3} - 2^{m+3})$ and $(n+4, m+2, n-m, 2^{n+3} - 2^{m+4})$. □

Some more constructions of resilient plateaued Boolean functions with disjoint spectra and high nonlinearity are provided below.

Theorem 3. *Suppose F and G are two functions such that (i) If $F = f \parallel f \parallel \overline{f} \parallel \overline{f} \parallel \overline{f} \parallel \overline{f} \parallel f \parallel f$ and $G = g \parallel \overline{g} \parallel \overline{h} \parallel h \parallel \overline{g} \parallel g \parallel \overline{h} \parallel h$, where f, g, h are same as defined in Theorem 1. Then F and G are disjoint spectra plateaued resilient functions of profile $(n+4, m+2, n-m, 2^{n+3} - 2^{m+4})$.*

By Proposition 1, we have the Walsh-Hadamard spectrum of F and G as $W_F(000\omega) = 0$, $W_F(001\omega) = 0$, $W_F(010\omega) = 0$, $W_F(011\omega) = 0$, $W_F(100\omega) = 0$, $W_F(101\omega) = 0$, $W_F(110\omega) = 8W_f(\omega)$, $W_F(111\omega) = 0$, and $W_G(000\omega) = 0$, $W_G(001\omega) = -4W_h(\omega)$, $W_G(010\omega) = 0$, $W_G(011\omega) = 4W_h(\omega)$, $W_G(100\omega) = 0$, $W_G(101\omega) = 4W_g(\omega)$, $W_G(110\omega) = 0$, $W_G(111\omega) = 4W_g(\omega)$.

The profile of F and G is same and equals $(n+4, m+2, n-m, 2^{n+3} - 2^{m+4})$. The following constructions are also have same profile

(ii) If $F = \overline{f} \parallel f \parallel \overline{f} \parallel f \parallel f \parallel \overline{f} \parallel f \parallel \overline{f}$ and $G = \overline{g} \parallel \overline{g} \parallel g \parallel g \parallel \overline{h} \parallel \overline{h} \parallel h \parallel h$.

(iii) If $F = g \parallel \overline{g} \parallel \overline{h} \parallel h \parallel \overline{g} \parallel g \parallel \overline{h} \parallel h$ and $G = \overline{g} \parallel \overline{g} \parallel g \parallel g \parallel \overline{h} \parallel \overline{h} \parallel h \parallel h$.

4   Conclusion

In this paper, we use Gao et al.'s technique to construct plateaued resilient functions. We provide some new constructions of highly nonlinear resilient Boolean functions on large number of variables with disjoint spectra by concatenating disjoint spectra functions on small number of variables. From Proposition 2 and Theorem 2, we observe that the nonlinearity of the concatenated functions (for some functions) as constructed in Theorem 1 has improved upon the bounds obtained in Proposition 2.

Acknowledgement. The work of the author was supported by NBHM (DAE), INDIA.

References

1. Zhang, W.G., Xiao, G.Z.: Constructions of Almost Optimal Resilient Boolean Functions on Large Even Number of Variables. IEEE Trans. Inf. Theory 55, 5822–5831 (2009)
2. Wang, Z.W., Zhang, W.: A New Construction of Leakage-Resilient Signature. J. Comput. Inf. Syst. 6, 387–393 (2010)
3. Carlet, C., Charpin, P.: Cubic Boolean Functions with Highest Resiliency. IEEE Trans. Inf. Theory 51, 562–571 (2005)

4. Sarkar, P., Maitra, S.: Nonlinearity Bounds and Constructions of Resilient Boolean Functions. In: Bellare, M. (ed.) CRYPTO 2000. LNCS, vol. 1880, pp. 515–532. Springer, Heidelberg (2000)
5. Tarannikov, Y.: On Resilient Boolean Functions with Maximal Possible Nonlinearity. In: Roy, B., Okamoto, E. (eds.) INDOCRYPT 2000. LNCS, vol. 1977, pp. 19–30. Springer, Heidelberg (2000)
6. Maitra, S., Pasalic, E.: Further Constructions of Resilient Boolean Functions with Very High Nonlinearity. IEEE Trans. Inf. Theory 48, 1825–1834 (2002)
7. Pasalic, E., Johansson, T.: Further Results on the Relation between Nonlinearity and Resiliency for Boolean Functions. In: Walker, M. (ed.) Cryptography and Coding 1999. LNCS, vol. 1746, pp. 35–45. Springer, Heidelberg (1999)
8. Pasalic, E., Maitra, S., Johannson, T., Sarkar, P.: New Constructions of Resilient and Correlation Immune Boolean Functions Achieving Upper Bound on Nonlinearity. In: Proc. WCC 2001, Int. Work. on Coding and Cryptography. ENDM, vol. 6, pp. 158–167 (2001)
9. Gao, S., Ma, W., Zhao, Y., Zhuo, Z.: Walsh Spectrum of Cryptographically Concatenating Functions and its Application in Constructing Resilient Boolean Functions. J. Comput. Inf. Syst. 7, 1074–1081 (2011)
10. Singh, D.: Second Order Nonlinearities of Some Classes of Cubic Boolean Functions Based on Secondary Constructions. Int. J. Comput. Sci. Inf. Tech. 2, 786–791 (2011)
11. Rothaus, O.S.: On Bent Functions. J. Comb. Theory, Series A 20, 300–305 (1976)
12. Maitra, S., Sarkar, P.: Highly Nonlinear Resilient Functions Optimizing Siegenthaler's Inequality. In: Wiener, M. (ed.) CRYPTO 1999. LNCS, vol. 1666, pp. 198–215. Springer, Heidelberg (1999)
13. Siegenthaler, T.: Correlation Immunity of Nonlinear Combining Functions for Cryptographic Applications. IEEE Trans. Inf. Theory 30, 776–780 (1984)
14. Iwata, T., Kurosawa, K.: Probabilistic Higher Order Differential Attack and Higher Order Bent Functions. In: Lam, K.-Y., Okamoto, E., Xing, C. (eds.) ASIACRYPT 1999. LNCS, vol. 1716, pp. 62–74. Springer, Heidelberg (1999)

An Extended RELIEF-DISC for Handling of Incomplete Data to Improve the Classifier Performance

E. Chandra Blessie[1,*], E. Karthikeyan[2,*], and V. Thavavel[3,**]

[1] Department of Computer Science,
D.J. Academy for Managerial Excellence, Coimbatore, Tamil Nadu, India
chandra_blessie@yahoo.co.in
[2] Department of Computer Science,
Government Arts College, Udumalpet, Tamil Nadu, India
e_karthi@yahoo.com
[3] Department of Computer Application,
School of Computer Science and Technology, Karunya University, Tamil Nadu, India

Abstract. Feature selection is one of the important issues in machine learning. Some of the RELIEF based algorithms are considered as the most successful algorithms for assessing the quality of features. RELIEF-DISC which was shown to be efficient in estimating features, cannot handle incomplete data for continuous features. In this paper, we propose an extended RELIEF-DISC algorithm by introducing a new approach to deal with the noisy and incomplete data sets. We investigate the performance of the Decision Tree classifier by imputing the missing values in RELIEF-DISC algorithm. The datasets are taken from the UCI ML Repository.

Keywords: Missing values, Machine Learning, C4.5, Discretization, Feature Selection and Decision Tree Classifier.

1 Introduction

Many learning algorithms perform poorly when the training data are incomplete [5]. Missing feature values commonly exist in real-world data set. They may come from the data collecting process or redundant diagnose tests, unknown data and so on. One standard approach involves imputing the missing values, then giving the completed data to the learning algorithm.

In general, the methods for treating the missing values can be divided into three categories [1]: 1) ignoring/discarding the data which are the easiest and most commonly applied. 2) Parameter estimation where maximum likelihood procedures are used to estimate the parameters of a model. 3) Imputation techniques, where missing values are replaced with estimated ones. The objective of this paper is to introduce a new approach to impute the missing values in the RELIEF-DISC [10] algorithm which can improve the classifier performance. While discretizing the intervals in RELIEF-DISC, the mean value of each initial interval is taken for each

---

* Assistant Professor.
** HOD and Assistant Professor (SG).

P. Balasubramaniam and R. Uthayakumar (Eds.): ICMMSC 2012, CCIS 283, pp. 530–536, 2012.
© Springer-Verlag Berlin Heidelberg 2012

class and the minimum value of the mean for each class is used for imputing the missing values for the corresponding class.

The rest of the paper is organized as follows. Section 2 discusses about the previous work in missing value imputation. Section 3 discusses about the RELIEF-DISC algorithm. Section 4 gives the proposed method. Experimental analysis and the comparison results are described in section 5. Conclusion and result discussion are described in section 6.

2   Review of the Previous Work

This section surveys [6] some commonly and widely used imputation methods.

A. **Mean Imputation (MI).** This is one of the most frequently used methods [8]. It consists of replacing the missing data for a given feature (attribute) by the mean of all known values of that attribute in the class where the instance with missing attribute belongs. Let us consider that the value x_{ij} of the k-th class, C_k, is missing then it will be replaced by

$$\text{Mean}(x_{ij}) = \sum_{x_{ij} \in C_k} x_{ij}/n_k \qquad (1)$$

where n_k represents the number of non-missing values in the j-th attribute of the k-th class.

B. **Ignoring the data.** There are two methods to discard the data having missing values. The first method is known as complete case analysis. This method discards all instances having missing values [7]. The second method determines the extents of missing values before deleting it.

C. **Most Common Attribute Value.** It is one of the simplest methods to deal with missing attribute values. The CN2 algorithm [4] uses this idea. The value of the attribute that occurs most often is selected to be the value for all the unknown values of the attribute.

D. **Concept Most Common Attribute Value.** The most common attribute value method does not pay any attention to the relationship between attributes and a decision. The concept most common attribute value method is a restriction of the first method to the concept, i.e., to all examples with the same value of the decision as an example with missing attribute vale.

E. **Method of Treating Missing Attribute Values as Special Values.** In this method [3], we deal with the unknown attribute values using a totally different approach: rather than trying to find some known attribute value as its value, we treat "unknown" itself as a new value for the attributes that contain missing values and treat it in the same way as other values.

F. **Imputation using Decision Trees Algorithms.** All the decision trees classifiers handle missing values by using built in approaches. CART replaces a missing value of a given attribute using the corresponding value of a surrogate attribute, which has the highest correlation with the original attribute. C4.5 uses a probabilistic approach to handle missing data in both the training and the test sample [9].

3   RELIEF-DISC

RELIEF-DISC [10] is one of the extended RELIEF algorithm which uses the discretization method for selecting of instance and nearest hit and miss values. It consists of two phases. In the first phase, entire dataset is sorted. Then Discretization [2] is applied on the sorted data set to partition the feature into finite number of intervals. The lowest value instance in each interval is selected as instance for finding the nearest hit and nearest miss instances. The number of instances selected is same as the number of intervals. So, there is no need for the expert to specify the sample size. In the second phase, only these instances are used for finding the nearest hit and nearest miss instances using RELIEF algorithm.

Pseudo-code of RELIEF-DISC

Let   D   be   the   training   data   set   with   continuous features F_i; S classes. For every F_i do:

Step 1

1.1 Find maximum (d_n) and minimum (d_o) values

1.2 sort all distinct values of F_i in ascending order

1.3 Initialize all possible interval boundaries, B, with the   minimum,   maximum   and   the   midpoints   where   the continuous   features   have   different   classes   in   the   set $B=\{[d_0,d_1][d_1,d_2],\ldots.,[d_{n-1},d_n]\}$

Step 2

2.1 For every features F_i , assign weight to 0.0

   $W(F_i)=0.0$

For every [d_i, d_{i+1},] in B_j , where i =1..n is the initial boundary points and j is the number of intervals

2.2 let the first instance x_i be the instance i in [d_i, d_{i+1},]

2.3 Take the next instance with the same class from the same interval as nearest hit H.

2.4 Take first instance from the next interval [d_{i+1}, d_{i+2}] as nearest miss M.

Step 3

3.1 Find the difference between the instance and the nearest hit and nearest miss instance diff(F,I,H) and diff(F,I,M)

3.2 Update the weight for the features

   $W(F_i)=W(F_i)-diff(F,I,H)/m+diff(F,I,M)/m$

Step 4

4.1 if W(F$_i$) > τ The features are selected as relevant features. End.

4   Proposed System

4.1   Imputation Using Discretization

Let D={d$_1$,d$_2$,d$_3$,........d$_n$} be the dataset and let the features be F={F$_1$,F$_2$,F$_3$,........F$_m$}where m is the number of features. For each feature, the data are sorted. Initial cutting points were found out between each pair of the instances in the attribute where the two consecutive values have different class value.

Next step is to find the mean value within each interval for each class instead of finding the mean value of the entire non missing values in the dataset. Then the

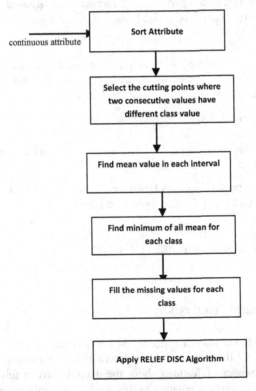

Fig. 1. Flow of extended RELIEF-DISC algorithm

minimum values of the mean for each class are used to fill the missing values corresponding to that class. This will also increase the relevancy between the instances and attributes.

4.2   Pseudocode for Extended RELIEF-DISC

```
Let D be the training data set with continuous features
Fᵢ; S classes.
```

For every F_i do:

Step 1

1.1 Find maximum (d_n) and minimum (d_o) values

1.2 sort all distinct values of F_i in ascending order

1.3 Initialize all possible interval boundaries, B, with the minimum, maximum and the midpoints where the continuous features have different classes in the set $B=\{[d_0,d_1][d_1,d_2],\ldots,[d_{n-1},d_n]\}$

Step 2

2.1 For every interval $[d_i,d_j]$ where I is the lower bound and j is the upper bound, find the mean value corresponding to a single class value

$$Mean(x_{ij}) = \sum_{x_{ij}\in C_k} x_{ij}/n_k \qquad (2)$$

2.2 Find the minimum value of all mean values corresponding to each class C_k.

2.3 Fill the missing values of each class C_k with the minimum mean value of the same class C_k.

Step 3

3.1. Step2 to step4 in RELIEF-DISC.

End.

5   Experimental Analysis

Our experiments were carried out using breast cancer datasets taken from the Machine Learning Database UCI Repository. Table 1 describes the information such as number of instances and the number of features about the dataset used in this paper. The main objective of the experiments conducted in this work is to analyze the efficiency of the C4.5 classification algorithm after selecting the relevant features. In this experiment, missing values are artificially imputed in different rates in different attributes.

Datasets without missing values are taken and few values are removed from it randomly. The rates of the missing values removed are from 2% to 4%.

Table 1. Dataset used for analysis

Datasets	Instances	Attributes
Breast Cancer	699	10

Performance Comparison of Breast Cancer Dataset

The original dataset without missing values yields the accurate classification rate of 94.56%. But the feature selection algorithm RELIEF-DISC yields an increase rate of 95.42%. RELIEF-DISC when applied on the data set with missing values decreases the rate. So, a comparative analysis is made using some imputation methods in RELIEF-DISC algorithm as shown in table 2. From the table, the proposed method (Discend) seems to increases the accuracy rate to 96.34%. The performance comparisons of 4 different imputation methods, RELIEF-DISC without filling the missing value and the Extended RELIEF-DISC by filling the missing value are shown in table 2. Figure 2 shows the comparison of the classification rate.

Table 2. Classification rate using imputation methods

Imputation Method	Features selected	Classification rate
RELIEF-DISC	1,3,5,6,7,8	95.422%
With Missing values	1,3,5	92.34%
Mean	1,3,5,6,7	93.4192%
Most often	3,5,6,7,8	94.7067%
Remove	1,3,5,6,7,8	94.1538%
Proposed (Discend)	1,3,5,6,7,8	96.34%

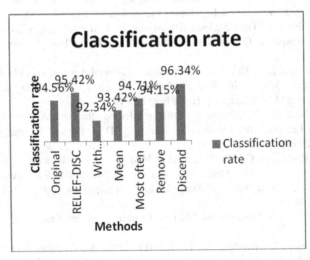

Fig. 2. Comparison result of C4.5 using Breast Cancer dataset

6   Conclusion and Discussion

RELIEF-DISC selects the features which are relevant, but does not handle the incomplete data. But the extended RELIEF-DISC algorithm fills the missing values in the initial stage of discretization and then selects the relevant features. The general method uses the mean of the entire non missing values. But in extended RELIEF-DISC, since the missing values are filled using the minimum value of all the mean values in each class, the classification rate is proved to be more and efficient.

Our experiment for extended RELIEF-DISC was conducted using MatLab 7.0.1 and the classifier performance was analyzed using Weka 3.6. Missing value problem must be solved before using the dataset as the incomplete data may lead to high misclassification rate. This work analyses the classification performance of the C4.5 classifier.

The proposed approach uses only the numerical attributes to impute the missing values. In further it can be extended to handle categorical attributes. From the above comparison, the proposed method seems to be better than the other methods as the accuracy rate is increased. Also, while filling the missing values, with the minimum value of all the mean values in each class, the relevancy between the instances and the attributes can be increased, which will give better result.

References

. 1. Mehala, B., Thangaiah, P.R.J., Vivekanandan, K.: Selecting Scalable Algorithms to Deal with Missing Values. J. Recent Trends. Engg. 1 (2009)
2. Chandra Blessie, E., Karthikeyan, E., Selvaraj, E.: NAD - A Discretization Approach for Improving Interdependency. J. Adv. Res. Comput. Sci. 2, 9–17 (2010)
3. Mundfrom, D.J., Whitcomb, A.: Imputing Missing Values: The Effect on the Accuracy of Classification. J. Mult. Linear Regression Viewpoints 25, 13–19 (1998)
4. Clark, P., Niblett, T.: The CN2 Induction Algorithm. J. Mach. Learn. 3, 261–283 (1989)
5. Kalton, G., Kasprzyk, D.: The Treatment of Missing Survey Data. J. Surv. Methodol. 12, 1–16 (1986)
6. Grzymala-Busse, J.W., Hu, M.: A Comparison of Several Approaches to Missing Attribute Values in Data Mining. In: Ziarko, W., Yao, Y. (eds.) RSCTC 2000. LNCS (LNAI), vol. 2005, pp. 378–385. Springer, Heidelberg (2001)
7. Tresp, V., Neuneier, R., Ahmad, S.: Efficient Methods for Dealing with Missing Data in Supervised Learning. In: Tesauro, G., Touretzky, D.S., Leen, T.K. (eds.) Advances in NIPS 7. MIT Press (1995)
8. Acuna, E., Rodriguez, C.: The Treatment of Missing Values and its Effect in the Classifier Accuracy. In: Gaul, W., Banks, D., House, L., McMorris, F.R., Arabie, P. (eds.) Classification, Clustering and Data Mining Applications, pp. 639–648. Springer, Heidelberg (2004)
9. Quinlan, J.R.: C4.5: Programs for Machine Learning. Morgan Kaufmann, San Mateo CA (1993)
10. Chandra Blessie, E., Karthikeyan, E.: RELIEF-DISC: An Extended RELIEF Algorithm Using Discretization Approach for Continuous Features. In: Second Int. Conf. Emerging Appl. Inf. Tech. (EAIT), pp. 161–164. IEEE Xplore (2011)

Prime Factorization without Using Any Approximations

P. Balasubramaniam[1,2,*,**], P. Muthukumar[1],
and Wan Ainun Binti Mior Othman[2]

[1] Department of Mathematics, Gandhigram Rural Institute - Deemed University
Gandhigram - 624 302, Tamilnadu, India
[2] Institute of Mathematical Sciences, Faculty of Science, University of Malaya,
50603 Kuala Lumpur, Malaysia
{balugru,muthukumardgl}@gmail.com, wanainun@um.edu.my

Abstract. Factoring number is a non-trivial operation, and that fact is the source of a lot of cryptographic algorithms. Many cryptosystems are based on the factorization of large integers. In this paper, factorization algorithm based on number theory is proposed and get the exact prime factorization without any approximations. The major advantages of the proposed method are listed and the disadvantages of the existing factorization algorithm based on square root approximation are highlighted. The time complexity of the proposed method is less because there is no recursive steps in this proposed algorithm.

Keywords: Prime factorization, Cryptography, RSA, Square root approximation.

1 Introduction

The factorization remains a challenging problem in cryptosystems while dealing the size of numbers are very large. In recent years, primality testing and factorization has become one of the most active areas of investigation in number theory and cryptography. RSA(Rivest-Shamir-Adleman)[1] public key cryptosystem is one of the best known cryptosystem. The necessary part of that system is to choose two largest distinct prime numbers p and q such that $n = pq$. Such cryptosystems are importance in industry or government concerned with safeguarding the transmission of data. In that sense, many factorization methods are existing with different manner. In this paper, a new method is proposed based on number theory concept for factoring integer into two large primes. Further the MATLAB program of the proposed method is given and drawbacks of existing factorization algorithm based on square root approximation are highlighted. It is proven that the proposed novel method consumes less time and space.

---

* Corresponding author.
** The work is carried out while he is working as a Visiting Professor, Institute of Mathematical Sciences, Faculty of Science, University of Malaya, 50603 Kuala Lumpur, Malaysia.

P. Balasubramaniam and R. Uthayakumar (Eds.): ICMMSC 2012, CCIS 283, pp. 537–541, 2012.
© Springer-Verlag Berlin Heidelberg 2012

2   Overview of Existing Prime Factorization Methods

In this section, we are briefly stating the existing factorization methods and their drawbacks.

2.1   Trail and Error Method

In this method, the prime factorization involves checking all the prime numbers starting with the smallest, and working to the larger ones. It is enough to check the largest prime number less than or equal to the square root of the given number because if a number is composite it has a factor less than or equal to its square root. Suppose if we use this method to find the prime factorization of a large number, then this method becomes inefficient.

2.2   Fermat's Method

This method of factorization is based on the numbers of special form that an integer can be written as the difference of two squares.

That is, every integer n such that $n = x^2 - y^2 = (x + y)(x - y)$. Then say that $n = ab$ with $b \geq a$, so setting $a = x - y$ and $b = x + y$. By solving for x and y we get $x = \frac{b+a}{2}, y = \frac{b-a}{2}$.

To factor n, a positive integer y such that $n + y^2$ is a square. So it is enough to compute $n + y^2$ for $y = 1, 2, \ldots, k$. Here the number of iterations needed to factor by Fermat's method is $\frac{b-a}{2}$, which is $O(b)$, thus greater than $O(\sqrt{n})$. Therefore, unless $b - a$ is small, Fermat's method is computationally infeasible.

2.3   Euler's Method

This method is applied only if the number can be represented in at least two different ways as the sum of two perfect squares. That is, $n = a^2 + b^2 = c^2 + d^2$ where a, b, c and d are the different positive integers.

Then $n = a^2 + b^2 = c^2 + d^2 \Rightarrow a^2 - c^2 = d^2 - b^2 \Rightarrow (a-c)(a+c) = (d-b)(d+b)$.

2.4   Recent Factorization Methods

After 1975, so many factorization methods are available in the existing literature. Recently, the prime factorization methods such as Pollard's rho method [3], Pollard's P-1 method [4], Elliptic curve method [5,6] and Square root approximation technique have been viewed by the authors in [2]. For more efficiency, they have modified the factorization algorithm based on square root approximation technique.

Even though their algorithm is good. It has so many drawbacks which are to be pointed out in the subsequent section 3.1. To over come the above drawbacks, the new technique is proposed in this paper to factor the positive integer into product of two primes without any approximation.

3   The Proposed Prime Factorization Method

Let n be an odd composite positive integer. If n is not a square, then $n = pq$ where p and q are primes. The proposed method includes the following steps.

1. Verify $n - 3 \equiv 0 \ (mod \ 4)$.

2. If step 1 is satisfied, then find l, m such that $4lm + 3l + m = N$ where $N = (n - 3)/4$ else goto step 4.

3. From the values of l, m in step 2, we get $p - 1 = 4l$ and $q - 3 = 4m$.

4. Verify $n - 1 \equiv 0 \ (mod \ 4)$.

5. If step 4 is satisfied, then find l, m such that $4lm + l + m = N$ where $N = (n - 1)/4$ else goto step 7 .

6. From the values of l, m in step 5, we get $p - 1 = 4l$ and $q - 1 = 4m$.

7. Verify $n - 9 \equiv 0 \ (mod \ 4)$.

8. If step 7 is satisfied, then find l, m such that $4lm + 3l + 3m = N$ where $N = (n - 9)/4$.

9. From the values of l, m in step 8, we get $p - 3 = 4l$ and $q - 3 = 4m$.

3.1   Program for Proposed Method

```
n = input ('Input a positive number except zero: ');
    if (mod(n-3,4)== 0)
    s=n-3;
    N=s/4;
        for l=1:a
            for m=1:b    //where a and b are some positive integers
                if (N==(4*l*m)+3*l+m)
                    disp ([l , m]);
                end
            end
        end

    if (mod(n-1,4)== 0)
    s=n-1 ;
    N=s/4 ;
        for l=1:a
            for m=1:b
                if (N==(4*l*m)+l+m)
                    disp ([l , m]);
                end
            end
        end
```

```
if (mod(n-9,4)==0)
s=n-9;
N=s/4;
    for l=1:a
        for m=1:b
            if (N==(4*l*m)+3*l+3*m)
                disp ([l , m]);
            end
        end
    end
```

3.2 Numerical Examples

Consider n as in the example of [2] in section 3. Let $n = 51923$.

Here step 1 is satisfied, using the above program we get $l = 34$ and $m = 94$. Then $p = 137$ and $q = 397$. Hence $n = 51923 = 137.397$.

Consider n as the running example of proposed algorithm in [2] and let $n = 14789166241$.

Here step 7 is satisfied, using the above program we get $l = 3757$ and $m = 245977$.

Then $p = 15031$ and $q = 983911$. Hence $n = 14789166241 = 15031.983911$.

3.3 Disadvantages of Factorization Algorithm Based on Square Root Approximation

Consider the Algorithm 1 and the section 4.1 of [2], we have

1. It is hard to find an approximate integer r such that the fractional part of $\sqrt{n} \times r$.
2. It is hard to find the adjacent prime for the approximate factor getting from the algorithm because so many prime numbers lie between them.

    For example, the authors [2] found out an integer $A = 985041$ in the running example. But A is not a divisor of n, then the next adjacent prime is 983911. The number of primes between 983911 and 985041 is 80. So it is necessary to check 80 times to justify the prime factor.
3. Suppose we cannot get any prime divisor, the process will be continued by the increment r.

3.4 The Major Advantages of the Proposed Method

1. One can get an exact prime factor of n.
2. It is not necessary to find any approximations.
3. It is not necessary to check l and m to be a divisor or not.
4. The algorithm holds for any number n that has more than two prime factors.
5. There is no recursive steps are adopted.

4 Conclusion

In this paper a new method has been proposed for factoring an integer into two large primes. The main advantage of this method is to get an exact prime factors of large number without any approximations or an initial value. It consumes less time and space complexity. Comparing the square root approximation technique, it is very easy to find the factors without any adjacent divisor or prime number. The disadvantages of the factorization method based on square root approximation have been analyzed through numerical examples. The lower and upper bound for integers l and m could be analyzed in the proposed method, that will be investigated in the near future.

References

1. Stallings, W.: Cryptography and Network Security, 4th edn. Prentice-Hall (2005)
2. Zalaket, J., Hajj-Boutros, J.: Prime factorization using square root approximations. Comput. Math. Appl. 61, 2463–2467 (2011)
3. Pollard, J.M.: A Monte Carlo method for factorization. BIT 15, 331–334 (1975)
4. Pollard, J.M.: Theorems on factorization and primality testing. Math. Proce. Cambridge Philoso. Soc. 76, 521–528 (1974)
5. Lenstrs, H.W.: Factoring integers with elliptic curves. Ann. Math. 126, 649–673 (1987)
6. Montgomery, P.L.: Speeding the Pollard and elliptic curve methods of factorization. Math. Comput. 48, 243–264 (1987)

Coding and Capacity Calculation for the $\mathcal{T}$-User F-Adder Channel

R.S. Raja Durai[1,*] and Meenakshi Devi[2]

[1] Department of Mathematics
Jaypee University of Information Technology
Waknaghat, Solan-173 234, Himachal Pradesh, India
rsraja.durai@juit.ac.in
[2] Department of Mathematics
Bahra University, Shimla Hills
Waknaghat, Solan-173 234, Himachal Pradesh, India
meenakshi_juit@yahoo.co.in

Abstract. The class of $\mathcal{T}$-*Direct* codes are a generalization to the class of linear codes with complementary duals (LCD codes) and are defined as the set of $\mathcal{T}$ F-ary linear codes $\Gamma_1, \Gamma_2, \ldots, \Gamma_{\mathcal{T}}$ such that $\Gamma_i \cap \Gamma_i^{\perp} = \{\mathbf{0}\}$, where $\Gamma_i^{\perp} = \Gamma_1 \oplus \Gamma_2 \oplus \cdots \oplus \Gamma_{i-1} \oplus \Gamma_{i+1} \oplus \cdots \oplus \Gamma_{\mathcal{T}}$ is the dual of Γ_i with respect to the direct sum $\Lambda = \Gamma_1 \oplus \Gamma_2 \oplus \cdots \oplus \Gamma_{\mathcal{T}}$ for each $i = 1, 2, \ldots, \mathcal{T}$. In this paper, an application of a class of $\mathcal{T}$-*Direct* codes over the $\mathcal{T}$-user F-Adder Channel is described. Further, the channel capacity limits for the two-user F-Adder Channel are derived both for the *noiseless* and *noisy* cases.

Keywords: LCD Codes, $\mathcal{T}$-*Direct* Codes, q-cyclic MRD Codes, $\mathcal{T}$-user F-Adder Channel, Channel Capacity.

1 Introduction

For an arbitrary finite field F, an F-ary linear code Γ is called a *linear code with complementary dual* (LCD code) if $\Gamma \cap \Gamma^{\perp} = \{\mathbf{0}\}$ [1]. The class of $\mathcal{T}$-*Direct* codes are a generalization to the class of LCD codes and are defined as the set of $\mathcal{T}$ F-ary linear codes $\Gamma_1, \Gamma_2, \ldots, \Gamma_{\mathcal{T}}$ such that $\Gamma_i \cap \Gamma_i^{\perp} = \{\mathbf{0}\}$, where $\Gamma_i^{\perp} = \Gamma_1 \oplus \Gamma_2 \oplus \cdots \oplus \Gamma_{i-1} \oplus \Gamma_{i+1} \oplus \cdots \oplus \Gamma_{\mathcal{T}}$ is the dual of Γ_i with respect to the direct sum $\Lambda = \Gamma_1 \oplus \Gamma_2 \oplus \cdots \oplus \Gamma_{\mathcal{T}}$ for each $i = 1, 2, \ldots, \mathcal{T}$ denoted by $(\Gamma_1, \Gamma_2, \ldots, \Gamma_{\mathcal{T}})$; each Γ_i is called the *constituent* code [2]. The $\mathcal{T}$-*Direct* codes can be effectively used over the *noiseless* $\mathcal{T}$-user *binary adder channel* [3] and consequently to the *noiseless* $\mathcal{T}$-user F-Adder Channel. A construction of a class of $\mathcal{T}$-*Direct* codes with constituent codes from the class of q-cyclic MRD codes, namely $\mathcal{T}$-*Direct* q-cyclic MRD codes, is given in [4]. Coding for the *noisy* $\mathcal{T}$-user F-Adder Channel is considered in this paper.

The study of multiple-access communication systems were first initiated by Shannon in 1961 - he studied the two-way communication channels and also

---

* Corresponding author.

P. Balasubramaniam and R. Uthayakumar (Eds.): ICMMSC 2012, CCIS 283, pp. 542–551, 2012.
© Springer-Verlag Berlin Heidelberg 2012

bounds on the capacity region were established [5]. In 1971, Ahlswede determined the *capacity regions* for the two-user and three-user multiple access channels with independent sources [6], and Van der Meulen put forward a limiting expression and simple inner and outer bounds on the capacity region for the two-user multiple-access channel [7]. In 1972, Liao [8] studied the general $\mathcal{T}$-user multiple-access channel with independent sources. In 1974, Ahlswede [9] determined the capacity region of a channel with two senders and two receivers. An extensive survey on the information-theoretic aspects of multiple-access channels until 1976 had been documented by van der Meulen [10]. In the literature, there has been an extensive research work on the coding schemes and capacity calculations for a multiple access channel known as the *binary adder channel* [11]-[14]. In the recent years, linear network coding has been a promising new approach to information dissemination over networks as discussed in [15], [16]. A study on the multi-user F-Adder Channel has been discussed in [17], [18].

Our main objective in this paper is to present a coding scheme for multi-user F-Adder Channel (for *noisy* case) by incorporating the ideas of the class of $\mathcal{T}$-*Direct* codes and to derive capacity limits for the multi-user F-Adder Channel. The class of 2-cyclic $\mathcal{T}$-Direct codes, defined in [4], are used in coding for the *noisy* $GF(2^n)$-user F-Adder Channel. Using the class of T-*Direct* codes with the constituent codes being Rank Distance (RD) codes, namely $\mathcal{T}$-Direct 2-cyclic MRD codes [4], a coding scheme for the *noisy* T-user F-Adder Channel is presented. The class of Rank Distance codes is a newer branch of Algebraic coding theory and is introduced by E. M. Gabidulin in 1985 [19]. Unlike the codes with Hamming metric, Rank Distance codes which are equipped with the rank metric are capable of correcting "crisscross" errors in memory chip arrays or in magnetic-tape recording [20], [21].

The paper is organized as follows. We present the basic definitions in the next section. Section III describes an application of the class of $\mathcal{T}$-*Direct* 2-cyclic MRD codes over the *noisy* $\mathcal{T}$-user $GF(2^n)$-Adder Channel. In section IV, capacity calculation for the two-user F-Adder Channel is given for *noiseless* and *noisy* cases. Final section draws the concluding remarks.

2   Preliminary Ideas

Let V^n be an n-dimensional *vector space* over the field $GF(q^N)$, where q is a prime power and $n \leq N$. Then any element $x_i \in GF(q^N)$ can be uniquely represented as $x_i = a_{1i}u_1 + a_{2i}u_2 + \cdots + a_{Ni}u_N$ for some fixed basis $u_1, u_2, \ldots,$ u_N of $GF(q^N)$, regarded as a *vector space* over $GF(q)$. Let $\mathbf{x} = (x_1, x_2, \ldots, x_n)$ $\in V^n$ and $\mathbf{A}_N^n$ denote the collection of all $N \times n$ matrices over $GF(q)$. Then the bijective map $\mathcal{A} : GF(q^N)^n \rightarrow \mathbf{A}_N^n$, defined by $\mathbf{x} \mapsto \mathcal{A}(\mathbf{x})$, associates each n-tuple $\mathbf{x}$ to an $(N \times n)$-array $\mathcal{A}(\mathbf{x}) = [a_{ij}]_{i,j=1}^{N,n}$. The rank of $\mathbf{x} \in V^n$ over $GF(q)$ is defined as the rank of $\mathcal{A}(\mathbf{x})$ and is denoted by $r(\mathbf{x}; q)$. The norm $r(\mathbf{x}; q)$ specifies a rank metric on V^n as $d(\mathbf{x}, \mathbf{y}) = r(\mathbf{x} - \mathbf{y}; q)$ for all $\mathbf{x}, \mathbf{y} \in V^n$.

Definition 2.1. [19] A linear (n, k, d) code which is a k-dimensional subspace of V^n is said to be a Rank Distance code if its metric is induced by the rank norm. An (n, k, d) RD code is said to be an MRD code if $d = n - k + 1$. Here d is the minimum distance of the code and is defined as the minimum rank any non-zero codeword can have.

Definition 2.2. [19] An (n, k, d) RD code Γ is called q-cyclic if for $(c_0, c_1, \ldots, c_{n-1})$ belongs to Γ, then $(c_{n-1}^{[1]}, c_0^{[1]}, \ldots, c_{n-2}^{[1]})$ also belongs to Γ, where and here after we follow the notation: $[m] = q^m$ for some positive integer m.

Definition 2.3. (Construction of q-cyclic MRD codes) [19] An (n, k, d) q-cyclic code with $d = n - k + 1$, termed as q-cyclic MRD code, is generated by the generator matrix G defined as follows:

$$G = \begin{bmatrix} \alpha^{[0]} & \alpha^{[1]} & \cdots & \alpha^{[n-1]} \\ \alpha^{[1]} & \alpha^{[2]} & \cdots & \alpha^{[n]} \\ \vdots & \vdots & \ddots & \vdots \\ \alpha^{[k-1]} & \alpha^{[k]} & \cdots & \alpha^{[k+n-2]} \end{bmatrix}$$

where $\{\alpha^{[0]}, \alpha^{[1]}, \ldots, \alpha^{[n-1]}\}$ forms a normal basis in $GF(q^n)$.

The paper considers the case when n and N are equal. The following theorem gives the necessary and sufficient condition for a set of $\mathcal{T}$ F-ary linear codes to constitute a $\mathcal{T}$-*Direct* code.

Theorem 2.4. [2] Let Γ_i be an (n, k_i) F-ary linear code with the generator matrix G_i such that $G_i G_j^{\mathsf{T}} = (0)$ for each $i = 1, 2, \ldots, \mathcal{T}$ with $i \neq j$. Then $(\Gamma_1, \Gamma_2, \ldots, \Gamma_{\mathcal{T}})$ is a $\mathcal{T}$-*Direct* code if and only if the $k_i \times k_i$ matrix $G_i G_i^{\mathsf{T}}$ is non-singular for every i. Further, if $(\Gamma_1, \Gamma_2, \ldots, \Gamma_{\mathcal{T}})$ is a $\mathcal{T}$-*Direct* code, then $\Pi_{\Gamma_i} = G_i^{\mathsf{T}} (G_i G_i^{\mathsf{T}})^{-1} G_i$ is the *orthogonal projector* from $\Lambda = \Gamma_1 \oplus \Gamma_2 \oplus \cdots \oplus \Gamma_{\mathcal{T}}$ onto Γ_i for each i.

Definition 2.5. [4] A $\mathcal{T}$-*Direct* q-cyclic MRD code is a $\mathcal{T}$-*Direct* code with the *constituent* codes being q-cyclic MRD codes.

$\mathcal{T}$-**user F-Adder Channel**: Given any finite field F, the F-Adder Channel is described as the channel whose inputs are elements of F and the output is the sum (over F) of the inputs. The reader is advised to refer [17] and [18] for

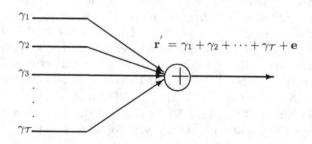

Fig. 1. *Noisy* $\mathcal{T}$-user F-Adder Channel

more details on F-Adder Channel. In this communication system, the $\mathcal{T}$ users send $\mathcal{T}$ n-tuples say γ_1, γ_2, $\ldots$, $\gamma_{\mathcal{T}}$ respectively from the F-ary linear codes $\Gamma_1, \Gamma_2, \ldots, \Gamma_{\mathcal{T}}$. For the *noisy* case, as depicted in Fig. 1, the received sequence $\mathbf{r}'$ is the componentwise sum $\mathbf{r}' = \gamma_1 + \gamma_2 + \cdots + \gamma_{\mathcal{T}} + \mathbf{e}$ over $GF(2^n)$ where the n-tuple $\mathbf{e}$ is an error induced by the noisy channel. In this case, the problem for the receiver is to employ an efficient error-correcting decoding technique to decode the received sequence into the actual codewords originally transmitted. The class of $\mathcal{T}$-*Direct* 2-cyclic MRD codes provides a solution to this problem - which is described in the next section.

3 Coding for the Noisy $\mathcal{T}$-User F-Adder Channel

Let F be a Galois field with characteristic 2, namely $F = GF(2^n)$. Consider the *noisy* $\mathcal{T}$-user F-Adder Channel. Let $\Gamma_1, \Gamma_2, \ldots, \Gamma_{\mathcal{T}}$ respectively denote (n, k_1, d_1), (n, k_2, d_2), $\ldots$, $(n, k_{\mathcal{T}}, d_{\mathcal{T}})$ 2-cyclic MRD codes with their respective generator matrices $G_{k_1}, G_{k_2}, \ldots, G_{k_{\mathcal{T}}}$ (defined as in section 3, [4]) such that $(\Gamma_1, \Gamma_2, \ldots, \Gamma_{\mathcal{T}})$ is a $\mathcal{T}$-*Direct* code. Let $K = k_1 + k_2 + \cdots + k_{\mathcal{T}}$ and $D = n - K + 1$.

$$\text{Let } G = \begin{bmatrix} G_{k_1} \\ G_{k_2} \\ \vdots \\ G_{k_{\mathcal{T}}} \end{bmatrix}$$

be the $K \times n$ matrix with the K row-vectors from the generator matrices G_{k_1}, $G_{k_2}, \ldots, G_{k_{\mathcal{T}}}$. Let H be such that $GH^{\mathrm{T}} = (\mathbf{0})$. By the construction of G, it defines an (n, K, D) 2-cyclic MRD code with complementary duals. Let Γ denote the (n, K, D) 2-cyclic MRD code defined by G. Being an LCD code, the *orthogonal projector* Π_{Γ} from $[GF(2^n)]^n$ onto Γ defined as $\mathbf{r}\Pi_{\Gamma} = \mathbf{r}G^{\mathrm{T}}(GG^{\mathrm{T}})^{-1}G$ for all $\mathbf{r} \in [GF(2^n)]^n$ exists [1]. Also note that if $\gamma_i \in \Gamma_i$ for each $i = 1, 2, \ldots, \mathcal{T}$, then $\gamma_1 + \gamma_2 + \cdots + \gamma_{\mathcal{T}} \in \Gamma$.

Suppose that the $\mathcal{T}$ users send the $\mathcal{T}$ codewords, say $\gamma_1, \gamma_2, \ldots, \gamma_{\mathcal{T}}$ respectively from the *constituent* codes $\Gamma_{k_1}, \Gamma_{k_2}, \ldots, \Gamma_{k_{\mathcal{T}}}$ of the $\mathcal{T}$-*Direct* 2-cyclic MRD code $(\Gamma_{k_1}, \Gamma_{k_2}, \ldots, \Gamma_{k_{\mathcal{T}}})$. As described, in this *noisy* channel, the received vector $\mathbf{r}'$ is $\gamma_1 + \gamma_2 + \cdots + \gamma_{\mathcal{T}} + \mathbf{e}$ over $GF(2^n)$, where $\mathbf{e} = (e_1, e_2, \ldots, e_n) \in [GF(2^n)]^n$ is an error-vector. Assume that the rank of the error-vector $\mathbf{e}$ is $m \leq \lfloor (D-1)/2 \rfloor$. The receiver recovers the codewords $\gamma_1, \gamma_2, \ldots, \gamma_{\mathcal{T}}$ from $\mathbf{r}'$ as described below. The decoder first computes the syndrome of the received vector. The syndrome S of the received vector $\mathbf{r}'$ is,

$$S = \mathbf{r}'H^{\mathrm{T}}$$
$$= \mathbf{e}H^{\mathrm{T}} \; [\textit{since } \gamma_i H^{\mathrm{T}} = (\mathbf{0}) \; \forall \; i]$$

The receiver then determines the error-vector $\mathbf{e}$ by applying an error-correcting decoding technique of the underlying code Γ. Then $\mathbf{r} = \mathbf{r}' - \mathbf{e} = \gamma_1 + \gamma_2 + \cdots + \gamma_{\mathcal{T}}$ is the sum of the codewords $\gamma_1, \gamma_2, \ldots, \gamma_{\mathcal{T}}$ over $GF(2^n)$. The decoder's problem now reduces to finding the codewords $\gamma_1, \gamma_2, \ldots, \gamma_T$ from $\mathbf{r}$, for which the receiver simply applies the *orthogonal projector* Π_{Γ_i} on $\mathbf{r}$ for each i: $\mathbf{r}\Pi_{\Gamma_i} = \gamma_i$, thus retrieving the transmitted codewords $\gamma_1, \gamma_2, \ldots, \gamma_T$ from $\mathbf{r}'$.

4   Capacity Calculation

The channel capacity establishes limits on the performance of practical communication systems. So, calculation of capacity region allows us to see, how much improvement is theoretically possible. In this section, we are investigating the *capacity region* for the *two*-user F-Adder Channel, both *noiseless* and *noisy* cases and derive their information capacity limits. Without loss of generality, we assume that the channel input and output alphabets to be the elements from $GF(q)$, i.e., $F = GF(q)$.

4.1   Capacity for the Noiseless 2-User F-Adder Channel

Consider a discrete memoryless *noiseless multiple access F-Adder channel* with two-independent sources. First the inputs, output and channel statistics are defined. Let Z_1 and Z_2 denote the channel input random variables with their respective marginal probabilities be denoted by $P_{Z_1}(z_1), P_{Z_2}(z_2)$. Denote Z as the channel output random variable so that

$$Z = Z_1 + Z_2$$

The channel is then specified by the conditional probability between the two-inputs and the output as,

$$P_{Z|Z_1,Z_2}(z|z_1, z_2) = \begin{cases} 1, & \text{for } z = z_1 + z_2 \\ 0, & \text{for } z \neq z_1 + z_2 \end{cases}$$

The *capacity region* is the closure of the set of all achievable rate pairs satisfying the following inequalities:

$$0 \leq R_1 + R_2 \leq I(Z_1, Z_2; Z)$$
$$0 \leq \quad R_1 \quad \leq I(Z_1; Z|Z_2)$$
$$0 \leq \quad R_2 \quad \leq I(Z_2; Z|Z_1)$$

where $I(Z_1; Z|Z_2)$, $I(Z_2; Z|Z_1)$, and $I(Z_1, Z_2; Z)$ are *conditional mutual informations*. The reader is encouraged to refer [22] for the *information theoretic approach* to multi-access channels including *additive white Gaussian noise* channel. Using the channel characteristics defined above, the conditional mutual information between Z_1 and Z given Z_2 is calculated as follows:

$$I(Z_1; Z|Z_2) = \sum_{z_1, z_2, z} P_{Z_1}(z_1) P_{Z_2}(z_2) \, P_{Z|Z_1,Z_2}(z|z_1, z_2) \, \log_q \left[\frac{P_{Z|Z_1,Z_2}(z|z_1, z_2)}{P_{Z|Z_2}(z|z_2)} \right]$$

$$= \sum_{z_1, z_2} P_{Z_1}(z_1) P_{Z_2}(z_2) \, \log_q \left[\frac{1}{P_{Z|Z_2}(z|z_2)} \right]$$

$$= \sum_{z_1} P_{Z_1}(z_1) \log_q \left[\frac{1}{P_{Z_1}(z_1)} \right]$$

$$= H(Z_1) \quad = \quad 1$$

where we have used the fact that the maximal value of the entropy $H(Z_1)$ is achieved when the input probabilities are equal:

$$P_{Z_1}(0) = P_{Z_1}(1) = \cdots = P_{Z_1}(\alpha^{p^{m-2}}) = \frac{1}{p^m}$$

here $0, 1, \alpha, \ldots, \alpha^{p^{m-2}}$ are the elements of $GF(q)$, where $q = p^m$ and m is the degree of the corresponding *primitive polynomial*. Similarly, the conditional mutual information between Z_2 and Z given Z_1 can be calculated as

$$I(Z_2; \, Z|Z_1) = H(Z_2) \quad = \quad 1$$

Finally, the mutual information between Z_1, Z_2 and Z can be calculated as

$$I(Z_1, Z_2; Z) = \sum_{z_1, z_2, z} P_{Z_1}(z_1) P_{Z_2}(z_2) P_{Z|Z_1, Z_2}(z|z_1, z_2) \, \log_q \left[\frac{P_{Z|Z_1, Z_2}(z|z_1, z_2)}{P_Z(z_1 + z_2)} \right]$$

$$= \sum_{z_1, z_2} P_{Z_1}(z_1) P_{Z_2}(z_2) \, \log_q \left[\frac{1}{P_{Z_1+Z_2}(z_1 + z_2)} \right]$$

$$= H(Z_1 + Z_2) \quad = \quad 1$$

where we have assumed again that the channel inputs to be *equally probable*. Thus we have the following theorem which states the capacity region of the *noiseless* two-user F-Adder Channel.

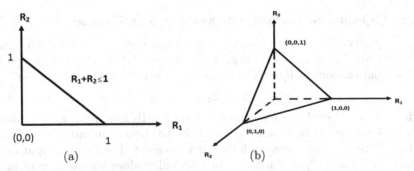

Fig. 2. Capacity region for *noiseless* (a) 2-user F-Adder Channel and (b) 3-user F-Adder Channel

Theorem 4.1. *Capacity region C of the noiseless two-user F-Adder Channel is*

$$C = \{ (R_1, R_2) \mid 0 \le R_1, R_2 \le 1, \, 0 \le R_1 + R_2 \le 1 \}$$

The *capacity region* C of the *noiseless* two-user F-Adder Channel depicted in Fig. 2 represents the set of all simultaneously achievable rate vectors. The *capacity region* for the *noiseless* three-user F-Adder Channel is also depicted in Fig. 2. If the users transmit data with a rate pair as a point inside the *capacity region*, encoders and decoder exist for which each user can communicate with the receiver with an arbitrary small probability of error. However, any rate pair not in the *capacity region* shown in Figs. 2(a) and 2(b) is not achievable i.e., transmission at a rate pair not in the region will lead to errors.

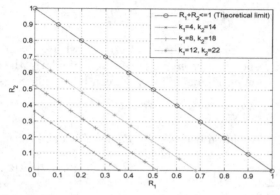

Fig. 3. Capacity bound for *noiseless* 2-user F-Adder Channel

As a justification of the theoretical capacity region for the *noiseless* case, we considered MRD codes of length $n = 50$ with varying code dimensions: $k_1 = 4, 8, 12$ and $k_2 = 14, 18, 22$. The rates for these codes are calculated and compared with the theoretical rate as shown in Fig. 3. It is observed that the calculated rate is within the theoretical limits as expected for the *noiseless* two-user F-Adder Channel.

4.2 Capacity for the Noisy 2-User F-Adder Channel

The capacity of a *noisy* two-user F-Adder Channel is calculated for *Additive White Gaussian Noise* conditions. A non-binary *AWGN* channel for two users with input random variables Z_1, Z_2 can be characterized by the expression

$$Y = Z_1 + Z_2 + N$$

where N is a Gaussian noise random variable with *zero mean* and *variance* σ_n^2 that is independent of Z_1 and Z_2, and Y is the channel output random variable with different sample value each time it is received. The channel input random variables Z_1 and Z_2 are assumed to take only values within a finite interval $[-A, +A]$, where A is an arbitrary positive real number. In practice, Y and N are samples of continuous random processes. As the amplitude probability density function of a Gaussian random variable X is given by

$$p(x) = \frac{1}{\sqrt{2\pi}\,\sigma}\, e^{-\frac{1}{2}\left(\frac{x-m}{\sigma}\right)^2}$$

with m, σ^2 are the mean and variance of X respectively, it follows that for a given sample value of $Z = z_i$, $Z = Z_1 + Z_2$, the channel output Y is Gaussian with mean z_i and variance σ_n^2:

$$p_{Y|Z}(Y|Z = z_i) = \frac{1}{\sqrt{2\pi}\,\sigma_n}\, e^{-\frac{1}{2}\left(\frac{y-z_i}{\sigma_n}\right)^2}, \quad -\infty < y < \infty$$

Define the achievable rate region as the convex hull of the set of rate pairs (R_1, R_2) which satisfy each of the inequalities:

$$0 \leq R_1 + R_2 \leq I(Z_1, Z_2; Y) \tag{1}$$
$$0 \leq \quad R_1 \quad \leq I(Z_1; Y|Z_2) \tag{2}$$
$$0 \leq \quad R_2 \quad \leq I(Z_2; Y|Z_1) \tag{3}$$

The average mutual information for the $AWGN$ channel under consideration has been shown to be [23]

$$I(Z; Y) = \sum_{i=1}^{q}\left\{\int_{-\infty}^{\infty} p_{Y|Z}(y|z_i)\mathbb{P}(Z = z_i)\,\log_q\left[\frac{p_{Y|Z}(y|z_i)}{p_Y(y)}\right]dy\right\}$$

where $p_Y(y)$ is the density function of the output random variable given by

$$p_Y(y) = \sum_{i=1}^{q} p_{Y|Z}(y|z_i)\,\mathbb{P}(Z = z_i)$$

and $\mathbb{P}(Z = z_i)$ is the probability of occurrence of the input level z_i.

Suppose that the inputs Z_1 and Z_2 are each constrained to have mean square values at most E_1 and E_2, respectively: $E\left[Z_1^2\right] \leq E_1$ and $E\left[Z_2^2\right] \leq E_2$. If the channel under consideration is considered as cascade of a *noiseless* channel adding Z_1 and Z_2 followed by a single input Gaussian channel, $I(Z_1, Z_2; Y)$ is at most the capacity of the single input non-binary channel with the input constrained to energy $E_1 + E_2$. Thus

$$I(Z_1, Z_2; Y) \leq \frac{1}{q}\log_q\left[1 + \frac{E_1 + E_2}{\sigma_n^2}\right]$$

with,

$$I(Z_1; Y|Z_2) \leq \frac{1}{q}\log_q\left[1 + \frac{E_1}{\sigma_n^2}\right]$$
$$I(Z_2; Y|Z_1) \leq \frac{1}{q}\log_q\left[1 + \frac{E_2}{\sigma_n^2}\right]$$

Since Z_1 and Z_2 are assumed to have independent distributions, the rate region for the *noisy* two-user F-Adder Channel for which the equations (1)-(3) are satisfied is given by

$$0 \leq R_1 + R_2 \leq \frac{1}{q} \log_q \left[1 + \frac{E_1 + E_2}{\sigma_n^2} \right] \tag{4}$$

$$0 \leq \quad R_1 \quad \leq \frac{1}{q} \log_q \left[1 + \frac{E_1}{\sigma_n^2} \right] \tag{5}$$

$$0 \leq \quad R_2 \quad \leq \frac{1}{q} \log_q \left[1 + \frac{E_2}{\sigma_n^2} \right] \tag{6}$$

where $E_1 = E_2 = \frac{A^2}{2}$ is considered. Fig. 4 illustrates the capacity C (in symbols per channel use) of the *noisy* 2-user F-Adder Channel, as a function of the *SNR*: $\frac{A^2}{2\sigma_n^2}$ for $\mathcal{T} = 1, 6, 12$ users.

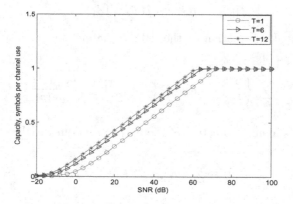

Fig. 4. Capacity as a function of SNR

It can be seen that the channel capacity C increases monotonically from 0 to 1 symbols per channel use as the SNR increases. The capacity calculations for the two-user F-Adder Channel given can be extended in a straight forward manner to calculate the information capacity limits for the general case of multi-user F-Adder Channel, for both the *noiseless* and *noisy* cases.

5   Conclusion

In this paper, coding problem for the *noisy* $\mathcal{T}$-user F-Adder Channel is considered. The class of $\mathcal{T}$-*Direct* 2-cyclic MRD codes are shown to be effective in coding for the *noisy* $\mathcal{T}$-user $GF(2^n)$-Adder Channel, in that they uniquely determine the transmitted codewords from the received *noisy* sequence. Further, capacity bounds for the two-user F-Adder Channel are obtained for both *noiseless* and *noisy* cases.

References

1. Massey, J.L.: Linear Codes with Complementary Duals. Discrete Mathematics 106 and 107, 337–342 (1992)

2. Vasantha, W.B., Raja Durai, R.S.: $\mathcal{T}$-Direct codes: An Application to $\mathcal{T}$-user BAC. In: 2002 IEEE Information Theory Workshop, Bangalore, p. 214 (2002)
3. Vasantha, W.B., Raja Durai, R.S.: Some Results on $\mathcal{T}$-Direct Codes. In: 3rd Asia-Europe Workshop on Information Theory, Japan, pp. 43–44 (2003)
4. Raja Durai, R.S.: Distributed Source Coding Using $\mathcal{T}$-Direct Codes. In: SympoTIC 2006, Slovakia, pp. 24–27 (2006)
5. Shannon, C.E.: Two-way Communication Channels. In: 4th Berkeley Symposium on Mathematical Statistics and Probability, Berkeley, vol. 1, pp. 611–644 (1961)
6. Ahlswede, R.: Multi-way Communication Channels. In: 2nd International Symposium on Information Theory, U.S.S.R., pp. 23–52 (1971)
7. Van der Meulen, E.C.: The Discrete Memoryless Channel with Two Senders and One Receiver. In: 2nd International Symposium on Information Theory, U.S.S.R., pp. 103–135 (1971)
8. Liao, H.: A Coding Theorem for Multiple Access Communications. In: International Symposium on Information Theory, Asilomar (1972)
9. Ahlswede, R.: The Capacity Region of a Channel with Two Senders and Two Receivers. The Annals of Probability 2(5), 805–814 (1974)
10. Van der Meulen, E.C.: A Survey of Multi-way Channels in Information Theory: 1961-1976. IEEE Transactions on Information Theory 23(1), 1–37 (1977)
11. Kasami, T., Lin, S.: Coding for a Multiple-Access Channel. IEEE Transactions on Information Theory 22(2), 129–137 (1976)
12. Chang, S.C., Weldon, E.J.: Coding for $\mathcal{T}$-user Multiple Access Channels. IEEE Transactions on Information Theory 25(6), 684–691 (1979)
13. Honary, B., Ali, F.: Capacity of $\mathcal{T}$-user Collaborative Coding Multiple-access Scheme Operating Over Noisy Channel. Electronics Letters 25(11), 742–744 (1989)
14. Honary, B., Ali, F., Darnell, M.: Information Capacity of Additive White Gaussian Noise Channel with Practical Constraints. In: IEE Proceedings, Part I: Communications, Speech and Vision, vol. 137(5), pp. 295–301 (1990)
15. Silva, D., Kshischang, F.R.: Using Rank-Metric Codes for Error Correction in Random Network Coding. In: IEEE International Symposium on Information Theory, France, pp. 796–800 (2007)
16. Silva, D., Kshischang, F.R., Koetter, R.: A Rank-Metric Approach to Error Control in Random Network Coding. IEEE Transactions on Information Theory 54(9), 3951–3967 (2008)
17. Urbanke, R., Rimoldi, B.: Coding for the F-Adder Channel: Two applications for Reed-Solomon Codes. In: IEEE International Symposium on Information Theory, San Antonio, p. 85 (1993)
18. Rimoldi, B.: Coding for the Gaussian Multiple-Access Channel: An Algebraic Approach. In: IEEE International Symposium on Information Theory, Austin, p. 81 (1993)
19. Gabidulin, E.M.: Theory of Codes with Maximum Rank Distance. Problems of Information Transmission 21, 1–12 (1985)
20. Blaum, M., McEliece, R.J.: Coding Protection for Magnetic Tapes: A Generalization of the Patel-Hong Code. IEEE Transactions on Information Theory 31(5), 690–693 (1985)
21. Levine, L., Meyers, W.: Semiconductor Memory Reliability with Error Detecting and Correcting Codes. Computer 9(10), 43–50 (1976)
22. Gallager, R.G.: A perspective on Multiaccess Channels. IEEE Transactions on Information Theory 31(2), 124–142 (1985)
23. Gallager, R.G.: Information Theory and Reliable Communication, pp. 71–97. Wiley, New York (1968)

Retraction Note to: Mathematical Modelling and Scientific Computation

P. Balasubramaniam and R. Uthayakumar

Department of Mathematics, Gandhigram Rural Institute - Deemed University,
Gandhigram 624302, Tamil Nadu, India
balugru@gmail.com, uthayagri@gmail.com

Retraction Note to:
P. Balasubramaniam and R. Uthayakumar (Eds.):
Mathematical Modelling and Scientific Computation, CCIS,
DOI: 10.1007/978-3-642-28926-2

The papers starting on pages 22, 32 and 47 of this volume have been retracted for reasons of plagiarism. In addition, the name of the second author was added without his knowledge or consent.

The updated original online version for these chapters can be found at
DOI: 10.1007/978-3-642-28926-2_3
DOI: 10.1007/978-3-642-28926-2_4
DOI: 10.1007/978-3-642-28926-2_5

P. Balasubramaniam and R. Uthayakumar (Eds.): ICMMSC 2012, CCIS 283, p. E1, 2012.
DOI: 10.1007/978-3-642-28926-2_63

Author Index